UNITARY SYMMETRY AND COMBINATORICS

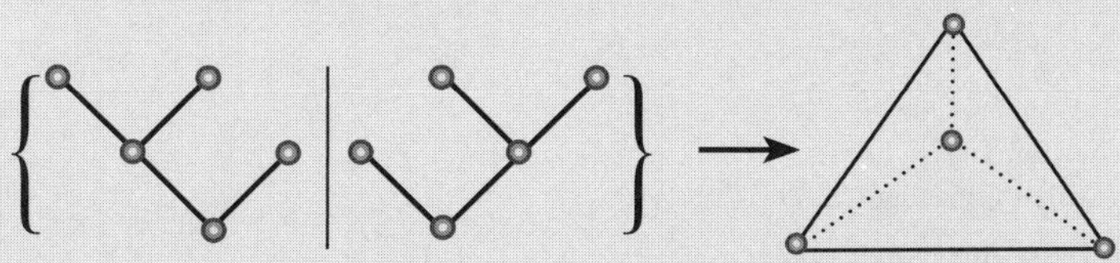

UNITARY SYMMETRY AND COMBINATORICS

James D. Louck

Los Alamos National Laboratory Fellow
Santa Fe, New Mexico, USA

World Scientific

NEW JERSEY • LONDON • SINGAPORE • BEIJING • SHANGHAI • HONG KONG • TAIPEI • CHENNAI

Published by

World Scientific Publishing Co. Pte. Ltd.

5 Toh Tuck Link, Singapore 596224

USA office: 27 Warren Street, Suite 401-402, Hackensack, NJ 07601

UK office: 57 Shelton Street, Covent Garden, London WC2H 9HE

British Library Cataloguing-in-Publication Data
A catalogue record for this book is available from the British Library.

UNITARY SYMMETRY AND COMBINATORICS

ISBN-13 978-981-281-472-2
ISBN-10 981-281-472-8

Printed in Singapore by B & JO Enterprise

Dedicated to the memory of my sons
Samuel Victor Louck
Joseph Patrick Louck

And to the courage of my son and wife
Thomas James Louck
Margaret Marsh Louck

Preface

This monograph is an outgrowth of the books "Angular Momentum in Quantum Physics" and "The Racah-Wigner Algebra in Quantum Theory," by L. C. Biedenharn and myself, published in 1981, originally by Addison-Wesley in the Gian-Carlo Rota series "Encyclopedia of Mathematics and Its Applications," and subsequently by Cambridge University Press. Biedenharn and I planned to extend the results for SU(2), which is the quantum mechanical rotation group of 2 x 2 unitary unimodular matrices, to the general unitary group $U(n)$, based on our research over thirty years of collaboration. The plan was to use the methods of the boson calculus because of its close relationship to the creation and annihilation operators associated with physical processes and the natural invariance of this calculus to unitary transformations. The broad outline of such a monograph on unitary symmetry based on the boson calculus was laid out some fifteen years ago, but was never implemented. Biedenharn became very interested in quantum groups and q-tensor operator theory, while I, under the influence of Gian-Carlo Rota and his student, William Y. C. Chen, became interested in the combinatorial basis of group representation and tensor operator theory. Biedenharn's death in 1996 ended any possibility of a rejoining of efforts, but our earlier collaborations have had a heavy bearing on the present work.

The role of combinatorics in the representation theory of groups is more encompassing than possibly could have been forseen. The fundamental role developed here evolved from research with William Y. C. Chen and Harold W. Galbraith, postdoctoral student of mine, and collaborator on a number of articles on symmetry in physics, all of which was tempered by Rota's global viewpoint of the pervasiveness of combinatorics. This monograph is about the discoveries made, as described by a algorithmic approach to enhance the computability of the complex objects encountered. It is against this background that the viewpoints advanced in this monograph emerged.

Boson polynomials are homogeneous polynomials defined over a collection of n^2 commuting boson creation operators. These polynomials give all the irreducible unitary representations of the general unitary group $U(n)$ by the simple device of replacing the boson operators by the n^2 elements of a unitary matrix. The multiplication property of these matrix group respresentations of $U(n)$ is preserved even by the boson polynomials. This suggests that the boson operators should be taken to be commuting indeterminates, and that the properties of these homogeneous polynomials should be developed in this context. The polynomials are themselves the basic objects, independent of any interpretation of the indeterminates over which they are defined. Then, not only are the irreducible representations of $U(n)$ (and the general linear group) obtained in one assignment of the indeterminates, but also in the original assignment the rich physical interpretation in terms of boson operators is regained.

But much more emerges. The group multiplication property of representations is a consequence of a new class of identities among multinomial coefficients, which themselves have a combinatorial origin and proof, and which hold for arbitrary interpretations of the n^2 indeterminates, including even singular matrices of order n. The structure is fully combinatorial. The study of these polynomials is thus brought under the purview of combinatorics and special functions, extended to many variables. These polynomials may be regarded as generalizations of the functions that arise in the study of the symmetric group, with its associated catalog of symmetric functions, such as the Schur functions, etc. Even more unexpected is that the famous MacMahon [129] Master Theorem, a classical result in combinatorics, is the basis for Schwinger's [160] famous generating function approach to angular momentum theory. Indeed, it is the MacMahon Master Theorem that unifies the angular momentum properties of composite systems in the binary build-up of such systems from more elementary constituents.

This monograph consists, essentially, of three distinct, but interrelated parts: Chapters 1-4, Chapters 5-9, and Chapters 10-11. The last two chapters are compendiums which define, develop, and summarize concepts used in the first nine chapters.

Chapters 1-4 deal with basic angular momentum theory and the properties of the famous Wigner $D-$functions, now extended to polynomial forms over four commuting indeterminates, and with the properties of arbitrary many multiple Kronecker products of these extended $D-$polynomials. These four chapters may be regarded as a summary of results that subsume all of standard angular momentum theory with a focus on the combinatorial underpinnings of these polynomials, as captured by the concept of $SU(2)$ solid harmonics. As examples, the famous Wigner-Clebsch-Gordan coefficients are shown to be objects that combinatorially come under the purview of the umbral calculus, while the binary coupling theory of angular momentum is intrinsically an application of the theory of graphs, specifically, binary trees, Cayley trivalent trees, and cubic graphs. This leads to a number of combinatorial interpretations of the well-known Racah sum rule and Biedenharn-Elliott identity, and the fundamental role of Racah coefficients in the binary recoupling theory of angular momenta.

Chapters 5-9 deal with the generalization of the solid harmonics to polynomials called $D^{\lambda}-$polynomials, where λ is a partition, and these polynomials are defined over n^2 commuting indeterminates, which when specialized to the elements of a complex matrix of order n give the integral irreducible representations of the general linear group of complex matrices of order n, and, in particular, all inequivalent irreducible representations of the general unitary group of matrices of order n. Again, the focus is on the combinatorial properties of the general polynomials themselves, such as their unique generation by shift operator actions, which involve diagraphs, Sylvester's identity, Schur functions, skew Schur functions, Kostka numbers, and Littlewood-Richardson numbers, all combinatorial concepts underlying modern treatments of the symmetric group S_n. It is the labeling of these polynomials by Gelfand-Tsetlin patterns, which are one-to-one with the semistandard Young-Weyl tableau, that underlies the relationship to the symmetric group. The reduction of the single Kronecker product $D^{\mu} \otimes D^{\nu} = \sum_{\lambda} c_{\mu\nu}^{\lambda} D^{\lambda}$ of two such irreducible polynomials into a direct sum of irreducible polynomials is extraordinarily rich in combinatorial structures.The $D^{\lambda}-$polynomials subsume many of the properties of classical Schur functions, and the matrix $D^{\lambda}(Z)$ might well be called a *matrix Schur function.*The complexity of these polynomials, although elegant in their structure, allows us to deal comprehensibly only with the Kronecker product of a pair of such polynomials. Multiple Kronecker products and the associated concepts of Racah coefficients, etc., and the relationship to graph theory is beyond our reach. New viewpoints of tensor operators as operator-valued $D^{\lambda}-$polynomials emerge. A comprehensive theory of (generalized) Racah coefficients must await further developments.

The Littlewood-Richardson numbers $c_{\mu\nu}^{\lambda}$ that occur in the reduction of the Kronecker product is so pervasive that we give a great deal of attention to their properties (Compendium B). These numbers express the number of repetitions of a given $D^{\lambda}-$polynomial in the Kronecker product reduction. They give the generalization to the general unitary group $U(n)$ of the familiar addition rule

$$j = j_1 + j_2, j_1 + j_2 - 1, \ldots, |j_1 - j_2|$$

of two interacting quantum-mechanical constituents with separate angular momenta j_1 and j_2, constituting a composite system of angular momentum j; the Littlewood-Richardson number is 0 or 1.

Three (at least) nontrivial combinatorial objects enter into the combinatorial interpretation of the Littlewood-Richardson numbers: Gelfand-Tsetlin patterns, semistandard skew tableaux, and the lattice permutations associated with these entities. The intricacies of such counting methods would appear to be a rather high price for obtaining the rule for the addition of two angular momenta, which was deduced by physicists from experimental spectroscopy and subsequently from algebraic techniques (see Condon and Shortley [45]) that involved neither Lie algebras nor combinatorics. But the new insights gained are well worth the effort.

These techniques underlie the development of the properties of the $D^{\lambda}-$polynomials over arbitrary commuting indeterminates. One of the principal purposes of this monograph is to demonstrate, by construction, the details and inter-relations of these concepts.

Chapters 10-11 comprise the third part of this monograph. They consist of two extensive Compendiums A and B of results from algebra, analysis, and combinatorics that relate to the first two parts. They have been included so as to be able to refer and use the results in the main parts of the monograph without having to interrupt the flow of presentation with technical asides. The presentation of the material in the Compendiums is very uneven: some is given in great detail and some is very brief, depending on their role in the main text.

There are a number of unsolved problems and unaddressed topics. Unsolved problems include the following, where further details can be found in the referenced sections:

1. Counting formula for the Clebsch-Gordon numbers that give the multiplicity of a given state of total angular momentum in the coupling of n angular momenta (Sect. 2.2).
2. The enumeration of the nonisomorphic unlabeled cubic graphs on $2n$ points that correspond to the coupling of n angular momenta (Sects. 3.3, 3.4, 4.5, 4.6).
3. Extension of the step-function formulas for Kostka numbers and Littlewood-Richardson numbers to $n \geq 4$ with a geometrical interpretation (Sects. 9.4.3, 9.6, 11.3.7, 11.3.8).
4. The geometrical meaning of operator patterns (Sects. 9.4, 9.6, 9.7.2).
5. A comprehensive theory of multiple Kronecker products of the D^λ−polynomials and of the associated recoupling matrices; that is, the generalization of $3n - j$ coefficients of $SU(2)$ and of the geometry of cubic graphs (p. 446).

Inadequately addressed and nonaddressed topics include the following:

(i). Full development of the properties of the skew-symmetric matrix associated with a standard labeled binary tree corresponding to the addition of angular momenta (Sect. 4.2).

(ii). Path formulation of recoupling matrices (Sect. 2.2.10).

(iii). Relation of D^λ−polynomials to special functions, such as a theory of multivariable Hermite polynomials (Sect. 11.9.4).

(iv). Formulation of a comprehensive umbral calculus and invariant theory approach to the D^λ−polynomials (Sect. 11.9.3).

(v). Extension of combinatorial foundations to other groups.

(vi). Applications to physical problems.

The very detailed Table of Contents serves as a summary of topics covered. The readership is intended to be graduate students and researchers interested in learning of the relation between symmetry and combinatorics and of challenging unsolved problems. The many examples serve partially as exercises. It is hoped that the topics presented promote further and more rigorous developments.

We mention some unconventional matters of style. We present significant result in italics, but do not grade and stylize them as lemmas and theorems. Such italicized statements serve as summaries of results, and often do not merit the title as theorems. Diagrams and figures are integrated into the text, and not set aside on nearby pages, so as to have a smooth flow of ideas. Our informality of presentation, including proofs, does not attain the status of rigor demanded in more formal approaches, but our purpose is better served, and our objectives met, by focusing on algorithmic, constructive methods, as illustrated by many examples. It is particularly encouraging to read in Andrews, Askey, and Roy [3] about the usefulness of algorithmic based, complex, mathematical relationships in today's computer oriented approach. Such relations encode information amenable to computer processing; perhaps, not to extent envisioned by Wolfram [187], but nonetheless naturally and innovatively.

This monograph is not democratically assembled. The enormous literature on physical applications of unitary symmetry are not amenable to a synthesis of technique, except in the broadest sense of Lie algebra and group representations. Moreover, the subject has received little attention from the combinatorial orientation presented here. Accordingly, the monograph is heavily biased toward the understanding I have been able to acquire over a fifteen year period of presenting lectures on these subjects at small conferences in Poland organized by Tadeusz and Barbara Lulek on Symmetry and Structural Properties of Condensed Matter, and also at Nankai University, PR China, at the invitation of William Y. C. Chen, Director, The Center for Combinatorics. The opportunity to address a sizeable number of students has been particularly rewarding. Important special contributions to the subject have come from my colleagues Bill Chen, Harold Galbraith, and Miguel Méndez. General encouragement from George Andrews, Bill Chen, Gordan Drake, Harold Galbraith, Brian Judd, Ron King, Tadeusz and Barbara Lulek, Steven Milne, Peter Paule, Gian-Carl Rota, and in earlier years, Larry Biedenharn, and in later years, my son Tom and wife Marge; all have helped to sustain the effort. I have also been inspired by the many lectures of Gian-Carlo Rota, the comprehensive book by Stanley [163], and the terse, but scholarly book by Macdonald [126].

James D. Louck

Contents

Preface vii

Notation xxi

1 Quantum Angular Momentum 1

1.1 Background and Viewpoint 1

1.1.1 Euclidean and Cartesian 3-space 1

1.1.2 Newtonian physics 9

1.1.3 Nonrelativistic quantum physics 10

1.1.4 Unitary frame rotations 17

1.2 Abstract Angular Momentum 28

1.2.1 Brief background and history 28

1.2.2 One angular momentum 30

1.2.3 Two angular momenta 35

1.3 $SO(3,\mathbb{R})$ and $SU(2)$ Solid Harmonics 43

1.3.1 Inner products 49

1.4 Combinatorial Features 51

1.4.1 Combinatorial definition of Wigner-Clebsch-Gordan coefficients 51

1.4.2 Magic square realization 59

1.5 Kronecker Product of Solid Harmonics 61

1.6 $SU(n)$ Solid Harmonics 64

1.6.1 Definition and properties of $SU(n)$ solid harmonics 64
1.6.2 MacMahon and Schwinger master theorems 66
1.6.3 Combinatorial proof of the multiplication property 67
1.6.4 Maclaurin monomials 71
1.6.5 Summary of relations 74
1.7 Generalization to $U(2)$ 75
1.7.1 Definition of $U(2)$ solid harmonics 75
1.7.2 Basic multiplication properties 78
1.7.3 Indeterminate and derivative actions on the $U(2)$ solid harmonics 81

2 Composite Systems **83**
2.1 General Setting 83
2.1.1 Angular momentum state vectors of a composite system 83
2.1.2 Group actions in a composite system 89
2.1.3 Standard form of the Kronecker direct sum 90
2.1.4 Reduction of Kronecker products 93
2.1.5 Recoupling matrices 94
2.2 Binary Coupling Theory 97
2.2.1 Binary trees 100
2.2.2 Standard labeling of binary trees 107
2.2.3 Generalized WCG coefficients defined in terms of binary trees 109
2.2.4 Binary coupled state vectors 112
2.2.5 Binary reduction of Kronecker products 114
2.2.6 Binary recoupling matrices 115
2.2.7 Triangle patterns and triangle coefficients 119
2.2.8 Racah coefficients 131
2.2.9 Recoupling matrices for $n = 3$ 137
2.2.10 Recoupling matrices for $n = 4$ 140

2.2.11 Structure of general triangle coefficients and recoupling matrices 152

2.3 Classification of Recoupling Matrices 156

3 Graphs and Adjacency Diagrams 163

3.1 Binary Trees and Trivalent Trees 163

3.2 Nonisomorphic Trivalent Trees 172

3.2.1 Shape labels of binary trees 184

3.3 Cubic Graphs and Trivalent Trees 189

3.4 Cubic Graphs 198

3.4.1 Factoring properties of cubic graphs 200

3.4.2 Join properties of cubic graphs 213

3.4.3 Cubic graph matrices 221

3.4.4 Labeled cubic graphs 226

3.4.5 Summary and unsolved problems 228

4 Generating Functions 229

4.1 Pfaffians and Double Pfaffians 230

4.2 Skew-Symmetric Matrix 232

4.3 Triangle Monomials 238

4.4 Coupled Wave Functions 239

4.5 Recoupling Coefficients 241

4.6 Special Cases 245

4.6.1 Cubic graph geometry of the $6-j$ and $9-j$ coefficients 255

4.7 Concluding Remarks 259

5 The D^λ–Polynomials: Form 261

5.1 Overview 261

5.2 Defining Relations 268

5.2.1 Another proof of the multiplication rule 277

5.3 Restriction to Fewer Variables 278
5.4 Vector Space Aspects 280
5.5 Fundamental Structural Relations 283
5.5.1 Structural relations 283
5.5.2 Principal objectives and tasks 284

6 Operator Actions in Hilbert Space 285
6.1 Introductory Remarks 285
6.2 Action of Fundamental Shift Operators 286
6.3 Digraph Interpretation 295
6.4 Algebra of Shift Operators 301
6.5 Hilbert Space and D^λ–Polynomials 304
6.6 Shift Operator Polynomials 308
6.7 Kronecker Product Reduction 312
6.8 More on Explicit Operator Actions 318

7 The D^λ–Polynomials: Structure 325
7.1 The $C^\lambda(A)$ Matrices 325
7.2 Reduction of $D^p(Z) \otimes D^\lambda(Z)$ 329
7.3 Binary Tree Structure: C^λ–Coefficients 335
7.3.1 The D^λ–polynomials for partitions with two nonzero parts 342
7.3.2 Recurrence relation for the D^λ–polynomials . . . 351

8 The General Linear and Unitary Groups 355
8.1 Background and Review 355
8.2 $GL(n,\mathbb{C})$ and its Unitary Subgroup $U(n)$ 358
8.3 Commuting Hermitian Observables 368
8.3.1 The Gelfand invariants 368
8.4 Differential Operator Actions 370

8.5 Eigenvalues of the Gelfand Invariants 371
8.5.1 A new class of symmetric functions 373
8.5.2 The general eigenvalues of the Gelfand invariants 375

9 Tensor Operator Theory **379**
9.1 Introduction 379
9.1.1 A basis of irreducible tensor operators 382
9.2 Unit Tensor Operators 386
9.2.1 Summary of properties of unit tensor operators . . 389
9.2.2 Explicit unit tensor operators 392
9.3 Canonical Tensor Operators 394
9.3.1 Application of the factorization lemma to canonical tensor operators 399
9.3.2 Subgroup conditions and reduced matrix elements 403
9.4 Properties of Reduced Matrix Elements 408
9.4.1 Unit projective operators 410
9.4.2 The pattern calculus 415
9.4.3 Shift invariance of Kostka and Littlewood-Richardson numbers 421
9.5 The Unitary Group $U(3)$ 422
9.5.1 The $U(3)$ canonical tensor operators 422
9.6 The $U(3)$ Characteristic Null Spaces 425
9.7 The $U(3) : U(2)$ Unit Projective Operators 432
9.7.1 Coupling rules and Racah coefficients 433
9.7.2 Limit relations 440

10 Compendium A. Basic Algebraic Objects **447**
10.1 Groups 447
10.1.1 Group actions 449
10.2 Rings 451
10.2.1 Rings of polynomials 452

10.2.2 Vector spaces of polynomials 455
10.3 Abstract Hilbert Spaces 456
10.3.1 Inner product spaces 456
10.3.2 Linear operators 458
10.3.3 Inner products and linear operators 460
10.3.4 Orthonormalization methods 461
10.3.5 Matrix representations of linear operators 464
10.4 Properties of Matrices 465
10.4.1 Properties of normal matrices 467
10.4.2 Inner product on the space of complex matrices . . 468
10.4.3 Exponentiated matrices 469
10.4.4 Lie bracket polynomials 476
10.5 Tensor Product Spaces 486
10.6 Vector Spaces of Polynomials 488
10.7 Group Representations 494
10.7.1 Irreducible representations of groups 499
10.7.2 Schur's lemmas 501

11 Compendium B: Combinatorial Objects 505
11.1 Partitions and Tableaux 506
11.1.1 Restricted compositions 512
11.1.2 Linear ordering of sequences and partitions 513
11.2 Young Frames and Tableaux 514
11.2.1 Skew tableaux 520
11.3 Gelfand-Tsetlin Patterns 523
11.3.1 Combinatorial origin of Gelfand-Tsetlin patterns . 523
11.3.2 Group theoretical origin of Gelfand-Tsetlin patterns 528
11.3.3 Skew Gelfand-Tsetlin patterns 528
11.3.4 Notations 532

11.3.5 Triangular and skew Gelfand-Tsetlin patterns . . . 533
11.3.6 Words and lattice permutations 535
11.3.7 Lattice permutations and Littlewood-Richardson numbers 538
11.3.8 Kostka and Littlewood-Richardson numbers 550
11.4 Generating Functions and Relations 556
11.4.1 Notations 557
11.4.2 Counting relations 558
11.4.3 Generating relations of functions 559
11.4.4 Operator generated functions 561
11.5 Multivariable Special Functions 563
11.5.1 Solid harmonics 564
11.5.2 Double harmonic functions in $\mathbb{R}^3$ 565
11.5.3 Hypergeometric functions 565
11.5.4 MacMahon's master theorem 568
11.5.5 Power of a determinant 568
11.6 Symmetric Functions 570
11.6.1 Introduction 570
11.6.2 Four basic symmetric functions 571
11.6.3 Vandermonde determinants 572
11.6.4 Schur functions 573
11.6.5 Dual bases for symmetric functions 578
11.7 Sylvester's Identity 580
11.8 Derivation of Weyl's Dimension Formula 582
11.8.1 Vandermonde determinants, Bernoulli polynomials, and Weyl's formula 582
11.9 Other Topics 585
11.9.1 Alternating sign matrices 585
11.9.2 Binary trees, graphs, and digraphs 595
11.9.3 Umbral calculus and double tableau calculus . . . 595

11.9.4 Special functions 596
11.9.5 Other generalizations 596

Bibliography **597**

Index **611**

Notation

General symbols

$\mathbb{R}$	real numbers
$\mathbb{C}$	complex numbers
$\mathbb{P}$	positive numbers
$\mathbb{Z}$	integers
$\mathbb{N}$	nonnegative integers
$\mathbb{R}^n$	Cartesian $n-$space
$\mathbb{C}^n$	complex $n-$space
$\mathbb{E}^n$	Euclidean $n-$space
$O(n, R)$	group of real orthogonal matrices of order n
$SO(n, \mathbf{R})$	group of real, proper orthogonal matrices of order n
$U(n)$	group of unitary matrices of order n
$SU(n)$	group of unimodular unitary matrices of order n
$GL(n, \mathbf{C})$	group of complex nonsingular matrices of order n
$O(n, R)$	group of real orthogonal matrices of order n
$\mathbf{M}^p_{n\times n}(\alpha, \alpha')$	set of $n \times n$ matrix arrays with nonnegative elements with row-sum α and column-sum α'
$\times$	ordinary multiplication in split product, direct product
$\oplus$	direct sum of matrices
$\otimes$	tensor product, Kronecker product of matrices
$\mathbb{P}\mathrm{ar}_n$	set of partitions having n parts, including 0 as a part
$\delta_{i,j}$	the Kronecker delta for integers i, j
$\delta_{A,B}$, $\delta(A, B)$	the Kroneker delta for sets A and B
$K(\lambda, \alpha)$	the Kostka numbers
$c^\lambda_{\mu\,\nu}$	the Littlewood-Richardson numbers

Specialized symbols

Lists of specialized symbols are found on pp. 532–533, 538, 557; others are introduced as needed in the text.

Chapter 1

Quantum Theory of Angular Momentum: Introduction

1.1 Background and Viewpoint

1.1.1 Euclidean and Cartesian 3-space

It is a basic tenet of Newtonian physics that physical phenomena take place in Euclidean 3-space, denoted $\mathbb{E}^3$, which is a collection of entities called "points," a point being a primitive undefined entity. Such points are used in Euclidean geometry to construct other objects such as lines, planes, triangles, etc., from a set of axioms that constitute a deductive structure called Euclidean geometry. Euclidean geometry in 2-space and 3-space is the formalized mathematical method for deducing "facts" about the intuitive space of our surroundings and the objects perceived to be part of those surroundings. Lines, planes, and other objects are constructed as subsets of points in $\mathbb{E}^3$. But the primitive object "point" has no properties on its own; it is usually thought of as having no "size;" a point cannot be taken apart in terms of still other points. Euclidean geometry does not use the concept of "distance between points" to characterize its objects, but rather the notion of congruence. Mac Lane [128, p.127]) argues, however, that the notion of *geometric magnitude* is present in Euclid's geometry and that together with congruence and geometric ratios constitutes a geometrical description of the modern concept of the real-number line.

Euclid's geometry allows its objects to be "moved" around in the $\mathbb{E}^3$ "space" in the modern sense of rigid body motions that includes translations and rotations. But no explanation for the "cause" of such motions is given; they are intrinsic properties of the "ambient" $\mathbb{E}^3$-space (Mac Lane [128, p.76]).

Euclidean $\mathbb{E}^3$-space contains, as subsets, collections of three lines that are mutually perpendicular and have a common unique point of intersection. The operation of perpendicular projection of a point $p \in \mathbb{E}^3$ to a point $L(p)$ on a given line $L \subset \mathbb{E}^3$ is defined for all points and all lines. It is these notions of three perpendicular lines intersecting at a point, and the operation of perpendicular projection onto these lines that is used to describe physical phenomena in a concrete manner, based on the use of the real numbers, and the notion of Cartesian 3-space, denoted $\mathbb{R}^3$.

Cartesian 3-space $\mathbb{R}^3$ is the real linear vector space whose elements consist of the collection of all sequences $x = (x_1, x_2, x_3), x_i \in \mathbb{R}, i = 1, 2, 3$, where $\mathbb{R}$ is the set of real numbers, where, of course, all the usual operations of addition, multiplication by a scalar, and the distributive rules that define a vector space are applied to such sequences. We refer to a sequence of three real numbers $(x_1, x_2, x_3) \in \mathbb{R}^3$ as a coordinate point, and often as coordinates of a point in $\mathbb{R}^3$. Cartesian 3-space $\mathbb{R}^3$ comes equipped naturally with two functions defined on all pairs of points $x = (x_1, x_2, x_3)$ and $x' = (x'_1, x'_2, x'_3)$ belonging to $\mathbb{R}^3$, namely, a distance function, denoted $d(x, x')$, and a *scalar product* denoted, $x \cdot x'$, given by

$$d(x, x') = \sqrt{(x_1 - x'_1)^2 + (x_2 - x'_2)^2 + (x_3 - x'_3)^2}, \tag{1.1}$$

$$x \cdot x' = x_1 x'_1 + x_2 x'_2 + x_3 x'_3. \tag{1.2}$$

The space $\mathbb{R}^3$ contains subsets called lines, for example, the set of coordinate points $\{x_1, a_2, a_3) \,|\, x_1 \in \mathbb{R}\}$, for fixed real numbers a_1 and a_2, defines a line. In particular, the subsets of $\mathbb{R}^3$ defined by

$$\begin{aligned} \mathbb{L}_1 &= \{(x_1, 0, 0) | x_1 \in \mathbb{R}\}, \\ \mathbb{L}_2 &= \{(0, x_2, 0) | x_2 \in \mathbb{R}\}, \\ \mathbb{L}_3 &= \{(0, 0, x_3) | x_3 \in \mathbb{R}\}, \end{aligned} \tag{1.3}$$

determine three mutually perpendicular lines in $\mathbb{R}^3$, where perpendicular means the scalar product of points belonging to the separate lines is 0.

To relate the geometrical methods of Euclidean $\mathbb{E}^3-$space geometry over points to the vector space methods of the analytical geometry of Cartesian $\mathbb{R}^3-$space over the real numbers, it is convenient make the strong assumption (from Mac Lane's [128] observations) that each line $L \subset \mathbb{E}^3$ is identified with the real-number line $L(\mathbb{R}) = \{p(x) \,|\, x \in \mathbb{R}\}$, where it is the custom to identity the point $p(x)$ on the real-number line

$L(\mathbb{R})$ with the real number x itself: $p(x) \mapsto x$. Thus, let $q \in \mathbb{E}^3$, and let the three perpendicular lines $L_1(q), L_2(q), L_3(q) \subset \mathbb{E}^3$, respectively, have q as their (unique) intersection point. Identify each of these three perpendicular geometrical lines with the real-number line $L(\mathbb{R}) = \{x|x \in \mathbb{R}\}$, so that the geometrical intersection point q is identified with a point

$$q \mapsto (q_1, q_2, q_3) \in \mathbb{R}^3 = \{(x_1, x_2, x_3)|\text{each } x_i \in \mathbb{R}\}. \tag{1.4}$$

Thus, the three perpendicular lines $L_1(q), L_2(q), L_3(q)$ in $\mathbb{E}^3$ are identified with the three perpendicular lines in $\mathbb{R}^3$ given by

$$\begin{aligned} L_1(q) &\to L_1(\mathbb{R}^3) = \{(x_1(q), q_2, q_3)|x_1(q) \in \mathbb{R}\}, \\ L_2(q) &\to L_2(\mathbb{R}^3) = \{(q_1, x_2(q), q_3)|x_2(q) \in \mathbb{R}\}, \\ L_3(q) &\to L_3(\mathbb{R}^3) = \{(q_1, q_2, x_3(q))|x_3(q) \in \mathbb{R}\}. \end{aligned} \tag{1.5}$$

A geometrical point $p \in \mathbb{E}^3$, which is obtained by geometric projection onto the three perpendicular lines $L_1(q), L_2(q), L_3(q) \subset \mathbb{E}^3$ is now mapped to a general coordinate point $(x_1(p), x_2(p), x_3(p)) \in \mathbb{R}^3$ given by

$$p \mapsto (x_1(p), x_2(p), x_3(p)) = (x_1(q) + q_1, x_2(q) + q_2, x_3(q) + q_3). \tag{1.6}$$

The set of three perpendicular axes $L_1(\mathbb{R}^3), L_2(\mathbb{R}^3), L_3(\mathbb{R}^3) \subset \mathbb{R}^3$ is called a *set-of-axes* or a *reference frame* for the Cartesian $\mathbb{R}^3$ vector space. The point $q \mapsto (q_1, q_2, q_3) \in \mathbb{R}^3$ is called the *origin* of the Cartesian reference frame. Thus, the arithmetic description of points in Euclidean space $\mathbb{E}^3$ is effected by two related objects in Cartesian space $\mathbb{R}^3$: a reference frame and a set of coordinates relative to the reference frame. It is convenient to denote this description of $\mathbb{E}^3$ by $\mathbb{R}^3(q)$: The underlying ambient Euclidean 3-space $\mathbb{E}^3$ is unchanged by different choices of $q \in \mathbb{E}^3$; the Cartesian 3-space $\mathbb{R}^3(q)$, with origin located at $q \mapsto (q_1, q_2, q_3)$ is a *redescription* of the (unchanged) points of $\mathbb{E}^3$. Equivalently, the coordinate points corresponding to a coordinate frame located at any origin (a_1, a_2, a_3) are a redescription of the coordinate points corresponding to a coordinate frame located at any other origin (a'_1, a'_2, a'_3), each being a different coordinate presentation of the same collection of points in the underlying Euclidean space $\mathbb{E}^3$.

Discussions of the axioms of Euclidean geometry that suit our needs and give a more complete picture than the brief discussion given here can be found in Weyl [176, 177] and Mac Lane [128]. A discussion of the origin and foundations of these concepts from the point of view of cognitive scientists can be found in Lakoff and Núñez [100].

We complete our description of the relation between the spaces $\mathbb{E}^3$ and $\mathbb{R}^3$ by introducing the concept of a *geometric vector* as defined by physicists in terms of directed line segments (see, for example, Page [139]). This notion is well-suited to Euclidean 3-space with its point entities. A geometrical vector is a directed line segment in $\mathbb{E}^3$ (a line segment with one point called the origin and a second point called the terminal point of the vector). Two geometrical vectors are defined as equal if they lie along parallel lines and have the same directional sense and length (geometrical magnitude); addition is defined in terms of the familiar parallelogram rule; multiplication by a scalar (real numbers) is defined as a scaling of geometrical magnitude, including the possibility (negative number) of the reversal of direction; and the vector $\mathbf{0}$, called the zero vector, is the vector in which the origin coincides with the terminal point. The set of geometrical vectors satisfy the following two rules for all points $a, b \in \mathbf{E}^3$: (i) $\mathbf{x}(a \to a) = \mathbf{0}$; (ii) $\mathbf{x}(a \to b) + \mathbf{x}(b \to c) = \mathbf{x}(a \to c)$. The second property is called the *transitive* property of addition. Together with the first rule, it implies that $\mathbf{x}(a \to b) + \mathbf{x}(b \to a) = \mathbf{0}$, which gives the negative geometrical vector, $\mathbf{x}(b \to a) = -\mathbf{x}(a \to b)$.

These definitions, together with the concept of *parallel transport,* which allows for the "motion" of equal geometrical vectors necessary to bring equal vectors into coincidence so that all geometrical vectors can be added, determines a linear vector space of geometrical vectors that offers an alternative means of implementing the ideas set forth by Mac Lane [128, p.75]).

The linear space of geometric vectors with origin at point $q \in \mathbf{E}^3$ is denoted $\mathbf{V}^3(q)$ and is isomorphic to the vector space $\mathbb{R}^3(q)$. For q identified as the origin $(0,0,0)$, geometrical vectors are described in terms of the corresponding Cartesian 3-space as follows: Unit vectors $\mathbf{e}_1, \mathbf{e}_2, \mathbf{e}_3$ are defined to be the directed line segments of unit length going from the origin $(0,0,0)$ to the respective points $(1,0,0), (0,1,0), (0,0,1)$. The general vector $\mathbf{x} = x_1\mathbf{e}_1 + x_2\mathbf{e}_2 + x_3\mathbf{e}_3$ is then the directed line segment going from the origin $(0,0,0)$ to the coordinate point (x_1, x_2, x_3), and its length is $|\mathbf{x}| = \sqrt{x_1^2 + x_2^2 + x_3^2}$.

It is convenient to think of the isomorphic vector spaces $\mathbb{R}^3(q)$ and $\mathbf{V}^3(q)$ with origin at $q \mapsto (q_1, q_2, q_3)$ as superposed over the point $q \in \mathbb{E}^3$. This merger of "pictures" of vector spaces makes quite clear the one-to-one relation between vectors and coordinate points. In this sense, we speak of a triad of unit vectors along the coordinate axes as a *unit reference frame* , or simply as a reference frame. Thus, we may characterize the Euclidean 3-space corresponding to each point $q \in \mathbb{E}^3$ as consisting of the two sets of objects: a reference frame $(\mathbf{e}_1(q), \mathbf{e}_2(q)), \mathbf{e}_3(q))$ located at the origin $q \mapsto (q_1, q_2, q_3)$, and the set of coordinate points $(x_1(q), x_2(q), x_3(q))$ projected onto the corresponding coordinate axes with directional and orientational sense defined by these unit geometrical vectors. This picture allows us to speak of the Cartesian 3-space located at the Euclidean point q and having an arbitrary orientation of its reference frame in Euclidean space, thus simplifying the description. We do not, however, identify all the isomorphic vector spaces $\mathbb{R}^3(q), q \in \mathbb{E}^3$, since the notion of the redescription of Euclidean 3−space by different reference frames and coordinate points has important implications for the description of the behavior of physical systems. (The same statement applies to $\mathbf{V}^3(q)$.)

The family of isomorphic Cartesian 3-spaces $\mathbb{R}^3(q), q \in \mathbb{E}^3$ can be transformed into one another by geometrical congruence and other operations that leave the geometrical objects in $\mathbb{E}^3$ "unchanged." For physics, the meaning of unchanged is to be decided by experimental observation. Physical theory must in some sense be invariant under such transformations, since the different Cartesian 3-spaces $\mathbb{R}^3(q)$ are just a redescription of the properties of the objects in $\mathbb{E}^3$ constituting a physical system. But how a specified choice of reference frame is used to "measure" the properties of a given physical system, and how such measurements might influence these properties, are questions that, as yet, have no universal answers: the answers depend on additional assumptions. For example, in Newtonian physics, it is assumed that "equations of motion" govern the motion of an object against the background Euclidean space, and that the observation of an object and its motion have negligible (or accountable) influence on the object and no influence on the space. In nonrelativistic quantum theory, it is assumed that "equations of motion" still govern the behavior of an object against the background Euclidean space, but the new equations of motion are such that the concept of a "property" of an object and the "measurement" of that property, such as "point" particle and measurement of the location of the point, become interwoven in ways that dramatically negate their classical (Newtonian) meaning. While the equations of motion themselves are deterministic in their time evolution, the transformation of the "properties" of the system during the measurement process are subject to various interpretations that need not be continuous against the Euclidean background. Special and general relativity "correct" Newtonian relativity in a different context by recognizing that objects and space do not have mutually exclusive properties, but relate to one another in ways that require abandonment of the Euclidean space and its assigned universal Newtonian time. Deterministic motions are maintained in a broader measurement process with rods and clocks, but all this requires a completely new formulation of "equations of motion," which are at odds with the quantum equations of motion. A general theory that explains all the observed properties of microscopic objects, using the successful quantum theory, and all those of macroscopic objects, using Einstein's successful relativity theory, remains elusive. For a recent discussion, we refer to Leggett [103].

Our concern in Chapters 1-4 of this monograph is with the properties of the angular momentum of many-particle systems in nonrelativistic quantum mechanics, and, especially, the role of combinatorial concepts.

We do not concern ourselves with the puzzles of the interpretation of quantum mechanics, but rather with the development of the properties of angular momentum from the viewpoints described above. The properties of angular momentum from the perspective of coordinate changes attributed to a redescription of a physical system in Euclidean 3-space $\mathbb{E}^3$ by the use of different Cartesian 3-spaces $\mathbb{R}^3(q)$ leads to important properties of the quantal physical system. We concern ourselves with three types of redescriptions: translations, inversions, and rotations.

The first kind of redescription is a consequence of what is called a *translation* of reference frames: The reference frame $(\mathbf{e}_1, \mathbf{e}_2, \mathbf{e}_3)$ located at point $q \in \mathbb{E}^3$, which is the origin of the Cartesian 3-space $\mathbb{R}^3(q)$, is moved by parallel transport to the point $q' \in \mathbb{E}^3$ and becomes the reference frame $(\mathbf{e}'_1, \mathbf{e}'_2, \mathbf{e}'_3)$ at the origin of the Cartesian 3-space $\mathbb{R}^3(q')$. The description of one and the same point $p \in \mathbb{E}^3$ in terms of this pair of reference frames is best expressed in terms of the three geometric vectors $\mathbf{x}(q \to p), \mathbf{x}(q \to q')$, and $\mathbf{x}(q' \to p)$, which satisfy the general transitive rule of addition:

$$\mathbf{x}(q \to p) = \mathbf{x}(q \to q') + \mathbf{x}(q' \to p). \tag{1.7}$$

Since we can also write $\mathbf{x}(q \to q') = -\mathbf{x}(0 \to q) + \mathbf{x}(0 \to q')$, relation (1.7) can also be written in the invariant form:

$$\mathbf{x}(q \to p) + \mathbf{x}(0 \to q) = \mathbf{x}(q' \to p) + \mathbf{x}(0 \to q'). \tag{1.8}$$

Here $0 \in \mathbb{E}^3$ is an arbitrary point $0 \mapsto (0,0,0) \in \mathbb{R}^3(0)$. Relations (1.7) and (1.8) hold for arbitrary geometric vectors. It is the positioning of reference frames by parallel transport at the points $q, q' \in \mathbb{E}^3$ that gives the operation of translation, which, in turn, gives the relations between the coordinates (x_1, x_2, x_3) relative to the reference frame $(\mathbf{e}_1, \mathbf{e}_2, \mathbf{e}_3)$ and the coordinates (x'_1, x'_2, x'_3) relative to the reference frame $(\mathbf{e}'_1, \mathbf{e}'_2, \mathbf{e}'_3)$ of the same point $p \in \mathbb{E}^3$ as follows:

$$x_i + q_i = x'_i + q'_i, \; i = 1, 2, 3. \tag{1.9}$$

Thus, the translation (parallel transport) of reference frames from a point $q \in \mathbb{E}^3$ to a point $q' \in \mathbb{E}^3$ gives the redescription of coordinates of a point $p \in \mathbb{E}^3$ given by (1.8); this relation is called the translation of coordinates effected by a translation of coordinate frame.

The second kind of redescription is a consequence of what is called an *inversion* of the reference frame $(\mathbf{e}_1, \mathbf{e}_2, \mathbf{e}_3)$ located at the point $q \in \mathbb{E}^3$. An inversion of the frame is defined by the transformation

$$(\mathbf{e}_1, \mathbf{e}_2, \mathbf{e}_3) \mapsto (-\mathbf{e}_1, -\mathbf{e}_2, -\mathbf{e}_3). \tag{1.10}$$

The coordinates (x_1, x_2, x_3) of each point $p \in \mathbb{E}^3$ as assigned in the Cartesian 3-space $\mathbb{R}^3(q)$ correspondingly undergo the redescription

$$(x_1, x_2, x_3) \mapsto (-x_1, -x_2, -x_3). \tag{1.11}$$

Reflections through planes may also defined. For example, the reflection of the coordinate frame at point $q \in \mathbb{E}^3$ through the plane containing the unit vectors $(\mathbf{e}_1, \mathbf{e}_2)$ is defined by $(\mathbf{e}_1, \mathbf{e}_2, \mathbf{e}_3) \mapsto (\mathbf{e}_1, \mathbf{e}_2, -\mathbf{e}_3)$ and the corresponding redescription of the coordinates (x_1, x_2, x_3) of each point $p \in \mathbb{E}^3$ as assigned in the Cartesian 3-space $\mathbb{R}^3(q)$ is given by $(x_1, x_2, x_3) \mapsto (x_1, x_2, -x_3)$.

The third kind of redescription is a consequence of what is called a *rotation* of the reference frame $(\mathbf{e}_1, \mathbf{e}_2, \mathbf{e}_3)$ located at the point $q \in \mathbb{E}^3$. A rotation of the frame is defined by the linear transformation

$$\mathbf{e}_i \mapsto \mathbf{e}'_i = \sum_{j=1}^{3} R_{ij}\, \mathbf{e}_j,\ i = 1, 2, 3, \tag{1.12}$$

where the matrix $R = (R_{ij})_{1 \le i,j \le 3}$ with element R_{ij} in row i and column j is a real, proper, orthogonal matrix; that is, a matrix with real elements such that $R^T R = R R^T = I_3$, where T denotes matrix transposition, I_3 the unit matrix of order 3, and $\det R = 1$. The necessary and sufficient conditions that a general point $p \in \mathbb{E}^3$ with coordinates $(x_1, x_2, x_3) \in \mathbb{R}^3(q)$ is the redescription of the same point p with coordinates (x'_1, x'_2, x'_3) with respect to the rotated reference frame $(\mathbf{e}'_1, \mathbf{e}'_2, \mathbf{e}'_3)$ is expressed by the invariance of form given by

$$x_1\mathbf{e}_1 + x_2\mathbf{e}_2 + x_3\mathbf{e}_3 = x'_1\mathbf{e}'_1 + x'_2\mathbf{e}'_2 + x'_3\mathbf{e}'_3. \tag{1.13}$$

The rotation (1.12) of the reference frame $(\mathbf{e}_1, \mathbf{e}_2, \mathbf{e}_3)$ to the reference frame $(\mathbf{e}'_1, \mathbf{e}'_2, \mathbf{e}'_3)$ now gives the redescription of one and the same point $p \in \mathbb{E}^3$ given by

$$x_i \mapsto x'_i = \sum_{j=1}^{3} R_{ij}\, x_j,\ \ i = 1, 2, 3, \tag{1.14}$$

where (x_1, x_2, x_3) are the coordinates of p referred to the coordinate frame $(\mathbf{e}_1, \mathbf{e}_2, \mathbf{e}_3)$ and (x'_1, x'_2, x'_3) are the coordinates of p referred to the coordinate frame $(\mathbf{e}'_1, \mathbf{e}'_2, \mathbf{e}'_3)$, where both coordinate frames are located at q and used for the description of coordinate points of the Cartesian 3−space $\mathbb{R}^3(q)$. Relations (1.14) are necessary and sufficient conditions that the redescription of the point p is effected by a rotation of frames located at q. We will, unless otherwise noted, always choose coordinate frames such that the triad of unit vectors defining the axes are *right-handed* triads, as defined by the familiar right-handed screw rule. Left-handed frames can be included as well by the operation of inversion.

Mathematicians at the end of the nineteenth century realized that the operations of translation and rotation can be effected by the action of exponential differential operators (see Crofton [46], Glaisher [64]), which in vector notation are expressed by

$$e^{\boldsymbol{a}\cdot\boldsymbol{\nabla}}F(x_1,x_2,x_3) = F(x_1+a_1,x_2+a_2,x_3+a_3), \qquad (1.15)$$

$$e^{-\phi\mathbf{n}\cdot\mathbf{R}}F(x_1,x_2,x_3) = F(x'_1,x'_2,x'_3). \qquad (1.16)$$

The operator $\boldsymbol{a}\cdot\boldsymbol{\nabla}$, where $\boldsymbol{\nabla} = \mathbf{e}_1\partial/\partial x_1 + \mathbf{e}_2\partial/\partial x_2 + \mathbf{e}_3\partial/\partial x_3$ is called the *generator of the translation* $\mathbf{a} = a_1\mathbf{e}_1 + a_2\mathbf{e}_2 + a_3\mathbf{e}_3$. In the second relation, the components of $\mathbf{R} = R_1\mathbf{e}_1 + R_2\mathbf{e}_2 + R_3\mathbf{e}_3$ are the differential operators defined by

$$\begin{aligned} R_1 &= x_2\partial/\partial x_3 - x_3\partial/\partial x_2, \\ R_2 &= x_3\partial/\partial x_1 - x_1\partial/\partial x_3, \\ R_3 &= x_1\partial/\partial x_2 - x_2\partial/\partial x_1. \end{aligned} \qquad (1.17)$$

The coordinates (x'_1,x'_2,x'_3) are the redescription of a point $p\in\mathbb{E}^3$ in terms of the coordinates (x_1,x_2,x_3) of the same point p as described by the rotation of the frame $(\mathbf{e}_1,\mathbf{e}_2,\mathbf{e}_3)\mapsto(\mathbf{e}'_1,\mathbf{e}'_2,\mathbf{e}'_3)$ in the positive sense about the direction corresponding to the unit vector $\mathbf{n} = n_1\mathbf{e}_1 + n_2\mathbf{e}_2 + n_3\mathbf{e}_3$, $\mathbf{n}\cdot\mathbf{n} = 1$, where the matrix elements $R_{ij} = R_{ij}(\phi,\mathbf{n})$ in (1.12) and (1.14) are those of

$$\begin{aligned} R(\phi,\mathbf{n}) &= e^{\phi N} = I_3 + N\sin\phi + N^2(1-\cos\phi), \\ N &= \begin{pmatrix} 0 & -n_3 & n_2 \\ n_3 & 0 & -n_1 \\ -n_2 & n_1 & 0 \end{pmatrix},\ N^3 = -N. \end{aligned} \qquad (1.18)$$

The proper, real, orthogonal matrix R has here been parametrized by the components of the unit vector $\mathbf{n}$ and the angle $0\le\phi<2\pi$. The gradient operator $\boldsymbol{\nabla}$ and the rotation operator $\mathbf{R} = R_1\mathbf{e}_1 + R_2\mathbf{e}_2 + R_3\mathbf{e}_3$ both satisfy the form invariant relation (1.13). The operator $-\phi\mathbf{n}\cdot\mathbf{R}$ is called the *generator of the rotation* about direction $\mathbf{n}$ by angle ϕ.

We anticipate some results from the brief discussion of quantum theory given below, and observe that relations (1.15)-(1.17) are precursors to similar transformations that were to occur later in quantum theory under the association of classical position $\mathbf{x}$ and linear momentum $\mathbf{p}$ to operators by the rule $\mathbf{x}\mapsto\mathbf{x}$ and $\mathbf{p}/\hbar\mapsto -i\boldsymbol{\nabla}$, which, when applied to the classical angular momentum $\mathbf{L} = \mathbf{x}\times\mathbf{p}$ of a single particle located at position $\mathbf{x}$ with linear momentum $\mathbf{p}$ relative to the origin of a coordinate frame $(\mathbf{e}_1, \mathbf{e}_2, \mathbf{e}_3)$, gives $\mathbf{L}/\hbar = -i\mathbf{R} = -i\mathbf{x}\times\boldsymbol{\nabla}$. Classical linear momentum and angular momentum of a point particle are identified in their quantum mechanical interpretation as the generators of translations and rotations, as described above. The founders of quantum mechanics rediscovered relations (1.15) and (1.16) in the guise of the unitary operators $\exp(i\boldsymbol{a}\cdot\boldsymbol{p}/\hbar)$ and $\exp(-i\phi\mathbf{n}\cdot\mathbf{L})/\hbar$ acting in the Hilbert space of states of a physical system. From the definition of the quantum angular momentum, $\mathbf{L}/\hbar = -i\mathbf{x}\times\boldsymbol{\nabla}$, the operations of translation, inversion, and rotation effect the following transformations of the quantal angular momentum:

$$\begin{aligned} \text{translation}: &\quad \mathbf{L}\mapsto\mathbf{L} - i\hbar\mathbf{a}\times\boldsymbol{\nabla},\\ \text{inversion}: &\quad \mathbf{L}\mapsto\mathbf{L},\\ \text{rotation}: &\quad \mathbf{L}\mapsto\mathbf{L}. \end{aligned} \tag{1.19}$$

1.1.2 Newtonian physics

A physical system is said to be isolated if it has no influence on its surroundings, and conversely. Such ideal systems do not exist in nature, but much of the progress in physics can be attributed to the approximate validity of the concept of an isolated physical system.

In Newtonian physics, the Euclidean 3-space $\mathbb{E}^3$ is taken as the background against which the changes in an isolated physical system take place: the space is considered to be void of (isolated from) all other physical objects, homogeneous (sameness at every point), and isotropic (sameness in every direction). The physical system occupies a collection (subset) of points in $\mathbb{E}^3$, which may be a single point, but this collection of points can *change* relative to the fixed Euclidean background and the fixed reference frames used to assign coordinates to the points of $\mathbb{E}^3$. The measure of this change requires a new concept, that of *time,* which itself is assumed to have any value on the real-line (ignoring units). The change of the configuration of points defining the physical system with the time parameter is called *motion,* where it is assumed that motion is described by increasing values of time. The change in time itself is measured by *clocks,* which themselves occupy points of the background Euclidean space $\mathbb{E}^3$, but which advance their readings in a uniform manner, independently of their own motion, as measured by still other clocks: There exists a universal time t for measuring all motions of physical systems against the fixed Euclidean background of space; this motion is governed by Newton's equations of motion, which determine how the coordinate points of the physical system change with time. Thus, enters the concept of *velocity,* and the recognition that the equations of motions must be form invariant with respect to the class of reference frames moving with constant velocity relative to one other. Such classes of frames are called *inertial frames,* and the invariance of the equations of motion is called the principle of Newtonian or Galilean relativity. The detailed mathematical description of all this is effected though the use of the Cartesian space $\mathbb{R}^3$ and its collection of reference frames and associated coordinate points. Thus, the reference frame $(\mathbf{e}_1, \mathbf{e}_2, \mathbf{e}_3)$ located at some point in $\mathbb{E}^3$ is now taken as a Newtonian reference frame that has the dual role of assigning coordinates to all the points of $\mathbb{R}^3$ and of assigning coordinates to the subset of points occupied by the physical system at each time t. Newton's equation of motion are identical for all reference frames moving with constant velocity $\mathbf{v}$ with respect to this chosen frame. The collection of all reference frames as parametrized by $\mathbf{v}$ have an important role in identifying the Galilean group as the invariance group of Newton's equations of motion.

There is another class of reference frames called *accelerated frames.* Such frames are often attached to part of the physical system or to other moving points. Accelerating reference frames are called *noninertial frames;* they are often used to simplify the description of the internal motions of a complex physical system, but such frames are on a different footing—they provide convenient transformations of coordinates that modify the form of Newton's equations and simplify the descriptions of the motions, as adapted to special situations.

1.1.3 Nonrelativistic quantum physics

In nonrelativistic quantum physics, we retain the Newtonian notion of an isolated physical system, as well as the Cartesian space $\mathbb{R}^3$ and the set of Newtonian inertial frames, and the relationship between frames and points in $\mathbb{R}^3$, just as described above. Now, however, the coordinates ascribed to the points occupied by the physical system do not enjoy a deterministic motion against the space $\mathbb{R}^3$ as is the case for the Newtonian equations of motion of its points. The classical dual role of reference frames is lost. The role of frames as a means of assigning coordinates to the points of $\mathbb{R}^3$ against which the properties of the system are to be measured is retained, but the equation of motion in quantum mechanics, the Schrödinger equation, does not allow the classical concept of point particle to propagate in time: the concept of point particle must be modified. The role of coordinates of the points occupied by the classical physical system becomes that of parameters that belong to the *domain of definition* of a function Ψ with complex values. Time is still Newton's universal time, which also now becomes a parameter belonging to the domain of definition of Ψ. The function Ψ with complex values $\Psi(\mathbf{X}, t)$, $\mathbf{X} = (\mathbf{x}^{(1)}, \mathbf{x}^{(2)}, \ldots, \mathbf{x}^{(n)})$, for a classical system consisting of n point particles with geometric position vectors $\mathbf{x}^{(1)}, \mathbf{x}^{(2)}, \ldots, \mathbf{x}^{(n)}$, is called the *wavefunction* or the *state function* of the system. The state function of the physical system, together with *a set of rules* that give the possible outcomes of the measurements on the system, determine the observable properties of the system. The Schrödinger equation is such that if the state function $\Psi(\mathbf{X}_0, t_0)$ of a physical system is known for some position configuration $\mathbf{X}_0$ at time t_0, it evolves in a unitary deterministic fashion to the state function $\Psi(\mathbf{X}, t)$ of the physical system at time t. The meaning of coordinates and time, however, are to be inferred from the rules of measurement: The mathematical properties of the equations of motion play back into the very meaning of the properties ascribed to the objects themselves; the classical concept of point particle is lost. Indeed, the quantum object called a point particle in its classical description exhibits in its quantum description the property of being "partially present" at every point in $\mathbb{R}^3$, in such a way that a measurement of its position at any arbitrary point $\mathbf{x}$ finds the object in its entirety at that point with some probability, as determined by its state function, and the object always possesses its mass, charge, and spin with certainty. Despite the holistic aspect of its position, the object can be counted in a measurement and appears as a single entity.

One of the most significant mathematical properties of the equations of motions is their *linear property,* as expressed by the *superposition of state functions:* If $\Psi(\mathbf{X}, t)$ and $\Phi(\mathbf{X}, t)$ are two solutions at time t, then so also is the linear combination

$$\alpha\Psi(\mathbf{X}, t) + \beta\Phi(\mathbf{X}, t), \tag{1.20}$$

where α and β are arbitrary complex numbers. It is this superposition property, together with the property of complex numbers given by

$$\begin{aligned}|\alpha\Psi(\mathbf{X}, t) + \beta\Phi(\mathbf{X}, t)|^2 &= |\alpha|^2|\Psi(\mathbf{X}, t)^2 + |\beta|^2|\Phi(\mathbf{X}, t)|^2 \\ &+\alpha\beta^*\Psi(\mathbf{X}, t)\Phi^*(\mathbf{X}, t) + \alpha^*\beta\Psi^*(\mathbf{X}, t)\Phi(\mathbf{X}, t),\end{aligned} \tag{1.21}$$

that underlies many nonintuitive behaviors of quantum systems. The quantum-mechanical description corresponding to the classical system of many point particles, ascribes a holistic existence to the system in which each part seems to relate to every other part, even when the parts are noninteracting at the time of measurement, provided they were interactive in the past, and provided the system has not been previously measured. The "properties" of the system becomes entangled in intricate ways with those of a measuring apparatus that is introduced into the space $\mathbb{R}^3$ in an interactive mode with the system to determine the "value" of a given quantity associated with the originally isolated system. This, in turn, leads to fundamental questions about how the state function of the originally isolated physical system and that of the apparatus determine the measured properties of the original system. The properties of the measuring apparatus are presumably subject to the rules of quantum theory, which, in turn, now apply to the interacting composite whole—the original isolated system and the measuring device. This new system then has its own collection of state functions governing the properties of the composite whole. How all of this is to be sorted out to obtain information about the original system is the *problem of measurement* (see Leggett [103]). There is no generally agreed on resolution of the problem of measurement.

The early standard book on the foundations of quantum mechanics is by Von Neumann [174] with a more modern exposition by Mackey [127]. Bohm [27] challenged the standard interpretation by an ingenious example, and Bell's [9] critical analysis reopened the entire subject of measurement. The large book by Wheeler and Zurek [179] reviews the history through 1980-81, with many informative comments on the Problem of Measurement and a superb list of references. Recent comments by Griffiths [71] illustrate a popular viewpoint. The possibility of quantum computers has intensified the interest in these problems and new experiments are confirming the reality of the counterintuitive quantum world. These important problems are, however, not the subject of this monograph, which makes its presentations in the standard interpretation with a focus on combinatorial aspects.

There are, fortunately, general invariance properties of the Schrödinger equation and its solution state functions for a complex composite physical systems that can be used to classify the quantum states of physical systems into substates available to the system.

Our focus here is on the properties of the total angular momentum of a physical system, which is a quantity $\mathbf{L}$ that has a vector expression $\mathbf{L} = L_1\mathbf{e}_1 + L_2\mathbf{e}_2 + L_3\mathbf{e}_3$ in the right-handed frame $(\mathbf{e}_1, \mathbf{e}_2, \mathbf{e}_3)$ and the expression $\mathbf{L}' = L'_1\mathbf{e}'_1 + L'_2\mathbf{e}'_2 + L'_3\mathbf{e}'_3$ in a second rotated right-handed frame $(\mathbf{e}'_1, \mathbf{e}'_2, \mathbf{e}'_3)$. At a given instant of time, necessarily $\mathbf{L} = \mathbf{L}'$, since these quantities are just redescriptions of the total angular momentum of the system at a given time. The total angular momentum is a conserved quantity; that is, $d\mathbf{L}/dt = \mathbf{0}$, for all time t, and this property makes the total angular momentum an important quantity for the study of the behavior of complex physical systems. For a system of n point particles, the total angular momentum relative to the origin of the reference frame $(\mathbf{e}_1, \mathbf{e}_2, \mathbf{e}_3)$ is obtained by vector addition of that of the individual particles by $\mathbf{L} = \sum_{i=1}^{n} \mathbf{L}_i$, where $\mathbf{L}_i$ is expressed by the vector cross product $\mathbf{L}_i = \mathbf{x}_i \times \mathbf{p}_i$ in terms of the vector position $\mathbf{x}_i = x_{1i}\mathbf{e}_1 + x_{2i}\mathbf{e}_2 + x_{3i}\mathbf{e}_3$ and the vector linear momentum $\mathbf{p}_i = p_{1i}\mathbf{e}_1 + p_{2i}\mathbf{e}_2 + p_{3i}\mathbf{e}_3$ of the particle labeled i. While angular momentum can be exchanged between interacting particles, the total angular momentum remains constant in time for an isolated physical system of n particles. The quantum-mechanical operator interpretation of such classical physical quantities is obtained by Schrödinger's rule $\mathbf{p}_i \mapsto -\mathrm{i}\hbar\boldsymbol{\nabla}_i, \hbar = h/2\pi$, where h is Planck's constant. The reference frame vectors $(\mathbf{e}_1, \mathbf{e}_2, \mathbf{e}_3)$ remain intact.

The viewpoints of Newtonian physics and quantum physics may be contrasted in many ways. From the viewpoint of angular momentum, for example, the classical angular momentum $\mathbf{L} = \mathbf{x} \times \mathbf{p}$ is associated with the actual **motion** of a particle, while in quantum theory, as noted above, it becomes the generator of rotations of the state vector that describes the properties of the object called "particle" (active viewpoint), or, in the viewpoint we have advanced, the generator of rotations of the coordinate frame (passive interpretation) that gives a redescription of the state vector that describes the properties of the particle. Neither the active nor passive viewpoint implies an actual "motion;" it is more in accord with the concept of congruence in Euclidean geometry. An even more contrasting feature is that the object called the electron is regarded as a "point particle," and whereas points in Euclidean space have no intrinsic properties, this point object, the electron, has an "internal" angular-momentum-like property called "spin," as we next describe.

Internal degrees of freedom

The discovery of objects having internal degrees of freedom going beyond motions in Newtonian space-time led to the introduction of the concept of the *spin* of a object (see Pauli [140]), even though the object may exhibit no measurable spatial extension. For example, the electron, the carrier of electrical conduction in metals under the influence of an electric field, was already recognized in 1897 (see Ref. [167]) to be one of the principal constituents of atoms, requires an intrinsic property called *spin,* to account for the observed properties of the emission and absorption of radiation by atoms. A single electron is characterized not only as being a point structure with a fixed charge and mass, but also by having a fixed spin $\frac{1}{2}$. As a structure at a single point, the electron is assigned a position $x = (x_1, x_2, x_3) \in \mathbb{R}^3$ at time t, which in quantum theory become the domain of definition of its space-time wavefunction $\psi(x,t)$; as a structure with spin, it is assigned a "state vector" $|\frac{1}{2}\rangle$ belonging to a Hilbert space $\mathcal{H}_{\frac{1}{2}}$ of dimension 2, which is assumed to be spanned by a pair of orthonormal basis vectors that are written $|\frac{1}{2},\frac{1}{2}\rangle$ and $|\frac{1}{2},-\frac{1}{2}\rangle$ in the Dirac bra-ket notation, with orthonormality relations $\langle\frac{1}{2},\mu\,|\,\frac{1}{2},\nu\rangle = \delta_{\mu,\nu}$, the inner product being unspecified beyond these properties. The general state vector of the electron in the spin space $\mathcal{H}_{\frac{1}{2}}$ is given by the ket-vector $|\frac{1}{2}\rangle = \alpha|\frac{1}{2},\frac{1}{2}\rangle + \beta|\frac{1}{2},-\frac{1}{2}\rangle$, where α and β are complex coefficients. The corresponding bra-vector in the dual space is then given by $\langle\frac{1}{2}| = \alpha^*\langle\frac{1}{2},\frac{1}{2}| + \beta^*\langle\frac{1}{2},-\frac{1}{2}|$ with inner product $\langle\frac{1}{2}|\frac{1}{2}\rangle = |\alpha|^2 + |\beta|^2$. In this conceptualization of a point particle with intrinsic spin $\frac{1}{2}$, its full state vector belongs to the tensor product space $\mathcal{H}\otimes\mathcal{H}_{\frac{1}{2}}$, which has state vectors of the form $\Psi_{\frac{1}{2}} = \psi\otimes|\frac{1}{2}\rangle$, where ψ is a wavefunction in the usual sense, depending only on the spatial coordinates. The action of a frame rotation $R : (\mathbf{e}_1, \mathbf{e}_2, \mathbf{e}_3) \mapsto (\mathbf{e}'_1, \mathbf{e}'_2, \mathbf{e}'_3)$, where $\mathbf{e}'_i = \sum_{j=1}^3 R_{ij}\mathbf{e}_j, i = 1, 2, 3,$ is not only to effect the transformation to new spatial coordinates of the particles given by $(x_1, x_2, x_3) \mapsto (x'_1, x'_2, x'_3)$, where $x'_i = \sum_{j=1}^3 R_{ij}x_j, i = 1, 2, 3$, but also to effect the transformation of the spin-space ket-vector basis of $\mathcal{H}_{\frac{1}{2}}$ by

$$\begin{aligned} |\tfrac{1}{2}, \tfrac{1}{2}\rangle &\mapsto |\tfrac{1}{2}, \tfrac{1}{2}\rangle' = u_{11}(R)|\tfrac{1}{2}, \tfrac{1}{2}\rangle + u_{21}(R)|\tfrac{1}{2}, -\tfrac{1}{2}\rangle, \\ |\tfrac{1}{2}, -\tfrac{1}{2}\rangle &\mapsto |\tfrac{1}{2}, -\tfrac{1}{2}\rangle' = u_{12}(R)|\tfrac{1}{2}, \tfrac{1}{2}\rangle + u_{22}(R)|\tfrac{1}{2}, -\tfrac{1}{2}\rangle. \end{aligned} \tag{1.22}$$

The action of the frame rotation R on the components (α, β) of the ket-vector $|\frac{1}{2}\rangle = \alpha|\frac{1}{2}, \frac{1}{2}\rangle + \beta|\frac{1}{2}, -\frac{1}{2}\rangle$ giving a general spin state is given by

$$\begin{aligned} \alpha &\mapsto \alpha' = u^*_{11}(R)\alpha + u^*_{21}(R)\beta, \\ \beta &\mapsto \beta' = u^*_{12}(R)\alpha + u^*_{22}(R)\beta, \end{aligned} \tag{1.23}$$

where

$$U(R) = \begin{pmatrix} u_{11}(R) & u_{12}(R) \\ u_{21}(R) & u_{22}(R) \end{pmatrix} \in SU(2) \tag{1.24}$$

is an element of the group of 2×2 unitary unimodular matrices $SU(2)$ that depends on the frame rotation matrix $R \in SO(3, \mathbb{R})$. This pair of transformations of basis vectors and components leaves invariant the relation

$$\alpha|\tfrac{1}{2}, \tfrac{1}{2}\rangle + \beta|\tfrac{1}{2}, -\tfrac{1}{2}\rangle = \alpha'|\tfrac{1}{2}, \tfrac{1}{2}\rangle' + \beta'|\tfrac{1}{2}, \tfrac{1}{2}\rangle', \tag{1.25}$$

as required for a redescription of the state vector of the spin by a rotation of the reference frame. (For a history of the discovery of the electron and its properties, see Whittaker [180] and The Physical Review: The First Hundred Years [167].

Generally, in the one-particle case above, we have in mind a single particle with an internal spin that is a constituent of a larger isolated physical system, with all constituents described in $\mathbb{R}^3$, that provides the interactions that act on the "marked" single particle. We often use Newtonian metaphors to identify this object, but its properties are to be inferred from the those of its quantum-mechanical state vector $\Psi = \psi \otimes |\frac{1}{2}\rangle$. Newtonian space-time still occurs in the domain of definition of the wavefunction ψ. It is also possible to assign "internal coordinates" to the new degrees of freedom of the internal space that describes the spin, as discussed below in (1.58).

It is somewhat remarkable that the symmetry group $SU(2)$ actually can be used to unify the action of frame rotations $R \in SO(3, \mathbb{R})$ on both points in $\mathbb{R}^3$ and on points used to describe the internal degrees of freedom associated with spin. From this viewpoint, the group $SU(2)$ is the basic group for the study of angular momentum of complex systems with spin. We show this by using the method of Cartan [35].

The method of Cartan

In the Cartan method (see also Wigner [181] and Biedenharn and Louck [21]), a point $x = (x_1, x_2, x_3) \in \mathbb{R}^3$ is presented as the entries in a 2×2 traceless, Hermitian matrix of the form:

$$H(x) = \begin{pmatrix} x_3 & x_1 - ix_2 \\ x_1 + ix_2 & -x_3 \end{pmatrix} = \sum_{i=1}^{3} x_i \sigma_i, \tag{1.26}$$

where the σ_i, $i = 1, 2, 3$, are the Pauli matrices:

$$\sigma_1 = \begin{pmatrix} 0 & 1 \\ 1 & 0 \end{pmatrix}, \sigma_2 = \begin{pmatrix} 0 & -i \\ i & 0 \end{pmatrix}, \sigma_3 = \begin{pmatrix} 1 & 0 \\ 0 & -1 \end{pmatrix}. \tag{1.27}$$

Thus, each point $x \in \mathbb{R}^3$ is mapped to a traceless Hermitian matrix: $x \mapsto H(x)$. Conversely, given a traceless Hermitian matrix H, it is mapped to a point $x \in \mathbb{R}^3$ by the rule $x_i = \frac{1}{2}\mathrm{Tr}(\sigma_i H)$, where $\mathrm{Tr}A$ denotes the trace of a matrix A. In obtaining this result, we use the following multiplication rules of the Pauli matrices:

$$\sigma_i^2 = I_2 = \sigma_0,\ i = 1, 2, 3;\ \sigma_1\sigma_2 = i\sigma_3,\ \sigma_2\sigma_3 = i\sigma_1,\ \sigma_3\sigma_1 = i\sigma_2. \tag{1.28}$$

Thus, with these conventions for the placement of the components x_i into matrices, the set of points $\mathbb{R}^3$ is bijective with the set of matrices:

$$\mathbb{H} = \{H \mid H \text{ is a } 2 \times 2 \text{ traceless Hermitian matrix}\}. \tag{1.29}$$

Traceless Hermitian matrices are mapped into traceless Hermitian matrices by unitary similarity transformations. Moreover, the determinant of a Hermitian matrix is invariant under such transformations. Accordingly, it must be the case that, for each $U \in SU(2)$, the transformation $x \mapsto x'$ defined by

$$U \begin{pmatrix} x_3 & x_1 - ix_2 \\ x_1 + ix_2 & -x_3 \end{pmatrix} U^\dagger = \begin{pmatrix} x_3' & x_1' - ix_2' \\ x_1' + ix_2' & -x_3' \end{pmatrix}, \tag{1.30}$$

is a real orthogonal transformation, since the determinant of this transformation is $-(x_1^2 + x_2^2 + x_3^2) = -(x_1'^2 + x_2'^2 + x_3'^2)$. The explicit transformation may be worked out to be

$$x_i' = \sum_{j=1}^{3} R_{ij}(U) x_j,\ R_{ij}(U) = \frac{1}{2}\mathrm{Tr}(\sigma_i U \sigma_j U^\dagger). \tag{1.31}$$

The proof that $\det R(U) = 1$ takes more effort, but is correct. We also note the easily proved property $(R(U))^T = R(U^\dagger) = R(U^{-1}) =$

$(R(U))^{-1}$. Also, because we have a homomorphism of groups, the multiplication property $R(U')R(U) = R(U'U)$ holds for all pairs $U', U \in SU(2)$.

Given $R \in SO(3,\mathbb{R})$, there are exactly two solutions $U(R)$ and $-U(R)$ that solve relation (1.30). The solution $U(R)$ is given by

$$\begin{aligned} U(R) &= \begin{pmatrix} \alpha_0 - i\alpha_3 & -i\alpha_1 - \alpha_2 \\ -i\alpha_1 + \alpha_2 & \alpha_0 + i\alpha_3 \end{pmatrix} \\ &= \alpha_0\sigma_0 - i\boldsymbol{\alpha}\cdot\boldsymbol{\sigma}, \end{aligned} \tag{1.32}$$

where the parameters $(\alpha_0, \boldsymbol{\alpha}) = (\alpha_0, \alpha_1, \alpha_2, \alpha_3)$ are given in terms of the elements R_{ij} of R by

$$\begin{aligned} &\alpha_0 = (1 + \mathrm{Tr}R)/d,\ \alpha_1 = (R_{32} - R_{23})/d, \\ &\alpha_2 = (R_{13} - R_{31})/d,\ \alpha_3 = (R_{21} - R_{12})/d, \\ &d = 2\sqrt{1 + \mathrm{Tr}R},\ \text{for } \mathrm{Tr}R \neq -1; \end{aligned} \tag{1.33}$$

$$\begin{aligned} &\alpha_0 = 0,\ \alpha_1 = \sqrt{(1 + R_{11})/2},\ \alpha_2 = s_2\sqrt{(1 + R_{22})/2}, \\ &\alpha_3 = s_3\sqrt{(1 + R_{33})/2},\ \text{for } \mathrm{Tr}R = -1, \end{aligned} \tag{1.34}$$

where, for $\mathrm{Tr}R = -1$, we have the following definitions: $s_2 = \mathrm{sign}(R_{12})$, $s_3 = \mathrm{sign}(R_{13})$, where $\mathrm{sign}(x)$ denotes the sign of a real number x with $\mathrm{sign}(0) = +$. The condition that $\det R = 1$ requires that the parameters $(\alpha_0, \boldsymbol{\alpha})$ constitute a point on the unit sphere $\mathbb{S}^3 \subset \mathbb{R}^4$ in Cartesian 4–space; that is, $\alpha_0^2 + \alpha_1^2 + \alpha_2^2 + \alpha_3^2 = 1$. Relation (1.32) then gives $\det U(R) = 1$; that is, $U(R)$ is unimodular for all $R \in SO(3,\mathbb{R})$. A real orthogonal matrix R with $\mathrm{Tr}R = -1$ is also a symmetric matrix, hence, $R^2 = I_3$, and the unitary unimodular matrix U is traceless and antihermitian; that is, $\mathrm{Tr}U = 0$ and $U^\dagger = -U$.

It is also useful to give the expression for $R(U)$ for $U = U(\alpha_0, \boldsymbol{\alpha})$, where $(\alpha_0, \boldsymbol{\alpha}) \in \mathbb{S}^3$:

$$R(\alpha_0, \boldsymbol{\alpha}) = \tag{1.35}$$
$$\begin{pmatrix} \alpha_0^2 + \alpha_1^2 - \alpha_2^2 - \alpha_3^2 & 2\alpha_1\alpha_2 - 2\alpha_0\alpha_3 & 2\alpha_1\alpha_3 + 2\alpha_0\alpha_2 \\ 2\alpha_1\alpha_2 + 2\alpha_0\alpha_3 & \alpha_0^2 + \alpha_2^2 - \alpha_3^2 - \alpha_1^2 & 2\alpha_2\alpha_3 - 2\alpha_0\alpha_1 \\ 2\alpha_1\alpha_3 - 2\alpha_0\alpha_2 & 2\alpha_2\alpha_3 + 2\alpha_0\alpha_1 & \alpha_0^2 + \alpha_3^2 - \alpha_1^2 - \alpha_2^2 \end{pmatrix}.$$

The parametrizations (1.32) and (1.35) of matrices $U \in SU(2)$ and $R \in SO(3,\mathbb{R})$ in terms of points on the unit sphere $\mathbb{S}^3$ are very useful for obtaining other parametrizations of these groups simply by parametrizing the points on the unit sphere $\mathbb{S}^3$, as we give below.

The multiplication of two unitary unimodular matrices in terms of these parameters is expressed by the quaternionic multiplication rule

$$U(\alpha_0, \boldsymbol{\alpha})U(\alpha_0', \boldsymbol{\alpha}') = U(\alpha_0'', \boldsymbol{\alpha}''), \tag{1.36}$$

$$\begin{aligned}(\alpha_0, \boldsymbol{\alpha})(\alpha_0', \boldsymbol{\alpha}') &= (\alpha_0'', \boldsymbol{\alpha}'')\\ &= \left(\alpha_0\alpha_0' - \boldsymbol{\alpha}\cdot\boldsymbol{\alpha}', \alpha_0\boldsymbol{\alpha}' + \alpha_0'\boldsymbol{\alpha} + \boldsymbol{\alpha}\times\boldsymbol{\alpha}'\right).\end{aligned} \tag{1.37}$$

These same relations hold, of course, upon replacing U by R.

We also note the following results for unitary matrices. The group $U(2)$ of unitary matrices is given in terms of the group of unitary unimodular matrices $SU(2)$ by

$$U(2) = \left\{U_\phi = e^{i\phi}U \mid U \in SU(2), 0 \leq \phi < 2\pi\right\}. \tag{1.38}$$

The 2–to–1 homomorphism (1.31) of $SU(2)$ to $SO(3,\mathbb{R})$ now defines an ∞–to–1 homomorphism of $U(2)$ to $SO(3,\mathbb{R})$: *Every unitary matrix* $U_\phi = e^{i\phi}U, 0 \leq \phi < 2\pi$, *corresponds to the same proper orthogonal matrix* $R(U)$.

1.1.4 Unitary frame rotations

Using the 2–to–1 homomorphism $R_{ij}(U) = \frac{1}{2}\mathrm{Tr}(\sigma_i U \sigma_j U^\dagger)$ between the groups $SU(2)$ and $SO(3,\mathbb{R})$, we can now describe in greater detail the redescription of a particle with spin 1/2 that is effected by a frame rotation, where we note that $(R(U))^T = R(U^\dagger)$. Using Cartan's method, the entire process can be described in terms of the unitary unimodular group $SU(2)$ and its action on the various sets that enter into the description of the particle and its quantum-mechanical states, which we summarize as follows:

1. Rotation action of $SU(2)$ on the set $\mathbb{F}$ of right-handed frames:

$$(\mathbf{e}_1, \mathbf{e}_2, \mathbf{e}_3) \mapsto (\mathbf{e}_1', \mathbf{e}_2', \mathbf{e}_3'), \tag{1.39}$$

$$\begin{pmatrix} \mathbf{e}_3' & \mathbf{e}_1' - i\mathbf{e}_2' \\ \mathbf{e}_1' + i\mathbf{e}_2' & -\mathbf{e}_3' \end{pmatrix} = U \begin{pmatrix} \mathbf{e}_3 & \mathbf{e}_1 - i\mathbf{e}_2 \\ \mathbf{e}_1 + i\mathbf{e}_2 & -\mathbf{e}_3 \end{pmatrix} U^\dagger, \tag{1.40}$$

$$\mathbf{e}_i' = \sum_{j=1}^{3} R_{ij}(U)\mathbf{e}_j, \ i = 1, 2, 3. \tag{1.41}$$

2. Redescription of coordinates effected by a frame rotation under the action of $SU(2)$:

$$(x_1, x_2, x_3) \mapsto (x'_1, x'_2, x'_3), \tag{1.42}$$

$$\begin{pmatrix} x'_3 & x'_1 - ix'_2 \\ x'_1 + ix'_2 & -x'_3 \end{pmatrix} = U \begin{pmatrix} x_3 & x_1 - ix_2 \\ x_1 + ix_2 & -x_3 \end{pmatrix} U^\dagger, \tag{1.43}$$

$$x'_i = \sum_{j=1}^{3} R_{ij}(U) x_j, \; i = 1, 2, 3. \tag{1.44}$$

3. Action of each $U \in SU(2)$ on the Hilbert space $\mathcal{H}$ of spatial wavefunctions in the tensor product space $\mathcal{H} \otimes \mathcal{H}_{\frac{1}{2}}$:

$$\mathcal{T}_U \psi = \psi', \; (\mathcal{T}_U \psi)(x) = \psi'(x'), \; \text{each } \psi \in \mathcal{H}, \tag{1.45}$$

$$\begin{pmatrix} x'_1 \\ x'_2 \\ x'_3 \end{pmatrix} = (R(U))^T \begin{pmatrix} x_1 \\ x_2 \\ x_3 \end{pmatrix}. \tag{1.46}$$

The coordinate transformation (1.44) between column matrices may be expressed in terms of group action as $(x'_1, x'_2, x'_3) = A_U(x_1, x_2, x_3)$, where $A_U x_i = \sum_j R_{ij}(U) x_j$, each $U \in SU(2)$. The group action (1.45) in function space is given by (1.45)-(1.46) in which the coordinate transformation uses $(R(U))^T$. This is not an error; it is dictated by the requirement that the action $A_g, g \in G$, of a group G on a set Y is to satisfy, by definition, for each pair of group elements $g', g \in G$, the product rule $A_{g'}(A_g y) = A_{g'g} y$, each $y \in Y$. This rule may be verified directly for A_U acting in the set of all coordinate points in accordance with (1.44), and for T_U acting in the set of all functions in the Hilbert space $\mathcal{H}$ in accordance with (1.45)-(1.46).

4. Action of each $U \in SU(2)$ on the basis of the spin space $\mathcal{H}_{\frac{1}{2}}$:

$$|\tfrac{1}{2}, \tfrac{1}{2}\rangle \mapsto |\tfrac{1}{2}, \tfrac{1}{2}\rangle' = \mathcal{S}_U |\tfrac{1}{2}, \tfrac{1}{2}\rangle = u_{11} |\tfrac{1}{2}, \tfrac{1}{2}\rangle + u_{21} |\tfrac{1}{2}, -\tfrac{1}{2}\rangle, \tag{1.47}$$

$$|\tfrac{1}{2}, -\tfrac{1}{2}\rangle \mapsto |\tfrac{1}{2}, -\tfrac{1}{2}\rangle' = \mathcal{S}_U |\tfrac{1}{2}, -\tfrac{1}{2}\rangle = u_{12} |\tfrac{1}{2}, \tfrac{1}{2}\rangle + u_{22} |\tfrac{1}{2}, -\tfrac{1}{2}\rangle. \tag{1.48}$$

Again, the product action rule $T_{U'}(T_U |\tfrac{1}{2}, \pm\tfrac{1}{2}\rangle) = T_{U'U} |\tfrac{1}{2}, \pm\tfrac{1}{2}\rangle$ can be verified.

5. Action of each $U \in SU(2)$ on a state vector $\Psi \in \mathcal{H} \otimes \mathcal{H}_{\frac{1}{2}}$:

$$\begin{aligned} T_U &= \mathcal{T}_U \otimes \mathcal{S}_U : \mathcal{H} \otimes \mathcal{H}_{\frac{1}{2}} \to \mathcal{H} \otimes \mathcal{H}_{\frac{1}{2}}, \\ T_U \Psi &= \mathcal{T}_U \psi \otimes \mathcal{S}_U |\tfrac{1}{2}\rangle, \\ |\tfrac{1}{2}\rangle &= \alpha |\tfrac{1}{2}, \tfrac{1}{2}\rangle + \beta |\tfrac{1}{2}, -\tfrac{1}{2}\rangle. \end{aligned} \tag{1.49}$$

6. Action of each $U \in SU(2)$ on a state vector $\Psi \in \mathcal{H} \otimes \mathcal{H}_j$:

$$\begin{aligned} T_U &= \mathcal{T}_U \otimes \mathcal{S}_U : \mathcal{H} \otimes \mathcal{H}_j \to \mathcal{H} \otimes \mathcal{H}_j, \\ T_U \Psi &= \mathcal{T}_U \psi \otimes \mathcal{S}_U |j\rangle, \\ |j\rangle &= \sum_m \alpha_{jm} |jm\rangle, \; \alpha_{jm} \in \mathbb{C}, \end{aligned} \tag{1.50}$$

$$\mathcal{S}_U |j\, m\rangle = \sum_{m'} D^j_{m'\, m}(U) |j\, m'\rangle. \tag{1.51}$$

In order to summarize relations (1.50)-(1.51) in one place, we have anticipated from Sect. 1.2 below the following notation for an orthonormal basis of the spin space $\mathcal{H}_j$ of an object of internal spin j given by the standard Dirac ket notation:

$$|j\, m\rangle,\ m = j, j-1, \ldots, -j;\ j \in \{0, 1/2, 1, 3/2, \ldots, \}. \tag{1.52}$$

Under the action of $\mathcal{S}_U$ these basis vectors undergo the transformation (1.51), where the functions $D^j_{m'\, m}(U)$ give a unitary matrix representation of order $2j+1$ of $SU(2)$. The notation $D^j_{m'\, m}(U)$ in this transformation denotes that these functions are defined over the elements $u_{ij}, 1 \leq i, j \leq 2$, of $U \in SU(2)$, and not over matrices U. These functions are arranged into a matrix of order $2j+1$ by the convention of enumerating the rows and columns in the order $m' = j, j-1, \ldots, -j$, as read across the columns from left-to-right and $m = j, j-1, \ldots, -j$, as read down the rows from top-to-bottom. The matrices $D^j(U)$ are then a unitary matrix representation of $SU(2)$; that is, they satisfy

$$D^j(U')D^j(U) = D^j(U'U) \text{ and } D^j(U)(D^j(U))^\dagger = I_{2j+1}, \tag{1.53}$$

for all pairs $U', U \in SU(2)$.

The group action (1.40) of $SU(2)$ on reference frames in Cartesian space $\mathbb{R}^3$ assigns the unitary group $SU(2)$ the **primary role** in the redescription of quantum states under the redescription of $\mathbb{R}^3$ by $SU(2)$ frame rotations, which we henceforth call simply *unitary frame rotations.*

Realizations of spin space

We give two explicit realizations of spin space. In the first realization, $\mathcal{H}_j$ is replaced by unit column vectors $\mathbb{C}_{2j+1}$ of complex numbers (row vectors could also be used). Thus, we simply make the replacement of abstract basis vectors by unit column vectors of length $2j+1$ as given by

$$|j\,m\rangle \mapsto \mathbf{s}_{j\,m} = \mathrm{col}(0\ldots,0,1,0,\ldots,0), \tag{1.54}$$

where the single 1 appears in row $j-m+1, m=j,j-1,\ldots,-j$. Thus, we have that the tensor product space is realized by $\mathcal{H}\otimes\mathbb{C}_{2j+1}$ with state vectors given by

$$\mathbf{\Psi}_j(x) = \begin{pmatrix} \psi_{j,j}(x) \\ \psi_{j,j-1}(x) \\ \vdots \\ \psi_{j,-j}(x) \end{pmatrix} = \sum_m \psi_{j\,m}(x)\mathbf{s}_{j\,m}. \tag{1.55}$$

Under the unitary rotations of frame given by (1.41), this tensor product space undergoes the transformation

$$\begin{aligned} T_U \begin{pmatrix} \psi_{j,j}(x) \\ \psi_{j,j-1}(x) \\ \vdots \\ \psi_{j,-j}(x) \end{pmatrix} &= D^j(U) \begin{pmatrix} (\mathcal{T}_U\psi_{j,j})(x) \\ (\mathcal{T}_U\psi_{j,j-1})(x) \\ \vdots \\ (\mathcal{T}_U\psi_{j,-j})(x) \end{pmatrix} \\ &= D^j(U) \begin{pmatrix} \psi_{j,j}(x') \\ \psi_{j,j-1}(x') \\ \vdots \\ \psi_{j,\,-j}(x') \end{pmatrix}, \end{aligned} \tag{1.56}$$

where the coordinate transformation $(x_1,x_2,x_2) \mapsto (x_1',x_2',x_2')$ is still given by

$$x_i' = \sum_{k=1}^{3} R_{ik}(U)x_k,\ i=1,2,3. \tag{1.57}$$

In the second realization of the abstract spin space, $\mathcal{H}_j$ is replaced by the polynomial space $\mathcal{P}_j$ of polynomials of degree $2j$ defined over two complex variables (z_1,z_2), and the abstract basis vector $|j\,m\rangle$ is taken to be the polynomial with values given by

$$\langle z_1,z_2|j\,m\rangle \mapsto P_{j\,m}(z_1,z_2) = \frac{z_1^{j+m}z_2^{j-m}}{\sqrt{(j+m)!(j-m)!}}, \tag{1.58}$$

for $m = j, j-1, \ldots . -j$. In this formulation, under the unitary frame rotation given by (1.41), the underlying spin-space coordinates (z_1, z_2) undergo the redescription $(z_1, z_2) \mapsto (z'_1, z'_2)$ given by

$$\begin{aligned} z'_1 &= u_{11}z_1 + u_{12}z_2, \\ z'_2 &= u_{21}z_1 + u_{22}z_2, \end{aligned} \quad U = (u_{ij})_{1 \le i \le j \le 2} \in SU(2), \tag{1.59}$$

while the basis polynomials undergo the redescription given by

$$(\mathcal{S}_U P_{j\,m})(z_1, z_2) = P_{j\,m}(z'_1, z'_2) = \sum_{m'} D^j_{m'\,m}(U) P_{j\,m'}(z_1, z_2), \tag{1.60}$$

in which the new coordinates are given in column matrix form by $\mathrm{col}(z'_1, z'_2) = U^T \mathrm{col}(z_1, z_2)$. (Compare with (1.44) and (1.46).)

The transformation coefficients in relation (1.60) are given by (see van der Waerden [171] and Ref. [21]):

$$\begin{aligned} D^j_{m'\,m}(U) &= \sqrt{(j+m')!(j-m')!(j+m)!(j-m)!} \\ &\times \sum_s \frac{(u_{11})^{j+m-s}(u_{12})^{m'-m+s}(u_{21})^s(u_{22})^{j-m'-s}}{(j+m-s)!(m'-m+s)!s!(j-m'-s)!}, \end{aligned} \tag{1.61}$$

where the summation is over all nonnegative values of s for which all factorials in the denominator are nonnegative.

The column matrix formulation is less specific about the character of internal space, which is not directly accessible to measurement, and is, perhaps, to be preferred. Mathematically, the two finite vector spaces of polynomials and unit vectors of length $2j+1$ described above are isomorphic, and it is no restriction to use the *spinor* polynomials $P_{j\,m}(z_1, z_2), m = j, j-1, \ldots, -j$, as basis vectors. Indeed, these polynomials have an inner product (,) such that $(P_{j\,m'}, P_{j\,m}) = \delta_{m'\,m}$, one-to-one with that of the unit vectors $\mathbf{s}_{j\,m}, m = j, j-1, \ldots, -j$, as discussed below in Sect. 1.3.1. There are many advantages to following this approach, especially from the vantage point of combinatorics. Moreover, it is well known that the transformation properties of the spinor basis functions already gives all inequivalent irreducible representations of $SU(2)$, as given above in relation (1.61), a result that we will prove later. It is important to recognize that the use of either of these special realizations of spin space in no way restricts the general theory of spin space, since they do not introduce constraining conditions.

Many-particle systems

The above description of a single particle in $\mathbb{R}^3$ of spin j identified as a constituent of an isolated physical system is easily generalized to a physical systems S in which n identical particles in $\mathbb{R}^3$, each of spin j, are constituents, all of which may be interacting. The $k-$th particle is assigned the points $x^{(k)} = (x_{1k}, x_{2k}, x_{3k}) \in \mathbb{R}^3, k = 1, 2, \ldots, n$ relative to the Newtonian frame $(\mathbf{e_1}, \mathbf{e_2}, \mathbf{e_3})$ used for the description of a single particle above. We use spinor basis functions for the description of the spin states. It is convenient now to present the spatial coordinates and spin coordinates as the following $3 \times n$ matrix X and the $2 \times n$ matrix Z defined by

$$X = \begin{pmatrix} x_{11} & x_{12} & \cdots & x_{1n} \\ x_{21} & x_{22} & \cdots & x_{2n} \\ x_{31} & x_{32} & \cdots & x_{3n} \end{pmatrix}, \tag{1.62}$$

$$Z = \begin{pmatrix} z_{11} & z_{12} & \cdots & z_{1n} \\ z_{21} & z_{22} & \cdots & z_{2n} \end{pmatrix}, \tag{1.63}$$

in which column k of these matrices gives the spatial coordinates $x^{(k)} = (x_{1k}, x_{2k}, x_{3k})$ and the spin coordinates $z^{(k)} = (z_{1k}, z_{2k})$ of the $k-$th particle. These coordinates then become the domain of definition of the state vector $\Psi \in \mathcal{H}$ of the physical system with values given by $\Psi(X, Z)$. The Hilbert space $\mathcal{H}$ is the $n-$fold tensor product space of the tensor product spaces $\mathcal{H}^{(k)} \otimes \mathcal{S}_j^{(k)}$ of each of the particles of spin j, which we present in the form

$$\mathcal{H} = (\mathcal{H}^{(1)} \otimes \mathcal{H}^{(2)} \otimes \cdots \otimes \mathcal{H}^{(n)}) \otimes (\mathcal{S}_j^{(1)} \otimes \mathcal{S}_j^{(2)} \otimes \cdots \otimes \mathcal{S}_j^{(n)}). \tag{1.64}$$

The value of each $\Psi \in \mathcal{H}$ is denoted $\Psi(X, Z)$ for each point $X \in \mathbb{R}^3 \otimes \mathbb{R}^3 \otimes \cdots \otimes \mathbb{R}^3$ and each point $Z \in \mathbb{C}^2 \otimes \mathbb{C}^2 \otimes \cdots \otimes \mathbb{C}^2$.

The properties of this $n-$particle system under the action of $SU(2)$ are summarized as follows:

1. Action of each $U \in SU(2)$ on the reference frame: This action is still given by

$$(\mathbf{e}_1, \mathbf{e}_2, \mathbf{e}_3) \mapsto (\mathbf{e}'_1, \mathbf{e}'_2, \mathbf{e}'_3), \tag{1.65}$$

$$\begin{pmatrix} \mathbf{e}'_3 & \mathbf{e}'_1 - i\mathbf{e}'_2 \\ \mathbf{e}'_1 + i\mathbf{e}'_2 & -\mathbf{e}'_3 \end{pmatrix} = U \begin{pmatrix} \mathbf{e}_3 & \mathbf{e}_1 - i\mathbf{e}_2 \\ \mathbf{e}_1 + i\mathbf{e}_2 & -\mathbf{e}_3 \end{pmatrix} U^\dagger \tag{1.66}$$

$$\mathbf{e}'_i = \sum_{l=1}^{3} R_{il}(U)\mathbf{e}_l, \ i = 1, 2, 3. \tag{1.67}$$

2. Action of each $U \in SU(2)$ on the state vector space $\mathcal{H}$: The state vector $\Psi \in \mathcal{H}$ undergoes the transformation $\Psi \to T_U\Psi$ given by

$$(T_U\Psi)(X, Z) = \Psi\left((R(U))^T X, U^T Z\right). \tag{1.68}$$

This formulation not only makes transparent the *left action* of the group $SU(2)$, but also clearly invites the possibility of further transformations of the state vector by using *right transformations* $X \mapsto XY$ and $Z \mapsto ZY$ by an arbitrary matrix Y of order n. Thus, for example, if we choose $Y = P_\pi$ to be a permutation matrix, such transformations permute the spatial and spin coordinates of the particles. Here P_π is any one of the $n!$ matrices obtained by a permutation of the columns of the identity matrix I_n (permutation matrices). Moreover, there is also the possibility of doing right transformations by choosing $Y \in U(n)$, the group of $n \times n$ unitary matrices: *The occurrence of the general unitary group $U(n)$ in the classification of the state vectors of $n-$particle systems is always implicit.*

The theory of angular momentum arises naturally from the above presentation of unitary frame rotations. Let us first show this for the case of a single particle with spin j for which we have

$$X = \begin{pmatrix} x_1 \\ x_2 \\ x_3 \end{pmatrix}, \ Z = \begin{pmatrix} z_1 \\ z_2 \end{pmatrix}. \tag{1.69}$$

The *generators* of a group are defined in terms of abelian subgroups. For the group $SU(2)$, the generator of the subgroup $SU(2;\phi,\mathbf{n}) \subset SU(2)$ defined by

$$SU(2;\phi,\mathbf{n}) = \{U(\phi,\mathbf{n}) = \exp(-i\phi\mathbf{n}\cdot\boldsymbol{\sigma}/2)\,|\,0 \le \phi < 2\pi\} \tag{1.70}$$

of a frame rotation about a fixed direction $\mathbf{n}$ is defined by

$$\frac{1}{2}\mathbf{n}\cdot\boldsymbol{\sigma} = i\frac{d}{d\phi}U(\phi,\mathbf{n})\Big|_{\phi=0}. \tag{1.71}$$

The generator $\mathbf{n}\cdot\mathbf{J} = n_1J_1 + n_2J_2 + n_3J_3$ of the transformation of the state vector $\Psi \mapsto T_{U(\phi,\mathbf{n})}\Psi$ corresponding to the redescription of the state vector under the unitary frame transformation (1.67) is defined in analogy to (1.71) by

$$((\mathbf{n}\cdot\mathbf{J})\Psi)(X,Z) = \left(i\frac{d}{d\phi}(T_{U(\phi,\mathbf{n})}\Psi)(X,Z)\right)_{\phi=0}$$

$$= i\frac{d}{d\phi}\Psi(X(\phi,\mathbf{n}), Z(\phi,\mathbf{n}))\Big|_{\phi=0}, \tag{1.72}$$

$$X(\phi,\mathbf{n}) = (R(\phi,\mathbf{n}))^T X,\ \ Z(\phi,\mathbf{n}) = (U(\phi,\mathbf{n}))^T Z. \tag{1.73}$$

The unitary unimodular matrix $U(\phi,\mathbf{n})$ is given explicitly by

$$U(\phi,\mathbf{n}) = \exp(-i\phi\mathbf{n}\cdot\boldsymbol{\sigma}/2) \tag{1.74}$$
$$= \begin{pmatrix} \cos(\phi/2) - in_3\sin(\phi/2) & (-in_1 - n_2)\sin(\phi/2) \\ (-in_1 + n_2)\sin(\phi/2) & \cos(\phi/2) + in_3\sin(\phi/2) \end{pmatrix}.$$

The real, proper, orthogonal matrix $R(\phi,\mathbf{n}) = R(U(\phi,\mathbf{n}))$ is obtained by setting $\alpha_0 = \cos(\phi/2)$ and $\boldsymbol{\alpha} = \boldsymbol{n}\sin(\phi/2)$ in relation (1.32). We now carry out the differentiation $d/d\phi$ in (1.72), using the chain rule from calculus, and the relations

$$i\frac{d}{d\phi}(U(\phi,\mathbf{n}))^T|_{\phi=0} = \frac{1}{2}\begin{pmatrix} n_3 & n_1 + in_2 \\ n_1 - in_2 & -n_3 \end{pmatrix}, \tag{1.75}$$

$$i\frac{d}{d\phi}(R(\phi,\mathbf{n}))^T|_{\phi=0} = i\begin{pmatrix} 0 & n_3 & -n_2 \\ -n_3 & 0 & n_1 \\ n_2 & -n_1 & 0 \end{pmatrix}. \tag{1.76}$$

We thus obtain the following results: The generator

$$\mathbf{n}\cdot\mathbf{J} = n_1J_1 + n_2J_2 + n_3J_3, \tag{1.77}$$

which acts on functions $\Psi \in \mathcal{H}$, is given in terms of the differential operator $\mathbf{n}\cdot\boldsymbol{\mathcal{J}}$, which acts in the set of values $\{\Psi(X,Z)\,|\,X \in \mathbb{R}^3; Z \in \mathbb{C}^2\}$ by the following relations:

$$((\mathbf{n}\cdot\mathbf{J})\Psi)(X,Z) = (\mathbf{n}\cdot\boldsymbol{\mathcal{J}})\Psi(\mathrm{X},\mathrm{Z}), \tag{1.78}$$

$$\mathbf{J} = \mathbf{L} + \mathbf{S},\ \ \boldsymbol{\mathcal{J}} = \boldsymbol{\mathcal{L}} + \boldsymbol{\mathcal{S}}, \tag{1.79}$$

where $\boldsymbol{\mathcal{L}}$ and $\boldsymbol{\mathcal{S}}$ are the differential operators defined by

$$\begin{aligned} \mathcal{L}_1 &= -i(x_2\partial/\partial x_3 - x_3\partial/\partial x_2), \\ \mathcal{L}_2 &= -i(x_3\partial/\partial x_1 - x_1\partial/\partial x_3), \\ \mathcal{L}_3 &= -i(x_1\partial/\partial x_2 - x_2\partial/\partial x_2); \end{aligned} \tag{1.80}$$

$$\begin{aligned} \mathcal{S}_1 &= (z_1\partial/\partial z_2 + z_2\partial/\partial z_1)/2, \\ \mathcal{S}_2 &= -i(z_1\partial/\partial z_2 - z_2\partial/\partial z_1)/2, \\ \mathcal{S}_1 &= (z_1\partial/\partial z_1 - z_2\partial/\partial z_2)/2. \end{aligned} \tag{1.81}$$

Here we use the notation

$$(T\Psi)(q) = \mathcal{T}\Psi(q), \tag{1.82}$$

where Ψ is an element of some function space $\mathcal{F}$ on which an operator T acts, $T : \Psi \mapsto \Psi' = T\Psi$; q is a point in the domain of definition $\mathbb{Q}$ of the functions in $\mathcal{F}$; and $\mathcal{T}$ is the differential operator, acting in the space of values of the functions in $\mathcal{F}$, that defines the values of $(T\Psi)(q)$. In the context at hand, such a relation applies to the operator pairs $(J_i, \mathcal{J}_i), (L_i, \mathcal{L}_i), (S_i, \mathcal{S}_i)$, where these are the generator pairs corresponding to $\mathbf{n} = \mathbf{e}_i, i = 1, 2, 3$. Similar relations apply to linear combinations of these linear operator. In particular, we have that the matrix generator $\mathbf{n} \cdot \boldsymbol{\sigma}/2$ of the unitary matrix representation $\exp(-i\phi\mathbf{n} \cdot \boldsymbol{\sigma}/2)$ of order 2 of $SU(2)$ is realized in the space of functions by

$$(T_{U(\phi,\mathbf{n})}\Psi)(X, Z) = \left(e^{-i\phi\mathbf{n}\cdot\mathbf{J}}\Psi\right)(X, Z) = e^{-i\phi\mathbf{n}\cdot\boldsymbol{\mathcal{J}}}\Psi(X, Z). \tag{1.83}$$

Although the differential operators $\mathcal{L}_i$ and $\mathcal{S}_i$ are quite different in structural form, they satisfy identical commutation relations, where the commutator $[A, B]$ of two operator A and B acting in a Hilbert space is defined by $[A, B] = AB - BA$:

$$[\mathcal{L}_1, \mathcal{L}_2] = i\mathcal{L}_3,\ [\mathcal{L}_2, \mathcal{L}_3] = i\mathcal{L}_1,\ [\mathcal{L}_3, \mathcal{L}_1] = i\mathcal{L}_2, \tag{1.84}$$

$$[\mathcal{S}_1, \mathcal{S}_2] = i\mathcal{S}_3,\ [\mathcal{S}_2, \mathcal{S}_3] = i\mathcal{S}_1,\ [\mathcal{S}_3, \mathcal{S}_1] = i\mathcal{S}_2, \tag{1.85}$$

$$[\mathcal{L}_i, \mathcal{S}_j] = 0,\ i, j = 1, 2, 3. \tag{1.86}$$

The operators L_i and S_i, with action in the space of state vectors, satisfy, of course, these same commutation relations. The important observation is: There can be many realizations of these commutation relations by quite different operators, but as we shall see in Sect. 1.2 below, under certain assumptions, they all give rise to the same matrix realizations.

Let us next relate the mathematical quantities given above to the physical quantity called angular momentum. The classical angular momentum $\mathbf{L}$ of a point particle relative to a frame $(\mathbf{e}_1, \mathbf{e}_2, \mathbf{e}_3)$ is defined by the cross product $\mathbf{L} = \mathbf{x}\times\mathbf{p}$, where $\mathbf{p}$ is the linear momentum of the particle. The quantum angular momentum $\boldsymbol{\mathcal{L}}$ is obtained from the classical quantity by making the replacement $\mathbf{p} \mapsto -i\hbar\boldsymbol{\nabla}$, where $\hbar$ is the Planck constant $h/2\pi$, and $\boldsymbol{\nabla}$ is the Laplace operator with components $(\partial/\partial x_1, \partial/\partial x_2, \partial/\partial x_3)$. Thus, the components of the dimensionless angular momentum operators $\boldsymbol{\mathcal{L}}/\hbar = -i\mathbf{x}\times\boldsymbol{\nabla}$ are exactly the generators of rotations given by relations (1.80). There is, of course, no such rule for obtaining the spin operators (1.81). As remarked earlier, spin angular momentum can either be incorporated into the quantum theory in terms of an unspecified abstract finite-dimensional Hilbert space or by using realizations of this space that do not restrict the mathematical content of the theory, as we have done above.

The generalizations of the above one-particle results to n particles is immediate: One simply copies relations (1.79)-(1.81) n times, adjoining an extra index k to designate the $k-$th particle with spin j. The infinitesimal operators associated with each particle are clearly additive, so we have that the generator of the transformation associated with the redescription of the state vector resulting from a unitary frame rotation by angle ϕ about direction $\mathbf{n}$ is given by $\mathbf{n}\cdot\mathbf{J}$ and have the properties summarized by the following relations:

$$\boldsymbol{\mathcal{J}} = \boldsymbol{\mathcal{L}} + \boldsymbol{\mathcal{S}}, \tag{1.87}$$

$$\boldsymbol{\mathcal{J}} = \sum_{k=1}^{n} \boldsymbol{\mathcal{J}}^{(k)}, \tag{1.88}$$

$$\boldsymbol{\mathcal{J}}^{(k)} = \boldsymbol{\mathcal{L}}^{(k)} + \boldsymbol{\mathcal{S}}^{(k)}; \tag{1.89}$$

$$\boldsymbol{\mathcal{L}}^{(k)} = -i\mathbf{x}^{(k)} \times \boldsymbol{\nabla}^{(\mathrm{k})},\ \mathbf{x}^{(\mathrm{k})} = \sum_{i=1}^{3} x_{ik}\mathbf{e}_i; \tag{1.90}$$

$$\begin{aligned} S_1^{(k)} &= (z_{1k}\partial/\partial z_{2k} + z_{2k}\partial/\partial z_{1k})/2, \\ S_2^{(k)} &= -i(z_{1k}\partial/\partial z_{2k} - z_{2k}\partial/\partial z_{1k})/2, \\ S_3^{(k)} &= (z_{1k}\partial/\partial z_{1k} - z_{2k}\partial/\partial z_{2k})/2. \end{aligned} \tag{1.91}$$

These operators are just n replicas of relations (1.79)-(1.81) with commutation relations that are k replicas of relations (1.84)-(1.86) with superscript (k) adjoined, where now operators having distinct values of particle index k all mutually commute. Relation (1.78) also holds, of course, in the interpretation as operators acting in the state space:

$$\left((\mathbf{n}\cdot\mathbf{J}^{(k)})\Psi\right)(X,Z) = (\mathbf{n}\cdot\boldsymbol{\mathcal{J}}^{(k)})\Psi(\mathrm{X},\mathrm{Z}), \tag{1.92}$$

with similar relations for $\mathbf{n}\cdot\mathbf{L}^{(k)}$ and $\mathbf{n}\cdot\mathbf{S}^{(k)}$. In particular, we still have the commutation relations for the total angular momentum:

$$[J_1, J_2] = iJ_3,\ [J_1, J_2] = iJ_3,\ [J_1, J_2] = iJ_3 \tag{1.93}$$

and the associated transformations of the state vectors:

$$(T_{U(\phi,\mathbf{n})}\Psi)(X,Z) = \left(e^{-i\phi\mathbf{n}\cdot\mathbf{J}}\Psi\right)(X,Z) = e^{-i\phi\mathbf{n}\cdot\boldsymbol{\mathcal{J}}}\Psi(X,Z). \tag{1.94}$$

This relatively simple $n-$particle physical system embodies the full content of angular momentum theory. The simplicity of these additive relations does not reveal the intricacies of the theory.

Remark. Each $U \in SU(2)$ effects a rotation $R(U)$ of the reference frame of a physical system in $\mathbb{R}^3$ as given by relations (1.65)-(1.67).

It may seem mysterious that a unitary rotation of the reference frame should effect also a unitary transformation of the internal coordinates that describe spin. This is so because a *redescription* of the quantum state corresponding to the reassignment of coordinates

$$X' = R(U)X, \; Z' = UZ, \tag{1.95}$$

as effected by a unitary frame rotation, treats the spin coordinates on a par with the spatial coordinates. It is this relation that defines the action of each $U \in SU(2)$ to be that of an operator T_U acting in the Hilbert space $\mathcal{H}$ of states by the rule $(T_U\Psi)(X,Z) = \Psi\left((R(U))^T X, U^T Z\right)$. We note that we have the option of redefining the redescription transformation (1.95) by replacing U by U^*, so that the action of the operator T_U on state vectors would be $(T_U\Psi)(X,Z) = \Psi\left((R(U^*))^T X, U^\dagger Z\right)$. But this would lead to replacing the matrices $D^j(U)$ defined in relation (1.16) by $D^j(U^*)$. This is an acceptable procedure, but we have **elected** to use the redescription given by (1.95).

The system described above leads us below to investigate abstractly all possible realizations of operators satisfying the commutation relations (1.93), subject to certain conditions, together with the problem of adding two or more angular momenta, each of which has three components that satisfy these commutation relations, and the components of the separate angular momenta mutually commute. This is the problem of constructing representations of two or more copies of the Lie algebra of a group.

The state vector $\Psi \in \mathcal{H}$ of every physical system in the space $\mathbb{R}^3$, with or without spin, possesses $SU(2)$ unitary symmetry; that is, for each $U \in SU(2)$, we have for each $\Psi \in \mathcal{H}$ that $T_U\Psi = \Psi' \in \mathcal{H}$, since this transformation in the space of states available to the system is just the redescription of $\mathcal{H}$ corresponding to the redescription of the physical system induced by a unitary frame rotation. Ideally, the state vector space is a separable Hilbert space, and T_U, each $U \in SU(2)$, is a unitary operator with respect to the inner product on $\mathcal{H}$. This is equivalent to requiring that the three generators $J_i, i = 1,2,3$, corresponding to rotations about the three basis vector $\mathbf{e}_i, i = 1,2,3$, are Hermitian operators. Such a separable Hilbert space $\mathcal{H}$ is said to possess $SU(2)$ symmetry.

A separable Hilbert space having $SU(2)$ symmetry can always be decomposed into a direct sum of various subspaces $\mathcal{H}_{SU(2)}$ on which the action of T_U, each $U \in SU(2)$, is irreducible, which means that $T_U : \mathcal{H}_{SU(2)} \mapsto \mathcal{H}_{SU(2)}$, each $U \in SU(2)$, and there exists no subspace of $\mathcal{H}_{SU(2)}$ with this property. The invariance of such subspaces $\mathcal{H}_{SU(2)}$ under the unitary action of each T_U is, however, insufficient, in general, to determine the decomposition of $\mathcal{H}$ into its irreducible $SU(2)$ subspaces. For an $n-$particle system in $\mathbb{R}^3$, there are $3n - 3$ degrees of spatial classical degrees of freedom, after accounting for the three degrees of freedom associated with the three independent translations in $\mathbb{R}^3$. There may also be internal nonclassical degrees of freedom associated with each particle. Each complex physical system must be considered on its own to determine those degrees of freedom that are significant for the explanation of various phenomena exhibited by the system. In some model cases, a collection of independent, mutually commuting Hermitian operators

$$\mathbb{H} = \{H_1, H_2, \ldots\} \tag{1.96}$$

can be found such that $[T_U, H_k] = 0, k = 1, 2, \ldots,$ for each $SU(2)$ frame rotation. Such a set of mutually commuting Hermitian operators can always be diagonalized on the Hilbert space of state vectors; that is, there exist vectors $\Psi_\lambda \in \mathcal{H}, \lambda = (\lambda_1, \lambda_2, \ldots)$, such that $H_k\Psi_\lambda = \lambda_k\Psi_\lambda$, where the λ_k are real numbers. Then, each subspace $\mathcal{H}_\lambda \subset \mathcal{H}$, for each set of fixed eigenvalues λ, is invariant under the action of T_U; that is, $T_U : \mathcal{H}_\lambda \to \mathcal{H}_\lambda$. If the set of $SU(2)$ invariant operators (1.96) and the total angular momentum squared, defined by $\mathbf{J}^2 = J_1^2 + J_2^2 + J_3^2$, which is also an $SU(2)$ invariant, determine a unique irreducible subspace $\mathcal{H}_{SU(2)}$, then the set of operators (1.96) is said to be *complete with respect to the group* $SU(2)$. If the entire state vector space $\mathcal{H}$ for the physical system in question can be written as a direct sum of such subspaces: $\mathcal{H} = \sum_\lambda \oplus \mathcal{H}_\lambda$, then the set of operators is a complete set for the physical system. This property is rarely the case for real systems. But the role of $SU(2)$ rotational symmetry is still significant, since an incomplete set of operators (1.96) can still be used to characterize partially the various $SU(2)$ irreducible subspaces of the full Hilbert space $\mathcal{H}$. It is such irreducible subspaces that concern us in the theory of angular momentum.

We note that as a matter of notation, it is customary to use the same symbol for an operator acting in the **space of functions** as for the operator acting in the **space of values** of the function. The difference in the notation $(T\Psi)(q) = \mathcal{T}\Psi(q)$ is in the extra parenthesis pair ( ) enclosing $T\Psi = \Psi'$. The set, function space or value space, in which an operator acts can usually be inferred from the context of usage.

1.2 Abstract Angular Momentum

1.2.1 Brief background and history

One viewpoint in physics is that complex systems are built up from interacting elementary constituents. More significantly, perhaps, is the converse position that complex physical systems may be taken apart to unveil their elementary constituents. While this build-up and take-apart viewpoint of the makeup of physical systems can be called into question, it is the basis for the angular momentum theory of composite systems. Unitary $SU(2)$ rotational symmetry of the constituents, first viewed as noninteracting, as implemented through unitary rotations of **individual frames of reference,** one for each system, and, then, the unitary rotational symmetry of the interacting composite whole, as implemented through the $SU(2)$ unitary rotation of a **single common reference frame,** is a defining attribute of such a viewpoint.

The simplest composite system $\mathcal{S}$ is one that may be considered as composed of two independent parts $\mathcal{S}_1$ and $\mathcal{S}_2$. The two separate systems, considered as noninteracting with state vector spaces $\mathcal{H}(1)$ and $\mathcal{H}(2)$, are each described in terms of their individual reference systems, $(\mathbf{e}_1^{(1)}, \mathbf{e}_2^{(1)}, \mathbf{e}_3^{(1)})$ and $(\mathbf{e}_1^{(2)}, \mathbf{e}_2^{(2)}, \mathbf{e}_3^{(2)})$; each system is separately invariant under the redescription induced by the independent $SU(2)$ unitary frame rotations $T_{U_1} : \mathcal{H}(1) \mapsto \mathcal{H}(1)$ and $T_{U_2} : \mathcal{H}(2) \mapsto \mathcal{H}(2)$, as given by relation (1.65)-(1.67). Systems $\mathcal{S}_1$ and $\mathcal{S}_2$ have angular momenta $\mathbf{J}(1)$ and $\mathbf{J}(2)$, respectively, each of which has three components that satisfy the commutation relations (1.93), and the components of the separate angular momenta mutually commute.

The quantum systems $\mathcal{S}_1$ and $\mathcal{S}_2$ are next taken as a single composite system, denoted $\mathcal{S}_1 \otimes \mathcal{S}_2$, and described in terms of a common reference system $(\mathbf{e}_1, \mathbf{e}_2, \mathbf{e}_3)$. The quantum states of the composite system belong to the tensor product space $\mathcal{H}(1) \otimes \mathcal{H}(2)$, which contains the angular momentum subspaces $\mathcal{H}_{j_1}(1) \otimes \mathcal{H}_{j_2}(2) \subset \mathcal{H}(1) \otimes \mathcal{H}(2)$, where $\mathcal{H}_{j_1}$ and $\mathcal{H}_{j_2}$ are irreducible subspaces of $\mathcal{H}(1)$ and $\mathcal{H}(2)$ under the action of $SU(2)$ rotations of the respective reference frames. The composite system $\mathcal{S}_1 \otimes \mathcal{S}_2$, even if the parts are interactive, is invariant under the redescription induced by a unitary action of the frame $(\mathbf{e}_1, \mathbf{e}_2, \mathbf{e}_3)$, and must possess the rotational symmetry described by the total angular momentum $\mathbf{J} = \mathbf{J}(1) + \mathbf{J}(2)$, the three components of which again must satisfy the commutation relations (1.93). This situation motivates not only the determination of all possible actions of a single angular momentum in the state space of a physical system, but also that for *compounding* two such angular momenta.

In the third foundational paper of quantum mechanics in 1926, Born, Heisenberg, and Jordan [31] (see also Refs. [30, 32, 82]) not only gave the basic commutation relations for the components (J_1, J_2, J_3) of the total angular momentum $\mathbf{J}$ of a physical system, but also determined all finite-dimensional Hermitian matrix representations of these operators. Dirac [51] simultaneously deduced these basic commutation relations using his algebraic approach to quantum mechanics, and later gave a marvelously brief derivation of their matrix representations in his book (Dirac [52]).

Much of this mathematics had been discovered and developed earlier by Lie and Cartan, work of which physicists of the day were apparently unaware. The invariant theory of Clebsch, Gordan, and Young, and others also had no influence in the early developments of quantum theory, although Weyl [177] corrects this oversight in his 1928 book "Gruppentheorie und Quantenmechanik," but his work was mostly ignored by physicists of the time. It was Wigner [181] and Racah [143, 144] who initiated the application of Lie algebraic and symmetry methods into quantum physics and chemistry in a style that was accessible and eventually adopted by physical scientists. Weyl's work, however, was and remains fundamental; his book already initiates combinatorial methods (Young tableaux) into group theory, and deduces the fundamental rule for reducing an irreducible representation of $U(n)$ into a direct sum of irreducible representations of $U(n-1)$, which is the underlying group theoretical origin of Gelfand-Tsetlin [62] patterns that play in Chapters 5-9 a prominent role in this monograph.

1.2.2 One angular momentum: First and Second Fundamental Results

Abstract angular momentum theory addresses the problem of constructing all finite Hermitian matrices that satisfy the commutation relations

$$[J_1, J_2] = iJ_3,\ [J_2, J_3] = iJ_1,\ [J_3, J_1] = iJ_2, \tag{1.97}$$

where $[A, B] = AB - BA$ denotes the *commutator* of two matrices of the same order. If J_1, J_2, J_3 is such a set of Hermitian matrices, then $AJ_1A^{-1}, AJ_2A^{-1}, AJ_3A^{-1}$, is another such set, where A is an arbitrary unitary matrix, $A^{-1} = A^\dagger = (A^*)^T$, where $*$ denotes complex conjugation and T transposition. Two matrix representation of the commutations relations (1.97) related by such a unitary transformation are said to be *equivalent*. Thus, abstract angular momentum theory determines all Hermitian matrix realizations of the commutation relations (1.97) up to unitary equivalence. The commutation relations (1.97) may also be formulated as

$$\begin{gathered} [J_3, J_\pm] = \pm J_\pm,\ [J_+, J_-] = 2J_3, \\ J_\pm = J_1 \pm iJ_2,\ J_+^\dagger = J_- \,. \end{gathered} \tag{1.98}$$

The squared angular momentum

$$\begin{aligned} \mathbf{J}^2 &= J_1^2 + J_2^2 + J_3^2 = J_-J_+ + J_3(J_3+1) \\ &= J_+J_- + J_3(J_3-1) \end{aligned} \tag{1.99}$$

commutes with each J_i, and J_3 is, by convention, taken with $\mathbf{J}^2$ as the pair of commuting Hermitian matrices to be diagonalized.

Examples. Two examples of matrices satisfying relations (1.97) are provided by $J_i = \sigma_i/2$, where the σ_i denote the three Pauli matrices:

$$\begin{aligned} J_1 &= \frac{1}{2}\begin{pmatrix} 0 & 1 \\ 1 & 0 \end{pmatrix}, \; J_2 = \frac{1}{2}\begin{pmatrix} 0 & -i \\ i & 0 \end{pmatrix}, \\ J_3 &= \frac{1}{2}\begin{pmatrix} 1 & 0 \\ 0 & -1 \end{pmatrix}, \; \mathbf{J}^2 = \frac{3}{4}\begin{pmatrix} 1 & 0 \\ 0 & 1 \end{pmatrix}; \end{aligned} \tag{1.100}$$

and by the following J_i :

$$\begin{aligned} J_1 &= \frac{1}{\sqrt{2}}\begin{pmatrix} 0 & 1 & 0 \\ 1 & 0 & 1 \\ 0 & 1 & 0 \end{pmatrix}, \; J_2 = \frac{i}{\sqrt{2}}\begin{pmatrix} 0 & -1 & 0 \\ 1 & 0 & -1 \\ 0 & 1 & 0 \end{pmatrix}, \\ J_3 &= \begin{pmatrix} 1 & 0 & 0 \\ 0 & 0 & 0 \\ 0 & 0 & -1 \end{pmatrix}, \; \mathbf{J}^2 = 2\begin{pmatrix} 1 & 0 & 0 \\ 0 & 1 & 0 \\ 0 & 0 & 1 \end{pmatrix}. \end{aligned} \tag{1.101}$$

All Hermitian matrices solving (1.97) can be determined by using only matrix theory, but it is customary in quantum mechanics to formulate the problem using Hilbert space concepts in the spirit of quantum theory. Thus, the viewpoint is adopted that the J_i are linear Hermitian operators with an action defined in a separable Hilbert space $\mathcal{H}$ such that $J_i : \mathcal{H} \to \mathcal{H}$. (We now use the notation $\mathcal{H}$ to denote a generic Hilbert space that undergoes a redescription under unitary frame transformations.) The Hilbert space $\mathcal{H}$ is then decomposed into a direct sum of subspaces that are irreducible with respect to this action; that is, subspaces that cannot be further decomposed as a direct sum of subspaces that all the J_i leave invariant (map vectors in the space into vectors in the space). In this section, the solution of this fundamental problem for angular momentum theory is summarized. These results set the notation and phase conventions for the irreducible action of angular momentum operators, in all of their varied realizations, and the relations given here are therefore referred to as *standard.* The standard solution of this problem is among the most important in quantum theory because of its generality and applicability to a wide range of problems. Since this problem is solved in many places, including the seminal references given above, the standard solution and its properties are given in summary form (see Ref. [21] for the notations used here and more details. These results are also summarized in Drake [53]).

It is convenient to formulate the problem and its solution in a broad context without being too specific, so as not to restrict applications to general physical systems. Every isolated physical system described in Cartesian 3−space $\mathbb{R}^3$ has unitary rotational symmetry under unitary frame rotations and corresponding unitary group actions in $\mathcal{H}$, as described in detail earlier. The quantum rotation group $SU(2)$ is a symmetry group of the quantum-mechancial system. The system possesses irreducible subspaces under the $SU(2)$ action $T_U : \mathcal{H} \rightarrow \mathcal{H}$ of $SU(2)$ and the associated generators of this group, which is a set of angular momentum operators satisfying the commutation relations (1.97). The irreducible subspaces of $\mathcal{H}$ are given by $\mathcal{H}_{SU(2)}(\lambda)$, where the angular momentum operators $J_i, i = 1, 2, 3$, have an irreducible action on each subspace; that is, $J_i : \mathcal{H}_{SU(2)}(\lambda) \mapsto \mathcal{H}_{SU(2)}(\lambda)$. The symbol $\lambda = (\lambda_1, \lambda_2, \ldots)$ denotes a sequence of labels of finite length in a set such that, together with the subspaces determined by the irreducible action of the angular momentum $\mathbf{J}$, it is possible to describe the entire space $\mathcal{H}$ as a direct sum of these irreducible subspaces:

$$\mathcal{H} = \sum_{\lambda} \oplus \mathcal{H}_{SU(2)}(\lambda). \tag{1.102}$$

The sequence λ could, for example, specify the eigenvalues of a set of commuting Hermitian operators that are invariant under $SU(2)$, as described in (1.96), such that these operators together with $\mathbf{J}^2$ uniquely determine each irreducible space $\mathcal{H}_{SU(2)}(\lambda)$. Our main concern here is with the characterization of these subspaces that are invariant and irreducible under the action of the group $SU(2)$ and not with the bigger picture of how the description of complex physical systems synthesizes such subspaces into the totality of state vectors available for a full description.

For the purpose of angular momentum theory itself, we now drop all such labels λ from consideration and give the characterization of the subspaces $\mathcal{H}_{SU(2)}$ by $\mathcal{H}_{SU(2)} = \mathcal{H}_j$, where j is an nonnegative integer or half-odd integer; that is, $j \in \{0, 1/2, 1, 3/2, \ldots\}$. We next summarize the angular momentum properties of this space, using the Dirac bra-ket notation for the basis vectors of the space $\mathcal{H}_j$:

1. Orthonormal basis of $\mathcal{H}_j$:
$$\{|j\, m\rangle \mid m = j, j-1, \ldots, -j\}, \quad \langle j\, m' \mid j\, m\rangle = \delta_{m',m}\,. \tag{1.103}$$
2. Standard action of the angular momentum operators:
$$\begin{aligned} \mathbf{J}^2\,|j\, m\rangle &= j(j+1)\,|j\, m\rangle, \quad J_3\,|j\, m\rangle = m\,|j\, m\rangle, \\ J_+\,|j\, m\rangle &= \sqrt{(j-m)(j+m+1)}\,|j\, m+1\rangle, \\ J_-|j\, m\rangle &= \sqrt{(j+m)(j-m+1)}\,|j\, m-1\rangle\,. \end{aligned} \tag{1.104}$$
3. Defining properties of highest weight vector $|j\, j\rangle$:
$$J_+\,|j\, j\rangle = 0, \quad J_3\,|j\, j\rangle = j\,|j\, j\rangle\,. \tag{1.105}$$

4. Generation of general vector from highest weight vector $|j\,j\rangle$:

$$|j\,m\rangle = \sqrt{\frac{(j+m)!}{(2j)!(j-m)!}}\,J_-^{j-m}\,|j\,j\rangle\,. \tag{1.106}$$

5. Defining properties of lowest weight vector $|j,-j\rangle$:

$$J_-\,|j,-j\rangle = 0, \quad J_3\,|j,-j\rangle = -j\,|j,-j\rangle\,. \tag{1.107}$$

6. Standard action of $T_U, U \in SU(2)$, on $\mathcal{H}_j$:

$$T_U\,|j\,m\rangle \;=\; \sum_{m'} D^j_{m'\,m}(U)\,|\,j\,m'\rangle, \tag{1.108}$$

$$\begin{aligned} D^j_{m'\,m}(U) &= N^j_{m'\,m} \sum_{A\in\mathbb{M}_2(\alpha',\alpha)} \frac{(u_{11})^{a_{11}}(u_{12})^{a_{12}}(u_{21})^{a_{21}}(u_{22})^{a_{22}}}{a_{11}!a_{12}!a_{21}!a_{22}!}, \\ N^j_{m'\,m} &= \sqrt{(j+m')!(j-m')!(j+m)!(j-m)!}, \end{aligned} \tag{1.109}$$

where $\mathbb{M}_2(\alpha',\alpha)$ denotes the set of 2×2 matrix arrays defined by

$$\begin{aligned} &\mathbb{M}_2(\alpha',\alpha) \\ &= \left\{ \begin{pmatrix} a_{11} & a_{12} \\ a_{21} & a_{22} \end{pmatrix} \,\middle|\, \begin{array}{l} a_{11}+a_{12}=\alpha'_1, a_{21}+a_{22}=\alpha'_2, \\ a_{11}+a_{21}=\alpha_1, a_{12}+a_{22}=\alpha_2 \end{array} \right\}, \end{aligned} \tag{1.110}$$

$$\begin{pmatrix} \alpha'_1 & \alpha'_2 \\ \alpha_1 & \alpha_2 \end{pmatrix} = \begin{pmatrix} j+m' & j-m' \\ j+m & j-m \end{pmatrix}.$$

Thus, the summation in (1.109) is carried out over all 2×2 matrix arrays $A \in \mathbb{M}_2(\alpha',\alpha)$ of nonnegative integral exponents such that the *row-sum vector* is given by $(\alpha'_1,\alpha'_2) = (j+m', j-m')$ and the *column-sum vector* is given by $(\alpha_1,\alpha_2) = (j+m, j-m)$. This notation for effecting the summation in (1.109) is highly redundant for $n=2$, since there is only one "free" summation index after eliminating the row and column sum constraints, as given, for example, by $s = a_{21}$ in (1.61). It is, however, a notation that extends naturally in subsequence generalizations to $U(n)$. Even at level $n=2$, it is very useful for determining symmetries of the functions $D^j_{m'\,m}(U)$. In relation (1.109), U denotes an arbitrary unitary unimodular matrix with row and columns enumerated by $U = (u_{ij})_{1\le i,j\le 2}$. The representation functions corresponding to the $(\phi,\hat{\mathbf{n}})$ parametrization are obtained by setting $U = U(\phi,\hat{\mathbf{n}})$, as given by (1.74).

7. Unitary action of T_U on $\mathcal{H}$:

$$\langle T_U\Psi \,|\, T_U\Psi\rangle = \langle \Psi \,|\, \Psi\rangle, \text{ each} \,|\, \Psi\rangle \in \mathcal{H}, \text{ each}\, U \in SU(2), \quad (1.111)$$

where $|\,\Psi\rangle$ is again the Dirac notation for a general state vector in $\mathcal{H}$.

8. Irreducible unitary matrix representations of $SU(2)$: By convention, the $(2j+1)^2$ functions $D^j_{m'\,m}(U)$ are arranged as the rows and columns of a matrix $D^j(U)$ of order $2j+1$ by the rule: The element in row $j - m' + 1$ and column $j - m + 1$ of $D^j(U)$ is $D^j_{m'\,m}, m' = j, j-1, \ldots, -j; m = j, j-1, \ldots, -j$. These matrices are then unitary; that is, $(D^j(U))^\dagger = (D^j(U))^{-1} = D^j(U^\dagger)$, and they satisfy the group representation multiplication rule:

$$D^j(U')D^j(U) = D^j(U'U), \quad U' \in SU(2), \quad U \in SU(2). \quad (1.112)$$

The matrices constitute, for $j \in \{0, 1/2, 1, 3/2, \ldots\}$, all inequivalent irreducible unitary matrix representations of the group $SU(2)$. Interestingly, there is such a representation for every natural integer $n = 1, 2, \cdots$.

Most parametrizations of $SU(2)$ can be obtained from the quaternion parametrization given in by relations (1.36)-(1.37) by making an appropriate choice of parameters for a point on the unit sphere. For example, for

$$\alpha_0 = \cos(\phi/2), \quad \boldsymbol{\alpha} = \mathbf{n}\,\sin(\phi/2),\ 0 \le \phi \le 2\pi, \quad (1.113)$$

we obtain the parametrization of $U \in SU(2)$ corresponding to a unitary frame rotation in $\mathbb{R}^3$ by angle ϕ about direction $\mathbf{n}$ with $\mathbf{n}\cdot\mathbf{n} = n_1^2 + n_2^2 + n_3^2 = 1$. The corresponding operator generating the transformation in the Hilbert space $\mathcal{H}_j$ is given by

$$T_{U(\phi,\,\mathbf{n})} = e^{-i\phi\mathbf{n}\cdot\mathbf{J}},\ \mathbf{n}\cdot\mathbf{J} = n_1J_1 + n_2J_2 + n_3J_3, \quad (1.114)$$

$$e^{-i\phi\mathbf{n}\cdot\mathbf{J}}|j\, m\rangle = \sum_{m'} D^j_{m'\,m}(U(\phi, \hat{\mathbf{n}}))|\, j\, m'\rangle. \quad (1.115)$$

Still other parametrizations are given in Ref. [21].

We refer to relations (1.104) giving the standard action of angular momenta components (J_1, J_2, J_3) in the abstract Hilbert space $\mathcal{H}_j$ as the First Fundamental Result, and the relations (1.108)-(1.110) giving the standard action of the unitary group $SU(2$ in this space as the Second Fundamental Result:

First Fundamental Result. Up to unitary equivalence, the action of Hermitian angular momentum operators on the basis of a finite-dimensional Hilbert space on which the action is irreducible can always be put in the standard form given by relations (1.104).

Second Fundamental Result. Up to unitary equivalence, the action of the unitary unimodular group $SU(2)$ on the basis of a finite-dimensional Hilbert space on which the action of the group is irreducible can always be put in the standard form given by relations (1.108)-(1.110).

We never depart from these forms of the fundamental relationships.

1.2.3 Two angular momenta: Third Fundamental Result

Two angular momenta can arise in somewhat different contexts. Two distinct systems S_1 and S_2 can be brought together as the constituents of a single system, or a single object can have a spatial angular momentum and an intrinsic spin-like angular momentum, as we have already considered in the previous sections. In each case, we encounter the problem of addition of two angular momenta. We show below how each fits into the abstract theory of addition of two angular momenta.

Two distinct physical systems S_1 and S_2 are each described in terms of their individual reference frames $(\mathbf{e}_1^{(1)}, \mathbf{e}_2^{(1)}, \mathbf{e}_3^{(1)})$ and $(\mathbf{e}_1^{(2)}, \mathbf{e}_2^{(2)}, \mathbf{e}_3^{(2)})$, each of which undergoes its own unitary rotation $U_1 \in SU(2)$ and $U_1 \in SU(2)$. The first system has angular momentum $\mathbf{J}(1) = J_1(1)\mathbf{e}_1^{(1)} + J_2(1)\mathbf{e}_2^{(1)} + J_3(1)\mathbf{e}_3^{(1)}$ and the second angular momentum $\mathbf{J}(2) = J_1(2)\mathbf{e}_1^{(2)} + J_2(2)\mathbf{e}_2^{(2)} + J_3(2)\mathbf{e}_3^{(2)}$. This corresponds to two copies of the same Lie algebra; that is, the angular momenta components satisfy the commutation relations:

$$\begin{aligned} &[J_1(1), J_2(1)] = iJ_3(1), \ [J_2(1), J_3(1)] = iJ_1(1), \\ &[J_3(1), J_1(1)] = iJ_2(1); \end{aligned} \tag{1.116}$$

$$\begin{aligned} &[J_1(2), J_2(2)] = iJ_3(2), \ [J_2(2), J_3(2)] = iJ_1(2), \\ &[J_3(2), J_1(2)] = iJ_2(2), \end{aligned} \tag{1.117}$$

and each component of $\mathbf{J}(1)$ commutes with each component of $\mathbf{J}(2)$. Thus, the components of the angular momenta $\mathbf{J}(1)$ and $\mathbf{J}(2)$ are Hermitian operators that act in their own irreducible spaces $\mathcal{H}_{j_1}(1)$ and $\mathcal{H}_{j_2}(2)$, each with the standard action:

1. System S_1. Standard action of the components of $\mathbf{J}(1)$ on the orthonormal basis of $\mathcal{H}_{j_1}(1)$:

$$\{|\, j_1\, m_1\rangle \mid m_1 = j_1, j_1 - 1, \ldots, -j_1\}\,. \tag{1.118}$$

$$\begin{aligned}
\mathbf{J}^2(1)\,|j_1\,m_1\rangle &= j_1(j_1+1)\,|j_1\,m_1\rangle, \quad J_3(1)\,|j_1\,m_1\rangle = m_1\,|j_1\,m_1\rangle, \\
J_+(1)\,|j_1\,m_1\rangle &= \sqrt{(j_1-m_1)(j_1+m_1+1)}\,|j_1\,m_1+1\rangle, \qquad (1.119) \\
J_-(1)\,|j_1\,m_1\rangle &= \sqrt{(j_1+m_1)(j_1-m_1+1)}\,|j_1\,m_1-1\rangle\,.
\end{aligned}$$

2. System S_2. Standard action of the components of $\mathbf{J}(2)$ on the orthonormal basis of $\mathcal{H}_{j_2}(2)$:

$$\{|\,j_2\,m_2\rangle\mid m_2=j_2,j_2-1,\ldots,-j_2\}\,. \qquad (1.120)$$

$$\begin{aligned}
\mathbf{J}^2(2)\,|j_2\,m_2\rangle &= j_2(j_2+1)\,|j_2\,m_2\rangle, \quad J_3(2)\,|j_2\,m_2\rangle = m_2\,|j_2\,m_2\rangle, \\
J_+(2)\,|j_2\,m_2\rangle &= \sqrt{(j_2-m_2)(j_2+m_2+1)}\,|j_2\,m_2+1\rangle, \qquad (1.121) \\
J_-(2)\,|j_2\,m_2\rangle &= \sqrt{(j_2+m_2)(j_2-m_2+1)}\,|j_2\,m_2-1\rangle\,.
\end{aligned}$$

The two systems S_1 and S_2 are now brought together as the parts of a composite system $S_1\otimes S_2$ in which they may be interacting. The two angular momenta are now described in terms of a single reference frame $(\mathbf{e}_1,\mathbf{e}_2,\mathbf{e}_3)$ by $\mathbf{J}(1)=J_1(1)\mathbf{e}_1+J_2(2)\mathbf{e}_2+J_3(2)\mathbf{e}_3$ and $\mathbf{J}(2)=J_1(2)\mathbf{e}_1+J_2(2)\mathbf{e}_2+J_3(2)\mathbf{e}_3$. The components of the two angular momenta still satisfy the commutation relations (1.116) -(1.117), with each component of $\mathbf{J}(1)$ commuting with each component of $\mathbf{J}(2)$. But now we can add components to obtain the total angular momentum

$$\mathbf{J}=\mathbf{J}(1)+\mathbf{J}(2)=(J_1(1)+J_1(2))\mathbf{e}_1+(J_2(1)+J_2(2))\mathbf{e}_2+(J_3(1)+J_3(2))\mathbf{e}_3, \qquad (1.122)$$

whose components satisfy the commutation relations

$$[J_1,J_2]=iJ_3,\ \ [J_2,J_3]=iJ_1,\ \ [J_3,J_1]=iJ_2. \qquad (1.123)$$

Here the unitary rotation of the frame $(\mathbf{e}_1,\mathbf{e}_2,\mathbf{e}_3)$ to which both systems are referred induces the redescription of the composite system $S_1\otimes S_2$ that leads directly to the addition of angular momenta. In case of a single system with internal spin, the system is "composite" from the outset—it cannot be taken apart or assembled: steps 1 and 2 above would not be introduced. But the abstract results from the addition of angular momenta applies to both truly composite and intrinsically composite systems. In both cases, we are led to two different ways of describing the angular momentum states of the composite system in terms of the tensor product space $\mathcal{H}_{j_1}(1)\otimes\mathcal{H}_{j_2}(2)$, which is the operative space for such systems, as we next describe:

1. Uncoupled basis of $\mathcal{H}_{j_1}(1)\otimes\mathcal{H}_{j_2}(2)$:

$$\left\{|j_1\,m_1\rangle\otimes|j_2\,m_2\rangle\,\middle|\,\begin{array}{l} m_1=j_1,j_1-1,\ldots,-j_1\,; \\ m_2=j_2,j_2-1,\ldots,-j_2 \end{array}\right\}. \qquad (1.124)$$

The actions of the angular momentum operators $\mathbf{J}(1)$ and $\mathbf{J}(2)$ on this basis are defined by

$$(J_i(1) \otimes \mathbb{I}_2)|j_1\, m_1\rangle \otimes |j_2\, m_2\rangle = (J_i(1) \mid j_1\, m_1\rangle) \otimes \mid j_2\, m_2\rangle, \quad (1.125)$$

$$(\mathbb{I}_1 \otimes J_i(1))|j_1\, m_1\rangle \otimes |j_2\, m_2\rangle = |j_1\, m_1\rangle) \otimes ((J_i(2)|j_2\, m_2\rangle), \quad (1.126)$$

where $\mathbb{I}_i, i = 1, 2$, is the identity operator on $\mathcal{H}_{j_i}$. In these relations, the actions of the angular momentum components $J_i(1)$ and $J_i(2)$ on the respective basis vectors $|j_1\, m_1\rangle$ and $|j_2\, m_2\rangle$ are given by relations (1.119) and (1.121).

2. Coupled basis of $\mathcal{H}_{j_1}(1) \otimes \mathcal{H}_{j_2}(2)$:

$$|(j_1\, j_2)j\, m\rangle = \sum_{\substack{m_1, m_2 \\ m_1 + m_2 = m}} C^{j_1\; j_2\; j}_{m_1\, m_2\, m} |j_1\, m_1\rangle \otimes |j_2\, m_2\rangle, \quad (1.127)$$

$$\left\{ |(j_1\, j_2)j\, m\rangle \;\middle|\; \begin{array}{l} j = j_1 + j_2, j_1 + j_2 - 1, \ldots, |j_1 - j_2|\, ; \\ m = j, j-1, \ldots, -j \end{array} \right\}. \quad (1.128)$$

The coefficients $C^{j_1\; j_2\; j}_{m_1\, m_2\, m}$ in the linear transformation (1.127) of the space $\mathcal{H}_{j_1}(1) \otimes \mathcal{H}_{j_2}(2)$ are called Wigner-Clebsch-Gordan (WCG) coefficients, which were first obtained by Clebsch [43] and Gordan [68], and then Wigner [184, 185]. These coefficients effect a real orthogonal transformation of the uncoupled basis to the coupled basis on which the total angular momentum has the standard action:

$$\begin{aligned}
\mathbf{J}^2|(j_1\, j_2)j\, m\rangle &= j(j+1)|(j_1\, j_2)j\, m\rangle, \\
J_3|(j_1\, j_2)j\, m\rangle &= m|(j_1\, j_2)j\, m\rangle, \qquad (1.129)\\
J_+|(j_1\, j_2)j\, m\rangle &= \sqrt{(j-m)(j+m+1)}\, |(j_1\, j_2)j\, m+1\rangle, \\
J_-|(j_1\, j_2)j\, m\rangle &= \sqrt{(j+m)(j-m+1)}\, |(j_1\, j_2)j\, m-1\rangle\, .
\end{aligned}$$

It is also the case that

$$\mathbf{J}^2(1)|(j_1\, j_2)j\, m\rangle = j_1(j_1+1)|(j_1\, j_2)j\, m\rangle, \quad (1.130)$$

$$\mathbf{J}^2(2)|(j_1\, j_2)j\, m\rangle = j_2(j_2+1)|(j_1\, j_2)j\, m\rangle. \quad (1.131)$$

The three Hermitian operators $\mathbf{J}^2(1), \mathbf{J}^2(2), \mathbf{J}^2$ are a complete set of $SU(2)$ invariant operators in the sense defined in (1.102); we denote the $SU(2)$ irreducible space $\mathcal{H}_{SU(2)}(\lambda_1, \lambda_2)$ by $\mathcal{H}_{(j_1 j_2)j}$. Relations (1.127)-(1.131) apply to arbitrary values of $j_1, j_2 \in \{0, 1/2, 1, \ldots\}$, but only to values of j that satisfy, for given j_1 and j_2, the rule

$$j \in \{j_1 + j_2, j_1 + j_2 - 1, \ldots, |j_1 - j_2|\}. \quad (1.132)$$

This relation between j_1, j_2, j is called the *triangle rule.* Also, because $J_3 = J_3(1) + J_3(2)$, the summation in relation (1.127) is over only such values of $m_1 \in \{j_1, j_1 - 1, \cdots - j_1\}$ and $m_2 \in \{j_2, j_2 - 1, \ldots, -j_2\}$ that satisfy

$$m_1 + m_2 = m, \tag{1.133}$$

where $m \in \{j, j-1, \cdots - j\}$. This relation between m_1, m_2, m is called the projection quantum number *sum rule.* (This is discussed further below in connection with the domain of definition of a WCG coefficient.) Relation (1.127) is the expression of the decomposition of the vector space $\mathcal{H}_{j_1}(1) \otimes \mathcal{H}_{j_2}(2)$ of dimension $(2j_1+1)(2j_2+1)$ into it $SU(2)$ irreducible subspaces $\mathcal{H}_{(j_1,j_2)j}$, hence,

$$\sum_{j=|j_1-j_2|}^{j_1+j_2} (2j+1) = (2j_1+1)(2j_2+1). \tag{1.134}$$

In both instances, we follow the practice of using the 3−components $J_3(1), J_3(2), J_3$ of the respective angular momenta to determine the orthonormal basis $|j_1 m_1\rangle \otimes |j_2 m_2\rangle$ of $\mathcal{H}_{j_1}(1) \otimes \mathcal{H}_{j_2}(2)$ and the orthonormal basis $|(j_1 j_2) j m\rangle$ of $\mathcal{H}_{(j_1,j_2)j}$.

3. Standard action of $U \in SU(2)$ on $\mathcal{H}_{j_1}(1)$:

$$T_U |j_1\, m_1\rangle = \sum_{m_1'=-j_1}^{j_1} D^{j_1}_{m_1'\, m_1}(U) |j_1\, m_1'\rangle, \tag{1.135}$$
$$m_1 = j_1, j_1 - 1, \ldots, -j_1.$$

4. Standard action of $U \in SU(2)$ on $\mathcal{H}_{j_2}(2)$:

$$T_U |j_2\, m_2\rangle = \sum_{m_2'=-j_2}^{j_2} D^{j}_{m_2'\, m_2}(U) |j_2\, m_2'\rangle, \tag{1.136}$$
$$m_2 = j_2, j_2 - 1, \ldots, -j_2.$$

5. Standard action of $(U, V) \in SU(2) \times SU(2)$ on $\mathcal{H}_{j_1}(1) \otimes \mathcal{H}_{j_2}(2)$:

$$T_{(U,V)}\, |j_1\, m_1\rangle \otimes |j_2\, m_2\rangle = T_U\, |j_1\, m_1\rangle \otimes T_V\, |j_2\, m_2\rangle$$
$$= \sum_{m_1', m_2'} D^{j_1}_{m_1'\, m_1}(U) D^{j_2}_{m_2'\, m_2}(V)\, |j_1\, m_1'\rangle \otimes |j_2\, m_2'\rangle, \tag{1.137}$$

where the multiplication of elements in the direct product group $SU(2) \times SU(2)$ is defined by

$$T_{(U',V')} T_{(U,V)} = T_{(U'U,V'V)}. \tag{1.138}$$

6. Standard action of $(U,U) \in SU(2) \times SU(2)$ on the coupled space $\mathcal{H}_{(j_1\,j_2)j}$:

$$T_{(U,U)}\,|(j_1\,j_2)j\,m\rangle = \sum_{m'=-j}^{j} D^j_{m'm}(U)\,|(j_1\,j_2)j\,m'\rangle, \quad (1.139)$$
$$m = j, j-1, \ldots, -j.$$

The pairs of unitary matrices in the set $\{(U,U)|U \in SU(2)\}$ define the *diagonal subgroup* of the direct product group $SU(2) \times SU(2)$. The unitary operators in the set $\{T_{(U,U)}|U \in SU(2)\}$ obey the multiplication rule

$$T_{(U',U')}T_{(U,U)} = T_{(U'U,U'U)}. \quad (1.140)$$

The new objects that make their appearance in the addition of two angular momenta, as outlined above, are the WCG coefficients, which relate the two bases, the uncoupled basis of the space $\mathcal{H}_{j_1}(1) \otimes \mathcal{H}_{j_2}(2)$ of dimension $(2j_1+1)(2j_2+1)$ and the so-called coupled basis of this same space, which is a direct sum of spaces $\sum_j \oplus\, \mathcal{H}_{(j_1\,j_2)j}$, each space in the direct sum having the orthonormal basis $\{|(j_1\,j_2)j\,m\rangle\,|\,m = j, j-1, \ldots, -j\}$ of dimension $(2j+1)$. Under the action $T_{(U,U)}$ of the diagonal subgroup, the first space gives a reducible unitary matrix representation of $SU(2)$ of dimension $(2j_1+1)(2j_2+1)$, the Kronecker product representation, defined by

$$\left(D^{j_1}(U) \otimes D^{j_2}(U)\right)_{(m_1'\,m_2')(m_1\,m_2)} = D^{j_1}_{m_1'\,m_1}(U) D^{j_2}_{m_2'\,m_2}(U). \quad (1.141)$$

Under this same action of $T_{(U,U)}$ of the diagonal subgroup, the subspace $\mathcal{H}_{(j_1\,j_2)j} \subset \mathcal{H}_{j_1}(1) \otimes \mathcal{H}_{j_2}(2)$ gives the standard irreducible unitary matrix representation of $SU(2)$ of dimension $2j+1$. Thus, the transformation (1.127) effects the reduction of the space $\mathcal{H}_{j_1}(1) \otimes \mathcal{H}_{j_2}(2)$ into the direct sum of irreducible subspaces $\sum_j \oplus \mathcal{H}_{(j_1\,j_2)j}$, where the summation is over all j that satisfy the triangle rule. The sum rule (1.133) on the projection quantum numbers expresses the additive property of the $U(1)$ subgroup of $SU(2)$.

A WCG coefficient $C^{j_1\;j_2\;j}_{m_1\,m_2\,m}$ is defined if and only if its labels fall into the domains of definition of the state vectors appearing in relation (1.124) and (1.128):

(i). angular momentum quantum numbers: $j_1, j_2, j \in \{0, 1/2, 1, 3/2, \ldots\}$.

(ii). projection quantum numbers: $m_1 \in \{j_1, j_1-1, \ldots, -j_1\}$, $m_2 \in \{j_2, j_1-1, \ldots, -j_2\}$, $m \in \{j, j-1, \ldots, -j\}$. (1.142) (iii). sum rule and triangle rule: $m_1 + m_2 = m$, $j \in \{j_1 + j_2, j_1 + j_2 - 1, \ldots, |j_1 - j_2|\}$.

For all other values of the angular momentum quantum labels $j_1, j_2, j, m_1, m_2, m_3$ *a WCG coefficient is undefined.*

The WCG coefficients satisfy orthogonality relations resulting from the orthonormality of the basis vectors in each side of relation (1.127), which are

$$\langle j_1 m_1' \,|\, j_1 m_1 \rangle = \delta_{m_1', m_1}, \; \langle j_2 m_2' \,|\, j_2 m_2 \rangle = \delta_{m_2', m_2}; \tag{1.143}$$

$$\langle (j_1 j_2) j' m' \,|\, (j_1 j_2) j m \rangle = \delta_{j', j}\, \delta_{m', m}. \tag{1.144}$$

The consequent orthogonality relations for the WCG coefficients are:

$$\sum_{m_1, m_2} C^{j_1\; j_2\; j}_{m_1\, m_2\, m} C^{j_1\; j_2\; j'}_{m_1\, m_2\, m} = \delta_{j, j'}, \tag{1.145}$$

$$\sum_{j, m} C^{j_1\; j_2\; j}_{m_1\, m_2\, m} C^{j_1\; j_2\; j}_{m_1'\, m_2'\, m} = \delta_{m_1, m_1'}\, \delta_{m_2, m_2'}, \tag{1.146}$$

where, in the first relation j_1, j_2, j, m, are any values for which the coefficients are defined, and the summation is over all pairs m_1, m_2 for which the coefficients are defined. Similarly, in the second relation, $j_1, m_1, m_1', j_2, m_2, m_2'$ are any values for which the coefficients are defined, and the summation over all pairs j, m for which the coefficients are defined. This convention of effecting the summations in the orthogonality relations avoids otherwise very awkward summation symbols and ranges.

There are a number of relations between WCG coefficients and representation functions that are consequences of the actions of the various unitary operators on the spaces $\mathcal{H}_{j_1}(1) \otimes \mathcal{H}_{j_2}(2)$, the subspace $\mathcal{H}_{(j_1\, j_2) j} \subset \mathcal{H}_{j_1}(1) \otimes \mathcal{H}_{j_2}(2)$, and of the orthogonality of the WCG coefficients, which are the elements of the matrix that reduces the reducible Kronecker product representation of $SU(2)$ into irreducible representations. These relations constitute the basic content of the theory of addition of two angular momenta. We summarize these basic relations as follows:

1. The substitution of the basis vector relation (1.127) into the group action relation (1.139), using the action (1.137) for $V = U$, and equating coefficients of orthonormal basis vectors on each side of the resulting relation gives the identity:

$$\sum_{m_1, m_2} C^{j_1\; j_2\; j}_{m_1\, m_2\, m} D^{j_1}_{m_1'\, m_1}(U) D^{j_2}_{m_2'\, m_2}(U) = D^{j}_{m'\, m}(U) C^{j_1\; j_2\; j}_{m_1'\, m_2'\, m'}\,. \tag{1.147}$$

2. Using the orthogonality relation (1.145) in this relation now gives the identity:

$$\sum_{\substack{m_1, m_2 \\ m'_1, m'_2}} C^{j_1\ j_2\ j'}_{m'_1\ m'_2\ m'} C^{j_1\ j_2\ j}_{m_1\ m_2\ m} D^{j_1}_{m'_1\ m_1}(U) D^{j_2}_{m'_2\ m_2}(U) = \delta_{j',j} D^{j}_{m'\ m}(U). \tag{1.148}$$

Relation (1.148) between WCG coefficients and representation functions can be written concisely in matrix form. We first define the matrix $C^{(j_1\, j_2)}$ of order $(2j_1 + 1)(2j_2 + 1)$ to have elements as follows:

$$C^{(j_1\, j_2)}_{(j\, m);(m_1\, m_2)} = \left(C^{(j_1\, j_2)}\right)_{(j\, m);(m_1\, m_2)} \tag{1.149}$$
$$= \begin{cases} C^{j_1\ j_2\ j}_{m_1\ m_2\ m}, & \text{all values of } j_1, j_2, j, m_1, m_2, m \text{ for which the coefficient is defined;} \\ 0, & \text{all other values of } j_1, j_2, j, m_1, m_2, m. \end{cases}$$

The sets defined as follows enumerate the rows and columns, respectively, of the matrix $C^{(j_1\, j_2)}$:

$$\mathbb{R}(j_1\, j_2) = \left\{ (j, m) \;\middle|\; \begin{array}{l} j = j_1 + j_2, \ldots, |j_1 - j_2|, \\ m = j, j-1, \ldots, -j \end{array} \right\},$$
$$\mathbb{C}(j_1\, j_2) = \{(m_1, m_2) \mid m_i = j_i, j_i - 1, \ldots, -j_i\}. \tag{1.150}$$

That the number of rows and columns of $C^{(j_1\, j_2)}$ are equal is a consequence of the identity:

$$\sum_{j=j_{\min}}^{j_{\max}} (2j+1) = (2j_1 + 1)(2j_2 + 1), \tag{1.151}$$

where $j_{\min} = |j_1 - j_2|$ and $j_{\max} = j_1 + j_2$. The orthogonality relations (1.145) and (1.146) are now expressed by the property that $C^{(j_1\, j_2)}$ is a real orthogonal matrix:

$$\left(C^{(j_1\, j_2)}\right)^T C^{(j_1\, j_2)} = C^{(j_1\, j_2)} \left(C^{(j_1\, j_2)}\right)^T = I_{(2j_1+1)(2j_2+1)}. \tag{1.152}$$

Since rows and columns of $C^{(j_1\, j_2)}$ are enumerated by different indexing pairs, both orders of expressing the orthogonality relations are important; they yield back relations (1.145) and (1.146).

Relation (1.148) is now expressed as the matrix relation

$$C^{(j_1\, j_2)}\left(D^{j_1}(U)\otimes D^{j_2}(U)\right)\left(C^{(j_1\, j_2)}\right)^T = \sum_{j=j_{\min}}^{j_{max}} \oplus D^j(U). \tag{1.153}$$

The real orthogonal matrix $C^{(j_1\, j_2)}$ effects by a similarity transformation the complete reduction of the reducible unitary representation of $SU(2)$ given by the Kronecker product $D^{j_1}(U)\otimes D^{j_2}(U)$ of the two irreducible unitary matrix representations $D^{j_1}(U)$ and $D^{j_2}(U)$ into a direct sum of irreducible unitary representations of $SU(2)$, where all representations are standard.

Relation (1.153) can be written in three equivalent forms by moving one or both of the orthogonal matrices to the right-hand side:

$$C^{(j_1\, j_2)}\left(D^{j_1}(U)\otimes D^{j_2}(U)\right) = \left(\sum_{j=j_{\min}}^{j_{max}} \oplus D^j(U)\right) C^{(j_1\, j_2)}, \tag{1.154}$$

$$\left(D^{j_1}(U)\otimes D^{j_2}(U)\right)\left(C^{(j_1\, j_2)}\right)^T = \left(C^{(j_1\, j_2)}\right)^T \sum_{j=j_{\min}}^{j_{max}} \oplus D^j(U), \tag{1.155}$$

$$D^{j_1}(U)\otimes D^{j_2}(U) = \left(C^{(j_1\, j_2)}\right)^T \left(\sum_{j=j_{\min}}^{j_{max}} \oplus D^j(U)\right) C^{(j_1\, j_2)}. \tag{1.156}$$

There is still an incompleteness in relations (1.153)-(1.156) in that we have not given the rule for assigning the rows and columns of the matrix $C(j_1\, j_2)$, nor that for ordering the matrices in the direct sum as block matrices along the diagonal. We return to this in Sect. 2.1.3, Chapter 2. It is often the case that rows and columns of matrices arising in angular momentum theory are indexed by sets of different structure, such as given by (1.150); caution must be exercised in taking matrix elements.

Third Fundamental Result. The basis of the tensor product space $\mathcal{H}_{j_1}(1)\otimes\mathcal{H}_{j_2}(2)$ on which two independent Hermitian angular momentum operators $\mathbf{J}(1)$ and $\mathbf{J}(2)$ have the standard irreducible action is related to the basis $\mathcal{H}_{(j_1\, j_2)j}$ on which the total angular momentum $\mathbf{J}=\mathbf{J}(1)+\mathbf{J}(2)$ has the standard irreducible action by a real, orthogonal transformation in which the transformation coefficients are the WCG coefficients. The WCG coefficients are then also the elements of a real orthogonal matrix that effects the reduction, by a similarity transformation, of the reducible unitary Kronecker product representation $D^{j_1}(U)\otimes D^{j_2}(U)$ of $SU(2)$ that arises from the action of the diagonal subgroup of the direct product group $SU(2) \times SU(2)$ in the tensor product space $\mathcal{H}_{j_1}(1) \otimes \mathcal{H}_{j_2}(2)$ into a direct sum of irreducible representations obtained from the action of $SU(2)$ in the irreducible subspace $\mathcal{H}_{(j_1\, j_2)j} \subset \mathcal{H}_{j_1}(1) \otimes \mathcal{H}_{j_2}(2)$.

In the following sections of this chapter, we discuss explicit realizations of the abstract results given above.

1.3 $SO(3,\mathbb{R})$ and $SU(2)$ Solid Harmonics

The form (1.109) giving all the inequivalent unitary representations of $SU(2)$ is a polynomial in the elements of U. When a mathematical quantity is polynomial in, say, some restricted variables that define it, it is often a useful generalization is to replace those variables by arbitrary indeterminates, and to consider the properties of the new object. A simple example of this is provided by the binomial coefficients defined by

$$\binom{n}{k} = \frac{n!}{(n-k)!k!} = \frac{n(n-1)\cdots(n-k+1)}{k!}, \tag{1.157}$$

which is polynomial in the nonnegative integer n. This suggests we consider the general polynomial $\binom{x}{k}$ for an arbitrary variable x defined by

$$\binom{x}{k} = \frac{x(x-1)\cdots(x-k+1)}{k!}, \tag{1.158}$$

which is then a polynomial of degree k with roots $x = 0, 1, \ldots, k-1$, which reduces at $x = n \geq k$ to the binomial coefficient (1.157). These polynomials then have a host of properties generalizing those of the numerical binomial coefficients, such as the sum rule

$$\binom{x+y}{c} = \sum_{a+b=c} \binom{x}{a}\binom{y}{b}. \tag{1.159}$$

The idea for obtaining the $SU(2)$ solid harmonics extends this simple example in the obvious way.

The $SU(2)$ *solid harmonics* are defined to be the polynomials homogeneous of degree $2j$ in four commuting indeterminates given by

$$D^j_{m\,m'}(Z) = \sqrt{\alpha!\,\alpha'!} \sum_{A \in \mathbb{M}_2(\alpha,\alpha')} \frac{Z^A}{A!}, \tag{1.160}$$

in which the nonnegative exponents $A = (a_{11}, a_{21}, a_{12}, a_{22})$ and the indeterminates $Z = (z_{11}, z_{21}, z_{12}, z_{22})$ are encoded in the matrix arrays defined by

$$A = \begin{pmatrix} a_{11} & a_{12} \\ a_{21} & a_{22} \end{pmatrix}, \quad Z = \begin{pmatrix} z_{11} & z_{12} \\ z_{21} & z_{22}. \end{pmatrix}. \tag{1.161}$$

We also use the space-saving notations

$$Z^A = \prod_{i,j=1}^{2} z_{ij}^{a_{ij}}, \quad A! = \prod_{i,j=1}^{2} (a_{ij})!, \tag{1.162}$$

$$\alpha! = \alpha_1!\,\alpha_2!, \quad \alpha'! = \alpha'_1!\,\alpha'_2!.$$

(It is convenient to write m first and m' second in (1.160) and make the appropriate adjustments in notation in the right-hand side.)

The $SU(2)$ solid harmonics defined by (1.160) are among the most important functions in angular momentum theory. Not only do they unify the irreducible representations of $SU(2)$ in any parametrization by making the appropriate definition of the indeterminates in terms of generalized coordinates, as we gave seen above, but they also include the popular boson calculus realization of state vectors for quantum-mechanical systems, as well as the state vectors for the symmetric rigid rotator (see Refs. [21, 38]). The parametrization given by $Z = \alpha_0\sigma_0 - i\boldsymbol{\alpha}\cdot\boldsymbol{\sigma}, \alpha_0^2+\alpha_1^2+\alpha_2^2+\alpha_3^2 = 1$, gives the $SU(2)$ representation functions in terms of points on the unit sphere $\mathbb{S}^3 \subset \mathbb{R}^4$, and the further specialization to $\alpha_0 = \cos(\phi/2), \boldsymbol{\alpha} = \mathbf{n}\sin(\phi/2), 0 \leq \phi \leq 2\pi$, the representation functions for a unitary frame rotation by angle ϕ about the direction $\mathbf{n}$. Other parametrizations are (1) spherical polar coordinates in $\mathbb{R}^4$, which replace ϕ by 2ψ to obtain $\alpha_0 = \cos\psi, \boldsymbol{\alpha} = \mathbf{n}\sin\psi, 0 \leq \psi \leq \pi$, where the components of $\mathbf{n}$ are given in terms of the usual spherical polar coordinates in $\mathbb{R}^3$ given by $(\cos\phi\,\sin\theta, \sin\phi\,\sin\theta, \cos\theta), 0 \leq \phi < 2\pi, 0 \leq \theta \leq \pi$; (2) Euler angles (α,β,γ) obtained by choosing $\alpha_0 = \cos(\alpha+\gamma)/2\,\cos\beta/2, \alpha_1 = -\sin(\alpha-\gamma)/2\,\sin\beta/2, \alpha_2 = \cos(\alpha-\gamma)/2\,\sin\beta/2, \alpha_3 = \sin(\alpha+\gamma)/2\,\cos\beta/2$, where $0 \leq \alpha < 2\pi, 0 \leq \gamma < 2\pi$, and β can be in either the domain $0 \leq \beta \leq \pi$ or $2\pi \leq \beta \leq 3\pi$. These domains of definitions of angles are necessary to cover the unit sphere $\mathbb{S}^3$ and to have the property that $R(-U) = R(U)$ for the proper, orthogonal matrices $R(U) \in SO(3,\mathbb{R})$. See Ref. [21] for more detailed discussions, as well as the relationship to the symmetric rotator. The relation to boson operators is given below.

The $SU(2)$ solid harmonics yield the spinor polynomials defined by (1.58) for special values of the indeterminates:

$$D^j_{m\,m'}\begin{pmatrix} z_1 & 0 \\ z_2 & 0 \end{pmatrix} = \delta_{j,m'}\sqrt{(2j)!}\,P_{j\,m}(z_1,z_2),$$

$$D^j_{m\,m'}\begin{pmatrix} 0 & z_1 \\ 0 & z_2 \end{pmatrix} = \delta_{j,-m'}\sqrt{(2j)!}\,P_{j\,m}(z_1,z_2), \tag{1.163}$$

$$D^j_{m\,m'}\begin{pmatrix} z_1 & 0 \\ 0 & z_2 \end{pmatrix} = \delta_{m,m'}\sqrt{(j+m)!(j-m)!}\,P_{j\,m}(z_1,z_2);$$

$$P_{j\,m}(z_1,z_2) = \frac{z_1^{j+m}z_2^{j-m}}{\sqrt{(j+m)!(j-m)!}}. \tag{1.164}$$

Perhaps the most significant property of the $SU(2)$ solid harmonics is the multiplication property of the corresponding matrices given by

$$D^j(X)D^j(Y) = D^j(XY), \tag{1.165}$$

where this property hold for arbitrary matrices X and Y of commuting indeterminates, including singular matrices. A combinatorial proof of this result is given below in a more general context. The transposition property expressed by

$$D^j(Z^T) = \left(D^j(Z)\right)^T \tag{1.166}$$

is also important.

We also can consider left and right transformations of $D^j(Z)$ as given by the operators L_U and R_V for $U, V \in SU(2)$:

$$(L_U D^j)(Z) = D^j(U^T Z) = D^j(U^T)D^j(Z), \tag{1.167}$$

$$(R_V D^j)(Z) = D^j(ZV) = D^j(Z)D^j(V), \tag{1.168}$$

which in terms of matrix elements take the forms:

$$(L_U D^j_{m\,m'})(Z) = \sum_{m''=-j}^{j} D^j_{m''\,m}(U)D^j_{m''\,m'}(Z), \tag{1.169}$$

$$(R_V D^j_{m\,m'})(Z) = \sum_{m''=-j}^{j} D^j_{m''\,m'}(V)D^j_{m\,m''}(Z). \tag{1.170}$$

Left and right transformations commute; that is, $L_U R_V = R_V L_U$, for all pairs $U, V \in SU(2)$, in fact, for all left and right transformations L_X and L_Y, for arbitrary matrices X and Y. The generators of these left and right group transformations may be calculated from (1.72) to obtain:

$$J_+ = \sum_{k=1}^{2} z_{1k}\partial/\partial z_{2k},\ J_- = \sum_{k=1}^{2} z_{2k}\partial/\partial z_{1k}, \tag{1.171}$$

$$J_3 = \frac{1}{2}\sum_{k=1}^{2}(z_{1k}\partial/\partial z_{1k} - z_{2k}\partial/\partial z_{2k});$$

$$J'_+ = \sum_{k=1}^{2} z_{k1}\partial/\partial z_{k2},\; J'_- = \sum_{k=1}^{2} z_{k2}\partial/\partial z_{k1}, \tag{1.172}$$
$$J'_3 = \frac{1}{2}\sum_{k=1}^{2}(z_{k1}\partial/\partial z_{k1} - z_{k2}\partial/\partial z_{k2}).$$

These two sets of operators then satisfy the standard commutation relations, and the two sets mutually commute. The action of these operators on the $SU(2)$ solid harmonics is exactly the standard action (1.104) for (J_1, J_2, J_3) for quantum numbers (j, m), while that for (J'_1, J'_2, J'_3) is for quantum numbers (j, m'). Transposed elements of Z enters into relations (1.171)-(1.172) because $(XU)^T = U^T X^T$ That the same j occurs in the standard action is a consequence of the equality of operators

$$\mathbf{J}^2 = J_1^2 + J_2^2 + J_3^2 = \mathbf{J}'^2 = J_1'^2 + J_2'^2 + J_3'^2. \tag{1.173}$$

This result is shown by direct calculation: It is a consequence of the left and right actions acting on the same coordinates, namely, the elements of Z. The equality of these two invariants is an important result for relating the $SU(2)$ solid harmonics to the symmetric rotator wavefunctions (see Ref. [21]), as well as for our subsequent generalization to the general unitary group $U(n)$.)

The nomenclature *$SU(2)$ solid harmonics* for the polynomials defined by (1.160) is by analogy with the term "$SO(3,\mathbb{R})$ solid (spherical) harmonics." These latter solid harmonics are defined over all points of Cartesian 3-space $\mathbb{R}^3$ by

$$\mathcal{Y}_{l,m}(x_1, x_2, x_3) = \left[\frac{2l+1}{4\pi}(l+m)!(l-m)!\right]^{1/2}$$
$$\times \sum_k \frac{(-x_1 - ix_2)^{m+k}(x_1 - ix_2)^k x_3^{l-m-2k}}{2^{m+2k}(m+k)!k!(l-m-2k)!}. \tag{1.174}$$

The components (L_1, L_2, L_3) of the orbital angular momentum operator $\mathbf{L} = -i\mathbf{x}\times\boldsymbol{\nabla}$ have the standard action on the solid harmonics, which are homogeneous polynomial solutions of Laplace's equation in $\mathbb{R}^3$:

$$\begin{gathered}\mathbf{L}^2\mathcal{Y}_{l\,m}(x) = l(l+1)\mathcal{Y}_{l\,m}(x),\; L_3\mathcal{Y}_{l\,m}(x) = m\mathcal{Y}_{l\,m}(x),\\ L_\pm\mathcal{Y}_{l\,m}(x) = \sqrt{(l\pm m)(l\mp m+1)}\mathcal{Y}_{l\,m\pm1}(x),\\ l \in \{0,1,2,\ldots\}, m = l, l-1, \ldots, -l.\end{gathered} \tag{1.175}$$

The eigenvalue relation $\mathbf{L}^2\mathcal{Y}_{l\,m}(x) = l(l+1)\mathcal{Y}_{l\,m}(x)$ is a consequence of the fact that the polynomials $\mathcal{Y}_{l\,m}(x)$ are homogeneous of degree l that solve Laplace's equation, since $\mathbf{L}^2$ can be written as

$$\mathbf{L}^2 = -(\mathbf{x}\cdot\mathbf{x})\boldsymbol{\nabla}^2 + (\mathbf{x}\cdot\boldsymbol{\nabla})^2 + (\mathbf{x}\cdot\boldsymbol{\nabla}), \tag{1.176}$$

which is a sum of three commuting operators, each of which is invariant under unitary frame rotations.

The solid harmonics in $\mathbb{R}^3$ are normalized to unity over the unit sphere, and the components L_i are Hermitian with respect to the inner product of square-integral functions over the unit sphere. That only integral values of the angular momentum quantum number l can occur is a consequence of the Hermitian property and the fact that both $L_+\mathcal{Y}_{l\,l} = 0$ and $L_-\mathcal{Y}_{l,-l} = 0$ must be satisfied (see Ref. [21] for more discussion of this result.)

The polynomials $D^j_{m\,m'}(Z), z = (z_{11}, z_{21}, z_{12}, z_{22}) \in \mathbf{C}^4$ are homogeneous of degree $2j$. The angular momentum operator $\mathbf{J}^2$, where $\mathbf{J}$ has components (J_1, J_2, J_3), is given by

$$\begin{aligned}\mathbf{J}^2 &= -(\det Z)(\det \frac{\partial}{\partial Z}) + J_0(J_0+1),\;\; J_0 = (z\cdot\partial)/2, \\ \partial &= (\partial/\partial z_{11}, \partial/\partial z_{21}, \partial/\partial z_{12}, \partial/\partial z_{22}),\end{aligned} \tag{1.177}$$

which is a sum of two commutating operators $-(\det Z)(\det \frac{\partial}{\partial Z})$ and $J_0(J_0+1)$, each of which is invariant under $SU(2)$ transformations induced by unitary frame rotations. The $SU(2)$ solid harmonics are homogeneous polynomials of degree $2j$ such that

$$\det \frac{\partial}{\partial Z} D^j_{m\,m'}(Z) = 0,\;\; \mathbf{J}^2 D^j_{m\,m'}(Z) = j(j+1) D^j_{m\,m'}(Z), \tag{1.178}$$

since the Euler operator J_0 has eigenvalue j. The components (J_1, J_2, J_3) of the angular momentum operators $\mathbf{J}$ corresponding to left transformations of Z and the components (J'_1, J'_2, J'_3) corresponding to right transformations as given by (1.171) and (1.172) then have the standard actions on these polynomials. Under either left or right $SU(2)$ transformations these polynomials give the standard unitary irreducible representations of the group $SU(2)$ given explictly by (1.169) and (1.170).

The relation between $SU(2)$ solid harmonics and $SO(3,\mathbb{R})$ solid harmonics is given by

$$\begin{aligned}&(-1)^{l+m} D^l_{-m,0}\begin{pmatrix} x_3 & x_1 - ix_2 \\ x_1 + ix_2 & -x_3 \end{pmatrix} \\ &= \sqrt{\frac{4\pi}{2l+1}}\,\mathcal{Y}_{l,m}(2x_1x_3, 2x_2x_3, x_3^2 - x_1^2 - x_2^2), x \in \mathbb{R}^3,\end{aligned} \tag{1.179}$$

for $l = 0, 1, 2, \ldots$. This identity may be proved directly from the definitions of the respective polynomials by expanding the factor $(x_3^2 - x_1^2 - x_2^2)^{l-m-2k}$ that appears in $\mathcal{Y}_{l,m}$ and using the binomial sum identity (1.159) to transform the right-hand side of (1.179) to the form given by the left-hand side from definition (1.160). The mapping $(x_1, x_2, x_3) \mapsto (2x_1x_3, 2x_2x_3, x_3^2 - x_1^2 - x_2^2)$ carries the points on the sphere of radius r onto the sphere of radius r^2. In particular, the unit sphere $\mathbb{S}^2$ is mapped onto itself. Moreover, since (1.179) is an identity in polynomials, it may be extended to complex numbers $x = (x_1, x_2, x_3) \in \mathbb{C}^3$. If we set

$$x_1 = (-z_1^2 + z_2^2)/2\sqrt{z_1z_2}, x_2 = i(z_1^2 + z_2^2)/2\sqrt{z_1z_2}, x_3 = \sqrt{z_1z_2}, \quad (1.180)$$

so that $x_1^2 + x_2^2 + x_3^2 = 0$ is a complex Cartan vector of zero length and

$$2x_1x_3 = -z_1^2 + z_2^2, 2x_1x_3 = iz_1^2 + iz_2^2, x_3^2 - x_1^2 - x_2^2 = 2z_1z_2, \quad (1.181)$$

we obtain from (1.179) the following relation between $SU(2)$ solid harmonics, $SO(3,\mathbb{R})$ solid harmonics, and $SU(2)$ spinor harmonics for *integer* $l \geq 0$:

$$\begin{aligned}
&(-1)^{l+m} D^l_{-m,0}\begin{pmatrix} \sqrt{z_1z_2} & z_2^2/\sqrt{z_1z_2} \\ -z_1^2/\sqrt{z_1z_2} & -\sqrt{z_1z_2} \end{pmatrix} \\
&= \sqrt{\frac{4\pi}{2l+1}}\mathcal{Y}_{l,m}(-z_1^2 + z_2^2, iz_1^2 + iz_2^2, 2z_1z_2) \\
&= \frac{(2l)!}{l!}\sqrt{\frac{2l+1}{4\pi}}\frac{z_1^{l+m}z_2^{l-m}}{\sqrt{(l+m)!(l-m)!}}. \qquad (1.182)
\end{aligned}$$

The $SU(2)$ spinor harmonics are defined for all angular momentum quantum numbers $j \in \{0, 1/2, 1, 3/2, \ldots\}$ by

$$P_{j\,m}(z_1, z_2) = \frac{z_1^{j+m}z_2^{j-m}}{\sqrt{(j+m)!(j-m)!}}, \quad m = j, j-1, \ldots, -j, \quad (1.183)$$

but only integer values of j arise in the relation (1.182) because of choosing $m' = 0$, which forces also l to be a nonnegative integer. The angular momentum operators whose standard action on the spinor harmonics generates the polynomials are given by

$$\begin{aligned}
&J_+ = z_1\partial/\partial z_2, J_- = z_2\partial/\partial z_1, \\
&J_3 = (z_1\partial/\partial z_1 - z_2\partial/\partial z_2)/2. \qquad (1.184)
\end{aligned}$$

Indeed, as noted earlier in (1.60), the transformation properties of the spinor harmonics given by (1.61) (replace U by Z) gives the simplest direct derivation of the $SU(2)$ solid harmonics themselves, which then through relations (1.163) reproduce the spinor harmonics.

1.3.1 Inner products

The properties of spinor harmonics "fit" into the standard theory of angular momentum by defining the inner product of two polynomials $P(z_1, z_2)$ and $Q(z_1, z_2)$ in the complex variables z_1, z_2 by

$$(P, Q) = P^*(\partial/\partial z_1, \partial/\partial z_2)\, Q(z_1, z_2)|_{z_1=z_2=0}, \tag{1.185}$$

where P^* is the complex conjugate polynomial to P, and $P^*(\partial/\partial z_1, \partial/\partial z_2)$ is obtained from $P^*(z_1, z_2)$ by replacing z_1 and z_2 by the derivatives $\partial/\partial z_1$ and $\partial/\partial z_2$. Thus, if $P(z_1, z_2) = z_1 z_2 + i z_2^2$, then $P^*(\partial/\partial z_1, \partial/\partial z_2) = (\partial/\partial z_1)(\partial/\partial z_2) - i(\partial/\partial z_2)^2$. The differential operator $P^*(\partial/\partial z_1, \partial/\partial z_2)$ then acts on the polynomial $Q(z_1, z_2)$ to produce a new polynomial that is evaluated at the origin $z_1 = z_2 = 0$. In this inner product, z_i and $\partial/\partial z_i$ are Hermitian conjugate operators in the space of complex polynomials in the complex variables (z_1, z_2), so that the angular momentum operators defined from (1.184) by

$$J_1 = (J_+ + J_-)/2,\ J_2 = (J_+ - J_-)/2i, \tag{1.186}$$

$$J_3 = (z_1\partial/\partial z_1 - z_2\partial/\partial z_2)/2 \tag{1.187}$$

are Hermitian operators. It is the Hermitian property of the commuting operators $\mathbf{J}^2$ and J_3 that assures the orthogonality $(P_{j'm'}, P_{jm}) = 0,\ j' \neq j,\ m' \neq m$, of the spinor polynomials.

The inner product (1.185), extended to n indeterminates, including $n = 1$, in the obvious way, is identical in structure and numerical value to the inner product associated with boson creation and annihilation operators. Thus, let $a = (a_1, a_2, \ldots, a_n)$ be a sequence of n commuting boson creation operators and $\overline{a} = (\overline{a}_1, \overline{a}_2, \ldots, \overline{a}_n)$ the corresponding sequence of n commuting annihilation operators, such that the commutation relations operators $[\overline{a}_i, a_j] = \delta_{i,j},\ i, j = 1, 2, \ldots, n$ are satisfied. The operator $\overline{a}_i$ is then the Hermitian conjugate to a_i in the inner product defined by $\langle 0|P^*(\overline{a})Q(a)|0\rangle$, where $P(a)$ and $Q(a)$ are arbitrary polynomials in the creation operators, with $P^*(\overline{a})$ the complex conjugate polynomial in the annihilation operators. The ket-vector $|0\rangle = |0, 0, \ldots, 0\rangle$ is the unique vector annihilated by each of the annihilation operators; that is, $\overline{a}_i|0\rangle = 0,\ i = 1, 2, \ldots, n$.

The complete set of commuting Hermitian operators $H_i = \overline{a}_i a_i, i = 1, 2, \ldots, n$ then have the eigenvalues given by $H_i|k\rangle = k_i|k\rangle$, where $k_i = 0, 1, 2, \ldots$, and the simultaneous ket-eigenvectors are $|k\rangle = P_k(a)|0\rangle$, where $k = (k_1, k_2, \ldots, k_n)$ and

$$P_k(a) = \frac{a_1^{k_1} a_2^{k_2} \cdots a_n^{k_n}}{\sqrt{k_1! k_2! \cdots k_n!}}. \tag{1.188}$$

The inner product of two polynomials $P(a)$ and $Q(a)$ is defined by

$$\langle P \,|\, Q \rangle = \langle\, 0 \,|\, P^*(\overline{a})\, Q(a) |\, 0 \,\rangle, \tag{1.189}$$

from which follows the identity of numerical value given by the two inner products:

$$\langle P \,|\, Q \rangle = (P, Q). \tag{1.190}$$

In the right-hand side the polynomials are $P(z)$ and $Q(z)$, where the indeterminates $z = (z_1, z_2, \ldots, z_n)$ replace the boson creation operators $a = (a_1, a_2, \ldots, a_n)$. We then also have the orthonormality of basis ket-vectors or basis polynomials as given by

$$\langle P_k \,|\, Q_{k'} \rangle = (P_k, Q_{k'}) = \delta_{k,k'} = \prod_{i=1}^{n} \delta_{k_i, k'_i}. \tag{1.191}$$

The creation-annihilation operator formalism is used frequently to model complicated composite physical systems. It is then natural to use this formalism for developing the unitary representation theory of the unitary group $U(n)$ because the commutation relations $[\overline{a}_i, a_j] = \delta_{i,j}, i, j = 1, 2, \ldots, n$ are invariant under unitary transformations; that is, the new operators and their Hermitian conjugates defined by

$$a'_i = \sum_{j=1}^{n} u_{ij} a_j, \; \overline{a}'_i = \sum_{j=1}^{n} u^*_{ij} \overline{a}_j \tag{1.192}$$

also satisfy the commutation relations $[\overline{a}'_i, a'_j] = \delta_{i,j}$, $i, j = 1, 2, \ldots, n$. The invariance of the *number operators* H_i, $i = 1, 2, \ldots, n$, requires that the transformation be restricted to the unimodular subgroup $SU(n) \subset U(n)$. Since $[\frac{\partial}{\partial x_i}, x_j] = \delta_{i,j}$, and $\frac{\partial}{\partial x_i}$ is Hermitian conjugate to x_i in the inner product (P, Q), the two structures, complex polynomials in creation-annihilation boson operators with inner product $\langle \;|\; \rangle$ and complex polynomials in indeterminate variables-derivatives with inner product $(\;,\;)$, are one-to-one in their properties.

The inner product in the Hilbert space of states of quantum theory is basic to its physical interpretation in terms of probabilities of the outcome of measurements of observables. Born's probabilistic interpretation of the solutions of the Schrödinger equation for a complex many-particle physical system imposes the condition that such wavefunctions be square integrable over the coordinate parameters used to describe the system. Physical observables are represented by Hermitian operators is such Hilbert spaces. It is the Hermitian property that is important for angular momentum theory. The condition that "angular momentum operators" be Hermitian with respect to the inner product for the spaces being used assures orthogonality of functions. Results from one such realization can be transferred to another with compatability of relations. (See Louck and Galbraith [124] for applications of this method.)

We conclude this section by noting the following orthogonality properties of the $SU(2)$ solid harmonics, the spinor solid harmonics, and the $SO(3,\mathbb{R})$ solid harmonics in the inner product $(\,,\,)$:

$$\left(D^j_{m\,m'}, D^{j'}_{m''\,m'''}\right) = \delta_{j,j'}\delta_{m,m''}\delta_{m',m'''}(2j)!, \tag{1.193}$$

$$(P_{j\,m}, P_{j'\,m'}) = \delta_{j,j'}\delta_{m,m'}, \tag{1.194}$$

$$(\mathcal{Y}_{l\,m}, \mathcal{Y}_{l'\,m'}) = \frac{(2l)!}{2^l l!}\frac{2l+1}{4\pi}\delta_{l,l'}\delta_{m,m'}. \tag{1.195}$$

The orthogonality of the $SO(3,\mathbb{R})$ solid harmonics is assured in either the inner product given by integration over the unit sphere or in the inner product $(\,,\,)$ because the angular momentum components (L_1, L_2, L_3) are Hermitian operators, but the normalization can be different. (Since the standard action of angular momentum operators preserves normalization, it is sufficient to verify the above relations for maximal projection quantum numbers.) Often, in combinatorial arguments, the inner product plays no direct role.

1.4 Combinatorial Features

1.4.1 Combinatorial definition of Wigner-Clebsch-Gordan coefficients

The $SU(2)$ solid harmonics have a basic role in the interpretation of WCG coefficients in combinatorial terms. The abstract Hilbert space coupling rule for compounding two independent angular momenta $\mathbf{J}(1)$ and $\mathbf{J}(2)$ with components $(J_1(1), J_2(1), J_3(1))$ and $(J_1(2), J_2(2), J_3(2))$ to a total angular momentum $\mathbf{J} = \mathbf{J}(1) + \mathbf{J}(2)$ with components $(J_1, J_2, J_3) = (J_1(1) + J_1(2), J_2(1) + J_2(2), J_3(1) + J_3(2))$ is

$$|(j_1\,j_2)j\,m\rangle = \sum_{\substack{m_1,m_2\\ m_1+m_2=m}} C^{\,j_1\;\,j_2\;\,j}_{\,m_1\,m_2\,m}\;|j_1\,m_1\rangle\otimes|j_2\,m_2\rangle. \tag{1.196}$$

This relation in abstract Hilbert space is realized explicitly by $SU(2)$ solid harmonics and spinor harmonics as follows:

$$\begin{aligned}\psi_{(j_1\,j_2)j\,m}(Z) \;&=\; \sum_{\substack{m_1,m_2\\ m_1+m_2=m}} C^{\,j_1\;\,j_2\;\,j}_{\,m_1\,m_2\,m}\\ &\quad\times P_{j_1\,m_1}(z_{11}, z_{21})P_{j_2\,m_2}(z_{12}, z_{22}),\end{aligned} \tag{1.197}$$

$$\psi_{(j_1\,j_2)j\,m}(Z) = \sqrt{\frac{2j+1}{(j_1+j_2-j)!(j_1+j_2+j+1)!}} \times (\det Z)^{j_1+j_2-j} D^j_{m,j_1-j_2}(Z), \qquad (1.198)$$

where $P_{j_1\,m_1}(z_{11}, z_{21})$ and $P_{j_2\,m_2}(z_{12}, z_{22})$ are the spinor harmonics defined by (1.183). Explicit knowledge of the WCG coefficients is not needed to prove these relationships, as we now demonstrate by validating the following relations (1.199)-(1.203).

Proof. Consider the angular momentum operators

$$J_+(1) = z_{11}\partial/\partial z_{21}, J_-(1) = z_{21}\partial/\partial z_{11}, \qquad (1.199)$$
$$J_3(1) = (z_{11}\partial/\partial z_{11} - z_{21}\partial/\partial z_{21})/2;$$

$$J_+(2) = z_{12}\partial/\partial z_{22}, J_-(2) = z_{22}\partial/\partial z_{12}, \qquad (1.200)$$
$$J_3(2) = (z_{12}\partial/\partial z_{12} - z_{22}\partial/\partial z_{22})/2.$$

The $J_3(i)$ operators are Hermitian in the inner product (,) and the $J_+(i), J_-(i)$ are Hermitian conjugates. These operators have the standard action on the polynomials $\psi_{j_1\,m_1}(z_{11}, z_{21})$ and $\psi_{j_2\,m_2}(z_{12}, z_{22})$, respectively, which are normalized to unity in the inner product (,). The components of total angular momentum operator $\mathbf{J} = \mathbf{J}(1) + \mathbf{J}(2)$ have the standard action on the polynomials $\psi_{(j_1\,j_2)j\,m}(Z)$, since they have the standard action on the factor $D^j_{m,j_1-j_2}(Z)$, and $[\mathbf{J}, \det\, Z] = [\mathbf{J}, \det\, \frac{\partial}{\partial Z}] = \mathbf{0}$. Thus, we have

$$\mathbf{J}^2\psi_{(j_1\,j_2)j\,m}(Z) = j(j+1)\psi_{(j_1\,j_2)j\,m}(Z),$$
$$J_3\psi_{(j_1\,j_2)j\,m}(Z) = m\psi_{(j_1\,j_2)j\,m}(Z), \qquad (1.201)$$
$$J_\pm\psi_{(j_1\,j_2)j\,m}(Z) = \sqrt{(j \mp m)(j \pm m+1)}\psi_{(j_1\,j_2)j\,m\pm 1}(Z).$$

We also note that the two commuting parts of $\mathbf{J}^2$, which are given by (1.177), are diagonal on these functions:

$$J_0(J_0+1)\psi_{(j_1\,j_2)j\,m}(Z)$$
$$= (j_1+j_2)(j_1+j_2+1)\psi_{(j_1\,j_2)j\,m}(Z), \qquad (1.202)$$
$$(\det Z)(\det \frac{\partial}{\partial Z})\psi_{(j_1\,j_2)j\,m}(Z)$$
$$= (j_1+j_2-j)(j_1+j_2+j+1)\psi_{(j_1\,j_2)j\,m}(Z). \qquad (1.203)$$

The first relation is a direct consequence of the fact that the polynomial (1.198) is homogeneous of degree $2j_1 + 2j_2$, hence, J_0 has eigenvalue

j_1+j_2. The second relation then follows from (1.177) and the eigenvalue $j(j+1)$ of $\mathbf{J}^2$. It may also be verified directly by the action of the operator on the highest weight polynomial (1.198) for $m=j$. Since the shift operators $J_\pm$ preserve normalization, the polynomials $\psi_{(j_1\,j_2)jm}(Z)$ are normalized to unity, since the highest weight polynomial can be shown to be so normalized. □

Since we have the standard action on the uncoupled orthonormal polynomials and the orthonormal coupled polynomials in (1.196), the linear relationship between these polynomials must define the WCG coefficients. They could differ by a $\pm$ sign from other means of defining them, but, in fact, they agree with the standard Wigner phase convention. The above derivation is also given in the context of boson polynomials in Ref. [21, p. 223]; here, we regard it as a fundamental relation in the context of $SU(2)$ solid harmonics and combinatorics.

We recall that there is a left and right action of the group $SU(2)$ on the solid harmonics. We complete the above analysis by observing that the angular momentum $\mathbf{J}'$ with components (J_1', J_2', J_3') defined by (1.172) and effecting the right transformation action in (1.170) with components that mutually commute with the components (J_1, J_2, J_3) of $\mathbf{J}$ and having $\mathbf{J}'^2=\mathbf{J}^2$ also have a well-defined action on the functions $\psi_{(j_1\,j_2)j\,m}(Z)$. The action of J_+', J_-', and J_3' on the quantum numbers (j_1,j_2) is to effect the shifts, respectively, to $(j_1+\frac{1}{2},j_2-\frac{1}{2})$, $(j_1-\frac{1}{2},j_2+\frac{1}{2})$, and (j_1,j_2). These actions of Hermitian angular momentum operators satisfying the standard commutation relations appear to be unusual in that they depend only on the angular momentum quantum numbers j_1,j_2,j themselves, which satisfying the triangle rule, and give further interesting properties of the modified $SU(2)$ solid harmonics $\psi_{(j_1\,j_2)j\,m}(Z)$. We note them in full:

$$\begin{aligned} \mathbf{J}'^2\psi_{(j_1\,j_2)j\,m}(Z) &= j(j+1)\psi_{(j_1\,j_2)j\,m}(Z), & (1.204)\\ J_3'\psi_{(j_1\,j_2)j\,m}(Z) &= (j_1-j_2)\psi_{(j_1\,j_2)j\,m}(Z); & (1.205)\end{aligned}$$

$$\begin{aligned} & J_+'\psi_{(j_1\,j_2)j\,m}(Z) & (1.206)\\ =\ & \sqrt{(j-j_1+j_2)(j+j_1-j_2+1)}\,\psi_{(j_1+\frac{1}{2}\,j_2-\frac{1}{2})j\,m}(Z), & \\ & J_-'\psi_{(j_1\,j_2)j\,m}(Z) & (1.207)\\ =\ & \sqrt{(j+j_1-j_2)(j-j_1+j_2+1)}\,\psi_{(j_1-\frac{1}{2}\,j_2+\frac{1}{2})j\,m}(Z). & \end{aligned}$$

If we identify $m'=j_1-j_2$ in these relations, their nonstandard appearance is revealed to be standard.

Rota's evaluation operation

Relations (1.197)-(1.198) may be used to derive the explicit expression for the WCG coefficients. The result is most succinctly expressed in terms of the umbral calculus (see Refs. [153, 154, 155, 156, 157]) and the *evaluation* operation, especially as summarized by Rota [155] in his article on Hopf algebras.

The evaluation at y of a divided power $x^k/k!$ of a single indeterminate x to an nonnegative integral power k is defined by

$$\text{eval}_y \frac{x^k}{k!} = \frac{[y]_k}{k!} = \frac{y(y-1)\cdots(y-k+1)}{k!} = \binom{y}{k}, \tag{1.208}$$

where $[y]_k$ is the falling factorial. This definition is extended to products by

$$\text{eval}_{(y_1,y_2,\ldots,y_n)} \prod_{i=1}^{n} \frac{x^{k_i}}{k_i!} = \prod_{i=1}^{n} \text{eval}_{y_i} \frac{x^{k_i}}{k_i!} = \prod_{i=1}^{n} \binom{y_i}{k_i}. \tag{1.209}$$

It is also extended by linearity to sums of such divided powers, multiplied by arbitrary numbers.

The basic relation underlying relation (1.197) is the following:

$$\begin{aligned} \frac{(\det Z)^n}{n!} & \sum_{A\in\mathbb{M}_2(\alpha,\alpha')} \frac{Z^A}{A!} \\ &= \sum_{B\in\mathbb{M}_2(\beta,\beta')} \left(\text{eval}_B \frac{(\det Z)^n}{n!}\right) \frac{X^B}{B!}, \end{aligned} \tag{1.210}$$

in which the line-sums β and β' of B are given in terms of the line-sums α and α' of A by a shift by n :

$$\beta = (\alpha_1 + n, \alpha_2 + n),\ \beta' = (\alpha'_1 + n, \alpha'_2 + n). \tag{1.211}$$

The evaluation operation in (1.210) gives

$$\begin{aligned} \text{eval}_B & \frac{(\det\ Z)^n}{n!} \\ &= \sum_{k_1+k_2=n} (-1)^{k_2} k_1! k_2! \binom{b_{11}}{k_1}\binom{b_{12}}{k_2}\binom{b_{21}}{k_2}\binom{b_{22}}{k_1}. \end{aligned} \tag{1.212}$$

This relation is proved by expanding the 2×2 determinant of Z, multiplying into the summation expression, changing the order of the summation, and expressing the result in terms of the evaluation operation for four indeterminates. This identity is a purely combinatorial, algebraic relation for arbitrary indeterminates and arbitrary row and column sum constraints on the array A as specified by $\alpha = (\alpha_1, \alpha_2)$ and $\alpha' = (\alpha'_1, \alpha'_2)$.

We now apply relations (1.210)-(2.212) to the case at hand, where we have $n = j_1 + j_2 - j, \alpha = (j + m, j - m), \alpha' = (j + j_1 - j_2, j - j_1 + j_2), \beta = (j_1 + j_2 + m, j_1 + j_2 - m), \beta' = (2j_1, 2j_2)$. This gives the following result for the WCG coefficients:

$$\begin{aligned} & C^{j_1\; j_2\; j}_{m_1\, m_2\, m} \\ & = \sqrt{\frac{(j_1 + j_2 - j)!(j_1 - j_2 + j)!(-j_1 + j_2 + j)!}{(j_1 + j_2 + j + 1)!}} \\ & \times \sqrt{\frac{(2j + 1)(j + m)!(j - m)!}{(j_1 + m_1)!(j_1 - m_1)!(j_2 + m_2)!(j_2 - m_2)!}} \\ & \qquad \times \mathrm{eval}_A \frac{(\det Z)^{j_1 + j_2 - j}}{(j_1 + j_2 - j)!}, \end{aligned} \tag{1.213}$$

$$\begin{aligned} & \mathrm{eval}_A \frac{(\det Z)^{j_1 + j_2 - j}}{(j_1 + j_2 - j)!} \\ & = \sum_{k_1 + k_2 = j_1 + j_2 - j} (-1)^{k_2} k_1! k_2! \binom{j_1 + m_1}{k_1} \binom{j_2 + m_2}{k_2} \\ & \qquad \times \binom{j_1 - m_1}{k_2} \binom{j_2 - m_2}{k_1}. \end{aligned} \tag{1.214}$$

In summary, we have that

Up to multiplicative square-root factors, the WCG coefficient is the integer obtained from the evaluation at the point $B = \begin{pmatrix} j_1 + m_1 & j_2 + m_2 \\ j_1 - m_1 & j_2 - m_2 \end{pmatrix}$ *of the divided power* $\frac{(\det Z)^{j_1 + j_2 - j}}{(j_1 + j_2 - j)!}$ *of a* 2×2 *determinant.*

Power of a determinant and WCG coefficients

There is still another basic relation in which the evaluation coefficients enter, which leads to a generating function for the WCG coefficients. Evaluation coefficients enter into the expansion of the power of a 3×3 determinant in terms of the monomiala X^A. The expansion goes as follows. Define the 3×3 matrix X of indeterminates by

$$X = \begin{pmatrix} x_{11} & x_{12} & x_{13} \\ x_{21} & x_{22} & x_{23} \\ x_{31} & x_{32} & x_{33} \end{pmatrix}, \tag{1.215}$$

and express the determinant in the standard way as a sum over permutations π in the symmetric group S_3 of order 3 and the signature ε_π of a permutation. We write the permutation π of the sequence $(1, 2, 3)$ given by $1 \mapsto \pi_1, 2 \mapsto \pi_2, 3 \mapsto \pi_3$ by the sequence notation $\pi = (\pi_1, \pi_2, \pi_3)$, where we sometimes drop the separating commas if no confusion can arise. Then, the $k-$th power of the determinant of X is given by

$$(\det X)^k = (\sum_{\pi \in S_3} \varepsilon_\pi x_{1\,\pi_1} x_{2\,\pi_2} x_{3\,\pi_3})^k = \sum_{A \in \mathbb{M}_3(k)} C_k(A) X^A, \tag{1.216}$$

where the summation is over all 3×3 matrices with nonnegative integer entries given by

$$A = \begin{pmatrix} a_{11} & a_{12} & a_{13} \\ a_{21} & a_{22} & a_{23} \\ a_{31} & a_{32} & a_{33} \end{pmatrix} \tag{1.217}$$

and having the single line-sum k, which defines the set $\mathbb{M}_3(k)$ of *magic squares* of order 3 and line-sum k. The coefficients $C_k(A)$ in this expansion are given as restricted summations over multinomial coefficients, which, in turn, are given in terms of Rota evaluation coefficients by

$$\begin{aligned} C_k(A) &= \sum_{k_\pi}^{\text{restr}} (-1)^{\sum k_{odd}} \binom{k}{k_{123}, k_{231}, k_{312}, k_{132}, k_{213}, k_{321}} \\ &= \frac{(-1)^{n_{ij}} (-1)^{\operatorname{Tr} A_{ij}} k!}{A_{ij}!} \operatorname{eval}_{A_{ij}} \frac{(\det X_{ij})^{a_{ij}}}{a_{ij}!}, \end{aligned} \tag{1.218}$$

where the quantities in this relation have the following definitions (see Sect. 11.5.5, Compendium B):

1. The summation over the nonnegative integers $k_\pi, \pi = (123), (231), (312), (132), (213), (321)$ not only satisfy the conditions $\sum_\pi k_\pi = k$ so that the multinomial coefficients are nonzero, but are also further restricted in a way that relates to the symmetric group S_3. For each $A \in \mathbb{M}_3(k)$, the six permutations are also to satisfy the relation

$$A = \sum_{\pi \in S_3} k_\pi P_\pi, \tag{1.219}$$

where P_π denotes a permutation matrix of order 3. The six permutation matrices $P_\pi, \pi \in S_3$, are the six matrices obtained by permuting the rows of the identity matrix I_3 by a permutation $\pi \in S_3$, so that the first, second, and third rows of $P_{\pi_1\pi_2\pi_3}$ are, respectively, e_{π_1}, e_{π_2}, and e_{π_3}, where e_i is the unit row matrix with 1 in column i and 0 in the remaining columns (unit column matrices may also be used). This gives each a_{ij} as the sum of the two $k_{\pi_1\pi_2\pi_3}$ in which $\pi_i = j$, for example, $a_{32} = k_{132} + k_{312}$ has $\pi_3 = 2$.

2. The symbol $\sum k_{\text{odd}}$ denotes the summation of the k_π over the odd permutations: $\sum k_{\text{odd}} = k_{132} + k_{213} + k_{321}$.

3. The matrices A_{ij} and X_{ij} are the 2×2 minors of A and X obtained, respectively, by striking row i and column j. There are nine choices of the pairs (i, j) given by $1 \le i \le j \le 3$, each choice giving the same result for the right-hand side of (1.218).

4. The factor $(-1)^{n_{ij}}$ is given by $(-1)^k$ for $i + j$ odd, and 1 for $i + j$ even.

Relation (1.216), with coefficients given by (1.218), is proved directly by carrying out the expansion of the k–th power of the determinant and rearranging terms so as to write the right-hand side as a sum over monomial terms X^A. Because of the constraints on the k_π in terms of the elements of A given by (1.219), there are nine ways of relating the sum over multinomial coefficients to the evaluation coefficients. This leads to the equality of all the different evaluation coefficients among the various minors of A. Relations (1.216) and (1.218) stand on their own as a basic combinatorial relationship for the power of a determinant in terms of evaluation coefficients.

We now choose $i = j = 3$ in relation (1.218). We also choose A in terms of angular momentum quantum numbers as

$$A = \begin{pmatrix} a_{11} & a_{12} & a_{13} \\ a_{21} & a_{22} & a_{23} \\ a_{31} & a_{32} & a_{33} \end{pmatrix} = \begin{pmatrix} j_1 + m_1 & j_2 + m_2 & j - m \\ j_1 - m_1 & j_2 - m_2 & j + m \\ j_2 - j_1 + j & j_1 - j_2 + j & j_1 + j_2 - j \end{pmatrix}, \tag{1.220}$$

select the case of the evaluation coefficient corresponding to element $a_{33} = j_1 + j_2 - j$, so that $A_{33} = \begin{pmatrix} j_1 + m_1 & j_2 + m_2 \\ j_1 - m_1 & j_2 - m_2 \end{pmatrix}$, and write the evaluation coefficient in terms of the WCG coefficient from (1.213). This gives the following expression for the Wigner coefficients:

$$C^{j_1\; j_2\; j}_{m_1\, m_2\, m} = (-1)^{j_1-j_2+m}\sqrt{\frac{(2j+1)A!}{J!(J+1)!}}C_J(A), \tag{1.221}$$

where $J = j_1 + j_2 + j$ is the common row-column line-sum in the magic square (1.220), and A, as given by relation (1.220), is to be substituted into this relation to express it entirely in terms of angular momentum quantum numbers. We have used the relation $(-1)^{TrA_{33}} = (-1)^{(j_1+m_1)+(j_2-m_2)} = (-1)^{(j_1+m_1)-(j_2-m_2)} = (-1)^{j_1-j_2+m}$ in obtaining the sign factor. The coefficients $C_J(A)$ are then the expansion coefficient in

$$(\det X)^J = \sum_{A\in\mathbb{M}_3(J)} C_J(A)X^A. \tag{1.222}$$

We thus obtain the famous generating function of Regge [147] for the Wigner coefficients, but now by using the combinatorial Rota evaluation coefficients, all without needing to use direct knowledge of the WCG coefficients.

It is the custom to define the so-called $3j-$coefficients by the relation

$$\begin{aligned}\begin{pmatrix} j_1 & j_2 & j \\ m_1 & m_2 & -m \end{pmatrix} &= \frac{(-1)^{j_1-j_2+m}}{\sqrt{2j+1}}C^{j_1\; j_2\; j}_{m_1\, m_2\, m} \\ &= \sqrt{\frac{A!}{J!(J+1)!}}C_J(A),\end{aligned} \tag{1.223}$$

which enhances the symmetry relations of these coefficients: The $3j-$coefficients are invariant under even permutations of the columns, are multiplied by the factor $(-1)^{j_1+j_2+j}$ by odd permutations, and by this same factor under sign reversal of the projection quantum numbers. These are the classical pre-Regge symmetries.

A *symmetry* of a WCG coefficient is a linear transformation of its labels that results in a multiple of the original coefficient. Because a determinant either changes its sign or remains invariant under permutations of its columns or row and under transposition, there are $3! \times 3! \times 2 = 72$ symmetries of the WCG coefficients and the associated $3-j$ coefficients, all of which may be deduced explictly from relation (1.222). These have been given in various places, and will not all be repeated here. (See Ref. [21] for a generating set of these relations in terms of the present notations.) We require, however, in the next chapter, the following classical symmetries of the WCG coefficients, corresponding to reversal of sign of the projection quantum numbers and to permutations of j_1, j_2, j_3, all of which can be verified directly from (1.222). We set $j_1 = a, j_2 = b, j = c$; $m_1 = \alpha, m_2 = \beta, m = \gamma = \alpha + \beta$ for clarity of presentation:

$$\begin{aligned}
C^{a\ b\ c}_{\alpha\ \beta\ \gamma} &= (-1)^{a+b-c} C^{a\ b\ c}_{-\alpha.\ -\beta,\ -\gamma},\\
C^{b\ a\ c}_{\alpha\ \beta\ \gamma} &= (-1)^{a+b-c} C^{a\ b\ c}_{\alpha\ \beta\ \gamma},\\
C^{a\ \ c\ \ b}_{\alpha,-\gamma,-\beta} &= (-1)^{a-\alpha}\sqrt{\frac{2b+1}{2c+1}}\,C^{a\ b\ c}_{\alpha\ \beta\ \gamma},\\
C^{\ c\ \ b\ \ a}_{-\gamma,\beta,-\alpha} &= (-1)^{b+\beta}\sqrt{\frac{2a+1}{2c+1}}\,C^{a\ b\ c}_{\alpha\ \beta\ \gamma},\\
C^{\ b\ \ c\ a}_{-\beta,\gamma,\alpha} &= (-1)^{b+\beta}\sqrt{\frac{2a+1}{2c+1}}\,C^{a\ b\ c}_{\alpha\ \beta\ \gamma},\\
C^{c\ \ a\ \ b}_{\gamma,-\alpha,\beta} &= (-1)^{a-\alpha}\sqrt{\frac{2b+1}{2c+1}}\,C^{a\ b\ c}_{\alpha\ \beta\ \gamma}.
\end{aligned} \tag{1.224}$$

We note also the relation of the ${}_3F_2$ hypergeometric functions of unit argument to evaluation coefficients. This result can be verified by direct calculation. An example of such a relation, which is suggestive of generalization, is given by

$$\begin{aligned}
&{}_3F_2\left(\begin{matrix} -a,\ -b,\ -c \\ d+1, e+1 \end{matrix};1\right)\\
&= (-1)^c \frac{\mathrm{eval}_A(Z)}{(d+1)_c(e+1)_c(a+b+c+d+e+1)!},
\end{aligned} \tag{1.225}$$

where $c \geq a, c \geq b$, with $a, b, c, a+e, b+e, c+e, a+d, b+d, c+d$ all non-negative integers, and we have defined $A = \begin{pmatrix} a & c+e \\ e+d & b \end{pmatrix}$; $(x)_k = x(x+1)\cdots(x+k-1)$. We also point out (see Ref. [21]) that the classical transformation properties of the ${}_3F_2$ hypergeometric functions account for all the 72 symmetries of the WCG coefficients, which were discovered independently by physicists in the context of the Regge result (see Sect. 2.2.8, Chapter 2, for details).

1.4.2 Magic square realization of the addition of two angular momenta

The relation between the two orthonormal bases of the Hilbert space $\mathcal{H}_{j_1} \otimes \mathcal{H}_{j_2}$ corresponding to the addition of two angular momentum is fully encoded in the properties of magic squares of order 3. The latter are purely combinatorial objects, whereas the origin of relation (1.196) is usually attributed to properties of the direct sum of two copies of the Lie algebra of $SU(2)$ and of the differential operators that realize these Lie algebras. The realization of the addition of angular momentum in terms of magic squares is encoded in the observation of Regge [147] that the restrictions on the domains of the quantum numbers j_1, m_1, j_2, m_2, j, m can be expressed in terms of a magic square A with line-sum $J = j_1+j_2+j$ given by (1.220). The angular momentum quantum numbers are given in terms of the elements of $A = (a_{ij})_{1\leq i,j\leq 3}$ by the invertible relations

$$\begin{aligned} &j_1 = (a_{11} + a_{21})/2, \;\; j_2 = (a_{12} + a_{22})/2, \\ &j = (a_{13} + a_{23})/2; \\ &m_1 = (a_{11} - a_{21})/2, \;\; m_2 = (a_{12} - a_{22})/2, \\ &m = (a_{23} - a_{13})/2. \end{aligned} \tag{1.226}$$

It follow from these relations and the fact that A is a magic square of line-sum J that the sum rule $m_1 + m_2 = m$, and the triangle condition $j = j_1 + j_2, j_1 + j_2 - 1, \ldots, |j_1 - j_2|$ are fulfilled.

We introduce the special symbol $\langle j_1, j_2, j\rangle$ to denote the set $\{j_1, j_2, j\}$ of angular momentum quantum numbers that satisfy the triangle conditions, where we note that, if a given triple satisfies the triangle conditions, then all permutations of the triple also satisfy the triangle conditions. The number of magic squares that have the fixed line-sum J is obtained as follows: Define the sets Δ_J and $M(j_1, j_2, j)$ by

$$\Delta_J = \{\langle j_1, j_2, j\rangle \mid j_1 + j_2 + j = J\}, \tag{1.227}$$

$$\begin{aligned} M(j_1, j_2, j) = \{(m_1, m_2) \mid &-j_1 \leq m_1 \leq j_1; \\ &-j_2 \leq m_2 \leq j_2; \; -j \leq m_1 + m_2 \leq j\}. \end{aligned} \tag{1.228}$$

Then, we have the following identity, which gives the number of angular momentum magic squares (1.220) with line-sum J :

$$\sum_{\langle j_1,j_2,j\rangle\in\Delta_J} |M(j_1, j_2, j)| = \binom{J+5}{5} - \binom{J+2}{5}. \tag{1.229}$$

It is nontrivial to effect the summation on the left-hand side of this relation to obtain the right-hand side, but this expression is known from the theory of magic squares (see Stanley [163, Vol. 1, p. 92]).

Not only can the addition of two angular momenta in quantum theory with its triangle rule for three angular momentum quantum numbers and its sum rule on the corresponding projection quantum numbers be codified in the structure of magic squares of order 3 and arbitrary line-sum, but also the content of the abstract state vector relation (1.196) itself can be so formulated, as follows:

$$\begin{aligned}&\Big|((a_{11}+a_{21})/2,(a_{12}+a_{22})/2)\ (a_{13}+a_{23})/2,(a_{23}-a_{13})/2\Big\rangle\\ &= \sum_{A_2\in\mathbb{M}_2(\alpha,\beta)} W_N(A) \qquad\qquad (1.230)\\ &\times\Big|(a_{11}+a_{21})/2,(a_{11}-a_{21})/2\Big\rangle\otimes\Big|(a_{12}+a_{22})/2,(a_{12}-a_{22})/2\Big\rangle.\end{aligned}$$

In this relation, $A=(a_{ij})_{1\leq i,j\leq 3}$ is a magic square of order 3 and arbitrary line-sum N (nonnegative integer), and the summation is over all matrix arrays of nonnegative integers $A_2=\begin{pmatrix} a_{11} & a_{12}\\ a_{21} & a_{22}\end{pmatrix}$ of order 2 with row and column sums given by $\alpha_1=N-a_{13},\alpha_2=N-a_{23},\beta_1=N-a_{31},\beta_2=N-a_{32}$, in which the elements (a_{31},a_{32},a_{33}) in row 3 and the elements (a_{13},a_{23},a_{33}) in column 3 are held fixed. The coefficients $W_N(A)$ in relation (1.230) are not determined by the magic square structure, but all domains of definition of the entries in the ket-vectors are determined. The coefficients $W_N(A)$ themselves are WCG coefficients, as given in terms of the elements of the magic square A by

$$W_N(A)=C^{(a_{11}+a_{21})/2\ (a_{12}+a_{22})/2\ (a_{13}+a_{23})/2}_{(a_{11}-a_{21})/2\ (a_{12}-a_{22})/2\ (a_{23}-a_{13})/2}. \qquad (1.231)$$

Thus, to each magic square A of order 3 there corresponds a unique WCG coefficient $W_N(A)$, and conversely. Thus, the full abstract structure of the addition of angular momentum is one-to-one with the structure of magic squares of order 3. These rich combinatorial footings of angular momentum theory are completed by the observation that the WCG coefficients themselves are obtained by the Regge generating function (1.222) for the expansion of a determinant of order 3, again there being no reference to Lie algebraic structures. *Addition of angular momentum in quantum theory can be presented in terms of magic squares and the Regge generating function for WCG coefficients without reference to Lie algebras and differential operators.*

1.5 Kronecker Product of Solid Harmonics

The Kronecker product, which is sometimes called the *direct product* of a matrix A of order m and a matrix B of order n, is defined by

$$A \otimes B = \begin{pmatrix} a_{11}B & a_{12}B & \cdots & a_{1m}B \\ a_{21}B & a_{22}B & \cdots & a_{2m}B \\ \vdots & \vdots & \cdots & \vdots \\ a_{m1}B & a_{m2}B & \cdots & a_{mm}B \end{pmatrix}. \tag{1.232}$$

From this definition, it may be verified that

$$(A \otimes B) \otimes C = A \otimes (B \otimes C), \tag{1.233}$$

where C is a matrix of order l. The placement of the parenthesis pair ( ) indicates how the Kronecker product is to be composed, but it is, in fact, independent of this pairing; hence,

$$A \otimes B \otimes C = (A \otimes B) \otimes C = A \otimes (B \otimes C). \tag{1.234}$$

The rows of the Kronecker product $A \otimes B$, where $A = (a_{ij})_{1 \leq i,j \leq m}$ and $B = (b_{kl})_{1 \leq k,l \leq n}$ are inherited from the row pairs (i,k) and column pairs (j,l) from A and from B, so that$(A \otimes B)_{ik;jl} = a_{ij}b_{kl}$. In (1.232), the $(ij)-$th *block matrix* is $a_{ij}B$. Kronecker products satisfy the multiplication rule

$$(A \otimes B)(C \otimes D) = (AC) \otimes (BD), \tag{1.235}$$

where A and C are of order m, and B and D are of order n. The Kronecker product extends to rectangular matrices as well, it only be required that the row and column order of the matrices in the products always match.

The elements in the Kronecker product of two matrices $D^{j_1}(X)$ and $D^{j_1}(Y)$ whose elements are the $SU(2)$ solid harmonic are given by

$$\left(D^{j_1}(X) \otimes D^{j_2}(Y)\right)_{m_1\, m_2; m_1'\, m_2'} = D^{j_1}_{m_1\, m_1'}(X) D^{j_2}_{m_2\, m_2'}(Y). \tag{1.236}$$

The product of two matrices of solid harmonics satisfies the relation

$$\begin{aligned} &\left(D^{j_1}(X') \otimes D^{j_2}(Y')\right)\left(D^{j_1}(X) \otimes D^{j_2}(Y)\right) \\ &\quad = D^{j_1}(X'X) \otimes D^{j_2}(Y'Y), \end{aligned} \tag{1.237}$$

where the elements of the 2×2 matrices X, Y, X', Y' are arbitrary commuting indeterminates. This relation has interesting consequences for special functions, but we give here only those properties for $X = Y$ and $X' = Y'$.

The basic property of the Kronecker product of matrices of solid harmonics defined over the same indeterminates Z is

$$C^{(j_1\, j_2)}\left(D^{j_1}(Z) \otimes D^{j_2}(Z)\right)\left(C^{(j_1\, j_2)}\right)^T = \sum_{j=j_{min}}^{j_{max}} (\det Z)^{j_1+j_2-j} D^j(Z), \tag{1.238}$$

in which $\oplus$ denotes the direct sum of matrices. The validity of this result follows from the fact that the implied double WCG coefficient coupling of the solid harmonics $D^{j_1}_{m_1\,m'_1}(Z)$ and $D^{j_2}_{m_2\,m'_2}(Z)$ given by (1.239) below is insensitive to the detailed properties of the polynomials: It depends only on the angular momentum properties of the coupling of angular momenta. The presence of $(\det Z)^{j_1+j_2-j}$ is required by the homogeneity properties of the solid harmonics.

The expresssion of relation (1.238) in terms of the elements of the matrices is the first result below, the other three being obtained from the first by using the orthogonality relations for the WCG coefficients (see (1.145)-(1.156)):

$$\sum_{\substack{m_1,m_2\\ m'_1,m'_2}} C^{j_1\;j_2\;j}_{m_1\,m_2\,m} C^{j_1\;j_2\;j'}_{m'_1\,m'_2\,m'} D^{j_1}_{m_1\,m'_1}(Z) D^{j_2}_{m_2\,m'_2}(Z) = \delta_{j,j'}(\det Z)^{j_1+j_2-j} D^{j}_{m\,m'}(Z), \tag{1.239}$$

$$\sum_{m'_1,m'_2} C^{j_1\;j_2\;j}_{m'_1\,m'_2\,m'} D^{j_1}_{m_1\,m'_1}(Z) D^{j_2}_{m_2\,m'_2}(Z) = C^{j_1\;j_2\;j}_{m_1\,m_2\,m}(\det Z)^{j_1+j_2-j} D^{j}_{m\,m'}(Z), \tag{1.240}$$

$$\sum_{m_1,m_2} C^{j_1\;j_2\;j}_{m_1\,m_2\,m} D^{j_1}_{m_1\,m'_1}(Z) D^{j_2}_{m_2\,m'_2}(Z) = C^{j_1\;j_2\;j}_{m'_1\,m'_2\,m'}(\det Z)^{j_1+j_2-j} D^{j}_{m\,m'}(Z), \tag{1.241}$$

$$D^{j_1}_{m_1\,m'_1}(Z) D^{j_2}_{m_2\,m'_2}(Z) = \sum_{j} C^{j_1\;j_2\;j}_{m_1\,m_2\,m} C^{j_1\;j_2\;j}_{m'_1\,m'_2\,m'}(\det Z)^{j_1+j_2-j} D^{j}_{m\,m'}(Z). \tag{1.242}$$

These relations apply for an arbitrary matrix Z, including singular ones. These relations can be specialized in many ways to obtain results for special functions (see Chen and Louck [40]). In particular, they hold for Z an element of any of the groups $SU(2), U(2), GL(2,\mathbb{C})$.

Angular momentum theory is for the most part an extension to n independent angular momenta of the results summarized in the preceding sections, which, then, is about the properties of the direct product group $SU(2)\times SU(2)\times\cdots\times SU(2)$ (n times) and the diagonal subgroup $(U,U,\ldots,U)$, which is isomorphic to $SU(2)$. This structure would not be very interesting if such an extension merely involved the replication of the relations of this section with the generation of complicated formulas involving summations over multiple WCG coefficients. But, it turns out that the invariant theory underlying such structures and their combinatorial underpinnings lead to a rich comprehensible theory of many-particle complex quantum systems, which is considered in Chapters 2-4. Already, in this Introduction, we make this point by introducing a special class of polynomials that we call $SU(n)$ solid harmonics because of their implicit occurrence in the fundamental work of Schwinger [160], which has its basis in a classical result in combinatorics by MacMahon [129].

1.6 $SU(n)$ Solid Harmonics

1.6.1 Definition and properties of $SU(n)$ solid harmonics

Generating functions codify the content of many mathematical entities in a unifying, comprehensive way. They are very popular in combinatorics, and Schwinger [160] used them extensively in his fundamental treatment of angular momentum theory. In this subsection and the next, we present a natural generalization of the $SU(2)$ solid harmonics to a class of polynomials, called $SU(n)$ solid harmonics, that are homogeneous in n^2 indeterminates. While these polynomials are of interest in their own right (see Gelfand and Graev [61]), it is their fundamental role in the addition of n independent angular momenta that motivates their introduction here in the first chapter. *The $SU(n)$ solid harmonics already occur at the level of multiple copies of $SU(2)$ and the binary theory of addition of angular moment;* they bring an unexpected unity and coherence to angular momentum coupling and recoupling theory through their relationship to MacMahon's [129] master theorem, as rediscovered and slightly generalized by Schwinger [160].

We list in a compendium format in this section some of the principal properties of $SU(n)$ solid harmonica, all of which can be proved directly from their definition, or as outlined below:

1. Definition:

$$D^{p}_{\alpha\,\beta}(Z) = \sqrt{\alpha!\beta!}\sum_{A\in \mathbb{M}^{p}_{n\times n}(\alpha,\beta)} \frac{Z^A}{A!}, \tag{1.243}$$

where we employ the following space-saving notations: A is the matrix $A = (a_{ij})_{1\leq i,j\leq n}$ of order n in the nonnegative integers a_{ij}; $A! = \prod_{i,j=1}^{n} a_{ij}!$, $Z^A = \prod_{i,j=1}^{n} z_{ij}^{a_{ij}}$; α is a sequence $\alpha = (\alpha_1, \alpha_2, \ldots, \alpha_n)$ of n nonnegative integers that sum to p and is called a *composition* of k into p nonnegative parts, this property being denoted by

$\alpha \vdash p$. We also write $x^\alpha = x_1^{\alpha_1} x_2^{\alpha_2} \cdots x_n^{\alpha_n}$ and $\alpha! = \alpha_1! \alpha_2! \cdots \alpha_n!$, for each pair (α, β) of such compositions $\alpha \vdash p$ and $\beta \vdash p$. The notation $\mathbb{M}^p_{n\times n}(\alpha, \beta)$ denotes the set of all matrices A such that the entries in row i sum to α_i and those in column j to β_j, for all $i, j = 1, 2, \ldots, n$. The significance of the row-sum vector α is that α_i is the degree of the polynomial $D^p_{\alpha\,\beta}(Z)$ in the variables $(z_{i1}, z_{i2}, \ldots, z_{in})$ in row i of Z, while the column-sum vector β has the significance that β_j is the degree of the polynomial $D^p_{\alpha\,\beta}(Z)$ in the variables $(z_{1j}, z_{2j}, \ldots, z_{nj})$ in column j of Z.

2. Matrix of the $D^p_{\alpha\,\beta}(Z)$ polynomials:

   The number of compositions of the integer k into n nonnegative parts is given by $\binom{n+p-1}{p}$. The compositions in this set may be linearly ordered by the lexicographical rule: if the first nonzero part of $\alpha - \beta$ is positive, we write $\alpha > \beta$; if negative. $\alpha < \beta$; if zero, $\alpha = \beta$. The polynomial $D^k_{\alpha\,\beta}(Z)$ is then the entry in row α and column β of the square matrix $D^p(Z)$ of order

$$\dim D^p(Z) = \binom{n+p-1}{p}, \tag{1.244}$$

   where, following the convention for $SU(2)$, the rows are labeled from top-to-bottom by the greatest to the least sequence, and the columns are labeled in the same manner as read from left-to-right. We give below the combinatorial proof by Chen and Louck [40] that these polynomials satisfy the following multiplication rule for arbitrary matrices X and Y :

$$D^p(X)D^p(Y) = D^p(XY). \tag{1.245}$$

3. Orthogonality in the inner product (,):

$$\left(D^p_{\alpha\,\beta}, D^{p'}_{\alpha'\,\beta'}\right) = \delta_{p,p'}\delta_{\alpha,\alpha'}\delta_{\beta,\beta'}p!\,. \tag{1.246}$$

4. Reduction to spinor harmonics:

$$D^p_{\alpha\,\beta}(\text{diag}(0,\ldots,0,z^{(j)},0,\ldots,0)) = \left(\textstyle\prod_{\substack{i=1\\ i\neq j}}^n \delta_{0,\beta_i}\right)\sqrt{\frac{p!}{\alpha!}}\,z^\alpha, \tag{1.247}$$

in which all columns of Z are $0 = \text{col}(0, 0, \ldots, 0)$, except column j, which is $z^{(j)} = \text{col}(z_1, z_2, \ldots, z_n)$, and $\alpha \vdash p$. It is also true that

$$D^p_{\alpha\,\beta}(\text{diag}(z_1, z_2, \ldots, z_n)) = \delta_{\alpha,\beta} z^\alpha. \tag{1.248}$$

5. Diagonal matrix:

$$D^p(I_n) = I_{\binom{n+p-1}{p}}. \tag{1.249}$$

6. Transposition property:

$$D^p(Z^T) = (D^p(Z))^T. \tag{1.250}$$

7. Special irreducible unitary representations of $SU(n)$:

$$D^p(U)D^p(V) = D^p(UV), \text{all } U, V \in SU(n). \tag{1.251}$$

Of course, the multiplication property (1.251) is much more general than this, applying as it does to matrices X and Y, including singular matrices, of arbitrary commuting indeteminates.

1.6.2 MacMahon and Schwinger master theorems

1. Schwinger's master theorem: For any two matrices X and Y of order n, the following identities hold:

$$\begin{aligned} e^{(\partial_x :X: \partial_y)} e^{(x:Y:y)}|_{x=y=0} &= \sum_{p=0}^{\infty} \sum_{\alpha,\beta \vdash p} D^p_{\alpha\,\beta}(X) D^p_{\beta\,\alpha}(Y) \\ &= \frac{1}{\det(I - XY)}, \end{aligned} \tag{1.252}$$

$$(x :Z: y) = xZy^T = \sum_{i,j=1}^{n} x_i z_{ij} y_j. \tag{1.253}$$

2. MacMahon's master theorem: Let X be the diagonal matrix $X = \text{diag}(x_1, x_2, \ldots, x_n)$ and Y a matrix of order n. Then, the coefficient of x^α in the expansion of $\frac{1}{\det(I-XY)}$ equals the coefficient of x^α in the product y^α, where $y_i = \sum_{j=1}^{n} y_{ij} x_j$; that is,

$$\frac{1}{\det(I - XY)} = \sum_{p=0}^{\infty} \sum_{\alpha \vdash p} D^p_{\alpha\,\alpha}(Y)\, x^\alpha. \tag{1.254}$$

3. Basic master theorem: Let Z be a matrix of order n, then

$$\frac{1}{\det(I - tZ)} = \sum_{p=0}^{\infty} t^p \sum_{\alpha \vdash p} D^p_{\alpha\,\alpha}(Z). \tag{1.255}$$

Schwinger's relation (1.252) follows from (1.255) by setting $Z = XY$ and using the multiplication property (1.245); MacMahon's relation (1.254) then follows from Schwinger's result by setting $X = \mathrm{diag}(x_1, x_2, \ldots, x_n)$ and using property (1.248). Of course, MacMahon's Master Theorem preceded Schwinger's result by many years (Schwinger's report is reprinted in Ref. [25]). The unification into the single form by using properties of the $D^p_{\alpha\,\beta}(Z)$ polynomials was pointed out by in Ref. [108] . More surprisingly, relation (1.252) was already discovered for the general linear group in 1897 by Molien [137]. The properties of this relation for groups are developed extensively in Michel and Zhilinskii [133].

For many purposes, it is better in combinatorics to avoid all square roots by using the polynomials

$$L^p_{\alpha\,\beta}(Z) = \sum_{A \in \mathbb{M}^p_{n\times n}(\alpha,\beta)} \frac{Z^A}{A!}. \tag{1.256}$$

1.6.3 Combinatorial proof of the multiplication property

The multiplication of two $SU(n)$ solid harmonics is expressed by

$$\sum_{\beta \vdash p} \sum_{A \in \mathbb{M}^p_{n\times n}(\alpha,\beta)} \sum_{B \in \mathbb{M}^p_{n\times n}(\beta,\gamma)} \frac{\gamma!}{A!\,B!} X^A Y^B = \sum_{C \in \mathbb{M}^p_{n\times n}(\alpha,\gamma)} \frac{(XY)^C}{C!}. \tag{1.257}$$

The following identities can be used to simplify this relation:

$$\frac{(XY)^C}{C!} = \sum_{\beta \vdash p} \sum_{A \in \mathbb{M}^p_{n\times n}(\alpha,\beta)} \sum_{B \in \mathbb{M}^p_{n\times n}(\beta,\gamma)} \begin{Bmatrix} C \\ A\,B \end{Bmatrix} X^A Y^B,$$

$$C \in \mathbb{M}^p_{n\times n}(\alpha,\gamma); \tag{1.258}$$

$$\begin{Bmatrix} C \\ A\,B \end{Bmatrix} = \sum_{H \in \mathbb{M}^p_{n\times n\times n}(A,B,C)} \frac{1}{H!}, \tag{1.259}$$

$$A \in \mathbb{M}^p_{n\times n}(\alpha,\beta),\; B \in \mathbb{M}^p_{n\times n}(\beta,\gamma),\; C \in \mathbb{M}^p_{n\times n}(\alpha,\gamma).$$

The planar matrix arrays in these relations are defined by

$$\begin{aligned}
\mathbb{M}^p_{n\times n}(\alpha,\beta) &= \{a_{ij}\,|\,\sum_{j=1}^n a_{ij}=\alpha_i, \sum_{i=1}^n a_{ij}=\beta_j\},\\
\mathbb{M}^p_{n\times n}(\beta,\gamma) &= \{b_{jk}\,|\,\sum_{k=1}^n b_{jk}=\beta_j, \sum_{j=1}^n b_{jk}=\gamma_k\}, \qquad (1.260)\\
\mathbb{M}^p_{n\times n}(\alpha,\gamma) &= \{c_{ik}\,|\,\sum_{k=1}^n c_{ik}=\alpha_i, \sum_{i=1}^n c_{ik}=\gamma_k\}.
\end{aligned}$$

The set $\mathbb{M}^p_{n\times n\times n}(A,B,C)$ over which the summation in (1.259) is taken is the collection of cubic arrays in which the lattice points (i,j,k), $1\leq i,j,k\leq n$, are assigned all possible nonnegative integers, as defined by

$$\mathbb{M}^p_{n\times n\times n}(A,B,C)=\left\{h_{ijk}\,\middle|\,\begin{array}{l}\sum_{k=1}^n h_{ijk}=a_{ij};\ \sum_{i=1}^n h_{ijk}=b_{jk};\\ \sum_{j=1}^n h_{ijk}=c_{ik}\end{array}\right\}. \tag{1.261}$$

These cubic arrays are natural generalizations of the 2×2 planar arrays (1.260); such arrays are unavoidable in the development of properties of the monomials X^A and of the general unitary group, despite their explosive counting qualities. Further properties are developed in the next subsection.

The proof of relations (1.258) and (1.259) is given by using the multinomial theorem to expand $(\sum_j x_{ij}y_{jk})^{c_{ik}}$, rearranging terms, and paying careful attention to details. Using these relations in (1.257), we find that necessary and sufficient conditions that the multiplication property (1.245) holds are that the following relation is valid:

$$\sum_{C\in\mathbb{M}^p_{n\times n}(\alpha,\gamma)}\left\{\begin{array}{c}C\\ A\,B\end{array}\right\}=\frac{\beta!}{A!\,B!}, \tag{1.262}$$

where $A\in\mathbb{M}^p_{n\times n}(\alpha,\beta)$ and $B\in\mathbb{M}^p_{n\times n}(\beta,\gamma)$. A combinatorial proof of (1.262) has been given by Chen and Louck [40]. Thus, the multiplication rule (1.245) is true for arbitrary commuting indeterminates.

We present here a somewhat different proof of relation (1.262), using the more basic identity:

$$L^p_{\alpha\,\beta}(J)=\sum_{A\in\mathbb{M}^p_{n\times n}(\alpha,\beta)}\frac{1}{A!}=\frac{p!}{\alpha!\,\beta!}, \tag{1.263}$$

where $L^p_{\alpha\,\beta}(J)$ is the polynomial (1.256) evaluated at all $z_{ij}=1$; that is, at $Z=J$, where J is the matrix of order n containing all $1's$.

To show that (1.263) implies (1.262), we require a preliminary result. It follows from the definition of the cubic array $\mathbb{M}^p_{n\times n\times n}(A,B,C)$ given by (1.261) that this set has the following decomposition in terms of planar matrix arrays perpendicular to $\mathbf{e}_2-$axis of a right-handed coordinate frame $(\mathbf{e}_1,\mathbf{e}_2,\mathbf{e}_3)$:

Plane perpendicular to the $\mathbf{e}_2-$axis at $x_2=j, 1\leq j\leq n$:

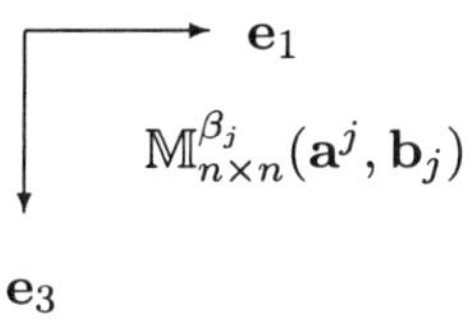

where $\mathbb{M}^{\beta_j}_{n\times n}(\mathbf{a}^j,\mathbf{b}_j)$ is the set of planar $n\times n$ matrix arrays of the form

$$\begin{pmatrix} h_{1j1} & h_{1j2} & \cdots & h_{1jn} \\ h_{2j1} & h_{2j2} & \cdots & h_{2jn} \\ \vdots & \vdots & \cdots & \vdots \\ h_{nj1} & h_{nj2} & \cdots & h_{njn} \end{pmatrix} \tag{1.264}$$

with row and column sums given by

$$\begin{aligned} \mathbf{a}^j &= (a_{1j},a_{2j},\ldots,a_{nj}) \vdash \beta_j, \\ \mathbf{b}_j &= (b_{j1},b_{j2},\ldots,b_{jn}) \vdash \beta_j. \end{aligned} \tag{1.265}$$

The cubic array $\mathbb{M}^p_{n\times n\times n}(A,B,C)$ is then expressed in terms of these planar arrays by

$$\mathbb{M}^p_{n\times n\times n}(A,B,C)=\mathbb{M}^{\beta_1}_{n\times n}(\mathbf{a}^1,\mathbf{b}_1)\boxtimes\mathbb{M}^{\beta_2}_{n\times n}(\mathbf{a}^2,\mathbf{b}_2)\boxtimes\cdots\boxtimes\mathbb{M}^{\beta_n}_{n\times n}(\mathbf{a}^n,\mathbf{b}_n), \tag{1.266}$$

where $\boxtimes$ denotes that the planes are stacked adjacent to one another (one unit between). Thus, the planar matrix arrays in (1.266) are located, respectively, at $x_2=1,2,\ldots,n$. The decomposition (1.266) assigns a unique nonnegative integer h_{ijk} to every point (i,j,k) in the cubic array such that $A\in\mathbb{M}^p_{n\times n}(\alpha,\beta)$, $B\in\mathbb{M}^p_{n\times n}(\beta,\gamma)$, $C\in\mathbb{M}^p_{n\times n}(\alpha,\gamma)$.

The cubic array $\mathbb{M}^p_{n\times n\times n}(A,B,C)$ can also be decomposed in terms of stacked planar arrays perpendicular to the $\mathbf{e}_1-$axis or to the $\mathbf{e}_3-$axis, but these decompositions are not required here. The general case is developed in the next section on Maclaurin monomials.

We can now use the decomposition (1.266) to show that (1.263) implies (1.262). Substitution of (1.259) into (1.262) gives

$$\sum_{C\in\mathbb{M}^p_{n\times n}(\alpha,\gamma)} \begin{Bmatrix} C \\ A\,B \end{Bmatrix} = \sum_{C\in\mathbb{M}^p_{n\times n}(\alpha,\gamma)} \; \sum_{H\in\mathbb{M}^p_{n\times n\times n}(A,B,C)} \frac{1}{H!}$$

$$= \prod_{j=1}^{n} \left(\sum_{H_j\in\mathbb{M}^{\beta_j}_{n\times n}(\mathbf{a}^j,\mathbf{b}_j)} \frac{1}{H_j!} \right) = \frac{\beta!}{A!\,B!}, \tag{1.267}$$

since relation (1.263) implies that

$$\sum_{H_j\in\mathbb{M}^{\beta_j}_{n\times n}(\mathbf{a}^j,\mathbf{b}_j)} \frac{1}{(H_j)!} = \frac{\beta_j!}{\prod_{i=1}^{n}(a_{ij})!\prod_{k=1}^{n}(b_{jk})!}. \tag{1.268}$$

In the middle relation in (1.267), the summation over $C \in \mathbb{M}^p_{n\times n}(\alpha,\gamma)$ becomes redundant when $\mathbb{M}^p_{n\times n\times n}(A,B,C)$ is written in the form (1.267), since it is automatically satisfied. We have thus shown that (1.263) implies (1.262), and it remains only to prove (1.263).

We now give a combinatorial proof of relation (1.263), following the method in Ref. [40]. Relation (1.263) may be rewritten in the multinomial coefficient form

$$\sum_{A\in\mathbb{M}^p_{n\times n}(\alpha,\beta)} \prod_{i=1}^{n} \binom{\alpha_i}{a_{i1}, a_{i2}, \ldots, a_{in}} = \binom{p}{\beta_1, \beta_2, \ldots, \beta_n}. \tag{1.269}$$

To prove this result we consider p distinct objects with α_i of the objects colored by color i, where $i = 1, 2, \ldots, n$, hence, $\alpha \vdash p$. Let these p objects be distributed into n cells $C_1, C_2, \ldots, C_n$, such that there are β_j objects in cell C_j with a_{ij} objects having color i. The left-hand side of (1.269) equals the number of ways of distributing the p objects into the n cells such that cell C_j contains β_j objects. The entries in the matrix $A \in \mathbb{M}^p_{n\times n}(\alpha,\beta)$ gives the configuration of these colored objects into the cells: The number of ways of distributing the α_i objects of color i among the n cells is given by $\binom{\alpha_i}{a_{i1},a_{i2},\ldots,a_{in}}$, and the sum of products of these coefficients is the number of ways $\binom{p}{\beta_1,\beta_2,\ldots,\beta_n}$ of distributing the p objects into the n cells. ☐

We remark again that there is also the direct combinatorial proof of relation (1.262) (Chen and Louck [40]) that does not require the intermediate step of showing that relation (1.263) implies (1.262). The above method reveals further properties of these relations.

1.6.4 Maclaurin monomials

The Maclaurin monomials are defined by

$$M_A(Z) = \frac{Z^A}{A!} = \prod_{i=1}^{l} \prod_{j=1}^{m} \frac{z_{ij}^{a_{ij}}}{a_{ij}!}, \tag{1.270}$$

where $A = (a_{ij})_{1 \le i \le l, 1 \le j \le m}$, is an $l \times m$ matrix array of nonnegative integers (exponents) a_{ij}, and $Z = (z_{ij})_{1 \le i \le l, 1 \le j \le m}$ is an $l \times m$ matrix of commuting indeterminates z_{ij}. These monomials are just the terms that occur in every formal power series expansion of a multivariable function depending on the $l \times m$ commuting indeterminates z_{ij}, as given by

$$F(Z) = \sum_A C(A) \frac{Z^A}{A!}, \tag{1.271}$$

where the $C(A)$ are numerical-valued coefficients, and the summation is over all $l \times m$ matrix arrays A. Since the Maclaurin monomials are orthogonal in the inner product (,), as given by

$$(M_{A'}, M_A) = \delta_{A',A} \frac{1}{A!}, \tag{1.272}$$

the coefficients $C(A)$ in the formal power series are

$$C(A) = A!(M_A, F) = \left(\frac{\partial}{\partial Z}\right)^A F(Z)\Big|_{Z=0_{l\times m}}. \tag{1.273}$$

We are interested in the product of two such functions (1.271), in particular, polynomials. The expansion of the product $(XY)^C$ into a sum over X^A and Y^B becomes a basic relation, which has already been given in (1.258)-(1.259) for the special case $l = m = n$, from which the extensions to the general case are evident, which we restate in detail for completeness:

$$\frac{(XY)^C}{C!} = \sum_{\beta \vdash p} \sum_{A \in \mathbb{M}^p_{l\times m}(\alpha,\beta)} \sum_{B \in \mathbb{M}^p_{m\times n}(\beta,\gamma)} \begin{Bmatrix} C \\ A\,B \end{Bmatrix} X^A Y^B,$$

$$C \in \mathbb{M}^p_{n\times l}(\alpha,\gamma); \tag{1.274}$$

$$\begin{Bmatrix} C \\ A\,B \end{Bmatrix} = \sum_{H \in \mathbb{M}^p_{l\times m\times n}(A,B,C)} \frac{1}{H!}, \tag{1.275}$$

$$A \in \mathbb{M}^p_{l\times m}(\alpha,\beta),\ B \in \mathbb{M}^p_{m\times n}(\beta,\gamma),\ C \in \mathbb{M}^p_{n\times l}(\alpha,\gamma).$$

$$\begin{aligned}
\mathbb{M}^p_{l\times m}(\alpha,\beta) &= \{a_{ij} \mid \sum_{j=1}^{m} a_{ij} = \alpha_i, \sum_{i=1}^{l} a_{ij} = \beta_j\}, \\
\mathbb{M}^p_{m\times n}(\beta,\gamma) &= \{b_{jk} \mid \sum_{k=1}^{n} b_{jk} = \beta_j, \sum_{j=1}^{m} b_{jk} = \gamma_k\}, \qquad (1.276) \\
\mathbb{M}^p_{n\times l}(\alpha,\gamma) &= \{c_{ik} \mid \sum_{k=1}^{n} c_{ik} = \alpha_i, \sum_{i=1}^{l} c_{ik} = \gamma_k\}.
\end{aligned}$$

The set $\mathbb{M}^p_{l\times m\times n}(A,B,C)$ over which the summation in (1.259) is taken is the collection of $l \times m \times n$ dimensional parallelepiped (six rectangular faces) in which the lattice points (i,j,k), $1 \le i \le l$, $1 \le j \le m$, $1 \le k \le n$, are assigned all possible nonnegative integers, as defined by

$$\mathbb{M}^p_{l\times m\times n}(A,B,C) = \left\{ h_{ijk} \;\middle|\; \begin{array}{l} \sum_{k=1}^{n} h_{ijk} = a_{ij};\ \sum_{i=1}^{l} h_{ijk} = b_{jk}; \\ \sum_{j=1}^{m} h_{ijk} = c_{ik} \end{array} \right\}. \qquad (1.277)$$

The set $\mathbb{M}^p_{l\times m\times n}(A,B,C)$ is described geometrically in Cartesian 3−space $\mathbb{R}^3$ by specifying the assignment of the h_{ijk} to points (i,j,k) referred to a right-handed triad of unit geometrical vectors $(\mathbf{e}_1,\mathbf{e}_2,\mathbf{e}_3)$ directed from the origin $(0,0,0)$ to the points $(1,0,0),(0,1,0),(0,0,1)$, respectively, as depicted by

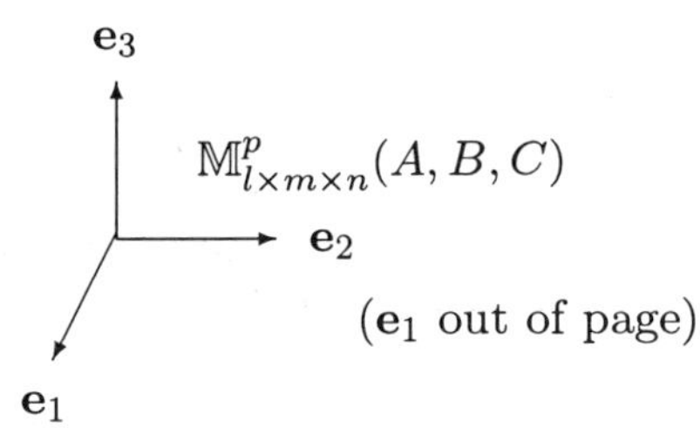

(1.278)

We require the sequences as follows, which for clarity we denote by bold letters:

$$\begin{aligned}
\mathbf{a}_i &= (a_{i1}, a_{i2}, \ldots, a_{im}),\ \mathbf{a}^j = (a_{1j}, a_{2j}, \ldots, a_{lj}), \\
\mathbf{b}_j &= (b_{j1}, b_{j2}, \ldots, b_{jn}),\ \mathbf{b}^k = (b_{1k}, b_{2k}, \ldots, b_{mk}), \qquad (1.279) \\
\mathbf{c}_i &= (c_{i1}, c_{i2}, \ldots, c_{in}),\ \mathbf{c}^k = (c_{1k}, c_{2k}, \ldots, c_{lk}).
\end{aligned}$$

These sequences give the row and column sums of various planar matrix arrays that are perpendicular to the respective unit vectors in (1.278) that determine the geometrical structure of the set $\mathbb{M}^{p}_{l\times m\times n}(A,B,C)$:

Plane perpendicular to the $\mathbf{e}_1$−axis at $x_1 = i, 1 \le i \le l$:

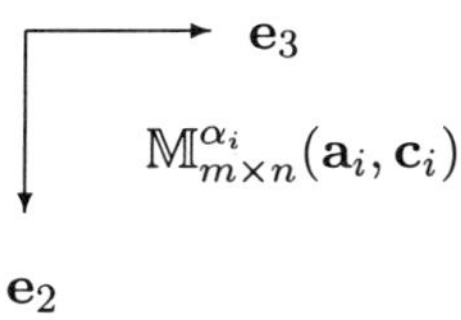

The entries in the set $\mathbb{M}^{\alpha_i}_{m\times n}(\mathbf{a}_i,\mathbf{c}_i)$ of $m \times n$ planar arrays can be any array given by

$$\begin{pmatrix} h_{i11} & h_{i12} & \cdots & h_{i1n} \\ h_{i21} & h_{i22} & \cdots & h_{i2n} \\ \vdots & \vdots & \cdots & \vdots \\ h_{im1} & h_{im2} & \cdots & h_{imn} \end{pmatrix}, \tag{1.280}$$

with row and column sums specified to be $\mathbf{a}_i \vdash \alpha_i$ and $\mathbf{c}_i \vdash \alpha_i$.

Plane perpendicular to the $\mathbf{e}_2$−axis at $x_2 = j, 1 \le j \le m$:

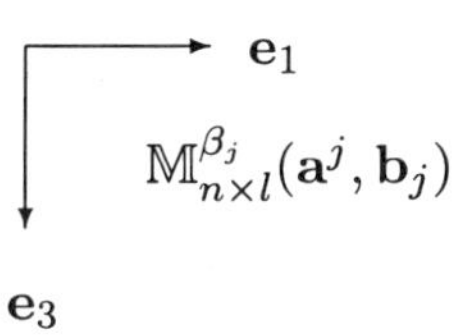

The entries in the set $\mathbb{M}^{\beta_j}_{n\times l}(\mathbf{a}^j,\mathbf{b}_j)$ of $n \times l$ planar arrays can be any array given by

$$\begin{pmatrix} h_{1j1} & h_{1j2} & \cdots & h_{1jl} \\ h_{2j1} & h_{2j2} & \cdots & h_{2jl} \\ \vdots & \vdots & \cdots & \vdots \\ h_{nj1} & h_{nj2} & \cdots & h_{njl} \end{pmatrix}, \tag{1.281}$$

with row and column sums specified to be $\mathbf{a}^j \vdash \beta_j$ and $\mathbf{b}_j \vdash \beta_j$.

Plane perpendicular to the $\mathbf{e}_3$−axis at $x_3 = k, 1 \leq k \leq n$:

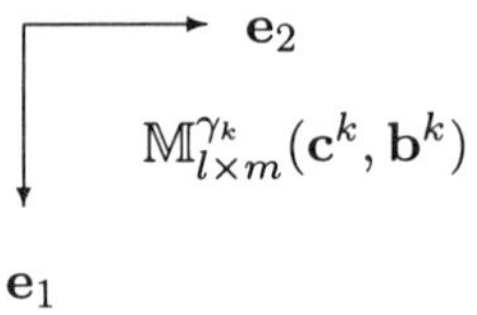

The entries in the set $\mathbb{M}^{\gamma_k}_{l\times m}(\mathbf{c}^k, \mathbf{b}^k)$ of $l \times m$ planar arrays can be any array given by

$$\begin{pmatrix} h_{11k} & h_{12k} & \cdots & h_{1mk} \\ h_{21k} & h_{22k} & \cdots & h_{2mk} \\ \vdots & \vdots & \cdots & \vdots \\ h_{l1k} & h_{l2k} & \cdots & h_{lmk} \end{pmatrix}, \tag{1.282}$$

with row and column sums specified to be $\mathbf{c}^k \vdash \gamma_k$ and $\mathbf{b}^k \vdash \gamma_k$.

We can write the set of lattice points $\mathbb{M}^p_{l\times m\times n}(A, B, C)$ in terms of these planes of lattice points perpendicular to the 1-axis, 2-axis, and 3-axis of the right-handed frame $(\mathbf{e}_1, \mathbf{e}_2, \mathbf{e}_3)$, respectively, by

$$\begin{aligned} & \mathbb{M}^p_{l\times m\times n}(A, B, C) \qquad (1.283) \\ = \; & \mathbb{M}^{\alpha_1}_{m\times n}(\mathbf{a}_1, \mathbf{c}_1) \boxtimes \mathbb{M}^{\alpha_2}_{m\times n}(\mathbf{a}_2, \mathbf{c}_2) \boxtimes \cdots \boxtimes \mathbb{M}^{\alpha_l}_{m\times n}(\mathbf{a}_l, \mathbf{c}_l) \\ = \; & \mathbb{M}^{\beta_1}_{n\times l}(\mathbf{a}^1, \mathbf{b}_1) \boxtimes \mathbb{M}^{\beta_2}_{n\times l}(\mathbf{a}^2, \mathbf{b}_2) \boxtimes \cdots \boxtimes \mathbb{M}^{\beta_m}_{n\times l}(\mathbf{a}^m, \mathbf{b}_m) \\ = \; & \mathbb{M}^{\gamma_1}_{l\times m}(\mathbf{c}^1, \mathbf{b}^1) \boxtimes \mathbb{M}^{\gamma_2}_{l\times m}(\mathbf{c}^2, \mathbf{b}^2) \boxtimes \cdots \boxtimes \mathbb{M}^{\gamma_n}_{l\times m}(\mathbf{c}^n, \mathbf{b}^n). \end{aligned}$$

The first line in (1.283) denotes the collection of stacked $m \times n$ planar matrix arrays located, respectively, at $x_1 = 1, 2, \ldots, l$, with similar interpretations for the second and third line.

1.6.5 Summary of relations

We summarize the following special results that originate from the general expansion (1.274) and the orthogonality of various polynomials with respect to the inner product (,) :

1. Solid harmonics in mn indeterminates:

$$L^p_{\beta\gamma}(Y) = \left(\frac{x^\beta}{\beta!}, \frac{(xY)^\gamma}{\gamma!}\right) = \sum_{B\in\mathbb{M}^p_{m\times n}(\beta,\gamma)} \frac{Y^B}{B!}, \tag{1.284}$$

which is a consequence of

$$\frac{z^\gamma}{\gamma!} = \frac{(xY)^\gamma}{\gamma!} = \sum_{\beta \vdash p} x^\beta L^p_{\beta\,\gamma}(Y). \tag{1.285}$$

This relation itself is obtained from (1.274) by choosing $l = 1$, so that we have the following simplifications:

$$\begin{aligned} C &= \gamma = (\gamma_1, \gamma_2, \dots, \gamma_n) \vdash p, A = (\beta_1, \beta_2, \dots, \beta_m) \vdash p, \\ B &\in \mathbb{M}^p_{m\times n}(\beta, \gamma); \end{aligned} \tag{1.286}$$

$$\begin{aligned} X &= (x_{11}, x_{12}, \dots, x_{1m}) = x = (x_1, x_2, \dots, x_m), \\ XY &= xY = z = (z_1, z_2, \dots, z_n), \\ z_k &= \sum_{j=1}^{m} x_j y_{jk},\ k = 1, 2, \cdots, n. \end{aligned} \tag{1.287}$$

Here, as will often be the case in this monograph, we abuse the notation by writing the inner product of two functions f with values $f(x)$ and g with values $g(x)$ as

$$(f, g) = (f(x), g(x)). \tag{1.288}$$

Despite the contradiction of notation, it presents the functions involved in the inner product in more comprehensible fashion.

2. Right and left action polynomials:

   The following polynomials, which are defined from the general relation (1.274), also occur in the sequel:

$$\begin{aligned} R^p_{A,C}(Y) &= \left(\frac{X^A}{A!}, \frac{(XY)^C}{C!}\right) = \sum_{B \in \mathbb{M}^p_{m\times n}(\beta,\gamma)} \left\{ \begin{array}{c} C \\ AB \end{array} \right\} Y^B, \\ L^p_{B,C}(X) &= \left(\frac{Y^B}{B!}, \frac{(XY)^C}{C!}\right) = \sum_{A \in \mathbb{M}^p_{l\times m}(\alpha,\beta)} \left\{ \begin{array}{c} C \\ AB \end{array} \right\} X^A, \\ A &\in \mathbb{M}^p_{l\times m}(\alpha, \beta),\ B \in \mathbb{M}^p_{m\times n}(\beta, \gamma),\ C \in \mathbb{M}^p_{n\times l}(\alpha, \gamma). \end{aligned} \tag{1.289}$$

1.7 Generalization to $U(2)$

1.7.1 Definition of $U(2)$ solid harmonics

A full list of properties of the $SU(2)$ solid harmonics $D^j_{m\,m'}(Z)$ defined by relation (1.160) must include the action of the indeterminates z_{ij} and the derivatives $\partial/\partial z_{ij}$ on these polynomials, as given below in Sect. 1.7.3. It is convenient to give these actions in terms of the extension of these polynomials to their $U(2)$ counterparts. The $U(2)$ solid harmonics are polynomials that are enumerated by double Gelfand-Tsetlin patterns that depend on partitions having two parts as defined by (λ_1, λ_2), where the λ_i are nonnegative integers satisfying $\lambda_1 \geq \lambda_2 \geq 0$. Just as it requires two pairs, $(j\,m)$ and $(j\,m')$, where j is the angular momentum shared by the two projection quantum numbers m and m', to define the $SU(2)$ solid harmonics, it requires two Gelfand-Tsetlin patterns

$$\begin{pmatrix} \lambda_1 & & \lambda_2 \\ & m_{11} & \end{pmatrix} \text{and} \begin{pmatrix} \lambda_1 & & \lambda_2 \\ & m'_{11} & \end{pmatrix} \tag{1.290}$$

to define the $U(2)$ solid harmonics. For each partition (λ_1, λ_2), the quantities m_{11} and m'_{11} are integers that can take on the $\lambda_1 - \lambda_2 + 1$ values given by

$$m_{11}, m'_{11} \in \{\lambda_2, \lambda_2 + 1, \ldots, \lambda_1\}. \tag{1.291}$$

These conditions are referred to as "betweenness conditions," and the labels m_{11} and m'_{11} are positioned between the partition labels to suggest this condition. To economize the notation, so as not to write the partition twice, it is convenient to invert the second pattern over the first in (1.290) and write the double pattern as

$$\begin{pmatrix} & m'_{11} & \\ \lambda_1 & & \lambda_2 \\ & m_{11} & \end{pmatrix} = \begin{pmatrix} & m'_{11} & \\ m_{12} & & m_{22} \\ & m_{11} & \end{pmatrix}, \tag{1.292}$$

where we often write the common partition as $(\lambda_1\, \lambda_2) = (m_{12}\, m_{22})$. The pattern (1.292) is referred to as a double Gelfand-Tsetlin pattern. General patterns of this form for the unitary group $U(n)$ are discussed in detail in Sect. 11.3, Compendium B. We use the notation (1.292) here for easy comparison with the general results given later for the general unitary group.

The $U(2)$ solid harmonics are defined in terms of the $SU(2)$ D^j- solid harmonics (1.160) by

$$\begin{aligned} D\begin{pmatrix} & m'_{11} & \\ \lambda_1 & & \lambda_2 \\ & m_{11} & \end{pmatrix}(Z) &= (\det Z)^{\lambda_2} D^j_{m\,m'}(Z) \\ &= \sum_{A \in \mathbb{M}^p_{2\times 2}(\alpha,\alpha')} C\begin{pmatrix} & m'_{11} & \\ \lambda_1 & & \lambda_2 \\ & m_{11} & \end{pmatrix}(A)\frac{Z^A}{A!}, \end{aligned} \tag{1.293}$$

where the relations between notations are given by

$$\begin{aligned} p &= \lambda_1+\lambda_2,\ j=(\lambda_1-\lambda_2)/2, \\ m &= m_{11}-(\lambda_1+\lambda_2)/2,\ m'=m'_{11}-(\lambda_1+\lambda_2)/2; \\ \alpha &= (\alpha_1,\alpha_2)=(m_{11},\lambda_1+\lambda_2-m_{11})\vdash p, \\ \alpha' &= (\alpha'_1,\alpha'_2)=(m'_{11},\lambda_1+\lambda_2-m'_{11})\vdash p. \end{aligned} \tag{1.294}$$

The compositions α and α' are also called the *weights* of the lower and upper Gelfand-Tsetlin patterns. The set $\mathbb{M}^p_{2\times2}(\alpha,\alpha')$ of 2×2 matrix arrays over which the summation is effected in the expansion on the right-hand side of (1.293) is defined by

$$\mathbb{M}^p_{2\times2}(\alpha,\alpha')=\left\{A=\begin{pmatrix} a_{11} & a_{12} \\ a_{21} & a_{22}\end{pmatrix}\middle|\begin{array}{l} a_{11}+a_{12}=\alpha_1, a_{21}+a_{22}=\alpha_2, \\ a_{11}+a_{21}=\alpha'_1, a_{12}+a_{22}=\alpha'_2\end{array}\right\}. \tag{1.295}$$

We have already met the special case of the polynomials (1.293) in the context of the addition of two angular momenta given by (1.198):

$$\begin{aligned} \psi_{(j_1\,j_2)j\,m}(Z) &= (M_{(j_1\,j_2)j})^{-1/2}D\begin{pmatrix} & 2j_1 & \\ j_1+j_2+j & & j_1+j_2-j \\ & j_1+j_2+m & \end{pmatrix}(Z), \\ M_{(j_1\,j_2)j} &= \frac{(j_1+j_2+j+1)!(j_1+j_2-j)!}{2j+1}. \end{aligned} \tag{1.296}$$

These coupled angular momentum basis functions are the special case $m'=j_1-j_2$ of the more general $U(2)$ solid harmonics defined by (1.293):

$$\begin{aligned} &D\begin{pmatrix} & j_1+j_2+m' & \\ j_1+j_2+j & & j_1+j_2-j \\ & j_1+j_2+m & \end{pmatrix}(Z)=(\det Z)^{j_1+j_2-j}D^j_{m\,m'}(Z) \\ &=\sum_{\substack{m_1,m_2\\ m'_1,m'_2}} C^{j_1\;j_2\;j}_{m_1\,m_2\,m}\,C^{j_1\;j_2\;j}_{m'_1\,m'_2\,m'} \\ &\times D\begin{pmatrix} & j_2+m'_2 & \\ 2j_2 & & 0 \\ & j_2+m_2 & \end{pmatrix}(Z)D\begin{pmatrix} & j_1+m'_1 & \\ 2j_1 & & 0 \\ & j_1+m_1 & \end{pmatrix}(Z). \end{aligned} \tag{1.297}$$

If we set $m'=j_1-j_2$, then this relation reduces to (1.198) and (1.296) in consequence of the relations:

$$C\begin{matrix} j_1 & j_2 & j \\ m_1' & m_2' & j_1-j_2 \end{matrix} = \delta_{m_1',j_1}\delta_{m_2',-j_2}\sqrt{\frac{(2j+1)(2j_1)!(2j_2)!}{(j_1+j_2-j)!(j_1+j_2+j+1)!}}, \tag{1.298}$$

$$\begin{aligned} D\begin{pmatrix} & 2j_1 & \\ 2j_1 & & 0 \\ & j_1+m_1 & \end{pmatrix}(Z) &= \sqrt{(2j_1)!}P_{j_1 m_1}(z_{11}, z_{21}), \\ D\begin{pmatrix} & 0 & \\ 2j_2 & & 0 \\ & j_1+m_1 & \end{pmatrix}(Z) &= \sqrt{(2j_2)!}P_{j_1 m_1}(z_{12}, z_{22}). \end{aligned} \tag{1.299}$$

(It is not particularly useful here to extract the C–coefficient in (1.293) from the above relations. See, however, Sect. 7.3.2, Chapter 7.)

The above results place the $U(2)$ solid harmonics in the notational context of general results that are presented for $U(n)$ in Chapters 5-7. Indeed, they are the model results for that generalization. We often refer to these polynomials as the $D^{(\lambda_1\lambda_2)}$–polynomials and the coefficients in (1.293) as the $C^{(\lambda_1\lambda_2)}$–coefficients. By design, the notation has been chosen such that the $D^{(\lambda_1\lambda_2)}$–polynomials and the $C^{(\lambda_1\lambda_2)}$–coefficients in (1.293) have similar forms, the first being defined over arbitrary indeterminates Z, the second over discrete nonnegative integers A. The polynomials are fully defined in terms of the monomials $Z^A/A!$, once the discrete $C^{(\lambda_1\lambda_2)}$–coefficients are known, as is the case above with $n = 2$. Indeed, the generalization to $U(n)$ is precisely that of determining the correspondence

$$D\begin{pmatrix} m' \\ \lambda \\ m \end{pmatrix}(Z) \leftrightarrow C\begin{pmatrix} m' \\ \lambda \\ m \end{pmatrix}(A) \tag{1.300}$$

between D^λ–polynomials and C^λ–coefficients for an arbitrary partition $\lambda = (\lambda_1, \lambda_2, \ldots, \lambda_n)$, where now Z and A are of order n, and the lower an upper patterns m and m' become general Gelfand-Tsetlin patterns.

1.7.2 Basic multiplication properties

It is useful to note several of the properties of the $D^{(\lambda_1\lambda_2)}$–polynomials and the $C^{(\lambda_1\lambda_2)}$–coefficients that generalize to arbitrary partitions. The general multiplication property of $D^{\lambda_1\lambda_2}-$polynomials follow from relation (1.245), the multiplication property of determinants, and definition (1.299):

$$\sum_{m''_{11}} D\begin{pmatrix} & m''_{11} & \\ \lambda_1 & & \lambda_2 \\ & m_{11} & \end{pmatrix}(X) D\begin{pmatrix} & m'_{11} & \\ \lambda_1 & & \lambda_2 \\ & m''_{11} & \end{pmatrix}(Y) = D\begin{pmatrix} & m'_{11} & \\ \lambda_1 & & \lambda_2 \\ & m_1 & \end{pmatrix}(XY). \tag{1.301}$$

This result implies that the multiplication of discretized $C^{(\lambda_1\lambda_2)}-$coefficients is given by

$$\begin{aligned} &\sum_{m''_{11}} C\begin{pmatrix} & m''_{11} & \\ \lambda_1 & & \lambda_2 \\ & m_{11} & \end{pmatrix}(A) C\begin{pmatrix} & m'_{11} & \\ \lambda_1 & & \lambda_2 \\ & m''_{11} & \end{pmatrix}(B) \\ &\quad = \sum_{C\in\mathbb{M}^p_{2\times 2}(\alpha,\alpha')} \left\{ \begin{matrix} C \\ AB \end{matrix} \right\} C\begin{pmatrix} & m'_{11} & \\ \lambda_1 & & \lambda_2 \\ & m_{11} & \end{pmatrix}(C), \end{aligned} \tag{1.302}$$

$$A \in \mathbb{M}^p_{2\times 2}(\alpha,\alpha''),\ B \in \mathbb{M}^p_{2\times 2}(\alpha'',\alpha'),$$

where the coefficients $\left\{ \begin{matrix} C \\ AB \end{matrix} \right\}$ are those that enter into the product of two Maclaurin polynomials for $n = 2$.

We also have the important Kronecker product relationship, as follows, from relation (1.242) (with similar relations corresponding to (1.239)-(1.241)):

$$\begin{aligned} &D\begin{pmatrix} & m'''_{11} & \\ \lambda'_1 & & \lambda'_2 \\ & m'_{11} & \end{pmatrix}(Z) D\begin{pmatrix} & m''_{11} & \\ \lambda_1 & & \lambda_2 \\ & m_{11} & \end{pmatrix}(Z) \\ &= \sum_{\lambda''_1,\lambda''_2} C\left[\begin{pmatrix} \lambda''_1 & & \lambda''_2 \\ & m_{11}+m'_{11} & \end{pmatrix}\begin{pmatrix} \lambda'_1 & & \lambda'_2 \\ & m'_{11} & \end{pmatrix}\begin{pmatrix} \lambda_1 & & \lambda_2 \\ & m_{11} & \end{pmatrix}\right] \\ &\quad \times C\left[\begin{pmatrix} \lambda''_1 & & \lambda''_2 \\ & m''_{11}+m'''_{11} & \end{pmatrix}\begin{pmatrix} \lambda'_1 & & \lambda'_2 \\ & m'''_{11} & \end{pmatrix}\begin{pmatrix} \lambda_1 & & \lambda_2 \\ & m''_{11} & \end{pmatrix}\right] \\ &\quad \times D\begin{pmatrix} & m''_{11}+m'''_{11} & \\ \lambda''_1 & & \lambda''_2 \\ & m_{11}+m'_{11} & \end{pmatrix}(Z). \end{aligned} \tag{1.303}$$

The $C-$coefficients in this relation are $U(2)$ WCG coefficients, which are related to $SU(2)$ WCG coefficients by associating each of the three GT patterns with an angular momentum quantum label and its projection by the rule (1.294). The three patterns in the $C-$coefficient are placed in the indicated positions, which corresponds to reading the two patterns of the $D-$polynomials on the left in (1.303) in the order right-to-left; if the order of the $D-$polynomials on the left is reversed, the patterns in the $C-$ coefficients are reversed. This rule is referred to as the *standard mapping rule* for WCG coefficients: Thus, the first $C-$coefficient is given by

$$C\left[\begin{pmatrix} \lambda''_1 & & \lambda''_2 \\ & m''_{11} & \end{pmatrix} \middle| \begin{pmatrix} \lambda'_1 & & \lambda'_2 \\ & m'_{11} & \end{pmatrix} \begin{pmatrix} \lambda_1 & & \lambda_2 \\ & m_{11} & \end{pmatrix}\right] = C^{j\;j'\;j''}_{m\,m'\,m''}\,; \tag{1.304}$$

$$\begin{aligned} j &= (\lambda_1-\lambda_2)/2, j'=(\lambda'_1-\lambda'_2)/2, j''=(\lambda''_1-\lambda''_2)/2, \\ m &= m_{11}-(\lambda_1+\lambda_2)/2, m'=m'_{11}-(\lambda'_1+\lambda'_2)/2, \\ m'' &= m''_{11}-(\lambda''_1+\lambda''_2)/2. \end{aligned} \tag{1.305}$$

The relation

$$\lambda_1+\lambda_2+\lambda'_1+\lambda'_1+\lambda'_2=\lambda''_1+\lambda''_2 \tag{1.306}$$

is a consequence of the homogeneity of the $D-$polynomials.

The reason for the ad hoc ordering rule in (1.303)-(1.304) is that it convenient to think of the patterns as being initial, intermediate, and final, a viewpoint that encompasses subsequent uses, when the initial and final $D-$polynomials are interpreted as state vectors, or the left-hand pair as noncommuting quantities (tensor operators). The order rule could have been done oppositely, as in relations (1.239)-(1.242), which is the usual presentation. Since the $C-$coefficients occur as pairs in such relations, no error is made because of the phase factor relation $C^{j_1\;j_2\;j_3}_{m_1\,m_2\,m_3}=(-1)^{j_1+j_2-j}C^{j_2\;j_1\;j_3}_{m_2\,m_1\,m_3}$. In all subsequent chapters, we use the "read backwards" rule (1.303)-(1.304).

All of the above relations can also be written in matrix form. For this, we order the Gelfand-Tsetlin patterns by

$$\begin{pmatrix} \lambda_1 & & \lambda_2 \\ & m_{11} & \end{pmatrix} > \begin{pmatrix} \lambda_1 & & \lambda_2 \\ & m'_{11} & \end{pmatrix}, \text{ for } m_{11}>m'_{11}, \tag{1.307}$$

where they are equal, of course, for $m_{11}=m'_{11}$. Then, the row and column elements of the matrix $D^{(\lambda_1\lambda_2)}(Z)$ are given by the pairs (m_{11},m'_{11}), where the elements in row m_{11} are ordered from left-to-right as read across the row by $\lambda_1,\lambda_1-1,\ldots,\lambda_2$; and the elements in column m'_{11} are ordered from top-to-bottom as read down the column by λ_1, $\lambda_1-1,\ldots,\lambda_2$. (This follows the convention used for $SU(2)$ representation functions.) In matrix form, the basic relations for the $D^{(\lambda_1\lambda_2)}-$

polynomials and the $C^{(\lambda_1\lambda_2)}-$coefficients are the following:

$$D^{(\lambda_1\lambda_2)}(Z) = \sum_A \frac{Z^A}{A!} C^{(\lambda_1\lambda_2)}(A),$$
$$D^{(\lambda_1\lambda_2)}(X)D^{(\lambda_1\lambda_2)}(Y) = D^{(\lambda_1\lambda_2)}(XY), \tag{1.308}$$

$$C^{(\lambda_1\,\lambda_2)}(A)C^{(\lambda_1\,\lambda_2)}(B) = \sum_C \left\{ \begin{matrix} C \\ A\,B \end{matrix} \right\} C^{(\lambda_1\,\lambda_2)}(C);$$

$$D^{(\lambda'_1\,\lambda'_2)}(Z) \otimes D^{(\lambda_1\,\lambda_2)}(Z) = (C^{(\lambda';\lambda)})^T \left(\sum_{\lambda''_1,\lambda''_2} \oplus D^{(\lambda''_1\,\lambda''_2)}(Z) \right) C^{(\lambda';\lambda)},$$
$$(\lambda;\lambda') = ((\lambda_1\,\lambda_2);(\lambda'_1\,\lambda'_2)). \tag{1.309}$$

1.7.3 Indeterminate and derivative actions on the $U(2)$ solid harmonics

The indeterminates z_{ij} and the derivatives $\partial/\partial z_{ij}$ have the action on the $D^{\lambda_1\lambda_2}-$polynomials listed below. We take these relations from the methods of the *pattern calculus* presented in Chapter 6 for $U(n)$. The results are presented in terms of a mixed notation that uses the partial hooks $p_{12} = \lambda_1 + 1, p_{22} = \lambda_2, p_{11} = m_{11}, p'_{11} = m'_{11}$ in place of the $\lambda_1, \lambda_2, m_{11}, m'_{11}$ that label the polynomials, because of the symmetry of coefficients shown by the partial hook notation, which is pervasive in $U(n)$: The below results apply also to the D^j-polynomials by making the conversion (1.293)-(1.294) of notations and canceling $\det Z$ factors, as appropriate:

$$z_{11}D\begin{pmatrix} & m'_{11} & \\ \lambda_1 & & \lambda_2 \\ & m_{11} & \end{pmatrix}(Z) = \sqrt{\frac{p_{11}-p_{22}+1}{p_{12}-p_{22}}}\sqrt{\frac{p'_{11}-p_{22}+1}{p_{12}-p_{22}}}D\begin{pmatrix} & m'_{11}+1 & \\ \lambda_1+1 & & \lambda_2 \\ & m_{11}+1 & \end{pmatrix}(Z)$$
$$+\sqrt{\frac{p_{12}-p_{11}-1}{p_{12}-p_{22}}}\sqrt{\frac{p_{12}-p'_{11}-1}{p_{12}-p_{22}}}D\begin{pmatrix} & m'_{11}+1 & \\ \lambda_1 & & \lambda_2+1 \\ & m_{11}+1 & \end{pmatrix}(Z),$$

$$z_{12}D\begin{pmatrix} & m'_{11} & \\ \lambda_1 & & \lambda_2 \\ & m_{11} & \end{pmatrix}(Z) = \sqrt{\frac{p_{11}-p_{22}+1}{p_{12}-p_{22}}}\sqrt{\frac{p_{12}-p'_{11}}{p_{12}-p_{22}}}D\begin{pmatrix} & m'_{11} & \\ \lambda_1+1 & & \lambda_2 \\ & m_{11}+1 & \end{pmatrix}(Z)$$
$$-\sqrt{\frac{p_{12}-p_{11}-1}{p_{12}-p_{22}}}\sqrt{\frac{p'_{11}-p_{22}}{p_{12}-p_{22}}}D\begin{pmatrix} & m'_{11} & \\ \lambda_1 & & \lambda_2+1 \\ & m_{11}+1 & \end{pmatrix}(Z),$$

$$z_{21}D\begin{pmatrix} & m'_{11} & \\ \lambda_1 & & \lambda_2 \\ & m_{11} & \end{pmatrix}(Z)=\sqrt{\tfrac{p_{12}-p_{11}}{p_{12}-p_{22}}}\sqrt{\tfrac{p'_{11}-p_{22}+1}{p_{12}-p_{22}}}D\begin{pmatrix} & m'_{11}+1 & \\ \lambda_1+1 & & \lambda_2 \\ & m_{11} & \end{pmatrix}(Z)$$
$$-\sqrt{\tfrac{p_{11}-p_{22}}{p_{12}-p_{22}}}\sqrt{\tfrac{p_{12}-p'_{11}-1}{p_{12}-p_{22}}}D\begin{pmatrix} & m'_{11}+1 & \\ \lambda_1 & & \lambda_2+1 \\ & m_{11} & \end{pmatrix}(Z),$$

$$z_{22}D\begin{pmatrix} & m'_{11} & \\ \lambda_1 & & \lambda_2 \\ & m_{11} & \end{pmatrix}(Z)=\sqrt{\tfrac{p_{12}-p_{11}}{p_{12}-p_{22}}}\sqrt{\tfrac{p_{12}-p'_{11}}{p_{12}-p_{22}}}D\begin{pmatrix} & m'_{11} & \\ \lambda_1+1 & & \lambda_2 \\ & m_{11} & \end{pmatrix}(Z)$$
$$+\sqrt{\tfrac{p_{11}-p_{22}}{p_{12}-p_{22}}}\sqrt{\tfrac{p'_{11}-p_{22}}{p_{12}-p_{22}}}D\begin{pmatrix} & m'_{11} & \\ \lambda_1 & & \lambda_2+1 \\ & m_{11} & \end{pmatrix}(Z);$$

$$\tfrac{\partial}{\partial z_{11}}D\begin{pmatrix} & m'_{11} & \\ \lambda_1 & & \lambda_2 \\ & m_{11} & \end{pmatrix}(Z)=\sqrt{\tfrac{p_{11}-p_{22}}{p_{12}-p_{22}-1}}\sqrt{\tfrac{p'_{11}-p_{22}}{p_{12}-p_{22}-1}}D\begin{pmatrix} & m'_{11}-1 & \\ \lambda_1-1 & & \lambda_2 \\ & m_{11}-1 & \end{pmatrix}(Z)$$
$$+\sqrt{\tfrac{p_{12}-p_{11}}{p_{12}-p_{22}+1}}\sqrt{\tfrac{p_{12}-p'_{11}}{p_{12}-p_{22}+1}}D\begin{pmatrix} & m'_{11}-1 & \\ \lambda_1 & & \lambda_2-1 \\ & m_{11}-1 & \end{pmatrix}(Z),$$

$$\tfrac{\partial}{\partial z_{12}}D\begin{pmatrix} & m'_{11} & \\ \lambda_1 & & \lambda_2 \\ & m_{11} & \end{pmatrix}(Z)=\sqrt{\tfrac{p_{11}-p_{22}}{p_{12}-p_{22}-1}}\sqrt{\tfrac{p_{12}-p'_{11}-1}{p_{12}-p_{22}-1}}D\begin{pmatrix} & m'_{11}-1 & \\ \lambda_1-1 & & \lambda_2 \\ & m_{11} & \end{pmatrix}(Z)$$
$$-\sqrt{\tfrac{p_{12}-p_{11}}{p_{12}-p_{22}+1}}\sqrt{\tfrac{p'_{11}-p_{22}+!}{p_{12}-p_{22}+1}}D\begin{pmatrix} & m'_{11} & \\ \lambda_1 & & \lambda_2-1 \\ & m_{11}-1 & \end{pmatrix}(Z),$$

$$\tfrac{\partial}{\partial z_{21}}D\begin{pmatrix} & m'_{11} & \\ \lambda_1 & & \lambda_2 \\ & m_{11} & \end{pmatrix}(Z)=\sqrt{\tfrac{p_{12}-p_{11}-1}{p_{12}-p_{22}-1}}\sqrt{\tfrac{p'_{11}-p_{22}}{p_{12}-p_{22}-1}}D\begin{pmatrix} & m'_{11}-1 & \\ \lambda_1-1 & & \lambda_2 \\ & m_{11} & \end{pmatrix}(Z)$$
$$-\sqrt{\tfrac{p_{11}-p_{22}+1}{p_{12}-p_{22}+1}}\sqrt{\tfrac{p_{12}-p'_{11}}{p_{12}-p_{22}+1}}D\begin{pmatrix} & m'_{11}-1 & \\ \lambda_1 & & \lambda_2-1 \\ & m_{11} & \end{pmatrix}(Z),$$

$$\tfrac{\partial}{\partial z_{22}}D\begin{pmatrix} & m'_{11} & \\ \lambda_1 & & \lambda_2 \\ & m_{11} & \end{pmatrix}(Z)=\sqrt{\tfrac{p_{12}-p_{11}-1}{p_{12}-p_{22}-1}}\sqrt{\tfrac{p_{12}-p'_{11}-1}{p_{12}-p_{22}-1}}D\begin{pmatrix} & m'_{11} & \\ \lambda_1-1 & & \lambda_2 \\ & m_{11} & \end{pmatrix}(Z)$$
$$+\sqrt{\tfrac{p_{11}-p_{22}+1}{p_{12}-p_{22}+1}}\sqrt{\tfrac{p'_{11}-p_{22}+1}{p_{12}-p_{22}+1}}D\begin{pmatrix} & m'_{11} & \\ \lambda_1 & & \lambda_2-1 \\ & m_{11} & \end{pmatrix}(Z).$$

Chapter 2

Abstract Angular Momentum Theory of Composite Systems

2.1 General Setting

We begin by describing the general setting for composite physical systems from the viewpoint of their angular momentum subspaces.

2.1.1 Angular momentum state vectors of a composite system

Every isolated quantum mechanical system with n constituent parts, each possessing angular momentum $\mathbf{J}(i), i = 1, 2, \ldots, n$, has total angular momentum

$$\begin{aligned} \mathbf{J} &= \mathbf{J}(1) + \mathbf{J}(2) + \cdots + \mathbf{J}(n), \\ \mathbf{J}(i) &= J_1(i)\mathbf{e}_1 + J_2(i)\mathbf{e}_2 + J_3(i)\mathbf{e}_3, \end{aligned} \tag{2.1}$$

where each angular momentum is referred to a common right-handed reference frame $(\mathbf{e}_1, \mathbf{e}_2, \mathbf{e}_3)$ in Cartesian 3−space $\mathbf{R}^3$. The components $J_k(i), k = 1, 2, 3,$ of $\mathbf{J}(i)$ have the standard action in a Hilbert space $\mathcal{H}_{j_i}$ of dimension $2j_i + 1$ (see (2.5) below).

Expression (2.1) for the sum of n angular momenta is not fully accurate in expressing the action of the total angular momentum $\mathbf{J}$ on the tensor product space $\mathcal{H}_{j_1} \otimes \mathcal{H}_{j_2} \otimes \cdots \otimes \mathcal{H}_{j_n}$ of dimension $\prod_{i=1}^{n}(2j_i + 1)$.

It is more carefully written as

$$\mathbf{J} = \sum_{i=1}^{n} \oplus \left(\mathbb{I}_{j_1} \otimes \cdots \otimes \mathbf{J}(i) \otimes \cdots \otimes \mathbb{I}_{j_n}\right), \tag{2.2}$$

where, in the factors in the direct sum, the identity operators $\mathbb{I}_{j_1}, \mathbb{I}_{j_2}, \ldots, \mathbb{I}_{j_n}$ appear in the corresponding positions $1, 2, \ldots, n$, except in position i, where $\mathbf{J}(i)$ stands. This notation correctly conveys that $\mathbf{J}(i)$ acts in the Hilbert space $\mathcal{H}_{j_i}$, and that the unit operators act in all other parts of the tensor product space. While we will often use the notation (2.1) for the sum of various angular momenta acting in the tensor product space, such expressions are always to be interpreted in the sense of a sum of tensor products of operators (see Sect. 10.5, Compendium A).

We introduce the following compact notations to describe the ket-vectors of the tensor product space:

$$\begin{aligned}
&\mathbf{j} = (j_1, j_2, \ldots, j_n),\ \text{each}\, j_i \in \{0, 1/2, 1, 3/2, \ldots\}, i = 1, 2, \ldots, n,\\
&\mathbf{m} = (m_1, m_2, \ldots, m_n),\ \text{each}\, m_i \in \{j_i, j_i - 1, \ldots, -j_i\},\\
&\qquad\qquad\qquad\qquad\qquad i = 1, 2, \ldots, n,\\
&\mathcal{H}_{\mathbf{j}} = \mathcal{H}_{j_1} \otimes \mathcal{H}_{j_2} \otimes \cdots \otimes \mathcal{H}_{j_n},\\
&|\mathbf{j}\,\mathbf{m}\rangle = |j_1\, m_1\rangle \otimes |j_2\, m_2\rangle \otimes \cdots \otimes |j_n\, m_n\rangle,\\
&\mathbb{C}(\mathbf{j}) = \{\mathbf{m} \,|\, m_i = j_i, j_i - 1, \ldots, -j_i;\ i = 1, 2, \ldots, n\}.
\end{aligned} \tag{2.3}$$

The set of $2n$ mutually commuting Hermitian operators

$$\mathbf{J}^2(1), J_3(1), \mathbf{J}^2(2), J_3(2), \ldots, \mathbf{J}^2(n), J_3(n) \tag{2.4}$$

is a complete set of operators in the tensor product space $\mathcal{H}_{\mathbf{j}}$, in that the set of vectors $|\mathbf{j}\,\mathbf{m}\rangle, \mathbf{m} \in \mathbb{C}(\mathbf{j})$ is an orthonormal basis. The action of the angular momentum operators $\mathbf{J}(i), i = 1, 2, \ldots, n$, is the standard action given by

$$\begin{aligned}
\mathbf{J}^2(i)|\mathbf{j}\,\mathbf{m}\rangle &= j_i(j_i + 1)|\mathbf{j}\,\mathbf{m}\rangle,\\
J_3(i)|\mathbf{j}\,\mathbf{m}\rangle &= m_i|\mathbf{j}\,\mathbf{m}\rangle,\\
J_+(i)|\mathbf{j}\,\mathbf{m}\rangle &= \sqrt{(j_i - m_i)(j_i + m_i + 1)}\, |\mathbf{j}\,\mathbf{m}_{+1}(i)\rangle,\\
J_-(i)|\mathbf{j}\,\mathbf{m}\rangle &= \sqrt{(j_i + m_i)(j_i - m_i + 1)}\, |\mathbf{j}\,\mathbf{m}_{-1}(i)\rangle,\\
\mathbf{m}_{\pm 1}(i) &= (m_1, \ldots, m_i \pm 1, \cdots, m_n).
\end{aligned} \tag{2.5}$$

The orthonormality of the basis functions is expressed by

$$\langle \mathbf{j}\,\mathbf{m} \,|\, \mathbf{j}\,\mathbf{m}'\rangle = \delta_{\mathbf{m},\mathbf{m}'},\ \text{each pair}\ \mathbf{m}, \mathbf{m}' \in \mathbb{C}(\mathbf{j}). \tag{2.6}$$

Since the collection of $2n$ commuting Hermitian operators (2.4) refers to the angular momenta of the individual constituents of a physical system, and the action of the angular momentum operators is on the basis vectors of each separate space, the basis $|\mathbf{j}\,\mathbf{m}\rangle, \mathbf{m} \in \mathbb{C}(\mathbf{j})$, is referred to as the *uncoupled basis* of the space $\mathcal{H}_{\mathbf{j}}$.

One of the most important observables for a composite system is the total angular momentum defined by (2.1). A set of $n+2$ mutually commuting Hermitian operators, which includes the square of the total angular momentum $\mathbf{J}$ and J_3 is the following:

$$\mathbf{J}^2(1), \mathbf{J}^2(2), \ldots, \mathbf{J}^2(n), \mathbf{J}^2, J_3. \tag{2.7}$$

This set of $n+2$ commuting Hermitian operators is an incomplete set with respect to the construction of the states of total angular momentum; that is, the simultaneous state vectors of the $n+2$ operators (2.7) do not determine a basis of the space $\mathcal{H}_{\mathbf{j}}$. There are many ways to complete such an incomplete basis. For example, an additional set of $n-2$ independent $SU(2)$ invariant Hermitian operators, commuting among themselves, as well as with each operator in the set (2.7), could serve this purpose. Other methods of labeling can also be used (see (2.44) below). We make the following assumptions:

Assumptions. The incomplete set of simultaneous eigenvectors of the $n+2$ angular momentum operators (2.7) has been extended to a basis of the space $\mathcal{H}_{\mathbf{j}}$ with properties as follows: A basis set of vectors can be enumerated in terms of an indexing set $\mathbb{R}(\mathbf{j})$ of the form

$$\mathbb{R}(\mathbf{j}) = \left\{ \boldsymbol{\alpha} = (\alpha_1, \alpha_2, \ldots, \alpha_{n-2}), j, m \;\middle|\; \begin{array}{l} j \in \mathbb{D}(\mathbf{j}); \alpha_i \in \mathbb{A}_i(\mathbf{j}, j); \\ m = j, j-1, \ldots, -j \end{array} \right\}, \tag{2.8}$$

where the domains of definition $\mathbb{D}(\mathbf{j})$ of j and $\mathbb{A}_i(\mathbf{j}, j)$ of α_i have the properties as follows. These domains of definition are to be such that for given quantum numbers $\mathbf{j}$ the cardinality of the set $\mathbb{R}(\mathbf{j})$ is given by

$$|\,\mathbb{R}(\mathbf{j})\,| = |\,\mathbb{C}(\mathbf{j})\,| = \prod_{i=1}^{n}(2j_1 + 1). \tag{2.9}$$

Moreover, these labels are to be such that the space $\mathcal{H}_{\mathbf{j}}$ has the orthonormal basis given by the ket-vectors

$$\begin{aligned} &|(\mathbf{j}\,\boldsymbol{\alpha})_{j\,m}\rangle, \;\; \boldsymbol{\alpha}, j, m \in \mathbb{R}(\mathbf{j}), \\ &\langle(\mathbf{j}\,\boldsymbol{\alpha})_{j,m}\,|\,(\mathbf{j}\,\boldsymbol{\alpha}')_{j',m'}\rangle = \delta_{j,j'}\delta_{m,m'}\delta_{\boldsymbol{\alpha},\boldsymbol{\alpha}'}, \\ &\boldsymbol{\alpha}, j, m \in \mathbb{R}(\mathbf{j});\; \boldsymbol{\alpha}', j', m' \in \mathbb{R}(\mathbf{j}). \end{aligned} \tag{2.10}$$

It is always the case that $\mathbb{D}(\mathbf{j})$ is independent of how the extension to a basis through the parameters $\boldsymbol{\alpha}$ is effected and that, for given j, the domain of m is $m = j, j-1, \ldots, -j$. Each of the quantum labels α_i belongs to some domain of definition $\mathbb{A}_i(\mathbf{j}, j)$ that can depend also on j.

Finally, the actions of the commuting angular momentum operators (2.7) and the total angular momentum $\mathbf{J}$ on the orthonormal basis set (2.10) are given by

$$\begin{aligned}
&\mathbf{J}^2(i)|(\mathbf{j}\,\boldsymbol{\alpha})_{j\,m}\rangle = j_i(j_i+1)|(\mathbf{j}\,\boldsymbol{\alpha})_{j\,m}\rangle, i = 1, 2, \ldots, n, \\
&\mathbf{J}^2|(\mathbf{j}\,\boldsymbol{\alpha})_{j\,m}\rangle = j(j+1)|(\mathbf{j}\,\boldsymbol{\alpha})_{j\,m}\rangle, \\
&J_3|(\mathbf{j}\,\boldsymbol{\alpha})_{j\,m}\rangle = m|(\mathbf{j}\,\boldsymbol{\alpha})_{j\,m+1}\rangle, \\
&J_+|(\mathbf{j}\,\boldsymbol{\alpha})_{j\,m}\rangle = \sqrt{(j-m)(j+m+1)}|(\mathbf{j}\,\boldsymbol{\alpha})_{j\,m+1}\rangle, \\
&J_-|(\mathbf{j}\,\boldsymbol{\alpha})_{j\,m}\rangle = \sqrt{(j+m)(j-m+1)}|(\mathbf{j}\,\boldsymbol{\alpha})_{j\,m-1}\rangle. \qquad \square
\end{aligned} \tag{2.11}$$

The notation for the ket-vectors in (2.10) and (2.11) places the total angular momentum quantum number j and its projection m in the subscript position to accentuate their privileged role. The set $R(\mathbf{j})$ enumerates an alternative unique orthonormal basis (2.10) of the space $\mathcal{H}_\mathbf{j}$ that contains the total angular momentum quantum numbers j, m; it is the analog of the set $C(\mathbf{j})$ that enumerates the orthonormal basis set (2.6). Any basis set with the properties (2.10)-(2.11) is called a *coupled basis* of $\mathcal{H}_\mathbf{j}$. For $n = 2$, the uncoupled basis set is $|j_1\, m_1\rangle \otimes |j_2\, m_2\rangle, C(j_1 j_2) = \{m_1, m_2 | m_i = j_i, j_i - 1, \ldots, -j_i, i = 1, 2\}$, and the coupled basis set is $|(j_1\, j_2) j\, m\rangle, R(j_1 j_2) = \{j, m | j \in \{j_1 + j_2, j_1 + j_2 - 1, \ldots, |j_1 - j_2|\}, m = j, j-1, \ldots, -j\}$. No extra $\boldsymbol{\alpha}$ labels are required. For $n = 3$, one extra label α_1 is required, and at this point, the domains of definition of $\alpha_1; j$ are left unspecified.

Angular momentum coupling theory is about the various ways of providing the extra set of $\boldsymbol{\alpha}$ labels and their domains of definition, together with the values of the total angular momentum quantum number j, such that the space $\mathcal{H}_\mathbf{j}$ is spanned by the vectors $|(\mathbf{j}\,\boldsymbol{\alpha})_{j\,m}\rangle$. It turns out, as shown below, that the set of values that the total angular momentum quantum number j can assume is independent of the α_i; the values of j being $j = j_{\min}, j_{\min} + 1, \ldots, j_{\max}$, for well-defined minimum and maximum values of j that are expressed in terms of $j_1, j_2, \ldots, j_n$. Thus, the burden of completing any basis is placed on assigning the labels α_i in the set

$$\mathbb{R}(\mathbf{j}, j) = \{\boldsymbol{\alpha} = (\alpha_1, \alpha_2, \ldots, \alpha_{n-2}) \,|\, \text{each}\, \alpha_i \in \mathbb{A}_i(\mathbf{j}, j)\}. \tag{2.12}$$

Such an assignment is called an $\boldsymbol{\alpha}-$*coupling scheme.* Since there are many ways of completing an incomplete basis of a finite vector space, there are also many coupling schemes. In this sense, the structure of the *coupling scheme set* $\mathbb{R}(\mathbf{j}, j)$ is the key object in angular momentum coupling theory; all the details of defining the coupling scheme are to be provided by the domains of definition $\mathbb{A}_i(\mathbf{j}, j)$.

The cardinality of the sets $\mathbb{R}(\mathbf{j})$ and $\mathbb{A}_j(\mathbf{j})$, and $\mathbb{C}(\mathbf{j})$ are related by

$$|\,\mathbb{R}(\mathbf{j})\,| = \sum_{j=j_{\min}}^{j_{\max}} (2j+1)N_j(\mathbf{j}) = N(\mathbf{j}) = \prod_{i=1}^{n}(2j_i+1) = |\,\mathbb{C}(\mathbf{j})\,|, \quad (2.13)$$

where we have defined $N_j(\mathbf{j}) = |\mathbb{R}(\mathbf{j}, j)|$. We give below a recursive method for generating the positive integers $N_j(\mathbf{j})$.

The purpose of introducing a coupling scheme is twofold: First, it encodes the instructions on effecting the reduction of the $n-$fold Kronecker product $D^{\mathbf{j}}(U) = D^{j_1}(U) \otimes D^{j_2}(U) \otimes \cdots \otimes D^{j_n}(U)$, $U \in SU(2)$, which is a reducible unitary representation of $SU(2)$ of dimension $N(\mathbf{j})$, into a direct sum of irreducible unitary representations $D^j(U)$. Second, it gives a new orthonormal basis of the space $\mathcal{H}_{\mathbf{j}}$ in which the total angular momentum quantum numbers (j, m) label the basis vectors. We discuss the first property in detail below, and next the second property.

The two orthonormal bases (2.6) and ((2.10) of the space $\mathcal{H}_{\mathbf{j}}$ must be related by a unitary transformation $A^{(\mathbf{j})}$ of order $N(\mathbf{j}) = \prod_{i=1}^{n}(2j_i+1)$ with rows and columns enumerated by

$$A^{(\mathbf{j})} = \left(A^{(\mathbf{j})}\right)_{\boldsymbol{\alpha},j,m\in\mathbb{R}(\mathbf{j});\,\mathbf{m}\in\mathbb{C}(\mathbf{j})}. \quad (2.14)$$

Thus, we must have the invertible relations between bases:

$$|(\mathbf{j}\,\boldsymbol{\alpha})_{j\,m}\rangle = \sum_{\mathbf{m}\in\mathbb{C}(\mathbf{j})} \left(A^{(\mathbf{j}}\right)_{\boldsymbol{\alpha},j,m;\,\mathbf{m}} |\mathbf{j}\,\mathbf{m}\rangle, \text{ each } \boldsymbol{\alpha}, j, m \in \mathbb{R}(\mathbf{j}),$$

$$(2.15)$$

$$|\mathbf{j}\,\mathbf{m}\rangle = \sum_{\boldsymbol{\alpha},j,m\in\mathbb{R}(\mathbf{j})} \left(A^{(\mathbf{j})*}\right)_{\boldsymbol{\alpha},j\,m;\,\mathbf{m}} |(\mathbf{j}\,\boldsymbol{\alpha})_{j\,m}\rangle, \text{ each } \mathbf{m} \in \mathbb{C}(\mathbf{j}).$$

Generation of the multiplicity numbers $N_j(\mathbf{j})$

The positive integers $N_j(\mathbf{j}) = |\,\mathbb{R}(\mathbf{j}, j)\,|$ in (2.13) give the number of occurrences of $D^j(U)$ in the reduction of the Kronecker product

$$D^{\mathbf{j}}(U) = D^{j_1}(U) \otimes D^{j_2}(U) \otimes \cdots \otimes D^{j_n}(U),\ U \in SU(2), \quad (2.16)$$

into a direct sum of $D^j(U)$, as given by

$$D^{\mathbf{j}}(U) = \sum_{j=j_{\min}}^{j_{\max}} \oplus \mathbb{D}^j(U),\ \ \mathbb{D}^j(U) = \underbrace{D^j(U) \oplus \cdots \oplus D^j(U)}_{N_j(\mathbf{j})}, \quad (2.17)$$

where all matrices are blocks along the diagonal, which, by convention, are arranged along the diagonal from the least to the greatest value of j. We refer to (2.17) as the *Kronecker direct sum* (see (2.25) below).

The Clebsch-Gordon (CG) numbers $N_j(\mathbf{j})$ may be generated by the following recursive procedure. Define the *multiset* $\langle j_1, j_2, \ldots, j_i \rangle, i = 2, 3, \ldots,$ for the addition of i angular momenta by

$$\begin{aligned} \langle j_1, j_2, \ldots, j_i \rangle &= \{\langle k, j_i \rangle | k \in \langle j_1, j_2, \ldots, j_{i-1} \rangle\}, \\ \langle j_1, j_2 \rangle &= \{|j_1 - j_2|, |j_1 - j_2| + 1, \ldots, j_1 + j_2\}. \end{aligned} \tag{2.18}$$

The second set then contains the set of values that the total angular momentum j can assume in the addition of two angular momenta $\mathbf{J}(1) + \mathbf{J}(2) = \mathbf{J}$. Starting with $i = 3$ and $k \in \langle j_1, j_2 \rangle$, the first relation can be iterated to obtain the multiset $\langle j_1, j_2, \ldots, j_n \rangle, n \geq 3$, from which the CG number $N_j(\mathbf{j})$ can be read out for every values of j : *Relation (2.18) is a method of generating all CG numbers.*

The cardinality of multiset $\langle j_1, j_2, \ldots, j_n \rangle$ is

$$| \langle j_1, j_2, \ldots, j_n \rangle | = \sum_{j=j_{\min}}^{j_{\max}} (2j+1) N_j(\mathbf{j}) = \prod_{i=1}^{n} (2j_i + 1). \tag{2.19}$$

The following properties of the multiset $\langle \mathbf{j} \rangle = \langle j_1, j_2, \ldots, j_n \rangle$ may be proved by induction on n :

1. Invariance under permutations:
$$\langle j_{\pi_1}, j_{\pi_2}, \ldots, j_{\pi_n} \rangle = \langle j_1, j_2, \ldots, j_n \rangle, \text{ each } \pi \in S_n. \tag{2.20}$$
2. Least and greatest elements, $j_{\min}$ and $j_{\max}$:
$$\begin{aligned} j_{\min} &= \min\{| j_1 \pm j_2 \pm \cdots \pm j_n |\}, \\ j_{\max} &= j_1 + j_2 + \cdots + j_n. \end{aligned} \tag{2.21}$$
All 2^{n-1} possible $\pm$ signs are to be considered in determining $j_{\min}$.
3. Values of the total angular momentum j :
$$j = j_{\min}, j_{\min} + 1, j_{\min} + 2, \ldots, j_{\max}. \tag{2.22}$$
4. Form of the set $\langle \mathbf{j} \rangle$:
$$\langle \mathbf{j} \rangle = \langle j_1, j_2, \ldots, j_n \rangle = \left\{ j_{\min}^{h_1}, (j_{\min} + 1)^{h_2}, \ldots, j_{\max}^{h_t} \right\}, \tag{2.23}$$
where $h_k = N_{j_{\min}+k-1}(\mathbf{j}),\ k = 1, 2, \ldots, t = j_{\max} - j_{\min} + 1$. It is always the case that $h_1 = h_t = 1$.

Example:

$$\begin{aligned}&\langle 1/2,1,5/2\rangle = \langle 1/2,5/2,1\rangle = \langle 1,5/2,1/2\rangle = \\ &\langle 1,1/2,5/2\rangle = \langle 5/2,1/2,1\rangle = \langle 5/2,1,1/2\rangle \\ &= \{1,2,2,3,3,4\}.\end{aligned} \tag{2.24}$$

The 36×36 block matrix $\mathbb{D}^{(1/2,1,5/2)}(U)$ is given by

$$\mathbb{D}^{(1/2,1,5/2)}(U) = \begin{pmatrix} \mathbb{D}^1(U) & 0 & 0 & 0 \\ 0 & \mathbb{D}^2(U) & 0 & 0 \\ 0 & 0 & \mathbb{D}^3(U) & 0 \\ 0 & 0 & 0 & \mathbb{D}^4(U) \end{pmatrix}, \tag{2.25}$$

$$\begin{aligned}&\mathbb{D}^1(U) = D^1(U),\ \mathbb{D}^4(U) = D^4(U), \\ &\mathbb{D}^2(U) = \begin{pmatrix} D^2(U) & 0 \\ 0 & D^2(U) \end{pmatrix}, \\ &\mathbb{D}^3(U) = \begin{pmatrix} D^3(U) & 0 \\ 0 & D^3(U) \end{pmatrix}.\end{aligned} \tag{2.26}$$

The multiset set $\langle \mathbf{j} \rangle$ is called the *generalized* CG series (or simply CG series), since it generalizes the Clebsch-Gordan series for $n = 2 : \langle j_1\, j_2\rangle = \{j_1 + j_2, j_1 + j_2 - 1, \ldots, |j_1 - j_2|\}$. Except for the number of multiple occurrences $N_j(\mathbf{j})$ of $D^j(U)$ in the direct sum $\mathbb{D}^j(U)$, the structure of the full direct sum is known. Indeed, since $N_j(\mathbf{j})$ can be generated recursively, as described above, this structure is uniquely determined. Thus, we know how the $n-$fold Kronecker product reduces into a direct sum without knowing the unitary similarity transformation that effects the reduction.

Since Clebsch and Gordan first found $C_j(j_1\, j_2) = 1$, for $j \in \langle j_1\, j_2\rangle$, we also call the numbers $N_j(\mathbf{j})$ the CG numbers. We give the combinatorial meaning of the CG numbers below in terms of counting compositions, after giving some additional structural properties of Kronecker products (see also relation (11.286), Compendium B).

2.1.2 Group actions in a composite system

Under the action of an $SU(2)$ unitary frame rotation of the common frame $(\mathbf{e}_1, \mathbf{e}_2, \mathbf{e}_3)$ used to describe the n constituents of a collection of physical systems in Cartesian space $\mathbb{R}^3$, where system i has angular momentum $\mathbf{J}(i) = J_1(i)\mathbf{e}_1 + J_2(i)\mathbf{e}_2 + J_3(i))\mathbf{e}_3$, the orthonormal basis of the angular momentum subspace

$$\mathcal{H}_{j_i} = \{|j_i\, m_i\rangle, m_i = j_i, j_i - 1, \ldots, -j_i\} \tag{2.27}$$

of systems i undergoes the standard unitary transformation

$$\mathcal{T}_U|j_i\, m_i'\rangle = \sum_{m_i} D^{j_i}_{m_i\, m_i'}(U)|j_i\, m_i\rangle, \text{ each } U \in SU(2). \tag{2.28}$$

The action of the components $(J_1(i), J_2(i), J_3(i))$ of $\mathbf{J}(i)$ on the basis $|j_i\, m_i\rangle$ of $\mathcal{H}_{j_i}$ is standard.

The uncoupled basis $\mathcal{H}_{j_1} \otimes \mathcal{H}_{j_2} \otimes \cdots \otimes \mathcal{H}_{j_n}$ of the angular momentum space $\mathcal{H}_{\mathbf{j}}$ of the collection of systems undergoes the unitary transformation given by

$$\begin{aligned}(\mathcal{T}_U \otimes \mathcal{T}_U \otimes \cdots \otimes \mathcal{T}_U)\left(|j_1\, m_1'\rangle \otimes |j_2\, m_2'\rangle \otimes \cdots \otimes |j_n\, m_n'\rangle\right. \\ = \sum_{\mathbf{m}} \left(D^{j_1}(U) \otimes D^{j_2}(U) \otimes \cdots \otimes D^{j_n}(U)\right)_{\mathbf{m}\,\mathbf{m}'} \\ \times(|j_1\, m_1\rangle \otimes |j_2\, m_2\rangle \otimes \cdots \otimes |j_n\, m_n\rangle),\end{aligned} \tag{2.29}$$

where $\mathbf{m} = (m_1, m_2, \ldots, m_n), \mathbf{m}' = (m_1', m_2', \ldots, m_n')$. This relation is described in the abbreviated notations (2.3) by

$$T_U|\mathbf{j}\,\mathbf{m}'\rangle = \sum_{\mathbf{m}} D^{\mathbf{j}}_{\mathbf{m}\,\mathbf{m}'}(U)\,|\mathbf{j}\,\mathbf{m}\rangle, \text{ each } U \in SU(2). \tag{2.30}$$

Similarly, the coupled basis (2.10) of $\mathcal{H}_{\mathbf{j}}$ undergoes the irreducible unitary transformation:

$$T_U|(\mathbf{j}\,\boldsymbol{\alpha})_{j\,m'}\rangle = \sum_{m} D^{j}_{m\,m'}(U)|(\mathbf{j}\,\boldsymbol{\alpha})_{j\,m}\rangle, \text{ each } U \in SU(2). \tag{2.31}$$

2.1.3 Standard form of the Kronecker direct sum

Schur's lemma (see Sect. 10.7.2, Compendium A) implies that the reducible unitary Kronecker product representation $D^{\mathbf{j}}(U)$ of $SU(2)$ defined by (2.16) and(2.30) is reducible into a direct sum of irreducible unitary representations $D^j(U)$ by a unitary matrix similarity transformation $U^{(\mathbf{j})}$ of order $N(\mathbf{j}) = \prod_{i=1}^{n}(2j_i + 1)$:

$$U^{(\mathbf{j})} D^{\mathbf{j}}(U) \left(U^{(\mathbf{j})}\right)^{\dagger} = \mathbb{D}^{\mathbf{j}}(U), \text{ each}\, U \in SU(2), \tag{2.32}$$

where the matrix $\mathbb{D}^{\mathbf{j}}(U)$, also of order $N(\mathbf{j})$, is the Kronecker direct sum defined in (2.17) with the block matrices ordered along the diagonal from the least value $j = j_{\min}$ to the greatest value $j = j_{max}$, as read down the diagonal (see the example given by (2.24)-(2.25)). The direct sum $\mathbb{D}^{j}(U)$ of $N_j(\mathbf{j})$ identical matrices $D^j(U)$ can then be written as the Kronecker product

$$\begin{aligned}\mathbb{D}^{j}(U) &= I_{N_j(\mathbf{j})} \otimes D^j(U)\\ &= \begin{pmatrix} D^j(U) & 0 & 0 & \cdots & 0 \\ 0 & D^j(U) & 0 & \ldots & 0 \\ \vdots & \vdots & \vdots & \ldots & \vdots \\ 0 & 0 & 0 & \cdots & D^j(U) \end{pmatrix},\end{aligned} \tag{2.33}$$

where $I_{N_j(\mathbf{j})}$ is the unit matrix of order $N_j(\mathbf{j})$. We give explicitly the following **standard form** of the Kronecker direct sum (2.32):

$$\sum_{j=j_{\min}}^{j_{\max}} \oplus \mathbb{D}^{\mathbf{j}}(U) = \mathbb{D}^{j_{\min}}(U) \oplus \mathbb{D}^{j_{min}+1}(U) \oplus \cdots \oplus \mathbb{D}^{j_{\max}}(U), \tag{2.34}$$

where each matrix $\mathbb{D}^{j}(U)$ is itself of the form (2.33). The reason for adopting a standard form (2.34) for the Kronecker direct sum is so that we can subsequently be very specific about the structure of the unitary matrix $U^{(\mathbf{j})}$ that effects the reduction in (2.32).

Combinatorial significance of the CG numbers

The CG numbers that give the number of times that the irreducible representation $D^j(U)$ occurs in the reduction of the direct product $D^{\mathbf{j}}(U)$ can be given in terms of the number of compositions of a certain form. We derive this result in this subsection. For each $j = j_{\min}, j_{\min} + 1, \ldots, j_{\max}$, that occurs in the reduction (2.32)-(2.33), we define the set of projection quantum numbers $\mathbf{m}$ by

$$\mathbb{M}_j(\mathbf{j}) = \left\{ (m_1, m_2, \ldots, m_n) \,\middle|\, \begin{array}{l} -j_i \leq m_i \leq j_i, i = 1, 2, \ldots, n; \\ m_1 + m_2 + \cdots + m_n = j \end{array} \right\}. \tag{2.35}$$

Let $M_j(\mathbf{j}) = |\,\mathbb{M}_j(\mathbf{j})\,|$ denote the cardinality of this set. Then, from this definition, and the standard form (2.34), it follows that

$$\sum_{k=j}^{j_{\max}} N_k(\mathbf{j}) = M_j(\mathbf{j}), \; j = j_{\min}, j_{\min} + 1, \ldots, j_{\max}. \tag{2.36}$$

Applying this relation, in turn, to $j = j_{\max}, j_{\max} - 1, \dots, j_{\min}$, we obtain the inverse relations

$$N_j(\mathbf{j}) = M_j(\mathbf{j}) - M_{j+1}(\mathbf{j}), \text{ for } j = j_{\max}, j_{\max} - 1, \dots, j_{\min}, \tag{2.37}$$

where $M_{j_{\max}+1}(\mathbf{j}) = 0, , M_{j_{\max}}(\mathbf{j}) = 1$. Thus, the CG numbers $N_j(\mathbf{j})$ are obtained by counting the number of elements in two adjacent sets in (2.35) and taking the difference.

The determination of the number of elements in the set $\mathbb{M}_j(\mathbf{j})$ can be formulated as a problem in counting the number of compositions of a certain type. Thus, we define the nonnegative integers $l_i = j_i - m_i, m_i = j_i, j_i - 1, \dots, -j_i$, so that $0 \leq l_i \leq 2j_i$. We now define, for each j such that $j_{\min} \leq j \leq j_{\max}$, the following set, which is just a shifted version of $\mathbb{M}_j(\mathbf{j})$, so that their cardinalities are equal:

$$\mathbb{L}_j(\mathbf{j}) = \left\{ (l_1, l_2, \dots, l_n)) \,\middle|\, \begin{array}{l} 0 \leq l_i \leq 2j_i, i = 1, 2, \dots, n; \\ l_1 + l_2 + \cdots + l_n = j_{\max} - j \end{array} \right\}, \tag{2.38}$$

with cardinality $L_j(\mathbf{j}) = |\,\mathbb{L}_j(\mathbf{j})\,| = M_j(\mathbf{j})$. Accordingly, we have the result: *$L_j(\mathbf{j})$ is the number of compositions of $j_{\max} - j$ into n nonnegative parts such that each part satisfies $0 \leq l_i \leq 2j_i$, and the cardinality $M_j(\mathbf{j})$ of the set (2.35) is given by*

$$M_j(\mathbf{j}) = L_j(\mathbf{j}), j = j_{\max}, j_{\max} - 1, \dots, j_{\min}. \tag{2.39}$$

For $j = j_{\max}$, this relation gives $M_{j_{\max}}(\mathbf{j}) = 1$, since there is only one composition of 0 into n nonnegative parts, namely, $(0, 0, \dots, 0)$. All CG numbers are obtained from the difference given by (2.37).

We illustrate the above relations for the case $n = 3$ and $(j_1, j_2, j_3) = (1/2, 1, 5/2)$ given by (2.24)-(2.25), so that $j_{\max} = 4, j_{\min} = 1$. For brevity, we drop the $\mathbf{j}$ arguments in the above relations, noting, however, that parts 1, 2, 3 of the composition must satisfy $0 \leq l_1 \leq 1, 0 \leq l_2 \leq 2, 0 \leq l_3 \leq 5$. By direct calculation of the compositions, we obtain:

$$\begin{aligned} &\mathbb{L}_4 = \{(0,0,0)\},\ \mathbb{L}_3 = \{(1,0,0),(0,1,0),(0,0,1)\}, \\ &\mathbb{L}_2 = \{(0,2,0),(0,0,2),(0,1,1),(1,1,0),(1,0,1)\}, \\ &\mathbb{L}_1 = \{(0,0,3),(0,1,2),(0,2,1),(1,1,1),(1,2,0),(1,0,2)\}, \\ &\mathbb{L}_0 = \{(0,0,4),(0,1,3),(0,2,2),(1,0,3),(1,1,2),(1,2,1)\}. \end{aligned} \tag{2.40}$$

Thus, we have $M_4 = 1, M_3 = 3, M_2 = 5, M_1 = 6, M_0 = 6$, and $N_4 = 1, N_3 = 2, N_2 = 2, N_1 = 1, N_0 = 0$, in agreement with (2.24).

When all the angular momenta j_i are equal, the CG numbers are related to Gaussian polynomials, which are treated in Andrews [1, Chapter 3]. This relation and further properties of the CG numbers are given in relations (11.24)-(11.26), Sect. 11.1.1, and relations (11.285)-(11.287), Compendium B. We learned of the Gaussian polynomial relation from a preprint by Sunko and Svrtan [168, publication place not found]).

2.1.4 Reduction of Kronecker products

The reduction of the Kronecker product $D^{\mathbf{j}}(U)$ into the standard form of the Kronecker direct sum by the unitary matrix similarity transformation in (2.32) is not unique. There are nondenumerably infinitely many unitary matrices $U^{(\mathbf{j})}$ of order $N(\mathbf{j}) = \prod_{i=1}^{n}(2j_i + 1)$ that effect the transformation

$$U^{(\mathbf{j})} D^{\mathbf{j}}(U) \left(U^{(\mathbf{j})}\right)^{\dagger} = \sum_{j=j_{\min}}^{j_{\max}} \oplus \mathbb{D}^{j}(U). \tag{2.41}$$

The rows and columns of $U^{(\mathbf{j})}$ are labeled by the indexing sets $\mathbb{R}(\mathbf{j})$ and $\mathbb{C}(\mathbf{j})$ as given by (2.8) and (2.3), respectively:

$$\left(U^{(\mathbf{j})}\right)_{\boldsymbol{\alpha},j,m;\mathbf{m}} = \langle \mathbf{j}\,\mathbf{m} \,|\, (\mathbf{j}\,\boldsymbol{\alpha})_{j\,m}\rangle, \tag{2.42}$$

where these matrix elements are also the transformation coefficients between the coupled and uncoupled basis vectors given by (2.15). The rows and columns can always be ordered such that $U^{(\mathbf{j})}$ effects the standard reduction given by (2.33)-(2.34); that is, given any coupling scheme, the transformation of the Kronecker product to the standard Kronecker direct sum can always be realized.

There is, however, an intrinsic non-uniqueness of the transformation (2.41) due to the multiplicity structure (2.33) of the standard reduction itself. Thus, define the matrix $\mathbb{W}_j^{(\mathbf{j})}$ of order $N(\mathbf{j})$ to be the direct product given by

$$\mathbb{W}_j^{(\mathbf{j})} = W_j^{(\mathbf{j})} \otimes I_{2j+1}, \tag{2.43}$$

where $W_j^{(\mathbf{j})}$ is an **arbitrary complex matrix** of order $N_j(\mathbf{j})$, the CG number. Then, the matrix $\mathbb{W}_j^{(\mathbf{j})}$ commutes with the direct sum matrix $\mathbb{D}^j$ defined by (2.33):

$$\mathbb{W}_j^{(\mathbf{j})} \mathbb{D}^j(U) = \mathbb{D}^j(U) \mathbb{W}_j^{(\mathbf{j})}, \text{ each } U \in SU(2). \tag{2.44}$$

The existence of the matrix $\mathbb{W}^{(\mathbf{j})}$ occurs because the set of $n+2$ commuting Hermitian operators (2.7) is incomplete for $n \geq 3$. *It is precisely the commuting property (2.44) of an arbitrary matrix of the form (2.43) that allows the additional degrees of freedom associated with interactions that break this degeneracy in physical systems.*

We may choose $W_j^{(\mathbf{j})}$ in (2.43) to be an arbitrary unitary matrix of order $N_j(\mathbf{j})$; that is, $W_j^{(\mathbf{j})} \in U(N_j(\mathbf{j}))$, the group of unitary matrices of order $N_j(\mathbf{j})$. Then, the direct sum matrix

$$W^{(\mathbf{j})} = \sum_{j=j_{\min}}^{j_{\max}} \oplus \mathbb{W}_j^{(\mathbf{j})} = \sum_{j=j_{\min}}^{j_{\max}} \oplus \left(W_j^{(\mathbf{j})} \otimes I_{2j+1} \right), W_j^{(\mathbf{j})} \in U(N_j(\mathbf{j})), \tag{2.45}$$

is a unitary matrix in the unitary group $U(N(\mathbf{j})$, which has the commuting property given by

$$W^{(\mathbf{j})} D^{\mathbf{j}}(U) = D^{\mathbf{j}}(U) W^{(\mathbf{j})}, \text{ each } U \in SU(2). \tag{2.46}$$

Thus, if we define the unitary matrix $V^{(\mathbf{j})}$ by $V^{(\mathbf{j})} = W^{(\mathbf{j})}\, U^{(\mathbf{j})}$, hence,

$$V^{(\mathbf{j})} U^{(\mathbf{j})\dagger} = W^{(\mathbf{j})}, \tag{2.47}$$

then $V^{(\mathbf{j})}$ also effects, for each $U \in SU(2)$, the transformation:

$$V^{(\mathbf{j})} D^{\mathbf{j}}(U) \left(V^{(\mathbf{j})}\right)^{\dagger} = U^{(\mathbf{j})} D^{\mathbf{j}}(U) \left(U^{(\mathbf{j})}\right)^{\dagger} = \sum_{j=j_{\min}}^{j_{\max}} \oplus \mathbb{D}^{j}(U). \tag{2.48}$$

Each unitary matrix $V^{(\mathbf{j})}$ effects exactly the same reduction of the Kronecker product representation $D^{\mathbf{j}}(U)$ of $SU(2)$ into standard Kronecker direct sum form as does $U^{(\mathbf{j})}$. We call all unitary similarity transformations with the property (2.48) *standard reductions.*

Summary: Define the subgroup $H(N(\mathbf{j}))$ of the unitary group $U(N(\mathbf{j}))$ of order $N(\mathbf{j}) = \prod_{i=1}^{n}(2j_i + 1)$ by

$$H(N(\mathbf{j})) = \left\{ \sum_{j=j_{\min}}^{j_{\max}} \oplus \left(W_j^{(\mathbf{j})} \otimes I_{2j+1} \right) \,\middle|\, W_j^{(\mathbf{j})} \in U(N_j(\mathbf{j})) \right\}. \tag{2.49}$$

Then, if the unitary matrix element $U^{(\mathbf{j})}$ effects the standard reduction, so does every unitary matrix $V^{(\mathbf{j})}$ such that

$$V^{(\mathbf{j})} U^{(\mathbf{j})\dagger} \in H(N(\mathbf{j})). \tag{2.50}$$

2.1.5 Recoupling matrices

Let $U^{(\mathbf{j})}$ and $V^{(\mathbf{j})}$ be unitary matrices of order $N(\mathbf{j})) = \prod_{i=1}^{n}(2j_i + 1)$ that effect the standard reduction (2.48). The unitary matrix $U^{(\mathbf{j})}$ corresponds to an $\boldsymbol{\alpha}-$coupling scheme and has its rows enumerated by the elements of the set

$$\mathbb{R}(\mathbf{j}) = \left\{ \boldsymbol{\alpha} \in \mathbb{R}(\mathbf{j}, j), j, m \,\middle|\, \begin{array}{l} j = j_{\min}, j_{\min} + 1, \ldots, j_{\max}; \\ m = j, j-1, \ldots, -j \end{array} \right\}, \tag{2.51}$$

where the domains of definition of the α_i quantum numbers are given by

$$\mathbb{R}(\mathbf{j}, j) = \{\boldsymbol{\alpha} = (\alpha_1, \alpha_2, \ldots, \alpha_{n-2}) \,|\, \alpha_i \in \mathbb{A}_i(\mathbf{j}, j)\}\,. \tag{2.52}$$

The unitary matrix $V^{(\mathbf{j})}$ corresponds to a $\boldsymbol{\beta}-$coupling scheme and has its rows enumerated by the elements of a set analogous to $\mathbb{R}(\mathbf{j})$ as defined by

$$\mathbb{S}(\mathbf{j}) = \left\{\boldsymbol{\beta} \in \mathbb{B}(\mathbf{j}, j), j, m \,\middle|\, \begin{array}{l} j = j_{\min}, j_{\min}+1, \ldots, j_{\max}; \\ m = j, j-1, \ldots, -j \end{array}\right\}. \tag{2.53}$$

where the domains of definition of the β_i quantum numbers are given by

$$\mathbb{S}(\mathbf{j}, j) = \{\boldsymbol{\beta} = (\beta_1, \beta_2, \ldots, \beta_{n-2}) \,|\, \beta_i \in \mathbb{B}_i(\mathbf{j}, j)\}\,. \tag{2.54}$$

The column indexing set for $U^{(\mathbf{j})}$ and $V^{(\mathbf{j})}$ is the same set of projection quantum numbers $\mathbb{C}(\mathbf{j})$.

There is a set of coupled state vectors associated with each of the unitary matrices $U^{(\mathbf{j})}$ and $V^{(\mathbf{j})}$ given by

$$\begin{aligned} |(\mathbf{j}\,\boldsymbol{\alpha})_{j\,m}\rangle &= \sum_{\mathbf{m}\in\mathbb{C}(\mathbf{j})} \left(U^{(\mathbf{j})}\right)_{(\boldsymbol{\alpha})_{j\,m};\,\mathbf{m}} |\mathbf{j}\,\mathbf{m}\rangle,\ \ \boldsymbol{\alpha}, j, m \in \mathbb{R}(\mathbf{j}), \\ |(\mathbf{j}\,\boldsymbol{\beta})_{j\,m}\rangle &= \sum_{\mathbf{m}\in\mathbb{C}(\mathbf{j})} \left(V^{(\mathbf{j})}\right)_{(\boldsymbol{\beta})_{j\,m};\,\mathbf{m}} |\mathbf{j}\,\mathbf{m}\rangle,\ \ \boldsymbol{\beta}, j, m \in \mathbb{S}(\mathbf{j}). \end{aligned} \tag{2.55}$$

Both the $\boldsymbol{\alpha}-$coupled basis and $\boldsymbol{\beta}-$coupled basis are orthonormal basis sets of the same tensor product space $\mathcal{H}_{\mathbf{j}}$ and satisfy all of the standard relations (2.10)-(2.12). Since these orthonormal basis sets span the same vector space, they are related by a unitary transformation of the form:

$$\begin{aligned} |(\mathbf{j}\,\boldsymbol{\beta})_{j\,m}\rangle &= \sum_{\boldsymbol{\alpha}\in\mathbb{R}(\mathbf{j},j)} U_{(\mathbf{j}\,\boldsymbol{\alpha})_j;(\mathbf{j}\,\boldsymbol{\beta})_j} |(\mathbf{j}\,\boldsymbol{\alpha})_{j\,m}\rangle, \\ U_{(\mathbf{j}\,\boldsymbol{\alpha})_j;(\mathbf{j}\,\boldsymbol{\beta})_j} &= \langle(\mathbf{j}\,\boldsymbol{\alpha})_{j\,m}\,|\,(\mathbf{j}\,\boldsymbol{\beta})_{j\,m}\rangle = \left(U^{(\mathbf{j})*}V^{(\mathbf{j})T}\right)_{(\boldsymbol{\alpha})_{j\,m};(\boldsymbol{\beta})_{j\,m}}, \\ &\qquad \boldsymbol{\alpha} \in \mathbb{R}(\mathbf{j}, j),\ \boldsymbol{\beta} \in \mathbb{S}(\mathbf{j}, j). \end{aligned} \tag{2.56}$$

It is the same value of j and m that appear in both sides of the first relation because the vectors in each basis set are eigenvectors of $\mathbf{J}^2$ and J_3. Moreover, the transformation coefficients $U_{(\mathbf{j}\,\boldsymbol{\alpha})_j;(\mathbf{j}\,\boldsymbol{\beta})_j}$ are independent of the value $m = j, j-1, \ldots, -j$ of the projection quantum number, as the notation indicates. This is true because the general relation (2.56) is generated from

$$|(\mathbf{j}\,\boldsymbol{\beta})_{j\,j}\rangle = \sum_{\boldsymbol{\alpha}\in\mathbb{R}_j(\mathbf{j})} U_{(\mathbf{j}\,\boldsymbol{\alpha})_j;(\mathbf{j}\,\boldsymbol{\beta})_j} |(\mathbf{j}\,\boldsymbol{\alpha})_{j\,j}\rangle \tag{2.57}$$

by the standard action of the lowering operator J_-, which does not affect the transformation coefficients.

The coefficients $U_{(\mathbf{j}\,\boldsymbol{\alpha})_j;(\mathbf{j}\,\boldsymbol{\beta})_j}$ *are called recoupling coefficients,* because they effect the transformation from one set of coupled state vectors to a second set. The unitary matrix $U^{(\mathbf{j})*}V^{(\mathbf{j})T} \in H(N(\mathbf{j}))$ of order $N(\mathbf{j}) = \prod_{i=1}^{n}(2j_i+1)$ is called a *recoupling matrix.* This recoupling matrix has the following expression of the form (2.49):

$$\begin{aligned} U^{(\mathbf{j})*}V^{(\mathbf{j})T} &= \sum_{j=j_{\min}}^{j_{\max}} \oplus \left(W_j^{(\mathbf{j})} \otimes I_{2j+1}\right),\; W_j^{(\mathbf{j})} \in U(N_j(\mathbf{j})), \\ & \left(W_j^{(\mathbf{j})}\right)_{\boldsymbol{\alpha};\boldsymbol{\beta}} = U_{(\mathbf{j}\,\boldsymbol{\alpha})_j;(\mathbf{j}\,\boldsymbol{\beta})_j}. \end{aligned} \tag{2.58}$$

The unitary matrix $W_j^{(\mathbf{j}} \in U(N_j(\mathbf{j}))$ is called a *reduced recoupling matrix;* its elements give the recoupling coefficients.

Recoupling matrices and reduced recoupling matrices for different pairs of coupling schemes are the principal objects of study in the theory of the addition of angular momenta. They bring an unsuspected unity to some of the classical relations in angular momentum theory, as developed subsequently in this monograph. Since the unitary matrix $W_j^{(\mathbf{j}}$ is of order $N_j(\mathbf{j})$, focus is again directed to the combinatorics of CG numbers.

We thus arrive at the Fourth Fundamental Result:

Fourth Fundamental Result: Let $\boldsymbol{\alpha}$ and $\boldsymbol{\beta}$ be quantum labels that give an $\boldsymbol{\alpha}$-coupling scheme and a $\boldsymbol{\beta}$-coupling scheme for the addition of n angular momenta. Then, the coupled state vectors for any two such coupling schemes are related by a unitary transformation, the coefficients in this transformation defining the recoupling coefficients for the two schemes. Moreover, there is a well-defined unitary recoupling matrix, which is fully defined by the pair of coupling schemes, such that this matrix is an element of the unitary group $H(N(\mathbf{j}))$ of matrices of order $N(\mathbf{j}) = \prod_{i=1}^{n}(2j_i+1)$, as defined by relation (2.58). This relation determines a reduced recoupling matrix $W_j^{\mathbf{j}}$, which is an element of the unitary group $U(N_j(\mathbf{j}))$, and the elements of this reduced recoupling matrix give the recoupling coefficients. Every pair of complete coupling schemes has this structure.

There is a general technique that provides $n-2$ Hermitian operators that completes the set (2.4) of commuting Hermitian operators by using the so-called intermediate angular momenta corresponding to different ways of associating pairs of angular momenta in the sum (2.1), so as to be able to apply at each step the coupling of two angular momenta, already described in Sect. 1.2.3. This method is known as the *binary theory* of the coupling of angular momentum. The procedure is simple, in principle: The reduction of the Kronecker product $D^{j_1}(U)\otimes D^{j_2}(U)\otimes\cdots\otimes D^{j_n}(U)$ is effected by repeated application of the reduction rule for pairs:

$$D^{a}_{\alpha\,\alpha'}(U)D^{b}_{\beta\,\beta'}(U)=\sum_{c} C^{a\,b\,c}_{\alpha\,\beta\,\alpha+\beta}C^{a\,b\,c}_{\alpha'\,\beta'\,\alpha'+\beta'}\,D^{c}_{\alpha+\beta\,\alpha'+\beta'}(U), \tag{2.59}$$

where we are now using (a,b,c) to represent triplets of the j_i-angular momentum quantum numbers, and Greek letters (α,β,γ) for their projections. After effecting the reduction for the selected pair $(a,b)\rightarrow c$, the same procedure is applied to the pair $(c,d)\rightarrow e$, etc. Despite the simplicity of the concept of binary couplings, the method is very intricate in its details. Surprisingly, perhaps, it is just these intricacies that lend themselves to the richness of combinatorial concepts. The first step in binary coupling theory, as noted, uses the four commuting Hermitian operators $\mathbf{J}^2(1),\mathbf{J}^2(2),\mathbf{J}^2,J_3$, which constitute a complete set of operators for determining the angular momentum states; the group $U(N_j(j_1,j_2))$ is the unitary group $U(1)$, the group of phase factors; and the CG number is $N_j(j_1,j_2)=1$. The phase factors may be chosen to be unity, and the elements of $U^{(j_1\,j_2)}=C^{(j_1\,j_2)}$ are the real WCG coefficients $\left(C^{(j_1\,j_2)}\right)_{(j\,m);(m_1\,m_2)}$. The remainder of this chapter and Chapters 3 and 4 develop binary coupling theory with a focus on combinatorial footings.

2.2 Binary Coupling Theory

Binary bracketings

The implementation of binary coupling theory leads to the problem of parentheses, namely, the problem of inserting pairs of parentheses into the sum $\mathbf{J}(1)+\mathbf{J}(2)+\cdots+\mathbf{J}(n)$, or, equivalently, into the product $D^{j_1}(U)\otimes D^{j_2}(U)\otimes\cdots\otimes D^{j_n}(U)$, in all possible ways such that objects are associated in pairs. For example, for $n=3$, this can be done in two ways as illustrated by the placement of ( ) in

$$\begin{gathered}(\,(\,\mathbf{J}(1)+\mathbf{J}(2))+\mathbf{J}(3)),\ \ (\,\mathbf{J}(1)+(\,\mathbf{J}(2)+\mathbf{J}(3))\,)\,,\\ (\,(\,D^{j_1}(U)\otimes D^{j_2}(U))\otimes D^{j_3}(U))\,,\\ (\,D^{j_1}(U)\otimes(\,D^{j_2}(U)\otimes D^{j_3}(U)))\,.\end{gathered} \tag{2.60}$$

Ignoring the $+$ and $\otimes$ operations, the placement of parentheses is represented symbolically as

$$((AB)C),\ (A(BC)). \tag{2.61}$$

Thus, the pair association is $(AB) = D$ and (DC) in the first case; and $(BC) = F$ and (AF) in the second case. A given insertion of parenthesis pairs into n such objects is called a *binary bracketing.*

In the case illustrated by (2.60) for three angular momenta, the two intermediate angular momenta are

$$\mathbf{J}(12) = \mathbf{J}(1) + \mathbf{J}(2),\ \mathbf{J}(23) = \mathbf{J}(2) + \mathbf{J}(3). \tag{2.62}$$

It may be verified that the squared angular momentum operator $\mathbf{J}^2(12)$ commutes with the five ($n = 3$) angular momentum operators (2.7), thus completing the set to the required six mutually commuting Hermitian operators. The squared angular momentum operator $\mathbf{J}^2(23)$ may similarly be adjoined to the set (2.7). The two operators $\mathbf{J}^2(12)$ and $\mathbf{J}^2(23)$ do not commute, so that we obtain by binary coupling exactly two complete sets of mutually commuting Hermitian operators, and, corresponding, two different coupling schemes. The case $n = 3$ already illustrates some fascinating, unexpected features that will be developed below, and which have far reaching implications for the general case.

The details of the similarity transformation that effects the explicit reduction of the Kronecker product depends strongly on the placement of parenthesis pairs in the Kronecker product, thus leading to many distinct coupling schemes. This structure might seem rather uninteresting, since the coupling matrices will always contain elements that are complicated multiple summations over products of WCG coefficients. It may come therefore as somewhat of a surprise that the combinatorial aspects of binary coupling theory leads deeply into the theory of binary trees, group invariants, the symmetric group, MacMahon's master theorem, intricate generating functions, cubic graphs, and more. The remainder of this chapter and the next two present these structures as they naturally unfold. We begin with the binary tree method of encoding the coupling schemes associated with the distinct placements of parenthesis pairs in a string of n objects $A_1 A_2 \cdots A_n$.

What is enclosed by a parenthesis pair is irrelevant to the problem of inserting parenthesis pairs into a string of n objects. It is purely a problem of placement, a combinatorial concept. We represent the objects by open circles on a line. Thus, for $n = 1, 2, 3, 4$, we have the following cases:

$$\begin{gathered} n = 1 : \circ;\ n = 2 : (\circ\ \circ)\,;\ n = 3 : ((\circ\ \circ)\,\circ)\,,\ (\circ\,(\circ\ \circ))\,; \\ n = 4 : (\circ\,((\circ\ \circ)\,\circ))\,,\ (\circ\,(\circ\,(\circ\ \circ)))\,, ((\circ\ \circ)\,(\circ\ \circ))\,, \\ (((\circ\ \circ)\,\circ)\,\circ)\,,\ ((\circ\,(\circ\ \circ))\,\circ)\,. \end{gathered} \tag{2.63}$$

We refer to any member of these arrangements of parenthesis pairs and n points as a *binary bracketing* B_n of order n. Thus, a binary bracketing of n $\circ$ points contains $n-1$ parenthesis pairs; there is a matched left (parenthesis and right) parenthesis that make up $n-1$ pairs. There is no parenthesis pair enclosing one point $\circ$. The term "binary" refers to the property that each parenthesis pair encloses exactly two "objects," where the object itself can be a binary bracketing of lower order. Each binary bracketing of order n has the property that under the mapping $(\circ\ \circ) \mapsto \circ$ it is transformed to a binary bracketing of order $n-1$. Performing this operation exactly $n-1$ times in a binary bracketing of order n reduces it to a single $\circ$ point. In general, there are many ways of doing this reduction, but there is always at least one way, since every binary bracketing of any order greater than 1 contains at least one pair $(\circ\ \circ)$.

A recursive procedure can be given for building-up all binary bracketings of order n. Let $\mathbb{B}_n, n = 1, 2, \ldots,$ denote the set of all binary bracketings of order n. Thus,

$$\mathbb{B}_1 = \{\circ\}, \mathbb{B}_2 = \{(\circ\ \circ)\}, \mathbb{B}_3 = \{((\circ\ \circ)\circ),\ (\circ(\circ\ \circ))\}. \tag{2.64}$$

Define the product of the set of binary bracketing of order j and order k by

$$\mathbb{B}_j * \mathbb{B}_k = \{(B_jB_k)\,|B_j \in \mathbb{B}_j, B_k \in \mathbb{B}_k\}. \tag{2.65}$$

Then, each binary bracketing in the set $\mathbb{B}_j * \mathbb{B}_k$ is of order $j+k$. This product in neither commutative nor associative. The bracketing $\mathbb{B}_n$ is built up recursively (see Comtet [44, p. 39]) by the rule

$$\mathbb{B}_n = (\mathbb{B}_1 * \mathbb{B}_{n-1}) \cup (\mathbb{B}_2 * \mathbb{B}_{n-2}) \cup \cdots \cup (\mathbb{B}_{n-1} * \mathbb{B}_1), \tag{2.66}$$

which contains binary bracketings of order less than n. For example,

$$\mathbb{B}_4 = (\mathbb{B}_1 * \mathbb{B}_3) \cup (\mathbb{B}_2 * \mathbb{B}_2) \cup (\mathbb{B}_3 * \mathbb{B}_1), \tag{2.67}$$

$$\begin{aligned} \mathbb{B}_1 * \mathbb{B}_3 &= \{(\circ((\circ\ \circ)\circ)),\ (\circ(\circ(\circ\ \circ)))\}, \\ \mathbb{B}_2 * \mathbb{B}_2 &= \{((\circ\ \circ)(\circ\ \circ))\}, \\ \mathbb{B}_3 * \mathbb{B}_1 &= \{(((\circ\ \circ)\circ)\circ),\ ((\circ(\circ\ \circ))\circ)\}. \end{aligned} \tag{2.68}$$

Relation (2.66) gives immediately the following recursion relation for the cardinality $a_n = |\mathbb{B}_n|$, $n = 1, 2 \ldots,$ of the set of binary bracketings of order n :

$$a_n = \sum_{k=1}^{n-1} a_k a_{n-k},\ a_1 = 1\,(\text{by convention}),\ n \geq 2, \tag{2.69}$$

with solution

$$a_n = \frac{1}{n}\binom{2n-2}{n-1} = \frac{(2n-2)!}{n!(n-1)!},\ n \geq 1. \tag{2.70}$$

The number a_n is called a *Catalan* number.

2.2.1 Binary trees

The set of binary bracketings of order n is bijective with the set of binary trees having n terminal points (see Stanton and White [165]). A binary tree is a graph of points and lines that is constructed recursively by a sequence of bifurcations and nonbifurcations of points. For our applications, it is convenient to keep account of points that bifurcate and those that do not. We start with a single point $\circ$ at *level* or *height* 0. The point $\circ$ either does not bifurcate or it bifurcates into a pair of points. If $\circ$ does not bifurcate, the graph contains the single point $\circ$; if the point $\circ$ bifurcates into a pair of points $(\circ\ \circ)$ at level 1, the bifurcating point $\circ$ is changed to a $\bullet$, which designates its change of status. These two cases are illustrated by the two pictures:

Each of the two points $\circ$ at level 1 in the second picture, then either does not bifurcate or bifurcates to give the three cases illustrated by the pictures:

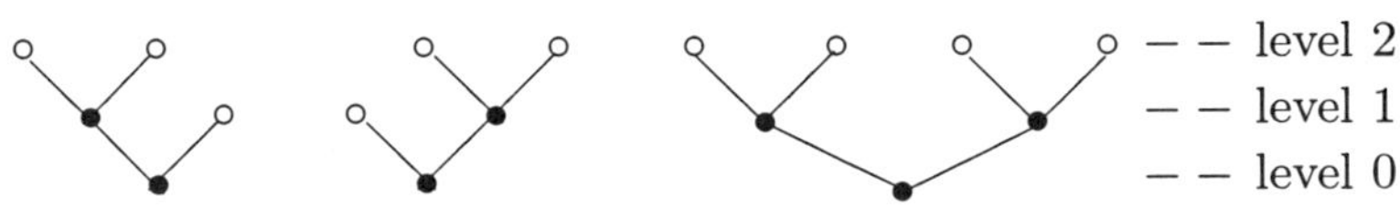

This procedure is continued to generate a binary tree upward to any level.

The points in a binary tree represented by $\bullet$ are called *internal points,* the point at level 0 being referred to as the root, since it is the source point of all the "branches." The points represented by $\circ$ are called *terminal* or *external points.* Binary trees are usually classified not by the number of levels, but by the number of external points $\circ$. We denote the set of trees having n external points by $\mathbb{T}_n$. As noted above, there is a bijective map between the set $\mathbb{B}_n$ of binary bracketings of order n and the set $\mathbb{T}_n$ of binary trees with n external $\circ$ points. It is also convenient to call the binary bracketing $B \in \mathbb{B}_n$ the *shape* of the tree $T \in \mathbb{T}_n$.

We list the binary trees of all shapes for $n = 2, 3, 4$:

$n = 2$:

$(\circ\ \circ) \quad \longrightarrow$  $\longrightarrow \quad \begin{pmatrix} \circ \\ \circ \\ \bullet \end{pmatrix}$

$n = 3$:

$((\circ\ \circ)\ \circ) \quad \longrightarrow$ 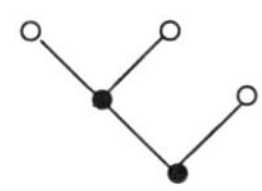 $\longrightarrow \quad \begin{pmatrix} \circ & \bullet \\ \circ & \circ \\ \bullet & \bullet \end{pmatrix}$

$(\circ\ (\circ\ \circ)) \quad \longrightarrow$ 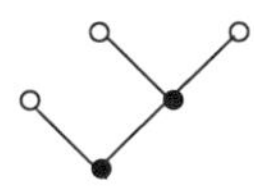 $\longrightarrow \quad \begin{pmatrix} \circ & \circ \\ \circ & \bullet \\ \bullet & \bullet \end{pmatrix}$

$n = 4$:

$(((\circ\ \circ)\ \circ)\ \circ) \quad \longrightarrow$ 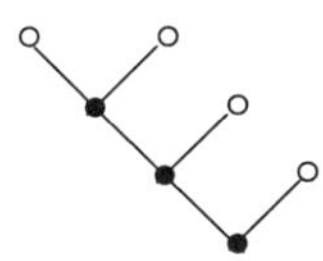 $\longrightarrow \quad \begin{pmatrix} \circ & \bullet & \bullet \\ \circ & \circ & \circ \\ \bullet & \bullet & \bullet \end{pmatrix}$

$(\circ\ (\circ\ (\circ\ \circ))) \quad \longrightarrow$ 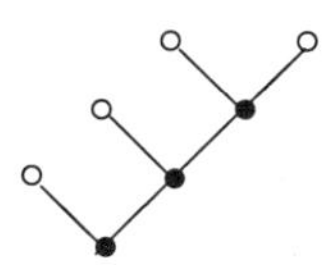 $\longrightarrow \quad \begin{pmatrix} \circ & \circ & \circ \\ \circ & \bullet & \bullet \\ \bullet & \bullet & \bullet \end{pmatrix}$

$((\circ\ \circ)\ (\circ\ \circ)) \quad \longrightarrow$ 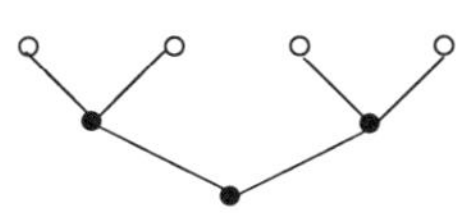 $\longrightarrow \quad \begin{pmatrix} \circ & \circ & \bullet \\ \circ & \circ & \bullet \\ \bullet & \bullet & \bullet \end{pmatrix}$

$((\circ\,(\circ\,\circ))\,\circ) \quad \rightarrow$ 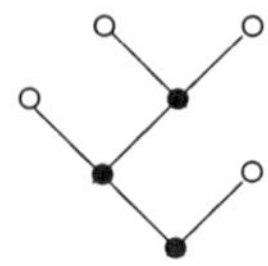 $\rightarrow \begin{pmatrix} \circ & \circ & \bullet \\ \circ & \bullet & \circ \\ \bullet & \bullet & \bullet \end{pmatrix}$

$(\circ\,((\circ\,\circ)\,\circ)) \quad \rightarrow$ 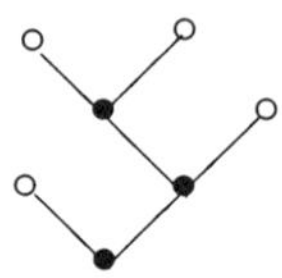 $\rightarrow \begin{pmatrix} \circ & \bullet & \circ \\ \circ & \circ & \bullet \\ \bullet & \bullet & \bullet \end{pmatrix}$

We have also shown in the pictures above a collection of matrices of three rows and $n-1$ columns containing elements $\bullet$ and $\circ$ that we next explain.

The basic structural element in a binary tree with internal points labeled by $\bullet$ and external points labeled by $\circ$ is a *fork,* which is the graph consisting of three points and two edges. There are four types of such forks as given by

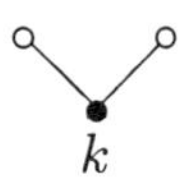 $\rightarrow \begin{pmatrix} \circ \\ \circ \\ \bullet \end{pmatrix}$ 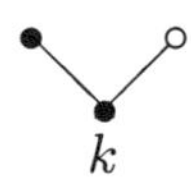 $\rightarrow \begin{pmatrix} \bullet \\ \circ \\ \bullet \end{pmatrix}$

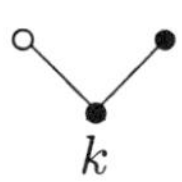 $\rightarrow \begin{pmatrix} \circ \\ \bullet \\ \bullet \end{pmatrix}$ 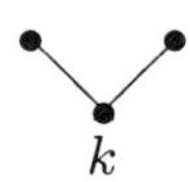 $\rightarrow \begin{pmatrix} \bullet \\ \bullet \\ \bullet \end{pmatrix}$

We use these correspondences of a fork to its column presentation to obtain the $3 \times (n-1)$ matrices in the above pictures. For a general binary tree, we start at the top level of the tree and read off the fork structure from left-to-right, then return to the next lower level and read the fork structure from left-to-right, ... The resulting $3 \times (n-1)$ matrix is called the *fork matrix* of the tree $T \in \mathbb{T}_n$, and is denoted $M_f(T)$.

We frequently invoke the notion of a fork as the elemental constituent in a binary tree. Our focus on assigning $\circ$ to external points and $\bullet$ to internal points is motivated by our application of binary trees to angular momentum coupling schemes, where the labeling of the two kinds of points by quantum numbers has a distinct physical significance, and different coupling schemes correspond to the assembly of forks into the various shapes constituting a binary tree.

The general bijection between the set of binary bracketings $\mathbb{B}_n$ of order n and the set of binary trees $\mathbb{T}_n$ with n external points can be shown as follows. We define the product of the set of trees $\mathbb{T}_j$ and $\mathbb{T}_k$ by

$$\mathbb{T}_j * \mathbb{T}_k = \{T_jT_k \mid T_j \in \mathbb{T}_j, T_k \in \mathbb{T}_k\}, \tag{2.71}$$

where the product T_jT_k of two binary trees is clear from the picture:

$T_j T_k \ =$ 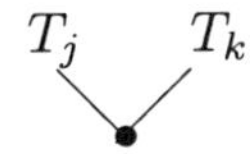

in which the root of T_j is attached to the upper end of the left edge, the root of the tree T_k is attached to the upper end of the right edge, and $T_1 = \circ$, but otherwise the root is a $\bullet$. It is apparent that

$$\mathbb{T}_n = (\mathbb{T}_1 * \mathbb{T}_{n-1}) \cup (\mathbb{T}_2 * \mathbb{T}_{n-2}) \cup \cdots \cup (\mathbb{T}_{n-1} * \mathbb{T}_1). \tag{2.72}$$

This gives exactly the recursion relation (2.69) for the cardinality $|\,\mathbb{T}_n\,|$, $n = 1, 2, \ldots,$ from which we conclude that $|\,\mathbb{T}_n\,| = a_n$, the Catalan number.

We next define an equivalence relation $\simeq$ in the set of binary trees $\mathbb{T}_n$. This will subsequently simplify greatly the number of binary trees that must be considered in our application to angular momentum theory. It is convenient to define the equivalence relation by using the fork presentation of the tree in terms of the corresponding $3 \times (n-1)$ fork matrix. The binary trees $T \in \mathbb{T}_n$ is said to be equivalent to the tree $T' \in \mathbb{T}_n$, written $T \simeq T'$, if the fork matrix $M_f(T')$ can be obtained from the fork matrix of $M_f(T)$ by any sequence of mappings of the columns of $M_f(T)$ to the columns of $M_f(T')$ of the form given by

$$\begin{pmatrix} \bullet \\ \circ \\ \bullet \end{pmatrix} \mapsto \begin{pmatrix} \circ \\ \bullet \\ \bullet \end{pmatrix}, \quad \begin{pmatrix} \circ \\ \bullet \\ \bullet \end{pmatrix} \mapsto \begin{pmatrix} \bullet \\ \circ \\ \bullet \end{pmatrix}. \tag{2.73}$$

If $T \simeq T'$, we also define the two fork matrices to be equivalent and write $M_f(T) \simeq M_f(T')$.

Examples.

$n = 4$: There are, from the pictures above, only two equivalence classes of binary trees under $\simeq$, which may be taken to be those with representatives given by

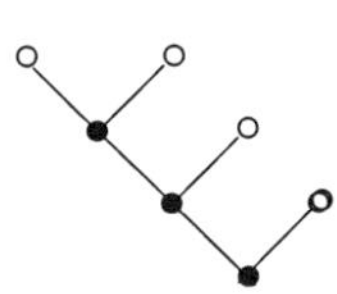 $\rightarrow \begin{pmatrix} \circ & \bullet & \bullet \\ \circ & \circ & \circ \\ \bullet & \bullet & \bullet \end{pmatrix}$

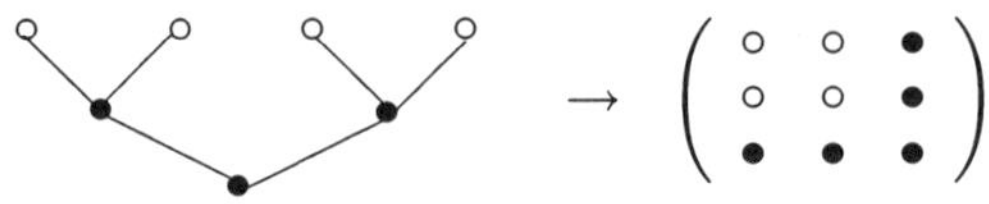

There are four trees (fork matrices) in the class with the first representative, and the second representative is in a class by itself.

$n = 5$:

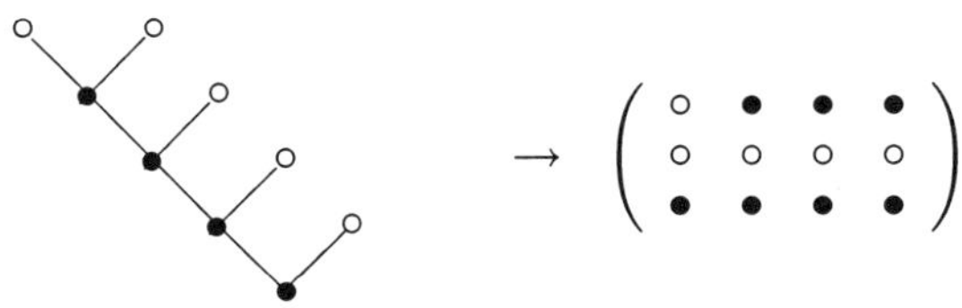

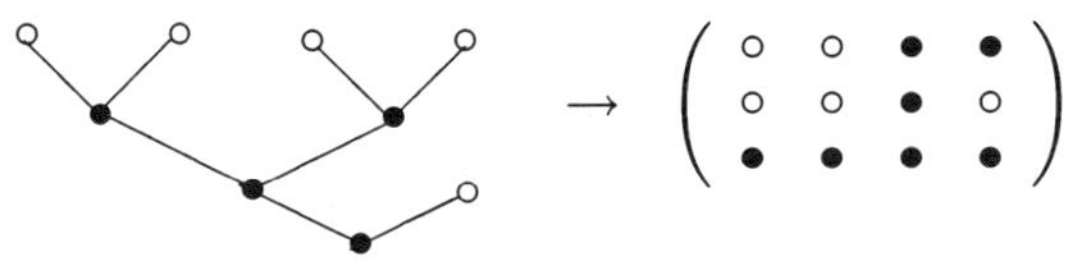

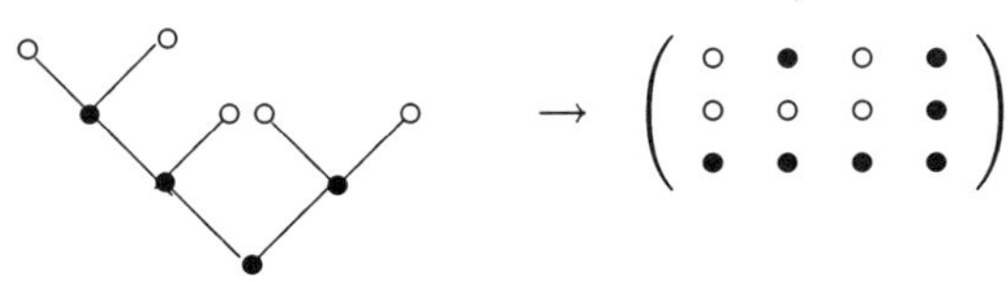

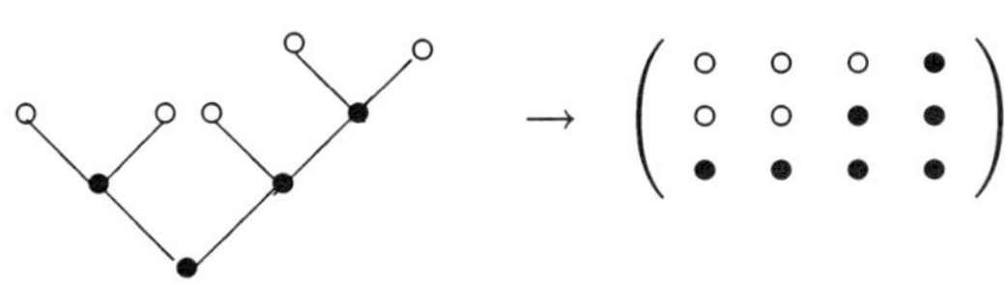

There are, respectively, $8, 2, 2, 2$ trees in $\mathbb{T}_5$ in the classes for which the above trees are the representatives, which gives altogether $a_5 = 14$ binary trees. $\square$

The number of equivalence classes in the quotient set $\mathbb{T}_n / \simeq$ can be given, recursively. We turn to this next, using the concept of a *binary tree polynomial.*

Binary tree polynomials

Let $\mathbb{T}_n(t)$ denote the subset of binary trees in $\mathbb{T}_n$ with n external $\circ$ points such that the fork matrix of each binary tree in $\mathbb{T}_n(t)$ contains exactly t columns of type col$(\circ\ \circ\ \bullet)$, and let $c_n(t) = |\,\mathbb{T}_n(t)\,|$ denote the number of such trees, so that

$$\sum_{t=1}^{[n/2]} c_n(t) = a_n = \frac{1}{n}\binom{2n-2}{n-1}, n \geq 2, \tag{2.74}$$

where $[n/2]$ equals $n/2$ for n even and $(n-1)/2$ for n odd. We define the *binary tree polynomial* $T_n(x)$ by

$$T_n(x) = \sum_{t=1}^{[n/2]} c_n(t)x^t,\ n = 1,2,3,\ldots;\ T_1(x) = 1, \tag{2.75}$$

where x is an indeterminate. The polynomial $T_n(x)$ is assigned to the set of trees $\mathbb{T}_n$. We may interpret the indeterminates $x, x^2, \ldots,$ as denoting the forks

$$x = \begin{pmatrix} \circ \\ \circ \\ \bullet \end{pmatrix},\ x^2 = \begin{pmatrix} \circ \\ \circ \\ \bullet \end{pmatrix}\begin{pmatrix} \circ \\ \circ \\ \bullet \end{pmatrix}, \cdots. \tag{2.76}$$

It follows from relation (2.72) that

$$T_n(x) = \sum_{k=1}^{n-1} T_k(x)T_{n-k}(x), n \geq 3. \tag{2.77}$$

The polynomials $T_n(x)$ are completely defined by this relation for $n \geq 3$ and the initial polynomials

$$T_1(x) = 1,\ T_2(x) = x. \tag{2.78}$$

Iteration of relation (2.77) gives the first few polynomials:

$$\begin{aligned} &T_3(x) = 2x,\ T_4(x) = 4x + x^2,\ T_5(x) = 8x + 6x^2, \\ &T_6(x) = 16x + 24x^2 + 2x^3,\ T_7(x) = 32x + 80x^2 + 20x^3, \\ &T_8(x) = 64x + 240x^2 + 120x^3 + 5x^4. \end{aligned} \tag{2.79}$$

The coefficient of x^t in these polynomials are the numbers $c_n(t)$ in the binary tree polynomial (2.75). Moreover, we have the following relation for the Catalan numbers:

$$a_n = \frac{1}{n}\binom{2n-2}{n-1} = T_n(1); \tag{2.80}$$

that is, *the Catalan number* a_n *is the value of the tree polynomial* $T_n(x)$ *at* $x = 1$.

Each $3 \times (n-1)$ fork matrix can be characterized as having t columns of type col($\circ\ \circ\ \bullet$); $n-2t$ total columns of type col($\bullet\ \circ\ \bullet$) and col($\circ\ \bullet\ \bullet$); and $t-1$ columns of type col($\bullet\ \bullet\ \bullet$), where $t = 1, 2, \ldots, [n/2]$. For counting purposes, we ignore the order of the columns in the fork matrix presentation of a binary tree and present it as the fork matrix given by

$$\underbrace{\begin{matrix} \circ & & \circ \\ \circ & \ldots & \circ \\ \bullet & & \bullet \end{matrix}}_{t} \quad \underbrace{\begin{matrix} \bullet & & \bullet & \circ & & \circ \\ \circ & \ldots & \circ & \bullet & \ldots & \bullet \\ \bullet & & \bullet & \bullet & & \bullet \end{matrix}}_{n-2t} \quad \underbrace{\begin{matrix} \bullet & & \bullet \\ \bullet & \ldots & \bullet \\ \bullet & & \bullet \end{matrix}}_{t-1}. \tag{2.81}$$

We have the following relations between equivalence classes of binary trees, under the equivalence relation $\simeq$, where we recall that $\mathbb{T}_n(t)$ denotes the number of binary trees containing t trees with forks of type col($\circ\ \circ\ \bullet$) :

$$\mathbb{T}_n = \bigcup_{t=1}^{[n/2]} \mathbb{T}_n(t), \ \mathbb{T}_n/\simeq\ = \bigcup_{t=1}^{[n/2]} \mathbb{T}_n(t)/\simeq,$$

$$\begin{matrix}\text{number of equivalence} \\ \text{classes contained in } \mathbb{T}_n\end{matrix} = b'_n = \sum_{t=1}^{[n/2]} \frac{c_n(t)}{2^{n-2t}}. \tag{2.82}$$

The quotient set $\mathbb{T}_n/\simeq$ is the union of $[n/2]$ disjoint quotient sets $\mathbb{T}_n(t)/\simeq$, one for each $t = 1, 2, \ldots, [n/2]$, where the quotient set $\mathbb{T}_n(t)/\simeq$ contains $c_n(t)/2^{n-2t}$ equivalence classes of binary trees induced on $\mathbb{T}_n(t)$ by the operation $\simeq$ defined by (2.73). The first few values of b'_n are given by

n	1	2	3	4	5	6	7	8
b'_n	1	1	1	2	4	8	21	51

The classification above of trees in accordance with the number of trees containing t trees of type col($\circ\ \circ\ \bullet$) is *not* the partitioning into types corresponding to the Wedderburn-Etherington number (see Comtet [44]), which admits the interchange of each pair of points at the endpoints of a fork. The mappings (2.73) are those that preserve the standard labeling of the internal points of a binary tree as defined below.

2.2.2 Standard labeling of binary trees

For the most part, we use only labeled binary trees, where the labels are quantum numbers or other objects that have a binary relationship to one another. Two examples illustrating this are

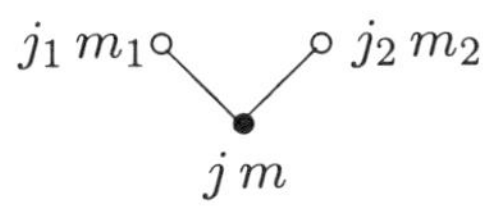

The first picture represents the addition of two angular momentum, $\mathbf{J} = \mathbf{J}(1) + \mathbf{J}(2)$, and the second the quantum numbers associated with the angular momenta. We often use the second type of labeled tree to specify numerical-valued objects, assigned to the labeled tree, for example, the WCG coefficient $C^{j_1\,j_2\,j}_{m_1\,m_2\,m}$ in the second picture. For our applications, we require *standard labeled trees;* that is, one set of labels attached to each tree, as assigned by a stated rule that is always to be followed. In this usage, the number of labeled trees is exactly the number a_n of trees. This is not the usual meaning of *labeled trees* used in combinatorics, as explained below.

In order to have some generality, let us assume that generic labels $\mathbf{u} = (u_1, u_2, \ldots, u_n)$ are assigned to the external points, and generic labels $\mathbf{v} = (v_1, v_2, \ldots, v_{n-2})$ are assigned to the internal points, excluding the root, which will be assigned the single generic label w.

The *standard rule* of assignment is the following: Select a binary tree in $T \in \mathbb{T}_n$, and label its points by the following two rules, as applied to internal and external points:

1. Internal labels: The internal points $\bullet$ are assigned the labels $\mathbf{v} = (v_1, v_2, \ldots, v_{n-2})$ by starting at the highest level and assigning the parts of $\mathbf{v}$, in order, to the internal points as read from left-to-right across the successive levels until level 1 is reached. The labels $\mathbf{v}$ are called *internal labels.* The root of the $i-$th fork is labeled by v_i, but the root at level 0 is assigned w. Internal labels may take on various meanings, but they are uniformly assigned to every tree $T \in \mathbb{T}_n$ by this rule.

2. External labels: The external points $\circ$ are assigned the labels $\mathbf{u} = (u_1, u_2, \ldots, u_n)$ in one-to-one correspondence with the binary bracketings B_n of the product $u_1 u_2 \cdots u_n$, as given by the bijection $\mathbb{B}_n \to \mathbb{T}_n$ (shown above for $n = 1, 2, 3, 4$.), and made clear by the examples below. The $\mathbf{u}$ are called *external labels.* External labels may take on various meanings, but they are uniformly assigned by this rule to every tree $T \in \mathbb{T}_n$. It will be necessary subsequently

to allow permutations of these standard labels, but it is always the standard assignment of labels to which the permutations are applied.

Examples:

$$(u_1\,((u_2\;u_3)\,u_4)) \qquad \longrightarrow$$

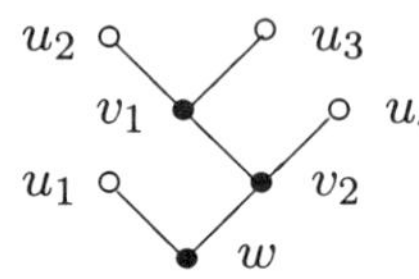

$$((u_1\;u_2)\,(u_3\;u_4)) \qquad \longrightarrow$$

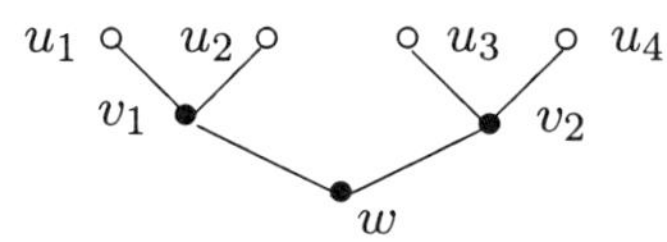

We often read off a labeled tree the triplets of letters associated with the forks, as read clockwise around the root of the fork. In these examples, we have the triplets

$$(u_2\,u_3\,v_1), (v_1\,u_4\,v_2), (u_1\,v_2\,w); (u_1\,u_2\,v_1), (u_3\,u_4\,v_2), (v_1\,v_2\,w). \qquad (2.83)$$

In accordance with the above rules, each $T \in \mathbb{T}_n$, is assigned a unique standard set of labels: $T \mapsto T(\mathbf{u};\mathbf{v})_w$, the number of such being the Catalan number a_n. Such standard assignments are used so that there will be no ambiguity in the association of numerical-valued objects with binary trees.

The term "labeled binary tree" above is *not* that of the technical definition of a connected labeled graph used in graph theory, which has the following definition: *The n points of the graph are labeled by permutations of 1,2,...,n, and two graphs are equivalent (isomorphic) if the labeling preserves adjacency of points. Labeled graphs are representatives of the equivalence classes. A graph is connected if there is a sequence of lines between each pair of points.* For example, the inequivalent labeled binary trees on three points may be taken to be

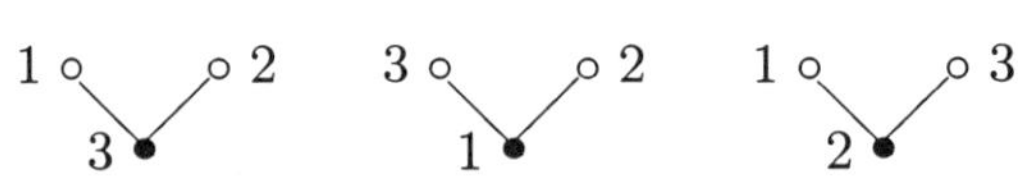

The assignment of numbers or other symbols to the points of a graph, as dictated in the standard rule above, does not qualify it as a labeled graph in the technical sense of this definition.

2.2.3 Generalized WCG coefficients defined in terms of binary trees

A basic object for angular momentum coupling theory is the labeled fork for $n = 2$, on which is defined a WCG coefficient:

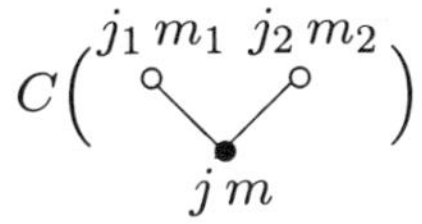

$$= \quad C^{j_1\ j_2\ j}_{m_1\, m_2\, m},\ m = m_1 + m_2, \tag{2.84}$$

where in terms of the $\mathbf{u}, \mathbf{v}, w$ notation above, we have $u_1 = (j_1\, m_1)$, $u_2 = (j_2\, m_2), (v_0) = (\emptyset)$, and $w = (j\, m)$.

The $i - th$ fork in a labeled tree corresponding to the coupling of n angular momenta is of the form:

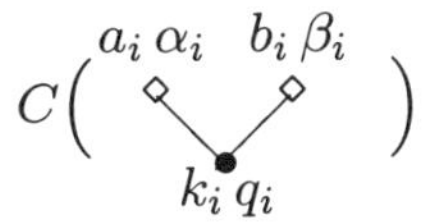

$$= \quad C^{a_i\ b_i\ k_i}_{\alpha_i\ \beta_i\ q_i},\ q_i = \alpha_i + \beta_i. \tag{2.85}$$

There are four types of forks and corresponding labels:

1. $(\diamond\ \diamond) = (\circ\ \circ) : (a_i\, \alpha_i)$ external; $(b_i\, \beta_i)$ external.
2. $(\diamond\ \diamond) = (\bullet\ \circ) : (a_i\, \alpha_i)$ internal; $(b_i\, \beta_i)$ external .
3. $(\diamond\ \diamond) = (\circ\ \bullet) : (a_i\, \alpha_i)$ external; $(b_i\, \beta_i)$ internal.
4. $(\diamond\ \diamond) = (\bullet\ \bullet) : (a_i\, \alpha_i)$ internal; $(b_i\, \beta_i)$ internal.

Thus, for each $T \in \mathbb{T}_n$, we have a labeled tree and associated coefficient:

1. Labeled tree:
$$T\,(\mathbf{j\, m}; \mathbf{k})_{j\, m}\,. \tag{2.86}$$

2. Generalized WCG coefficient:
$$C_{T(\mathbf{j\, m};\mathbf{k})_{j\, m}} = \prod_{i=1}^{n-1} C^{a_i\ b_i\ \ k_i}_{\alpha_i\ \beta_i\ \alpha_i+\beta_i}\,. \tag{2.87}$$

   The product is over all $n-1$ forks in the tree $T \in \mathbb{T}_n$, where the roots of the forks are labeled successively by $k_1\, q_1, k_2\, q_2, \ldots, k_{n-2}\, q_{n-2}$, $k_{n-1}\, q_{n-1}$, with $k_{n-1}\, q_{n-1} = j\, m$. The standard labeling $\mathbf{u} = (j_1\, m_1, j_2\, m_2, \ldots, j_n\, m_n) = \mathbf{j\, m}; \mathbf{v} = (k_1\, q_1, k_2\, q_2, \ldots, k_{n-2}\, q_{n-2}) = \mathbf{k\, q}$; $w = j\, m$ assigns unique quantum numbers to each pair of points $(\diamond, \diamond)$ labeled by $(a_i\, \alpha_i, b_i\, \beta_i)$ in (2.85). The coefficient (2.87) is called a *generalized* WCG coefficient; each term in the product of $n-1$ ordinary WCG coefficients is distinct; and each standard labeled tree $T\,(\mathbf{j}\,\mathbf{m};\mathbf{k})_{j\,m}$ has associated with it a unique generalized WCG coefficient $C_{T(\mathbf{j}\,\mathbf{m};\mathbf{k})_{j\,m}}$.

We call attention to the notations (2.86) for labeled trees and (2.87) for generalized WCG coefficients, which do not contain the labels $\mathbf{q} = (q_1, q_2, \ldots, q_{n-2})$ that appear in (2.85). This is because the $q-$labels are always completely determined in terms of the projection quantum numbers $\mathbf{m} = (m_1, m_2, \ldots, m_n)$ from the sum rule on projection quantum numbers, as illustrated by the examples (2.88) below. It is very important that the quantum labels appearing in the WCG coefficient associated with each labeled fork in (2.85), as well as in the full collection of products of coefficients in (2.87) belong to the domain of definition of the respective WCG coefficients. These quantum labels are interdependent, as illustrated for $n = 4$ by the following examples of standard labeled binary trees and their associated generalized WCG coefficients.

Examples. Generalized WCG coefficients of labeled binary trees:

$C\Big($ 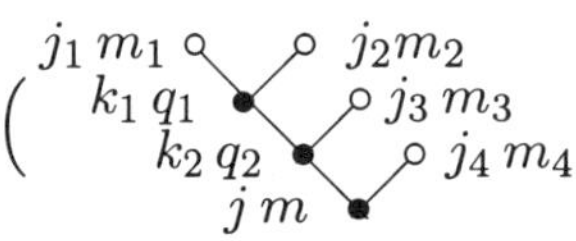 $\Big)$

$$= C^{j_1\;j_2\;k_1}_{m_1\,m_2\,q_1}\,C^{k_1\,j_3\;k_2}_{q_1\,m_3\,q_2}\,C^{k_2\,j_4\;j}_{q_2\,m_4\,m}$$

$C\Big($ 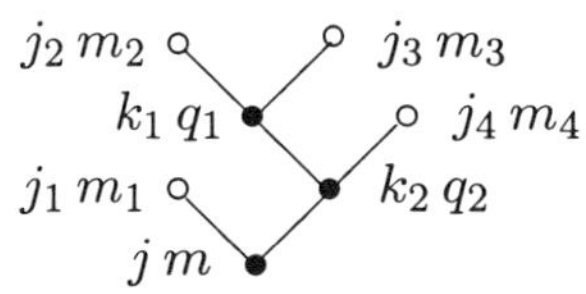 $\Big)$

$$= C^{j_2\;j_3\;k_1}_{m_2\,m_3\,q_1}\,C^{k_1\,j_4\;k_2}_{q_1\,m_4\,q_2}\,C^{j_1\;k_2\;j}_{m_1\,q_2\,m}$$

$C\Big($ 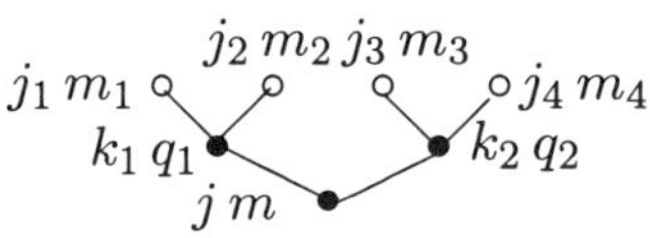 $\Big)$

$$= C^{j_1\;j_2\;k_1}_{m_1\,m_2\,q_1}\,C^{j_3\;j_4\;k_2}_{m_3\,m_4\,q_2}\,C^{k_1\,k_2\;j}_{q_1\,q_2\,m} \tag{2.88}$$

It is required that, for each given set of $\mathbf{j}; j\,m$, the triangle rule and the projection sum rule be enforced on the labels in each $C-$coefficients appearing in the product. In the middle coefficient, for example, it is required that $k_1 \in \langle j_2\, j_3\rangle, k_2 \in \langle k_1\, j_4\rangle, k_2 \in \langle j_1\, j\rangle; q_1 = m_2 + m_3, q_2 = m_2 + m_3 + m_4, m_1 + m_2 + m_3 + m_4 = m.$ $\square$

We next give the domain of definition of the quantum numbers appearing in the generalized WCG coefficient (2.87). These domains are uniquely determined by the tree $T \in \mathbb{T}_n$, the rules for the standard assignment of the quantum numbers giving the labeled tree $T\left(\mathbf{j}\,\mathbf{m};\mathbf{k}\right)_{j\,m}$, and the domains of each of the $n-1$ WCG-coefficients in the product (2.87), as given by (1.142), applied now to the quantum numbers in the factors $C^{a_i\ b_i\ k_i}_{\alpha_i\ \beta_i\ q_i}$. Thus, the domains of definition of the quantum numbers in the generalized WCG coefficient $C_{T(\mathbf{j}\,\mathbf{m};\mathbf{k})_{j\,m}}$ for specified $\mathbf{j}$ are the following:

1. Domain of definition of column indices—projection quantum numbers $\mathbf{m}$ (uncoupled states):
$$\mathbb{C}(\mathbf{j}) = \{\mathbf{m}\,|\, m_i = j_i, j_1 - 1, \ldots, -j_i; i = 1, 2, \ldots, n\} \qquad (2.89)$$
This is exactly the same as in (2.3), and is the same for all binary trees $T \in \mathbb{T}_n$.

2. Domain of definition of the internal quantum numbers $\mathbf{k}$:
$$\mathbb{R}_T(\mathbf{j}, j) = \{\mathbf{k} = (k_1, k_2, \ldots, k_{n-2})\,|\,\text{each}\ \langle a_i, b_i, k_i\rangle\ \text{is a triangle}\}, \qquad (2.90)$$
where $\langle a_i, b_i, k_i\rangle$ is the triangle associated with the $i-$th fork of the tree $T \in \mathbb{T}_n$, and the domains of definition of all k_i are to be expressed in terms of the $\mathbf{j}, j$ by use of the triangle rule.

3. Domain of definition of row indices—the internal angular momenta $\mathbf{k}$ and $j\,m$ (coupled states):
$$\mathbb{R}_T(\mathbf{j}) = \left\{\mathbf{k} \in \mathbb{R}_T(\mathbf{j}, j), j, m \,\middle|\, \begin{array}{l} \jmath = j_{\min}, j_{\min}+1, \ldots, j_{\max}; \\ m = j, j-1, \ldots, -j \end{array}\right\}. \qquad (2.91)$$

These sets are called column and row indexing sets because they are used to enumerate the orthogonality relations for the generalized WCG coefficients in relations (2.101)-(2.102) below.

The same symbols $\mathbf{k} = (k_1, k_2, \ldots, k_{n-1})$ are used in the standard assignment of labels to the internal points of each binary tree $T \in \mathbb{T}_n$, despite the fact that their placement depends on T, as illustrated in the examples (2.88). Thus, for each $T \in \mathbb{T}_n$, we have a $T-$coupling scheme for the addition of n angular momenta, each of which should be labeled by $\mathbf{k}(T)$, but which we abbreviate as $\mathbf{k} = \mathbf{k}(T)$ to avoid awkward notations. Altogether, we have defined a_n $T-$coupling schemes, one for each binary tree of order n. We emphasize again that the q_i projection numbers of the k_i are all expressible in terms of the m_i in the generalized WCG coefficients by enforcing the sum rule for projection quantum numbers.

2.2.4 Binary coupled state vectors

The labels $\mathbf{k}$ in the coupling of pairs of angular momenta associated with each fork in the labeled binary tree $T(\mathbf{j}\,\mathbf{m};\mathbf{k})_{j\,m}$ originate from what are called the *intermediate* angular momenta of a coupling scheme. The full set of mutually commuting angular momentum operators whose eigenvalues determine uniquely (up to phase factors) the normalized coupled state vectors corresponding to the labeled binary tree $T(\mathbf{j}\,\mathbf{m};\mathbf{k})_{j\,m}$ is the set of $2n$ operators given by

$$\begin{aligned}&\mathbf{J}^2(1), \mathbf{J}^2(2), \ldots, \mathbf{J}^2(n),\\ &\mathbf{K}_T^2(1), \mathbf{K}_T^2(2), \ldots, \mathbf{K}_T(n-2), \mathbf{J}^2, \mathbf{J}_3.\end{aligned} \tag{2.92}$$

The intermediate angular momentum $\mathbf{K}_T(i)$ has components $\left(K_T^{(1)}(i), K_T^{(2)}(i), K_T^{(3)}(i)\right)$, and $\mathbf{K}_T^2(i)$ is the sum of squares of these three components. In the standard labeling $T(\mathbf{u};\mathbf{v})_w$ of a tree $T \in \mathbb{T}_n$, the angular momentum assignment for the coupling scheme corresponding to the set of operators (2.92) is the following:

$$\begin{aligned}&\text{external points: } u_i = \mathbf{J}(i),\ i = 1, 2, \ldots, n,\\ &\text{internal points: } v_i = \mathbf{K}_T(i),\ i = 1, 2, \ldots, n-2,\\ &\text{root: } w = \mathbf{J}.\end{aligned} \tag{2.93}$$

Example: For $n = 4$, in the five trees given on pp. 101-102, the four external points $\circ$ are assigned the four angular momenta $\mathbf{J}(i), i = 1, 2, 3, 4$, in accordance with the binary bracketing B_4 and the rule for assigning (u_1, u_2, u_3, u_4), the root is assigned $\mathbf{J}$, and the two internal points are assigned the intermediate angular momenta $\mathbf{K}(i), i = 1, 2$, for each of the five cases, in order, as given by the following (We have dropped the T label, for brevity):

$$\begin{aligned}&\mathbf{K}(1) = \mathbf{J}(1) + \mathbf{J}(2),\ \mathbf{K}(2) = \mathbf{K}(1) + \mathbf{J}(3),\ \mathbf{J} = \mathbf{K}(2) + \mathbf{J}(4),\\ &\mathbf{K}(1) = \mathbf{J}(3) + \mathbf{J}(4),\ \mathbf{K}(2) = \mathbf{J}(2) + \mathbf{K}(1),\ \mathbf{J} = \mathbf{J}(1) + \mathbf{K}(2),\\ &\mathbf{K}(1) = \mathbf{J}(1) + \mathbf{J}(2),\ \mathbf{K}(2) = \mathbf{J}(3) + \mathbf{J}(4),\ \mathbf{J} = \mathbf{K}(1) + \mathbf{K}(2),\\ &\mathbf{K}(1) = \mathbf{J}(2) + \mathbf{J}(3),\ \mathbf{K}(2) = \mathbf{J}(1) + \mathbf{K}(1),\ \mathbf{J} = \mathbf{K}(2) + \mathbf{J}(4),\\ &\mathbf{K}(1) = \mathbf{J}(2) + \mathbf{J}(3),\ \mathbf{K}(2) = \mathbf{K}(1) + \mathbf{J}(4),\ \mathbf{J} = \mathbf{J}(1) + \mathbf{K}(2).\end{aligned} \tag{2.94}$$

In each of the five cases, the total angular momentum $\mathbf{J}$ at the right, which is assigned to the root, is the sum of the four angular momenta assigned to the external points: $\mathbf{J} = \mathbf{J}(1) + \mathbf{J}(2) + \mathbf{J}(3) + \mathbf{J}(4)$. $\square$

We next give the general coupled state vector corresponding to the labeled binary tree $T\,(\mathbf{j}\,\mathbf{m};\mathbf{k})_{j\,m}$, these coupled state vectors being simultaneous eigenvectors of the complete set of $2n$ commuting Hermitian operators (2.92) for specified $\mathbf{j}, j, m$:

$$|T(\mathbf{j}\,\mathbf{k})_{j\,m}\rangle = \sum_{\mathbf{m}\in\mathbb{C}(\mathbf{j})} C_{T(\mathbf{j}\,\mathbf{m};\mathbf{k})_{j\,m}}\,|\mathbf{j}\,\mathbf{m}\rangle, \text{ each } \mathbf{k}, j, m \in \mathbb{R}_T(\mathbf{j}), \tag{2.95}$$

with the inversion

$$|\mathbf{j}\,\mathbf{m}\rangle = \sum_{\mathbf{k},j,m\in\mathbb{R}_T(\mathbf{j})} C_{T(\mathbf{j}\,\mathbf{m};\mathbf{k})_{j\,m}}\,|T(\mathbf{j}\,\mathbf{k})_{j\,m}\rangle\,, \text{ each } \mathbf{m} \in \mathbb{C}(\mathbf{j}). \tag{2.96}$$

Each of these sets of vectors is an orthonormal basis of the vector space $\mathcal{H}_{\mathbf{j}} = \mathcal{H}_{j_1} \otimes \mathcal{H}_{j_2} \otimes \cdots \otimes \mathcal{H}_{j_n}$ of dimension $N(\mathbf{j}) = \prod_{i=1}^{n}(2j_i + 1)$. The row and column indexing sets (2.89)-(2.91) satisfy the same dimensionality relations as given earlier by (2.9) and (2.13):

$$\begin{aligned} |\mathbb{R}_T(\mathbf{j})| &= |\mathbb{C}(\mathbf{j})| = \prod_{i=1}^{n}(2j_i+1), \\ |\mathbb{R}_T(\mathbf{j})| &= \sum_{j=j_{\min}}^{j_{\max}} (2j+1)N_j(\mathbf{j}). \end{aligned} \tag{2.97}$$

These relations now hold for each labeled binary tree $T\,(\mathbf{j}\,\mathbf{m};\mathbf{k})_{j\,m}\,, T \in \mathbb{T}_n$. Each such labeled binary tree gives a $T-$coupling scheme, there being a_n (Catalan number) such coupling schemes in all.

The eigenvalues of the set of $2n$ Hermitian operators (2.92) on the coupled state vectors (2.95) are given by

$$\begin{aligned} &\mathbf{J}^2(i)|T(\mathbf{j}\,\mathbf{k})_{j\,m}\rangle = j_i(j_i+1)|T(\mathbf{j}\,\mathbf{k})_{j\,m}\rangle, i = 1,2,\ldots,n, \\ &\mathbf{K}_T^2(i)|T(\mathbf{j}\,\mathbf{k})_{j\,m}\rangle = k_i(k_i+1)|T(\mathbf{j}\,\mathbf{k})_{j\,m}\rangle, \\ &\qquad\qquad\qquad\qquad\qquad i = 1,2,\ldots,n-2, \\ &\mathbf{J}^2|T(\mathbf{j}\,\mathbf{k})_{j\,m}\rangle = j(j+1)|T(\mathbf{j}\,\mathbf{k})_{j\,m}\rangle, \end{aligned} \tag{2.98}$$

where, in addition, the action of the components of the total angular momentum $\mathbf{J}$ on these coupled states is standard:

$$\begin{aligned} J_+|T(\mathbf{j}\,\mathbf{k})_{j\,m}\rangle &= \sqrt{(j-m)(j+m+1)}\,|T(\mathbf{j}\,\mathbf{k})_{j\,m+1}\rangle, \\ J_-|T(\mathbf{j}\,\mathbf{k})_{j\,m}\rangle &= \sqrt{(j+m)(j-m+1)}\,|T(\mathbf{j}\,\mathbf{k})_{j\,m-1}\rangle, \\ J_3|T(\mathbf{j}\,\mathbf{k})_{j\,m}\rangle &= m|T(\mathbf{j}\,\mathbf{k})_{j\,m}\rangle. \end{aligned} \tag{2.99}$$

It follows from the construction of the coupled states (2.95) in terms of generalized WCG coefficients, and also from the Hermitian properties of the eigenvalue relations (2.98)-(2.99), that the following orthonormality of coupled state vectors (2.95) holds:

$$\langle T(\mathbf{j}\,\mathbf{k})_{j\,m} \,|\, T(\mathbf{j}\,\mathbf{k}')_{j'\,m'}\rangle = \delta_{\mathbf{k},\mathbf{k}'}\delta_{j,j'}\delta_{m,m'}. \tag{2.100}$$

It is important to observe in this relation that it is the same binary tree T that occurs in the bra-ket vectors of the coupled states: $\mathbf{j}\,\mathbf{k}$ is the standard labeling of T, and $\mathbf{j}\,\mathbf{k}'$ is the same standard labeling with primes on the parts of $\mathbf{k} = (k_1, k_2, \ldots, k_{n-2})$.

The row and column orthogonality of the generalized WCG coefficients are given, respectively, by

$$\sum_{\mathbf{m}\in\mathbb{C}(\mathbf{j})} C_{T(\mathbf{j}\,\mathbf{m};\mathbf{k})_{j\,m}} C_{T(\mathbf{j}\,\mathbf{m};\mathbf{k}')_{j'\,m'}} = \delta_{\mathbf{k},\mathbf{k}'}\delta_{j,j'}\delta_{m,m'}, \tag{2.101}$$

$$\sum_{\mathbf{k},j,m\in\mathbb{R}_T(\mathbf{j})} C_{T(\mathbf{j}\,\mathbf{m};\mathbf{k})_{j\,m}} C_{T(\mathbf{j}\,\mathbf{m}';\mathbf{k})_{j\,m}} = \delta_{\mathbf{m},\mathbf{m}'}. \tag{2.102}$$

The direct proof of, say, the orthogonality relation (2.101) matches together the same a_i, b_i pair in (2.87), uses the sum rule $m = m_1 + m_2 + \cdots + m_n$ for fixed m to express $\alpha_i + \beta_i = q_i$ in terms of other projection quantum numbers, and then uses the orthogonality relation

$$\sum_{\alpha_i,\beta_i} C^{a_i\,b_i\,k_i}_{\alpha_i\,\beta_i\,q_i} C^{a_i\,b_i\,k'_i}_{\alpha_i\,\beta_i\,q_i} = \delta_{k_i,k'_i}. \tag{2.103}$$

2.2.5 Binary reduction of Kronecker products

Corresponding to each labeled binary tree and the corresponding set of generalized WCG coefficients, we have the reduction of the $n-$fold Kronecker product representation of $SU(2)$ to standard form given by

$$C^{(\mathbf{j})}_T\, D^{\mathbf{j}}(U)\, C^{(\mathbf{j})tr}_T = \sum_{j=j_{\min}}^{j_{\max}} \oplus \mathbb{D}^j(U), \tag{2.104}$$

where we now use tr to denote transposition to avoid conflict with the tree symbol T. In this relation, $C^{(\mathbf{j})}_T$ is the real orthogonal matrix of order $N(\mathbf{j}) = \prod_{i=1}^n (2j_i + 1)$ with its matrix elements given in terms of the generalized WCG coefficients with rows enumerated by $\mathbf{k}, j, m \in \mathbb{R}_T(\mathbf{j})$

and columns by $\mathbf{m} \in \mathbb{C}(\mathbf{j})$, as defined by

$$\left(C_T^{(\mathbf{j})}\right)_{\mathbf{k},j,m;\mathbf{m}} = \begin{cases} C_{T(\mathbf{jm};\mathbf{k})_{j\,m}}, \text{for all } \mathbf{k}, j\,m \in \mathbb{R}_T(\mathbf{j}) \\ \text{and } \mathbf{m} \in \mathbb{C}(\mathbf{j}); \\ 0, \text{ otherwise} \end{cases} \tag{2.105}$$

In terms of matrix elements, relation (2.104) is given by

$$\sum_{\mathbf{m},\,\mathbf{m}'} C_{T(\mathbf{j}\,\mathbf{m};\mathbf{k})_{j\,m}} C_{T(\mathbf{j}\,\mathbf{m}';\mathbf{k}')_{j'\,m'}} D^{\mathbf{j}}_{\mathbf{m}\,\mathbf{m}'}(U) = \delta_{j,j'}\delta_{\mathbf{k},\mathbf{k}'} D^{j}_{m\,m'}(U). \tag{2.106}$$

The matrix element form of (2.104) obtained by moving $C_T^{(\mathbf{j})}$ and $C_T^{(\mathbf{j})\mathrm{tr}}$ to the right-hand side is

$$D^{\mathbf{j}}_{\mathbf{m}\,\mathbf{m}'}(U) = \sum_{\mathbf{k},\,j} C_{T(\mathbf{j}\,\mathbf{m};\mathbf{k})_{j\,m}} C_{T(\mathbf{j}\,\mathbf{m}';\mathbf{k})_{j\,m'}} D^{j}_{m\,m'}(U), \tag{2.107}$$

in which $m = m_1 + m_2 + \cdots + m_n$, $m' = m'_1 + m'_2 + \cdots + m'_n$.

2.2.6 Binary recoupling matrices

In terms of the notation (2.95) for the ket-vector $|T(\mathbf{j}\,\mathbf{k})_{j\,m}\rangle$ corresponding to the coupling scheme defined by the standard labeled binary tree $T(\mathbf{j}\,\mathbf{m};\mathbf{k})_{j\,m}$, we have the relation

$$\begin{aligned} |T(\mathbf{j}\,\mathbf{k})_{j\,m}\rangle &= \sum_{\mathbf{m}\in\mathbb{C}(\mathbf{j})} C_{T(\mathbf{j}\,\mathbf{m};\mathbf{k})_{j\,m}} |\mathbf{j}\,\mathbf{m}\rangle \\ &= \sum_{\mathbf{m}\in\mathbb{C}(\mathbf{j})} \left(C_T^{(\mathbf{j})}\right)_{\mathbf{k},j,m;\,\mathbf{m}} |\mathbf{j}\,\mathbf{m}\rangle. \end{aligned} \tag{2.108}$$

The ket-vector $|T'(\mathbf{j}\,\mathbf{k}')_{j\,m}\rangle$ corresponding to a second coupling scheme defined by the standard labeled binary tree $T'(\mathbf{jm};\mathbf{k}')_{j\,m}$ (primes are attached to the standard labels of the roots of the forks in the binary tree T') is given by

$$\begin{aligned} |T'(\mathbf{j}\,\mathbf{k}')_{j\,m}\rangle &= \sum_{\mathbf{m}\in\mathbb{C}(\mathbf{j})} C_{T'(\mathbf{j}\,\mathbf{m};\mathbf{k}')_{j\,m}} |\mathbf{j}\,\mathbf{m}\rangle \\ &= \sum_{\mathbf{m}\in\mathbb{C}(\mathbf{j})} \left(C_{T'}^{(\mathbf{j})}\right)_{\mathbf{k}',j,m;\mathbf{m}} |\mathbf{j}\,\mathbf{m}\rangle. \end{aligned} \tag{2.109}$$

The two sets of orthonormal basis vectors (2.108) and (2.109) span the same vector space $\mathcal{H}_{\mathbf{j}}$; hence, they are related by a unitary transformation of the form

$$|T'(\mathbf{j}\,\mathbf{k}')_{j\,m}\rangle = \sum_{\mathbf{k}\in\mathbb{R}_T(\mathbf{j},j)} R_{T(\mathbf{j}\,\mathbf{k})_j;\,T'(\mathbf{j}\,\mathbf{k}')_j} |T(\mathbf{j}\,\mathbf{k})_{j\,m}\rangle. \tag{2.110}$$

From the inner product of the two coupled ket-vectors in this relation and also between (2.108)-(2.109), we obtain the following relations:

$$\begin{aligned} \langle T(\mathbf{j}\,\mathbf{k})_{j\,m}|T'(\mathbf{j}\,\mathbf{k}')_{j\,m}\rangle &= R_{T(\mathbf{j}\,\mathbf{k})_j;\,T'(\mathbf{j}\,\mathbf{k}')_j} \qquad (2.111)\\ = \left(C_T^{(\mathbf{j})}C_{T'}^{(\mathbf{j})\mathrm{tr}}\right)_{\mathbf{k},j,m;\,\mathbf{k}',j,m} &= \sum_{\mathbf{m}\in\mathbb{C}(\mathbf{j})} C_{T(\mathbf{j}\,\mathbf{m};\mathbf{k})_{j\,m}} C_{T'(\mathbf{j}\,\mathbf{m};\,\mathbf{k}')_{j\,m}}. \end{aligned}$$

Relations (2.108)-(2.110) are the binary coupling analogs of the general relations (2.55)-(2.56), where $\boldsymbol{\alpha} = \mathbf{k}, U^{(\mathbf{j})} = C_T^{(\mathbf{j})}, \boldsymbol{\beta} = \mathbf{k}', V^{(\mathbf{j})} = C_{T'}^{(\mathbf{j})}$.

The real orthogonal matrix

$$R_{T,T'}^{(\mathbf{j})} = C_T^{(\mathbf{j})}C_{T'}^{(\mathbf{j})\mathrm{tr}} \in H(N(\mathbf{j})) \tag{2.112}$$

of order $N(\mathbf{j}) = \prod_{i=1}^{n}(2j_i+1)$ is called a *binary recoupling matrix* for the pair of standard labeled trees $T(\mathbf{j}\,\mathbf{k})_{j\,m}$ and $T'(\mathbf{j}\,\mathbf{k}')_{j\,m}$. This binary recoupling matrix has the following expression of the form (2.58):

$$R_{T,T'}^{(\mathbf{j})} = \sum_{j=j_{\min}}^{j_{\max}} \oplus \left(R_{T,T'}^{(\mathbf{j},j)} \otimes I_{2j+1}\right), \tag{2.113}$$

where the real orthogonal matrix $R_{T,T'}^{(\mathbf{j},j)} \in U(N_j(\mathbf{j}))$ is called a *reduced binary recoupling matrix.* The elements of the matrix (2.113) give the recoupling coefficients for the pair of standard labeled trees $T(\mathbf{j}\,\mathbf{k})_{j\,m}$ and $T'(\mathbf{j}\,\mathbf{k}')_{j\,m}$:

$$\left(R_{T,T'}^{(\mathbf{j})}\right)_{\mathbf{k},j;\mathbf{k}',j} = R_{T(\mathbf{j}\,\mathbf{k})_j;\,T'(\mathbf{j}\,\mathbf{k}')_j} = \left(C_T^{(\mathbf{j})}C_{T'}^{(\mathbf{j})\mathrm{tr}}\right)_{\mathbf{k},j,m;\,\mathbf{k}',j,m}. \tag{2.114}$$

As shown earlier in (2.56)-(2.57), as well as in the derivation of (2.111), these matrix elements are independent of the projection quantum number m. The binary recoupling matrix $R_{T,T'}^{(\mathbf{j})}$ is, of course, an element of the group $H(N(\mathbf{j}))$ of unitary matrices (see (2.49)) that commutes with the standard Kronecker direct sum.

Action of the symmetric group on the standard labels of binary trees

To include all ways of coupling pairs of angular momenta (all binary coupling schemes), we must allow permutations of the labels $(\mathbf{j}\,\mathbf{m})$ that are assigned to the external points of every tree $T \in \mathbb{T}_n$ in accordance with the bracket symbol $B_n \in \mathbb{B}_n$. We have chosen not to introduce permutations earlier so as to leave the notations less encumbered. But we must now consider permutations of the external labels. Thus, for $\pi : 1 \mapsto \pi_1, 2 \mapsto \pi_2, \ldots, n \mapsto \pi_n$, we permute the standard angular momentum $\mathbf{j}\,\mathbf{m}$ labels of the external points to $\pi : (j_1, j_2, \ldots, j_n) \mapsto (j_{\pi_1}, j_{\pi_2}, \ldots, j_{\pi_n})$ and $(m_1, m_2, \ldots, m_n) \mapsto (m_{\pi_1}, m_{\pi_2}, \ldots, m_{\pi_n})$, which we write as $\pi(\mathbf{j}\,\mathbf{m})$. Thus, the standard labeled tree $T(\mathbf{j}\,\mathbf{m}; \mathbf{k})_{j\,m}$ is now labeled by $T\,(\pi(\mathbf{j}\,\mathbf{m})\,; \mathbf{k})_{j\,m}$, where we keep the standard labels $\mathbf{k}$ of the internal points and $j\,m$ of the root. The angular momentum labels $(\mathbf{j}\,\mathbf{m})$ initially assigned to the external points in the standard labeling is the reference set for all permutations $\pi \in S_n$.

The generalized WCG coefficient corresponding to the permuted external labels $\pi(\mathbf{j}\,\mathbf{m})$ is

$$C_{T(\pi(\mathbf{j}\,\mathbf{m});\mathbf{k})_{j\,m}} = \text{product of } n-1 \text{ WCG coefficients obtained by replacing } (\mathbf{j}\,\mathbf{m}) \text{ in definition (2.87) by } \pi(\mathbf{j}\,\mathbf{m}). \quad (2.115)$$

Relations (2.108)-(2.109), with $\mathbf{j}$ replaced by $\pi(\mathbf{j})$ in the T' state vectors, transcribe to

$$\begin{aligned} |T(\mathbf{j}\,\mathbf{k})_{j\,m}\rangle &= \sum_{\mathbf{m}\in\mathbb{C}(\mathbf{j})} C_{T(\mathbf{j}\,\mathbf{m};\mathbf{k})_{j\,m}} |\mathbf{j}\,\mathbf{m}\rangle \\ &= \sum_{\mathbf{m}\in\mathbb{C}(\mathbf{j})} \left(C_T^{(\mathbf{j})}\right)_{\mathbf{k},j,m;\,\mathbf{m}} |\mathbf{j}\,\mathbf{m}\rangle, \end{aligned} \quad (2.116)$$

$$\begin{aligned} |T'(\pi(\mathbf{j})\,\mathbf{k}')_{j\,m}\rangle &= \sum_{\mathbf{m}\in\mathbb{C}(\mathbf{j})} C_{T'(\pi(\mathbf{j})\,\mathbf{m};\mathbf{k}')_{j\,m}} |\mathbf{j}\,\mathbf{m}\rangle \\ &= \sum_{\mathbf{m}\in\mathbb{C}(\mathbf{j})} \left(C_{T'}^{\pi(\mathbf{j})}\right)_{\mathbf{k}',j,m;\mathbf{m}} |\mathbf{j}\,\mathbf{m}\rangle. \end{aligned} \quad (2.117)$$

Similarly, relations (2.110) and (2.111) transcribe to

$$|T'(\pi(\mathbf{j})\,\mathbf{k}')_{j\,m}\rangle = \sum_{\mathbf{k}\in\mathbb{R}_T(\mathbf{j},j)} R_{T(\mathbf{j}\,\mathbf{k})_j;\,T'(\pi(\mathbf{j})\,\mathbf{k}')_j} |T(\mathbf{j}\,\mathbf{k})_{j\,m}\rangle, \quad (2.118)$$

with inner product given by all the following relations:

$$\begin{aligned}\langle T(\mathbf{j}\,\mathbf{k})_{j\,m}|T'(\pi(\mathbf{j})\,\mathbf{k}')_{j\,m}\rangle &= \sum_{\mathbf{m}\in\mathbb{C}(\mathbf{j})} C_{T(\mathbf{j}\,\mathbf{m};\mathbf{k})_{j\,m}}\, C_{T'(\pi(\mathbf{j})\,\mathbf{m};\mathbf{k}')_{j\,m}} \\ = R_{T(\mathbf{j}\,\mathbf{k})_j;\,T'(\pi(\mathbf{j})\,\mathbf{k}')_j} &= \left(C_T^{(\mathbf{j})} C_{T'}^{\pi(\mathbf{j})\mathrm{tr}}\right)_{\mathbf{k},j,m;\,\mathbf{k}',j,m}. \qquad (2.119)\end{aligned}$$

(The relation $\sum_{\pi(\mathbf{m})} = \sum_{\mathbf{m}}$ has been used in the above equations.)

The real orthogonal matrix defined by

$$R_{T,T'}^{(\mathbf{j});\,\pi(\mathbf{j})} = C_T^{(\mathbf{j})} C_{T'}^{\pi(\mathbf{j})\mathrm{tr}} \in H(N(\mathbf{j})) \qquad (2.120)$$

is the binary recoupling matrix for the pair of labeled trees $T(\mathbf{j}\,\mathbf{k})_{j\,m}$ and $T'(\pi(\mathbf{j})\,\mathbf{k}')_{j\,m}$. The matrix elements of this binary recoupling matrix are

$$\left(R_{T,T'}^{(\mathbf{j});\,\pi(\mathbf{j})}\right)_{\mathbf{k};\mathbf{k}'} = \left(C_T^{(\mathbf{j})} C_{T'}^{\pi(\mathbf{j})\mathrm{tr}}\right)_{\mathbf{k},j,m;\,\mathbf{k}',j,m}, \qquad (2.121)$$

which are independent of the projection quantum number m. Since the matrix $R_{T,T'}^{(\mathbf{j});\pi(\mathbf{j})}$ belongs to the unitary subgroup $H(N(\mathbf{j}))$, it can also be written in the form described in relations (2.49)-(2.50):

$$R_{T,T'}^{(\mathbf{j});\pi(\mathbf{j})} = \sum_{j=j_{\min}}^{j_{\max}} \oplus \left(R_{T,T'}^{(\mathbf{j},j);(\pi(\mathbf{j}),j)} \otimes I_{2j+1}\right), \qquad (2.122)$$

where the real orthogonal matrix $R_{T,T'}^{(\mathbf{j},j);(\pi(\mathbf{j}),j)} \in U(N_j(\mathbf{j}))$ is the reduced recoupling matrix for the pair of labeled trees $T(\mathbf{j}\,\mathbf{k})_{j\,m}$ and $T'(\pi(\mathbf{j})\,\mathbf{k}')_{j\,m}$.

For $\mathbf{j} \mapsto \pi(\mathbf{j})$, the first set of eigenvalue relations in (2.98) becomes

$$\mathbf{J}^2(\pi_i)|T(\pi(\mathbf{j})\,\mathbf{k})_{jm}\rangle = j_{\pi_i}(j_{\pi_i}+1)|T(\pi(\mathbf{j})\,\mathbf{k})_{jm}\rangle, \qquad (2.123)$$

which is identical to the original set, since it is just a rearrangement of those relations. Thus, the coupled state vectors $|T'(\pi(\mathbf{j})\,\mathbf{k}')_{j\,m}\rangle$ satisfy relations (2.98)-(2.99) with $\mathbf{k}$ replaced by $\mathbf{k}'$ and T by T'.

All permutations $\pi(\mathbf{j})$, $\pi'(\mathbf{j})$ can also be considered to obtain the labeled binary trees $T(\pi(\mathbf{j})\,\mathbf{k})_{j\,m}$ and $T'(\pi'(\mathbf{j})\,\mathbf{k}')_{j\,m}$. The modifications of relations (2.116)-(2.122) are evident: replace first π by π', then $(\mathbf{j}, j)$ by $(\pi(\mathbf{j}), j)$. We now obtain the real orthogonal recoupling matrix given by

$$R_{T,T'}^{\pi(\mathbf{j});\,\pi'(\mathbf{j})} = C_T^{\pi(\mathbf{j})} C_{T'}^{\pi'(\mathbf{j})\mathrm{tr}} \in H(N(\mathbf{j})). \qquad (2.124)$$

The real orthogonal matrix $R_{T,T'}^{(\pi(\mathbf{j}),j);(\pi'(\mathbf{j}),j)} \in U(N_j(\mathbf{j}))$ now appears in the form (2.122) and is the reduced binary recoupling matrix. *We now refer to all binary trees that are labeled by a permutation of the standard labels* $\mathbf{j}$ *as standard.The key property for being standard is the rule for assigning the internal* $\mathbf{k}$ *labels to the roots of the forks, including* j *to the root of full the binary tree.* (See the examples (2.127) below in which all permutations of j_1, j_2, j_3, j_4 are allowed.)

The product of two binary recoupling matrices is, in general, not again a binary recoupling matrix. Instead, we have the following trivial multiplication property of these real orthogonal matrices

$$R_{T,T'}^{\pi(\mathbf{j});\,\pi'(\mathbf{j})} R_{T',T''}^{\pi'(\mathbf{j});\,\pi''(\mathbf{j})} = R_{T,T''}^{\pi(\mathbf{j});\,\pi''(\mathbf{j})}, \tag{2.125}$$

for all $\pi, \pi', \pi'' \in S_n$ and all $T, T', T'' \in \mathbb{T}_n$. *This simple transitive multiplication property is the source for many well-known relations in angular momentum theory,* as will be shown subsequently.

We restate the Fourth Fundamental Result for binary coupling schemes:

Fifth Fundamental Result: Binary recoupling matrices are the fundamental quantities that determine the binary coupled angular momentum basis vectors for the addition of n angular momenta associated with the placement of parenthesis pairs into the sum (2.1) and all permutations thereof: All binary $T-$coupling schemes can be obtained from a selected one by using recoupling matrices.

2.2.7 Triangle patterns and triangle coefficients

Triangle patterns

The *standard triangle pattern* $\Delta_T(\mathbf{j}\,\mathbf{k})_j$ corresponding to the standard labeled tree $T(\mathbf{j}\,\mathbf{k})_j$ is a $3 \times (n-1)$ matrix array with columns entries that are the labeled forks in a standard labeled binary tree with n external points (no projection quantum numbers appear in the labeled trees $T(\mathbf{j}\,\mathbf{k})_j$). Thus, we have the matrix array defined by

$$\Delta_{T(\mathbf{j}\,\mathbf{k})_j} = \begin{pmatrix} a_1 & a_2 & \cdots & a_{n-1} \\ b_1 & b_2 & \cdots & b_{n-1} \\ k_1 & k_2 & \cdots & k_{n-1} \end{pmatrix}, \tag{2.126}$$

in which column i defined by $t_i = \mathrm{col}(a_i\, b_i\, k_i)$ is the triangle of labels of the $i-$th fork in the standard labeled tree $T(\mathbf{j}\,\mathbf{k})_j$. We refer to each column of this matrix array as a triangle, since its entries obey the triangle rule for addition of angular momenta. We write $k_{n-1} = j$, and always distinguish the root label j of the labeled tree T in the notation.

We call such an array of triangles a *triangle pattern of order* $n-1$. The triangles in the array are exactly those that appear in the product of WCG coefficients in (2.87). *Moreover, the columns in the triangle pattern are ordered from left-to-right in exactly the order corresponding to the root labeling* $k_1, k_2, \ldots, k_{n-1}$. This notation encodes exactly the same triangle information as exhibited by the labeled tree presentation, but is typographically more tractable.

The enumeration of standard labeled triangle patterns is essential to our subsequent developments, and at the risk of being repetitious, we give this standard labeling in complete detail for $n = 4$:

$$(((j_1\, j_2))\, j_3\,)\, j_4 \quad \rightarrow$$

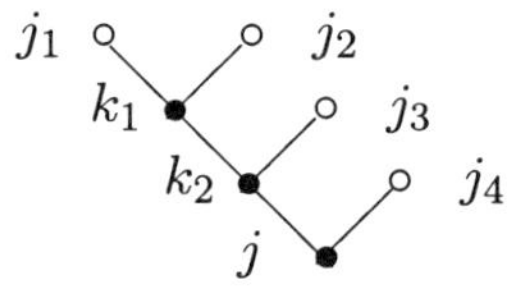

$$\rightarrow \begin{pmatrix} j_1 & k_1 & k_2 \\ j_2 & j_3 & j_4 \\ k_1 & k_2 & j \end{pmatrix}$$

$$(j_1\,(j_2\,(j_3\, j_4))) \quad \rightarrow$$

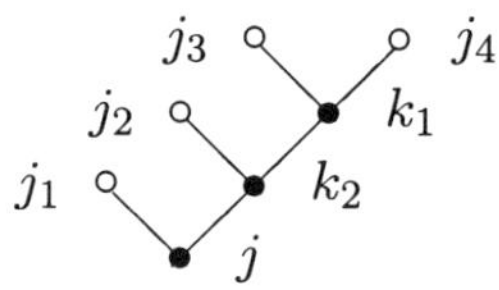

$$\rightarrow \begin{pmatrix} j_3 & j_2 & j_1 \\ j_4 & k_1 & k_2 \\ k_1 & k_2 & j \end{pmatrix}$$

$$((j_1\; j_2)\,(j_3\; j_4)) \quad \rightarrow$$

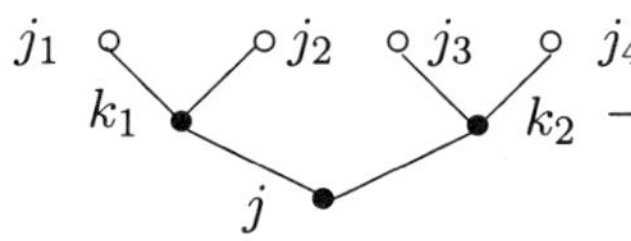

$$\rightarrow \begin{pmatrix} j_1 & j_3 & k_1 \\ j_2 & j_4 & k_2 \\ k_1 & k_2 & j \end{pmatrix}$$

$$((j_1\,(j_2\; j_3))\, j_4) \quad \rightarrow$$

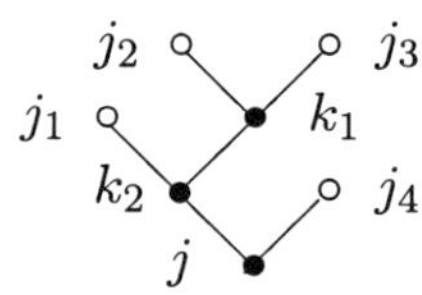

$$\rightarrow \begin{pmatrix} j_2 & j_1 & k_2 \\ j_3 & k_1 & j_4 \\ k_1 & k_2 & j \end{pmatrix}$$

$$(j_1\,((j_2\; j_3)\, j_4)) \quad \rightarrow$$

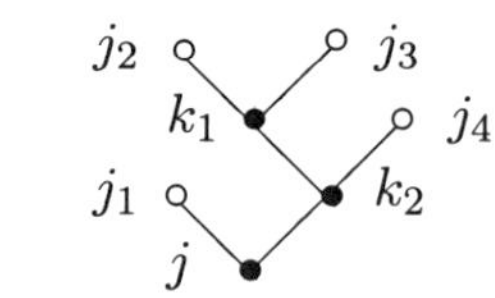

$$\rightarrow \begin{pmatrix} j_2 & k_1 & j_1 \\ j_3 & j_4 & k_2 \\ k_1 & k_2 & j \end{pmatrix} \tag{2.127}$$

As defined above, we also consider as standard the binary trees and the triangle patterns that correspond to the permutations $\pi(j_1\, j_2\, j_3\, j_4), \pi \in S_4$, of the labels of the external points. The key property for the standard labels of a binary tree and the corresponding triangle pattern is that the internal labels k_1, k_2, j are fixed in their assignment.

The *standard triangle pattern* of order $n-1$ corresponding to the standard labeled tree $T(\pi(\mathbf{j})\,\mathbf{k})_j$ is given by

$$\Delta_{T(\pi(\mathbf{j})\,\mathbf{k})_j} = \begin{pmatrix} a_1^\pi & a_2^\pi & \cdots & a_{n-1}^\pi \\ b_1^\pi & b_2^\pi & \cdots & b_{n-1}^\pi \\ k_1 & k_2 & \cdots & k_{n-1} \end{pmatrix}. \tag{2.128}$$

Equivalence classes of triangle patterns

The operation $\simeq$ defined by (2.73) can be applied to the columns of the triangle patterns defined by (2.127)-(1.128), where, by definition, the labels attached to the $\circ$ and $\bullet$ points are to be **carried along** with the interchange of external and internal points. We extend this operation to include labeled forks of type col$(\circ\ \circ\ \bullet) \mapsto$ col$(\circ\ \circ\ \bullet)$, which are now distinguished by their angular momentum (j_i, j_l) labels. We do not, however, include columns of type col$(\bullet\ \bullet\ \bullet)$, since the operation $\simeq$ destroys the standard labeling. Applied to a standard triangle pattern, the $\simeq$ operation $\Delta_{T(\pi(\mathbf{j})\,\mathbf{k})_j}$ maps the pattern to a new standard triangle pattern $\Delta_{T'(\pi'(\mathbf{j})\,\mathbf{k})_j}$, where it may happen that the binary tree is unchanged; that is, $T' = T$. We repeat: *No rearrangements of the entries in columns of the form col$(\bullet\ \bullet\ \bullet)$ are admitted, since such an operation destroys the standard ordering.* Thus, the operation $\simeq$ is applied to each column of a standard triangle pattern (2.128) as the interchange of the labels in row 1 (top) and row 2 (middle), except for those columns corresponding to forks of type col$(\bullet\ \bullet\ \bullet)$.

It is convenient to effect the equivalence relation $\simeq$ on a triangle pattern to introduce the transposition operators $\tau_i, i = 1, 2, \ldots, n-1$, which act in the set of triangle patterns

$$\mathbb{P}(\mathbf{j}\,\mathbf{k})_j = \{\Delta_{T(\pi(\mathbf{j})\,\mathbf{k})_j} \,|\, T \in \mathbb{T}_n, \pi \in S_n\}. \tag{2.129}$$

The action of the exchange operator τ_i is defined by

$$\tau_i \begin{pmatrix} a_1^\pi & \cdots & a_i^\pi & \cdots & a_{n-1}^\pi \\ b_1^\pi & \cdots & b_i^\pi & \cdots & b_{n-1}^\pi \\ k_1 & \cdots & k_i & \cdots & k_{n-1} \end{pmatrix} = \begin{pmatrix} a_1^\pi & \cdots & b_i^\pi & \cdots & a_{n-1}^\pi \\ b_1^\pi & \cdots & a_i^\pi & \cdots & b_{n-1}^\pi \\ k_1 & \cdots & k_i & \cdots & k_{n-1} \end{pmatrix}, \tag{2.130}$$

where, for each given triangle pattern $\Delta_{T(\pi(\mathbf{j})\,\mathbf{k})_j}$ the index i can assume only the subset of values in $\{1, 2, \ldots, n-1\}$ for which the columns of $\Delta_{T(\pi(\mathbf{j})\,\mathbf{k})_j}$ correspond to fork triangles of the binary tree $T(\pi(\mathbf{j})\,\mathbf{k})_j$ of the types col$(\circ\ \circ\ \bullet)$, col$(\circ\ \bullet\ \bullet)$, col$(\bullet\ \circ\ \bullet)$. The τ_i for an index i corresponding to columns of type col$(\bullet\ \bullet\ \bullet)$ give a triangle pattern not in the set $\mathbb{P}(\mathbf{j}\,\mathbf{k})_j$.

Two triangle patterns in the set $\mathbb{P}(\mathbf{j}\,\mathbf{k})_j$ that are related by any of the $n-t$ allowed exchange operations (see (2.81)) are called $\simeq$ equivalent.

The collection of all $\simeq$ equivalent triangle patterns in $\mathbb{P}(\mathbf{j}\,\mathbf{k})_j$ constitutes an equivalence class, and the full set $\mathbb{P}(\mathbf{j}\,\mathbf{k})_j$ of $n!a_n$ triangle patterns can, accordingly, be partitioned into disjoint equivalence classes. This partitioning is not difficult to carry out, using the results from Sect. 2.2.1, where it is shown that the number of triangle patterns having n columns and containing t columns of type $\mathrm{col}(\circ\ \circ\ \bullet)$ is $c_n(t)$ for which we have a generating function. The number of $\simeq$ equivalent triangle patterns in each subset $\mathbb{P}_t(\mathbf{j}\,\mathbf{k})_j$ of triangle patterns containing t columns of type $\mathrm{col}(\circ\ \circ\ \bullet)$ is 2^{n-t}, since there are $n-t$ exchange operations that can be applied to each triangle pattern. Hence, the number of $\simeq$ equivalence classes in the quotient set $\mathbb{P}_t(\mathbf{j}\,\mathbf{k})_j/\simeq$ is $|\mathbb{P}_t(\mathbf{j}\,\mathbf{k})_j/\simeq| = n!c_n(t)/2^{n-t}$. This gives the number of disjoint $\simeq$ equivalence classes in the quotient set $\mathbb{P}(\mathbf{j}\,\mathbf{k})_j/\simeq$ as the integer d_n defined by

$$d_n = n! \sum_{t=1}^{[n/2]} \frac{c_n(t)}{2^{n-t}}. \tag{2.131}$$

Examples. Equivalence classes of triangle patterns for $n = 2, 3, 4, 5:$

$n = 2:$ We have from Sect. 2.2.1 that there is one fork matrix, and it is given by

$$\begin{matrix} \circ \\ \circ \\ \bullet \end{matrix}\ . \tag{2.132}$$

Thus, there is one $\simeq$ equivalence class containing two triangle patterns, namely,

$$\begin{pmatrix} j_1 \\ j_2 \\ j \end{pmatrix},\ \begin{pmatrix} j_2 \\ j_1 \\ j \end{pmatrix}, \tag{2.133}$$

and the first triangle pattern can be taken as the representative.

$n = 3:$ We have from Sect. 2.2.1 that there are two fork matrices, each of type $\mathrm{col}(\circ\ \circ\ \bullet)$; namely,

$$\begin{matrix} \circ & \bullet \\ \circ & \circ \\ \bullet & \bullet \end{matrix}\ ,\quad \begin{matrix} \circ & \circ \\ \circ & \bullet \\ \bullet & \bullet \end{matrix}\ . \tag{2.134}$$

Thus, $t = 1$ and $c_3(1) = 2$. There are $3! = 6$ triangle patterns associated with each of these fork matrices, obtained by inserting a permutation of j_1, j_2, j_3 in each of the $\circ$ positions, k, j in row 3, and k in the remaining

• position. Thus, there are 12 triangle patterns in all. Relation (2.131) gives $d_3 = 3$ as the number of $\simeq$ inequivalent classes of triangle patterns. Representatives of these inequivalent classes can be chosen as

$$\begin{pmatrix} j_1 & k \\ j_2 & j_3 \\ k & j \end{pmatrix}, \begin{pmatrix} j_1 & k \\ j_3 & j_2 \\ k & j \end{pmatrix}, \begin{pmatrix} j_2 & k \\ j_3 & j_1 \\ k & j \end{pmatrix}. \tag{2.135}$$

Each of these triangle patterns then defines a family of four $\simeq$ triangle patterns under the operations of interchanging the symbols in row 1 and row 2 in either of the two columns. Thus, from these three patterns and their $\simeq$ equivalents, we regain the $n!a_n = 3!2 = 12$ patterns corresponding to the all standard labeled binary trees of order 3.

$n = 4$: We have $t = 1$ and $t = 2$, with $c_4(1) = 4$ fork matrices in the first class, and $c_4(2) = 1$ fork matrices in the second class (see Sect. 2.2.1), as given by

$$t = 1 : \begin{matrix} \circ & \bullet & \bullet \\ \circ & \circ & \circ \\ \bullet & \bullet & \bullet \end{matrix}, \quad \begin{matrix} \circ & \circ & \circ \\ \circ & \bullet & \bullet \\ \bullet & \bullet & \bullet \end{matrix}, \quad \begin{matrix} \circ & \circ & \bullet \\ \circ & \bullet & \circ \\ \bullet & \bullet & \bullet \end{matrix}, \quad \begin{matrix} \circ & \bullet & \circ \\ \circ & \circ & \bullet \\ \bullet & \bullet & \bullet \end{matrix};$$

$$\tag{2.136}$$

$$t = 2 : \begin{matrix} \circ & \circ & \bullet \\ \circ & \circ & \bullet \\ \bullet & \bullet & \bullet \end{matrix}.$$

There are $4! = 24$ triangle patterns associated with each of these five fork matrices, obtained by inserting any permutation of j_1, j_2, j_3, j_4 in each of the $\circ$ positions and the entries k_1, k_2, j in row 3; ; then, for $t = 1, k_1$ and k_2 in the remaining $\bullet$ positions in columns 2 and 3, respectively; then, for $t = 2$, the entries k_1, k_2 in the $\bullet$ positions in row 1 and row 2. Thus, there are 120 triangle patterns in all. Relation (2.131) gives $d_3 = 18$ as the total number of $\simeq$ equivalence classes of triangle patterns, twelve originating from the four $t = 1$ type of fork matrices and six from the single $t = 2$ type of fork matrix. Representatives of each of these $\simeq$ equivalent classes can be chosen as

$$t = 1 : \begin{pmatrix} j_p & k_1 & k_2 \\ j_q & j_r & j_s \\ k_1 & k_2 & j \end{pmatrix}; \; t = 2 : \begin{pmatrix} j_p & j_r & k_1 \\ j_q & j_s & k_2 \\ k_1 & k_2 & j \end{pmatrix}, \tag{2.137}$$

where p, q, r, s is any permutation of $1, 2, 3, 4$ such that $p < q$ in the pattern to the left, which gives twelve representative triangle patterns; and $p < q, r < s$ in the pattern to the right, which gives six representative triangle patterns. Each of the $t = 1$ classes contains eight $\simeq$ equivalent members, and each of the $t = 2$ classes contains four $\simeq$ equivalent members. This distribution into $\simeq$ equivalence classes accounts for all $(12)(8) + (6)(4) = 120 = n!a_n = (24)(5) = 120$ standard triangle patterns of order 3 corresponding to the set of standard labeled binary trees $\{T(\pi(\mathbf{j})\,\mathbf{k})_j \,|\, T \in \mathbb{T}_4, \pi \in S_4\}$.

$n = 5$: A less detailed accounting goes as follows. We have $t = 1$ and $t = 2$, with $c_5(1) = 8$ fork matrices in the first class, and $c_5(2) = 6$ fork matrices in the second class (see Sect. 2.2.1). There are $5! = 120$ triangle patterns associated with each of these fourteen fork patterns, obtained by inserting any permutation of j_1, j_2, j_3, j_4, j_5 in each of the $\circ$ positions and the entries k_1, k_2, k_3, j in row 3, thus giving a total of $n!a_n = (120)(14) = 1680$ triangle patterns. For $t = 1$, there are $5!c_5(1)/16 = 60$ equivalence classes, and, for $t = 2$, there are $5!c_5(2)/8 = 90$ equivalence classes. Each $t = 1$ triangle pattern has sixteen $\simeq$ equivalent triangle patterns, and the $t = 2$ triangle pattern has eight $\simeq$ equivalent triangle patterns. This distribution into $\simeq$ equivalence classes thus accounts for the total of $(60)(16) + (90)(8) = 1680$ standard triangle patterns associated with the standard labeled binary trees in the set $\{T(\pi(\mathbf{j})\,\mathbf{k})_j \,|\, T \in \mathbb{T}_5, \pi \in S_5\}$. $\square$

The above examples suggest how a set of representative triangle patterns, one from each of the $\simeq$ equivalence classes in the quotient set $\mathbb{P}(\mathbf{j}\,\mathbf{k})_j/\simeq$ can be chosen. We introduce the following $[n/2]$ subsets of permutations:

$$\begin{aligned} &S_n^{\mathrm{rep}}(t) \\ &= \{\pi = (\pi_1, \pi_2, \ldots, \pi_n) \in S_n \mid \pi_1 < \pi_2, \pi_3 < \pi_4, \ldots, \pi_{2t-1} < \pi_{2t}\}, \end{aligned} \tag{2.138}$$

where $t = 1, 2, \ldots, [n/2]$. The number of permutations in each set is

$$|\, S_n^{\mathrm{rep}}(t)\, | = n!\frac{c_n(t)}{2^{n-t}}. \tag{2.139}$$

A set of representative triangle patterns of the equivalence classes belonging to the quotient set $\mathbb{P}_t(\mathbf{j}\,\mathbf{k})_j/\simeq$ (see (2.131)) is given by

$$\mathbb{P}_t^{\mathrm{rep}}(\mathbf{j}\,\mathbf{k})_j/\simeq\; = \{\Delta_{T(\pi(\mathbf{j})\,\mathbf{k})_j} \,|\, \pi \in S_n^{\mathrm{rep}}(t)\}. \tag{2.140}$$

The union of this set of representative triangle patterns for all t then gives a set of d_n representative standard triangle patterns of the full set $\mathbb{P}(\mathbf{j}\,\mathbf{k})_j$ of triangle patterns defined by (2.129); that is, a set of representatives of the quotient set $\mathbb{P}(\mathbf{j}\,\mathbf{k})_j/\simeq$ is given by

$$\mathbb{P}^{\mathrm{rep}}(\mathbf{j}\,\mathbf{k})_j/\simeq\; = \bigcup_{t=1}^{[n/2]} \{\Delta_{T(\pi(\mathbf{j})\,\mathbf{k})_j} \,|\, \pi \in S_n^{\mathrm{rep}}(t)\}. \tag{2.141}$$

Triangle coefficients

We recall (see (2.124)) that the matrix elements of the real orthogonal binary recoupling matrix

$$R_{T,T'}^{\pi(\mathbf{j});\,\pi'(\mathbf{j})} = C_T^{\pi(\mathbf{j})} C_{T'}^{\pi'(\mathbf{j})\mathrm{tr}} \in H(N(\mathbf{j})) \tag{2.142}$$

are given by the inner product of coupled state vectors:

$$\begin{aligned}\langle T(\pi(\mathbf{j})\,\mathbf{k})_{j\,m} | T'(\pi'(\mathbf{j})\,\mathbf{k}')_{j\,m}\rangle &= R_{T(\pi(\mathbf{j})\,\mathbf{k})_j;\,T'(\pi'(\mathbf{j})\,\mathbf{k}')_j} \\ &= \left(C_T^{\pi(\mathbf{j})} C_{T'}^{\pi'(\mathbf{j})\mathrm{tr}}\right)_{\mathbf{k},j,m;\,\mathbf{k}',j,m} \\ &= \sum_{\mathbf{m}\in\mathbb{C}(\pi(\mathbf{j}))} C_{T(\pi(\mathbf{j})\,\mathbf{m};\mathbf{k})_{j\,m}} C_{T'(\pi'(\mathbf{j})\,\mathbf{m};\mathbf{k}')_{j\,m}},\end{aligned} \tag{2.143}$$

$$\begin{aligned}|T(\pi(\mathbf{j})\,\mathbf{k})_{j\,m}\rangle &= \sum_{\mathbf{m}\in\mathbb{C}(\mathbf{j})} C_{T(\pi(\mathbf{j})\,\mathbf{m};\mathbf{k})_{j\,m}} |\mathbf{j}\,\mathbf{m}\rangle \\ &= \sum_{\mathbf{m}\in\mathbb{C}(\mathbf{j})} \left(C_T^{(\mathbf{j})}\right)_{\mathbf{k},j,m;\,\mathbf{m}} |\mathbf{j}\,\mathbf{m}\rangle,\end{aligned} \tag{2.144}$$

$$\begin{aligned}|T'(\pi'(\mathbf{j})\,\mathbf{k}')_{j\,m}\rangle &= \sum_{\mathbf{m}\in\mathbb{C}(\mathbf{j})} C_{T'(\pi'(\mathbf{j})\,\mathbf{m};\mathbf{k}')_{j\,m}} |\mathbf{j}\,\mathbf{m}\rangle \\ &= \sum_{\mathbf{m}\in\mathbb{C}(\mathbf{j})} \left(C_{T'}^{\pi'(\mathbf{j})}\right)_{\mathbf{k}',j,m;\mathbf{m}} |\mathbf{j}\,\mathbf{m}\rangle.\end{aligned} \tag{2.145}$$

We now define the *triangle coefficient* in terms of the pair of standard triangle patterns $\Delta_{T(\pi(\mathbf{j})\,\mathbf{k})_j}$ and $\Delta_{T'(\pi'(\mathbf{j})\,\mathbf{k}')_j}$ corresponding to the pair of standard labeled trees $T(\pi(\mathbf{j})\,\mathbf{k})_j$ and $T'(\pi'(\mathbf{j})\,\mathbf{k}')_j$ by the inner product (2.143):

$$\begin{aligned}\left\{\Delta_{T(\pi(\mathbf{j})\,\mathbf{k})_j} \middle| \Delta_{T'(\pi'(\mathbf{j})\,\mathbf{k}')_j}\right\} &= \langle T(\pi(\mathbf{j})\,\mathbf{k})_{j\,m} | T'(\pi'(\mathbf{j})\,\mathbf{k}')_{j\,m}\rangle \\ &= \left(C_T^{\pi(\mathbf{j})} C_{T'}^{\pi'(\mathbf{j})\mathrm{tr}}\right)_{\mathbf{k},j,m;\,\mathbf{k}',j,m}.\end{aligned} \tag{2.146}$$

Thus, the inner product in this result, as well as the binary recoupling matrix, are vividly presented in terms of the triangle patterns of the form

(2.129) associated with the fork matrices of a pair of binary trees. Each triangle coefficient has the explicit definition in terms of $SU(2)$ WCG coefficients (see (2.87)) given by

$$\left\{\Delta_{T(\pi(\mathbf{j})\,\mathbf{k})_j}\middle|\Delta_{T'(\pi'(\mathbf{j})\,\mathbf{k}')_j}\right\} = \left\{\Delta_{T'(\pi'(\mathbf{j})\,\mathbf{k}')_j}\middle|\Delta_{T(\pi(\mathbf{j})\,\mathbf{k})_j}\right\}$$

$$= \sum_{\mathbf{m}\in\mathbb{C}(\mathbf{j})} \prod_{i=1}^{n-1} C^{a_i^\pi \;\; b_i^\pi \;\; k_i}_{\alpha_i^\pi \;\; \beta_i^\pi \;\; \alpha_i^\pi+\beta_i^\pi} \prod_{i=1}^{n-1} C^{a_i^{\pi'} \;\; b_i^{\pi'} \;\; k_i'}_{\alpha_i^{\pi'} \;\; \beta_i^{\pi'} \;\; \alpha_i^{\pi'}+\beta_i^{\pi'}}. \tag{2.147}$$

We refer to $\Delta_{T(\pi(\mathbf{j})\,\mathbf{k})_j}$ and $\Delta_{T'(\pi'(\mathbf{j})\,\mathbf{k}')_{j'}}$ as the *left-triangle pattern* and the *right-triangle pattern;* the interchange of these patterns gives the same triangle coefficient. Because the number of triangles appearing in the columns of each of the two triangle patterns of the triangle coefficient (2.147) is $n-1$, we refer to the triangle coefficient as a *triangle coefficient of order* $2n-2$. The triangle coefficient (2.147) encodes in its notation exactly the triangles of the forks of the pair of labeled binary trees underlying the triangle patterns entering into its definition; it takes on the numerical value of a sum over all projection quantum number $\mathbf{m}$ of the products of the $SU(2)$ WCG coefficients associated with these forks.

The columns of a triangle coefficients are **always** to be obtained directly from the labeled forks of the labeled binary trees $T(\pi(\mathbf{j})\,\mathbf{k})_j$ and $T(\pi'(\mathbf{j})\,\mathbf{k}')_j$. This is because not all $3 \times (n-1)$ matrices in which each column is an angular momentum triangle originate from a standard labeled binary tree, and the determination of the subset that has this property is nontrivial. Thus, triangle coefficients are always regarded as a notationally convenient way of presenting the fork triangles of a pair of labeled binary trees; row 3 of a triangle pattern is always the standard fork root labels $(k_1, k_2, \ldots, k_{n-1})$ with $k_{n-1} = j$.

Triangle coefficients of order $2n-2$ are very complicated objects, but are fully defined numerical-valued coefficients (2.147). It might appear that nothing remains to be done. The fact, however, that these objects depend only on collections of geometrical objects called triangles and their interconnections through binary trees suggests a deeper combinatorial setting than is evident from this preliminary definition in terms of WCG coefficients.

Many triangle coefficients are related by phase factors in consequence of the classical symmetries (1.224) of the $SU(2)$ WCG coefficients. The τ_i exchange operators (2.130), which may now act on both the left-triangle pattern and the right-triangle pattern give sets of *phase-equivalent* triangle coefficients, reduce greatly the number of triangle coefficients that need be considered. The determination of the phase-equivalence classes of triangle coefficients is simply an adaptation of the $\simeq$ equivalent classes of a single triangle pattern to pairs of triangle patterns, accounting now for the WCG symmetries (1.224). Thus, we define left-triangle pattern and right-triangle exchange operators as follows:

$$\tau_i^{(l)}\left\{\begin{pmatrix} a_1^\pi & \cdots & a_i^\pi & \cdots & a_{n-1}^\pi \\ b_1^\pi & \cdots & b_i^\pi & \cdots & b_{n-1}^\pi \\ k_1 & \cdots & k_i & \cdots & k_{n-1} \end{pmatrix} \middle| \begin{pmatrix} a_1^{\pi'} & a_2^{\pi'} & \cdots & a_{n-1}^{\pi'} \\ b_1^{\pi'} & b_2^{\pi'} & \cdots & b_{n-1}^{\pi'} \\ k_1 & k_2 & \cdots & k_{n-1} \end{pmatrix}\right\}$$
$$= (-1)^{a_i^\pi + b_i^\pi - k_i}$$
$$\times \left\{\begin{pmatrix} a_1^\pi & \cdots & b_i^\pi & \cdots & a_{n-1}^\pi \\ b_1^\pi & \cdots & a_i^\pi & \cdots & b_{n-1}^\pi \\ k_1 & \cdots & k_i & \cdots & k_{n-1} \end{pmatrix} \middle| \begin{pmatrix} a_1^{\pi'} & a_2^{\pi'} & \cdots & a_{n-1}^{\pi'} \\ b_1^{\pi'} & b_2^{\pi'} & \cdots & b_{n-1}^{\pi'} \\ k'_1 & k'_2 & \cdots & k'_{n-1} \end{pmatrix}\right\}, \tag{2.148}$$

$$\tau_i^{(r)}\left\{\begin{pmatrix} a_1^\pi & a_2^\pi & \cdots & a_{n-1}^\pi \\ b_1^\pi & b_2^\pi & \cdots & b_{n-1}^\pi \\ k_1 & k_2 & \cdots & k_{n-1} \end{pmatrix} \middle| \begin{pmatrix} a_1^{\pi'} & \cdots & a_i^{\pi'} & \cdots & a_{n-1}^{\pi'} \\ b_1^{\pi'} & \cdots & b_i^{\pi'} & \cdots & b_{n-1}^{\pi'} \\ k'_1 & \cdots & k'_i & \cdots & k'_{n-1} \end{pmatrix}\right\}$$
$$= (-1)^{a_i^{\pi'} + b_i^{\pi'} - k'_i}$$
$$\times \left\{\begin{pmatrix} a_1^\pi & a_2^\pi & \cdots & a_{n-1}^\pi \\ b_1^\pi & b_2^\pi & \cdots & b_{n-1}^\pi \\ k_1 & k_2 & \cdots & k_{n-1} \end{pmatrix} \middle| \begin{pmatrix} a_1^{\pi'} & \cdots & b_i^{\pi'} & \cdots & a_{n-1}^{\pi'} \\ b_1^{\pi'} & \cdots & a_i^{\pi'} & \cdots & b_{n-1}^{\pi'} \\ k'_1 & \cdots & k'_i & \cdots & k'_{n-1} \end{pmatrix}\right\}. \tag{2.149}$$

The values that the index i can assume are exactly those described below (2.130), applied now separately to the left-triangle pattern and the right-triangle pattern.

The set of triangle coefficients of order $2n-2$ can be partitioned into $\simeq$ equivalence classes by application of the exchange operators (2.148)-(2.149). We prefer now to call each $\simeq$ equivalence class of triangle coefficients a *phase-equivalent class.* Before determining the distribution of the set of triangle coefficients of order $2n-2$ into phase-equivalent classes, it is convenient to consider how this process is effected for small values of n. We return to the general problem in Sect. 2.2.11.

Examples: Triangle coefficients for $n = 2, 3$.

It is convenient to set $j_1 = a, j_2 = b, j_3 = c, j_4 = d$ in the following examples. Moreover, because of our focus on the properties of triangle coefficients in binary coupling theory, we present these examples in their entirely, giving the binary trees and triangle patterns, as well.

$n = 2$: There is but a single standard labeled binary tree and corresponding triangle pattern, as given by

$$T(ab)_j = $$ 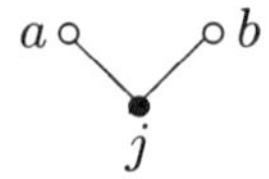 $$\rightarrow \begin{pmatrix} a \\ b \\ j \end{pmatrix}. \tag{2.150}$$

There are four triangle coefficients of order 2 corresponding to all permutations $\pi, \pi' \in S_2$:

$$\begin{aligned}
\left\{ \begin{array}{c|c} a & a \\ b & b \\ j & j \end{array} \right\} &= \sum_{\alpha,\beta} C^{a\ b\ j}_{\alpha\ \beta\ m} C^{a\ b\ j}_{\alpha\ \beta\ m} = 1, \\
\left\{ \begin{array}{c|c} a & b \\ b & a \\ j & j \end{array} \right\} = \left\{ \begin{array}{c|c} b & a \\ a & b \\ j & j \end{array} \right\} &= \sum_{\alpha,\beta} C^{a\ b\ j}_{\alpha\ \beta\ m} C^{b\ a\ j}_{\beta\ \alpha\ m} = (-1)^{a+b-j}, \\
\left\{ \begin{array}{c|c} b & b \\ a & a \\ j & j \end{array} \right\} &= \sum_{\alpha,\beta} C^{b\ a\ j}_{\beta\ \alpha\ m} C^{b\ a\ j}_{\beta\ \alpha\ m} = 1.
\end{aligned} \tag{2.151}$$

The middle relation is a consequence of the identity of triangle coefficients under the interchange of left- and right-triangle patterns.

$n = 3$: There are two standard labeled binary trees with labels $j_1 = a, j_2 = b, j_3 = c$ given by

$$T((a\,b\,c);\,k)_j \;=$$ 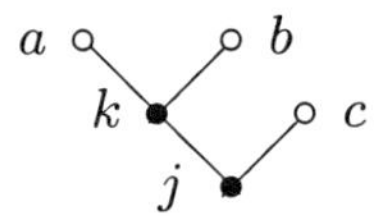 $$\rightarrow \begin{pmatrix} a & k \\ b & c \\ k & j \end{pmatrix},$$

$$(2.152)$$

$$T'((a\,b\,c);\,k)_j \;=$$ 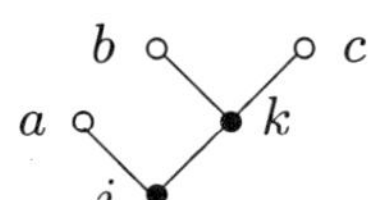 $$\rightarrow \begin{pmatrix} b & a \\ c & k \\ k & j \end{pmatrix}.$$

These two standard labeled triangle patterns, together with all permutations of a, b, c give the $3!2 = 12$ labeled triangle patterns. Any of the twelve can be chosen as the left-triangle pattern and any of the twelve as the right-triangle pattern, thus giving 144 triangle coefficients. This set of 144 triangle coefficients is then partitioned into nine equivalence classes by application of the exchange operators (2.148)-(2.149), which gives sixteen phase-equivalent triangle coefficients in each of the nine classes. The representatives of the nine equivalence classes can be obtained as follows: There are three $\simeq$ equivalence classes with representative triangle patterns given by (2.135):

$$\begin{pmatrix} a & k \\ b & c \\ k & j \end{pmatrix}, \begin{pmatrix} a & k \\ c & b \\ k & j \end{pmatrix}, \begin{pmatrix} b & k \\ c & a \\ k & j \end{pmatrix}. \tag{2.153}$$

There are four $\simeq$ equivalent patterns in each equivalence class.

We now select any of the three representatives from (2.153) to be the left-triangle pattern or to be the right-triangle pattern (with k replaced by k') to obtain nine equivalence classes of triangle coefficients, each of which contains sixteen phase-equivalence triangle coefficients by application of the exchange operators (2.148)-(2.149). But now a new phenomenon enters, again due to the properties of the WGC coefficients that enter into the expression of a triangle coefficient in terms of a summation over $2n-2$ such $SU(2)$ WCG coefficients. We list first these results and then outline the proof. We have the following three identities originating from the property that the entries in the first two rows of the first column in the left-triangle pattern match those in the first column in the right-triangle pattern, and this gives the three relations:

$$\begin{aligned}
\left\{\begin{array}{cc|cc} a & k & a & k' \\ b & c & b & c \\ k & j & k' & j \end{array}\right\} &= \delta_{k,k'}, \\
\left\{\begin{array}{cc|cc} a & k & a & k' \\ c & b & c & b \\ k & j & k' & j \end{array}\right\} &= \delta_{k,k'}, \\
\left\{\begin{array}{cc|cc} b & k & b & k' \\ c & a & c & a \\ k & j & k' & j \end{array}\right\} &= \delta_{k,k'}.
\end{aligned} \tag{2.154}$$

We list the remaining six in terms of the following reference coefficient:

$$\sqrt{(2k+1)(2k'+1)}\,W(abjc;kk') = \left\{\begin{array}{cc|cc} a & k & b & a \\ b & c & c & k' \\ k & j & k' & j \end{array}\right\}. \tag{2.155}$$

The notation $W(abjc;kk')$ was introduced by Racah [143]; hence, this coefficient is called a *Racah coefficient.* This expression of a Racah coefficient in terms of a summation over four WCG coefficient is, by convention, standard (see Ref. [21] for the notations used here). We discuss at length below the structure of Racah coefficients. The remaining six representative triangle coefficients are expressed in terms of the Racah coefficient by

$$\left\{\begin{array}{cc|cc} a & k & b & k' \\ b & c & c & a \\ k & j & k' & j \end{array}\right\} = (-1)^{a+k'-j}\sqrt{(2k+1)(2k'+1)}\,W(abjc;kk'),$$
$$\left\{\begin{array}{cc|cc} a & k & a & k' \\ b & c & c & b \\ k & j & k' & j \end{array}\right\} = (-1)^{a-k-k'+j}\sqrt{(2k+1)(2k'+1)}\,W(bajc;kk'),$$
$$\left\{\begin{array}{cc|cc} a & k & b & k' \\ c & b & c & a \\ k & j & k' & j \end{array}\right\} = (-1)^{a+b+c-j}\sqrt{(2k+1)(2k'+1)}\,W(acjb;kk'),$$
$$\left\{\begin{array}{cc|cc} b & k & a & k' \\ c & a & b & c \\ k & j & k' & j \end{array}\right\} = (-1)^{b+k-j}\sqrt{(2k+1)(2k'+1)}\,W(cbja;kk'),$$
$$\left\{\begin{array}{cc|cc} a & k & a & k' \\ c & b & b & c \\ k & j & k' & j \end{array}\right\} = (-1)^{a-k-k'+j}\sqrt{(2k+1)(2k'+1)}\,W(cajb;kk'),$$
$$\left\{\begin{array}{cc|cc} b & k & a & k' \\ c & a & c & b \\ k & j & k' & j \end{array}\right\} = (-1)^{a+b+c-j}\sqrt{(2k+1)(2k'+1)}\,W(bcja;kk'). \tag{2.156}$$

Proof. The verification of the validity of relations (2.154) and (2.156) goes as follows: First, each triangle coefficient of order four is expressed as a summation over the projection quantum numbers of four $SU(2)$ WCG coefficients as given explicitly in terms of the fork triangles by (2.147). Relations (2.154) are consequences of the orthogonality relations (1.145)-(1.146). Relation (2.155) is the definition given by Racah (see Ref. [21] for the notations used here) of the Racah W-coefficient. Relations (2.156) are then rearrangements of the order of the four WCG coefficients in relation (2.155), where the classical symmetries (1.224) are used to bring the rearrangement into the form that gives these relations. □

It is important to note that relations (2.156) are not applications of the exchange relations (2.148)-(2.149) alone. Moreover, we could, with slight changes of the representatives (2.153) (see (2.157 below), have placed the definition (2.155) among a representative set of six relations similar to (2.156). As the representative set of the equivalence classes stand in (2.156), relation (2.155) is phase-equivalent to the first-listed triangle coefficient. Which member we choose for the representative of an equivalence class is, of course, arbitrary.

It is also the case that the last three relations in (2.156) are obtained from the first three, respectively, by the interchange of the left-triangle pattern and the right-triangle pattern, followed by the interchange of k and k'. This gives the following three symmetries of the Racah coefficient: $W(abjc;k'k) = W(cbja;kk'), W(bajc;k'k) = W(cajb;kk')$, $W(acjb;k'k) = W(bcja;kk')$. These symmetries are special cases of more general symmetries that will be taken up later.

Remarkable simplifications have occurred. The set of $(n!a_n)^2 = ((6)(2))^2 = 144$ triangle coefficients of order four corresponding to all possible selections of the left- and right-triangle patterns reduces to 48 that are equal to phase factors or to zero, and 96 that are of the form (phase factors) $\times$ (square-root dimension factors associated with the internal angular momenta) $\times$ (**one basic coefficient**), the latter being the Racah coefficient $W(abjc;kk')$ with the six permutations of its external angular momentum parameters a, b, c.) This result itself is quite striking, but still would be quite complex, since a triangle coefficient of order four is a summation over four WCG coefficients. The fact that this complex object can be brought to a comprehensible form is also due to Racah [143]. The Racah coefficient, in turn, is related to a standard object in mathematics, the ${}_4F_3$ hypergeometric function of unit argument. Because Racah coefficients are the basic entities out of which all triangle coefficients are built (see the Sixth Fundamental Result below), we present their properties before returning to the general counting problem of equivalent classes of general triangle coefficients of order $2n-2$ in Sect. 2.2.11.

2.2.8 Racah coefficients

We present the definition of the Racah coefficient in terms of all three notations used above, the labeled tree diagram, the triangle coefficient of order four, and as matrix elements of the recoupling matrix:

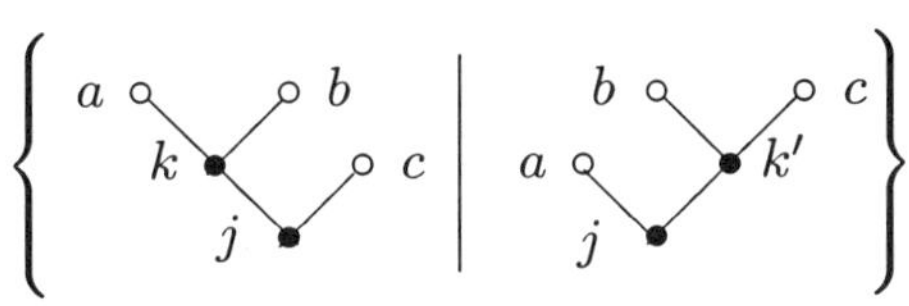

$$= \left\{ \begin{array}{cc|cc} a & k & b & a \\ b & c & c & k' \\ k & j & k' & j \end{array} \right\} = \left(R^{(ab)c;\, a(bc)}_{T,T'} \right)_{k,j,m;\, k',j,m}$$

$$= \sqrt{(2k+1)(2k'+1)}\, W(abjc;kk'). \qquad (2.157)$$

The orthogonal matrices appearing in the recoupling matrix

$$R^{(ab)c;\, a(bc)}_{T,T'} = C^{(ab)c}_{T} \left(C^{a(bc)}_{T'} \right)^{\mathrm{tr}} \qquad (2.158)$$

are the orthogonal matrices with elements given in terms of the WCG coefficients associated with the respective standard labeled binary trees by

$$\begin{aligned}\left(C_T^{(ab)c}\right)_{k,j,m;\alpha,\beta,\gamma} &= C^{a\ b\ \ k}_{\alpha\ \beta\ \alpha+\beta}C^{\ \ k\ \ \ c\ j}_{\alpha+\beta\ \gamma\ m},\\ \left(C_{T'}^{a(bc)}\right)_{k',j,m;\alpha,\beta,\gamma} &= C^{b\ c\ \ k'}_{\beta\ \gamma\ \beta+\gamma}C^{a\ \ k'\ \ j}_{\alpha\ \beta+\gamma\ m}.\end{aligned} \tag{2.159}$$

Thus, the explicit expression for the Racah coefficient in terms of four WCG coefficients is

$$\begin{aligned}&\sqrt{(2k+1)(2k'+1)}W(abjc;kk')\\ &= \sum_{\substack{(\alpha,\beta,\gamma)\in\mathbb{C}(a,b,c)\\ \alpha+\beta+\gamma=m}} C^{a\ b\ \ k}_{\alpha\ \beta\ \alpha+\beta}C^{\ \ k\ \ \ c\ j}_{\alpha+\beta\ \gamma\ m}C^{b\ c\ \ k'}_{\beta\ \gamma\ \beta+\gamma}C^{a\ \ k'\ \ j}_{\alpha\ \beta+\gamma\ m}.\end{aligned} \tag{2.160}$$

As noted several times, the summation over WCG coefficients gives a numerical value that is independent of $m = j, j-1, \ldots, -j$. It is also convenient to denote the elements of the real orthogonal matrix $C_T^{(\mathbf{j})} = U_T^{(\mathbf{j})}$ that brings the Kronecker product $D^{\mathbf{j}}$ to Kronecker direct sum form (see (2.32)) by a slightly modified notation:

$$C_T^{(\mathbf{j})} = C_T^{\mathrm{sh}(\mathbf{j})}, \tag{2.161}$$

where sh($\mathbf{j}$) denotes the actual bracketing or *shape* of the angular momenta $\mathbf{j}$ assigned to the external points of the tree T.

The summation in (2.160) can be carried out to give

$$\begin{aligned}&W(abjc;kk') = (-1)^{a+b+c+j}\Delta(abk)\Delta(kcj))\Delta(ak'j)\Delta(bck')\\ &\times \sum_{h=h_{\min}}^{h_{\max}} \frac{(-1)^h(h+1)!}{(h-a-b-k)!(h-k-c-j)!(h-a-k'-j)!}\\ &\times\frac{1}{(h-b-c-k')!(a+b+c+j-h)!}\\ &\times\frac{1}{(a+c+k+k'-h)!(b+j+k+k'-h)!},\end{aligned} \tag{2.162}$$

where $h_{\min}$ and $h_{\max}$ are defined by

$$\begin{aligned}h_{\min} &= \max\{a+b+k, k+c+j, a+k'+j, b+c+k'\},\\ h_{\max} &= \min\{a+b+c+j, a+c+k+k', b+j+k+k'\}.\end{aligned} \tag{2.163}$$

We refer to the Δ-factors as delta-triangle coefficients. They are defined on all triplets of angular momentum labels a, b, c such that the triangle rule $c \in \langle a\, b\rangle$ holds:

$$\Delta(abc) = \sqrt{\frac{(a+b-c)!(a-b+c)!(-a+b+c)!}{(a+b+c+1)!}}. \tag{2.164}$$

The delta-triangle coefficients $\Delta(abk), \Delta(kcj)), \Delta(ak'j), \Delta(bck')$ occurring in the Racah coefficient (2.162) are intrinsic characteristics of a Racah coefficient as inherited from the WCG coefficients occurring in (2.160). Thus, from the definition $\langle j_1\, j_2\rangle = j_1+j_2, j_1+j_2-1, \ldots |j_1-j_2|$, we have, for specified (a, b, c, j), that

$$k \in \langle a\, b\rangle \text{ and } k \in \langle c\, j\rangle;\;\; k' \in \langle b\, c\rangle \text{ and } k' \in \langle a\, j\rangle. \tag{2.165}$$

It is quite remarkable that Racah was able to carry out the summation over four WCG coefficients occurring in (2.160) to obtain the one-parameter summation expression for the W-coefficient given by (2.162). Since each WCG coefficient is itself a one-parameter summation expression, the summation in (2.160) is over six parameters. We comment below in the Remarks on Racah's method.

Racah coefficients satisfy the row and column orthogonality relations given by

$$\begin{gathered}\sum_k (2k+1)(2k'+1)W(abjc;kk')W(abjc;kk'') = \delta_{k',k''},\\ \sum_{k'} (2k+1)(2k'+1)W(abjc;kk')W(abjc;k''k') = \delta_{k,k''}.\end{gathered} \tag{2.166}$$

We prove this relationship and others below in the context of the properties of recoupling matrices.

It is customary to introduce the so-called $6-j$ coefficient in place of the Racah W-coefficient because of the enhancement of symmetries that the $6-j$ coefficient exhibits. It is defined by

$$\left\{ \begin{matrix} a & b & k \\ c & j & k' \end{matrix} \right\} = (-1)^{a+b+c+j} W(abjc;kk'). \tag{2.167}$$

The $6-j$ coefficient is then invariant under all permutations of its columns, and under the interchange of any pair of elements in the top row with the corresponding elements in the bottom row. The triangles in the $6-j$ coefficient are $(a\, b\, k), (c\, j\, k), (a\, j\, k'), (c\, b\, k')$, and the symmetry relations preserve these delta-triangles. There are actually 144 symmetries of the $6-j$ coefficient, or the W-coefficient, as derived from their generating function in Sect. 4.6, Chapter 4. All these symmetries can also be verified directly from the explicit form (2.162).

Remarks.

(a). The summation on the right-hand side of (2.160) is over four WCG coefficients, each of which is itself a one-parameter summation expression, this giving rise to a six-parameter summation over factorial terms. Racah's method of summation was to locate a term $\binom{x+y}{c}$ in this expression and apply the binomial sum rule (1.159), thus "lifting" the summation to a seven-parameter summation. After repeating this process four times, a ten-parameter summation expression was reached in which the factors were organized in such a way that all but one of the internal summations could be carried out, thus giving (2.162). This process is given in detail in Ref. [21], but it has never been analyzed structurally in a way that reveals why it works. Such an analysis might be quite useful, especially, in view of the relationship of WCG coefficients and Racah coefficients to hypergeometric series, as we next point out.

(b). The WCG coefficients and Racah coefficients are expressible in terms of terminating ${}_3F_2$ and ${}_4F_3$ hypergeometric coefficients, respectively, evaluated at unit argument. These formulas are complicated by the fact that neither WCG coefficients nor Racah coefficients start off in their summation expressions as given by (1.213)-(1.214) and (2.162) with the first term in the sum being equal to 1, as is the case for terminating hypergeometric series. Adjustments for this must be made by shifting the summation indices. But the shift itself depends on the magnitude of the various denominator factors of the form $(\alpha_i - k)!$ of which there are several. Taking this into account leads to the following forms of the WCG coefficients and Racah coefficients in which a shift factor δ occurs in the parameters of the hypergeometric coefficients:

$$\begin{aligned} C^{a\ b\ c}_{\alpha\ \beta\ \gamma} &= (-1)^{a-b+\gamma}\sqrt{2c+1}\begin{pmatrix} a & b & c \\ \alpha & \beta & -\gamma \end{pmatrix} \\ &= N_1 \begin{pmatrix} a & b & c \\ \alpha & \beta & \gamma \end{pmatrix} {}_3F_2\left(\begin{matrix} -a_1,\ -a_2,\ -a_3 \\ b_1+1,\ b_2+1 \end{matrix} ;1 \right), \end{aligned} \tag{2.168}$$

where the multiplying factor N_1 and the parameters a_1, a_2, a_3 have the following definitions:

$$N_1\begin{pmatrix} a\ b\ c \\ \alpha\,\beta\ \gamma \end{pmatrix} = (-1)^{a+b+\gamma-\delta}\delta_{\alpha+\beta,\gamma}\frac{\Delta(abc)}{a_1!a_2!a_3!b_1!b_2!}\times$$
$$[(2c+1)(a+\alpha)!(a-\alpha)!(b+\beta)!(b-\beta)!(c+\gamma)!(c-\gamma)!]^{1/2}, \tag{2.169}$$
$$-a_1 = a+\alpha-\delta,\ -a_2 = b+\alpha+\gamma-\delta,\ -a_3 = c+\gamma-\delta,$$
$$b_1 = \delta_1-\delta,\ b_2 = \delta_2-\delta.$$

The shift factor δ is the minimum value in the sequence

$$(a+\alpha+b+\beta,,b-\beta+c+\gamma,a+\alpha+c+\gamma), \tag{2.170}$$

and δ_1 and δ_2 are the two quantities remaining in this sequence, in either order, after striking out the minimum value δ. The ${}_3F_2$ series (2.168) is always well-defined: It terminates at the summation index value $k = \min(a_1, a_2, a_3)$.

The W coefficient has a similar relation to the ${}_4F_3$ hypergeometric function of unit argument:

$$W(abdc;ef) = (-1)^{a+b+c+d}\begin{Bmatrix} a & b & e \\ c & d & f \end{Bmatrix}, \tag{2.171}$$
$$= N_2\begin{pmatrix} a & b & e \\ c & d & f \end{pmatrix} {}_4F_3\begin{pmatrix} -a_1,\ -a_2,\ -a_3,\ -a_4 & \\ & ;1 \\ -\delta-1,\ b_1+1,\ b_2+1 & \end{pmatrix},$$

where the multiplying factor N_2 has the following definition:

$$N_2\begin{pmatrix} a & b & e \\ c & d & f \end{pmatrix}$$
$$= \frac{(-1)^{a+b+c+d-\delta}\Delta(abe)\Delta(ecd))\Delta(afd)\Delta(bcf)(\delta+1)!}{(a_1)!(a_2)!(a_3)!(a_4)!(b_1)!(b_2)!},$$
$$-a_1 = a+b+e-\delta,\ -a_2 = c+d+e-\delta, \tag{2.172}$$
$$-a_3 = a+d+f-\delta,\ -a_4 = b+c+f-\delta,$$
$$b_1 = \delta_1-\delta,\ b_2 = \delta_2-\delta.$$

The shift factor δ is the minimum value in the sequence

$$(a+b+c+d, a+c+e+f, b+d+e+f), \tag{2.173}$$

and δ_1 and δ_2 are the two quantities remaining in this sequence, in either order, after striking out the minimum value δ. The ${}_4F_3$ series (2.171) is always well-defined: It terminates at the summation index value $k = \min(a_1, a_2, a_3, a_4)$, before the 0 in the denominator is reached because of the relation $a_i > -\delta - 1, i = 1, 2, 3, 4$. The ${}_4F_3$ hypergeometric series is Saalschützian; that is, its parameters satisfy $-(a_1 + a_2 + a_3 + a_4) + 1 = -\delta_1 - 1 + \delta_2 + \delta_2$, in consequence of $2(a+b+c+d+e+f) = \delta + \delta_1 + \delta_2$.

If the definition (2.160) of the Racah coefficient is written in terms of ${}_3F_2$ and ${}_4F_3$ hypergeometric series of unit arguments, the square-root delta-triangle factors cancel from each side of the relation, so that an intricate identity remains with no square roots, which involves a quadruple summation over four ${}_3F_2$ hypergeometric series, with multiplying factors, being equal to a Saalschützian ${}_4F_3$ hypergeometric series with multiplying factors. This is the relation that Racah proved. To our knowledge, the underlying structure of this identity has not been put in the context of hypergeometric function theory, especially from a combinatorial perspective.

(c). Hypergeometric series in which one or more numerator parameters is a negative integer $-n$ are finite sums that are essentially polynomials, as shown by the following identity:

$$ {}_pF_q\left(\begin{array}{c} x_1, x_2, \ldots, x_{p-1}, -n \\ \\ y_1, y_2, \ldots, y_q \end{array} ; z \right) \tag{2.174}$$

$$= \sum_{k=0}^{n} \frac{(x_1)_k (x_2)_k \cdots (x_{p-1})_k (-n)_k \, z^k}{(y_1)_k (y_2)_k \cdots (y_q)_k \, k!} = \frac{{}_pP_q(\mathbf{x}; \mathbf{y};\, z)}{\prod_{j=1}^{q} (y_j)_n},$$

$${}_pP_q(\mathbf{x}; \mathbf{y}; z) = \sum_{k=0}^{n} (-1)^k \binom{n}{k} z^k \prod_{i=1}^{p-1} (x_i)_k \prod_{j=1}^{q} (y_j + k)_{n-k}, \tag{2.175}$$

where $\mathbf{x} = (x_1, x_2, \ldots, x_{p-1})$, $\mathbf{y} = (y_1, y_2, \ldots, y_q)$. Relation (2.174) is valid for all values of these parameters such that no zeros are introduced into the denominator. (See Sect. 11.5.3, Compendium B.)

(d). A fascinating aspect of the polynomials ${}_pP_q$ is that they can have integral zeros that give rise to zeros of a WCG or Racah coefficient that would otherwise have no reason to be zero: The polynomial ${}_pP_q$ may be viewed as a Diophantine equation that gives what are called nontrivial zeros of WCG and Racah coefficients. While such zeros are well-known (see Ref. [113] for a review), this subject of zeros has not been examined in the context of terminating hypergeometric series, to our knowledge.

We return now to the properties of recoupling matrices and their relationship to triangle coefficients and Racah coefficients.

2.2.9 Recoupling matrices for $n = 3$

All 144 recoupling matrices for $n = 3$ given by

$$R_{T,T'}^{\mathrm{sh}(\pi(abc));\,\mathrm{sh}'(\pi'(abc))} = C_T^{\mathrm{sh}(\pi(abc))}\left(C_{T'}^{\mathrm{sh}'(\pi'(abc))}\right)^{\mathrm{tr}}, \quad \pi, \pi' \in S_3, \tag{2.176}$$

can be obtained from the 9 representative triangle coefficients defined by (2.154) and (2.156). where we now use the shape notation (2.161) for the enumeration of these recoupling matrices. For example, for $\pi(abc) = bca$ and $\mathrm{sh} = ((\circ\,\circ)\circ)$, we have $\mathrm{sh}(\pi(abc)) = \mathrm{sh}(bca) = ((bc)a) = (bc)a$, where we drop the outside parenthesis pair. The representative recoupling matrices are those with elements given by the following relations:

$$\begin{aligned}
\left(R_{T,T}^{(ab)c;\,(ab)c}\right)_{k,j,m;\,k',j,m} &= \left\{\begin{array}{cc|cc} a & k & a & k' \\ b & c & b & c \\ k & j & k' & j \end{array}\right\} = \delta_{k,k'}, \\
\left(R_{T,T}^{(ac)b;\,(ac)b}\right)_{k,j,m;\,k',j,m} &= \left\{\begin{array}{cc|cc} a & k & a & k' \\ c & b & c & b \\ k & j & k' & j \end{array}\right\} = \delta_{k,k'}, \\
\left(R_{T,T}^{(bc)a;\,(bc)a}\right)_{k,j,m;\,k',j,m} &= \left\{\begin{array}{cc|cc} b & k & b & k' \\ c & a & c & a \\ k & j & k' & j \end{array}\right\} = \delta_{k,k'};
\end{aligned} \tag{2.177}$$

$$\begin{aligned}
\left(R_{T,T}^{(ab)c;\,(bc)a}\right)_{k,j,m;\,k',j,m} &= \left\{\begin{array}{cc|cc} a & k & b & k' \\ b & c & c & a \\ k & j & k' & j \end{array}\right\}, \\
\left(R_{T,T}^{(ab)c;\,(ac)b}\right)_{k,j,m;\,k',j,m} &= \left\{\begin{array}{cc|cc} a & k & a & k' \\ b & c & c & b \\ k & j & k' & j \end{array}\right\}, \\
\left(R_{T,T}^{(ac)b;\,(bc)a}\right)_{k,j,m;\,k',j,m} &= \left\{\begin{array}{cc|cc} a & k & b & k' \\ c & b & c & a \\ k & j & k' & j \end{array}\right\}, \\
\left(R_{T,T}^{(bc)a;\,(ab)c}\right)_{k,j,m;\,k',j,m} &= \left\{\begin{array}{cc|cc} b & k & a & k' \\ c & a & b & c \\ k & j & k' & j \end{array}\right\}, \\
\left(R_{T,T}^{(ac)b;\,(ab)c}\right)_{k,j,m;\,k',j,m} &= \left\{\begin{array}{cc|cc} a & k & a & k' \\ c & b & b & c \\ k & j & k' & j \end{array}\right\}, \\
\left(R_{T,T}^{(bc)a;\,(ac)b}\right)_{k,j,m;\,k',j,m} &= \left\{\begin{array}{cc|cc} b & k & a & k' \\ c & a & c & b \\ k & j & k' & j \end{array}\right\}.
\end{aligned} \tag{2.178}$$

While the matrix elements of these representative recoupling matrices are independent of m, this projection quantum number is retained so as to enumerate all matrix elements for $m = j, j-1, \ldots, -j$ in accordance with the order of these matrices:

$$\sum_{j=j_{\min}}^{j_{\max}} (2j+1)C_j(abc) = (2a+1)(2b+1)(2c+1) = N(abc), \tag{2.179}$$

where $j_{\min} = \min\{|\, a \pm b \pm c\,|\}$ and $j_{\max} = a+b+c$ (see (2.13) and (2.21)).

Each recoupling matrix $R_{T,T'}^{\mathrm{sh}(\pi(abc));\,\mathrm{sh}'(\pi'(abc))}$ is an element of the group of unitary matrices $H(N(a\,b\,c))$, as given by (2.49). For the case at hand, we have:

$$H(N(a\,b\,c)) = \left\{ \sum_{j=j_{\min}}^{j_{\max}} \oplus \left(W_{T,T'}^{(\mathrm{sh}(\pi(abc)),j);\,(\mathrm{sh}'(\pi'(abc)),j)} \otimes I_{2j+1} \right) \right.$$
$$\left. \middle| W_{T,T'}^{(\mathrm{sh}(\pi(abc)),j);\,(\mathrm{sh}'(\pi'(abc)),j)} \in U(N_j(abc) \right\}. \tag{2.180}$$

Accordingly, the recoupling matrix $R_{T,T'}^{\mathrm{sh}(\pi(abc));\,\mathrm{sh}'(\pi'(abc))}$ has the direct sum expression given by

$$R_{T,T'}^{\mathrm{sh}(\pi(abc));\,\mathrm{sh}'(\pi'(abc))} = \sum_{j=j_{\min}}^{j_{\max}} \oplus \left(R_{T,T'}^{\mathrm{sh}(\pi(abc)),j);\,\mathrm{sh}'(\pi'(abc)),j)} \otimes I_{2j+1} \right), \tag{2.181}$$
$$R_{T,T'}^{\mathrm{sh}(\pi(abc)),j);\,\mathrm{sh}'(\pi'(abc)),j)} \in U(N_j(abc)).$$

We recall that the CG number $N_j(abc)$ is determined by relations (2.37)-(2.38), where $M_j(abc)$ is the number of compositions (l_1, l_2, l_3) of $a+b+c-j$ into three nonnegative integer parts such that $0 \le l_1 \le 2a, 0 \le l_2 \le 2b, 0 \le l_3 \le 2c$; it is also determined by the multiset $\langle abc\rangle$.

All 144 reduced recoupling matrices are members of the unitary group $U(N_j(abc))$. For example, we have that $W_{T,T'}^{(ab)c,j;\,c(ab),j}$ is the matrix with nonzero elements given by

$$\left(W_{T,T'}^{(ab)c,j;\,c(ab),j} \right)_{k,j;k',j} = (-1)^{c+k-j}\delta_{k,k'}. \tag{2.182}$$

Applied to the Racah coefficient defined by (2.155), we have that

$$W_{T,T'}^{(ab)c,j;\,a(bc),j} \in U(N_j(abc)), \tag{2.183}$$

which signifies that the matrix elements given by

$$\left(W_{T,T'}^{(ab)c,j;\,a(bc),j}\right)_{k,j;k',j} = \sqrt{(2k+1)(2k'+1)}W(abjc;kk') \tag{2.184}$$

are the elements of a real orthogonal matrix of order $N_j(abc)$; that is, they satisfy the orthogonality relations given by

$$\begin{gathered}\sum_k (2k+1)(2k'+1)W(abjc;kk')W(abjc;kk'') = \delta_{k',k''},\\ \sum_{k'} (2k+1)(2k'+1)W(abjc;kk')W(abjc;k''k') = \delta_{k,k''}.\end{gathered} \tag{2.185}$$

All other nontrivial recoupling matrices give these same relations, with permutations of a, b, c.

The set of reduced recoupling matrices does not constitute a subgroup of $U(N_j(abc))$, as noted earlier for the general case. There is, however, an interesting result originating from relation (2.125). We apply this relation for $n = 3$ to the following special case of labeled trees, using the shape notation above:

$$R_{T,T}^{(ab)c;(bc)a}\, R_{T,T}^{(bc)a;(ac)b} = R_{T,T}^{(ab)c;(ac)b}. \tag{2.186}$$

This relation is expressed in terms of triangle coefficients by

$$\begin{aligned}\sum_{k''} &\left\{\begin{array}{cc|cc} a & k & b & k'' \\ b & c & c & a \\ k & j & k'' & j \end{array}\right\}\left\{\begin{array}{cc|cc} b & k'' & a & k' \\ c & a & c & b \\ k'' & j & k' & j \end{array}\right\}\\ &= \left\{\begin{array}{cc|cc} a & k & a & k' \\ b & c & c & b \\ k & j & k' & j \end{array}\right\}.\end{aligned} \tag{2.187}$$

These triangle coefficients are, in turn, related to Racah coefficients by relations (2.178) and their phase-equivalents. Making the appropriate substitutions gives the following relation between Racah coefficients:

$$\begin{aligned}\sum_{k''} &(-1)^{a+k''-j}(2k''+1)\, W(abjc;kk'')W(bcja;k''k')\\ &= W(abjc;kk').\end{aligned} \tag{2.188}$$

This relation is an identity known as Racah's *sum rule:*

The trivial relation (2.186) for recoupling matrices expresses the famous Racah sum rule.

The use of $(bc)a$ as the middle factor in relation (2.186) is arbitrary, and we could just have well used some other shape and permutation, but no new relations emerge.

2.2.10 Recoupling matrices for $n = 4$

There are $(4!(5))^2 = 14,400$ recoupling matrices

$$R_{T,T'}^{\pi(a\,b\,c\,d);\pi'(a\,b\,c\,d)} = C_T^{\pi(a\,b\,c\,d)}\, C_{T'}^{\pi'(a\,b\,c\,d)tr}, \tag{2.189}$$

corresponding to all $\pi, \pi' \in S_4$, and all $T, T' \in \mathbb{T}_4$, there being $a_4 = 5$ binary trees in the set $\mathbb{T}_4$. These are given in Sect. 2.2, and we denote them in the ordered listed by T_1, T_2, T_3, T_4, T_5. The corresponding standard labeled trees are given in the shape notation by

$$\begin{aligned}&T_1([(ab)c]d), T_2(a[b(cd)]), T_3((ab)(cd)),\\ &T_4(a[(bc)d]), T_5(a[(bc)d]),\end{aligned} \tag{2.190}$$

where, for clarity, we have also used $[\,]$ as a parenthesis pair. Thus, considering all twenty-four permutations $\pi \in S_4$ of (a, b, c, d), there are 120 triangle patterns. There are $d_4 = 18$ representative binary trees (and corresponding triangle patterns) given by (2.141).

Twelve representatives correspond to $t = 1$ and are given by

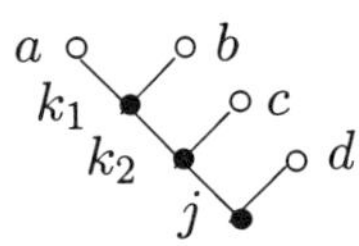

$$= T_1([(ab)c]\,d) \rightarrow \begin{pmatrix} a & k_1 & k_2 \\ b & c & d \\ k_1 & k_2 & j \end{pmatrix}, \tag{2.191}$$

together with those with external $\circ$ labels given by the permutations

$$\begin{aligned}[(ab)c]\,d \quad \mapsto \quad &[(ab)d]\,c, [(ac)b]\,d, [(ac)d]\,b,\\ &[(ad)b]\,c, [(ad)c]\,b, [(bc)a]\,d, [(bc)d]\,a,\\ &[(bd)a]\,c, [(bd)c]\,a, [(cd)a]\,b, [(cd)b]\,a.\end{aligned} \tag{2.192}$$

There are eight $\simeq$ equivalent triangle patterns in each of the twelve classes with these representatives, giving 96 triangle patterns in all for $t = 1$.

Four representatives correspond to $t = 2$ and are given by

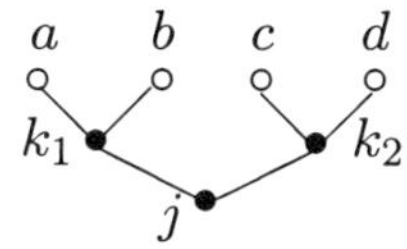

$$= T_3((ab)(cd)) \rightarrow \begin{pmatrix} a & c & k_1 \\ b & d & k_2 \\ k_1 & k_2 & j \end{pmatrix}, \qquad (2.193)$$

together together with those with external $\circ$ labels given by the permutations

$$(ab)(cd) \mapsto (ac)(bd), (ad)(bc), (bc)(ad), (bd)(ac), (cd)(ab). \qquad (2.194)$$

There are four $\simeq$ equivalent triangle patterns in each of the six classes with these representatives, giving 24 triangle patterns in all for $t = 2$.

The distribution (2.191)-(2.194) into $\simeq$ equivalence classes accounts for all $(12)(8) + (6)(4) = 120 = n!a_n = (24)(5) = 120$ triangle patterns $\mathbb{P}(\mathbf{j}\,\mathbf{k})_j$ of order 4 (see (2.129)) corresponding to the set of labeled binary trees $\{T(\pi(\mathbf{j})\,\mathbf{k})_j \,|\, T \in \mathbb{T}_4, \pi \in S_4\}$.

The left-triangle pattern and the right-triangle pattern in a triangle coefficient of order $2(4) - 2 = 6$ can be any of the 18 representatives of a $\simeq$ equivalence class of triangle patterns of order 4. Thus, there are $(18)(18) = 324$ disjoint equivalent classes of triangle coefficients, representatives of which are given in terms of the representatives of left-triangle patterns and right-triangle patterns under the $\simeq$ operation as follows:

(i). left: 12 of type $t = 1$, right: 12 of type $t = 1$;

(ii). left: 12 of type $t = 1$, right: 6 of type $t = 2$;

(iii). left: 6 of type $t = 2$, right: 12 of type $t = 1$;

(iv). left: 6 of type $t = 2$, right: 6 of type $t = 2$.

Set (i) of representative triangle coefficients contains $(12)(8)(12)(8) = 9,216$ phase-equivalent triangle coefficients; set (ii) contains $(12)(8)(6)(4) = 2,304$ phase-equivalent triangle coefficients; set (iii) contains $(6)(4)(12)(8) = 2,304$ phase-equivalent triangle coefficients; and set (iv) contains $(6)(4)(6)(4) = 576$ phase-equivalent triangle coefficients. This distribution of triangle coefficients into 324 disjoint phase-equivalent classes thus accounts for all $(4!(5))^2 = 14,400$ triangle coefficients (2.147) corresponding to all labeled trees for $T, T' \in \mathbb{T}_4$ and $\pi, \pi' \in S_4$.

The above analysis gives 324 representative triangle coefficients with six columns to consider, three columns in the left-triangle patterns and three columns in the right-triangle pattern. If any column in the left-triangle pattern and any column in the right-triangle pattern have the same entries in row 1 and row 2, we say that the two patterns have a *common fork.* If the left-triangle pattern and the right-triangle pattern have a common fork, then it follows from the expression (2.147) of the triangle coefficient as a summation over a product of six WCG coefficients that the common forks give the factor

$$\left\{ \begin{array}{c|c} x & x \\ y & y \\ k & k' \end{array} \right\} = \sum_{\alpha,\beta} C^{x\ y\ k}_{\alpha\ \beta\ m} C^{x\ y\ k'}_{\alpha\ \beta\ m} = \delta_{k,k'}. \tag{2.195}$$

Thus, whenever the left-triangle pattern and the right-triangle pattern have a common fork, a triangle coefficient with six columns reduces to a triangle coefficient with four columns; that is, to a the Kronecker delta factor $\delta_{k,k'}$ times a triangle coefficient of order four.

A pair of the eighteen representative left-triangle patterns enumerated in (2.191)-(2.194) have a common fork, if and only if the first column in each pattern is a common fork. This gives $(18)(3) = 54$ representative triangle coefficients that reduce to triangle coefficients of lower order. We are left with 270 triangle coefficients of order 6 to consider. All other triangle coefficients of order 6 are related to these by phase factors. Correspondingly, there are 270 recoupling matrices to consider. Further progress on the classification of triangle coefficients and recoupling matrices depends on finding additional elements of structure that bring deeper insight to the problem.

Such progress can be made by considering further special examples that suggest a more general conceptual framework. We consider three examples for $n = 4$ of recoupling matrices taken from Ref. [21] that illustrate such possibilities:

It is the simple multiplication property (2.125) of recoupling matrices that is the basis for relations among triangle coefficients of order $2n-2$.

We need to consider relations such as

$$\begin{aligned} R^{[(ac)(bd);[(ab)c]d}_{T_3,T_1} &= R^{[(ac)(bd);[(ac)b]d}_{T_3,T_1} R^{[(ac)b]d;[b(ac)]d}_{T_1,T_4} \\ &\quad \times R^{[b(ac)]d;[(ba)c]d}_{T_4,T_1} R^{[(ba)c]d;[(ab)c]d}_{T_1,T_1}. \end{aligned} \tag{2.196}$$

This notation is still too encumbered for such products with many factors, expecially when matrix elements are taken.

We introduce some abbreviated notations in order to display the relations of interest by the (arbitrary) assignment of integers as follows:

$$\begin{aligned} 1 &= T_1([(ab)c]d), \quad 2 = T_1([(ab)d]c), \quad 3 = T_1([(ac)b]d), \\ 4 &= T_1([(ba)c]d), \quad 5 = T_2(c[a(bd)]), \quad 6 = T_3((ab)(cd)), \\ 7 &= T_3((ac)(bd)), \quad 8 = T_3((ad)(bc)), \quad 9 = T_3((ca)(bd)), \end{aligned}$$

$$\begin{aligned} 10 &= T_4([b(ac)]d),\ 11 = T_4(c[(ab)d]),\ 12 = T_4([d(ab)]c), \\ 13 &= T_5(d[(ab)c]). \end{aligned} \tag{2.197}$$

We then have the following relations from (2.196)-(2.197) and (2.127):

$$\begin{aligned} W^{7;1} &= W^{7;3}W^{3;10}W^{10;4}W^{4;1} \\ &= W^{7;9}W^{9;5}W^{5;11}W^{11;2}W^{2;12}W^{12;13}W^{13;1}; \end{aligned} \tag{2.198}$$

$$W^{7;6} = W^{7;3}W^{3;10}W^{10;4}W^{4;1}W^{1;6}.$$

The two distinct expressions for $W^{7;1}$ give the Biedenharn-Elliott identity (see Ref. [21]) and that for $W^{7;6}$ gives Wigner's [183] definition of the so-called $9-j$ coefficient. Remarkably, as noted above, these results are expressions of the simple multiplicative property of recoupling matrices given by (2.125).

The evaluation of the matrix elements occurring in the products (2.198) is quite tedious. We illustrate, by two examples, how the calculations are effected, and how the triangle coefficients of order 6 reduce to lower order triangle coefficients of order 4 and 2. The underlying reason for these reductions is, in every instance, that summations can be effected over subsets of WCG coefficients in the explicit expression of triangle coefficients of order 6. It is never required to examine explicitly these sums over WCG coefficients, since it is always evidenced directly in terms of the triangles patterns that occur in the triangle coefficient. There are, however, some subtleties in the standard labelings that occur and to which careful attention must be paid. This is best illustrated by an example:

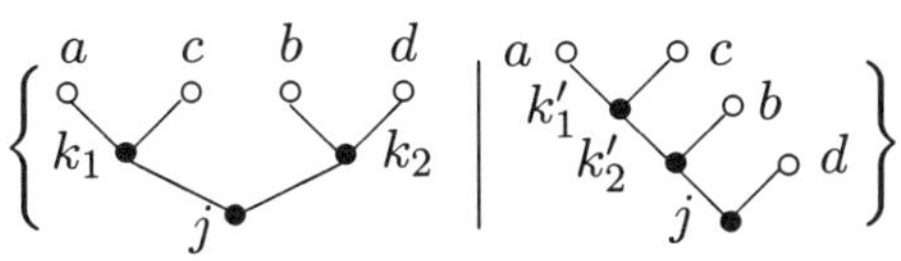

$= \delta_{k_1,k_1'}$ 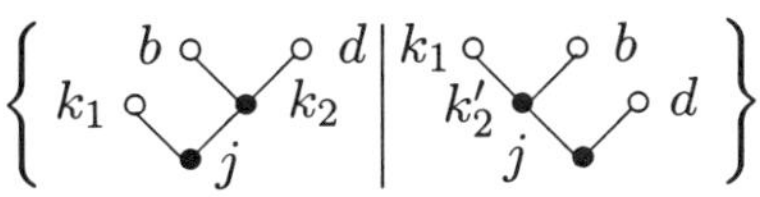

$$= \delta_{k_1,k_1'} \left\{ \begin{matrix} b & k_1\circ \\ d & k_2 \\ k_2 & j \end{matrix} \middle| \begin{matrix} k_1\circ & k_2' \\ b & d \\ k_2' & j \end{matrix} \right\} \tag{2.199}$$

$$= \delta_{k_1,k_1'} \left\{ \begin{matrix} d & k_2 \\ b & k_1\circ \\ k_2 & j \end{matrix} \middle| \begin{matrix} b & d \\ k_1\circ & k_2' \\ k_2' & j \end{matrix} \right\}$$

$$= \delta_{k_1,k_1'} \sqrt{(2k_2+1)(2k_2'+1)}\, W(dbjk_1;k_2k_2').$$

This result shows that when a fork of type col($\circ\ \circ\ \bullet$) is removed by the orthogonality relation (2.195), which in terms of labeled forks is expressed by

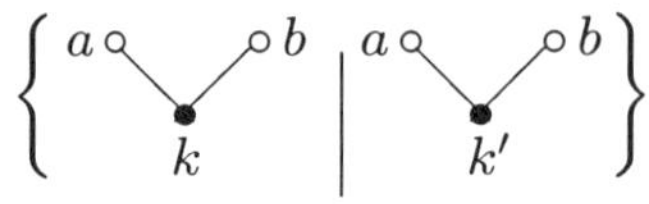

$$= \delta_{k,k'},$$

(2.200)

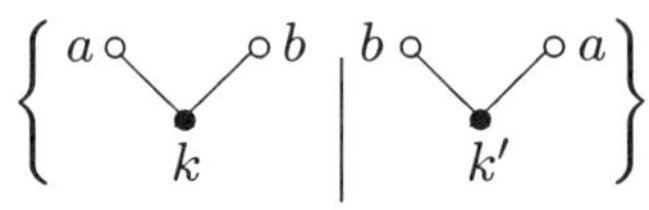

$$= (-1)^{a+b-k}\delta_{k,k'},$$

then, the root of the fork must be replaced by $\circ$. Indeed, this is the inverse process to

$$\circ \mapsto$$  , (2.201)

which represents the bifurcation of an external point $\circ$ into another fork that is adjoined to the tree. We indicate this property by writing a $\circ$ to the right of k_1 in the triangle coefficient in (2.199), as illustrated by col($k_1\circ, k_2, j$) and col($k_1\circ, b, k_2'$), which indicate that these triangles are of type col($\circ\ \bullet\ \bullet$) and col($\circ\ \circ\ \bullet$). The adjacent $\circ$ to k_1 does not change its numerical value; it is only a reminder that this k_1, which was previously a root label, has now become the label $\circ$ of an external point.

We emphasize again that triangle coefficients are a convenient presentation of pairs of standard labeled binary tree coefficients, as defined in terms of generalized WCG coefficients by (2.146)-(2.147). It is the labeled binary trees that are basic, since we have **not** solved the inverse mapping problem of defining a triangle coefficient such that its triangles always define a pair of standard labeled binary trees.

The evaluation (2.199) is for the factor $(W^{7;3}))_{k_1,k_2,j;\,k_1',k_2',j}$ in relation (2.198). The remaining factors are similarly evaluated:

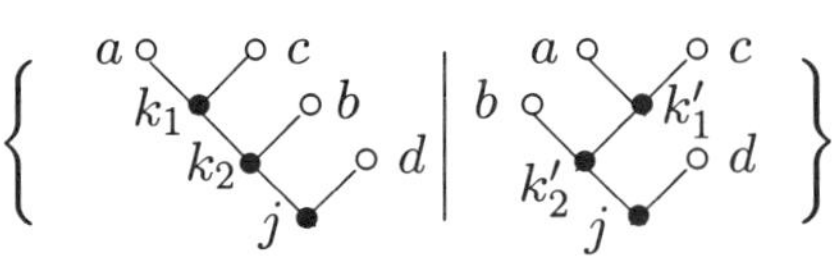

$$= \left(W^{3;10}\right)_{k_1,k_2,j;\,k_1',k_2',j} = \left\{\begin{array}{ccc|ccc} a & k_1 & k_2 & a & b & k_2' \\ c & b & d & c & k_1' & d \\ k_1 & k_2 & j & k_1' & k_2' & j \end{array}\right\}$$

$$= \delta_{k_1,k_1'}\delta_{k_2,k_2'}(-1)^{b+k_1-k_2}; \tag{2.202}$$

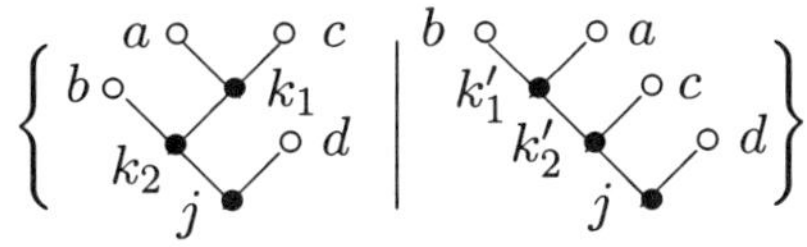

$$= \left(W^{10;4}\right)_{k_1,k_2,j;\,k_1',k_2',j} = \left\{ \begin{array}{ccc|ccc} a & b & k_2 & b & k_1' & k_2' \\ c & k_1 & d & a & c & d \\ k_1 & k_2 & j & k_1' & k_2' & j \end{array} \right\} \qquad (2.203)$$

$$= \delta_{k_2,k_2'} \left\{ \begin{array}{cc|cc} a & b & b & k_1' \\ c & k_1 & a & c \\ k_1 & k_2 & k_1' & k_2 \end{array} \right\} = \delta_{k_2,k_2'} \left\{ \begin{array}{cc|cc} b & k_1' & a & b \\ a & c & c & k_1 \\ k_1' & k_2 & k_1 & k_2 \end{array} \right\}$$

$$= \delta_{k_2,k_2'} \sqrt{(2k_1+1)(2k_1'+1)}\, W(bak_2c;k_1'k_1);$$

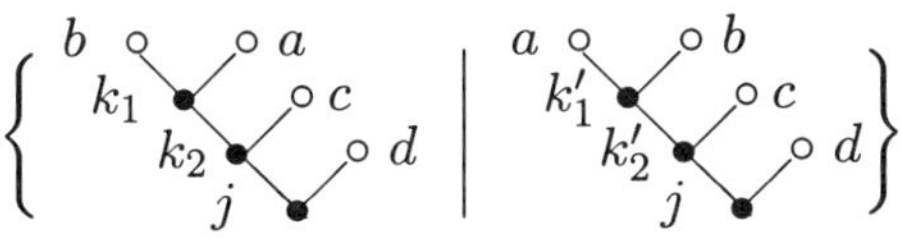

$$= \left(W^{4;1}\right)_{k_1,k_2,j;\,k_1',k_2',j} = \left\{ \begin{array}{ccc|ccc} b & k_1 & k_2 & a & k_1' & k_2' \\ a & c & d & b & c & d \\ k_1 & k_2 & j & k_1' & k_2' & j \end{array} \right\}$$

$$= \delta_{k_1,k_1'}\delta_{k_2,k_2'}(-1)^{a+b-k_1}. \qquad (2.204)$$

Finally, taking matrix elements of the four products in (2.198), and using the above relations (making the adjustments in the intermediate row and column indices that are summed over), we obtain:

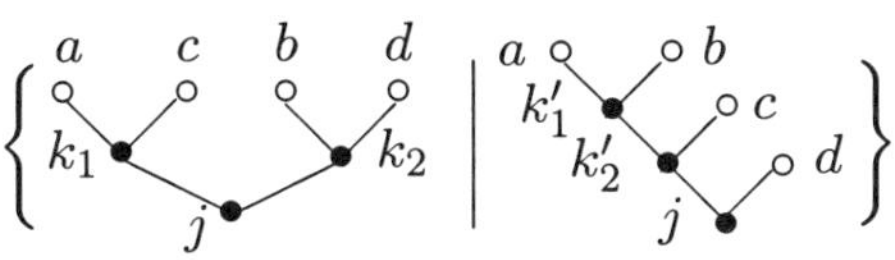

$$= \left(R^{[(ac)(bd);[(ab)c]d}_{T_3,\,T_1}\right)_{k_1,k_2,j;\,k_1',k_2',j} = \left\{ \begin{array}{ccc|ccc} a & b & k_1 & a & k_1' & k_2' \\ c & d & k_2 & b & c & d \\ k_1 & k_2 & j & k_1' & k_2' & j \end{array} \right\}$$

$$= (-1)^{a+k_2'-k_1-k_1'} \left\{ \begin{array}{cc|cc} b & k_1' & a & b \\ a & c & c & k_1 \\ k_1' & k_2' & k_1 & k_2' \end{array} \right\} \left\{ \begin{array}{cc|cc} d & k_2 & b & d \\ b & k_1\circ & k_1\circ & k_2' \\ k_2 & j & k_2' & j \end{array} \right\}$$

$$= (-1)^{a+k_2'-k_1-k_1'} \sqrt{(2k_1+1)(2k_1'+1)(2k_2+)(2k_2'+1)}$$

$$\times W(bak_2'c;k_1'k_1)W(dbjk_1;k_2k_2'). \qquad (2.205)$$

Similar calculations for the second form of $W^{7:1}$ in (2.198) give:

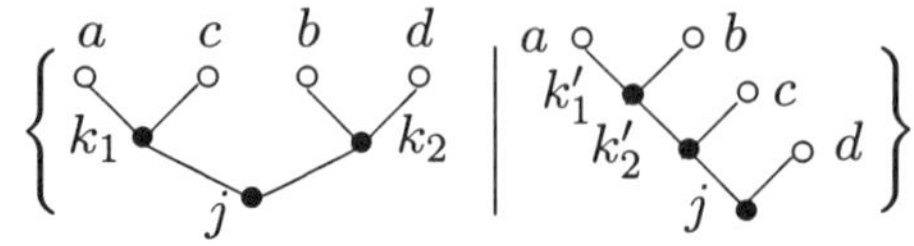

$$= \left(R^{[(ac)(bd);[(ab)c]d}_{T_3,\,T_1}\right)_{k_1,k_2,j;\,k_1',k_2',j} = \left\{\begin{array}{ccc|ccc} a & b & k_1 & a & k_1' & k_2' \\ c & d & k_2 & b & c & d \\ k_1 & k_2 & j & k_1' & k_2' & j \end{array}\right\}$$

$$= (-1)^{a+k_2'-k_1-k_1'} \sum_k \left\{\begin{array}{cc|cc} a & k_1' & b & a \\ b & d & d & k_2 \\ k_1' & k & k_2 & k \end{array}\right\} \tag{2.206}$$

$$\times \left\{\begin{array}{cc|cc} c & k_1 & a & c \\ a & k_2\circ & k_2\circ & k \\ k_1 & j & k & j \end{array}\right\} \left\{\begin{array}{cc|cc} d & k & k_1'\circ & d \\ k_1'\circ & c & c & k_2' \\ k & j & k_2' & j \end{array}\right\}$$

$$= (-1)^{a+k_2'-k_1-k_1'} \sqrt{(2k_1+1)(2k_1'+1)(2k_2+)(2k_2'+1)}$$

$$\times \sum_k (2k+1) W(abkd;k_1'k_2) W(cajk_2;k_1k) W(dk_1'jc;kk_2').$$

The B-E identity in the three notations, paired-tree coefficients, triangle coefficients, and Racah coefficients is expressed by (2.207)-(2.209):

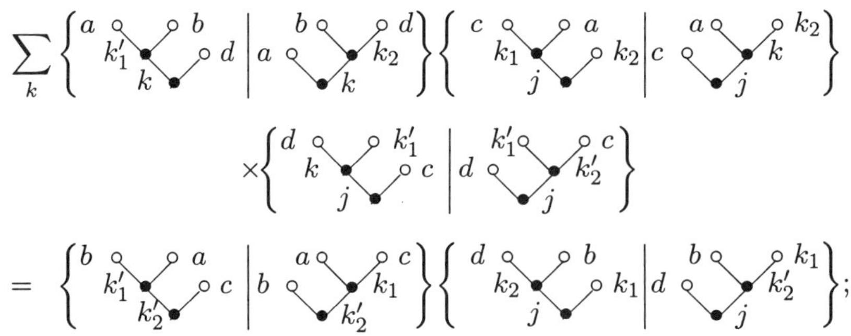

(2.207)

$$\sum_k \left\{\begin{array}{cc|cc} a & k_1' & b & a \\ b & d & d & k_2 \\ k_1' & k & k_2 & k \end{array}\right\} \left\{\begin{array}{cc|cc} c & k_1 & a & c \\ a & k_2\circ & k_2\circ & k \\ k_1 & j & k & j \end{array}\right\} \left\{\begin{array}{cc|cc} d & k & k_1'\circ & d \\ k_1'\circ & c & c & k_2' \\ k & j & k_2' & j \end{array}\right\}$$

$$= \left\{\begin{array}{cc|cc} b & k_1' & a & b \\ a & c & c & k_1 \\ k_1' & k_2' & k_1 & k_2' \end{array}\right\} \left\{\begin{array}{cc|cc} d & k_2 & b & d \\ b & k_1\circ & k_1\circ & k_2' \\ k_2 & j & k_2' & j \end{array}\right\}; \tag{2.208}$$

$$\sum_{k}(2k+1)W(abkd;k_1'k_2)W(cajk_2;k_1k)W(dk_1'jc;kk_2')$$
$$= W(bak_2'c;k_1'k_1)W(dbjk_1;k_2k_2'). \tag{2.209}$$

Wigner's notation for the $9-j$ coefficient, which occurs in the second line below, encodes the labeled forks of the pair of binary trees as rows and columns of a 3×3 array enclosed within brackets in the manner shown. Thus, the $9-j$ coefficient is defined in terms of the triangle coefficent associated with the left-triangle pattern and right-triangle patterns shown, multiplied by the square roots of the root labels of the forks. It is expressed in terms of the matrix elements of the recoupling matrix $W^{7;6}$ defined in (2.198) by adjoining the recoupling matrix $W^{1;6}$ to the right of $W^{7;1}$. The matrix element calculation gives

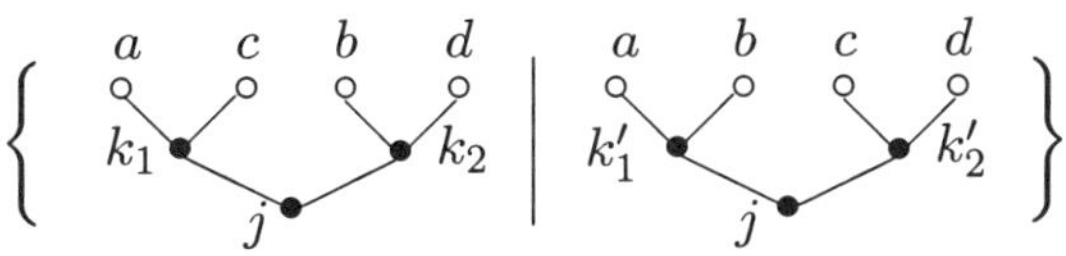

$$= \sqrt{(2k_1+1)(2k_1'+1)(2k_2+1)(2k_2'+1)}\left\{\begin{array}{ccc} a & b & k_1' \\ c & d & k_2' \\ k_1 & k_2 & j \end{array}\right\}$$

$$= \left(W^{7;6}\right)_{k_1,k_2,j;\,k_1',k_2',j} = \left(W^{7:1}W^{1;6}\right)_{k_1,k_2,j;\,k_1',k_2',j}$$

$$= \left(R^{[(ac)(bd);(ab)(cd)]}_{T_3,\,T_3}\right)_{k_1,k_2,j;\,k_1',k_2',j} = \left\{\begin{array}{ccc|ccc} a & b & k_1 & a & c & k_1' \\ c & d & k_2 & b & d & k_2' \\ k_1 & k_2 & j & k_1' & k_2' & j \end{array}\right\}$$

$$= \sum_{k}(-1)^{a+k-k_1-k_1'}\left\{\begin{array}{cc|cc} b & k_1\circ & k_1\circ & k \\ d & k_2 & b & d \\ k_2 & j & k & j \end{array}\right\}$$

$$\times\left\{\begin{array}{cc|cc} b & k_1' & a & b \\ a & c & c & k_1 \\ k_1' & k & k_1 & k \end{array}\right\}\left\{\begin{array}{cc|cc} k_1'\circ & k & c & k_1'\circ \\ c & d & d & k_2' \\ k & j & k_2' & j \end{array}\right\}$$

$$= \sqrt{(2k_1+1)(2k_1'+1)(2k_2+1)(2k_2'+1)}\sum_{k}(-1)^{a+k-k_1-k_1'}$$

$$\times(2k+1)W(dbjk_1;k_2k)W(bakc;k_1'k_1)W(k_1'cjd;kk_2'). \tag{2.210}$$

The computations given above are very tedious, but the underlying recoupling matrix multiplication structure (2.198) of such relations is elementary. *Relations between Racah coefficients can be constructed almost at will simply by writing out such relations.* Summations of products of Racah coefficients can be obtained that can't be reduced to a sum over products of a fewer number, which is the case of the Wigner $9-j$ coefficient, as will be shown. Such relations are encoded by the shapes of the trees appearing in the sequence of multiplications of the recoupling matrices. Let us illustrate this for the cases considered above, including the Racah sum rule, and comment on their significance afterwards:

1. Racah sum rule:
   (a) first path:
$$(ab)c \xrightarrow{A} a(bc) \xrightarrow{C} (bc)a \xrightarrow{A} b(ca) \xrightarrow{C} b(ac) \xrightarrow{C} (ac)b. \tag{2.211}$$
   (b) second path:
$$(ab)c \xrightarrow{C} c(ab) \xrightarrow{A} (ca)b \xrightarrow{C} (ac)b. \tag{2.212}$$
2. Biedenharn-Elliott identity:
   (a) first path:
$$(ac)(bd) \xrightarrow{A} [(ac)b]d \xrightarrow{C} [b(ac)]d \xrightarrow{A} [(ba)c]d \xrightarrow{C} [(ab)c]d. \tag{2.213}$$
   (b) second path:
$$(ac)(bd) \xrightarrow{C} (ca)(bd) \xrightarrow{A} c[a(bd)] \xrightarrow{A} c[(ab)d] \xrightarrow{C} [(ab)d]c \xrightarrow{C} [d(ab)]c \xrightarrow{A} d[(ab)c] \xrightarrow{C} [(ab)c]d. \tag{2.214}$$
3. Wigner $9-j$ coefficient:
   (a) first path:
$$(ac)(bd) \xrightarrow{A} [(ac)b]d \xrightarrow{C} [b(ac)]d \xrightarrow{A} [(ba)c]d \xrightarrow{C} [(ab)c]d \xrightarrow{A} (ab)(cd). \tag{2.215}$$
   (b) second path:
$$[(ab)c]d \xrightarrow{A} [a(bc)]d \xrightarrow{A} a[(bc)d] \xrightarrow{C} a[d(bc)] \xrightarrow{A} a[(db)c] \xrightarrow{C} [(db)c]a \xrightarrow{C} [(bd)c]a. \tag{2.216}$$

These sequences of commutation transformations C and association transformations A from one binary tree shape to another encode various *paths* whereby the final bracketing scheme can be reached from the initial one by using only these elementary operations. The corresponding product of recoupling matrices is then written out directly from these paths. For example, the application of the two paths (2.211) and (2.212) gives

$$\begin{aligned} R^{(ab)c;(ac)b}_{T,T} &= R^{(ab)c;a(bc)}_{T,T'} R^{a(bc);(bc)a}_{T',T} R^{(bc)a;b(ca)}_{T,T'} R^{b(ca);b(ac)}_{T,T'} R^{b(ac);(ac)b}_{T',T} \\ &= R^{(ab)c;c(ab)}_{T,T'} R^{c(ab);(ca)b}_{T',T} R^{(ca)b;(ac)b}_{T,T'}. \end{aligned} \tag{2.217}$$

This relation is just (2.186), rewritten now in terms of recoupling matrices of the form

$$R^{\mathrm{sh}(\pi(\mathbf{j}));\,\mathrm{sh}'(\pi'(\mathbf{j}))}_{T,T'}, \qquad \mathrm{sh}(\pi(\mathbf{j})) \xrightarrow{C} \mathrm{sh}'(\pi'(\mathbf{j})), \text{ or} \tag{2.218}$$

$$\mathrm{sh}(\pi(\mathbf{j})) \xrightarrow{A} \mathrm{sh}'(\pi'(\mathbf{j})),$$

in which the two shapes are related by a commutation or an association operation. Exactly the same analysis applies to the B-E identity. The Wigner $9-j$ expression (2.210) is, of course, a definition of this coefficient, but, just as was the case with the $6-j$ coefficient, there are other ways of writing this coefficient in terms of paths connecting quite different initial and final shapes. The calculation for the second path given by (2.216) gives

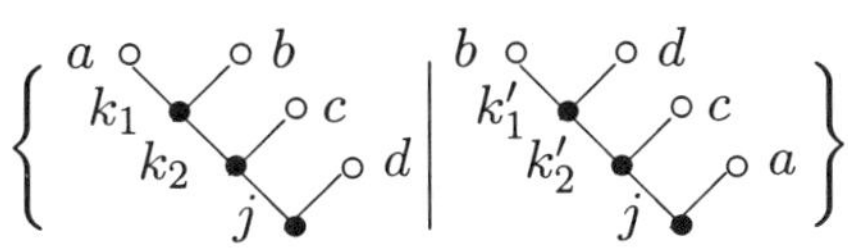

$$= \left(R^{[(ab)c]d;[(bd)c]a}_{T_1,T_1}\right)_{k_1,k_2,j;\,k'_1,k'_2,j}$$

$$= \left\{ \begin{array}{ccc|ccc} a & k_1 & k_2 & b & k'_1 & k'_2 \\ b & c & d & d & c & a \\ k_1 & k_2 & j & k'_1 & k'_2 & j \end{array} \right\} = (-1)^{a+b-k'_1-k_2+k'_2} \tag{2.219}$$

$$\times \sqrt{(2k_1+1)(2k'_1+1)(2k_2+1)(2k'_2+1)} \left\{ \begin{array}{ccc} b & a & k_1 \\ d & j & k_2 \\ k'_1 & k'_2 & c \end{array} \right\}.$$

We must, of course, prove that this is the same $9-j$ coefficient as defined by (2.210), which is shown in Sect. 2.3 (see (2.240)).

It is important to be precise about the meaning of commutations and associations in the above relations: The operations $(xy) = (yx)$ and

$(xy)z = z(xy)$ are commutations, no matter how intricate x, y, and z themselves are, as bracketings. For example, $(ca)x \xrightarrow{A} c(ax)$ for $x = (bd)$ is an association: $(ca)(bd) \xrightarrow{A} c[a(bd)]$.

The results for $n = 4$ above on recoupling matrices demonstrate that new elements of structure are entering the problem, elements of great simplicity: Commutations, marked by C, are always accompanied by a phase-factor change in the associated reoupling matrix, and associations, marked A, are always accompanied by a Racah coefficient transformation in the associated recoupling matrix, including possibly a phase-factor. Because Racah coefficients are not defined as elements of an orthogonal matrix, such transformations are always accompanied by the square root of a pair of dimension factors. A sequence of transformations as indicated by expressions such as (2.211) encode completely the transformation between pairs of labeled binary trees, from which the matrix elements of the product of recoupling matrices can be written out. We can construct recoupling matrix transformations from one standard labeled tree to another almost at will by supplying arbitrary intermediate recoupling matrices both in form and in number. Before addressing this property further in the next section, we complete our preliminary results on the Wigner $9-j$ coefficient.

Let $\mathbf{j} = (j_1, j_2, j_3, j_4) = (a, b, c, d)$ in the following relations. Each recoupling matrix $R_{T,T'}^{\mathrm{sh}(\pi(\mathbf{j}));\,\mathrm{sh}'(\pi'(\mathbf{j}))}$ is an element of the group of unitary matrices $H(N(\mathbf{j}))$, as given by (2.49). For the case at hand, we have:

$$H(N(\mathbf{j})) = \left\{ \sum_{j=j_{\min}}^{j_{\max}} \oplus \left(W_{T,T'}^{(\mathrm{sh}(\pi(\mathbf{j})),j);\,(\mathrm{sh}'(\pi'(\mathbf{j})),j)} \otimes I_{2j+1} \right) \;\middle|\; W_{T,T'}^{(\mathrm{sh}(\pi(\mathbf{j})),j);\,(\mathrm{sh}'(\pi'(\mathbf{j})),j)} \in U(N_j(\mathbf{j})) \right\}. \tag{2.220}$$

Accordingly, the recoupling matrix $R_{T,T'}^{\mathrm{sh}(\pi(\mathbf{j}));\,\mathrm{sh}'(\pi'(\mathbf{j}))}$ has the direct sum expression given by

$$R_{T,T'}^{\mathrm{sh}(\pi(j));\,\mathrm{sh}'(\pi'(j))} = \sum_{j=j_{\min}}^{j_{\max}} \oplus \left(R_{T,T'}^{\mathrm{sh}(\pi(\mathbf{j})),j);\,\mathrm{sh}'(\pi'(\mathbf{j})),j)} \otimes I_{2j+1} \right), \tag{2.221}$$

$$R_{T,T'}^{\mathrm{sh}(\pi(\mathbf{j})),j);\,\mathrm{sh}'(\pi'(\mathbf{j})),j)} \in U(N_j(\mathbf{j})).$$

We recall that the CG number $N_j(\mathbf{j}) = N_j(a, b, c, d)$ is determined by relation (2.37), where $M_j(a, b, c, d)$ is the number of compositions $(l_1, l_2,$

$l_3, l_4)$ of $a+b+c+d-j$ into four nonnegative integer parts such that $0 \leq l_1 \leq 2a, 0 \leq l_2 \leq 2b, 0 \leq l_3 \leq 2c, 0 \leq l_4 \leq 2d$; it is also determined by the multiset $\langle a, b, c, d \rangle$.

In particular, we have that the $9-j$ coefficient, which is defined in terms of the elements of the recoupling matrix by relation (2.210) determines uniquely a real orthogonal matrix $W^{(ac)(bd),j;(ab)(cd),j})_{T,T}$ belonging to the group $U(N_j(a\,b\,c\,d))$ with rows and columns given by

$$\left(W_{T,T}^{(ac)(bd),j;(ab)(cd),j}\right)_{k_1\,k_2;\,k_1'\,k_2'} \tag{2.222}$$
$$= \left\{ \begin{array}{ccc|ccc} a & b & k_1 & a & c & k_1' \\ c & d & k_2 & b & d & k_2' \\ k_1 & k_2 & j & k_1' & k_2' & j \end{array} \right\}$$
$$= \sqrt{(2k_1+1)(2k_1'+1)(2k_2+)(2k_2'+1)} \left\{ \begin{array}{ccc} a & b & k_1' \\ c & d & k_2' \\ k_1 & k_2 & j \end{array} \right\},$$

for all angular momenta such that the six triangle conditions are fulfilled; otherwise, the elements of the matrix are defined to be 0. Thus, the $9-j$ coefficients satisfy the row and column orthogonality relations given by

$$\sum_{k_1,\,k_2} (2k_1+1)(2k_1'+1)(2k_2+)(2k_2'+1) \tag{2.223}$$
$$\times \left\{ \begin{array}{ccc} a & b & k_1' \\ c & d & k_2' \\ k_1 & k_2 & j \end{array} \right\} \left\{ \begin{array}{ccc} a & b & k_1'' \\ c & d & k_2'' \\ k_1 & k_2 & j \end{array} \right\} = \delta_{k_1',k_1''}\delta_{k_2',k_2''},$$

$$\sum_{k_1',\,k_2'} (2k_1+1)(2k_1'+1)(2k_2+)(2k_2'+1) \tag{2.224}$$
$$\times \left\{ \begin{array}{ccc} a & b & k_1' \\ c & d & k_2' \\ k_1 & k_2 & j \end{array} \right\} \left\{ \begin{array}{ccc} a & b & k_1' \\ c & d & k_2' \\ k_1'' & k_2'' & j \end{array} \right\} = \delta_{k_1,k_1''}\delta_{k_2,k_2''}.$$

The $9-j$ coefficient inherits its symmetries from its expression in terms of WCG coefficients, or, equivalently, from its expression in terms of Racah coefficients and their symmetries: the 9−j coefficient

$$\left\{ \begin{array}{ccc} a & b & k_1' \\ c & d & k_2' \\ k_1 & k_2 & j \end{array} \right\} \tag{2.225}$$

is invariant under all even permutations of its rows and columns, under transposition of the 3×3 array, and it is multiplied by the factor

$$(-1)^{a+b+c+d+k_1+k_2+k_1'+k_2'} \tag{2.226}$$

under odd permutations of its rows or columns. Accounting for the symmetry under interchange of rows and column, the orthogonality relations (2.223) and (2.224) are identical.

We turn next to the development of the general properties of recoupling matrices building on the elements of structure observed above for $n = 4$.

2.2.11 Structure of general triangle coefficients and recoupling matrices

We first consider properties of the general triangle coefficients of order $2n-2$ as defined by relation (2.147), which we repeat in terms of the shapes sh and sh$'$ determined by the trees $T \in \mathbb{T}_n$ and $T' \in \mathbb{T}_n$:

$$\left\{\Delta_{(\mathrm{sh}(\pi(\mathbf{j}))\,\mathbf{k})_j} \middle| \Delta_{(\mathrm{sh}'(\pi'(\mathbf{j}))\,\mathbf{k}')_j}\right\} = \left\{\Delta_{(\mathrm{sh}'(\pi'(\mathbf{j}))\,\mathbf{k}')_j} \middle| \Delta_{(\mathrm{sh}(\pi(\mathbf{j}))\,\mathbf{k})_j}\right\}$$

$$= \sum_{\mathbf{m}\in\mathbb{C}(\mathbf{j})} \prod_{i=1}^{n-1} C^{a_i^\pi\ b_i^\pi\ k_i}_{\alpha_i^\pi\ \beta_i^\pi\ \alpha_i^\pi+\beta_i^\pi} \prod_{i=1}^{n-1} C^{a_i^{\pi'}\ b_i^{\pi'}\ k_i'}_{\alpha_i^{\pi'}\ \beta_i^{\pi'}\ \alpha_i^{\pi'}+\beta_i^{\pi'}}. \tag{2.227}$$

The collection of $(n!a_n)^2$ triangle coefficients of order $2n-2$ given by (2.227) can be organized into classes of phase-equivalent triangle coefficients as determined by the exchange operations defined by (2.148)-(2.149) and from properties (2.139)-(2.141) of left-triangle and right-triangle patterns: There are $2^{n-t} \cdot 2^{n-t'}$ phase-equivalent triangle coefficients in the class having a representative of the left-triangle pattern labeled by any of the angular momenta $(j_{\pi_1}, j_{\pi_2}, \ldots, j_{\pi_n})$, $\pi \in S_n^{\mathrm{rep}}(t)$, and a representative the right-triangle pattern labeled by any of the angular momenta $(j_{\pi_1'}, j_{\pi_2'}, \ldots, j_{\pi_n'}), \pi' \in S_n^{\mathrm{rep}}(t')$. Altogether there are $(d_n)^2$ representatives among the full set of $(n!a_n)^2$ triangle coefficients, as given by

$$(d_n)^2 = \left(n! \sum_{t=1}^{[n/2]} \frac{c_n(t)}{2^{n-t}}\right)\left(n! \sum_{t'=1}^{[n/2]} \frac{c_n(t')}{2^{n-t'}}\right). \tag{2.228}$$

If the left-triangle and right-triangle patterns in a triangle coefficient of order $2n-2$ contain a common fork, then the triangle coefficient reduces to one of order $2n-4$ in consequence of the relation:

$$\left\{\begin{array}{c|c} x & x \\ y & y \\ k & k' \end{array}\right\} = (-1)^{x+y-k'} \left\{\begin{array}{c|c} x & y \\ y & x \\ k & k' \end{array}\right\} = \delta_{k,k'}. \tag{2.229}$$

Thus, the new triangle coefficient of order $2n-2$ is the Kronecker delta factor $\delta_{k,k'}$, or this factor with a phase, times the triangle coefficient of order $2n-4$ obtained by striking the pair of columns in (2.229) from the original. This result follows directly from the right-hand side of relation (2.227) in which the sum over the projection quantum numbers of the pair of $SU(2)$ WCG coefficients having the triangles in (2.229) can be carried out to give (2.229), and this also effects the removal of the pair of WCG coefficients. We refer to this property as the *common fork reduction rule.* A *fundamental triangle coefficient* defined below in Sect. 2.3 has no such common fork.

The structure of recouping matrices is intrinsically related to the general operations of commutation and association of symbols, two of the most basic operations in mathematics. These operations are implemented into the bracketing scheme or shape sh of a tree as follows: Let x, y, and z denote any bracketing of p, q, and r letters, respectively, in which the total of $p+q+r=n$ letters are all distinct, and where for a single letter a, the notation (a) denotes a itself. Then, the operations of commutation and association of these bracketings are defined by

$$\begin{aligned} &\text{commutation: } (xy) \rightarrow (yx);\ (x)(yz) \rightarrow (yz)x; \\ &\text{association: } (x)(yz) \rightarrow (xy)(z). \end{aligned} \tag{2.230}$$

The importance of these operations for the set of labeled binary trees in $\mathbb{T}_n$ with n external points is the following:

Shape transformation rule between labeled binary trees: There exists a sequence $S_{sh(\pi(\mathbf{j}));sh'(\pi'(\mathbf{j}))}(C,A)$ of commutations C and associations A that transforms the labeled shape $sh(\pi(\mathbf{j}))$ of the external points of a standard labeled tree $T \in \mathbb{T}_n$ to the labeled shape $sh'(\pi'(\mathbf{j}))$ of the external points of a standard labeled tree $T' \in \mathbb{T}_n$:

$$S_{sh(\pi(\mathbf{j}));sh'(\pi'(\mathbf{j}))}(C,A):\ sh(\pi(\mathbf{j})) \rightarrow sh'(\pi'(\mathbf{j})), \tag{2.231}$$

this result being true for all pairs of tree $T, T' \in \mathbb{T}_n$.

Proof. The proof of this result uses the build-up rule (2.72) and induction on n. Each term in the union (2.72) has the form

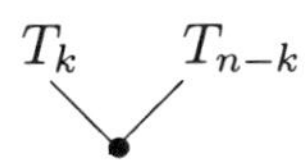

in which $T_k \in \mathbb{T}_k$ and $T_{n-k} \in \mathbb{T}_{n-k}, k = 1, 2, \ldots, n-1$ are labeled trees with a number of external points k and $n - k$. By the induction hypothesis, there exist sequences of commutations and associations such that

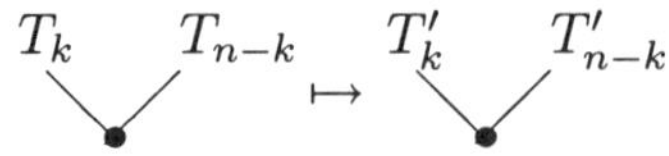

where T_k and T_{n-k} are labeled trees $T_k \in \mathbb{T}_k$ and $T_{n-k} \in \mathbb{T}_{n-k}$ (for brevity we omit the labels). Since this result is true for every each $k = 1, 2, \ldots, n-1$, we obtain a sequence of commutations and associations giving the stated result (2.231). Since the result is true for $n = 2, 3$, it is true for arbitrary n. $\square$

The mapping between shapes and permutations of the external labels of binary trees expressed by (2.231) can be carried one step further. Repeated application gives:

For each pair of labeled binary trees $T_1 = T(\pi(\mathbf{j})\mathbf{k})_j$ *and* $T'_{1'} = T'(\pi'(\mathbf{j})\mathbf{k}')_j$, *the corresponding recoupling matrix can be written as a product of recoupling matrices in which each recoupling matrix in the product has an initial and final shape related by a single commutation operation or by a single association operation. We use the abbreviated notations* $1 = sh(\pi(\mathbf{j}))$ *and* $1' = sh'(\pi'(\mathbf{j}))$. *Then, the recoupling matrix* $R^{1;1'}_{T_1,T'_{1'}}$ *can be written as the following product of recoupling matrices for some positive integer* p :

$$R^{(1;1')}_{T_1,T'_{1'}} = R^{(1;2)}_{T_1;T_2} R^{(2;3)}_{T_2,T_3} \cdots R^{(p-2;p-1)}_{T_{p-2},T_{p-1}} R^{(p-1;1')}_{T_{p-1},T'_{1'}}. \tag{2.232}$$

The abbreviated shapes denoted $(h;k)$ *in the recoupling matrix* $R^{(h;k)}_{T_h,T_k}$ *for* $(h;k) = (1;2), (2;3), \ldots, (p-2;p-1), (p-1;1')$ *are such that the shapes h and k are related by either a single commutation or by a single association of the form (2.230).*

This decomposition rule for an arbitrary recoupling matrix leads immediately to the Sixth Fundamental Result of angular momentum theory:

Sixth Fundamental Result: *Each triangle coefficient of order* $2n-2$ *is either a phase factor times a triangle coefficient of lower order, or it is a summation over a product of triangle coefficients of order four, each of the latter having the form (phase factor)* $\times$ *(Racah coefficient)* $\times$ *(square root of two dimension factors).*

This result is one of the most fundamental results of the binary coupling theory of the addition of angular momentum. It elevates the role of the operations of commutation and association in the algebra and geometry of recoupling matrices to a basic structural principle.

Example. It is useful to give a nontrivial example of the reduction that takes place for a triangle coefficient in which the left-triangle pattern and the right-triangle pattern are related by an association, which we take to be $(xy)z \to x(yz)$, for $x = (ab), y = (cd), x = (ef)g$:

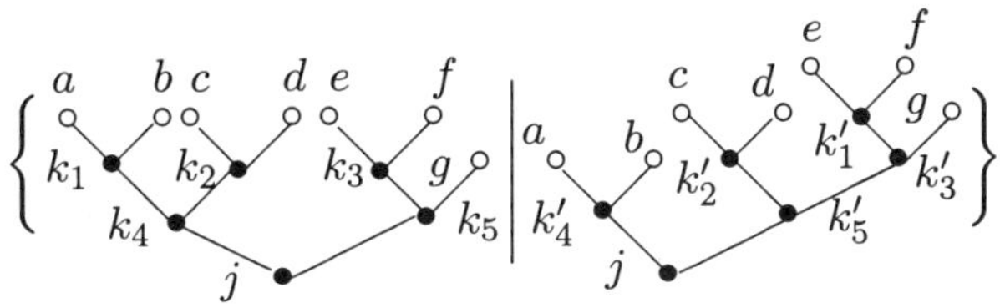

$$= \left(R_{T,T'}^{(xy)z;\,x(yz)} \right)_{\mathbf{k},j;\,\mathbf{k}',j}$$

$$= \left\{ \begin{array}{cccccc|cccccc} a & c & e & k_1 & k_3 & k_4 & e & c & k_1' & a & k_2' & k_4' \\ b & d & f & k_2 & g & k_5 & f & d & g & b & k_3' & k_5' \\ k_1 & k_2 & k_3 & k_4 & k_5 & j & k_1' & k_2' & k_3' & k_4' & k_5' & j \end{array} \right\}$$

$$= \delta_{k_1,k_4'}\delta_{k_2,k_2'}\delta_{k_3,k_1'} \left\{ \begin{array}{ccc|ccc} k_1\circ & k_1'\circ & k_4 & k_1'\circ & k_2\circ & k_1\circ \\ k_2\circ & g & k_5 & g & k_3' & k_5' \\ k_4 & k_5 & j & k_3' & k_5' & j \end{array} \right\}$$

$$= \delta_{k_1,k_4'}\delta_{k_2,k_2'}\delta_{k_3,k_1'}\delta_{k_5,k_3'} \left\{ \begin{array}{cc|cc} k_1\circ & k_4 & k_2\circ & k_1\circ \\ k_2\circ & k_3'\circ & k_3'\circ & k_5' \\ k_4 & j & k_5' & j \end{array} \right\}$$

$$= \delta_{k_1,k_4'}\delta_{k_2,k_2'}\delta_{k_4,k_1'}\delta_{k_5,k_3'}$$

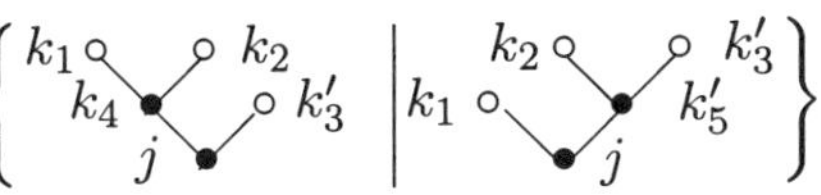

$$= \delta_{k_1,k_4'}\delta_{k_2,k_2'}\delta_{k_4,k_1'}\delta_{k_5,k_3'} \sqrt{(2k_4+1)(2k_5'+1)}\, W(k_1k_2jk_3';k_4k_5'). \quad (2.233)$$

This example illustrates nicely all aspects of the reduction rule that occurs for an association of the form $(xy)z \to x(yz)$, including the convention of writing a $\circ$ adjacent to a fork label upon removal of that fork (see (2.201)). Indeed, the pair of labeled binary trees in the coefficient on the left-hand side of (2.233) with which we began the reduction is obtained from the pair of binary trees in the coefficient on the right-side by adjoining the labeled subtrees

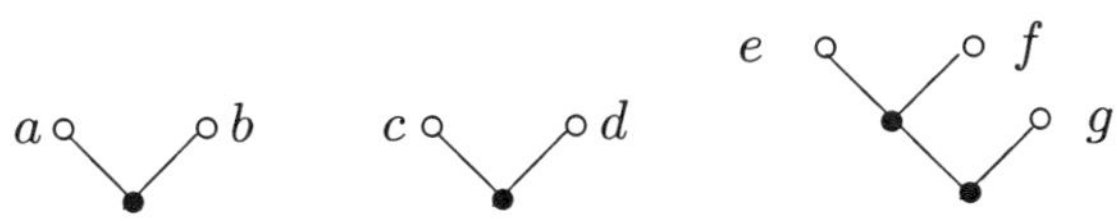

to the ∘ points marked k_1, k_2, k_3', respectively, and changing these ∘ points to • points (see (2.201)). The root labels of the forks in the resulting expanded labeled binary tree are then adjusted, as necessary, so that they are all standard (this accounts for the Kronecker delta factors). The process given by (2.233) is, of course, just the inverse of this. This shows how to go directly from $(xy)z \to x(yz)$, for $x = (ab), y = (cd), x = (ef)g$, without the necessity of providing all the intermediate steps given in the example.

2.3 Classification of Recoupling Matrices

As shown in the prevous section, there abounds an unlimited number of ways of writing a given recoupling matrix in terms of products of other recoupling matrices. In consequence of the Sixth Fundamental Result, this result is now expressed in terms of identities between triangle coefficients of order four, or, equivalently, between Racah coefficients with adjoined dimension factors and phases. This raises the question of how to classify such relations, and, indeed, what is meant by the term "classification," issues to which we now turn.

Recoupling matrices are matrices of order $N(\mathbf{j}) = \prod_{i=1}^{n}(2j_i + 1)$ defined by

$$R_{T,T'}^{\mathrm{sh}(\pi(\mathbf{j}));\,\mathrm{sh}'(\pi'(\mathbf{j}))} = C_T^{\mathrm{sh}(\pi(\mathbf{j}))}\left(C_{T'}^{\mathrm{sh}'(\pi'(\mathbf{j}))}\right)^{\mathrm{tr}}. \tag{2.234}$$

Here, $C_T^{\mathrm{sh}(\pi(\mathbf{j}))}$ is a real orthogonal matrix that brings the Kronecker product $D^{(\mathbf{j})}(U)$ to standard Kronecker direct sum form (2.33)-(2.34) for the coupling scheme corresponding to the labeled binary tree $(\mathrm{sh}(\pi(\mathbf{j}))\,\mathbf{k})_j$, and $C_{T'}^{\mathrm{sh}'(\pi'(\mathbf{j}))}$ is a second such real orthogonal matrix for the coupling scheme corresponding to the labeled binary tree $(\mathrm{sh}'(\pi'(\mathbf{j}))\,\mathbf{k}')_j$. The matrix elements of this recoupling matrix are given in terms of the triangle coefficient defined over the left-triangle pattern $\Delta_{(\mathrm{sh}(\pi(\mathbf{j}))\,\mathbf{k})_j}$ and the right-triangle pattern $\Delta_{(\mathrm{sh}'(\pi'(\mathbf{j}))\,\mathbf{k}')_j}$ by

$$\begin{aligned}\left(R_{T,T'}^{\mathrm{sh}(\pi(\mathbf{j}));\,\mathrm{sh}'(\pi'(\mathbf{j}))}\right)_{\mathbf{k},j;\mathbf{k}',j} &= \langle(\mathrm{sh}(\pi(\mathbf{j}))\,\mathbf{k})_j \mid (\mathrm{sh}'(\pi'(\mathbf{j}))\,\mathbf{k}')_j\rangle \\ &= \{\Delta_{(\mathrm{sh}(\pi(\mathbf{j}))\,\mathbf{k})_j} \mid \Delta_{(\mathrm{sh}'(\pi'(\mathbf{j}))\,\mathbf{k}')_j}\},\end{aligned} \tag{2.235}$$

where the middle term is the bra-ket inner product of the coupled state vectors $|(\mathrm{sh}(\pi(\mathbf{j}))\,\mathbf{k})_j\rangle$ and $|(\mathrm{sh}'(\pi'(\mathbf{j}))\,\mathbf{k}')_j\rangle$.

We next summarize the principal properties of general recoupling matrices of order $N(\mathbf{j}) = \prod_{i=1}^{n}(2j_i + 1)$ in Items 1-4, followed by some new terminology in Items 5-11 to describe these properties:

1. There are $(n!a_n)^2$ recoupling matrices of order $N(\mathbf{j})$ whose matrix elements, the triangle coefficients of order $2n-2$, are partitioned into d_n^2 equivalence classes consisting of phase-equivalent triangle coefficients, there being $2^{n-t}2^{n-t'}$ such phase-equivalent triangle coefficients in the same class for a left-triangle pattern of type t and a right-triangle pattern of type t'.

2. We take the representatives of the equivalence classes in Item 1 to be those having left-triangle patterns and right-triangle patterns with the external angular momenta j_{π_i} indexed by the sets $S_n^{\mathrm{rep}}(t)$ and $S_n^{\mathrm{rep}}(t')$, as described in (2.138)-(2.141).

3. The subgroup properties of recoupling matrices are those stated in relations (2.220)-(2.221), now applied to general $\mathbf{j} = (j_1, j_2, \ldots, j_n)$.

4. Each recoupling matrix can be factored into a product such that each recoupling matrix in the product is that for a pair of labeled trees whose labeled shapes are related by either a commutation operation or by an association operation. This gives the result that the matrix elements of every recoupling matrix can be written as a summation over a phase factor times a product of triangle coefficients of order four.

5. A recoupling matrix $R_{T,T'}^{\mathrm{sh}(\pi(\mathbf{j}));\,\mathrm{sh}'(\pi'(\mathbf{j}))}$ is called *fundamental* if the labeled trees $T((\pi(\mathbf{j})\,\mathbf{k})_j$ and $T'((\pi'(\mathbf{j})\,\mathbf{k}')_j$ are representatives of their respective $\simeq$ equivalence classes as described in Items 1 and 2, and if, in addition, they contain no common forks; that is, no labeled forks with triangles of the form $\mathrm{col}(x\,y\,k)$ and $\mathrm{col}(x\,y\,k')$ (or $\mathrm{col}(y\,x\,k')$), respectively. The matrix elements of fundamental recoupling matrices are necessarily *fundamental triangle coefficients;* that is, triangle coefficients in which the left-tree pattern and the right-tree pattern are those described in Item 2, together with the property that the left-tree pattern and the right-tree pattern contain no common fork columns.

6. Two labeled forks in a pair of labeled binary trees of order n are said to be *adjacent* if the triangles associated with the forks contain a common symbol. The same terminology is applied to the corresponding triangle coefficient of order $2n-2$. Adjacent forks may or may not be contiguous columns in the triangle coefficient.

7. The *adjacency diagram* of a fundamental triangle coefficient is a labeled graph that has $2n-2$ labeled points and $3n-3$ labeled

lines in the Cartesian plane $\mathbb{R}^2$, obtained by the following mapping rule: The labeled points are those of the $2n-2$ distinct triangles, which are sequences of length three. Each pair of adjacent triangles is joined by a labeled line, where the label of the line is the one common to the adjacent triangles. Since there are $n-1$ triangle columns in the left-triangle pattern and in the right-triangle pattern, and adjacent triangles only occur between left-triangle and right-triangle patterns in a fundamental triangle coefficient, there are $3n-3$ such labeled lines. Moreover, there are exactly 3 lines incident on each point. The points of this graph can be placed at arbitrarily positions in the plane, but it is convenient to organize the $n-1$ points associated with the left-triangle pattern into a *top configuration,* and the $n-1$ points associated with the right-triangle pattern into a *bottom configuration,* (see examples (2.237), (2.240), (2.243) below). Such configurations of points and lines are *cubic graphs,* which are defined more precisely in the next chapter.

8. A *path* of a fundamental recoupling matrix $R_{T,T'}^{\mathrm{sh}(\pi(\mathbf{j}));\,\mathrm{sh}'(\pi'(\mathbf{j}))}$ is any sequence of commutations and associations mapping the left shape to the right shape:

$$S_{\mathrm{sh}(\pi(\mathbf{j}));\mathrm{sh}'(\pi'(\mathbf{j}))}(C,A):\ \mathrm{sh}(\pi(\mathbf{j}))\rightarrow \mathrm{sh}'(\pi'(\mathbf{j})). \tag{2.236}$$

   Examples of paths are given in (2.211)-(2.216).

9. The *length of the path* $S_{\mathrm{sh}(\pi(\mathbf{j}));\mathrm{sh}'(\pi'(\mathbf{j}))}(C,A)$ is the number of associations A that occurs in the sequence in Item 8, and is denoted $L_{\mathrm{sh}(\pi(\mathbf{j})),\mathrm{sh}'(\pi'(\mathbf{j}))}$. The path of minimal length is denoted $\underline{S}_{\mathrm{sh}(\pi(\mathbf{j})),\mathrm{sh}'(\pi'(\mathbf{j}))}$ and has length $\underline{L}_{\mathrm{sh}(\pi(\mathbf{j})),\mathrm{sh}'(\pi'(\mathbf{j}))}$.

10. A pair of fundamental recoupling matrices is said to be an *isometric pair* if they possess paths of the same minimal length, this terminology being also applied to fundamental triangle coefficients.

11. An isometric pair of recoupling matrices is said to be an *isomorphic pair* if the $2n-2$ triangles in each of the corresponding isometric pair of triangle coefficients can be put into one-to-one correspondence such that the adjacency relations between the $2n-2$ triangles is preserved; that is, an isometric pair of recoupling matrices is isomorphic if and only if they have the same adjacency diagram.

Examples. We illustrate the concept of isomorphic pairs of recoupling coefficients for $n = 3, 4$:

$n = 3$: There is a single adjacency diagram, as given by

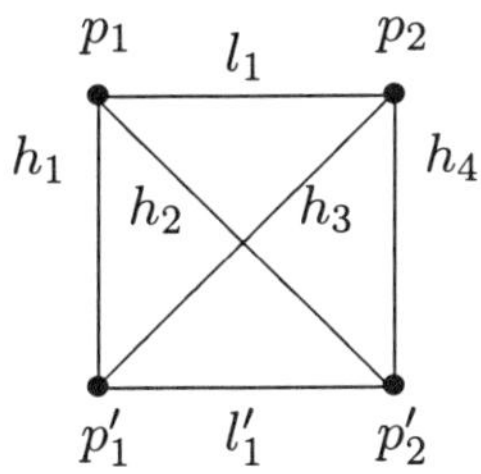

(2.237)

If we give the points in this diagram the labels associated with the recoupling matrix $R^{(ab)c;a(bc)}_{T,T'}$ in relation (2.155); that is, the assignment

$$\begin{aligned} &p_1 = (a\,b\,k),\ p_2 = (k\,c\,j);\ p'_1 = (b\,c\,k'),\ p'_2 = (a\,k'\,j), \\ &l_1 = k,\ l'_1 = k';\ h_1 = b,\ h_2 = a,\ h_3 = c,\ h_4 = j, \end{aligned} \tag{2.238}$$

then we obtain the adjacency diagram shown. If we select any other recoupling coefficient, such as $R^{(ab)c;(ac)b}_{T,T}$, and make the assignment

$$\begin{aligned} &p_1 = (a\,b\,k),\ p_2 = (k\,c\,j);\ p'_1 = (a\,c\,k'),\ p'_2 = (k'\,b\,j) \\ &l_1 = k,\ l_2 = k';\ h_1 = a,\ h_2 = b,\ h_3 = c,\ h_4 = j, \end{aligned} \tag{2.239}$$

then we obtain the same adjacency diagram. These two fundamental recoupling coefficients are isometric and isomorphic. Indeed, for $n = 3$, all fundamental recoupling coefficients are isometric and isomorphic, which explains why there is only one Racah coefficient.

$n = 4$: The concept of isomorphic pairs of recoupling matrices is nicely illustrated by the two expressions (2.210) and (2.219) for the $9 - j$ coefficient. The recoupling matrices $R^{(ac)(bd):(ab)(cd)}$ and $R^{[(ab)c]d;[(bd)c]a}$ are not only isometric, as shown by the sequences of commutations and associations (2.215) and (2.216), which are minimal, but also their sets of triangles can also be put into one-to-one correspondence such that the adjacency relations are preserved. This latter property is most easily shown from the following adjacency diagram:

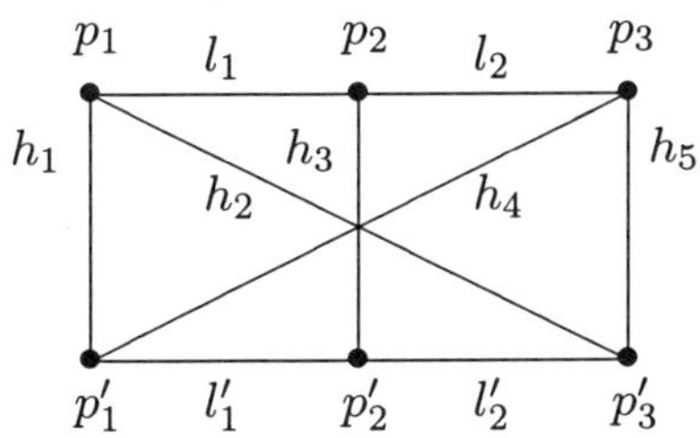

(2.240)

where the point and line labels in these two diagrams for the recoupling matrix $R^{[(ab)c]d;[(bd)c]a}$ (see (2.219)) and $R^{(ac)(bd):(ab)(cd)}$ (see(2.210)) are:

$$\begin{aligned} &p_1 = (a\, b\, k_1),\ p_2 = (k_1\, c\, k_2),\ p_3 = (k_2\, d\, j),\\ &p'_1 = (b\, d\, k'_1),\ p'_2 = (k'_1\, c\, k'_2),\ p'_3 = (k'_2\, a\, j),\\ &l_1 = k_1,\ l_2 = k_2;\ l'_1 = k'_1,\ l'_2 = k'_2,\\ &h_1 = b,\ h_2 = a,\ h_3 = c,\ h_4 = d,\ h_5 = j; \end{aligned} \tag{2.241}$$

$$\begin{aligned} &p_1 = (a\, c\, k_1),\ p_2 = (k_1\, k_2\, j),\ p_3 = (b\, d\, k_2),\\ &p'_1 = (c\, d\, k'_2),\ p'_2 = (k'_1\, k'_2\, j),\ p'_3 = (a\, b\, k'_1),\\ &l_1 = k_1,\ l_2 = k_2;\ l'_1 = k'_2,\ l'_2 = k'_1,\\ &h_1 = c,\ h_2 = a,\ h_3 = j,\ h_4 = d,\ h_5 = b. \end{aligned} \tag{2.242}$$

Thus, the recoupling matrices $R^{[(ab)c]d;\,[(db)c]a}_{T_1,\,T_1}$ and $R^{(ac)(bd);\,[(ab)(cd)}_{T_3,\,T_3}$ are isomorphic, since they are fundamental, isometric, and have the same adjacency diagram. This result suggests that isomorphic pairs of recoupling matrices define the same angular momentum coefficient, up to a phase factor. This is pursued, in depth, in the next chapter.

The structural difference between the recoupling matrix $R^{(ac)(bd);[(ab)c)d}_{T_3,T_1}$ giving rise to the B-E identity and those recoupling matrices giving rise to different expressions for the $9-j$ coefficient in now made apparent from their distinct adjacency diagrams. The adjacency diagram for the recoupling matrix recoupling matrix $R^{(ac)(bd);[(ab)c)d}_{T_3,T_1}$ is verified from (2.206) to be

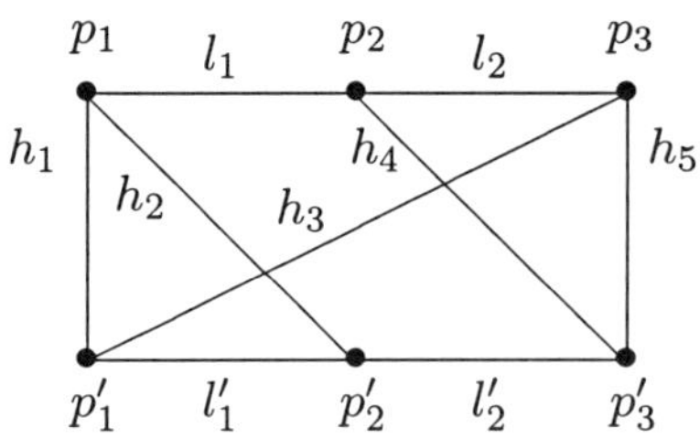

(2.243)

with the point-line assignment

$$\begin{aligned} &p_1 = (a\, c\, k_1),\ p_2 = (k_1\, k_2\, j),\ p_3 = (b\, d\, k_2),\\ &p'_1 = (a\, b\, k'_1),\ p'_2 = (k'_1\, c\, k'_2),\ p'_3 = (k'_2\, d\, j),\\ &l_1 = k_1,\ l_2 = k_2;\ l'_1 = k'_1,\ l'_2 = k'_2,\\ &h_1 = a,\ h_2 = c,\ h_3 = b,\ h_4 = j,\ h_5 = d. \end{aligned} \tag{2.244}$$

The nonisomorphism of the adjacency diagrams (2.240) and (2.243) for the $9-j$ coefficient and the Biedenharn-Elliott identity is a geometric realization of the fact that there are exactly two nonisomorphic cubic graphs on six points. each of which is realized by these two diagrams. Every fundamental recoupling matrix for the coupling of four angular momenta has an adjacency diagram that is isomorphic to one of these two diagrams. We will appeal in the next chapter to the theory of cubic graphs to show this, as well as to explain the factoring property on the right-hand side of the B-E identity (2.209). □

Examples can be misleading as well as revealing. The top and bottom lines in the adjacency diagrams above originating from the left-triangle patterns and the right-triangle patterns, respectively, do not generalize in the obvious way as a single line of points with adjoining incident lines. This is illustrated by the following example from $n = 6$:

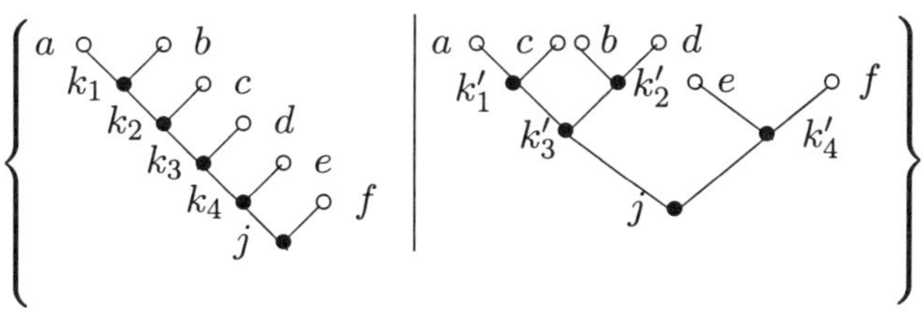

(2.245)

→

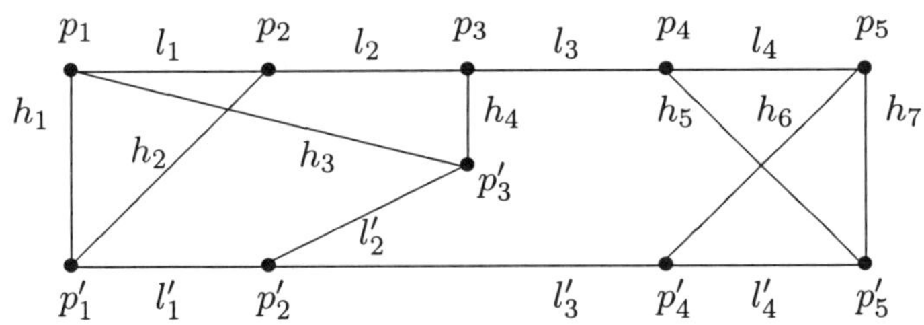

where we have the following definition of points and lines:

$$
\begin{aligned}
&p_1 = (a\,b\,k_1),\; p_2 = (k_1\,c\,k_2),\; p_3 = (k_2\,d\,k_3),\; p_4 = (k_3\,e\,k_4),\\
&p_5 = (k_4\,f\,j);\\
&p'_1 = (a\,c\,k'_1),\; p'_2 = (k'_1\,k'_2\,k'_3),\; p'_3 = (b\,d\,k'_2),\; p'_4 = (k'_3\,k'_4\,j),\\
&p'_5 = (e\,f\,k'_4); \qquad (2.246)\\
&l_1 = k_1,\; l_2 = k_2,\; l_3 = k_3,\; l_4 = k_4;\; l'_1 = k'_1,\; l'_2 = k'_2,\\
&l'_3 = k'_3,\; l'_4 = k'_4;\\
&h_1 = a,\; h_2 = c,\; h_3 = b,\; h_4 = d,\; h_5 = e,\; h_6 = j,\; h_7 = f.
\end{aligned}
$$

The subset of points $\{p_1, p_2, p_3, p_4, p_5\}$ and lines $\{l_1, l_2, l_3, l_4, l_5\}$ and the subset of points $\{p'_1, p'_2, p'_3, p'_4, p'_5\}$ and lines $\{l'_1, l'_2, l'_3, l'_4, l'_5\}$ are each graphs known as *trivalent trees;* that is, the points and lines associated with the forks of the left-triangle pattern and the points and lines associated with the forks of the right-triangle pattern, each define a trivalent tree. Moreover, the full graph is a labeled cubic graph on ten points. *It is this pair of inter-related graph structures that generalizes and allows us to continue with the analysis of the properties of recoupling coefficients.*

General classes of isomorphic fundamental recoupling matrices can be identified. We introduce the symbol $\cong$ to denote that two fundamental recoupling matrices are isomorphic. Then, we have the isomorphism between fundamental recoupling matrices given by

$$R_{T,T'}^{\mathrm{sh}(\pi(\mathbf{j}));\,\mathrm{sh}'(\pi'(\mathbf{j}))} \cong R_{T,T'}^{\mathrm{sh}(\mathbf{j});\,\mathrm{sh}'(\pi''(\mathbf{j})}, \quad \pi'' = \pi'\pi^{-1}, \tag{2.247}$$

since the two recoupling matrices are isometric, and the relationship between the pair of labels $\mathrm{sh}(\pi(\mathbf{j}))$ and $\mathrm{sh}'(\pi'(\mathbf{j}))$ is one-to-one with the relationship between the pair of labels $\mathrm{sh}((\mathbf{j}))$ and $\mathrm{sh}'(\pi''(\mathbf{j})$; hence, the point-to-line relationship in the two adjacency diagrams is preserved. From the isomorphism given by (2.247), it follows that the set of fundamental recoupling matrices given by

$$R_{T,T'}^{\mathrm{sh}(\mathbf{j});\,\mathrm{sh}'(\pi'(\mathbf{j}))}, \text{ all } \pi' \in S_n; \text{ all } T, T' \in \mathbb{T}_n \tag{2.248}$$

includes all possible nonisomorphic recoupling matrices. This does not preclude the possibility that there are still isomorphic recoupling matrices in the set (2.248) such as demonstrated for the $9-j$ coefficient. It is also true that

$$R_{T,T'}^{\mathrm{sh}(\pi(\mathbf{j}));\,\mathrm{sh}'(\pi'(\mathbf{j}))} \cong R_{T',T}^{\mathrm{sh}'(\pi'(\mathbf{j}));\,\mathrm{sh}(\pi(\mathbf{j}))}, \tag{2.249}$$

since the interchange of left-triangle patterns and right-triangle patterns simply interchanges the pair of trivalent trees, thus preserving all line-to-point relationships in the corresponding adjacency diagrams.

We can now state what we mean by the classification of fundamental recoupling matrices:

Partition the set of fundamental recoupling matrices into equivalence classes of isomorphic recoupling matrices.

This is a very difficult classification scheme to implement: We have no algorithm for determining when a pair of fundamental recoupling matrix is isometric, and no algorithm for determining when two isometric recoupling matrices have the same adjacency diagram; that is, are isomorphic. But further progress can still be made by appealing to the theory of trivalent trees and cubic graphs.

Chapter 3

Binary Trees, Trivalent Trees, Cubic Graphs, and Adjacency Diagrams

3.1 Binary Trees and Trivalent Trees

We have given in the last section a mapping of representatives of phase-equivalent classes of pairs of fundamental (labeled) binary trees associated with binary angular momentum coupling schemes to what we have called adjacency diagrams. Underlying this pair mapping is the mapping of a single unlabeled binary tree to a subgraph called a trivalent tree. The rule for this mapping is very simple: We perform the following operation on each binary tree of order n.

Binary tree $\mapsto$ *trivalent tree mapping rule: Remove all external* $\circ$ *points and their incident lines from each* $T \in \mathbb{T}_n$.

This rule gives a graph containing $n-1$ points and $n-2$ lines, where each point has either $1, 2$, or 3 incident lines. Such a graph is called a *trivalent tree.* Their properties and enumeration were already studied by Cayley [39]. If the configuration of points in the original trivalent tree is left undisturbed in effecting the mapping, various arrays of 'bent' lines occur, which we straighten, maintaining the adjacency of the $\bullet$ points. Trivalent trees that preserve adjacency of points in applying this rule are taken as equal. Quite generally, *two points in a graph are called adjacent if there is a line joining them, and two graphs are called isomorphic if their points can be put into one-to-one correspondence such that the adjacency of points is preserved.* The arrangement of the points may be diagrammed in any manner whatsoever that is convenient for visual description. Thus, the binary trees given in Sect. 2.2.1 map, under the binary tree mapping rule, to the following nonisomorphic trivalent trees, where all points are arranged into diagrams that "grow" from left-to-right with n. This orientation rule is to obtain agreement with the adjacency diagrams in Sect. 2.3. The number n is the number of points in the trivalent tree, hence, the parent binary tree belongs to $\mathbb{T}_{n+1}$:

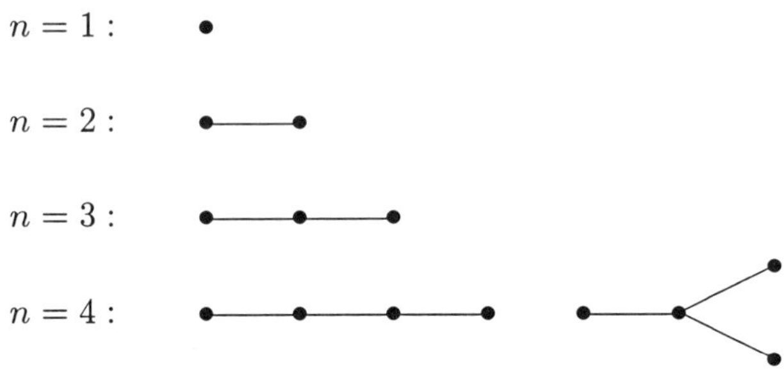

(3.1)

If T and T' are isomorphic trivalent trees of order n, we write $T \cong T'$. We denote the set of nonisomorphic trivalent trees on n points by $\mathbb{V}_n$, and their number by $v_n = |\mathbb{V}_n|$.

The unlabeled trivalent trees having n points for $n = 2 - 11$ may be found in Harary [77] and Harary and Palmer [78] (see also Ref. [21]). The number v_n of such trivalent trees is obtained from Table A3 in Harary by striking from the listed set for $p = n$ all those trees that have four or more incident lines on any point. The numbers so obtained are:

n	2	3	4	5	6	7	8	9	10	11
v_n	1	1	2	2	4	6	11	18	37	66

(3.2)

It is important to observe that we are counting **unlabeled** trivalent trees. While we may label the n points of a given trivalent tree by $1, 2, \ldots, n$ to keep account of adjacency, this is followed by the determination of the equivalence classes (isomorphic) trivalent trees corresponding to this labeling, and then the labels $1, 2, \ldots, n$ are removed. This process is not that of determining the subset of permutations of $1, 2, \ldots, n$ that give inequivalent labeled trivalent trees in the sense of labeled graphs defined on p. 108. Our graphs carry labels, but the classification is not into sets of inequivalent labeled graphs. To keep the distinction clear, we henceforth use the underline <u>labeled</u> to designate labeled graphs in the sense of the technical definition.

Every trivalent tree in the set $\mathbb{V}_n$ has the general configuration given by the following diagram with points labeled by the rules described below in (3.3)-(3.8):

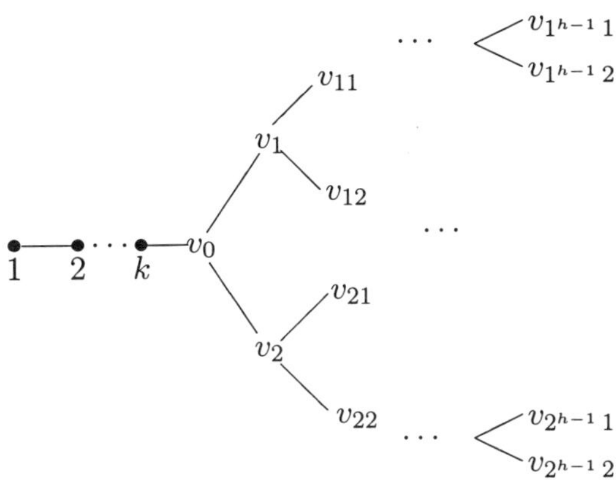

(3.3)

The subscript w attached to each symbol v_w is a word on the two letters 1 and 2, described in detail in (3.4) below. Each symbol v_w appearing in this diagram is either a vertex point of degree 3, an end point of degree 1 from which no lines emanate to the right, or an empty point $\emptyset$ of degree 0, which is not to appear in the diagram. The points marked $v_w, w \in \mathbb{W}_h$, where $\mathbb{W}_h$ is the set of words of length h on the letters 1 and 2 are distributed such that there are d internal points of degree 3; $d+1$ external points of degree 1; and $2^{h+1}-2d-2$ empty points. The collection of $2d+1\,(d \geq 1)$ nonempty points is to have the configuration of a binary tree of order $d+1$ (number of external points) that grows to the right with increasing d. Also, the set of $2d$ lines incident on this collection $\{v_w\}$ of $2d+1$ nonempty points contains $n-k-2d-1 \geq 0$ additional points (not shown in (3.3)), p of which are to belong to the $d-1$ internal lines (lines between internal points) and q to the $d+1$ external lines (lines with an end point), where $p+q = n-k-2d-1$.

The subtree of (3.3) from which the k $\bullet$ points $1, 2, \ldots, k$ and associated k lines to the left of v_0 are removed, and which includes no additional points on its internal and external lines, is called the *binary tree skeleton* T_{d+1} of the trivalent tree V_n. We write $\mathbb{T}_{d+1} \subseteq \mathbb{V}_n$.

The vertex points v_w in (3.3) are listed by natural page ordering:

$$\begin{gathered}0<1<2<11<12<21<22<111<112<121< \\ 122<211<212<221<222<1111<1112<1121< \\ 1122<1211<1212<1221<1222<2111<2112< \\ 2121<2122<2211<2212<2221<2222<\cdots\end{gathered} \tag{3.4}$$

Thus, column $m = 0, m = 1, m = 2, \ldots, d$ in diagram (3.3) use, respectively, the word 0, the words 1 and 2, the words 11, 12, 21, 22, etc. to index the points v_w, where the least word is at the top and the greatest word at the bottom. We always label the point in column $m = 0$ by $v_0 = 0$. We denote by $\mathbb{W}_m$ the set of words of length m on the letters (integers) 1 and 2, so that the cardinality of $\mathbb{W}_m$ is $|\,\mathbb{W}_m\,| = 2^m$. The points $v_w, w \in \mathbb{W}_m$, are those appearing in column m, where $m = 0, 1, 2, \ldots, h$. The notations 1^s and 2^s denote that the integers 1 and 2 are repeated s times with $s = 0$ meaning no occurrence. Since there are 2^m points appearing in column m of the diagram, there are $2^0 + 2^1 + 2^2 + \cdots + 2^h = 2^{h+1} - 1$ such points in the diagram (3.3), but $2^{h+1} - 2d - 2$ of them are the empty point, which do not appear in the diagram. We call the integer $h \geq 1$, the *height* of the binary tree skeleton. While many of the points v_w in (3.3) are the empty point, each of the columns $m = 0, 1, 2, \ldots, h$ must contain at least one nonempty point for $n \geq k + 3$ to have a binary tree skeleton present. The distribution of the additional points on the lines of the skeleton $\mathbb{T}_{d+1} \subseteq \mathbb{V}_n$ is described in detail below in (3.8), using the concept of a compound fork

The k • points belonging to the horizontal line to the left of v_0 in diagram (3.3) define the *base line* and are called *base points*.The baseline parameter k can be any value $k \geq 1$ (see (3.1)).

The parameters n, d, and h are inter-related through their domains of definition, as follows: (i). The domain of definition of h for given d is

$$h = \underline{h}, \underline{h} + 1, \ldots, d, \tag{3.5}$$

in which the least value $\underline{h}$ of h is the greatest integer $\underline{h}$ such that $2^{\underline{h}-1} \leq d$. This result follows from the property that the greatest value $\overline{h} = d$ occurs for one nonempty compound fork in each column; the least value $\underline{h}$ of h occurs when the compound forks are stacked according to 2^0 in column $0; 2^1$ in column $1; 2^2$ in column $2; \ldots$, remainder $= r < 2^{\underline{h}}$ in column $\underline{h}$, which gives the stated value. (ii).The domain of definition of n for given $k \geq 1$ and given $d \geq 1$ is easily seen to be

$$n \geq 2d + k + 1. \tag{3.6}$$

Examples. We repeat from the list of binary trees given in Sect. 2.2.1, Chapter 2, the binary tree skeletons for $d = 1, 2$:

(1) : 

(1, 1) :

(1, 2) :

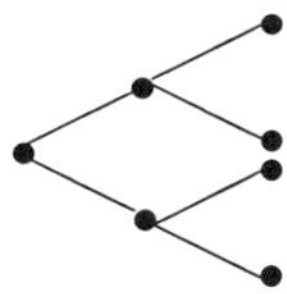

(3.7)

The sequences standing to the left specify the number of forks appearing in the columns $0, 1, \ldots, h-1$ of the binary trees. We have omitted the base line containing k points in these diagrams to focus on that part of the diagram (3.3) emanating to the right from the point v_0. These diagrams are binary tree skeletons of a trivalent tree having k base points and containing $d = 1$ forks, $d = 2$ forks, and $d = 3$ forks, respectively. We no longer use $\circ$ to label the end points of a binary tree as done in Sect. 2.2.1, Chapter 2.

We introduce the concept of a *compound fork* to describe the points that are placed on the lines of the skeleton $\mathbb{T}_{d+1}$ to complete the trivalent tree graph $\mathbb{V}_n$. A compound fork is a simple fork of a binary tree to which points have been added to the upper and lower branches. Just as in diagrams in Sect. 2.2.1, Chapter 2, there are four types of compound forks, which have the following four diagrams:

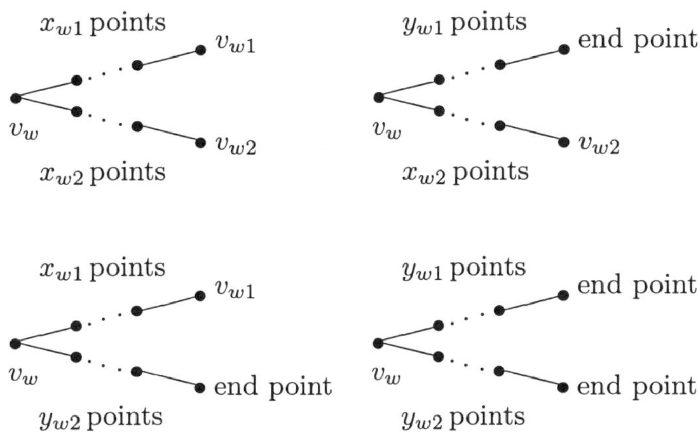

(3.8)

The distinction in these diagrams is in the labeling of the points, depending on whether the end point of a fork in the binary tree skeleton is that of another fork or an end point.

We use the following nomenclature for describing the compound forks (3.8), the binary tree skeleton, and the full trivalent tree $\mathbb{V}_n$. An internal line of $\mathbb{V}_n$ is a line incident on two (simple) roots of a fork in the binary tree skeleton, which are points of degree 3; all point belonging to an internal line are called internal points; these points are x_{w1} and x_{w2} in number, as shown in the compound forks (3.8). An external line of $\mathbb{V}_n$ is a line incident to a root of a fork in the binary tree skeleton and to an end point; all points belonging to an external line are called external points; these points are y_{w1} and y_{w2} in number, as shown in the compound forks (3.8). All points v_w are called *root points* of the binary tree skeleton, and the k points to the left of v_0 are called base line points. By convention, the nonnegative integers $x_{w1}, x_{w2}, y_{w1}, y_{w2}$ in the compound fork diagram (3.8) do **not** count a point labeled with a v or an end point; these numbers count the points added to each fork of the skeleton binary tree, and can be 0. This nomenclature, applied to the compound forks (3.8), gives the following descriptions:

1. In the first diagram, the upper and lower branches contain $x_{w1} \geq 0$ and $x_{w2} \geq 0$ internal points, respectively, all added to a fork of the binary tree skeleton whose lines are incident on fork roots.

2. In the second diagram, the upper branch contains $y_{w1} \geq 0$ external points, all added to a fork of the binary tree skeleton whose upper line is incident on an end point; the lower branch contains $x_{w2} \geq 0$ internal points, all added to the same fork whose lower line is incident on a fork root.

3. In the third diagram, the upper branch contains $x_{w1} \geq 0$ internal points, all added to a fork of the binary tree skeleton, whose upper line is incident on another fork root; the lower branch contains $y_{w1} \geq 0$ external points, all added to the same fork whose lower line is incident on an end point.

4. In the fourth diagram, the upper and lower branches contain $y_{w1} \geq 0$ and $y_{w2} \geq 0$ external points, respectively, all added to a fork of the binary tree skeleton whose lines are incident on end points.

The description of the full trivalent tree $\mathbb{V}_n$ in diagram (3.3) is now completed by replacing each fork in that diagram by a compound fork in accordance with the four types (3.8), where the structure of the binary tree skeleton determines uniquely which of the four compound forks is to replace each simple fork of the skeleton. We denote by $T_{n,k,d}$ a trivalent tree containing n total points, k base line points, and a binary tree skeleton T_{d+1} with d forks, where each simple fork is now replaced by a compound fork of the appropriate form (3.8). The set of all such trivalent trees is denoted by $\mathcal{T}_{n,k,d}$. By definition, *the set* $\mathcal{T}_{n,k,d}$ *contains no trivalent tree with an end branch containing more than* $k-1$ *added*

y_w*−type points.* This is no restriction because a trivalent tree having an end branch containing $k' \geq k+1$ points (counting now the end point) can be oriented into a position such that the end branch containing k' points becomes the base line—such a trivalent tree will be included in the set $\mathcal{T}_{n,k',d}$ of trivalent trees.

The family of linear Diophantine equations that must be solved to obtain all trivalent trees in the set $\mathcal{T}_{n,k,d}$ can be formulated. There are $k \geq 1$ base line points, and each trivalent tree $T_{n,k,d} \in \mathcal{T}_{n,k,d}$ contains a binary tree skeleton $T_{d+1} \in \mathbb{T}_{d+1}$ that has d forks. Thus, there are $n-2d-k-1 \geq 0$ additional points belonging to the $2d$ lines constituting the compound forks as described in diagram (3.8). Let $\mathbb{I}_{n,k,d}$ denote the set of words labeling the root points of internal lines, **not including** v_0, and let $\mathbb{E}_{n,k,d}$ denote the set of words labeling the end points of the external lines. Then, if $p \geq 0$ points belong to the internal lines and $q \geq 0$ points belong to the external lines, the following Diophantine equations must hold for the number of points constituting the compound forks:

$$\begin{aligned} p+q &= n-2d-k-1, \\ \sum_{w \in \mathbb{I}_{n,k,d}} x_w &= p; \sum_{w \in \mathbb{E}_{n,k,d}} y_w = q, \text{ each } y_w \leq k-1, \end{aligned} \tag{3.9}$$

where the cardinalities of the sets $\mathbb{I}_{n,k,d}$ and $\mathbb{E}_{n,k,d}$ are

$$|\,\mathbb{I}_{n,k,d}\,| = d-1, \; |\,\mathbb{E}_{n,k,d}\,| = d+1. \tag{3.10}$$

We denote the set of solutions of this family of Diophantine equations by $\mathbb{D}_{n,k,d}$. Each solution gives a member of the set of trivalent trees $\mathcal{T}_{n,k,d}$ having a given binary tree $T_{d+1} \in \mathbb{T}_{d+1}$ as its skeleton, and conversely.

Counting formulas for the number of solutions $|\mathbb{D}_{n,k,d}|$ of the three Diophantine relations (3.9) can be given. For the purpose of counting, we can rewrite these relations in the simplified form given by

$$\begin{aligned} p+q &= n-2d-k-1, \\ x_1+x_2+\cdots+x_{d-1} &= p, \\ y_1+y_2+\cdots+y_{d+1} &= q, \text{ each } y_i \leq k-1. \end{aligned} \tag{3.11}$$

Thus, the sequence $(x_1, x_2, \ldots, x_{d-1})$ is a composition of p into $d-1$ nonnegative parts. The number of such compositions is given by the binomial coefficient

$$\binom{p+d-2}{d-2}, \; d \geq 2. \tag{3.12}$$

Similarly, the sequence $(y_1, y_2, \ldots, y_{d+1})$ is a composition of q into $d+1$ nonnegative parts, where now each part must satisfy the restriction $y_i \leq k-1$. The counting of such restricted compositions has already been encountered in the discussion of the values of the Clebsch-Gordan numbers for the coupling of equal angular momenta, as described in Sect. 2.1.3, Chapter 2.

The set (2.38) of compositions $\mathbb{L}_j(\mathbf{j})$ transcribes in terms of the present notation to the following set:

$$\mathbb{L}_{\frac{1}{2}(d+1)(k-1)-q}\left(\tfrac{1}{2}(k-1),\ldots,\tfrac{1}{2}(k-1)\right)$$
$$= \left\{ y=(y_1,y_2,\ldots,y_{d+1}) \,\middle|\, \begin{array}{l} 0\le y_i\le k-1, i=1,\ldots,d+1, \\ y_1+y_2+\ldots+y_{d+1}=q \end{array} \right\}, \tag{3.13}$$

where the sequence $(\frac{1}{2}(k-1),\ldots,\frac{1}{2}(k-1))$ is $\frac{1}{2}(k-1)$ repeated $d+1$ times. The parameter transformation from (2.38) to (3.13) is given by

$$\begin{aligned} &n=d+1,\ \mathrm{j_{min}}=0 \text{ for } d \text{ odd},\ j_{min}=\ \tfrac{1}{2}(k-1) \text{ for } d \text{ even}, \\ &j_{\max}=\tfrac{1}{2}(d+1)(k-1),\ q=j_{max}-j, \\ &q=0,1,\ldots,q_{\max},\ q_{\max}=\begin{cases} \tfrac{1}{2}(d+1)(k-1), & d \text{ odd}, \\ \tfrac{1}{2}d(k-1), & d \text{ even}. \end{cases} \end{aligned} \tag{3.14}$$

If the CG numbers have already been calculated by some method, then the cardinality of the set (3.13) is given by relation (2.36):

$$\begin{aligned} &\sum_{h=j_{\max}-q}^{j_{\max}} N_h\left(\tfrac{1}{2}(k-1),\ldots,\tfrac{1}{2}(k-1)\right)=l_{n,k,q}, \\ &l_{n,k,q}=|\mathbb{L}_{\frac{1}{2}(d+1)(k-1)-q}\left(\tfrac{1}{2}(k-1),,\ldots,\tfrac{1}{2}(k-1)\right)|. \end{aligned} \tag{3.15}$$

We have two methods for determining the CG coefficients that enter into (3.15). The first method described by relation (2.18), Chapter 2 generates the multiset

$$\begin{aligned} &\langle\tfrac{1}{2}(k-1),\ldots,\tfrac{1}{2}(k-1)\rangle=\left\{0^{k_0},1^{k_1},\ldots,h^{k_h},\ldots,(q_{\max})^{k_{q_{\max}}}\right\}, \\ &\qquad k_h=N_h(\tfrac{1}{2}(k-1),\ldots,\tfrac{1}{2}(k-1)). \end{aligned} \tag{3.16}$$

The second method uses the $n-$fold product of Schur functions given by

$$\begin{aligned} &\prod_{i=1}^{n} s_{(2j_i,0)}(z_1,z_2)=\sum_{j=j_{\min}}^{j_{\max}} N_j(\mathbf{j})\, s_{(2j,0)}(z_1,z_2), \\ &s_{(2a,0)}(z_1,z_2)=\sum_{\alpha=-a}^{a} z_1^{a+\alpha}z_2^{a-\alpha},\ a=0,1/2,1,\ldots \end{aligned} \tag{3.17}$$

to generate the CG numbers $C_j(\mathbf{j})$ (see (11.286)-(11.287)), Compendium B). These Schur functions for $n = 2$ are also obtained directly from the $D^a(Z)$ matrices as given by $\mathrm{Tr}D^a(\mathrm{diag}(z_1, z_2))$. The transcription of notation (3.14) needs, of course, to be effected in (3.17) to obtain the CG numbers in (3.15).

The number of elements in the solution set $\mathbb{D}_{n,k,d}$ of Diophantine equations (3.9) is given by

$$|\,\mathbb{D}_{n,k,d}\,| = \sum_{p+q=n-2d-k-1} \binom{p+d-2}{d-2} l_{n,k,q},\ d \geq 2. \tag{3.18}$$

Since there are $\frac{1}{d+1}\binom{2d}{d}$ skeleton binary trees to include, we obtain the following formula for the cardinality of the set $\mathcal{T}_{n,k,d}$:

$$|\,\mathcal{T}_{n,k,d}\,| = \frac{1}{d+1}\binom{2d}{d} |\,\mathbb{D}_{n,k,d}\,|. \tag{3.19}$$

The construction of the trivalent trees is the set $\mathcal{T}_{n,k,d}$ for small values of d can be carried out directly by solving the set of Diophantine equations (3.9). We illustrate this by two examples:

Examples. For $d = 2$, we have the following two binary tree skeletons, where we also use the notations in relations (3.9) to denote the number of points that belong to the internal and external lines:

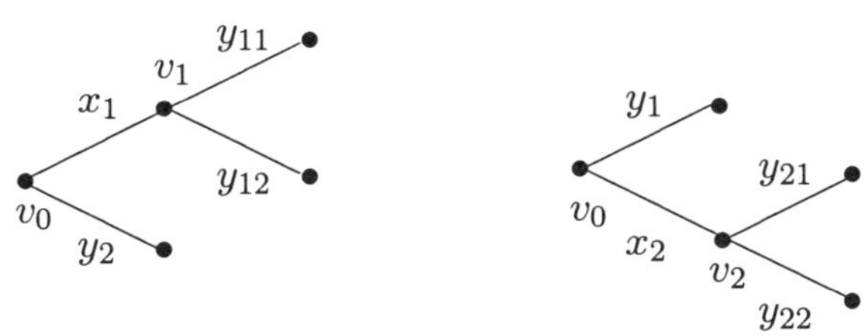

(3.20)

We consider the trivalent trees in the set $\mathcal{T}_{10,3,2}$. Thus, 2 points are to be adjoined as the base line to the left of each vertex v_0 in each diagram. This leaves 2 points to be distributed onto the one internal line and three external lines in all possible ways. This gives the following Diophantine relations for the respective binary tree skeletons in (3.20):

$$\begin{gathered} p+q=2;\ x_1=p;\ y_2+y_{11}+y_{12}=q,\ \text{each } y_w \leq 2, \\ p+q=2;\ x_2=p;\ y_1+y_{21}+y_{22}=q,\ \text{each } y_w \leq 2. \end{gathered} \tag{3.21}$$

The solution set of each Diophantine relation (3.21) is

$$\begin{aligned}\mathbb{D}_{10,3,2} = \{&(0,2,0,0),(0,0,2,0),(0,0,0,2),(0,0,1,1),(0,1,0,1),\\&(0,1,1,0),(1,1,0,0),(1,0,1,0),(1,0,0,1),(2,0,0,0)\}\,,\end{aligned} \tag{3.22}$$

Thus, we obtain

$$\{(x_1,y_2,y_{11},y_{12})\} = \mathbb{D}_{10,3,2},\ \{(x_2,y_1,y_{21},y_{22})\} = \mathbb{D}_{10,3,2}. \tag{3.23}$$

While the numerical sequences in these two sets are equal, the corresponding trivalent trees are not: *their shapes are different.* There are twenty trivalent trees in the set $\mathcal{T}_{10,3,2}$.

We consider also the trivalent trees in the set $\mathcal{T}_{9,2,2}$. Thus, 2 points are to be adjoined as the base line to left of each vertex v_0 in each diagram (3.20). This leaves 2 points to be distributed onto the one internal line and three external lines in all possible ways. This gives the Diophantine relations (3.21) in which the restriction in the second relation is now $y_w \leq 1$. The solution set of each of the Diophantine relations is

$$\begin{aligned}\mathbb{D}_{9,2,2} = \{&(0,0,1,1)(0,1,0,1),(0,1,1,0)\\&(1,1,0,0)(1,0,1,0),(1,0,0,1),(2,0,0,0)\}\,,\end{aligned} \tag{3.24}$$

Thus, we obtain

$$\{(x_1,y_2,y_{11},y_{12})\} = \mathbb{D}_{9,2,2},\ \{(x_2,y_1,y_{21},y_{22})\} = \mathbb{D}_{9,2,2}. \tag{3.25}$$

There are fourteen trivalent trees in the set $\mathcal{T}_{9,2,2}$. □

The set $\mathcal{T}_{n,k,d}$ contains all possible nonisomorphic trivalent trees containing a total of n points such that the base line contains k points and the skeleton binary tree contains d forks. But the set $\mathcal{T}_{n,k,d}$ also contains, in general, a number of isomorphic trivalent trees.

3.2 Nonisomorphic Trivalent Trees

It is useful to give examples of some sets of nonisomorphic trivalent trees before considering the general case:

Examples. For $d = 0$, the diagram (3.3) reduces to a trivalent tree with $n = k + 1$ points on a line:

$\mathcal{T}_{n,n-1,0} =$

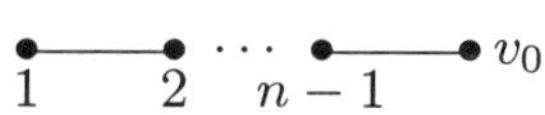

(3.26)

For $n = 2, 3$, this is the only trivalent tree.

For $d = 1$, diagram (3.3) reduces to the following diagram, which has $n = k + y_1 + y_2 + 3$:

$$\mathcal{T}_{k+y_1+y_2+3,k,1} =$$

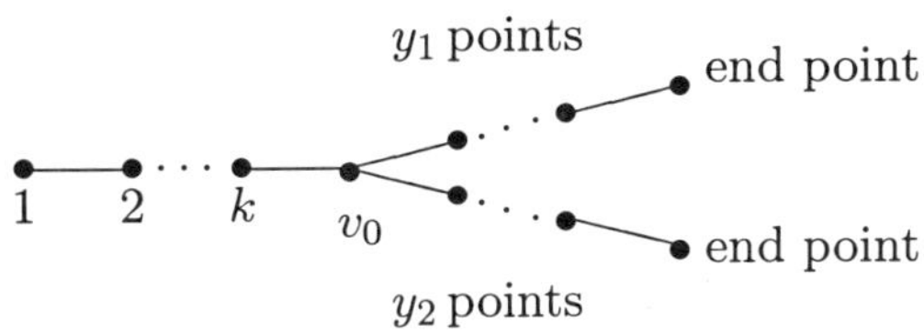

(3.27)

The conditions $0 \le y_1 \le k - 1$, $0 \le y_2 \le k - 1$ are imposed in order to satisfy the base line rule that no end line can contain more than k points. This example illustrates the occurrence of isomorphic trivalent trees. The operation of reflection through the horizonal line containing the vertex (root) v_0 clearly preserves adjacency: *Only those trivalent trees in diagram (3.27) for which, say, $y_1 \ge y_2$ can be nonisomorphic.* But there are still adjacency preserving operations that must be effected should $y_1 = k - 1$ in order to conclude that we have found all nonisomorphic trivalent trees for $d = 1$. The first is a counterclockwise rigid-body rotation by 120° (first adjust the angles between the three lines in (3.27) to be 120°) about the point v_0, which brings the upper end line to the horizontal base line position, followed by a reflection through the horizontal line containing the new base line, which also has k points. These two operations leave the trivalent tree (3.27) invariant. Should also $y_2 = k - 1$, we rotate counterclockwise by 240° about the point v_0 to bring the lower end line to the horizontal position base line position, and find that the trivalent tree (3.27) again remains invariant. Thus, these operations do not reduce the number of nonisomorphic trivalent trees. We conclude that each member of the family of trivalent trees in diagram (3.27) is nonisomorphic for all pairs (y_1, y_2) for which the condition $y_1 \ge y_2$ is fulfilled; and all nonisomorphic trivalent trees for $d = 1$ are so obtained. $\square$

We introduce the notation $[n, k, d]$ to denote a complete set of nonisomorphic trivalent trees containing n points with k base line points and d forks in the skeleton binary tree. *Complete* means that every trivalent tree in the set $\mathcal{T}_{n,k,d}$ is isomorphic to a trivalent tree in the set $[n, k, d]$, and there are no isomorphisms between trivalent trees in the set $[n, k, d]$. The examples above then give the following families of complete nonisomorphic trivalent trees:

$$\begin{aligned} [n, n-1, 0] &= \mathcal{T}_{n,n-1,0}, \quad |\,[n, n-1, 0]\,| = 1,\ n \geq 2, \\ [n, k, 1] &= \left\{ \mathcal{T}_{k+y_1+y_2+3,k,1} \,\middle|\, \begin{array}{c} y_1 + y_2 = n - k - 3, \\ 0 \leq y_2 \leq y_1 \leq k - 1 \end{array} \right\}, \\ & \qquad k \geq 1,\ n \geq k + 3. \end{aligned} \tag{3.28}$$

The second relation in (3.28) shows that (y_1, y_2) is a partition of $n - k - 3$ into two nonnegative parts with each part $\leq k - 1$. It follows that the complete set of nonisomorphic trivalent trees $[n, k, 1]$ is not only determined by the elements of the set $\mathbb{D}_{n,k,1}$, it is also determined by the restricted set of partitions (see Sect. 11.1.1, Compendium B) given by

$$[n, k, 1] = \mathbb{P}\mathrm{ar}(2, k, n - k - 3). \tag{3.29}$$

Thus, the cardinality of $[n, k, 1]$ is given by

$$|\,[n, k, 1]\,| = p(2, k, n - k - 3), \tag{3.30}$$

where the coefficient $p(2, k, n - k - 3)$ is obtained from the expansion of the Gaussian polynomial:

$$\begin{bmatrix} k+2 \\ k \end{bmatrix}_t = \frac{(1 - t^{k+2})(1 - t^{k+1})}{(1 - t^2)(1 - t)} = \sum_{s=0}^{2k} p(2, k, s) t^s. \tag{3.31}$$

It is always the case that $[n, k, d] \subseteq \mathcal{T}_{n,k,d}$, but it is, in general, nontrivial to sort out a complete set of nonisomorphic trivalent trees $[n, k, d]$ from $\mathcal{T}_{n,k,d}$. This is because a trivalent tree can be presented in many configurations of points and lines. Picturesquely, a trivalent tree may be viewed as a collection of solid balls (points) joined by flexible, unbreakable threads (lines); any entanglement of this object preserves adjacency of points. This is why we present trivalent trees in a standard configuration given by diagrams (3.3) and (3.8) with the order relation on sites given by (3.4), where the general rule for drawing such pictures of trivalent trees is that the points of degree 3—the vertices (roots) of the compound forks (3.8)— should always be aligned in vertical columns $0, 1, \ldots, h$, and in the word order (3.4). This requires that the points belonging to each compound fork (3.8) be compressed in their spacing. We refer to this as *local scaling.* This is a convenience of description; it is irrelevant for the point-line relations in the graph. While we are not always careful to make this adjustment in all diagrams, it is essential for applying the local reflections described below. We assume without notice that such minor adjustments are effected.

Symmetries of figures in the plane have an important role in the determination of nonisomorphic trivalent trees. It is clear that overall rotations, translations, reflections through any line, scaling, and general entanglement, as described above, all preserve adjacency of points and give isomorphic trivalent trees. All symmetries described here refer to the standard configuration (3.3) with compound forks given by (3.8), where we use local scalings within a compound fork (3.8) to bring a pair of trivalent trees into coincidence through a symmetry operation. Trivalent trees whose points and lines are brought into coincidence by such operations are take to be equal. In a standard configuration, the base line is placed horizontally and to the left of the underlying binary tree skeleton.

There are two types of reflections that are important for sorting out the nonisomorphic trivalent trees in the set $\mathcal{T}_{n,k,d}$.

The first type of reflection operation is called a *global reflection* and denoted by τ_w. It is an active reflection of all point and lines of the entire trivalent tree $T_{n,k,d} \in \mathcal{T}_{n,k,d}$ through the horizontal line containing the vertex point v_w. There is such a global reflection operation τ_w for each vertex contained in $T_{n,k,d}$; that is, one for each word w assigned to the root of a compound fork, or what is the same thing, to the root of the fork of the binary tree skeleton of $T_{n,k,d}$. Accordingly, there are d such reflection operations τ_w. These reflection operations all give isomorphic trivalent trees. We denote these types of isomorphisms by $\cong_\tau$, and refer to the trivalent trees in an equivalence class in the quotient set $\mathcal{T}_{n,k,d}/\cong_\tau$ as being $\tau-$equivalent. The number of equivalence classes in the quotient set $\mathcal{T}_{n,k,d}/\cong_\tau$ is given by the Wedderburn-Etherington numbers, as presented below.

The second type of reflection operation is called a *local reflection* and denoted by σ_w. It is defined as follows. Let T_w denote the subtrivalent tree consisting of all points and lines of $T_{n,k,d} \in \mathcal{T}_{n,k,d}$ for which the vertex (root) point v_w serves as root. The trivalent tree T_w thus has as its binary tree skeleton all points and lines that emerge from the root v_w by the bifurcation process corresponding to the subbinary tree of the binary tree skeleton of $T_{n,k,d} \in \mathcal{T}_{n,k,d}$. The local reflection σ_w is an active reflection of all point and lines of T_w through the horizontal line containing the vertex point v_w; all other points and lines of $T_{n,k,d}$ are to remain unchanged. There is such a local reflection operation σ_w for each compound fork contained in $T_{n,k,d}$; that is, one for each word w assigned to the root of a compound fork; that is, to the root of the fork of the binary tree skeleton of $T_{n,k,d}$. Accordingly, there are d such local reflection operations σ_w. These local reflection operations all give isomorphic trivalent trees. We denote these types of isomorphisms by $\cong_\sigma$, and refer to the trivalent trees in an equivalence class in the quotient set $\mathcal{T}_{n,k,d}/\cong_\sigma$ as being $\sigma-$equivalent. In particular, $\sigma_0 = \tau_0$ is the reflection of all points of $T_{n,k,d}$ through the horizontal line containing the base line.

The global and local reflection operations τ_w and σ_w that can be applied to a given $T_{n,k,d}$ depend on the fork structure of the binary tree skeleton of $T_{n,k,d}$. Thus, the sets of global and local reflection operators may be defined by

$$\mathbb{S}_{n,k,d} = \{\tau_w \,|\, w \in \mathbb{F}_{n,k,d}\},\ \mathbb{R}_{n,k,d} = \{\sigma_w \,|\, w \in \mathbb{F}_{n,k,d}\}, \tag{3.32}$$

where $\mathbb{F}_{n,k,d}$ is the set of words located at the roots of the forks in the binary tree skeleton of a given binary tree $T_{n,k,d} \in \mathcal{T}_{n,k,d}$. Thus, we have the cardinality of sets given by

$$|\,\mathbb{S}_{n,k,d}\,| = |\,\mathbb{R}_{n,k,d}\,| = |\,\mathbb{F}_{n,k,d}\,| = d. \tag{3.33}$$

The reflections operations τ_w and σ_w do not exhaust all operations that need to be considered in isolating the nonisomorphic trivalent trees in the set $\mathcal{T}_{n,k,d}$. This is because some of these sets can contain trivalent trees that have end branches containing exactly k points. Such a trivalent tree can then be rotated by a rigid-body rotation and translated such that this end branch is horizontal with all of its k points to the left of the point v_0 of the transformed trivalent tree. Further individual *local rotations* of subtrivalent trees T_w and local reflections can then be applied to bring the transformed trivalent tree to the standard form (3.3); that is, to a trivalent tree in the set $\mathcal{T}_{n,k,d}$. Such operations must be effected on every end branch containing k points of each trivalent tree $T_{n,k,d} \in \mathcal{T}_{n,k,d}$ to determine the nonisomorphic trivalent trees in $\mathcal{T}_{n,k,d}$. We refer to these kinds of operations as Ω equivalences, but do not specify the details of such operations. The isomorphism of two Ω−equivalent trivalent trees is denoted by $\cong_\Omega$.

We have introduced three operations $\cong_\tau$, $\cong_\sigma$, and $\cong_\Omega$ above that give isomorphisms $\cong$ between trivalent trees. We could dispense with the $\cong_\tau$ isomorphisms, since *all trivalent trees obtain from a given one by a single* $\cong_\tau$ *reflection operation can also be obtained by a sequential application of* σ_w *reflections.* Nonetheless, it is sometimes convenient to use τ reflections in place of the more basic σ_w reflections.

The Ω isomorphisms, while vague in their definition, are very important. It is useful to illustrate this operation:

Example: An Ω equivalence of trivalent trees:

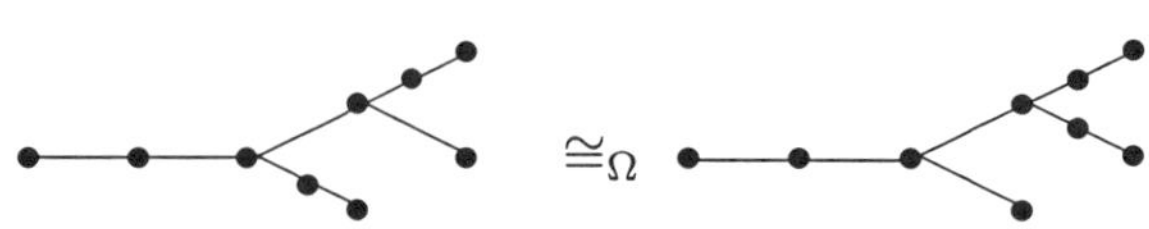

(3.34)

The isomorphism of these two trivalent trees is shown as follows: By a rigid-body counterclockwise rotation of the left-hand figure about the point v_0 (not marked) the uppermost branch is brought to horizontal position. This gives the trivalent tree on the left below:

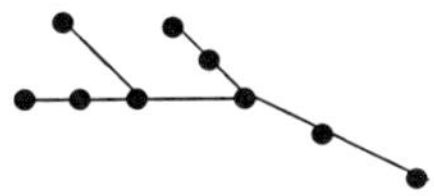 $\cong_\Omega$ 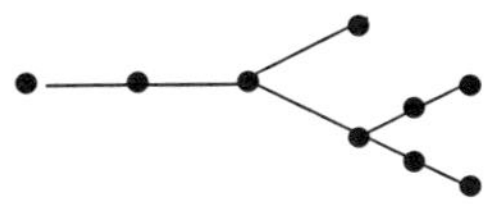 (3.35)

By local rotations of branches, the configuration on the left is brought to the configuration on the right. Application of the reflection operation $\tau_0 = \sigma_0$ to the trivalent tree on the right in (3.35) now gives the trivalent tree on the right in (3.34). *But no reflection operation can bring the two trivalent trees in (3.34) into coincidence.* □

Trivalent trees that are $\cong_\Omega$ isomorphic in consequence of an end branch that contains the same number of points k as the base line are the most difficult to recognize. Such trivalent trees require a rigid-body rotation, rotations of individual end branches, and reflection operations of the form σ_w to bring them to the standard form (3.3). The defining characteristic of an $\cong_\Omega$ isomorphism is that it contains a rigid-body rotation that transforms an end branch containing k points into a new horizontal base line containing k points. Such isomorphisms must be considered in every set $\mathcal{T}_{n,k,d}$ in which some trivalent trees have end branches containing k points.

The Ω operations are applicable only to those trivalent trees in $\mathcal{T}_{n,k,d}$ that admit k points on an end branch. Thus, we have two domains of definition of n to consider:

$$\begin{aligned} 2d+k+1 \;\leq\; & n \leq 2d+2k-1, \text{ no end branch contains } k \text{ points,} \\ n \;\geq\; & 2d+2k, \text{ some end branches contain } k \text{ points.} \end{aligned} \tag{3.36}$$

These inequalities must hold because the base line and binary tree skeleton already contain $2d + k + 1$ points. If all the extra $n - 2d - k - 1$ points are placed on a single one of the $d + 1$ end branches, then for $n \leq 2d + 2k - 1$ there are $k - 2 + 1 = k - 1$ points on the end branch, while for $n \geq 2d + 2k$ there are $k - 1 + 1 = k$ points on the end branch.

Examples (See (3.20)-(3.25) for notations). It is useful to illustrate the effect of $\cong_\Omega$ equivalences for $n \geq 2d+2k$. The set of trivalent trees $\mathcal{T}_{10,3,2}$ contains seven nonisomorphic trivalent trees:

$$\begin{aligned} [10,3,2] = \; & \{T_{10}(0,2,0,0), T_{10}(0,0,2,0), T_{10}(0,0,1,1), T_{10}(0,1,1,0), \\ & T_{10}(1,1,0,0), T_{10}(1,0,1,0), T_{10}(2,0,0,0)\}, \end{aligned} \tag{3.37}$$

where the notation $T_{10}(x_1, y_2, y_{11}, y_{12})$ denotes a trivalent tree corresponding to the indicated sequences $\{(x_1, y_2, y_{11}, y_{12})\}$, which is a subset of the ten solutions (3.22) of the Diophantine equations (3.21). This result follows from the fact that the trivalent trees $T_{10}(x_1, y_2, y_{11}, y_{12})$ and $T'(y_1, x_2, y_{21}, y_{22})$ corresponding to the solution sets (3.22) and (3.23) are reflection-equivalent by the reflection operation σ_0. While the numerical sequences in (3.22) and (3.23) are the same, the shapes T and T' of the trivalent trees are different. Application of the local reflection operation σ_1 to the remaining three trivalent trees $T_{10}(1,0,0,1), T_{10}(0,1,0,1)$, and $T_{10}(0,0,0,2)$ corresponding to the solution set (3.22) gives the following reflection equivalences: $T_{10}(1,0,0,1) \cong_\sigma T_{10}(1,0,1,0), T_{10}(0,1,0,1) \cong_\sigma T_{10}((0,1,1,0), T_{10}(0,0,0,2) \cong_\sigma T_{10}(1,0,0,2,0)$. This gives the set (3.37) above, which contains only nonisomorphic trivalent trees. We still must examine the trivalent trees with end branches containing $k = 3$ points, which are $T_{10}(0,2,0,0), T_{10}(0,0,2,0), T_{10}(0,0,0,2)$. All possible $\Omega-$type operations map this subset into itself. Thus, the seven nonisomorphic trivalent trees in the set (3.37) is not reduced by the Ω operations. The set $[10,3,2]$ of trivalent trees (3.22) is complete; the trivalent trees in the set have been chosen to have the property $y_{11} \geq y_{12}$.

The set of trivalent trees $\mathcal{T}_{9,2,2}$ contains four nonisomorphic trivalent trees:

$$[9,2,2] = \{T_9(0,0,1,1), T_9(1,1,0,0), T_9(1,0,1,0), T_9(2,0,0,0)\}. \quad (3.38)$$

This result follows by application of the same σ_0 and σ_1 reflection operations as in the example $\mathcal{T}_{10,3,2}$ above, which gives the following two $\cong_\sigma$ isomorphisms: $T_9(1,0,0,1) \cong_\sigma T_9(1,0,1,0), T_9(0,1,0,1) \cong_\sigma T_9(0,1,1,0)$. These reflection-equivalent isomorphisms reduce the set of fourteen trivalent trees having the parameters (3.24) to the four trivalent trees in $[9,2,2]$ given by (3.38), and one more given by $T_9(0,1,1,0)$. This extra one cannot be transformed into a trivalent tree in the set $[9,2,2]$ by a reflection operation. But from the example (3.34) above, we have that $T_9(0;1,1,0) \cong_\Omega T_9(0;0,1,1)$, so that all of $[9,2,2]$ is correctly given by (3.38). The set $[9,2,2]$ of nonisomorphic trivalent trees (3.38) is complete, *but it requires an Ω operation to effect the reduction of the set $\mathcal{T}_{9,2,2}$ to $[9,2,2]$*. The trivalent trees in this set $[9,2,2]$ have been chosen to have the property $y_{11} \geq y_{12}$ and $x_1 \geq y_2$.

It is an important feature, exhibited by these examples, $\mathcal{T}_{10,3,2}$ and $\mathcal{T}_{9,2,2}$, that once these sets have been reduced to the trivalent trees of different shapes corresponding to the solution sets (3.22) and (3.23) of the Diophantine relations (3.21), then the only σ_w operations that can be applied are those corresponding to reflections of a pair of end branches, since all other reflection operations change the shape of the trivalent tree. This is a general feature, as developed below. □

It is useful to present the general case for the set of nonisomorphic trivalent trees for $[n,k,2]$, which has 2 forks in the binary tree skeleton. There are two classes of trivalent trees to consider corresponding to the two binary tree skeletons. These are given by the diagrams:

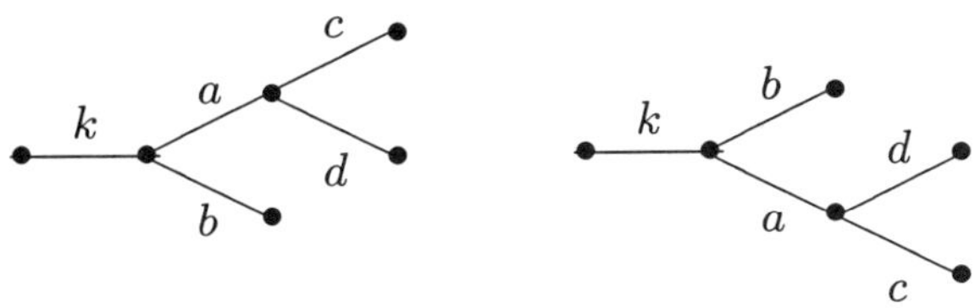

(3.39)

Here we have introduced a simplified notation in place of the general word notation: The integer parameters k, a, b, c, d next to the lines denote the number of • points that belong to the line, not counting the roots of the forks nor the end points (see (3.20)). Thus, the total number of points in the trivalent trees (3.39) is

$$n = k + a + b + c + d + 5, \tag{3.40}$$

where the integers k, a, b, c, d must satisfy the restrictions

$$k \geq 1, , a \geq 0,\ 0 \leq b \leq k-1,\ 0 \leq c \leq k-1,\ 0 \leq d \leq k-1. \tag{3.41}$$

The set $\mathcal{T}_{n,k,2}$ of trivalent trees is obtained by enumerating all parameters a, b, c, d such that conditions (3.41) are satisfied for given n and k. (Note that d in the above relations is not the number of forks in (3.39), which is fixed at 2). The Diophantine relations that must be satisfied (see (3.11)) are given by

$$\begin{aligned} a + q &= n - k - 5 \geq 0, \\ b + c + d &= q \geq 0,\ 0 \leq b \leq k-1,\ 0 \leq c \leq k-1,\ 0 \leq d \leq k-1. \end{aligned} \tag{3.42}$$

The number of trivalent trees in $\mathcal{T}_{n,k,2}$ has already been given by relation (3.19) in which we set the fork parameter $d = 2$. Our purpose here in not a rediscussion of that result, but rather directed toward finding the parameters a, b, c, d that characterize a complete subset $[n, k, 2] \subset \mathcal{T}_{n,k,2}$ of nonisomorphic trivalent trees.

Application of the reflection operation σ_0 to the trivalent tree on the right in (3.39) shows that it is reflection-equivalent to the trivalent tree on the left. Application of the local reflection operator σ_1 to the trivalent tree on the right preserves its shape, and also shows that it is always possible to choose $c \geq d$ in the set of trivalent trees $[n, k, 2]$. Any possible further restrictions can only occur in the case where one or more of the parameters b, c, d is equal to $k - 1$, in which case at least one end branch contains k points, the number of points in the base line.

We introduce the notation (as in the $[10, 3, 2]$ and $[9, 2, 2]$ examples above)

$$T_n(a, b, c, d),\ a + b + c + d = n - k - 5, \tag{3.43}$$

for the trivalent tree on the left-hand side in (3.39). Then, we obtain the following trivalent trees in the complete nonisomorphic set $[n,k,2]$ of trivalent trees, where the restrictions on the nonnegative parameters a,b,c,d depend on the relation between n and k :

$$[n,k,2] = \{T_n(a,b,c,d)\,|\,0 \leq a+b+c+d \leq k-2, c \geq d\}\,, \quad \text{for}\, k+5 \leq n \leq 2k+3. \tag{3.44}$$

$$[n,k,2] = \{T_n(a,b,c,d)\,|\,a+b+c+d \geq k-1, c \geq d\}\,, \quad \text{for}\, n \geq 2k+4. \tag{3.45}$$

In the set $[n,k,2]$ defined by (3.44), no end branch can contain k points, while in the set $[n,k,2]$ defined by (3.45), at least one end branch can contain k points (see(3.36)).

There are no reflection operations that can be applied to the sets of trivalent trees defined by (3.44) and (3.45) to reduce the number of trivalent trees in these sets, since σ_1 is the only shape preserving reflection operation. Any further restrictions on the parameters a,b,c,d can only come from Ω isomorphisms, which only apply to the set (3.45) when one or more of the parameters b,c,d is equal to $k-1$, so that there are end branches containing k points. It is not difficult to verify the following isomorphisms, where $n \geq 2k+4$:

$$\begin{aligned}
b &= k-1 : T_n(a,k-1,c,d) \cong_\Omega T_n(a,k-1,c,d), \\
c &\geq d,\ a+c+d = n-2k-4; && (3.46)\\
c &= k-1 : T_n(a,b,k-1,d) \cong_\Omega T_n(a,d,k-1,b), \\
k-1 &\geq d,\ a+b+d = n-2k-4; && (3.47)\\
d &= k-1 : T_n(a,b,c,k-1) \cong_\Omega T_n(a,c,k-1,b), \\
c &\geq k-1,\ a+b+c = n-2k-4. && (3.48)
\end{aligned}$$

The self-isomorphism (3.46) effects no conditions on the parameters b,c,d; the isomorphism (3.47) implies it is always possible to choose $b \geq d$, and the isomorphism (3.48), together with $T_n(a,b,c,k-1) \cong_{\sigma_1} T_n(a,b,k-1,c)$, implies it is always possible to choose $b \geq c$. We conclude that the parameters a,b,c,d in the complete set of trivalent trees $[n,k,2]$ defined by (3.45) can always be chosen such that

$$c \geq d, \text{ all cases; } b \geq d, \text{ if}\, c = k-1; \ b \geq c, \text{ if}\, d = k-1. \tag{3.49}$$

The examples given for $[10,3,2]$ and $[9,2,2]$, which both fall with the general case (3.45), validate the conditions (3.49).

Further information on the determination of a complete nonisomorphic set $[n,k,d] \subset \mathcal{T}_{n,k,d}$ can be obtained from the Diophantine relations

(3.9) by application of the $\cong_\sigma$ and $\cong_\Omega$ operations to all the elements of $\mathcal{T}_{n,k,d}$. We require more notation to elucidate these steps. We first introduce the composition

$$\mathbf{p} = (p_0, p_1, \ldots, p_{h-1}), p_0 = 1, \tag{3.50}$$

where p_m denotes the number of forks in column m in the binary tree skeleton of the trivalent tree (3.3) with compound forks given by (3.8), for each $m = 0, 1, \ldots, h-1$ (all points occurring in column h in (3.3) are end points). The complete specification of such a trivalent tree then requires also the *occupation numbers* $\{x_w\}$ and $\{y_w\}$ of the compound forks giving the number of points belonging to the various lines of the binary tree skeleton, as detailed in (3.8). These occupation numbers are denoted by the notation (see (3.9)):

$$(\mathbf{x}; \mathbf{y}),\ \ \mathbf{x} = \{x_w \,|\, w \in \mathbb{I}_{n,k,d}\}\,,\ \ \mathbf{y} = \{y_w \,|\, w \in \mathbb{E}_{n,k,d}\}\,. \tag{3.51}$$

The notation $T^{\mathbf{p}}_{n,k,d}(\mathbf{x}; \mathbf{y})$ completely specifies each binary tree $T^{\mathbf{p}}_{n,k,d}(\mathbf{x}; \mathbf{y}) \in \mathcal{T}_{n,k,d}$.

Example. The following example for $d = 3$ illustrates the above notations:

$$T^{(1,2)}_{n,k,3}(x_1, x_2; y_{11}, y_{12}, y_{21}, y_{22}) =$$

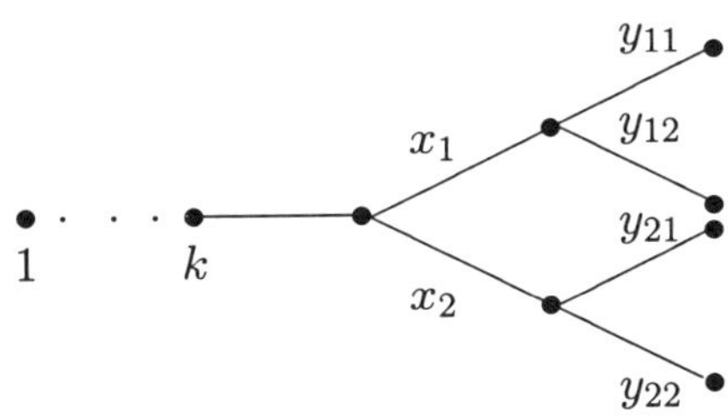

(3.52) □

Table 3.1 on page 183 lists all eighteen nonisomorphic trivalent trees of order 9 in terms of the above notations. The thirty-seven nonisomorphic trivalent trees of order 10 in the subsets $[10, k, d]$ can be constructed by the same methods. These calculations give the following results, where the number in parentheses ( ) following $[10, k, d]$ is the number of trivalent trees in that set:

$$\begin{aligned}
&[10, 9, 0](1)\\
&[10, 3, 1](1), [10, 4, 1](2), [10, 5, 1](2),\\
&[10, 6, 1](1), [10, 7, 1](1);
\end{aligned} \tag{3.53}$$

$$[10,3,2](7), [10,4,2](4), [10,5,2](2);$$
$$[10,1,3](5), [10,2,3](7), [10,3,3](3);$$
$$[10,1,4](1).$$

With more effort, the detailed structure $T^{\mathbf{P}}_{10,k,d}(\mathbf{x};\mathbf{y})$ of each nonisomorphic trivalent tree in each of the above subsets can also be given.

We next show how the general counting formula (3.19) can be refined by taking advantage of the τ_w global reflection operations. These formulas utilize the Wedderburn-Etherington numbers. The Wedderburn-Etherington number b_{d+1} is the number of binary trees in the quotient set defined by $\mathbb{T}_{d+1}/\cong_\tau$, where $\mathbb{T}_{d+1}$ is the set of binary trees of order $d+1$, which always contains d forks. The numbers b_{d+1} enter into a relation or the form (3.19) in place of the Catalan numbers because the reflection operations τ_w can just as well be applied to the trivalent tree skeleton of each trivalent tree in $\mathcal{T}_{n,k,d}$ first, and then the distributions of points corresponding to the solutions of the Diophantine equations (3.9) is effected on just a set of representative binary trees of the equivalence classes in the quotient set $\mathbb{T}_{d+1}/\cong_\tau$.

The recurrence relations for the Wedderburn-Etherington numbers are given by

$$\begin{aligned} b_{2p-1} &= \sum_{s=1}^{p-1} b_s b_{2p-s-1}, \\ b_{2p} &= \sum_{s=1}^{p-1} b_s b_{2p-s} + \binom{b_p+1}{2}, \end{aligned} \tag{3.54}$$

where $p \geq 2$, and, by definition, $b_0 = b_1 = b_2 = 1$. We refer to Comtet [44, p. 55] and Stanley [163, pp. 245, 278, Vol. 2] for their derivation and references to the original literature. The first few Wedderburn-Etherington numbers are given by

d	0	1	2	3	4	5	6	7	8	9
b_{d+1}	1	1	1	2	3	6	11	23	46	98

We next define the quotient set $\mathcal{C}_{n,k,d}$ by

$$\mathcal{C}_{n,k,d} = \mathcal{T}_{n,k,d}/\cong_\tau, \tag{3.55}$$

Table 3.1. The $[9, k, d]$ Nonisomorphic Trivalent Trees.*

subset	$\mathbf{p}$	$\mathbf{x}; \mathbf{y}$
$[9, 8, 0]:$	$\emptyset$	$---$
$[9, 3, 1]:$	(1)	$(2, 1)$
$[9, 4, 1]:$	(1)	$(3; 1)$
	(1)	$(1; 1)$
$[9, 5, 1]:$	(1)	$(1; 0)$
$[9, 6, 1]:$	(1)	$(0; 0)$
$[9, 1, 2]:$	$(1, 1)$	$(2; 0, 0, 0)$
$[9, 2, 2]:$	$(1, 1)$	$(2; 0, 0, 0)$
	$(1, 1)$	$(1; 0, 1, 0)$
	$(1, 1)$	$(1; 1, 0, 0)$
	$(1, 1)$	$(0; 0, 1, 1)$
$[9, 3, 2]:$	$(1, 1)$	$(1; 0, 0, 0)$
	$(1, 1)$	$(0; 0, 1, 0)$
	$(1, 1)$	$(0; 1, 0, 0)$
$[9, 4, 2]:$	$(1, 1)$	$(0; 0, 0, 0)$
$[9, 1, 3]:$	$(1, 2)$	$(1, 0; 0, 0, 0, 0)$
$[9, 2, 3]:$	$(1, 2)$	$(0, 0; 0, 0, 0, 0)$
	$(1, 1, 1)$	$(0, 0; 0, 0, 0, 0; 1, 1, 0^6)$

* The first entry $[9, 8, 0]$ is the trivalent tree (3.26) with no forks; the entries under $[9, k, 2]$ are of the form given by the left-most diagram (3.39); and the first two under $[9, k, 3]$ of the form of diagram (3.52). Addition of all trivalent trees in the listed sets $[9, k, d]$ gives the eighteen nonisomorphic trivalent trees of order 9 (see (3.2)). The explicit trivalent trees corresponding to the members of these sets are easily read off in terms of the notation $T^{\mathbf{p}}_{n,k,d}(\mathbf{x}; \mathbf{y})$ from the listed sequences $\mathbf{p}$ and the occupation numbers $(\mathbf{x}; \mathbf{y})$.

and denote a set of representative trivalent trees of the equivalence classes $\mathcal{C}_{n,k,d}$ by $\mathcal{R}_{n,k,d}$. Since one trivalent tree is selected from each equivalence class, we have that $|\mathcal{R}_{n,k,d}| = b_{d+1}$. The Diophantine relations (3.9) are now solved just for each representative trivalent tree $R_{n,k,d} \in \mathcal{R}_{n,k,d}$. The number of such solutions is obtained from relation (3.19) by replacing the multiplicative Catalan number by the Wedderburn-Etheringon number, which gives

$$b_{n,k,d} = b_{d+1}|\mathbb{D}_{n,k,d}| \tag{3.56}$$

as the number of trivalent trees in the set $\mathcal{T}_{n,k,d}$ that are $\tau-$inequivalent. (This does not take into account σ_w-type operations. For example, for $\mathcal{T}_{10,3,2}$ and $\mathcal{T}_{9,2,2}$, this is just the reduction, respectively, from twenty trivalent trees to ten trivalent trees, and from fourteen to seven.)

3.2.1 Shape labels of binary trees

The notation $T^{\mathbf{p}}_{n,k,d}(\mathbf{x};\mathbf{y})$ fully specifies a trivalent tree in the set $\mathcal{T}_{n,k,d}$. But additional parameters that specify the shape of the binary tree skeleton of $\mathcal{T}_{n,k,d}$ are needed for further classification into nonisomorphic subsets. For this purpose, we introduce the concept of a *fork vector.* Recall that the positive integer p_m of the composition $\mathbf{p} = (p_0, p_1, \ldots, p_{h-1})$ defined by (3.50) gives the number of forks in column m of the binary tree skeleton of height h. We now go one step further and define the distribution of forks in the binary tree by the following rule: A 1 is assigned to each position v_w where a fork occurs, and a 0 to every other point, where all 2^m points in each column $m = 0, 1, \ldots, h-1$ are included in this assignment, but not column h in (3.3), which is all $0's$:

$$\text{column } m \text{ contains } p_m \ 1's \text{ and } 2^m - p_m \ 0's,\ m = 0, 1, \ldots, h-1. \tag{3.57}$$

The collection of $1's$ and $0'1$ obtained by this rule are now arranged in a row vector obtained by using the order rule (3.4) to place a 1 in each position w where a fork v_w occurs, and 0 at every other place. This vector is called the fork vector $\mathbf{f}_T$ of the binary tree skeleton T; it is of length $2^h - 1$.

Examples. The definition of a fork vector $\mathbf{f}_T$ of a binary tree T is made clear from the following set of eight binary trees corresponding to the composition $\mathbf{p} = (1, 1, 2, 1)$:

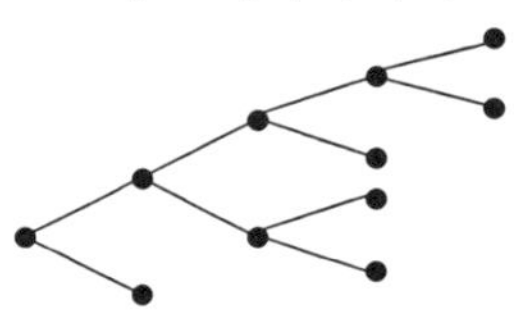

$\rightarrow (1; 1, 0; 1, 1, 0, 0; 1, 0, 0, 0, 0, 0, 0, 0)$

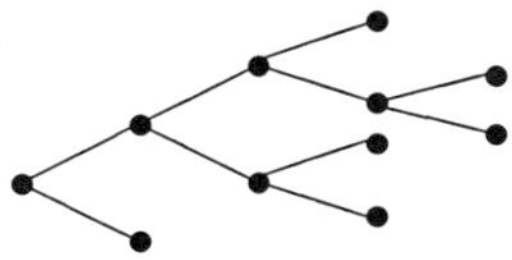

$$\to (1;1,0;1,1,0,0;0,1,0,0,0,0,0,0)$$

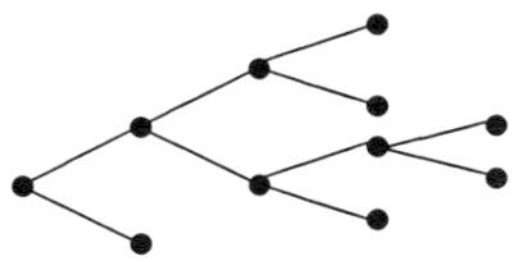

$$\to (1;1,0;1,1,0,0;0,0,1,0,0,0,0,0)$$

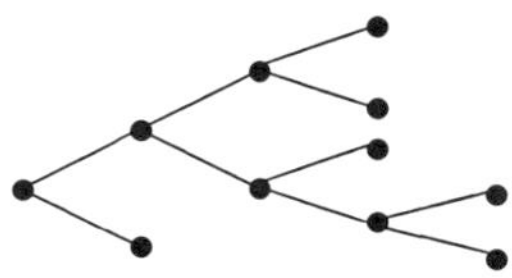

$$\to (1;1,0;1,1,0,0;0,0,0,1,0,0,0,0)$$

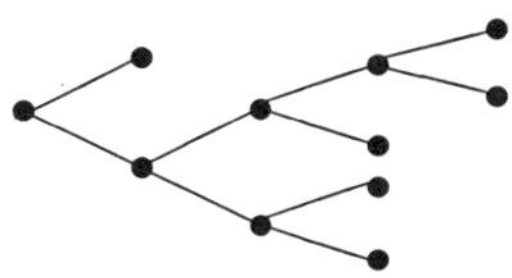

$$\to (1;0,1;0,0,1,1;0,0,0,0,1,0,0,0)$$

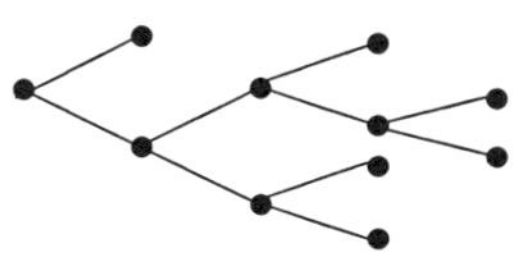

$$\to (1;0,1;0,0,1,1;0,0,0,0,0,1,0,0)$$

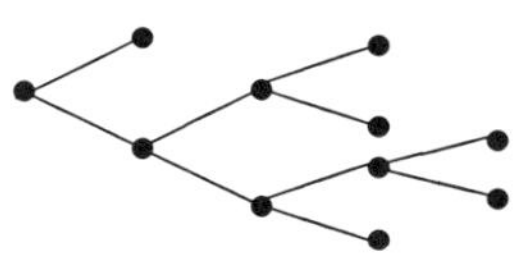

$$\to (1;0,1;0,0,1,1;0,0,0,0,0,0,1,0)$$

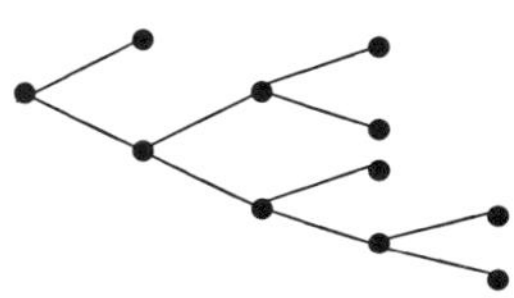

$$\to (1;0,1;0,0,1,1;0,0,0,0,0,0,0,1) \tag{3.58}$$

The fork vector $\mathbf{f}_T$ uniquely determines the shape of the binary tree T, including the composition $\mathbf{p}$. We now modify the notation $T^{\mathbf{p}}_{n,k,d}(\mathbf{x};\mathbf{y})$

to $T^{\mathbf{f}_T}_{n,k,d}(\mathbf{x};\mathbf{y}), T \in \mathbb{T}_{d+1}$. The new notation enumerates all trivalent trees in the set $\mathcal{T}_{n,k,d}$, but now in a manner that admits the application of the local reflection operations σ_w to effect further classification into nonisomorphic subsets.

The *incidence matrix* $\mathbb{I}_{\mathbf{f}_T}$ is the $h \times (2^h - 1)$ matrix of $0's$ and $1's$ having row sums given by the parts of the fork vector $\mathbf{f}_T$ and column sums by the parts of the composition $\mathbf{p}$, as illustrated by the following examples:

Examples. The fork vectors, respectively, of the four binary trees T_1, T_2, T_3, T_4 in (3.7) are $\mathbf{f}_{T_1} = (1), \mathbf{f}_{T_2} = (1;1,0), \mathbf{f}_{T_3} = (1;0,1), \mathbf{f}_{T_4} = (1;1,1)$ with corresponding incidence matrices as follows:

$$\begin{aligned} \mathbb{I}_{f_{T_1}} &= (1),\ \mathbb{I}_{f_{T_2}} = \begin{pmatrix} 1 & 0 & 0 \\ 0 & 1 & 0 \end{pmatrix}, \\ \mathbb{I}_{f_{T_3}} &= \begin{pmatrix} 1 & 0 & 0 \\ 0 & 0 & 1 \end{pmatrix},\ \mathbb{I}_{f_{T_4}} = \begin{pmatrix} 1 & 0 & 0 \\ 0 & 1 & 1 \end{pmatrix}. \quad \square \end{aligned} \tag{3.59}$$

Since the fork vector $\mathbf{f}_T, T \in \mathbb{T}_{d+1}$ fully determines the incidence matrix, we make no use of the latter.

A representative trivalent tree $R_{n,k,d} \in \mathcal{R}_{n,k,d}$ (see (3.56)) is fully characterized by the fork vector $\mathbf{f}_T, T = R_{n,k,d} \in \mathbb{T}_{d+1}$ of the binary tree skeleton and the occupation parameters $(\mathbf{x};\mathbf{y})$: The fork vector $\mathbf{f}_T$ specifies the shape of the binary root skeleton and the composition $\mathbf{p}$; and the occupation parameters $(\mathbf{x};\mathbf{y})$ specify the distribution of points onto the internal and external lines of the binary tree skeleton. Thus, the notation

$$R^{\mathbf{f}_T}_{n,k,d}(\mathbf{x};\mathbf{y}),\ T \in \mathcal{R}_{n,k,d}\,. \tag{3.60}$$

fully specifies each trivalent tree in the set of representatives $\mathcal{R}_{n,k,d}$.

We summarize the present status of the classification problem of trivalent trees in $\mathcal{T}_{n,k,d}$ into nonisomporhic trivalent trees:

Construction of complete sets of nonisomorphic trivalent trees: *(i) Select a set of representative trivalent trees $\mathcal{R}_{n,k,d}$ from the equivalence classes $\mathcal{C}_{n,k,d}$ (3.55), as described above. (ii) Describe each representative $R_{n,k,d} \in \mathcal{R}_{n,k,d}$ by the detailed notation $R_{n,k,d} = R^{\mathbf{f}_T}_{n,k,d}(\mathbf{x};\mathbf{y}), T \in \mathcal{R}_{n,k,d}$. (iii) Apply to each $R^{\mathbf{f}_T}_{n,k,d}(\mathbf{x};\mathbf{y})$ the subset of reflection operators σ_w that preserve its shape, and use the resulting pairs of reflection-equivalent relations to identify a set $\widehat{\mathcal{R}}_{n,k,d} \subseteq \mathcal{R}_{n,k,d}$ of trivalent trees, such that each trivalent tree in $\widehat{\mathcal{R}}_{n,k,d}$ is characterized by the property that no pair of trivalent trees in the set is reflection-equivalent. (iv) Obtain a complete* *set* $[n,k,d] \subseteq \widehat{\mathcal{R}}_{n,k,d}$ *of nonisomorphic trivalent trees by taking into account, as necessary, for* $n \geq 2d+2k$, *all* $\cong_\Omega$ *isomorphic trivalent trees.*

More details for implementing step (iii) are available. Step (i) classifies trivalent trees into a set $\mathcal{R}_{n,k,d}$ of representatives of the equivalence classes $\mathcal{C}_{n,k,d}$; step (ii) takes into account the shape of each of the binary tree skeletons T of the representative trivalent trees in $\mathcal{R}_{n,k,d}$ by assigning a unique fork vector $\mathbf{f}_T$ to each, and a distribution $(\mathbf{x};\mathbf{y}) \in \mathbb{D}_{n,k,d}$ of solution points onto the lines of the binary tree skeletons. But no additional constraints are placed on the distribution $(\mathbf{x};\mathbf{y})$. Thus, each trivalent tree $R_{n,k,d} \in \mathcal{R}_{n,k,d}$ is fully labeled by $R_{n,k,d} = R^{\mathbf{f}_T}_{n,k,d}(\mathbf{x};\mathbf{y})$ and, in step (iii), we exercise the freedom of assigning restricted solution points $\mathbf{x};\mathbf{y} \in \mathbb{D}_{n,k,d}$ to the trivalent tree

$$R^{\mathbf{f}_T}_{n,k,d}(\mathbf{x};\mathbf{y}) \in \mathcal{R}_{n,k,d} \tag{3.61}$$

by application of the relevant local reflection operations. It is this freedom that we next implement.

Relations (2.74)-(2.81), Chapter 2, enumerate the number $c_{d+1}(t)$ of binary trees having t forks (type t) for which both end points are external, which we call *free forks*. These numbers are related to the Catalan numbers a_{d+1} by

$$a_{d+1} = \sum_{t=1}^{[(d+1)/2]} c_{d+1}(t)\,. \tag{3.62}$$

Since a_{d+1} counts all binary trees with d forks, all binary trees skeletons that are distributed among the b_{d+1} representatives trivalent trees in $\mathcal{R}_{n,k,d}$ are accounted for, as described in the preceding paragraph: Each such binary tree skeleton has $c_{d+1}(t)$ forks of type t for some $t \in \{1,2,\ldots,[(d+1)/2]$. (Different members of $\mathcal{R}_{n,k,d}$ can have the same number $c_{d+1}(t)$ of forks of type t.) We now select a trivalent tree $R_{n,k,d} = R^{\mathbf{f}_T}_{n,k,d}(\mathbf{x};\mathbf{y})$, identify the type t to which it belongs, enumerate the t free fork roots of the t binary tree skeleton by $v_{w_1}, v_{w_2}, \ldots, v_{w_t}$, and the corresponding local reflection operators by $\sigma_{w_1}, \sigma_{w_2}, \ldots, \sigma_{w_t}$. The application of these t shape-preserving local reflection operators to the set of all trivalent trees

$$R^{\mathbf{f}_T}_{n,k,d}(\mathbf{x};\mathbf{y}),\ \ (\mathbf{x};\mathbf{y}) \in \mathbb{D}_{n,k,d} \tag{3.63}$$

corresponding to the solution set $\mathbb{D}_{n,k,d}$ of the Diophantine relations (3.9)-(3.10) allows us to choose exactly the subset with the property

$$R^{\mathbf{f}_T}_{n,k,d}(\mathbf{x};\mathbf{y}),\ \ y_{w_i1} \geq y_{w_i2},\ i = 1,2,\ldots t\,. \tag{3.64}$$

This gives a set of trivalent trees that we denote by

$$\widehat{\mathcal{R}}_{n,k,d} \subseteq \mathcal{R}_{n,k,d}. \tag{3.65}$$

If there are no $\Omega-$type isomorphisms among the trivalent trees in the set $\widehat{\mathcal{R}}_{n,k,d}$, then this set is the subset of nonisomorphic trivalent trees contained in $\mathcal{T}_{n,k,d}$. Thus, we have from (3.36) that

$$\begin{aligned} |\,[n,k,d]\,| &= |\,\widehat{\mathcal{R}}_{n,k,d}\,|,\ 2d+k+1 \leq 2d+2k-1, \\ |\,[n,k,d]\,| &\leq |\,\widehat{\mathcal{R}}_{n,k,d}\,|,\ n \geq 2d+2k. \end{aligned} \tag{3.66}$$

A counting formula in the second instance, where Ω operations necessarily enter, appears to be quite difficult. As examples, we have the following: The steps leading to the set $\widehat{\mathcal{R}}_{n,k,d}$ is illustrated for $\mathcal{T}_{10,3,2}$ by $[10,3,2] = \widehat{\mathcal{R}}_{10,3,2}$ given by (3.37). It is also the case that the set $\widehat{\mathcal{R}}_{9,2,2}$ is given by $\widehat{\mathcal{R}}_{9,2,2} = \{[9,2,2], T_9(0,1,1,0)\}$, where $[9,2,2]$ is given by (3.38), and where $T_9(0,1,1,0)$ is the trivalent tree given by the left-hand side of (3.34); it can only be removed from $\widehat{\mathcal{R}}_{9,2,2}$ by an $\Omega-$type isomorphism to obtain the set of nonisomorphic trivalent trees $[9,2,2]$.

Counting formulas can be given for the following related problems. Let $\mathbb{B}_{d,h}(\mathbf{p})$ denote the set of binary trees in the subset of $\mathbb{T}_{d+1}$ containing d forks in which the number of forks in column m is given by the $m-$th part p_m of the composition $\mathbf{p} = (p_0, p_1, \ldots, p_{h-1}) \vdash d, m = 0,1,2,\ldots,h-1$. Then, since these parts must satisfy the conditions $1 \leq p_m \leq 2p_{m-1}$, $m = 1,2,\ldots,h-1; p_0 = 1$, there are

$$|\,\mathbb{B}_{d,h}(\mathbf{p})\,| = \prod_{m=1}^{h-1} \binom{2p_{m-1}}{p_m} \tag{3.67}$$

trivalent trees in the set $\mathbb{B}_{d,h}(\mathbf{p})$. It follows from this result that the Catalan number $a_{d+1} = |\,\mathbb{T}_{d+1}\,|$ is obtained as the sum:

$$\sum_{h=\underline{h}}^{d} \sum_{\mathbf{p}\vdash d} \prod_{m=1}^{h-1} \binom{2p_{m-1}}{p_m} = a_{d+1} = \frac{1}{d+1}\binom{2d}{d}, \tag{3.68}$$

where $\underline{h}$ is defined by (3.5). We encounter again a solvable problem of counting the number of compositions of a given positive integer into d positive parts, where there are restrictions on each of the parts (see Sect. 11.1.1, Compendium B). The results (3.67)-(3.68) are, of course, valid for binary trees, in general, independently of their occurrence as skeletons of trivalent trees.

3.3 Cubic Graphs Associated with Pairs of Trivalent Trees

We have seen in relations (2.237)-(2.246) how pairs of trivalent trees lead to adjacency diagrams. We examine this structure again in the context of unlabeled trivalent trees. Let $V_n, V'_n \in \mathbb{V}_n$ be any pair of trivalent trees of order n. Adjoin $n+2$ lines that go **between** the two trivalent trees such that the resulting graph contains only points of degree 3. This is always possible, since the $d+2$ points of degree 1 and $n-2d-2$ points of degree 2 belonging to V_n can always be matched with the $d'+2$ points of degree 1 and $n-2d'-2$ points of degree 2 belonging to $V'n$ by exactly $2(d+2)+(n-2d-2) = 2(d'+2)+(n-2d'-2) = n+2$ *joining lines* such that every point in the conjoined graph has degree 3. We call such a conjoining of trivalent trees graph a *regular union* of V_n and V'_n. Each possible regular union of V_n and V'_n gives a new graph containing $2n$ points and $2(n-1)+n+2 = 3n$ lines, with each point of degree 3. Such a graph is called a *cubic graph*. There are, of course, many regular unions of two given trivalent trees V_n and V'_n, each of which gives a cubic graph. We denote by $\mathbb{C}(V_n, V'_n)$ the set of all cubic graphs obtained by all possible regular unions of a pair of trivalent trees $V_n, V'_n \in \mathbb{V}_n$ and an element of this set by $C(V_n, V'_n) \in \mathbb{C}(V_n, V'_n)$.

Examples. It is useful to look at the number of regular unions for $n = 2, 3, 4$:

1. $n = 2$: There is but a single trivalent tree, which is •—•, and the conjoining by four lines going between two such trivalent trees gives uniquely the cubic graph with diagram:

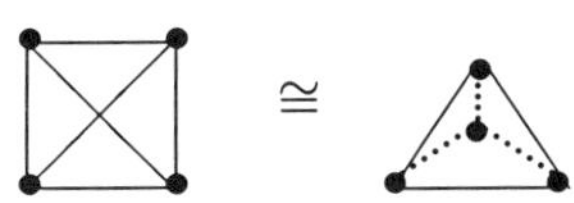 (3.69)

There is only one cubic graph for $n = 2$, and it is characterized by the property that each of its points is adjacent to each of the remaining three points. Shown is the planar $\mathbb{R}^2$ diagram and the corresponding tetrahedral diagram in $\mathbb{R}^3$. There is only one cubic graph on four points, which accounts for there being only one Racah coefficient.

2. $n = 3$: There is but a single trivalent tree, which is •—•—•, and the conjoining by five lines going between two such trivalent trees gives five cubic graphs with the following diagrams:

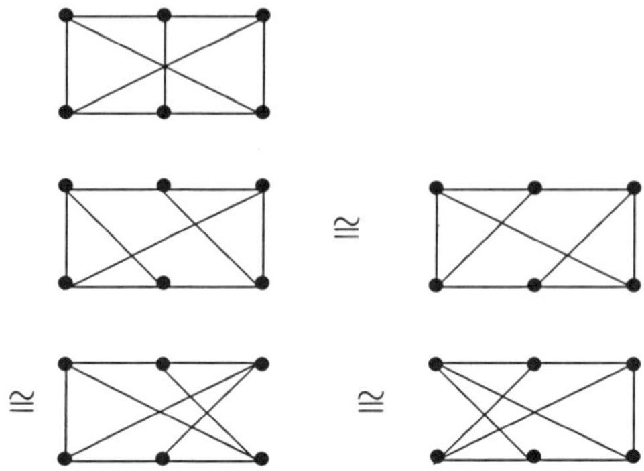

(3.70)

But applying the criterion that two graphs are isomorphic if there is a one-to-one correspondence between their points that preserves adjacency, we find that there are only two nonisomorphic cubic graphs on six points. The equivalences shown are evident from the diagrams, except for that between the first and third one in (3.70). This equivalence is shown by labeling the points in the first one by $1, 2, 3$ from left-to-right across the top horizontal line and by $4, 5, 6$ from left-to-right across the bottom horizontal line, and labeling the points in the third one by $1, 5, 6$ from left-to-right across the top horizontal line and by $2, 3, 4$ from left-to-right across the bottom horizontal line. The two nonequivalent planar cubic graphs on six points can also be presented as diagrams in $\mathbb{R}^3$:

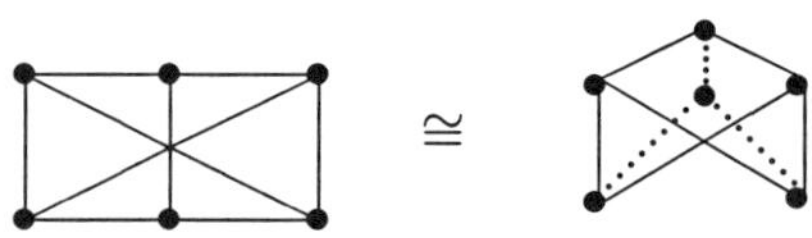

(3.71)

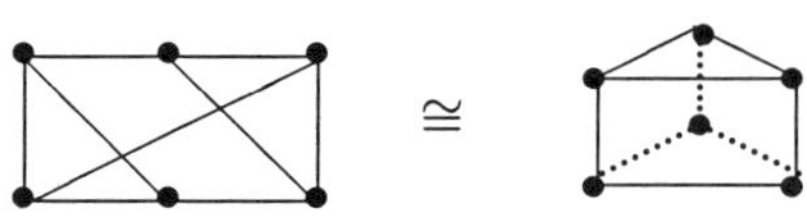

(3.72)

Thus, only two nonisomorphic cubic graphs can arise in the coupling of four angular momenta, the first one being the geometrical cubic graph presentation of the Wigner $9-j$ coefficient; the second that of the B-E identity.

The basis for identifying the cubic graph (3.72) with the B-E identity can be given on purely geometrical grounds by presenting this cubic graph in the following form in which it has been partitioned into two subgraphs joined by three dotted lines:

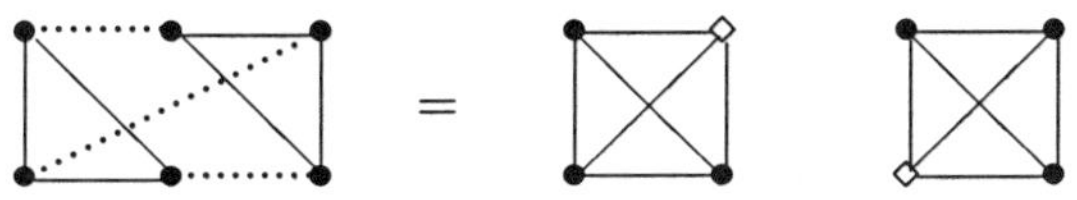

(3.73)

The diagram on the right is obtained from the regular union of trivalent trees on the left by cutting the three ⋯ lines and joining the ends together to form a point $\diamond$, which is shown as two points on the right of the equal sign. The diagram on the right is then the product of two tetrahedra in which the two *diamond* points are to be identified as the same point. (We write the product as simple juxtaposition, but will subsequently need to refine the concept of product in Sect. 3.4.2.) This is the content of the right-hand side of the B-E identity as expressed in the three forms (2.207)-(2.209), where the point $\diamond$ receives the same label $(b\, k_1\, k_2')$. This extra triangle does not appear in the left-hand side of any of the forms (2.207)-(2.209); its appearance in the B-E identity may be attributed to the multiplication properties of recoupling matrices, which is the source of the B-E identity, as noted earlier. This geometrical property of the joining of a pair of trivalent trees into a cubic graph for $n = 3$ is the fundamental reason for the existence of the B-E identity, and for the existence of only one Wigner $9-j$ coefficient.

3. $n = 4$: The trivalent trees in $\mathbb{V}_4$ are

(3.74)

Since the pairing is independent of order of the trivalent trees, there are three paired sets of trivalent trees. Each pair is joined by six lines in all possible ways such that the degree of each point is three, thus obtaining a large collection of cubic graphs. But each cubic graph in this collection is isomorphic to one in the following set of five, each one in this set originating from the pairing of the left-most trivalent tree in (3.74) with itself:

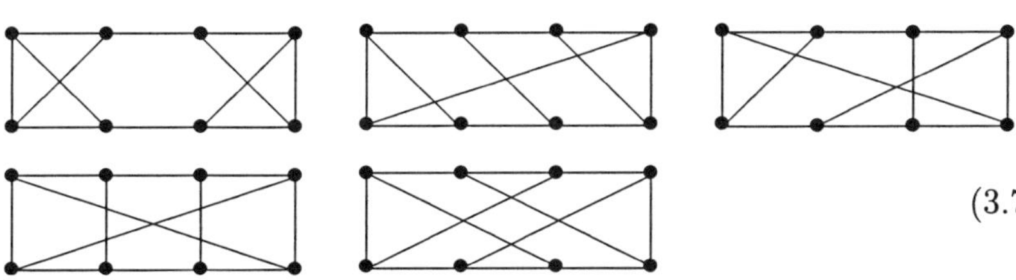

(3.75)

Several of the isomorphisms between cubic graphs of order eight are illustrated by

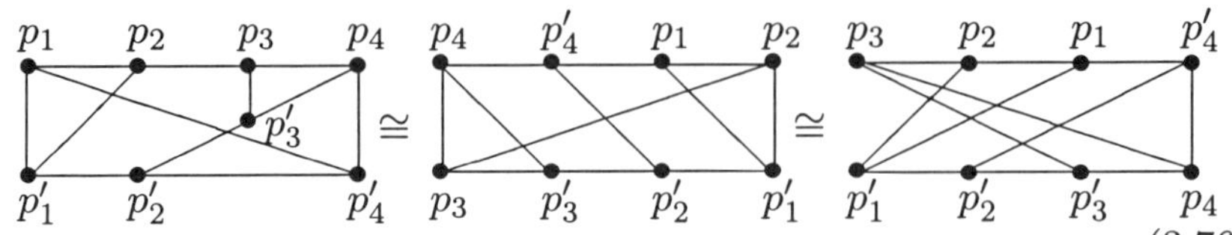

(3.76)

The first isomorphism shows that the cubic graph obtained by the regular union of the separate pair of inequivalent trivalent trees (3.74) can be isomorphic to a cubic graph obtained by the regular union of the left-most trivalent tree with itself.

It is known from graph theory that the five cubic graphs on eight points given by (3.75) constitute all such cubic graphs. It turns out that all cubic graphs on four, six, and eight points arise in the binary theory of the coupling of angular momenta, but this property fails for ten points. *Not all cubic graphs containing* $2n$ *points can be obtained by the regular union of two trivalent trees of order* n *by* $n+2$ *lines* (see (3.87) below). We require more detailed analysis of this property, as well as the properties of those cubic graphs that do occur and give rise to the factoring of the associated fundamental triangle coefficients such as occurs for the B-E identity. We return to this below, but let us first note some geometrical aspects of the coupling of three and four angular momenta.

The geometrical content of the Racah sum rule, the B-E identity, and the Wigner $9-j$ coefficient can be described in terms of cubic graphs, but the labels associated with the vertices of the cubic graphs must be assigned in accordance with the rules already set forth. This point needs further discussion. We have chosen in relation (2.155) to associate the Racah coefficient with a special pair of standard labeled binary trees and the corresponding fundamental triangle coefficient of order four:

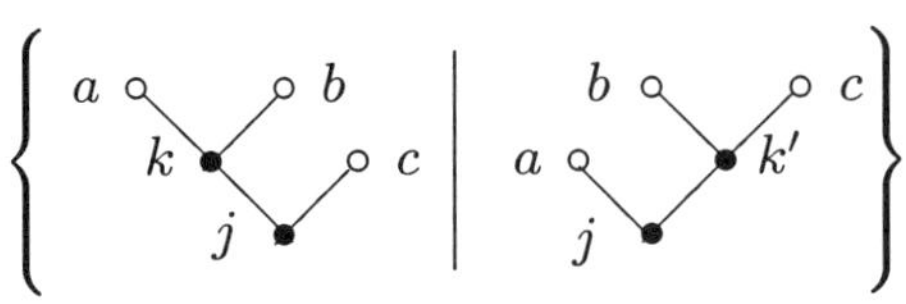

(3.77)

$$= \left\{ \begin{array}{cc|cc} a & k & b & a \\ b & c & c & k' \\ k & j & k' & j \end{array} \right\} = \sqrt{(2k+1)(2k'+1)}\,W(abjc;kk').$$

We have earlier remarked that a tetrahedron has the unique property that each of its points is adjacent to every other point. This property can be realized by assigning four arbitrary triangles, say, $(abc), (dec), (dbf)$, (aef), each of which shares exactly one entry with each of the other triangles, to any four distinct points and joining by lines each pair of points that shares a common letter. But the only labelings of cubic graphs that we allow in the treatment of angular momentum are those that arise from the standard labeling of a pair of binary trees of order 2—there are four external points to be assigned angular momentum labels, since the two internal points and the roots have the standard k_1, k_2, j and k_1', k_2', j assignment. If we keep the order of the parts in each of $(abc), (dec), (dbf), (aef)$ fixed, then there is no assignment of these triangles to the pair of standard labeled binary trees that gives the coupling rules for three angular momenta. By design, all standard labelings leave invariant the set of eigenvalue relations associated with the coupling scheme. Thus, extra conditions are imposed on the structure of angular momentum triangles that label the vertices of the tetrahedron. Exactly these same restrictions apply to the associated fundamental triangle coefficients with four columns of angular momentum triangles—they are always read off the pair of labeled binary trees by the stated rule (2.126), as illustrated by the examples (2.127). It is, of course, the case that if permutations among the parts of the four triangles $(abc), (dec), (dbf), (aef)$, are allowed, then they can always be brought to the form of two distinct angular momentum coupling schemes. In our usage, it is always the coupling schemes themselves that determine the relations between the angular momentum triangles, and, most significantly, the adjacency of points in the corresponding labeled cubic graph. It is useful to discuss these rules in relation to the Racah coefficient.

The assignment of a tetrahedron to a Racah coefficient is made on the basis of relation (2.155), which is (3.77) above, with phase factors in accordance with the properties of the triangle coefficient of order four as presented in (2.155)-(2.156):

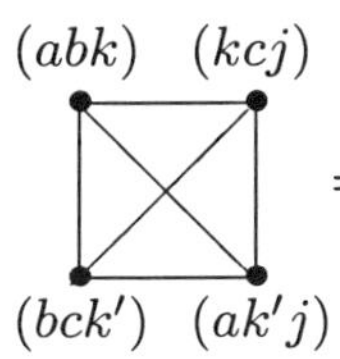

$$= \left\{ \begin{array}{cc|cc} a & k & b & a \\ b & c & c & k' \\ k & j & k' & j \end{array} \right\}. \qquad (3.78)$$

Given the point assignment $p_1 = (abk), p_2 = (kcj), p_1' = (bck'), p_2' = (ak'j)$ in terms of the angular momentum quantum numbers a, b, c, j, k, k', as read off the pair of binary trees in (3.77), the fundamental triangle coefficient of order four is assigned exactly the labels shown. This provides the labeling of the four points of the tetrahedron on the left. We can think of this relation as providing a symbolic identity between the labeled tetrahedron and the fundamental triangle coefficient, as shown. But now we can apply all six permutations of the three angular momenta a, b, c

labeling the external points to obtain *signed* labeled tetrahedra. For example, the interchange of a and b in (3.78) gives the signed tetrahedron:

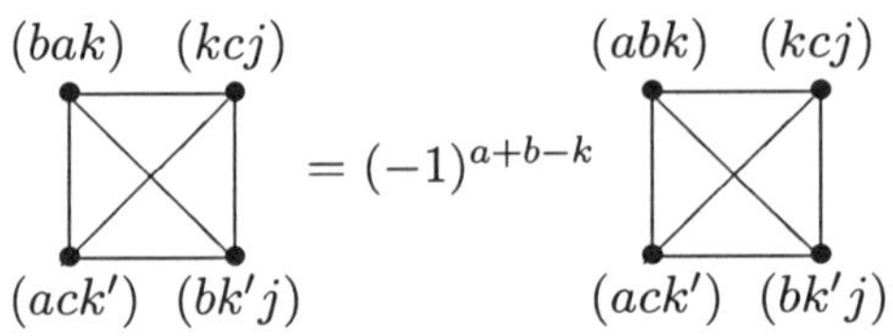

(3.79)

The interchange of the top and bottom labels in (3.78) is also admitted:

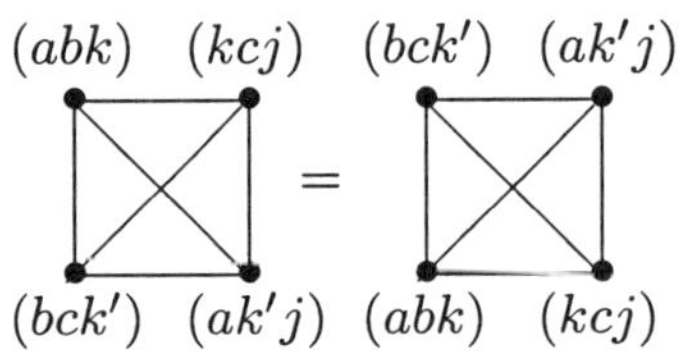

(3.80)

The general rule is: A labeled (possibly signed) tetrahedron is a symbolic graphical presentation of the numerical-valued triangle coefficient of order 4 corresponding to a standard labeled pair of binary trees, each of order 2. The content of such relations is exactly the same as their triangle coefficient presentation, no more and no less.

The entire content of angular momentum recoupling theory can be expressed by symbolic relations between signed tetrahedra, as defined above. We illustrate this by writing relations (2.185),(2.187),(2.208), and (2.210) for the orthogonality relations, the Racah sum rule, the Biedenharn-Elliott identity, and the Wigner $9-j$ coefficient as symbolic rules for composing tetrahedra:

1. Orthogonality relations: :

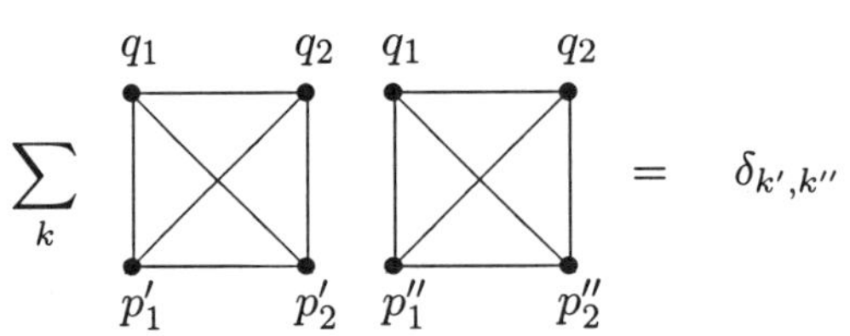

(3.81)

where all the triangles q_i have a part equal to the summation parameter k and are given in terms of angular momentum quantum numbers by $q_1 = (abk), q_2 = (kcj)$; and the 'free' triangles are given by $p'_1 = (bck'), p'_2 = (ak'j), p''_1 = (bck''), p''_2 = (ak''j)$.

2. Racah sum rule:

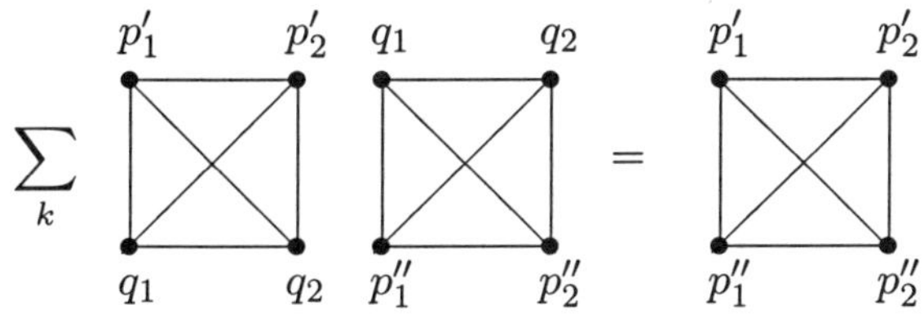

(3.82)

where all the triangles q_i have a part equal to the summation parameter k and are given by in terms of angular momentum quantum numbers: $q_1 = (bck), q_2 = (kaj)$; and the free triangles are given by $p'_1 = (abk'), p'_2 = (k'cj), p''_1 = (ack''), p''_2 = (k''bj)$.

3. Biedenharn-Elliott identity:

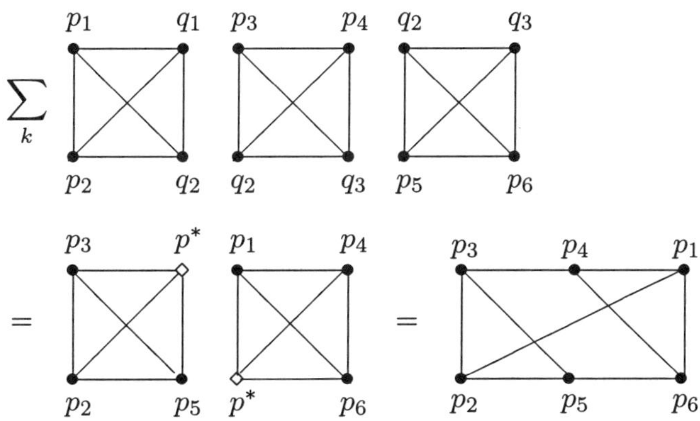

(3.83)

where all the triangles q_i have a part equal to the summation parameter k and are given by $q_1 = (ak_2k), q_2 = (k'_1dk), q_3 = (ckj)$; the free triangle p_i by $p_1 = (bdk_2), p_2 = (abk'_1), p_3 = (cak_1), p_4 = (k_1k_2j), p_5 = (k'_1ck'_2), p_6 = (dk'_2j)$; and the new, created triangle in the middle term only by $p^* = (bk_1k'_2)$. Expression (3.83) conceals the fact that the B-E identity has a row-column matrix element form similar to the Racah sum rule. This is because of the reductions that take place in consequence of Kronecker delta factors.

4. Wigner $9-j$ defining relation:

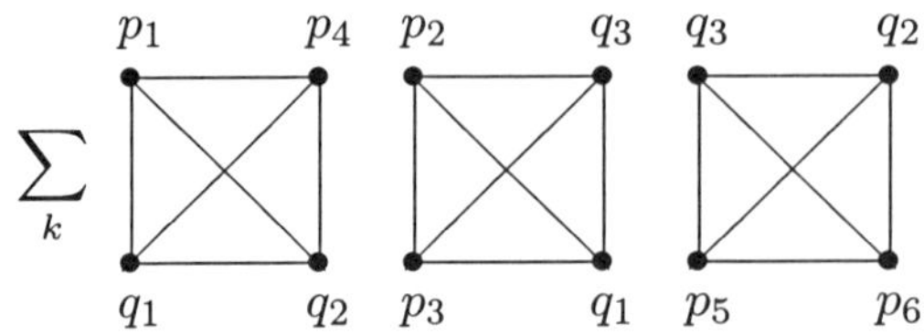

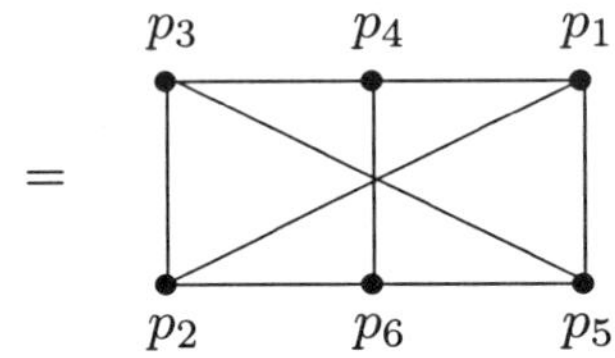

(3.84)

where all the triangles q_i contain a part equal to the summation parameter k and are given by $q_1 = (bk_1k), q_2 = (kdj), q_3 = (k_1'ck)$; and the p_i by $p_1 = (bdk_2), p_2 = (abk_1'), p_3 = (cak_1), p_4 = (k_1k_2j), p_5 = (cdk_2'), p_6 = (k_1'k_2'j)$.

There are denumerably infinitely many such relations between recoupling coefficients of the sort given above. We will eventually understand in the next chapter on generating functions that all such relations are expressions between **integers** because all recoupling coefficients are integers times square-root factors, and all such square-root factors either get squared or can be canceled from the two sides of relations between recoupling coefficients, leaving behind integer relationships. In particular, this is true for the expressions of recoupling coefficients in terms of the basic $6-j$ coefficients. The "tetrahedral integers" constitute, in some sense, a "basis." Angular momentum theory is about the properties of integers in which Diophantine theory eventually gets entangled.

Adjacency properties of triangles have a vital role in the structure of relations between recoupling coefficients. In expressing a recoupling coefficient in terms of labeled tetrahedra, we connect only the four points with six lines within each of the individual tetrahedra. No lines go between different tetrahedra. This is the content of the symbolic representation (3.83) of the relation (2.208) for the B-E identity. In particular, all $q-$triangle labeling points have a part equal to the summation index in the above examples and are adjacent, but only those belonging to a given tetrahedron are connected. The other triangles that label points that belong to different tetrahedra under the summation may likewise be adjacent, but they are not connected.

The symbolic identity

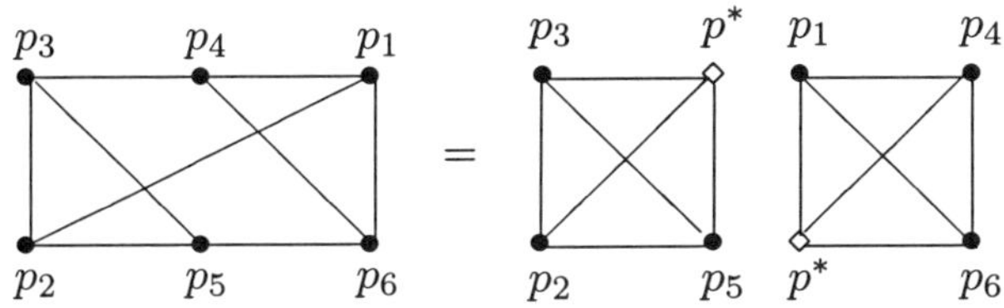

(3.85)

is crucial in the B-E identity. Relation (3.85) is rather remarkable in that it shows that a labeled cubic graph on six points can be constructed from the product of two labeled tetrahedra and a point p^* that appears as a *virtual* point, where we use the term virtual to describe a point or triangle that appears in one side of a relation between labeled cubic graphs, but not in the other. It is an intrinsic property of the cubic graph on the left in (3.85).

From the viewpoint of angular momentum theory, the factorizing property of the particular cubic graph in (3.85) is a consequence of the simple relation between orthogonal recoupling matrices expressed by relation (2.125), as implemented through the matrix elements of this identity in relations (2.196)-(2.198), and using properties of triangle coefficients. The virtual triangle made its appearance automatically in the process of taking matrix elements. But there is nothing special about this procedure, which implies that we can expect virtual triangles to enter in many relations between recoupling coefficients.

A general property of all relations between recoupling coefficients is that there must be *conservation of free triangles and of the adjacency of these triangles.* This phrase is intended to mean that the free triangles and their adjacency in the left- and right-hand sides of an identity between two cubic graphs be invariant (the same); "free" applies to those triangles not containing a summation index that are not virtual triangles, where we observe that virtual triangles come in identical pairs, should they appear. This property of such relations must be true or the relation is self-contradictory. It is interesting how this property comes about under a summation over labeled tetrahedra, as in the above examples. It is directly verified that when two triangles are free and adjacent in a given tetrahedron, they are adjacent in both sides of the identity. If they belong to different tetrahedra, the adjacent or non-adjacent property is verified by comparison of the triangles, and is carried forward to the other side of the relation. The transitive property of adjacent points belonging to different tetrahedra can also be important. Adjacency of triangles is reflective, but it is not, in general, transitive. But there can be subsets of triangles that are transitive. For example, in the expression for the Wigner $9-j$ coefficient, we have $p_1 \equiv p_4, p_1 \equiv q_1 \equiv p_2$ implies

$p_1 \equiv p_2$ and $p_1 \equiv q_2 \equiv p_5$ implies $p_1 \equiv p_5$ in the left-hand side, which accounts for the three lines adjacent to p_1 in the cubic graph symbol for the $9-j$ coefficient. Similarly, in the expression for the B-E identity, we have $p_1 \equiv p_2, p_1 \equiv q_2 \equiv p_4$ implies $p_1 \equiv p_4$ and $p_1 \equiv q_2 \equiv p_6$ implies $p_1 \equiv p_6$ in the left-hand side, which accounts for the three lines adjacent to p_1 in the cubic graph symbol of the B-E identity.

We next address the general relationship between pairs of unlabeled nonisomorphic trivalent trees and unlabeled cubic graphs. We have defined on p. 189 the notion of the regular union of two trivalent trees $V_n, V'_n \in \mathbb{V}_n$ of order n : all points of degree 1 and 2 of belonging to the separate trivalent trees are joined together to obtain a graph having $2n$ points and $3n$ lines such that each point of the composite graph has degree 3. Each such joining gives a cubic graph, and the set of all such cubic graphs is denoted $\mathbb{C}(V_n, V'_n)$. The required $n+2$ joining lines always exists, as shown in the first paragraph of this section. Unfortunately, the regular union of two trivalent trees, one in the set of complete nonisomorphic $[n, k, d]$, the other in $[n, k', d']$, does not, in general, give a set of nonisomorphic cubic graphs: The identification of a set of nonisomorphic cubic graphs with $2n$ points is not resolved. The set of cubic graphs $\mathbb{C}(V_n, V'_n)$ does contain all cubic graphs that can arise in recoupling matrices associated with the binary coupling of angular momenta, but further refinement of the structure of the set $\mathbb{C}(V_n, V'_n)$ is needed to determine the nonisomorphic cubic graphs contained therein, a task to which we next turn.

3.4 Cubic Graphs

As an abstract graph, a cubic graph may be defined as follows:

A cubic graph C_{2n} of order $2n$ is a collection of $2n$ distinct points $p_i, i = 1, 2, \ldots, 2n$, and $3n$ lines $l_j, j = 1, 2, \ldots, 3n$, such that three lines are incident on each point, there being exactly one line going between each pair of points.

Cubic graphs can be presented in various forms for visualizing their properties: A collection of points in Cartesian $\mathbb{R}^3$ space; a collection of points (planar graphs) in Cartesian $\mathbb{R}^2$ space; and as arrangements of triangles as columns in a matrix having 3 rows and $3n$ columns (see Sect. 3.4.3). The previous definitions of adjacency of points and of isomorphic graphs apply also to cubic graphs:

1. Two points in a cubic graph are *adjacent* if there exists a line incident on each point.
2. Two cubic graphs are *isomorphic* if there exists a one-to-one correspondence between their points that preserves adjacency.

We introduce the following notations for the sets of cubic graphs of order $2n$ encountered:

$$\begin{aligned} \mathbb{C}_{2n} &= \{\text{all nonisomorphic cubic graphs of order } 2n\}, \\ \mathbb{C}_{n,n} &= \{\mathbb{C}(V_n, V'_n) \,|\, V_n \in [n,k,d],\ V'_n \in [n,k',d']\}. \end{aligned} \qquad (3.86)$$

Examples. It is useful to illustrate some unusual planar cubic graphs:

1. A cubic graph with one line joining two (noncubic) subgraphs:

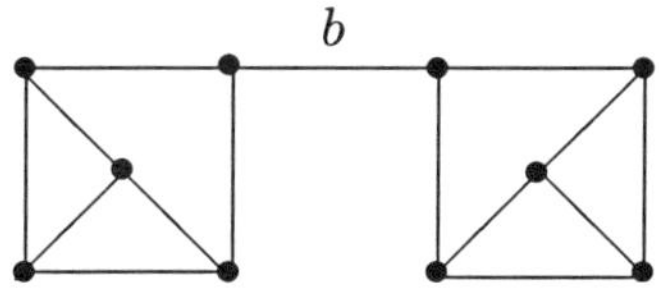

(3.87)

   This cubic graph on ten points is a member of $\mathbb{C}_{10}$, but not of $\mathbb{C}_{5,5}$; it cannot be assembled from two trivalent trees of order five joined by seven lines, because, by symmetry one of the joining lines must be b, and if this line is removed, there is no end point of degree 1 in either of the two disjoint subgraphs that remains to serve as base point of a trivalent tree. This illustrates the property: *Not all cubic graphs of order* $2n$ *are contained in* $\mathbb{C}_{n,n}$.

2. The following cubic graph on sixteen points factors into four cubic subgraphs of order four:

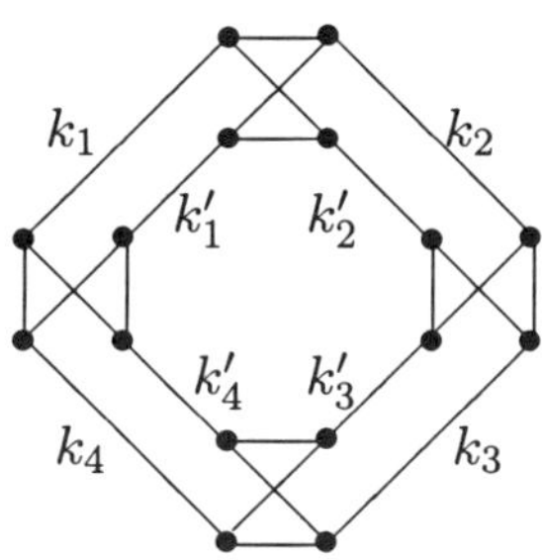

$$= \prod_{i=1}^{4} \delta_{k_i,k'_i}$$

(3.88)

$\times$

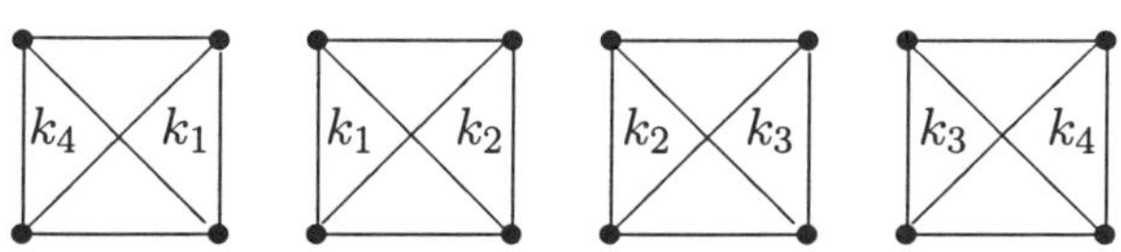

We can cut the lines k_2, k_2', k_4, k_4' and join together smoothly (no new • point) the end points k_2 and k_2' and those of k_4 and k_2' thus obtaining a product of two cubic subgraphs on eight points. This procedure can be repeated on each of these subgraphs in the product thus obtaining the product of the four tetrahedra shown on the right-hand side having the shared vertical line labels shown. This planar cubic graph of order sixteen generalizes in the obvious way to arbitrarily many subunits containing the four • points shown, with pairs of points on two concentric circles. Thus, there are denumerably infinitely many cubic graphs that can be viewed as factoring into cubic subgraphs under the cutting and joining of lines. *None of these cubic graphs belongs to* $\mathbb{C}_{n,n}$. *They do not occur in angular momentum binary recoupling theory.* This is proved below in the Yutsis factoring theorem for cubic graphs in the set $\mathbb{C}_{n,n}$ by showing that the line labels in the product on the right-hand side of (3.88) cannot occur in cubic graphs belonging to the set $\mathbb{C}_{n,n}$.

These examples illustrate the subtleties of the role of general cubic graphs in the binary recoupling theory of angular momenta. The first example (3.87) shows that a cubic graph may fail to be in the set $\mathbb{C}_{n,n}$ because the lines joining subgraphs do have the required properties; the second example (3.88) fails, as shown below, because the structure of the factors is incompatible with a cubic graph belonging to $\mathbb{C}_{n,n}$. The cutting and joining of lines and the associated factoring of cubic graphs in the set $\mathbb{C}_{n,n}$ is an important refinement for determining the subset $\mathbb{C}_{n,n} \subset \mathbb{C}_{2n}$ of nonisomorphic cubic graphs that occur in binary recoupling theory of angular momenta.

3.4.1 Factoring properties of cubic graphs

A significant property of cubic graphs for the classification of recoupling matrices is the following:

Yutsis factoring theorem for cubic graphs (sufficiency version). *If a cubic graph is constituted of two subgraphs joined by two or three lines, and each of the subgraphs contains at least three points, then this cubic graph factors into two cubic graphs of lower order by the cut and join of lines. In the case of two lines labeled by* k *and* k', *there is a Kronecker delta contraction* $\delta_{k,k'}$ *that gives two disjoint cubic graphs that share the common line labeled* k. *In the case of three lines labeled by* k, q, k', *a new virtual point* $p^* = (kqk')$ *is created that gives two disjoint cubic graphs that share the common point* p^*, *provided the cut and join produces no loops (a line that closes on the same point) or double lines (two lines between the same point).*

Proof. We give the proof by examples, since it is clear that the illustrated properties generalize. The two cubic graph with diagrams given by the first and second lines of (3.89) below illustrate this factoring. Note, however, that the cutting of the three lines in the first graph gives a double line; these lines cannot be cut and rejoined to obtain a pair of cubic graphs. Moreover, the graph on the third line shows that a cubic graph with four joining lines can also be partitioned into two cubic subgraphs by the cut and join operation.

Examples. Partitioning of cubic graphs into cubic subgraphs by cutting and rejoining of lines:

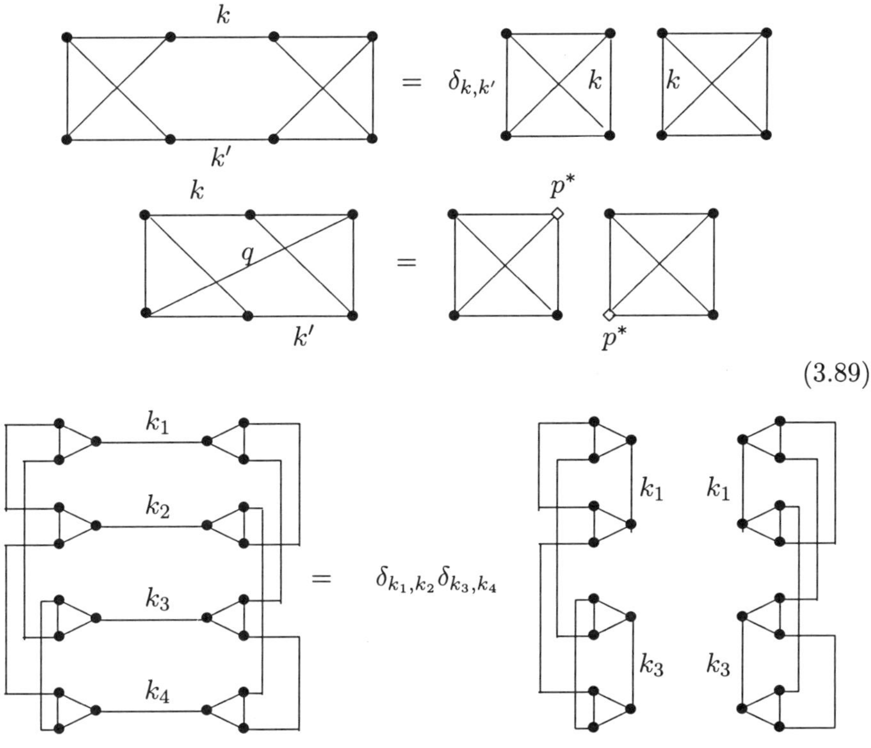

(3.89)

If one, or both, of the cubic subgraphs that occurs in the factoring theorem is itself constituted of two subgraphs joined by two or three lines, then the factoring process can be continued. The cubic graph on the left in the third line of (3.89) has $2 \cdot 12 = 24$ points and $3 \cdot 12 = 36$ lines, while each of the two cubic subgraphs on the right has $2 \cdot 6 = 12$ points and $3 \cdot 6 = 18$ lines. (Each right-angle line incident on a pair of • points represents the same smooth line, with no points of the graph.)

The number of points and lines is conserved in these factorings for pairs of cut lines that are rejoined into lines, as in the first and third examples in (3.89), two lines are replace by two lines, one each in the two disjoint cubic subgraphs. But when three cut lines are joined into a point, as in the second example, the number of points is increased by two because a single new virtual point has been created, and it belongs to each of the two disjoint cubic subgraphs—points are not conserved in this process. □

Application of the Yutsis factoring theorem to the five nonisomorphic cubic graphs on eight points given by (3.75) gives the following: The first one in the top row factors into a product of two tetrahedra, the second into the product of three tetrahedra, and the third into a tetrahedra times a cubic graph on six points, the cubic graph on six points being the one that corresponds to the Wigner $9-j$ coefficient. Thus, these three nonisomorphic cubic graphs give no coefficients not already encountered in the binary coupling of four angular momenta or less. The Yutsis factoring theorem, as stated, does not apply to the two cubic graphs given in the second row of (3.75), but it is the case that these two cubic graphs cannot be factored (see 3.90 below) by cut and join operations. These two give rise to two $12-j$ coefficients. We have not yet proved this, nor have we yet given a definition of a $3n-j$ coefficient, results we supply below. The factoring theorem is due to the Yutsis *et al.* [192, 193] school and repeated in the present notation in Ref. [21].

We have discussed in some detail how binary recoupling theory leads to cubic graphs in the sets $\mathbb{C}_{n,n}, n \geq 2$. These cubic graphs are always obtained from a pair of binary trees which present the coupling of $n+1$ angular momenta $j_1, j+2, \ldots, j_{n+1}$ to a final angular momentum j, as exhibited in the standard labeling of two binary trees without common forks, as explained in detail in Chapter 2. It is convenient to refer to these cubic graphs as being of *angular momentum type*. We denote this set of cubic graphs by $\mathbb{G}(T,T')$. Only cubic graphs $G(T,T') \in \mathbb{G}(T,T')$ of angular momentum type can occur in the binary recoupling theory of angular momenta, including those associated with the cubic graphs that occur in a factoring. The five nonisomorphic cubic graphs on eight points given by (3.75) are all obtained from the regular union of a pair of trivalent trees of order 4, which all are members of $\mathbb{G}(T,T')$, where T,T' are binary trees of order five. The three in the top line factor into products of the required angular momentum type, while the two in the bottom line cannot be factored: they correspond to two $12-j$ coefficients, as noted above. *Being of angular momentum type (including factors) severely restricts the cubic graphs that can arise in binary recoupling theory for* $n \geq 5$.

The results discussed above pose the following question: Given a cubic graph $C_{n,n} \in \mathbb{C}_{n,n}$ that is the regular union of two trivalent trees of order n, what are the conditions that it factor into a product of two cubic subgraphs $C_{r,r} \in \mathbb{C}_{r,r}$ and $C_{s,s} \in \mathbb{C}_{s,s}$, by the cut and join process such that each subgraph is of angular momentum type? This is a paramount question that we next address.

We first observe that each cubic graph $G_{p,q,r,s} \in \mathbb{C}_{2n}$ satisfies:

The cut and join of lines in a cubic graph $G_{p,q,r,s} \in \mathbb{C}_{2n}$ can effect a product into two cubic graphs $C_{2r} \in \mathbb{C}_{2r}$ and $C_{2s} \in \mathbb{C}_{2s}$, if and only if $G_{p,q,r,s}$ has the form presented by the diagram:

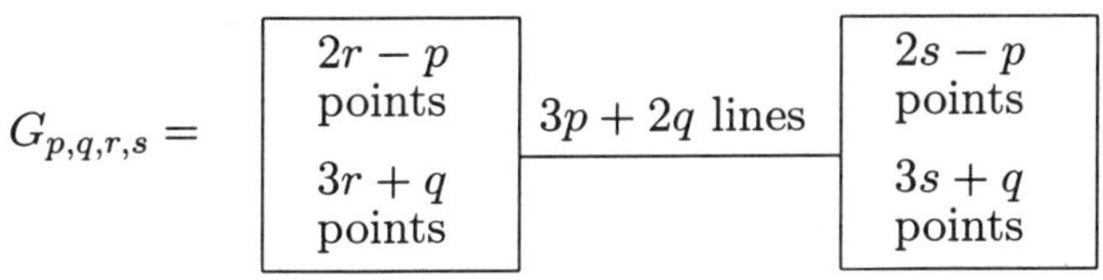

(3.90)

The diagram on the right represents two subgraphs joined by $t = 3p+2q \geq 2$ lines, the first containing $2r - p \geq 3$ points and $3r + q$ lines, the second $2s - p \geq 3$ points and $3s + q$ lines, where p and q are some pair of nonnegative integers, not both 0, and we require the graphs concealed in the boxes to have exactly three lines incident on each point (degree 3).

Proof. Suppose there are $t \geq 2$ joining lines. Then, since every positive integer $t \geq 2$ can be written in the form $t = 3p + 2q$ for some pair of nonnegative integers p and q, we can collect the lines together into p subsets of three lines and q subsets of two lines. After cutting all the t lines, each identified subset of three lines is joined to a point, thereby creating p new pairs of points, each of degree 3; each identified subset of two lines is joined smoothly to a line, thus preserving the number 2q of lines. One set of p points of degree three goes back into the left-hand box and one set into the right-hand box, thus increasing the number of points of degree three in each box to $2r$ and $2s$, respectively; similarly, q lines go back into each of the boxes, thus decreasing the number of lines by q. Thus, the number of points and lines is $2r$ and $3r$ in the left-hand box, and $2s$ and $3s$ in the right-hand box, each point being of degree 3. Thus, diagram (3.90) is necessary; it is clearly sufficient. □

We can also derive conditions for a cubic graph to be realized as a regular union of two trivalent trees. Consider the following diagram:

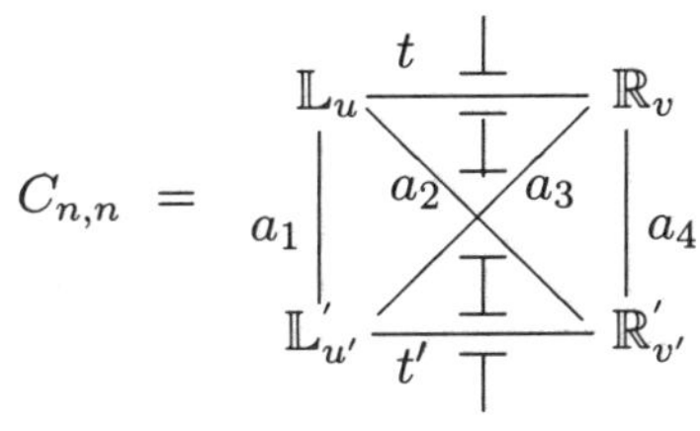

(3.91)

This diagram presents a cubic graph $C_{n,n}$ in terms of points and lines that taken together constitute a cubic graph that belongs to the regular union $\mathbb{C}_{n,n}$ of any pair of trivalent trees in the set $\mathbb{V}_n$. This is proved below, after defining the symbols in diagram (3.91), where we use the language of angular momentum adjacency diagrams to describe the collection of points and lines. The nomenclature a $k-$line, a $k'-$line, and a $j-$line refers to lines in the adjacency diagram that are labeled, respectively, by $k_1, k_2, \ldots, k_{n-1}$, in the upper trivalent tree, by $k'_1, k'_2, \ldots, k'_{n-1}$ in the lower trivalent tree, and by $j_1, j_2, \ldots, j_{n+1}, j$ for the $n+2$ joining lines.

Diagram (3.91) has the following description:

1. The top trivalent tree is the union of a left and right subgraph denoted $\mathbb{L}_u$ and $\mathbb{R}_v$ that are joined by a number t of $k-$lines.
   (a) The subgraph $\mathbb{L}_u$ consists of u points $l_1, l_2, \ldots, l_u$ and a number $t+x$ of $k-$lines.
   (b) The subgraph $\mathbb{R}_v$ consists of v points $r_1, r_2, \ldots, r_v$ and a number $t+y$ of $k-$lines.
2. The bottom trivalent tree is the union of a left and right subgraph denoted $\mathbb{L}'_{u'}$ and $\mathbb{R}'_{v'}$ that are joined by a number t' of $k'-$lines.
   (a) The subgraph $\mathbb{L}'_{u'}$ consists of u' points $l'_1, l'_2, \ldots, l'_{u'}$ and a number $t'+x'$ of $k'-$lines.
   (b) The subgraph $\mathbb{R}'_{v'}$ consists of v' points $r'_1, r'_2, \ldots, r'_{v'}$ and a number $t'+y'$ of $k'-$lines.
3. The lines joining these four subgraphs are the following:
$$\begin{aligned} &a_1\ j-\text{lines join } \mathbb{L}_u \text{ and } \mathbb{L}'_{u'},\ a_2\ j-\text{lines join } \mathbb{L}_u \text{ and } \mathbb{R}'_{v'}, \\ &a_3\ j-\text{lines join } \mathbb{R}_v \text{ and } \mathbb{L}'_{u'},\ a_4\ j-\text{lines join } \mathbb{R}_v \text{ and } \mathbb{R}'_{v'}. \end{aligned} \tag{3.92}$$
   The total number of $j-$lines in (3.91) is
$$a_1 + a_2 + a_3 + a_4 = n + 2. \tag{3.93}$$
4. The parameters $u, v, t, x, y; u', v', t', x', y'$ entering into the definitions in Items 1 and 2 are nonnegative integers that satisfy the following conditions:
$$u + v = n,\ u' + v' = n, \tag{3.94}$$
$$\begin{aligned} &t + x + y = n - 1,\ t' + x' + y' = n - 1,\ t \geq 1,\ t' \geq 1; \\ &x + x' \geq 1,\ x \geq 0,\ x' \geq 0;\ y + y' \geq 1,\ y \geq 0,\ y' \geq 0. \end{aligned} \tag{3.95}$$
   These conditions are necessary and sufficient for a regular union of two trivalent trees of order n to give a cubic graph (3.91) that has $2n$ points, $3n$ lines, with each point of degree 3.

5. The vertical line partitioning the cubic graph into the left sets of points $\mathbb{L}_u$ and $\mathbb{L}'_{u'}$ and the right sets of points $\mathbb{R}_v$ and $\mathbb{R}'_{v'}$ is called the channel line; it contains three channels, the upper one called the $k-$channel, the middle one called the $j-$channel, and the lower one called the $k'-$channel. There are t lines going through the $k-$channel, t' lines going through the $k'-$channel, and $a_{23} = a_2 + a_3$ lines going through the $j-$channel. The total number of lines going through these three channels is $t + a_{23} + t' = 3p + 2q$. The channel line is not part of the cubic graph $C_{n,n}$; it is placed in the diagram over the graph to account for the left-right partitioning of the adjacency diagram into subgraphs and joining lines.

We can now provide the proof of the result above, which we state as follows:

The graph, denoted $C_{n,n}$ in diagram (3.91), which satisfies all the properties set forth in Items 1-5, is a cubic graph obtained by the regular union of two trivalent trees in the set $\mathbb{V}_n$, where the joining lines can be any of the compositions of nonnegative integers $(a_1, a_2, a_3, a_4) \vdash n + 2$; that is $C_{n,n} \in \mathbb{C}_{n,n}$. Every cubic graph $C_{n,n} \in \mathbb{C}_{n,n}$ has this structure.

Proof. The proof is given by showing that the sets of points $\mathbb{R}_v$ and $\mathbb{L}'_{u'}$ can be transfered in diagram (3.91) in such a way as required in the definition of a regular union, as shown in the following picture:

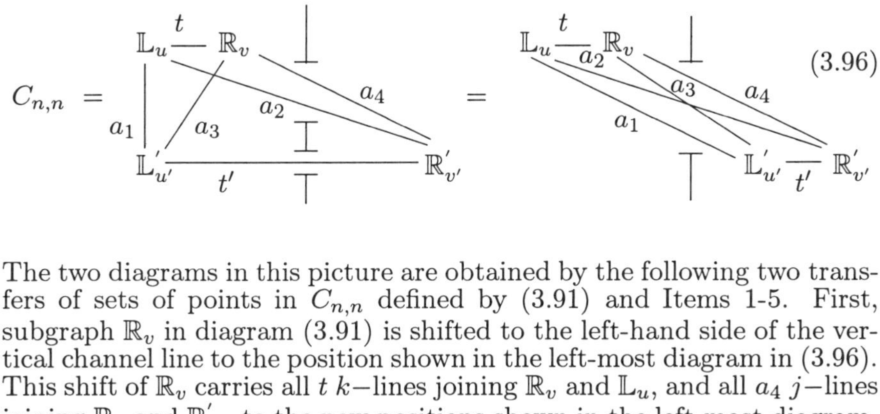

(3.96)

The two diagrams in this picture are obtained by the following two transfers of sets of points in $C_{n,n}$ defined by (3.91) and Items 1-5. First, subgraph $\mathbb{R}_v$ in diagram (3.91) is shifted to the left-hand side of the vertical channel line to the position shown in the left-most diagram in (3.96). This shift of $\mathbb{R}_v$ carries all t $k-$lines joining $\mathbb{R}_v$ and $\mathbb{L}_u$, and all a_4 $j-$lines joining $\mathbb{R}_v$ and $\mathbb{R}'_{v'}$, to the new positions shown in the left-most diagram, leaving the other subgraphs and lines unchanged; in particular, the upper channel is now closed, and the middle channel gains a_4 new $j-$lines. Second, the subgraph $\mathbb{L}'_{u'}$ in the left-most diagram (3.96) is shifted to the right-hand side of the vertical channel line to the new position shown. This shift carries all t' $k'-$lines joining $\mathbb{L}'_{u'}$ and $\mathbb{R}'_{v'}$, and all a_3 $j-$lines joining $\mathbb{L}_{u'}^{'}$ and $\mathbb{R}_v$, to the new positions shown, leaving the other subgraphs and lines on the left-hand side unchanged; in particular, the lower channel is now closed, and the middle channel gains $a_1 + a_3$ new $j-$lines. The final result of these two positional shifts of subgraphs and lines is the right-hand diagram in (3.96). But this is just the configuration of points for the joining of two trivalent trees in $\mathbb{V}_n$, since the subgraphs $\mathbb{L}_u$ and $\mathbb{R}_v$, joined by t $k-$lines, and the subgraphs $\mathbb{L}_{u'}^{'}$ $\mathbb{R}_{v'}^{'}$, joined by t' $k'-$lines, constitute the top and bottom trivalent trees, respectively. Moreover, the number of $j-$lines is $a_1 + a_2 + a_3 + a_4 = n + 2$. Since the two shifts of subgraphs and the "dragging along" of lines is just a reconfiguration of the subgraphs and points in (3.91) that preserves all features of the graph from which we started, diagram (3.91) is that of a cubic graph $C_{n,n} \in \mathbb{C}_{n,n}$. Since the entire procedure just described is reversible, every cubic graph $C_{n,n} \in \mathbb{C}_{n,n}$ can be presented as a diagram of the form (3.91). □

Examples. By a slight rearrangement of points and lines, the cubic graph (3.88) can be presented as the following diagram:

$$C_{16} =$$

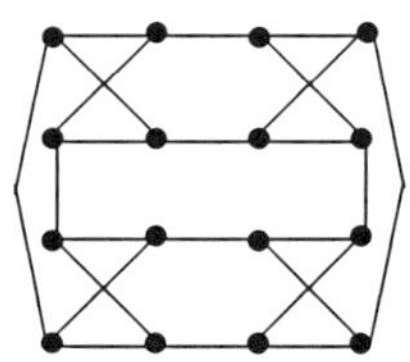

(3.97)

(The bent lines in this diagram are smooth lines containing no • points.) Comparing this diagram with (3.91), we obtain the parameter values given by $n = 8; p = 0, q = 2; u = v = u' = v' = 4; a_1 = 2, a_2 = 0, a_3 = 0, a_4 = 2; t = t' = 2, a_{23} = 0$; hence, $a_1 + a_2 + a_3 + a_3 = 4; t + a_{23} + t' = 4$. Since the condition $a_1 + a_2 + a_3 + a_3 = n + 2$ is violated, the cubic graph (3.88) cannot be presented as the joining of two trivalent trees of order 8—it cannot occur in the addition of $n + 2 = 10$ angular momenta. Similarly, it can be shown that the third cubic graph in (3.89) cannot be presented as the joining of two trivalent trees of order 12—it cannot occur in the addition of $n + 2 = 14$ angular momenta. □

The results obtained above on trivalent trees and cubic graphs can now be used to derive conditions under which a cubic graph $\mathbb{C}_{n,n}$ can be presented as two cubic subgraphs $C_{r,r} \in \mathbb{C}_{r,r}$ and $C_{s,s} \in \mathbb{C}_{s,s}$ joined by $3p + 2q$ lines. These conditions come in the form of a family of Diophantine relations that must be satisfied. Thus, we can prove the following result:

Under the cut and join operations of the $3p + 2q$ joining lines in diagram (3.90), applied now to $C_{n,n}$, each cubic graph $C_{n,n} \in \mathbb{C}_{n,n}$ factors into a *cubic graph $C_{r,r} \in \mathbb{C}_{r,r}$ and a cubic graph $C_{s,s} \in \mathbb{C}_{s,s}$:*

$$C_{n,n} = C_{r,r} \times C_{s,s}, \tag{3.98}$$

where the positive integral parameters n, r, s, p, q satisfy the relations

$$n = r + s - p \geq 3,\ 2r \geq p + 3,\ 2s \geq p + 3,\ p + q \geq 1. \tag{3.99}$$

In addition, the nonnegative parameters $(a_1, a_2, a_3, a_4), t, x, y, t', x', y'$ entering into the regular union of the two trivalent trees in the left-most diagram (3.96), which is an alternative presentation of the same diagram (3.90), must satisfy the Diophantine relations (3.100)-(3.102) below.

Proof. The following relations are just a systematic reassembly of conditions set forth in Items 1-5 in the description of the subgraphs and lines in diagram (3.91), which, via diagram (3.96), express diagram (3.90) as a regular union of two trivalent trees:

1. It follows from (3.99) that the integers n, r, s, p, q satisfy:

$$\begin{aligned} &r \geq 2,\ s \geq 2, |r - s| \leq n - 3, \\ &p = r + s - n,\ p + q \geq 1. \end{aligned} \tag{3.100}$$

2. The composition (a_1, a_2, a_3, a_4) of nonnegative integers satisfies:

$$a_1 + a_2 + a_3 + a_4 = n + 2. \tag{3.101}$$

3. The parameters t, x, y, t', x', y' satisfy:

$$\begin{aligned} t + t' &= (3p + 2q) - (a_2 + a_3), \\ t + x + y &= n - 1, \\ t' + x' + y' &= n - 1, \\ t + t' + x + x' &= (3r + q) - (a_1 + a_2 + a_3), \\ t + t' + y + y' &= (3s + q) - (a_2 + a_3 + a_4). \end{aligned} \tag{3.102}$$

The method of solving the Diophantine relations (3.100)-(3.102) is by the following four steps, where $n \geq 3$ is specified: (i) Any pair of integers (r, s) satisfying (3.100) is selected; (ii) p is defined from step (i) as given by (3.101), and q is selected to satisfy $p + q \geq 1$; (iii) any composition (a_1, a_2, a_3, a_4) satisfying (3.101) is selected; (iii) all solutions in nonnegative integers of relations (3.102) are constructed, using the selected values of the parameters from (i), (ii), and (iii). The family of all such solutions obtained in this way gives a factoring of cubic graphs of the form (3.98). □

The five linear relations are not independent, but they are always compatible—relations (3.102) presents the conditions in the most symmetric form. It will also be noticed that no use has been made in (3.98)-(3.102) of the conditions $u+v = u'+v' = n$. These are restrictions on the internal structure of the three cubic graphs entering into relation (3.98) that have not yet been enforced.

The factoring property $C_{n,n} = C_{r,r} \times C_{s,s}$ is, however, too general for the coupling of angular momenta: *all cubic graphs in this relation must be of angular momentum type* $\mathbb{G}(T, T')$. The requirement of being of angular momentum type places severe restrictions on these three cubic graphs. It is useful to illustrate the occurrence of a cubic graph that is not of angular momentum type and one that is, both of which factor. (We have introduced $\times$ between the factors in the above relations and below to keep the factors clear; it is a generic separation symbol with no specific properties.)

Example 1. A factoring not of angular momentum type occurs for $n = 5, p = q = 1, r = 2, s = 4$:

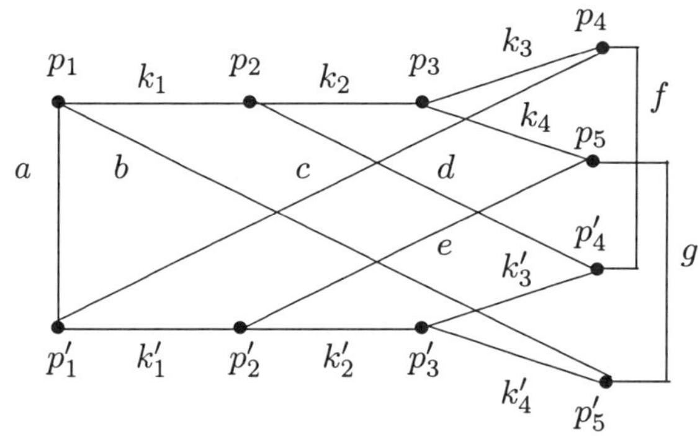

(3.103)

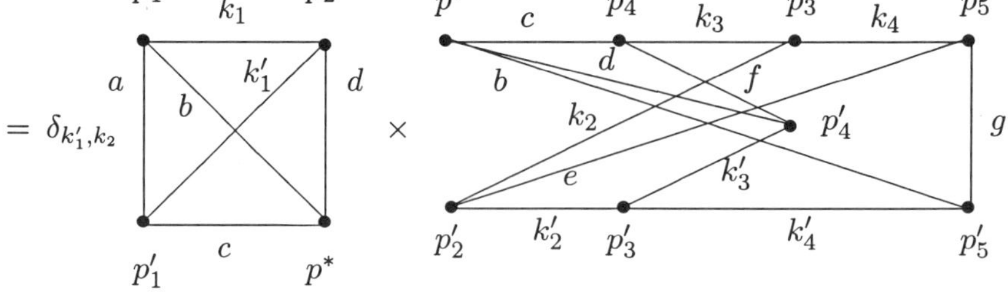

The factoring illustrated in this diagram is the result of the cut of the five lines k_2, d, b, c, k_1' and join of d, b, c to form the virtual point p^*, and then the smooth join of the lines k_1', k_2, which are now identified as the same line and introduces the factor δ_{k_1',k_2}. It is quite easy to read off the second cubic graph in the second line of (3.103) directly from the left-hand side of the relation. It is, however, the case that this graph is isomorphic to the second one in the top row of (3.75), as shown by the adjacency of points in the following two diagrams:

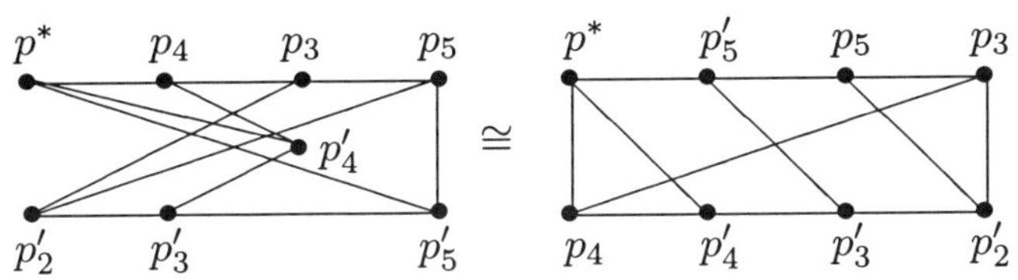

(3.104)

The parameters (3.91) of the graphs in (3.103) are:

$$\begin{aligned}
&\mathbb{L}_u :\ u = 2, t = 1, x = 1, a_1 = 1, a_2 = 2;\\
&\mathbb{L}'_{u'} :\ u' = 1, t' = 1, x' = 0, a_1 = 1, a_3 = 1;\\
&\mathbb{R}_v :\ v = 3, t = 1, y = 2, a_3 = 1, a_4 = 3 :\\
&\mathbb{R}'_{v'} :\ v' = 4, t' = 1, y' = 3, a_2 = 1, a_4.
\end{aligned} \tag{3.105}$$

These parameters satisfy the Diophantine relations (3.100)-(3.102), including $u + v = u' + v' = n$.

The diagram on the left in (3.103) with parameters (3.105) cannot be realized as a member of $\mathbb{G}(T, T')$, despite the factoring property: The joining lines a, b, c, d, e, f, g of external angular momenta going between the pair of trivalent trees of order five on the left must all remain joining lines of the factored trivalent trees on the right, and this is clearly violated by the virtual point $p^{*.}$ The cubic graph (3.103) factors, but it is not of angular momentum type.

Example 2. A factoring of angular momentum type occurs for $n = 5, p = 1, q = 0, r = 2, s = 4$. These parameters also satisfy the Diophantine relations (3.100)-(3.102). The structure of $C_{n,n}$ is now assured because we start with a pair of binary trees with no common forks. The three joining lines have the proper form to be cut and joined into a single point p^* of the proper form. We give all the details of the continued factoring because it is a nice representative of the methods advanced here, using binary trees and the associated triangle coefficients:

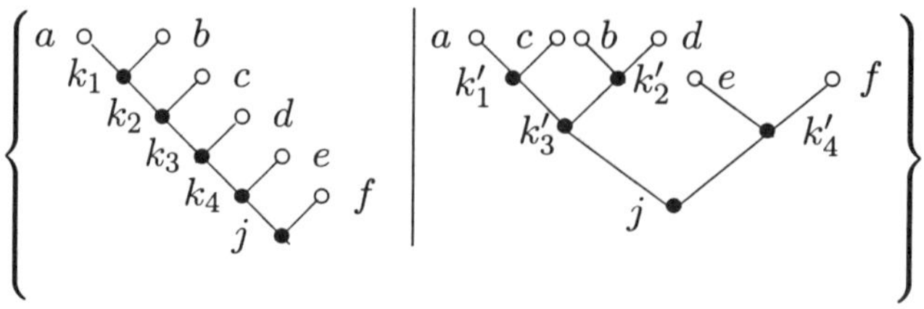

$\cong$

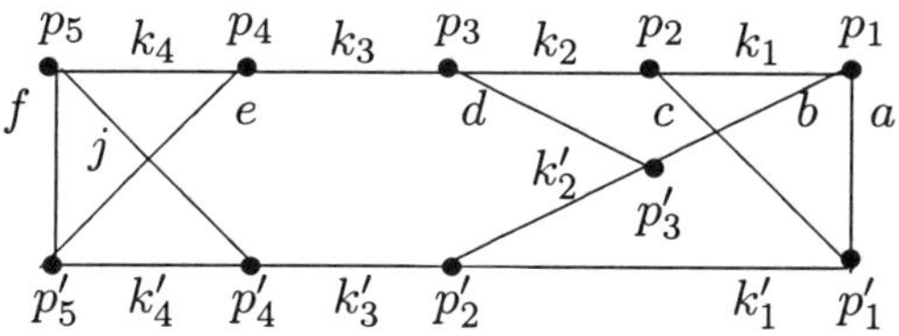

$= \delta_{k_3,k_3'}$

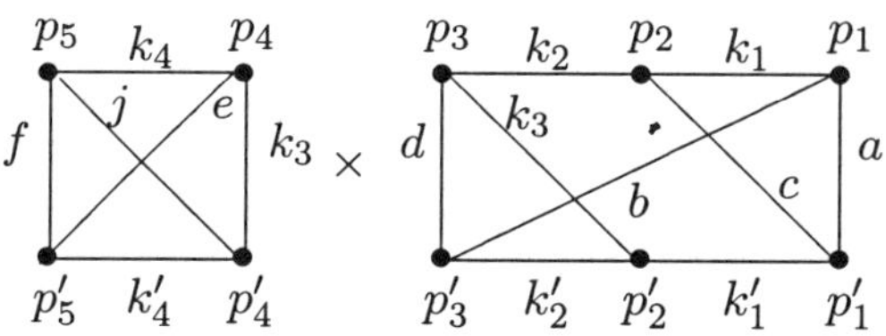

$= \delta_{k_3,k_3'}$

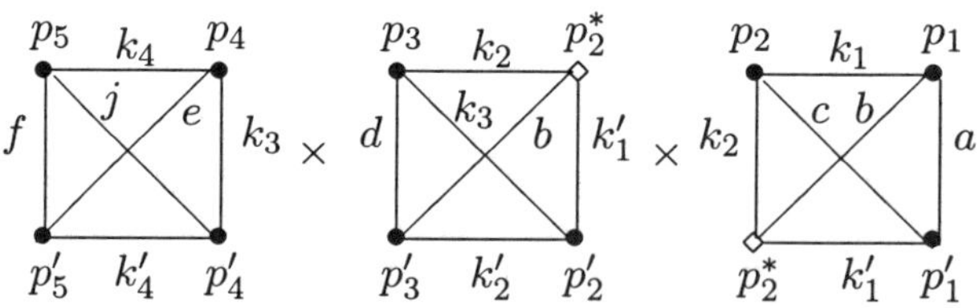

(3.106)

The set of factors $C_{2,2} \times C_{3,3}$, as well as $C_{2,2} \times (C_{2,2} \times C_{2,2})$ are all seen to be of angular momentum type. Points in this diagram are labeled by

$$\begin{aligned} &p_1 = (a\,b\,k_1),\ p_2 = (k_1\,c\,k_2),\ p_3 = (k_2\,d\,k_3),\ p_4 = (k_3\,e\,k_4),\\ &p_5 = (k_4\,f\,j), p_2^* = (k_2\,b\,k_1');\ p_1' = (a\,c\,k_1'),\ p_2' = (k_1'\,k_2'\,k_3),\\ &p_3' = (b\,d\,k_2'),\ p_4' = (k_3\,k_4'\,j),\ p_5' = (e\,f\,k_4'). \end{aligned} \qquad (3.107)$$

This symbolic identity (3.106), of course, implies relations between the various corresponding fundamental triangle coefficients by the rules given in Sect. 2.2.11, Chapter 2.

Further refinements of the family of Diophantine relations (3.100)-(3.102) requires that we enforce the requirement that $C_{r,r}$ and $C_{s,s}$ in (3.98) be of angular momentum type, which we now consider. The general situation for the cut and join of the $3p + 2q$ lines lines may be described by the following diagram:

$$\left.\begin{array}{ll} t\ k\text{-lines} & \text{———} \\ a_2 + a_3\ j\text{-lines} & \text{———} \\ t'\ k'\text{-lines} & \text{———} \end{array}\right\} 3p + 2q \ \ \text{total lines} \tag{3.108}$$

The $3p+2q = t + a_2 + a_3 + t'$ lines must be cut and joined in a way that effects a factoring into the product $C_{r,r} \times C_{s,s}$, where each of these cubic graphs is of angular momentum type.

It is useful to examine first the special case $t = t' = 1$: There are $3p + 2q - 2$ join lines of j–type. For $p + q \geq 2$, there are at least $p + 2$ j–type join lines. Thus, for $p \geq 1$, at least three j–lines must be joined, which is prohibited. If $p = 0$, there are at least two j–lines that must be joined. This is also prohibited: The smooth join of two j–lines into a line introduces a Kronecker delta factor on the labels of the two j–lines, which identifies two j–lines as having identical labels, and this creates a double line in either $C_{r,r}$ or $C_{s,s}$. We conclude that it is necessary that $p + q \leq 1$ to obtain a factoring into cubic graphs of angular momentum type.

It is, of course, possible to make a cut for which $t+t' \geq 2$, as diagram (3.104) shows: the result for $t = t' = 1$ is too restrictive, unless proved otherwise.

A general result obtained by considering each case is the following:

Under the cut of the $3p + 2q$ join lines in (3.108), those lines with the following properties cannot be joined: three lines with j–labels to a point; pairs of j–lines smoothly to a line; pairs of k–lines smoothly to a line; pairs of k'–lines smoothly to a line; a k–line and a k'–line smoothly to a line; and lines that create loops (a line that joins a point to itself) or double lines (two lines that join the same pair of points).

These ancillary conditions imply that the only possible matchings in diagram (3.108) are those that give q smooth joins of a k–line and a k'–lines, and those that give p joins to a virtual point p^* of a k–line, a k'–line, and a j–line. By adding the number of k–lines, j–lines, and k'– lines that appear in the cut and join, we obtain the relations $t = t' = p + q$ and $a_2 + a_3 = p$.

These conditions on the cut and join of lines are now used in relations (3.101)-(3.102) to obtain a refined version of those Diophantine equations given by

1. The positive integers n, r, s and nonnegative integers p, q satisfy:
$$\begin{aligned} &r \geq 2,\ s \geq 2, |r-s| \leq n-3, \\ &p = r+s-n,\ p+q \geq 1. \end{aligned} \tag{3.109}$$
2. The composition (a_1, a_2, a_3, a_4) of nonnegative integers satisfies:
$$a_1 + a_2 + a_3 + a_4 = n+2,\ a_2 + a_3 = p. \tag{3.110}$$
3. The parameters t, x, y, t', x', y' satisfy
$$\begin{aligned} &t = t' = q, \\ &x + y = n - 1 - (p+q), \\ &x' + y' = n - 1 - (p+q), \\ &x + x' = 3r - (3p+q) - a_1, \\ &y + y' = 3s - (3p+q) - a_4. \end{aligned} \tag{3.111}$$
The last four relations are always compatible in consequence of the first relation and of $n = r + s - p$.

Relations (3.110) still do not take into account the requirements that $C_{r,r}$ and $C_{s,s}$ be cubic graphs of angular momentum type; that is, $C_{r,r}, C_{s,s} \in \mathbb{G}(T, T')$. These conditions are the following: When the q lines obtained by the smooth joining of the q pairs consisting of a $k-$line and a $k'-$line and the p points obtained by the joining of three lines to a virtual point containing a $k-$line, a $j-$line, and a $k'-$line are all adjoined to the subgraph $\{\mathbb{L}_u, \mathbb{L}'_{u'}\}$, and separately to the subgraph $\{\mathbb{R}_u, \mathbb{R}'_{u'}\}$, the cubic graphs $C_{r,r}$ and $C_{s,s}$ of angular momentum type must be created. These requirements impose further conditions on the parameters. In particular, it must be the case that $a_1 + 2p + q = r + 2$, so that the number of lines joining the upper trivalent tree and the lower trivalent tree in $C_{r,r}$ is $r + 2$, with a similar result $a_4 + 2p + q = s + 2$ for $C_{s,s}$. This result and (3.110) now gives $p + q = 1$. This proves:

Every cubic graph $C_{n,n} \in \mathbb{G}(T, T')$ of angular momentum type that factors into a product $C_{n,n} = C_{r,r} \times C_{s,s}$ of cubic graphs $C_{r,r}$ and $C_{s,s}$, each of angular momentum type, has parameter values that satisfy the following family of Diophantine relations:

Two-line case $(p = 0,\ q = 1)$:
$$\begin{aligned} &n = r + s \geq 4,\ r \geq 2,\ s \geq 2, \\ &a_2 = a_3 = 0,\ \ a_1 + a_4 = n + 2, \\ &x + y = x' + y' = n - 2, \\ &x + x' = 3r - a_1 - 1,\ \ y + y' = 3s - a_4 - 1. \end{aligned} \tag{3.112}$$

Three-line case $(p = 1,\ q = 0)$:

$$\begin{aligned}
&n = r + s - 1 \geq 3,\ r \geq 2,\ s \geq 2,\\
&a_2 + a_3 = 1,\ a_1 + a_4 = n + 1,\\
&x + y = x' + y' = n - 2,\\
&x + x' = 3r - a_1 - 3,\ y + y' = 3s - a_4 - 3.
\end{aligned} \tag{3.113}$$

The Yutsis factoring theorem is also necessary for cubic graphs of angular momentum type (necessary version).

The general family of Diophantine relations (3.100)-(3.102) is invariant under the interchange of the primed and unprimed parameters, and under the simultaneous interchange of the parts of the three pairs of parameters $(r, s), (a_1, a_4), (a_2, a_3)$. The first operation interchanges the upper and lower trivalent trees in $C_{n,n}, C_{r,r}, C_{s,s}$; the second the subgraphs to the left and right of the $3p + 2q$ joining lines. These symmery operations give isomorphic cubic graphs. These results carry over to relations (3.112) and (3.113).

3.4.2 Join properties of cubic graphs

The factoring of cubic graphs obtained by the cut and join of two or three lines depends entirely on the local configuration of points and lines. The three possible local pictures have the following diagrams:

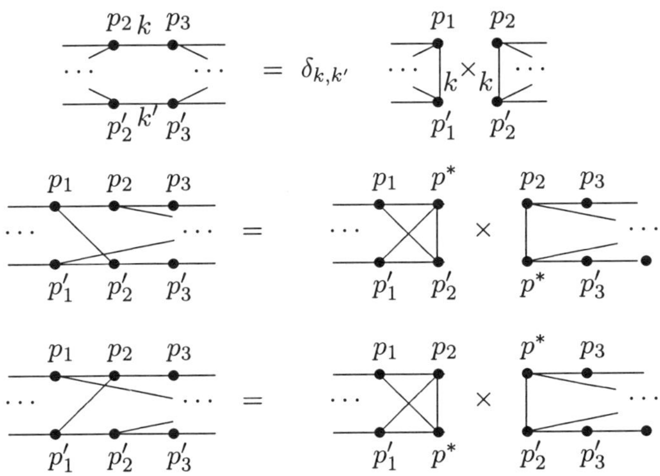

(3.114)

The diagram on the right is the result of the cut and join operation on the diagram to the left. The symbol $\cdots$ denotes that the points and lines to the right and left of the depicted local configuration complete the picture to a full cubic graph.

The joins in these diagrams are of different types: The first is a smooth-line join, and the two three-line joins are of a down-diagonal type and an up-diagonal type as represented by the diagrams:

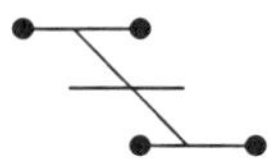 and 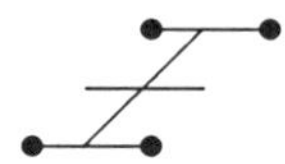 (3.115)

The neutral $\times$ symbol in (3.114) needs to be refined: We introduce the wedge symbol $\wedge$ to denote the cut and smooth-line join of two lines; the join symbol $\vee_*$ to denote the cut and join of three lines to a point in the down-diagonal case; and the join symbol $\vee^*$ to denote the cut and join of three lines to a point in the up-diagonal case. The joins down-diagonal three-line join $\vee_*$ and the up-diagonal three-line join $\vee^*$ are also distinguished by the positions of the two virtual points created.

A characteristic of the $\times = \wedge$ operation in the first diagram in (3.114) is: it is immediately preceded and immediately followed by a three-line join. But such three-line joins have the property:

A three-line cut is not allowed in the pair of contiguous three-line joins to a two-line join; such a cut creates a double line. Only the two-line cut and join is allowed.

A full cubic graph $C_{n,n}$ of which (3.114) is just a local part may exhibit several two-line or three-line joins. We require further nomenclature for their description. With each cubic graph $C_{n,n} \in \mathbb{C}_{n,n}$, we associate a sequence consisting of all the two-line joins, the down-diagonal three-line joins, and the up-diagonal three-line joins, ordered in the sequence as read from left-to-right across the upper horizontal line. We call this sequence the *characteristic join sequence* of $C_{n,n}$. The characteristic join sequence of $C_{n,n}$, in turn, uniquely determines a *maximal product* of cubic graphs of the form $\prod_t^\times C_{t,t}$, which is obtained by the cut and join of every two-line join and every **allowed** three-line join in its characteristic join sequence. We define two cubic graphs $C_{n,n}$ and $C'_{n,n}$, both in $\mathbb{C}_{n,n}$, to be *sequentially isomorphic* if their characteristic join sequences are the same; otherwise, sequentially nonisomorphic.

The characteristic join sequence of a cubic graph may be presented as a word in the three-letter alphabet N, L, R defined in terms of the line-join operations by making the mappings

$$\wedge \mapsto N,\ \vee_* \mapsto L,\ \vee^* \mapsto R, \tag{3.116}$$

where the mapping of each join subsequence $\vee_* \wedge \vee^* \mapsto \wedge$ and, correspondingly, $LNR \mapsto N$ take into account that three-line cuts of the 3-line joins that always precede and follow a two-line join are not allowed.

It is useful to illustrate these concepts. For this purpose, we introduce the following space-saving notations:

$W =$ 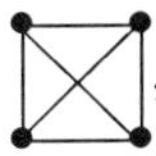 , $V =$ 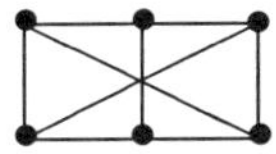 . (3.117)

Example 1. There are two nonisomorphic cubic graphs on six points. They can be taken to be those in diagram (3.83) and (3.84). The first one has the maximal factors given by

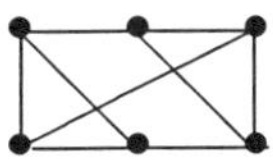 $= W \vee_* W.$ (3.118)

The second one is V, given by (3.117); it has no two-line or three-line joins and is the Wigner $9-j$ symmetric cubic graph (3.84). The characteristic join sequence of (3.18) is $\vee_* \mapsto L$.

Example 2. There are five nonisomorphic cubic graphs on eight points. They can be taken to be those in diagram (3.75). Three have the maximal factor cubic graphs as follows:

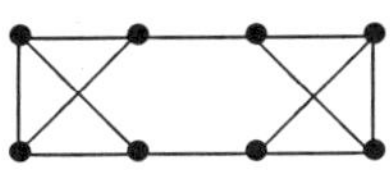 $= W \wedge W\,,$

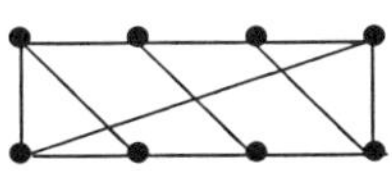 $= W \vee_* (W \vee_* W)\,,$ (3.119)

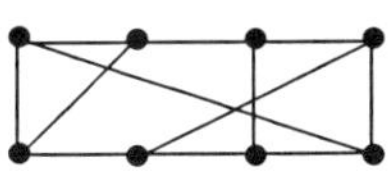 $= W \vee^* V\,.$

The remaining two are given by the following symmetric cubic graphs:

$U =$ 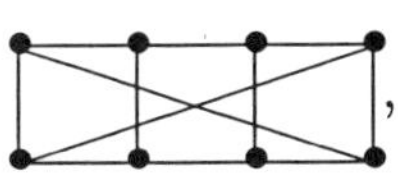 , $X =$ 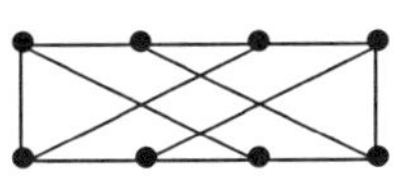 . (3.120)

The characteristic join sequences of the cubic graphs (3.119) and the corresponding words are given, respectively, by

$$\vee_* \wedge \vee^* \mapsto \wedge \mapsto N,\ \vee_* \vee_* \mapsto LL,\ V^* \mapsto R. \tag{3.121}$$

The two cubic graphs in (3.120) that do not factor may be regarded as corresponding to the empty word $\emptyset$. All five cubic graphs are of angular momentum type, since each can be realized in terms of a pair of standard labeled binary trees, as can each of the factors in the maximal factorings (3.119) (see Ref. [21, Vol. 9]). Sequentially nonisomorphic cubic graphs on eight points are nonisomorphic; there are two $12-j$ coefficients. □

The determination of a set of nonisomorphic cubic graphs on ten points is more complicated; it is the first case where features representative of many of the complex relations between general cubic graphs and cubic graphs of angular momentum type appear, so it is worth examining in some detail.

There are nineteen nonisomorphic cubic graphs on ten points (Korflage [93, p. 53] and Harary [77]. Two members of the set $\mathbb{C}_{10}$ can be chosen to be the cubic graph given by (3.87) and the one given by (3.103), where we now replace the second factor by the isomorphic cubic graph (3.104) and remove all labels, since they serve no purpose here. These two cubic graphs, which are not of angular momentum type, illustrate the nontrivial nature of the problem of determining whether a given cubic graph belongs to the set $\mathbb{G}(T,T')$.

Additional features of cubic graphs that must be taken into account are the following:

(i). The isomorphism of a pair of cubic graphs may or may not preserve its $3p+2q$ join-line structure. This is illustrated by examples (3.76),(3.104), and (3.123)-(3.124) below.

(ii). The balance of points between the number $2n$ of points in $C_{n,n}$ and the number of points $2t$ that appear in the factors $C_{t,t}$ of any product relation for $C_{n,n}$ must satisfy the relation

$$\sum_t n_t = n + h, \tag{3.122}$$

where h is the number of $\vee$ operations (either type) that appears in the factored product, $n_t = |\mathbb{C}_{t,t}|$ is the number of cubic graphs on $2t$ points that occurs in the factoring, which is $h+1$, and the summation is over all such factor. This relation must hold because the $\wedge$ operation does not change the number of points, and the $\vee$ operation (either type) increases the number of points by two.

We can now assemble maximal join diagrams of the form (3.118) and (3.119) to realize a set of eighteen nonisomorphic cubic graphs of order ten, example (3.87) excepted, by implementing the following build-up rule, which is a summary of properties of products developed above.

Build-up rule:

1. Form all products $\wedge, \vee_*, \vee^*$ of lower order cubic graphs of order (number of factors) $f = 1, 2, \ldots, 2h+1$, where h is the number of join operations $\vee$ (either type) in the product, such that the balance of products relation (3.122) is satisfied.

2. Insert parenthesis pairs $(A\,(B\,(C\,\cdots(X\,(Y\,Z))))\cdots)$ into each such product, and read off the characteristic join sequence.

3. Take into account the symmetry that interchanges the upper and lower trivalent trees (and reverses the slopes of the join lines), the left-right reflection symmetry through any vertical line, and the symmetry that interchanges factors in any parenthesis pair, for example, $(Y\,Z) = (Z\,Y), (X\,(Y\,Z)) = ((Y\,Z)\,X), \ldots,$ and keep only those with the standard parenthesis pairs in Item 2 that have characteristic join sequences that are nonisomorphic. (Such parenthesis pair bracketings are called commutative, and their number is given by the Wedderburn-Etherington numbers—see Comtet [44, p. 54].)

4. Replace each subsequence $\vee_* \wedge \vee^*$ appearing in the characteristic join sequence by $\wedge \mapsto N$ to account for the special feature (3.121) of all two-line joins.

5. Reverse the direction of the factor process in relations (3.114) to obtain the (unique) cubic graph represented by the factors obtained in Items 1-4; if the cubic graph so obtained contains a two-line or three-line join, continue the factor process until a maximal factoring is effected, where we note that the maximal factoring of each cubic graph in the set $\mathbb{C}_{n,n}$ gives a unique characteristic join sequence.

We implement the build-up rule to obtain a set of eighteen nonisomorphic cubic graphs on ten points, each of which factors. For this, we require two additional cubic graphs on eight points that are not already listed in Example 2 above:

$Y =$ 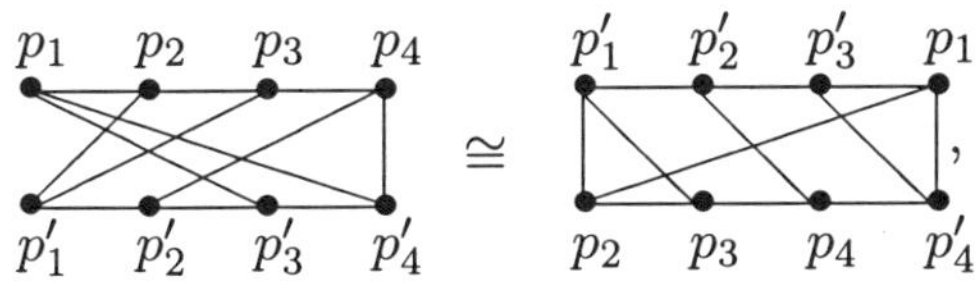

, (3.123)

$Z =$ 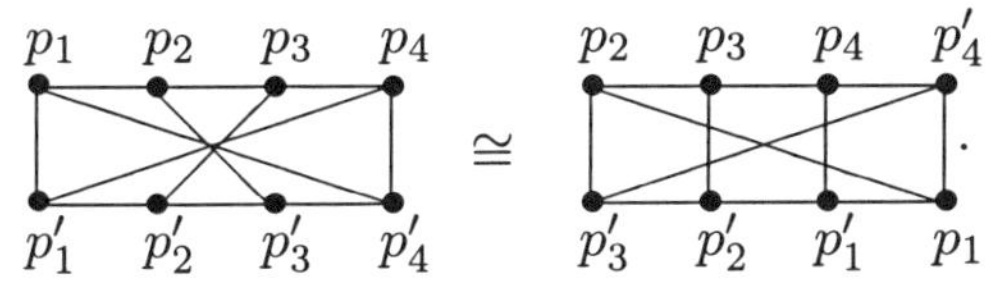

. (3.124)

The eighteen nonisomorphic cubic graphs may be chosen to be those given by the notation U, V, W, X, Y, Z by the following maximal factors:

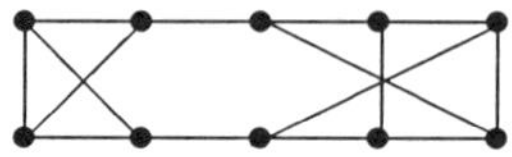

$$= W \wedge V \,,$$

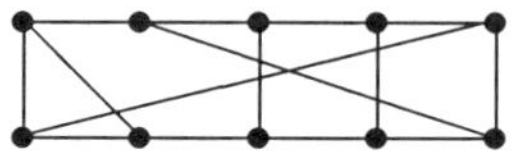

$$= W \vee_* U \,,$$

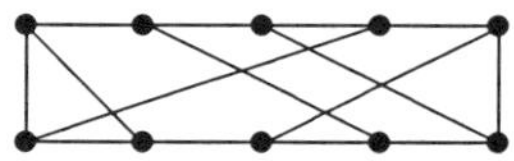

$$= W \vee_* X \,,$$

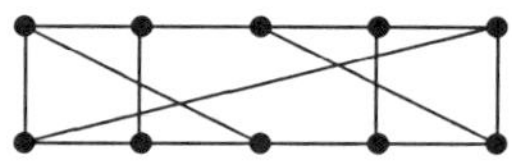

$$= V \vee_* V \,,$$

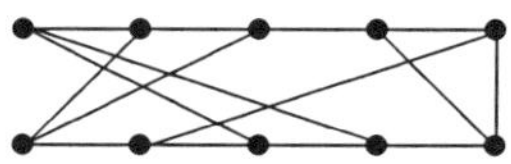

$$= Y \vee_* W \,,$$

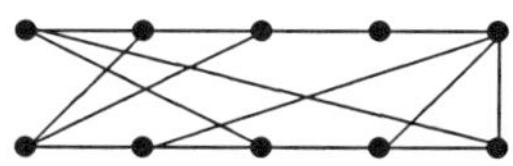

$$= Y \vee^* W \,,$$

(3.125)

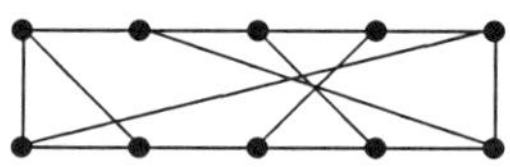

$$= W \vee_* Z \,,$$

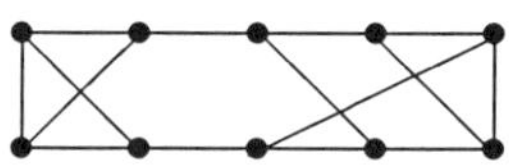

$$= W \wedge (W \wedge W) \,,$$

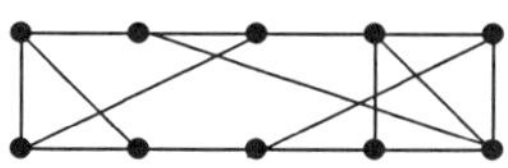

$$= W \vee_* (W \vee^* V) \,,$$

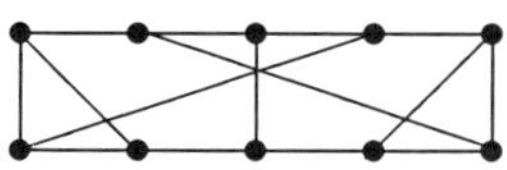

$$= W \vee_* (W \vee^* Z) \,,$$

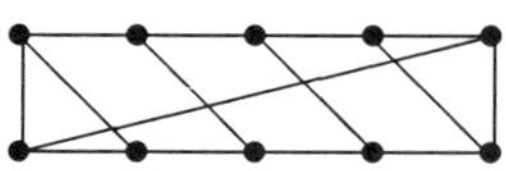

$$= W \vee_* (W \vee_* (W \vee_* W)) \,,$$

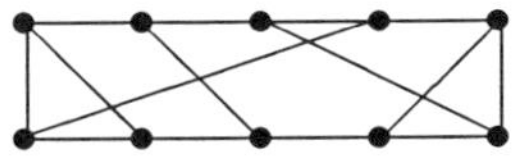

$$= W \vee_* (W \vee_* (W \vee^* W)),$$

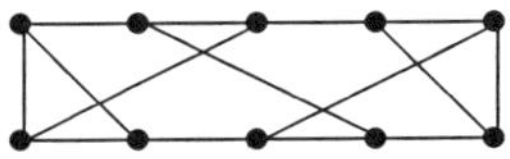

$$= W \vee_* (W \vee^* (W \vee_* W)).$$

This leaves the following five nonisomorphic cubic graphs that do not factor on two or three lines:

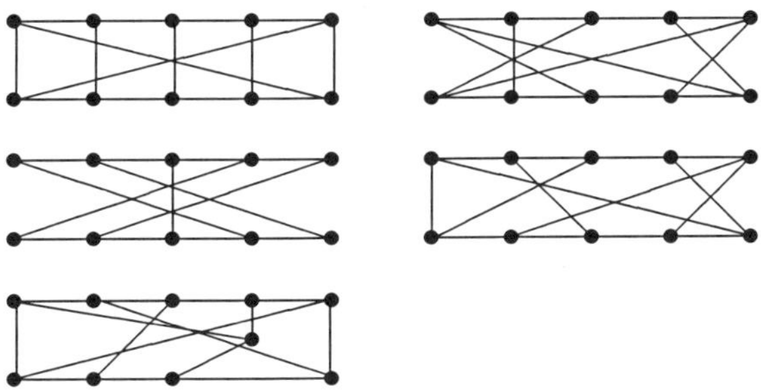

(3.126)

Sequentially nonisomorpic cubic graphs are nonisomorphic, and the cubic graphs in (3.126) determine five $15-j$ coefficients.

Three of the nonisomorphic cubic graphs on ten points are not of angular momentum type $\mathbb{G}(T, T')$. These three are the one-line join given by (3.87) and the fifth and sixth one in the collection of thirteen give by (3.125). To verify this fully, it is necessary to show that each of the five cubic graphs given by (3.126) can be realized as the standard labeling of a pair of binary trees selected from

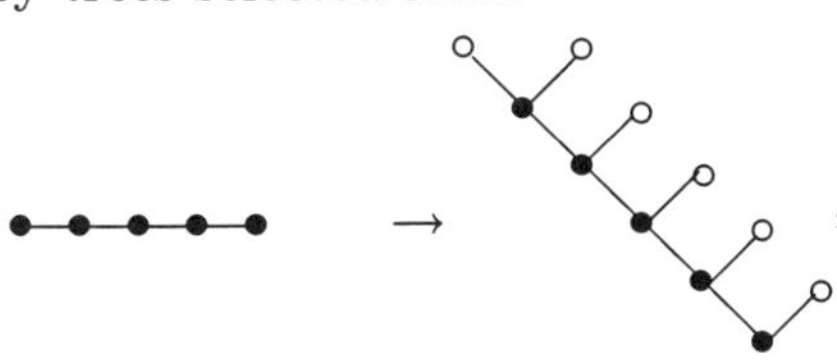

$$= T$$

(3.127)

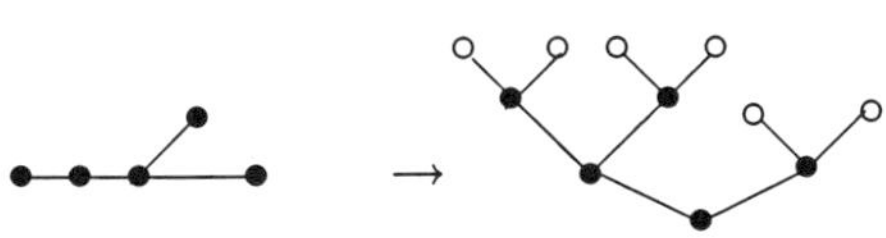

$$= T'$$

such that the pair of binary trees contains no common forks.

It is quite interesting that all members of the class of nonisomorphic cubic graphs that factor into a product of $h+1$ tetrahedral $W-$factors can be given. The class is determined by the set of corresponding characteristic join sequences for which the words of length h in the two letters L and R have certain symmetry properties (see (3.122) for the definition of h), as follows:

Select any cubic graph that factors into $h+1$ tetrahedra and determine its corresponding word w of length h. Determine the set of words with the following properties: (i) w is in the set; (ii) the reversal of every word in the set is in the set; and (iii) each word obtained by interchanging L and R in every word in the set is in the set. This collection of words uniquely determines a family of isomorphic cubic graphs with the required factoring properties. A word not in the set with w is then selected, and a second class of isomorphic cubic graphs is constructed. This process is continued in the standard way to partition the set of all cubic graphs in question into disjoint equivalence classes. Then, any set of representatives of these equivalence classes gives a complete set of nonisomorphic cubic graphs that factor into a product of $h+1$ tetrahedral $W-$factors.

Example. The last three in the list (3.125) corresponding, respectively, to the words LLL, LLR, LRL are nonisomorphic cubic graphs for which a generalized B-E identity holds. This is because the five cubic graphs corresponding to the words RRR; RRL, LRR, RLL; RLR are isomorphic to the original three; hence, we have the word isomorphisms given by

$$RRR \cong LLL;\ RRL \cong LRR \cong RLL \cong LLR;\ RLR \cong LRL. \quad \square \tag{3.128}$$

When applied to cubic graphs of angular momentum type, so that the product into $h+1$ W-factors gives the product of $h+1$ Racah coefficients (with adjoined dimension and phase factors), each member of the family of relations described above generalizes the B-E identity (3.83).

The build-up rule is a guide for constructing nonisomorphic cubic graphs, but it is not yet an algorithm. For example, the selection of the additional cubic graphs on $2n$ points (Y and Z in (3.123)-(3.124) for $n=4$) in going to the next higher level $2n+2$ has not been given. The property that the class of characteristic join sequences corresponding to the set of words having the symmetries in the letters N, R, L defined above, and in Item 3, are an isomorphic set is a powerful tool. But to effect this, even at the next level of twelve points, is nontrivial. It has not been attempted. Alternative constructive methods should also be developed, keeping the approach fully under the purview of combinatorial methods.

Alternative approaches

Abstract triangles (abc) of symbols with $a \neq b \neq c$ have a vital role in the theory of cubic graphs from the point of view of angular momentum theory. The points of a cubic graph can always be represented by a collection of $2n$ such triangles $p_i = \{a_i\, b_i\, c_i\}, i = 1, 2, \ldots, 2n$, with the property that each of the three parts of the triangle is shared by exactly one other triangle. Lines incident on points triangles) are then defined by the pairs of triangles that share a common letter. This gives a total of $3n$ lines with three lines incident on each point; that is, a cubic graph. Each selection of such triangles gives a unique cubic graph. All cubic graphs of order $2n$ can be constructed in this manner, but many cubic graphs obtained in this manner are isomorphic. We address this problem in the next section.

3.4.3 Cubic graph matrices

The enumeration of the full set of cubic graphs on $2n$ points can be approached by using matrix arrays similar to the fundamental triangle coefficients of order $2n$. Thus, we consider a $3 \times 2n$ matrix of $6n$ indeterminates $x = (x_1, x_2, \ldots, x_{3n})$ and $y = (y_1, y_2, \ldots, y_{3n})$ that are distributed among the rows and columns of the matrix by certain rules given below. We denote such an array by $A_{3\times 2n}(x; y)$, refer to the indeterminates as coordinates, and call the three coordinates in each column the *internal coordinates* of a point. Thus, a matrix array $A_{3\times 2n}(x; y)$ consists of $2n$ points, $p_1, p_2, \ldots, p_{2n}$, as read from left-to-right across the columns; and each point (column) p_i has three distinct internal coordinates, which can be interpreted as generic points in Cartesian 3−space $\mathbb{R}^3$, or as triples of labels of planar points, but with no significance as Cartesian coordinates in $\mathbb{R}^2$. The rules for the distribution of the $6n$ coordinates $x = (x_1, x_2, \ldots, x_{3n})$ and $y = (y_1, y_2, \ldots, y_{3n})$ into the matrix $A_{3\times 2n}(x; y)$ are called *adjacency conditions,* and are as follows:

(1). Adjacency conditions: Let the three internal coordinates of a given column of $A_{3\times 2n}(x; y)$ be denoted by (z_a, z_b, z_c), where each component is selected from the $6n$ coordinates $x = (x_1, x_2, \ldots, x_{3n})$ and $y = (y_1, y_2, \ldots, y_{3n})$ by the following rules:

- (i). All three subscripts a, b, c of the internal coordinates (z_a, z_b, z_c) are distinct.
- (ii). Subscript a appears in $\mathrm{col}(z_a, z_b, z_c)$ and in exactly one other column, distinct from $\mathrm{col}(z_a, z_b, z_c)$; subscript b appears in $\mathrm{col}(z_a, z_b, z_c)$, and in exactly one other column, distinct from $\mathrm{col}(z_a, z_b, z_c)$, and distinct from the second column in which a appears; and subscript c appears in $\mathrm{col}(z_a, z_b, z_c)$ and in exactly one other column, distinct from $\mathrm{col}(z_a, z_b, z_c)$, and distinct from the second column in which a appears and from the second column in which b appears. These relations between columns are to hold for each of the $2n$ columns of $A_{3\times 2n}(x; y)$.

We introduce the following notation for these matrix arrays:

$$\mathbb{A}_{3\times 2n}(x;y) = \{\text{all matrix arrays satisfying conditions (1)}\}. \quad (3.129)$$

We call each matrix in this set a *cubic graph matrix.* Each cubic graph matrix maps to a cubic graph whose points p_i, $i = 1, 2, \ldots, 2n$, are presented as the columns of the matrix (3.129) with incident lines given by the coordinates that share a common subscript. Conversely, each cubic graph whose points are presented by 3-tuples of points satisfying the adjacency rules with incident lines given by the 3-tuples that share a common subscript maps to a cubic graph matrix (3.129). This bijection, however, does not differentiate between isomorphic cubic graphs.

Permutations may be used to reduce the number of cubic graph matrices $A_{3\times 2n}(x;y)$ that give nonisomorphic cubic graphs. The following three permutation groups leave the adjacency conditions invariant:

$$\begin{aligned} S_3(2n) &= S_3 \times S_3 \times \cdots \times S_3 \ (2n \text{ times}), \\ S_{2n} &= \text{standard permutation group on 1,2, } \ldots\text{, 2n,} \quad (3.130) \\ S_2 &= \text{interchange of symbols } x \text{ and } y. \end{aligned}$$

The action of each $\rho \in S_3(2n)$ (direct product group) on each $A_{3\times 2n}(x;y) \in \mathbb{A}_{3\times 2n}(x;y)$ is defined to be a permutation of the three row entries in each of its columns, while the action of each $\pi \in S_{2n}$ is defined to be a permutation of the $2n$ columns themselves. Thus, we have that $|\,S_3(2n)\,| = 3!(2n)!$ and $|\,S_{2n}\,| = (2n)!$.

The action of these invariance groups can be used to bring each cubic matrix $A_{3\times 2n}(x;y) \in \mathbb{A}_{3\times 2n}(x;y)$ to the following form:

$$A^*_{3\times 2n}(x;y) = \begin{pmatrix} x_1 & y_3 & y_5 \cdots y_{2n-1} & y_{i_1} & y_{i_3} \cdots y_{i_{2n-3}} & y_{i_{2n-1}} \\ x_2 & x_4 & x_6 \cdots x_{2n} & y_{i_2} & y_{i_4} \cdots y_{i_{2n-2}} & y_{i_{2n}} \\ x_3 & x_5 & x_7 \cdots x_{2n+1} & x_{2n+2} & x_{2n+3} \cdots x_{3n} & y_{3n} \end{pmatrix}, \quad (3.131)$$

where the entry $y_{i_{2n+1}}$ in row 3 and column $2n$ has subscript $i_{2n+1} = 3n$, and the remaining subscripts $i_1, i_2, \ldots, i_{2n}$ are a permutation of the $2n$ integers in the set $\mathbb{H}_{2n} = \{1, 2, 4, \ldots, 2n-2, 2n, 2n+1, 2n+2, \ldots, 3n-1\}$ such that following three conditions hold:

(i). Adjacency conditions (1i) and (1ii) are preserved.

(ii). The column subscripts $\{i_1, i_2\}, \{i_3, i_4\}, \ldots, \{i_{2n-2}, i_{2n}\}$ are disjoint 2−subsets of $\mathbb{H}_{2n}$ such that $i_1 < i_2, i_3 < i_4, \ldots, i_{2n-1} < i_{2n}$.

(iii). The set of integer subscripts $i_1, i_2, \ldots, i_{2n}$, which are permutations of the integers in the set $\mathbb{H}_{2n}$, must preserve the adjacency conditions (i), the 2-subset conditions (ii), and yield nonisomorphic cubic graph matrices.

The term *nonisomorphic cubic graph matrices* designates that the set of cubic graphs with their $2n$ points presented as the columns of the set of cubic graph matrices are nonisomorphic. We can now prove:

Each cubic graph matrix $A_{3\times 2n}(x;y) \in \mathbb{A}_{3\times 2n}(x;y)$ *is isomorphic to a cubic graph matrix in the set of cubic graph matrices* $\mathbb{A}^*_{3\times 2n}(x;y)$ *defined by (3.131) and conditions (i)-(iii);* $\mathbb{A}^*_{3\times 2n}(x;y)$ *is a complete set of nonisomorphic cubic matrices.*

Examples. It is useful to see some examples of cubic graph matrices before proceeding to the proof.

$n = 2$: The following cubic graph matrix represents the tetrahedron (3.69):

$$\begin{pmatrix} x_1 & y_3 & y_1 & y_2 \\ x_2 & x_4 & y_4 & y_5 \\ x_3 & x_5 & x_6 & y_6 \end{pmatrix}. \tag{3.132}$$

$n = 3$: The following two cubic graph matrices represent the two nonisomorphic cubic graphs (3.83)-(3.84) on six points:

$$\begin{pmatrix} x_1 & y_3 & y_5 & y_2 & y_4 & y_1 \\ x_2 & x_4 & x_6 & y_6 & y_8 & y_7 \\ x_3 & x_5 & x_7 & x_8 & x_9 & y_9 \end{pmatrix}, \begin{pmatrix} x_1 & y_3 & y_5 & y_1 & y_2 & y_4 \\ x_2 & x_4 & x_6 & y_6 & y_8 & y_7 \\ x_3 & x_5 & x_7 & x_8 & x_9 & y_9 \end{pmatrix}. \tag{3.133}$$

$n = 4$: The following five cubic graph matrices represent the five nonisomorphic cubic graphs (3.75) on eight points:

$$\begin{gathered}
\begin{pmatrix} x_1 & y_3 & y_5 & y_7 & y_1 & y_2 & y_8 & y_6 \\ x_2 & x_4 & x_6 & x_8 & y_4 & y_{10} & y_{11} & y_9 \\ x_3 & x_5 & x_7 & x_9 & x_{10} & x_{11} & x_{12} & y_{12} \end{pmatrix}, \\
\begin{pmatrix} x_1 & y_3 & y_5 & y_7 & y_1 & y_2 & y_4 & y_6 \\ x_2 & x_4 & x_6 & x_8 & y_8 & y_{10} & y_{11} & y_9 \\ x_3 & x_5 & x_7 & x_9 & x_{10} & x_{11} & x_{12} & y_{12} \end{pmatrix}, \\
\begin{pmatrix} x_1 & y_3 & y_5 & y_7 & y_1 & y_8 & y_6 & y_2 \\ x_2 & x_4 & x_6 & x_8 & y_4 & y_{10} & y_{11} & y_9 \\ x_3 & x_5 & x_7 & x_9 & x_{10} & x_{11} & x_{12} & y_{12} \end{pmatrix}, \\
\begin{pmatrix} x_1 & y_3 & y_5 & y_7 & y_2 & y_4 & y_6 & y_1 \\ x_2 & x_4 & x_6 & x_8 & y_8 & y_{10} & y_{11} & y_9 \\ x_3 & x_5 & x_7 & x_9 & x_{10} & x_{11} & x_{12} & y_{12} \end{pmatrix}, \\
\begin{pmatrix} x_1 & y_3 & y_5 & y_7 & y_1 & y_8 & y_2 & y_4 \\ x_2 & x_4 & x_6 & x_8 & y_6 & y_{10} & y_{11} & y_9 \\ x_3 & x_5 & x_7 & x_9 & x_{10} & x_{11} & x_{12} & y_{12} \end{pmatrix}.
\end{gathered} \tag{3.134}$$

The cubic graph (3.87) is represented by the cubic graph matrix:

$$\begin{pmatrix} x_1 & y_3 & y_5 & y_7 & y_9 & y_1 & y_4 & y_2 & y_6 & y_8 \\ x_2 & x_4 & x_6 & x_8 & x_{10} & y_{14} & y_{12} & y_{13} & y_{10} & y_{11} \\ x_3 & x_5 & x_7 & x_9 & x_{11} & x_{12} & x_{13} & x_{14} & x_{15} & y_{15} \end{pmatrix}$$

$\rightarrow$ 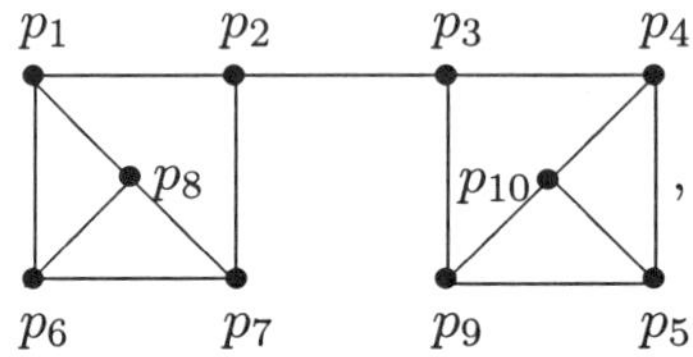

, (3.135)

where p_i denotes column $i = 1, \ldots, 10$ of the matrix. □

Proof that $A^*_{3\times 2n}(x;y)$ is a complete set of nonisomorphic cubic graph matrices. We first show that the permutation operations (3.130) can be used to bring every cubic graph matrix $A_{3\times 2n}(x;y)$ to the form $A^*_{3\times 2n}(x;y)$ in which the indices $i_1, i_2, \ldots, i_{2n+1}$ have not yet been specified, except that they must be a permutation of the integers in the set $\{\mathbb{H}_{2n}, 3n\}$. Let n_1, n_2, n_3, n_4 denote the number of columns, respectively, containing three $x's$, two $x's$ and one y, two $y's$ and one x, three $y's$. Then, since the number of $x's$ and the number of $y's$ is equal, we have that $3n_1 + 2n_2 + n_3 = n_2 + 2n_3 + 3n_4 = 3n$. Since adjacency is preserved by the interchange of $x's$ and $y's$, we must have $n_1 = n_4, n_2 = n_3$, hence, $n_1 + n_2 = n$. But, since an x_i can only share the subscript i with y_i, the adjacency conditions imply that $n_1 = 1$ and $n_2 = n - 1$. But each of the $(2n)!$ arrangements of the columns and the 3! arrangements of the three coordinates of a point give a trivial isomorphism (such rearrangements of coordinates leave the point-to-line relation trivially invariant). Thus, the columns of each $A_{3\times 2n}(x;y)$ can be brought to the form in which columns 1 through $2n$ have the column with three $x's$ occuring as column 1, followed by the $n-1$ columns containing two $x's$, followed by the $n-1$ columns containing one x, followed by the column containing the three $y's$, where the coordinate assignment within each column is still not specified . But, again by application of permutations in the group $S_3(2n)$, we can bring the $x's$ to the positions shown in (3.131), and this is a trivial isomorphism. Thus, by some permutation $\pi \in S_{3n}$, each cubic graph matrix $A_{3\times 2n}(x;y)$ can be brought to a form in which the x_i occur in the positions shown in (3.131). This still leaves open the assignment of the subscripts of the y_i, for all $i = 1, 2, \ldots, 3n$. Without exception, every cubic graph on $2n$ points has a subgraph of the form

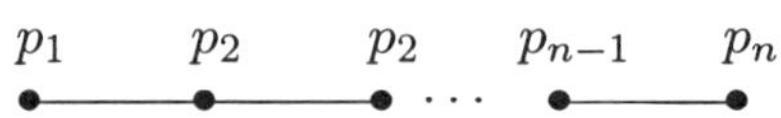

(3.136)

This implies that we can choose the entries in row 1 of columns 2 through $n-1$ of each $A_{3\times 2n}(x;y)$ to be $y_3, y_5, \cdots, y_{2n-1}$, since the first n columns correspond to a subgraph of the form (3.136). We conclude that the subscripts of the entries $y_{i_1}, y_{i_2}, \ldots, y_{i_{2n+1}}$ must be a permutation of $\{1, 2, 4, \ldots, 2n-2, 2n, 2n+1, \ldots, 3n\}$ such that those in each column are distinct (condition (1i), p. 221). But, again by application of permutations in $S_3(2n)$ that leave the first n columns unchanged, the subscript entries in rows 1 and 2 in columns $n+1, n+2, \ldots, 2n$ can be brought to the form in which the following conditions are satisfied: $i_1 < i_2, i_3 < i_4, \ldots, i_{2n-3} < i_{2n-2}, i_{2n-1} < i_{2n} < i_{2n+1}$, where $y_{i_{2n+1}}$ is the entry in row 3 and column $2n$, since all permutations that effect this ordering are trivial isomorphisms. The choice of the entry $y_{i_{2n+1}}$ in row 3 and column $2n$ to have subscript $i_{2n+1} = 3n$ assigns the last two columns a common subscript, hence, a line incident on each of the points in the corresponding cubic graph. But there is always the freedom, in any cubic graph, of choosing any two points to have a common incident line without restricting the remaining point-line relationships. Correspondingly, there is no restriction in realizing all nonisomorphic cubic graph matrices by making the choice $i_{2n+1} = 3n$ in (3.131). Thus, the integer $3n$ can be omitted from further consideration, and the subscripts of the entries $y_{i_1}, y_{i_2}, \ldots, y_{i_{2n}}$ can chosen to be the 2−subsets of $\mathbb{H}_{2n}$ given by

$$i_1 < i_2,\ i_3 < i_4,\ \ldots\ i_{2n-1} < i_{2n};\ i_1 < i_3 < \cdots < i_{2n-1}. \tag{3.137}$$

Condition (iii) is imposed to ensure that no isomorphic cubic graphs are admitted into the collection of cubic graph matrices of the form (3.131). These conditions are, however, very difficult to implement, since this requires that the set of cubic graph matrices corresponding to all solutions that satisfy condition (i) and (ii) be partitioned into equivalence classes of nonisomorphic cubic matrices. Completeness of the set $\mathbb{A}^*_{3\times 2n}(x;y)$, with condition (iii) enforced, follows because every cubic graph matrix in the set (3.129) is isomorphic to one in this set, and, by definition, the set (3.129) includes all possible cubic graphs. □

The necessity of enforcing condition (iii)—the set of cubic graph matrices $\mathbb{A}^*_{3\times 2n}(x;y)$ must contain only nonismorphic matrices—is already illustrated by the tetrahedron; it can also be represented by

$$\begin{pmatrix} x_1 & y_3 & y_1 & y_2 \\ x_2 & x_4 & y_5 & y_4 \\ x_3 & x_5 & x_6 & y_6 \end{pmatrix}. \tag{3.138}$$

This cubic graph matrix satisfies conditions (i) and (ii), and is isomorphic to the one given by (3.132).

Conditions (ii) in (3.131) need to be restricted further to determine a set of nonisomorphic cubic graphs, which is the purpose of condition (iii). The situation is similar to that involved in the concept of *matchings,* where the 2−subset conditions (ii) are further augmented by the requirement that $i_1 < i_3 < \cdots < i_{2n-1}$ (see Brualdi and Ryser [34, p. 318] and Sect. 4.1 of the next chapter). While there must exist a set of permutations of the set of integers in the set $\mathbb{H}_{2n}$ that give exactly a set $\mathbb{A}^*_{3\times 2n}(x;y)$ of nonisomorphic cubic graphs, such permutations of $\mathbb{H}_{2n}$ must not only respect the adjacency conditions, they must also account for isomorphisms. This procedure can always be effected by computing the full set of indices $\{i_1, i_2, \ldots, i_{2n}\}$ that satisfy conditions (i) and (ii) and selecting a subset of nonisomorphic cubic matrices by partitioning this set into equivalence classes of isomorphic cubic matrices—those that preserve adjacency of points. This is very difficult to implement algebraically, especially as a statement of conditions on the set of indices $\{i_1, i_2, \ldots, i_{2n}\}$. Of course, given such a set of nonisomorphic cubic graph matrices, the corresponding set of nonisomorphic cubic graphs is read off the matrix by assigning the columns to be the points of the graph.

Cubic graph matrices can be adapted by appropriate notational adjustments to the set $\mathbb{G}(T, T')$ of cubic graphs of angular momentum type. In this case, the cubic graph matrix $\mathbb{A}^*_{3\times 2n}(x;y)$ has for its columns exactly the columns of the fundamental triangle coefficient of order $2n$, which itself is read off the pair of binary trees of order $n+1$ without common forks—now seen to be nontrivial conditions—corresponding to a given coupling scheme. Many of the properties of such triangle coefficients can be transcribed to cubic graph matrices of angular momentum type.

3.4.4 Labeled cubic graphs

It is quite interesting that the counting formula giving the number of labeled cubic graphs on $2n$ points is known. We recall the definition of a labeled cubic graph on $2n$ points. We assign to the points of a given cubic graph the integers $1, 2, \ldots, 2n$ in all possible $(2n)!$ ways. Two such cubic graphs with such permuted labels are then defined to be isomorphic if the adjacency of the points is preserved (Harary and Palmer [78]); otherwise, they are nonisomorphic (see Sect. 2.2.2 for an example of labeled forks). We denote the set of nonisomorphic labeled cubic graphs by $L\mathbb{C}_{2n}$, and the cardinality by

$$L_{2n} = \mid L\mathbb{C}_{2n} \mid . \tag{3.139}$$

A formula giving the number L_{2n} has been derived by Read [146], and a rederivation of that result, using the principle of inclusion-exclusion, has been given by Chen and Louck [42]. The formula for L_{2n} can be put in the form

$$L_{2n} = \frac{(2n)!}{6^n}\sum_{i=0}^{n}\frac{A_i}{(n-i)!}, \tag{3.140}$$

where the quantity A_i is independent of n. It is given by

$$A_i = \frac{3^i}{2^i}\sum_{j=0}^{i}\frac{(-1)^j}{j!}\sum_{k=0}^{2i-2j}\frac{(-1)^k(2i+2k-1)!!}{3^k\, k!(2i-2j-k)!}. \tag{3.141}$$

Thus, in evaluating L_{2n}, the quantities $A_i, i = 0, 1, 2, \ldots,$ can be evaluated separately, and then used back in (3.140). For example, we find that

$$\begin{aligned} A_0 &= (-1)!! = 1, A_1 = -1, A_2 = 2, A_3 = 58/3, \\ L_0 &= 1, L_2 = 0, L_4 = 1, L_6 = 70. \end{aligned} \tag{3.142}$$

That A_i is independent of n suggests that this number has a combinatorial interpretation beyond the context of cubic graphs.

The enumeration of the set of labeled graphs is often taken as the starting point for the enumeration of unlabeled graphs by application of Polya's enumeration theorem (see Harary and Palmer [78]). In the case of cubic graphs on six points, the relationship is the following. We label the points of the nonisomorphic cubic graphs (3.84) and (3.83) as follows:

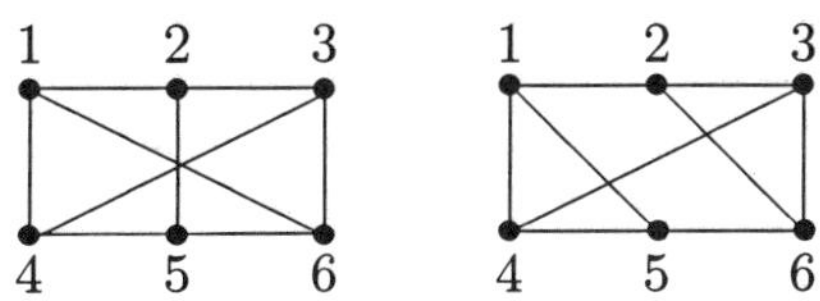

(3.143)

Then, because of the symmetries of these cubic graphs, the following permutations of (1,2,3,4,5) give all the isomorphic labeled cubic graphs, respectively:

1. The labelings given by $(1,2,3,4,5,6), (1,2,3,6,5,4),$ $(1,2,5,4,3,6), (1,2,5,6,3,4), (1,4,3,2,5,6), (1,4,3,6,5,2),$ $(1,4,5,2,3,6), (1,4,5,6,3,2), (1,6,3,2,5,4), (1,6,3,4,5,2),$ $(1,6,5,2,3,4), (1,6,5,4,3,2)$, and all six cyclic permutations of each of these permutations, give altogether seventy-two labeled cubic graphs that are isomorphic to the first labeled cubic graph in (3.143). Thus, there are $720/72 = 10$ nonisomorphic labeled cubic graphs corresponding to the first cubic graph in (3.143).
2. The labelings given by $(1,2,3,4,5,6), (1,2,6,5,4,3)$, and all six cyclic permutations of each of these permutations, give altogether

twelve labeled cubic graphs that are isomorphic to the second labeled cubic graph in (3.143). Thus, there are $720/12 = 60$ nonisomorphic labeled cubic graphs corresponding to the second cubic graph in (3.143).

Thus, the number of nonisomorphic labeled cubic graphs is $10 + 60 = 70$, in agreement with Read's formula. The implementation of Polya's enumeration theorem, or other methods, to obtain a formula for $\mid \mathbb{C}_{2n} \mid$ remains unsolved to our knowledge.

3.4.5 Summary and unsolved problems

A principal method of this monograph for obtaining recoupling coefficients is the use of pairs of standard labeled binary trees that share no common fork and of the corresponding fundamental triangle coefficients. These objects, in turn, define cubic graphs of angular momentum type. It can now be recognized that the property "no common fork" is nontrivial to implement: it is equivalent to fulfilling the adjacency conditions stated at the beginning of Sect. 3.4.3. The methods developed in Chapter 2 are now supplemented by those of this chapter by effecting the factoring of each cubic graph of angular momentum type that has a two-line join or a three-line join into maximal factors. This can be carried out for all such pairs of binary trees with no common fork, thus expressing all fundamental triangle coefficients as maximal factored forms. For all those that do not factor, there is the deeper problem of relating each such fundamental triangle coefficient to one that is associated with a set of nonisomorphic cubic graphs. It is the latter nonisomorphic cubic graphs of angular momentum type that define the so-called $3n-j$ coefficients. It is for this reason that the study of cubic graphs and the subset of cubic graphs of angular momentum type is important. Some of the properties of these two classes of cubic graphs have been presented, but the structural details that identify, in general, the subset that is of angular momentum type is still to be found. In the context of cubic graph matrices, these issues may be viewed as a combinatorial problem of distinguishing between distributions of subscript indices in a $3 \times 3n$ matrix. The discovery of structural principles that allow such an identification of cubic graphs of angular momentum type is one of the most important unsolved problems in angular momentum theory, as are counting formulas for their number (and the number of cubic graphs, as well).

Chapter 4

Generating Functions in Angular Momentum Theory

In the three previous chapters, we have developed the rich interplay between angular momentum theory and combinatorics in the setting of binary trees, trivalent trees, and cubic graphs. This is a very intricate relationship in which the binary coupling theory of angular momentum selects its own pathway into these mathematical subjects by utilizing pairs of binary and trivalent trees, culminating in the study of the properties of a subset of all cubic graphs of $2n$ points, namely, those that can be realized as the joining of pairs of trivalent trees. It seems unlikely that such a subset of cubic graphs would be studied on its own were it not for this relationship to angular momentum theory, which itself may be regarded as the study of the properties of copies of the group $SU(2)$. But the inter-relations do not stop here. Physics demands explicit formulas for its complex entities before the subject is complete; combinatorics affords the possibility for understanding the structure of complex relationships through the concept of generating functions. It is here that again the binary coupling theory of angular momentum and combinatorics continue their relationship. The purpose of this chapter is to set this forth in some detail. New concepts entering these developments include that of a double Pfaffian, the skew-symmetric matrix of a binary tree coupling scheme, and triangle monomials, as well as MacMahon's Master Theorem in its general form given by (1.255) in Chapter 1. We repeat it for easy reference:

$$\frac{1}{\det(I - tZ)} = \sum_{p=0}^{\infty} t^p \sum_{\alpha \vdash p} D^{p}_{\alpha\,\alpha}(Z), \tag{4.1}$$

in which $Z = (z_{ij})_{1\leq i,j\leq n}$ is an arbitrary matrix of commuting indeterminates, and the $D^{p}_{\alpha,\beta}(Z)$ are $SU(n)$ solid harmonics defined over these indeterminates. Schwinger's Master Theorem (1.252) follows from this relation by setting $Z = XY$ and using the multiplication property (1.245); MacMahon's relation (1.254) follows from Schwinger's result by setting $X = \mathrm{diag}(x_1, x_2, \ldots, x_n)$.

4.1 Pfaffians and Double Pfaffians

Schwinger [160] observed that the calculation of $3n-j$ coefficients involves taking the square root $\sqrt{(I-AB)}$, where A and B are skew-symmetric (antisymmetric) matrices of order n, but the procedure is rather obscure. The appropriate concepts for taking the square root $\sqrt{A}$ of a skew-symmetric matrix A and of $\sqrt{(I-AB)}$ is that of a Pfaffian and a double Pfaffian, which we denote by $\mathrm{Pf}(A)$ and $\mathrm{Pf}(A,B)$. Since $\det A = 0$ for skew-symmetric matrices of odd order, we define the Pfaffian only for matrices of even order $2n$, but the double Pfaffian is defined for matrices A and B of all orders $n \geq 2$. The definitions use the concept of a *matching* of the set of integers $\{1, 2, \ldots, 2n\}$ (see Brualdi and Ryser [34, p. 318] for definitions and references to the literature, and Chen and Louck [40] for further references and the relation to unitary group theory). A matching of the set of integers $\{1, 2, \ldots, 2n\}$ is an unordered set of n disjoint 2–subsets given by

$$\mathbb{I}_{2n} = \{\{i_1, i_2\}, \{i_3, i_4\}, \ldots, \{i_{2n-1}, i_{2n}\}\}, \tag{4.2}$$

where the notation is standardized by choosing

$$i_1 < i_2,\; i_3 < i_4,\; \ldots\; i_{2n-1} < i_{2n};\; i_1 < i_3 < \cdots < i_{2n-1}. \tag{4.3}$$

The cardinality of the set $\mathbb{I}_{2n}$ is $|\,\mathbb{I}_{2n}\,| = (2n-1)!! = (1)(3)\cdots(2n-1)$.

The definition of the Pfaffian is given in terms of the matchings of $\{1, 2, \ldots, 2n\}$ by

$$\mathrm{Pf}(A) = \sum_{\substack{\text{all matchings of}\\ \{1,\ldots,2n\}}} \varepsilon(i_1, i_2, \ldots, i_{2n}) a_{i_1,i_2} a_{i_3,i_4} \cdots a_{i_{2n-1},i_{2n}}, \tag{4.4}$$

where $\varepsilon(i_1, i_2, \ldots, i_{2n})$ is the sign of the permutation $(i_1, i_2, \ldots, i_{2n})$.

We standardize the form of the elements in a skew-symmetric matrix A of order n by writing the elements in row i and column j that are above the diagonal of $0's$ as $a_{ij}, i < j$; hence, the element in row i and column j below the diagonal is $a_{ij} = -a_{ji}, i > j$. We abbreviate this convention of indexing the elements of a skew-symmetric matrix $A = -A^T$ by the notation:

$$A = (a_{ij})_{1 \leq i < j \leq n} \,. \tag{4.5}$$

Examples. The Pfaffians of skew-symmetric matrices of order 2 and 4 are obtained from the single 2–set matching $\{\{1,2\}\}$ of the set $\{1,2\}$, and the three 2–set matchings of the set $\{1,2,3,4\}$ given by the three sets of 2–sets $\{\{1,2\},\{3,4\}\};\{\{1,3\},\{2,4\}\};\{\{1,4\},\{2,3\}\}$:

$$\begin{aligned} n = 1 &: \ \mathrm{Pf}(A) = a_{12}, \qquad (4.6)\\ n = 2 &: \ \mathrm{Pf}(A) = a_{12}a_{34} - a_{13}a_{24} + a_{14}a_{23}. \quad \square \end{aligned}$$

For the definition of the double Pfaffian, we need the set of all subsets of $\{1,2,\ldots,n\}$ of even length. We denote the set of all such subsets of $\{1,2,\ldots,n\}$ by the notation $\mathbb{I}_{2k}$, and a $2k$–subset member of this set by $\{i_1,i_2,\ldots,i_{2k}\} \in \mathbb{I}_{2k}$. The notation for the $2k$–subset $\{i_1,i_2,\ldots,i_{2k}\}$ is standardized by writing $i_1 < i_2 < \cdots < i_{2k}$. For example, there are three 2-sets of $\{1,2,3\}$ given by $\mathbb{I}_2 = \{\{1,2\},\{1,3\},\{2,3\}\}$. The definition of the double Pfaffian for two skew-symmetric matrices A and B of order n may be given in terms of a summation over ordinary Pfaffians:

$$\mathrm{Pf}(A,B) = 1 + \sum_{k\geq 1} \sum_{\{i_1,i_2,\ldots,i_{2k}\}\in\mathbb{I}_{2k}} \mathrm{Pf}(A_{i_1,i_2,\ldots,i_{2k}})\mathrm{Pf}(B_{i_1,i_2,\ldots,i_{2k}}). \tag{4.7}$$

The skew-symmetric matrix $A_{i_1,i_2,\ldots,i_{2k}}$, with $i_1 < i_2 < \cdots < i_{2k}$ of order $2k$ is obtained by selecting its elements from a subset of elements $\{a_{ij}\}$ of the skew-symmetric matrix A given by (4.5) and placing them in row r and column s by the following rule:

$$(A_{i_1,i_2,\ldots,i_{2k}})_{r,s} = a_{i_r,i_s}, \text{ each pair } 1 \leq r < s \leq 2k. \tag{4.8}$$

The elements of the skew-matrix $B_{i_1,i_2,\ldots,i_{2k}}$ are similarly defined.

Examples. The details of the construction of the double Pfaffian for $n = 3,4$ are as follows: The standardized subsets of $\{1,2,3\}$ of even length are $\mathbb{I}_2 = \{\{1,2\},\{1,3\},\{2,3\}\}$, while those of $\{1,2,3,4\}$ are $\mathbb{I}_2 = \{\{1,2\},\{1,3\},\{1,4\},\{2,3\},\{2,4\},\{3,4\}\}$ and $\mathbb{I}_4 = \{\{1,2,3,4\}\}$. For $n = 3$, there is only one $k = 1$ term in (4.7), which is followed by a sum over the three terms corresponding to the three subsets in $\mathbb{I}_2$, which are $a_{12}b_{12}, a_{13}b_{13}, a_{23}b_{23}$. For $n = 4$, the $k = 1$ term in (4.7) gives the sum of the six terms $a_{i_1i_2}b_{i_1i_2}$ corresponding to the standard matchings, while the $k = 2$ term gives a product of the order 2 Pfaffian in (4.6) and the same term with the letter a replaced by b. Thus, we obtain for $n = 2$ and $n = 3$, respectively, the double Pfaffians:

$$\mathrm{Pf}(A,B) = \mathrm{Pf}(B,A) = 1 + a_{12}b_{12} + a_{13}b_{13} + a_{23}b_{23}, \tag{4.9}$$

$$\begin{aligned}\mathrm{Pf}(A,B) = \mathrm{Pf}(B,A) &= 1 + a_{12}b_{12} + a_{13}b_{13} + a_{14}b_{14} \\ &+ a_{23}b_{23} + a_{24}b_{24} + a_{34}b_{34} \\ &+ (a_{12}a_{34} - a_{13}a_{24} + a_{23}a_{14})(b_{12}b_{34} - b_{13}b_{24} + b_{23}b_{14}). \quad \square\end{aligned} \tag{4.10}$$

The general relations of determinants of skew-symmetric matrices A, B to Pfaffians are given by

$$\det A = (\mathrm{Pf}(A))^2, \tag{4.11}$$

$$\det(I - AB) = (\mathrm{Pf}(A,B))^2,\ \mathrm{Pf}(A,B) = \mathrm{Pf}(B,A). \tag{4.12}$$

It turns out that the double Pfaffian is, in fact, the Pfaffian of a skew-symmetric matrix of order $2n$, which is given by

$$\det(I - AB) = Pf\begin{pmatrix} A & I \\ -I & -B \end{pmatrix} = (Pf(A,B))^2,\ n \geq 1. \tag{4.13}$$

(Private communication from John Stembridge as noted in Chen and Louck [40]). These relations can be verified by direct expansion for the cases $n = 1, 2, 3$.

Pfaffians have a very important role in the theory of generating functions for all of the fundamental triangle coefficients. This is because a skew-symmetric matrix, hence, a Pfaffian, can be associated with each standard labeled binary tree, then a double Pfaffian with each pair of such binary trees without common forks, hence, with each fundamental triangle coefficient.

4.2 The Skew-Symmetric Matrix of a Standard Labeled Binary Tree

A skew-symmetric matrix A of order $n+2$ can be associated with each binary tree $T \in \mathbb{T}_{n+1}$. Each element a_{ij} of the matrix A depends on the binary tree T, a permutation $\pi = (\pi_1, \pi_2, \ldots, \pi_{n+1}) \in S_{n+1}$ of a standard labeling $(u_1, u_2, \ldots, u_{n+1})$ of the $n+1$ external $\circ$ points, and on the indeterminates x_{kl} that belong to a $3 \times n$ matrix $X = (x_{kl})_{1\leq k\leq 3, 1\leq l\leq n}$; each a_{ij} is a monomial in the x_{kl}. (See Sect. 2.2, pp. 101-102, for examples of the shape of a binary tree and the corresponding external $\circ$ points to which the standard labeling $(u_1, \ldots, u_{n+1})$ is assigned). Each element a_{ij} is labeled by the shape of T, filled in with the parts of the permutation π, and the variables (indeterminates) X—the a_{ij} do not depend on the dummy variables $(u_1, \ldots, u_{n+1})$, only on the permutation thereof). We introduce the following notation for the matrix A and its elements:

$$A = A_{\mathrm{sh}_T(\pi)}(X), \; a_{ij} = (A_{\mathrm{sh}_T(\pi)}(X))_{ij}. \tag{4.14}$$

It is useful to illustrate this notation for the skew-symmetric matrix of a binary tree before giving the rules for its determination:

Example. $n+1=3 : \mathrm{sh}_T = ((\circ\circ)\circ), \pi(u_1,u_2,u_3) = (u_{\pi_1}, u_{\pi_2}, u_{\pi_3}) = (u_2,u_3,u_1), \pi = (\pi_1,\pi_2,\pi_3) = (2,3,1), \mathrm{sh}_T(\pi) = ((\pi_1\pi_2)\pi_3)) = ((23)1)$:

$$A = A_{((2\,3)\,1)}(X) = \begin{pmatrix} 0 & a_{12} & a_{13} & a_{14} \\ -a_{12} & 0 & a_{23} & a_{24} \\ -a_{13} & -a_{23} & 0 & a_{34} \\ -a_{14} & -a_{24} & -a_{34} & 0 \end{pmatrix}, \tag{4.15}$$

in which each $a_{ij} = a_{ij}(X)$ is a monomial in the indeterminate elements x_{ij} of a 2×3 matrix

$$X = \begin{pmatrix} x_{11} \; x_{12} \\ x_{21} \; x_{22} \\ x_{31} \; x_{32} \end{pmatrix}. \tag{4.16}$$

The explicit monomials x_{ij} are given by

$$\begin{array}{lll} a_{12} = -x_{21}x_{23} & a_{13} = -x_{11}x_{21} & a_{14} = x_{21} \\ & a_{23} = x_{31} & a_{24} = x_{21}x_{22} \\ & & a_{34} = x_{11}x_{12} \end{array} \qquad \square \tag{4.17}$$

It is the rules for obtaining the elements of this matrix for the general $3 \times n$ case (4.14) in terms of the structure of a binary tree of order $n+1$ and general shape that are developed in this section.

The rules for writing out the elements of the skew-symmetric matrix $A = A_{\mathrm{sh}_T(\pi)}(X)$ for a given tree $T \in \mathbb{T}_{n+1}$ are quite intricate; they utilize still another set of indeterminates assigned to the forks of a binary tree, which carry the full structure of the binary tree, but these indeterminates are generating function variables and do not appear in the final result for $A_{\mathrm{sh}_T(\pi)}(X)$.

We begin with a binary tree $T \in \mathbb{T}_{n+1}$ and consider the standard labeled binary tree $T(\pi(\mathbf{j})\,\mathbf{k})_j$, where for $n+1$ angular momenta, we have $\mathbf{k} = (k_1, k_2, \ldots, k_{n-1}), \pi(\mathbf{j}) = (j_{\pi_1}, j_{\pi_2}, \ldots, j_{\pi_{n+1}}), \pi \in S_{n+1}$. We now introduce new generic labels in the spirit of the u, v notation in Sect. 2.2.2, because the labels required need not be angular momentum labels; they need only reflect the structure of the underlying standard labeled binary tree. The adjustments of notation are: Replace each j_i by

z_i, $i = 1, 2, \ldots, n+1$; each k_i by w_i, $i = 1, 2, \ldots, n-1$; and j by w_n, and write $\mathbf{z} = (z_1, z_2, \ldots, z_{n+1})$ and $\mathbf{w} = (w_1, w_2, \ldots, w_n)$. We also consider all permutations $\pi(\mathbf{z})$. The standard labeled binary tree $T(\pi(\mathbf{j})\,\mathbf{k})_j$, under this renaming of parameters, is now denoted by $T(\pi(\mathbf{z})\,\mathbf{w}), T \in \mathbb{T}_{n+1}$.

We next take the unusual step, motivated by the work of Schwinger [160], of writing each root symbol $w_i, i = 1, 2, \ldots, n$, of the $i-$th fork as a linear combination of the symbols that appear at the end points of the fork. There are four cases to consider, as given below (2.85), Chapter 2:

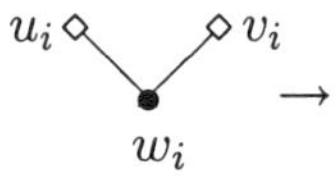

$$\rightarrow \begin{pmatrix} u_i \\ v_i \\ w_i \end{pmatrix} \quad : \quad w_i = x_{2i}u_i + x_{1i}v_i, \tag{4.18}$$

where the labels u_i and v_i are, respectively, the labels of the points that occur in the $i-$th fork in the standard labeled tree $T(\pi(\mathbf{z})\,\mathbf{w})$:

1. For $(\diamond, \diamond) = (\circ, \circ)$, the labels u_i and v_i are both external $z-$labels.
2. For $(\diamond, \diamond) = (\bullet, \circ)$, the label u_i is an internal $w-$label, and v_i is an external $z-$label.
3. For $(\diamond, \diamond) = (\circ, \bullet)$, the label u_i is an external $z-$label, and v_i is an internal $w-$label.
4. For $(\diamond, \diamond) = (\bullet, \bullet)$, the labels u_i and v_i are both internal $w-$labels.

The important point is that for a given standard labeled binary tree $T(\pi(\mathbf{z})\,\mathbf{w})$, the labels of the $i-$th fork are uniquely assigned and fall into one of these four classes, so that the labels u_i and v_i are uniquely identified in terms of the parts of $\mathbf{z}$ and $\mathbf{w}$. Since there are n such forks, we have n such relations of the form (4.18):

$$\begin{aligned} w_1 &= x_{21}u_1 + x_{11}v_1, \\ w_2 &= x_{22}u_2 + x_{12}v_2, \\ &\vdots \\ w_n &= x_{2n}u_n + x_{1n}v_n. \end{aligned} \tag{4.19}$$

In these relations, the $x_{ij}, 1 \leq i \leq 2,\ 1 \leq j \leq n$ are taken as the elements in row i and column j of a matrix $X_{2\times n}$, which constitutes rows 1 and 2 of an extended $3 \times n$ matrix $X = (x_{ij})_{1\leq i\leq 3, 1\leq j\leq n}$ of indeterminates—this extension is used in (4.21) below. The pairs $(u_i, v_i), i = 1, 2, \ldots, n$, in these relations consist of $n+1$ symbols z_i and n symbols w_i. (See (2.81) with n replaced by $n+1$ and examples (2.83).) The assignment of the

z_i and w_i to a standard labeled binary tree is such that the first relation in (4.19) always gives w_1 in terms of a pair (u_1, v_1) of $z-$symbols, and in all relations above the $i-$th relation, $w_i = x_{2i}u_i + x_{1i}v_i$, only those w_j with $j < i$ occur in the right-hand side of (4.19). This distribution of the n symbols w_i, $i = 1, 2, \ldots, n$, among the variables on the right-hand side of (4.19) allows each w_i to be expressed as a linear combination of the z_j, $i = 1, 2, \ldots, n$, with coefficients $M_{ij}(X_{2\times n})$ that are monomials in the $2n$ indeterminates $X_{2\times n} = (x_{ij})_{1\leq i\leq 2, 1\leq j\leq n}$:

$$w_i = \sum_j M_{ij}(X_{2\times n}) z_j. \tag{4.20}$$

The structure of relations (4.19)-(4.20) is uniquely determined by the underlying binary tree T, and this leads to the following rule for associating a unique skew-symmetric matrix with each binary tree $T \in \mathbb{T}_{n+1}$:

Skew-symmetric matrix of a binary tree. *The skew-symmetric matrix* $A_{sh_T(\pi)}(X)$ *of a standard labeled binary tree* $T(\pi(\mathbf{z})\,\mathbf{w})$ *is determined by equating coefficients of* $\det(z_i, z_j)$ *in the following relation:*

$$\sum_{1\leq i<j\leq n+1} a_{ij} \det(z_i, z_j) = \sum_{j=1}^{n} x_{3j} \det(u_j, v_j) + \det(w_n, z_{n+1})$$
$$= \sum_{1\leq i<j\leq n+1} \left(A_{sh_T(\pi)}(X)\right)_{ij} \det(z_i, z_j). \tag{4.21}$$

This gives uniquely the elements a_{ij} *of*

$$A = A_{sh_T(\pi)}(X). \tag{4.22}$$

Relation (4.21) is called the generating function for the skew-symmetric matrix of a binary tree. It is useful to review how the structure of relations (4.19)-(4.20) leads to the determination of A from (4.21): The $2n$ indeterminates $(u_j, v_j), j = 1, 2, \ldots, n$, are identified, for each standard labeled tree $T(\pi(\mathbf{z})\,\mathbf{w})$ in terms of the n indeterminates $z_1, \ldots, z_n$ by the rules (1)-(4) following (4.18); the extra indeterminates x_{3j} (row 3 of X) mark these terms; the quantity w_n is defined by the last expression in (4.19); and each w_i has the form (4.20) that can be substituted into the middle term in (4.21), whenever a w_i occurs. Thus, there are $(n+1)(n+2)/2$ terms $\det(z_i, z_j)$ in the left-most and middle expressions in relation (4.21). This leads to the right-most expression in (4.21), in which the elements of the matrix $A_{\mathrm{sh}_T(\pi)}(X)$ are independent of all variables z_i and w_i, but still encode the shape of the tree and the permutation π. The examples below illustrate the implementation of relation (4.21).

Examples:

$n+1=2:$

$$: \begin{array}{l} \mathrm{sh}_T(\pi) = (1\,2), \\ w_1 = x_{21}z_1 + x_{11}z_2 \end{array} ;$$

$$\begin{aligned} &\sum_{1\le i<j\le 3} a_{ij}\det(z_i,z_j) = x_{31}\det(u_1,v_1) + \det(w_1,z_3) \\ &= x_{31}\det(z_1,z_2) + x_{21}\det(z_1,z_3) + x_{11}\det(z_2,z_3); \\ &A = A_{(1\,2)}(X) = \begin{pmatrix} 0 & x_{31} & x_{21} \\ -x_{31} & 0 & x_{11} \\ -x_{21} & -x_{11} & 0 \end{pmatrix}. \end{aligned} \tag{4.23}$$

$n+1=3:$

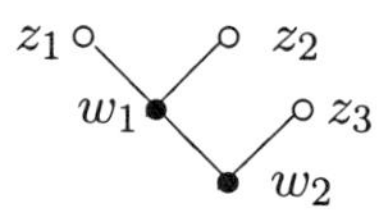

$$: \begin{array}{l} \mathrm{sh}_T(\pi) = ((1\,2)\,3); \\ w_1 = x_{21}z_1 + x_{11}z_2, \\ w_2 = x_{22}w_1 + x_{12}z_3 \\ \quad = x_{22}x_{21}z_1 + x_{22}x_{11}z_2 + x_{12}z_3; \end{array}$$

$$\begin{aligned} &\sum_{1\le i<j\le 4} a_{ij}\det(z_i,z_j) \\ &= x_{31}\det(u_1,v_1) + x_{32}\det(u_2,v_2) + \det(w_2,z_4) \\ &= x_{31}\det(z_1,z_1) + x_{32}\det(w_1,z_3) + \det(w_2,z_4) \\ &= x_{31}\det(z_1,z_2) + x_{21}x_{32}\det(z_1,z_3) + x_{21}x_{22}\det(z_1,z_4) \\ &+x_{11}x_{32}\det(z_2,z_3) + x_{11}x_{22}\det(z_2,z_4) + x_{12}\det(z_3,z_4); \end{aligned}$$

$$A = A_{((1\,2)\,3)}(X) = \begin{pmatrix} 0 & x_{31} & x_{21}x_{32} & x_{21}x_{22} \\ -x_{31} & 0 & x_{11}x_{32} & x_{11}x_{22} \\ -x_{21}x_{32} & -x_{11}x_{32} & 0 & x_{12} \\ -x_{21}x_{22} & -x_{11}x_{22} & -x_{12} & 0 \end{pmatrix}. \tag{4.24}$$

$n+1=4:$

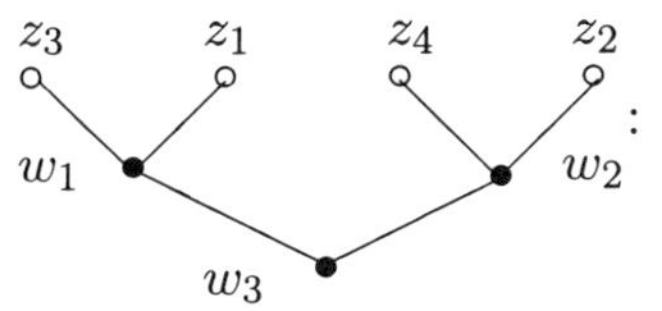

$$
\begin{aligned}
&\mathrm{sh}_T(\pi) = ((3\,1)(4\,2));\\
&w_1 = x_{21}z_3 + x_{11}z_1,\\
&w_2 = x_{22}z_4 + x_{12}z_2,\\
&w_3 = x_{23}w_1 + x_{13}w_2\\
&\quad = x_{11}x_{23}z_1 + x_{12}x_{13}z_2\\
&\quad +x_{21}x_{23}z_3 + x_{13}x_{22}z_4;
\end{aligned}
$$

$$
\begin{aligned}
&\sum_{1\le i<j\le 5} a_{ij}\det(z_i,z_j)\\
&= x_{31}\det(z_3,z_1) + x_{32}\det(z_4,z_2) + x_{33}\det(w_1,w_2) + \det(w_3,z_5)\\
&= x_{11}x_{12}x_{33}\det(z_1,z_2) - x_{31}\det(z_1,z_3) + x_{11}x_{22}x_{33}\det(z_1,z_4)\\
&+x_{11}x_{23}\det(z_1,z_5) - x_{12}x_{21}x_{33}\det(z_2,z_3) - x_{32}\det(z_2,z_4)\\
&+x_{12}x_{13}\det(z_2,z_5) + x_{21}x_{22}x_{33}\det(z_3,z_4) + x_{21}x_{23}\det(z_3,z_5)\\
&+x_{13}x_{22}\det(z_4,z_5);
\end{aligned}
\tag{4.25}
$$

$$
\begin{array}{llll}
a_{12} = x_{11}x_{12}x_{33} & a_{13} = -x_{31} & a_{14} = x_{11}x_{22}x_{33} & a_{15} = x_{11}x_{23}\\
 & a_{23} = -x_{12}x_{21}x_{33} & a_{24} = -x_{32} & a_{25} = x_{12}x_{13}\\
 & & a_{34} = x_{21}x_{22}x_{33} & a_{35} = x_{21}x_{23}\\
 & & & a_{45} = x_{13}x_{22}
\end{array}
$$

where these are the elements above the diagonal of the skew-symmetric matrix $A = A_{\mathrm{sh}_T(\pi)}(X) = A_{((3\,1)(4\,2))}(X)$.

It is not necessary to calculate the elements of $A_{\mathrm{sh}_T(\pi)}(X)$ for every $\pi \in S_{n+1}$, since these can be obtained by a rearrangement of the elements of $A_{\mathrm{sh}_T(\pi)}(e)$, $e = (1\,2\,\cdots\,n+1)$:

$$
\begin{aligned}
\left(A_{\mathrm{sh}_T(\pi)}(X)\right)_{i,j} &= \left(A_{\mathrm{sh}_T(\pi)}(e)\right)_{\pi_i^{-1},\pi_j^{-1}},\ 1\le i<j\le n+1,\\
\left(A_{\mathrm{sh}_T(\pi)}(X)\right)_{i,n+2} &= \left(A_{\mathrm{sh}_T(\pi)}(e)\right)_{\pi_i^{-1},n+2},\ 1\le i\le n+1.
\end{aligned}
\tag{4.26}
$$

The skew-symmetric matrix $A_{\mathrm{sh}_T(\pi)}(X)$ of a standard labeled binary tree $T(\pi(\mathbf{z})\,\mathbf{w})$ occurs in the generating functions for coupled wave functions given below in Sect. 4.4.

4.3 Triangle Monomials

The definition of a triangle monomial associated with a standard labeled binary tree is an essential step in developing generating functions.

Let $\langle a,b,c\rangle$ be a triangle of quantum numbers (a,b,c), and let (x,y,z) be three indeterminates. The *elementary triangle monomial* $\Phi_{\langle a,b,c\rangle}(x,y,z)$ is defined by

$$\Phi_{\langle a,b,c\rangle}(x,y,z) = \{abc\}^{-1}x^{b+c-a}y^{a+c-b}z^{a+b-c}, \tag{4.27}$$

where $\{abc\}$ is $\sqrt{2c+1}$ times the $\delta-$triangle coefficient $\Delta(abc)$ defined by (2.164):

$$\{abc\} = \left[\frac{(2c+1)(b+c-a)(a+c-b)!(a+b-c)!}{(a+b+c+1)!}\right]^{1/2}. \tag{4.28}$$

These monomials satisfy the following orthogonality relations in the inner product $(\ ,\ )$:

$$\left(\Phi_{\langle a',b',c'\rangle},\Phi_{\langle a,b,c\rangle}\right) = \delta_{a',a}\delta_{b',b}\delta_{c',c}(a+b+c+1)!/(2c+1). \tag{4.29}$$

Using the definition (4.27) of an elementary triangle monomial, we now define the triangle monomial associated with a standard labeled binary tree $T(\mathbf{j}\,\mathbf{k})_j$ of order $n+1$, where now

$$\mathbf{j} = (j_1,j_2,\ldots,j_{n+1}),\ \mathbf{k} = (k_1,k_2,\ldots,k_{n-1}), \tag{4.30}$$

by taking the product of all elementary triangle monomials over the n triangles that label the n forks of the standard labeled binary tree. This gives a triangle monomial for each binary tree $T\in\mathbb{T}_{n+1}$:

$$\Phi_{T(\mathbf{j}\,\mathbf{k})_j}(X) = \prod_{i=1}^{n}\Phi_{\langle a_i,b_i,k_i\rangle}(x_{1i},x_{2i},x_{3i}), \tag{4.31}$$

where (a_i,b_i,k_i) are the labels of the $i-$th fork in the standard labeled tree $T(\mathbf{j}\,\mathbf{k})_j$, as defined in detail in relations (2.84)-(2.87), with n replaced by $n+1$. These triangle monomials satisfy the orthogonality relations given by

$$\left(\Phi_{T'(\mathbf{j}\,\mathbf{k}')_j},\Phi_{T(\mathbf{j}\,\mathbf{k})_j}\right) = \delta_{T',T}\,\delta_{\mathbf{k}',\mathbf{k}}\prod_{i=1}^{n}\frac{(a_i+b_i+k_i+1)!}{2k_i+1}. \tag{4.32}$$

4.4 Generating Functions for Coupled Wave Functions

The defining relation (4.21) for the skew-symmetric matrix associated with a standard labeled binary tree $T(\pi(\mathbf{j})\,\mathbf{k})_j, T \in \mathbb{T}_{n+1}$, can be written in matrix form (tr now denotes transposition):

$$\begin{aligned}
&\sum_{i=1}^{n} x_{3i} \det(u_i, v_i) + \det(w_n, z_{n+1}) = xAy^{\mathrm{tr}}, \\
&A = A_{\mathrm{sh}_T(\pi)}(X), \;\; z_i = \mathrm{col}(x_i, y_i), i = 1, 2, \ldots, n+1, \qquad (4.33) \\
&x = (x_1, x_2, \ldots, x_{n+1}), \;\; y^{\mathrm{tr}} = \mathrm{col}(y_1, y_2, \ldots, y_{n+1}).
\end{aligned}$$

The abstract relation (2.116) for the coupled state vectors $|T(\pi(\mathbf{j})\,\mathbf{k})_{jm}\rangle$ is realized in terms of coupled spinor wave functions for $n+1$ angular momenta as

$$\begin{aligned}
\Psi_{T(\pi(\mathbf{j})\,\mathbf{k})_{jm}}(x, y) &= \sum_{\mathbf{m}} C_{T(\pi(\mathbf{j}\,\mathbf{m})\,\mathbf{k})_{jm}} \prod_{i=1}^{n+1} P_{j_i m_i}(x_i, y_i), \\
&\qquad\qquad\qquad\qquad\qquad\qquad\qquad (4.34) \\
P_{j_i m_i}(x_i, y_i) &= \frac{x_i^{j_i+m_i} y_i^{j_i-m_i}}{\sqrt{(j_i+m_i)!(j_i-m_i)!}}, i = 1, 2, \ldots, n+1,
\end{aligned}$$

in which now the standard labeled tree $T(\pi(\mathbf{j})\,\mathbf{k})_{jm}$ has $\mathbf{j}$ and $\mathbf{k}$ extended to $n+1$ angular momenta as given by (4.30), and also $\mathbf{m} = (m_1, m_2, \ldots, m_{n+1})$. The coefficients $C_{T(\pi(\mathbf{j}\,\mathbf{m})\,\mathbf{k})_{jm}}$ are the generalized WCG coefficients defined by the product (2.115), extended to n.

Schwinger's generating function for the coupled spinor wave functions (4.34) is given by

$$\begin{aligned}
\exp\left(x\, A_{\mathrm{sh}_T(\pi)}(X)\, y^{\mathrm{tr}}\right) &= \sum_{\mathbf{j}, \mathbf{k}, j} \Phi_{T(\pi(\mathbf{j})\,\mathbf{k})_j}(X) \\
&\times \sum_{m} (-1)^{j-m} P_{j,-m}(x_{n+1}, y_{n+1}) \Psi_{T(\pi(\mathbf{j})\,\mathbf{k})_{jm}}(x, y). \quad (4.35)
\end{aligned}$$

The summation in this formal generating relation is over the denumerably infinite set of values $\mathbf{j}, \mathbf{k}, j$ for which the WCG coefficients in (4.34) are defined, as explained in detail in (2.89)-(2.91). The proof of Schwinger's result is by exponentiating relation (4.34), and using the definition of the triangle monomials.

We also have from relations (1.252)-(1.253) that the exponential in (4.35) is given by

$$\exp\left(x\,A_{\mathrm{sh}_T(\pi)}(X)\,y^{\mathrm{tr}}\right) = \sum_{p=0}^{\infty}\sum_{\alpha,\beta\vdash k}\frac{x^{\alpha}}{\sqrt{\alpha!}}D^{p}_{\alpha,\beta}\left(A_{\mathrm{sh}_T(\pi)}(X)\right)\frac{y^{\beta}}{\sqrt{\beta!}}. \tag{4.36}$$

Using this relation in (4.35) gives the relation between D^p-functions of argument $A_{\mathrm{sh}_T(\pi)}(X)$ and triangle monomials as follows:

$$D^{p}_{\alpha,\beta}\left(A_{\mathrm{sh}_T(\pi)}(X)\right) = (-1)^{j-m}\sum_{\mathbf{k}} C_{T(\pi(\mathbf{j}\,\mathbf{m});\mathbf{k})_{jm}}\Phi_{T(\pi(\mathbf{j})\,\mathbf{k})_j}(X), \tag{4.37}$$

where the relations between parameters are given by

$$\begin{aligned}&\alpha_i = j_{\pi_i} + m_{\pi_i},\ \beta_i = j_{\pi_i} - m_{\pi_i} \quad i = 1,2,\ldots,n+1;\\ &\alpha_{n+2} = j - m,\ \beta_{n+2} = j + m;\\ &p = \sum_{i=1}^{n+2}\alpha_i = \sum_{i=1}^{n+2}\beta_i = j_1 + j_2 + \cdots + j_{n+1} + j.\end{aligned} \tag{4.38}$$

Relation (4.37) gives directly the Van der Waerden form for the presentation of the $SU(2)$ WCG coefficients. Thus, for $n = 1$, the skew-symmetric matrix of order 3 is given by (4.23). For this case, there is no summation over $\mathbf{k}$ in the right-hand side of (4.37), since $k_1 = j$ and there is no k_0; the right-hand side reduces to

$$(-1)^{j-m}C^{j_1\ j_2\ j}_{m_1 m_2 m}\Phi_{\langle j_1,j_2,j\rangle}(x_{11},x_{21},x_{31}), \tag{4.39}$$

since $C_{T(j_1\,m_1\,j_2\,m_2)_{j\,m}} = C^{j_1\ j_2\ j}_{m_1 m_2 m}$. The direct substitution of the skew-symmetric matrix $A = A_{(12)}(X)$ of order 3 into (1.243) for the $SU(3)$ solid harmonics gives an expression that is a multiple of the triangle monomial $x_{11}^{j_2+j-j_1}x_{21}^{j_1+j-j_2}x_{31}^{j_1+j_2-j}$. This monomial cancels from each side of relation (4.37) for $n = 1$, leaving behind the following formula for the $SU(2)$ WCG coefficients:

$$C^{j_1\ j_2\ j}_{m_1 m_2 m} = \{j_1 j_2 j\}\sqrt{\alpha!\beta!}\sum_{B\in\mathbb{M}'(\alpha,\beta)}\frac{(-1)^{b_{21}}}{B!}, \tag{4.40}$$

$$\alpha = (j_1+m_1, j_2+m_2, j-m), \beta = (j_1-m_1, j_2-m_2, j+m).$$

The set of 3×3 matrix arrays $\mathbb{M}'(\alpha,\beta)$ is the subset of $\mathbb{M}^p_{3\times 3}(\alpha,\beta), p = j_1 + j_2 + j$, with diagonal elements $b_{ii} = 0$. This result is just Van der Waerden's form (1.213) (when written conventionally) of the WCG coefficients, now written in terms of the redundant summation over all elements of the matrix array $\mathbb{M}'(\alpha,\beta)$ (put $k_2 = b_{12}$ in (1.214)).

4.5 Generating Functions for Recoupling Coefficients

MacMahon's Master Theorem in the form (4.1) and relation (4.37) above for the U_{n+1} solid harmonics defined on the skew-symmetric matrix $A_{\mathrm{sh}_T(\pi)}(X)$ of a standard labeled binary tree $T(\pi(\mathbf{j})\,\mathbf{k})_j$ can now be used to obtain generating function for all recoupling coefficients for the binary coupling of $n+1$ angular momenta. For economy of notation, we write

$$A = A_{\mathrm{sh}_T(\pi)}(X), \quad A' = A_{\mathrm{sh}_{T'}(\pi')}(X'). \tag{4.41}$$

Then, setting $t=1$ and $Z = AA'$ in (4.1) and using (4.7) for the double Pfaffian, we obtain the relation:

$$\frac{1}{(Pf(A,A'))^2} = \frac{1}{\det(I-AA')} = \sum_{p=0}^{\infty}(-1)^p \sum_{\alpha,\beta\vdash p} D^p_{\alpha,\beta}(A)D^p_{\alpha,\beta}(A'), \tag{4.42}$$

where we have used the multiplication property (1.245) and the transposition symmetry property (1.250) to obtain

$$D^p_{\alpha,\alpha}(AA') = \sum_{\beta\vdash p}(-1)^p D^p_{\alpha,\beta}(A)D^p_{\alpha,\beta}(A'). \tag{4.43}$$

We next use relation (4.31) to obtain the following form of (4.42):

$$\frac{1}{(Pf(A,A'))^2} = \frac{1}{\det(I-AA')} = \sum_{p=0}^{\infty}(-1)^p \sum_{\substack{\mathbf{j},\,j \\ j_1+j_2+\cdots+j_{n+1}+j=p}} (2j+1)^{-1}$$
$$\times \sum_{\mathbf{k},\,\mathbf{k}'} \Phi_{T(\pi(\mathbf{j})\,\mathbf{k})_j}(X)\Phi_{T'(\pi'(\mathbf{j})\,\mathbf{k}')_j}(X') \left(R_{T,T'}^{\mathrm{sh}(\pi(\mathbf{j}));\mathrm{sh}'(\pi'(\mathbf{j}))}\right)_{\mathbf{k}\,j;\mathbf{k}'\,j}. \tag{4.44}$$

In obtaining this result, we have used the inner product relation

$$\left(\Psi_{T(\pi(\mathbf{j})\,\mathbf{k})_{j\,m}}(x,y)\,,\ \Psi_{T'(\pi'(\mathbf{j})\,\mathbf{k}')_{j\,m}}(x,y)\right) = \langle T(\pi(\mathbf{j})\,\mathbf{k})_j\,|\,T'(\pi'(\mathbf{j})\,\mathbf{k}')_j\rangle$$
$$= \left(C_T^{\mathrm{sh}(\pi(\mathbf{j}))}C_{T'}^{\mathrm{sh}'(\pi'(\mathbf{j}))tr}\right)_{\mathbf{k}\,j;\mathbf{k}'\,j} = \left(R_{T,T'}^{\mathrm{sh}(\pi(\mathbf{j}))\,;\,\mathrm{sh}'(\pi'(\mathbf{j}))}\right)_{\mathbf{k}\,j;\mathbf{k}'\,j}. \tag{4.45}$$

These relations are discussed in detail in Chapter 2. We have also used the property that the inner product is independent of m, which introduces the factor $(2j+1)^{-1}$, since m has been summed over. Moreover, the relations between the α_i, β_i and the j_i, m_i and j, m are invertible so that the summation over all $\alpha, \beta \vdash p$ can be replaced by the summation over all $\mathbf{j}, j, m$, which gives the slightly modified form:

$$\frac{1}{(Pf(A,A'))^2} = \frac{1}{\det(I - AA')} = 1 + \sum_{\mathbf{j},\mathbf{k},\mathbf{k}',j} \frac{(-1)^{j_1+\cdots+j_{n+1}+j}}{(2j+1)} \\ \times \Phi_{T(\pi(\mathbf{j})\,\mathbf{k})_j}(X)\Phi_{T'(\pi'(\mathbf{j})\,\mathbf{k}')_j}(X') \left(R^{\mathrm{sh}(\pi(\mathbf{j}))\,;\,\mathrm{sh}'(\pi'(\mathbf{j}))}_{T,T'}\right)_{\mathbf{k}\,j;\mathbf{k}'\,j}. \tag{4.46}$$

The summation is over the denumerably many sets of angular momentum quantum numbers $\mathbf{j}, \mathbf{k}, \mathbf{k}', j$ such that all triangle relations are satisfied among the triangles $\langle a_i\, b_i\, k_i\rangle$, $\langle a'_i\, b'_i\, k'_i\rangle$ associated with the forks of the standard labeled binary trees $T(\pi(\mathbf{j})\,\mathbf{k})_j$ and $T'(\pi'(\mathbf{j})\,\mathbf{k}')_j, T, T' \in \mathbb{T}_{n+1}$. The form (4.46) is useful for comparing with a second expansion (4.53) below that gives the recoupling coefficients themselves.

The reciprocal of the double Pfaffian in (4.46) can be expanded in a second way, using

$$\mathrm{Pf}(A, A') = 1 + G(X, X'), \tag{4.47}$$

where $G(X, X')$ is *a polynomial with integral coefficients* in the elements of the $3 \times (n+1)$ matrices X and X', as described in detail in Sect. 4.2, and determined by $(1 + G(X, X'))^2 = \det(I - AA')$, where A and A' are given by (4.41). We now effect the following expansion:

$$\begin{aligned}\frac{1}{(\mathrm{Pf}(A,A'))^2} &= (1 + G(X,X'))^{-2} \\ &= 1 + \sum_{p=1}^{\infty}(-1)^p(p+1)(G(X,X'))^p. \end{aligned}\tag{4.48}$$

But each $G(X, X')$ in this result is a sum of monomial terms of the form

$$G(X, X') = \sum_{q=1}^{h} G_q(X, X'), \tag{4.49}$$

where each $G_q(X, X')$ is a monomial in the variables x_{ij} and x'_{ij}, multiplied by an integer, and h is the number of distinct monomials occurring in the double Pfaffian $\mathrm{Pf}(A, A') = 1 + G(X, X')$. Thus, using the multinomial expansion theorem, we obtain

$$(G(X,X'))^p = \sum_{\substack{p_1,p_2,\ldots,p_h \\ p_1+p_2+\cdots+p_h=p}} \binom{p}{p_1, p_2, \ldots, p_h} \\ \times (G_1(X,X'))^{p_1}(G_2(X,X'))^{p_2}\cdots(G_h(X,X'))^{p_h}. \tag{4.50}$$

But, since each $G_q(X, X')$ is a monomial in the variables x_{ij} and x'_{ij}, multiplied by an integer, it must be the case that

$$\begin{aligned}
&(G_1(X,X'))^{p_1}(G_2(X,X'))^{p_2}\cdots(G_h(X,X'))^{p_h} \\
&= I^{(p)}_{p_1,p_2,\ldots,p_h}\prod_{i=1}^{n}(x_{1i})^{b_i+k_i-a_i}(x_{2i})^{a_i+k_i-b_i}(x_{3i})^{a_i+b_i-k_i} \\
&\quad\times\prod_{i=1}^{n}(x'_{1i})^{b'_i+k'_i-a'_i}(x'_{2i})^{a'_i+k'_i-b'_i}(x'_{3i})^{a'_i+b'_i-k'_i},
\end{aligned} \tag{4.51}$$

where $I^{(p)}_{p_1,p_2,\ldots,p_h}$ is an integer. The triangles $\langle a_i\, b_i\, k_i\rangle$ and $\langle a'_i\, b'_i\, k'_i\rangle$ are those read off the i–th forks of the standard labeled binary trees $T(\pi(\mathbf{j})\,\mathbf{k})_j$ and $T'(\pi'(\mathbf{j})\,\mathbf{k}')_j, T, T' \in \mathbb{T}_{n+1}$. Relation (4.51) giving the product of powers of the $G_q(X, X')$ in terms of the triangle monomials (without normalizing factors) must hold because of the orthogonality of the triangle monomials.

Because of the monomial property of each $G_q(X, X')$ in (4.51), the product of the powers of these monomials must give $6n$ linear relations of the following form, where each of the L_{li}, L'_{li} is a linear combination of the $p_1, p_2, \ldots, p_h$ with coefficients 0 or 1.

$$\begin{aligned}
L_{1i}(p_1,p_2,\ldots,p_h) &= b_i + k_i - a_i\,, \\
L_{2i}(p_1,p_2,\ldots,p_h) &= a_i + k_i - b_i\,, \\
L_{3i}(p_1,p_2,\ldots,p_h) &= a_i + b_i - k_i\,; \\
L'_{1i}(p_1,p_2,\ldots,p_h) &= b'_i + k'_i - a'_i\,, \\
L'_{2i}(p_1,p_2,\ldots,p_h) &= a'_i + k'_i - b'_i\,, \\
L'_{3i}(p_1,p_2,\ldots,p_h) &= a'_i + b'_i - k'_i\,.
\end{aligned} \tag{4.52}$$

The effect of different orderings of the factors in (4.51) is to permute the $p_1, p_2, \ldots, p_h$ in these relations. Relations (4.52) are quite significant: The solutions for the p_i in terms of an independent subset of $\{p_1, p_2, \ldots, p_h\}$, and in terms of the $3n$ quantum numbers consisting of the angular momentum quantum numbers $\mathbf{j}$, the intermediate angular momentum quantum numbers $\mathbf{k}, \mathbf{k}'$, and the total angular momentum j determine the form of the recoupling matrices. These relations must always be solvable because of the form of the expansion (4.44). This is illustrated by the examples below in Sect. 4.6.

We next substitute (4.51) into (4.50), and the resulting relation back into (4.48), to obtain the following expression for the expansion of the reciprocal of the double Pfaffian:

$$\frac{1}{(\mathrm{Pf}(A,A'))^2} = (1+G(X,X'))^{-2} = 1 +$$
$$\sum_{\mathbf{j},\mathbf{k},\mathbf{k}',j} \sum_{p=1}^{\infty} \sum_{\substack{p_1,p_2,\ldots,p_h \\ p_1+p_2+\cdots+p_h=p}}^{\text{restricted}} (-1)^p (p+1) \begin{pmatrix} p \\ p_1,p_2,\ldots,p_h \end{pmatrix} I^{(p)}_{p_1,p_2,\ldots,p_h}$$
$$\times \prod_{i=1}^{n} (x_{1i})^{b_i+k_i-a_i} (x_{2i})^{a_i+k_i-b_i} (x_{3i})^{a_i+b_i-k_i}$$
$$\times \prod_{i=1}^{n} (x'_{1i})^{b'_i+k'_i-a'_i} (x'_{2i})^{a'_i+k'_i-b'_i} (x'_{3i})^{a'_i+b'_i-k'_i}. \tag{4.53}$$

The summation is over the denumerably many sets of angular momentum quantum numbers $\mathbf{j},\mathbf{k},\mathbf{k}',j$ such that all triangle relations are satisfied among the triangles $\langle a_i\, b_i\, k_i\rangle$, $\langle a'_i\, b'_i\, k'_i\rangle$ associated with the forks of the standard labeled binary trees $T(\pi(\mathbf{j})\,\mathbf{k})_j$ and $T'(\pi'(\mathbf{j})\,\mathbf{k}')_j$, $T, T' \in \mathbb{T}_{n+1}$. The summation over the p_i is a finite summation that is **restricted** in a way that takes into account all of the linear relations (4.52). It is important to take the summation in (4.53) in this manner because it is the coefficients of the triangle monomials having the same set of triangles that are to be compared between (4.53) and (4.46).

The equating of the coefficients of like triangle monomials in (4.46) and (4.53) now gives the matrix elements of the recoupling matrix:

$$\langle T(\pi(\mathbf{j})\,\mathbf{k})_j \,|\, T'(\pi'(\mathbf{j})\,\mathbf{k}')_j\rangle = \left(R^{\mathrm{sh}(\pi(\mathbf{j}));\mathrm{sh}'(\pi'(\mathbf{j}))}_{T,T'}\right)_{\mathbf{k}\,j;\mathbf{k}'\,j} \tag{4.54}$$
$$= \frac{(-1)^{j_1+\cdots+j_{n+1}+j}}{2j+1} \prod_{i=1}^{n} \sqrt{(2k_i+1)(2k'_i+1)}\; \Delta(a_i\, b_i\, k_i)\Delta(a'_i\, b'_i\, k'_i)$$
$$\times \sum_{p}^{\text{restricted}} \sum_{\substack{p_1,p_2,\ldots,p_h \\ p_1+p_2+\cdots+p_h=p}}^{\text{restricted}} (-1)^p (p+1) \begin{pmatrix} p \\ p_1,p_2,\ldots,p_h \end{pmatrix} I^{(p)}_{p_1,p_2,\ldots,p_h}.$$

We have used the relation $\{abc\} = \sqrt{2c+1}\,\Delta(abc)$ (compare (4.28) and (2.164)). We recall also that the $\delta-$triangle factors $\Delta(a_i\, b_i\, k_i)$ and $\Delta(a'_i\, b'_i\, k'_i)$ are known in terms of the angular momentum quantum numbers $j_1, j_2, \ldots, j_{n+1}$, the intermediate angular momentum quantum numbers, $k_1, k_2, \ldots, k_{n-1}; k'_1, k'_2, \ldots, k'_{n-1}$, and the total angular momentum j. These $\delta-$triangle factors contain all the triangles coming from the forks of the associated pair of labeled binary trees; they assure in (4.54) that all triangles are carried over to the matrix elements of the associated recoupling matrix.

We conclude:

The matrix elements of each recoupling matrix are given by an integer multiplied by the δ−triangle factors determined by the forks of the pair of binary trees that define the two coupling schemes and the dimensional factors associated with the angular momenta labeling the roots of the forks. The integer itself is a restricted sum over multinomial coefficients, weighted by an integral factor that depends on the skew-symmetric matrices associated with each binary tree.

This integer property can be very useful in the development of algorithms for the calculation of $3n-j$ coefficients (see Wei [175]).

There is no direct reference in (4.54) to Racah coefficients or to cubic graphs. Since the matrix elements of each nontrivial recoupling matrix can also be expressed in terms of Racah coefficients, the generating function gives an entirely different expression of the fundamental triangle coefficients.

Relation (4.54) depends on the explicit construction of the skew-symmetric matrices in (4.41) and the matrices of indeterminates X and X', as illustrated by (4.15) and(4.23)-(4.25), which, in turn, determine the integral weight factor $I^{(p)}_{p_1,p_2,\ldots,p_h}$. Since we only have a recursive method for the generation of the matrices X and X', these weight factors have not been determined. It is useful to see how the preceding method works for $6-j$ and $9-j$ coefficients, and how it compares to published results.

4.6 Special Cases of the Generating Function

The methods developed above are illustrated in this section for $n+1 = 2,3,4$. To avoid excessive use of the prime $'$ and for comparison with earlier results, we introduce the notations $B = A'$ and $Y = X'$ for the the second skew-symmetric matrix in (4.41):

$$A = A_{\mathrm{sh}_T(\pi)}(X), \quad B = A_{\mathrm{sh}_{T'}(\pi')}(Y). \tag{4.55}$$

$n+1 = 2$: This case corresponds to the triangle coefficient given by (2.151), so that we must find $\langle T(ab)_j \mid T(ab)_j \rangle = 1$. There are no intermediate angular momenta for this case so that the symbols $\mathbf{k}$ and $\mathbf{k}'$ are empty, and the summation over these quantum numbers does not occur in the general formulas above. From (4.9), (4.23), and (4.48), we obtain the following relations:

$$\mathrm{Pf}(A,B) = 1 + x_{11}y_{11} + x_{21}y_{21} + x_{31}y_{31}, \tag{4.56}$$

$$\frac{1}{(\mathrm{Pf}(A,B))^2} = 1 + \sum_p^\infty (-1)^p\,(p+1)(x_{11}y_{11} + x_{21}y_{21} + x_{31}y_{31})^p, \quad (4.57)$$

$$\begin{aligned}&(x_{11}y_{11} + x_{21}y_{21} + x_{31}y_{31})^p \\ &= \sum_{\substack{a,b,j\\ a+b+j=p}} \binom{p}{b+j-a, a+j-b, a+b-j} \quad (4.58)\\ &\times (x_{11})^{b+j-a}(x_{21})^{a+j-b}(x_{31})^{a+b-j}(y_{11})^{b+j-a}(y_{21})^{a+j-b}(y_{31})^{a+b-j}.\end{aligned}$$

The restriction relations (4.52) give uniquely $p_1 = b+j-a, p_2 = a+j-b, p_3 = a+b-j$, so that there is no summation over p in (4.54) and $I^{(p)}_{p_1,p_2,p_3} = 1$. Thus, we obtain

$$\begin{aligned}\langle T(ab)_j \mid T(ab)_j\rangle &= (C_T^{(ab)} C_T^{(ab)\mathrm{tr}})_{j;j} = R_{T,T}^{(ab);\,(ab)} = 1 \quad (4.59)\\ &= (a+b+j+1)(\Delta(abj))^2 \binom{a+b+j}{b+j-a, a+j-b, a+b-j}.\end{aligned}$$

The $3-j$ coefficient is unity (the Wigner coefficients (4.40) are more properly called $3-jm$ coefficients).

$n+1 = 3$: The generating function (4.46) for $n = 2$ (in an equivalent form) is often called the Schwinger-Bargmann (Bargmann [8]) generating function for the $6-j$ coefficients. Historically, it is an important result in angular momentum theory because it was the first derivation giving the Racah coefficients directly in the form (2.162) without the necessity of summing over WCG coefficients, as carried out by Racah. Regge [147] also found this formula from the generating function and pointed out the 144 symmetries of the $6-j$ coefficient, which were overlooked by Schwinger. We give the derivation in some detail, because it confirms the details of relations (4.46) and (4.54), based on the double Pfaffian. This derivation starts with the pair of binary trees given by (2.157) and arrives at the same result:

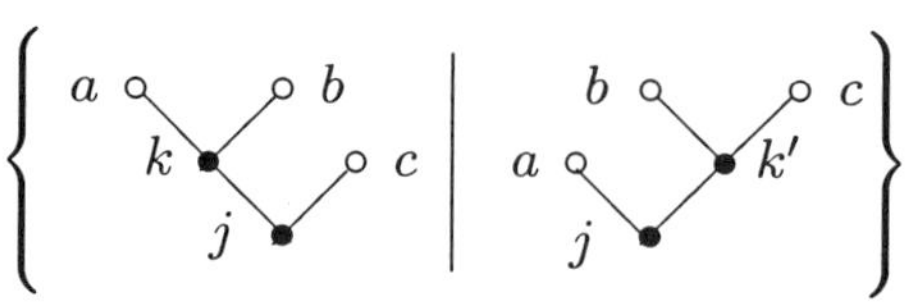

$$= \left\{ \begin{array}{cc|cc} a & k & b & a \\ b & c & c & k' \\ k & j & k' & j \end{array} \right\} = \left(R_{T,T'}^{(ab)c;\,a(bc)} \right)_{k,j;\,k',j}$$

$$= \sqrt{(2k+1)(2k'+1)}\,W(abjc;kk'). \quad (4.60)$$

The two skew-symmetric matrices that enter the double Pfaffian are

$$A = \begin{pmatrix} 0 & x_{31} & x_{21}x_{32} & x_{21}x_{22} \\ -x_{31} & 0 & x_{11}x_{32} & x_{11}x_{22} \\ -x_{21}x_{32} & -x_{11}x_{32} & 0 & x_{12} \\ -x_{21}x_{22} & -x_{11}x_{22} & -x_{12} & 0 \end{pmatrix}; \tag{4.61}$$

$$B = \begin{pmatrix} 0 & y_{21}y_{32} & y_{11}y_{32} & y_{22} \\ -y_{21}y_{32} & 0 & y_{31} & y_{12}y_{21} \\ -y_{11}y_{32} & -y_{31} & 0 & y_{11}y_{12} \\ -y_{22} & -y_{12}y_{21} & -y_{11}y_{12} & 0 \end{pmatrix}. \tag{4.62}$$

The result for B is obtained by application of the method of Sect. 4.2.

The double Pfaffian is given by (4.10). Because of the identities $a_{13}a_{24} = a_{23}a_{14} = x_{21}x_{32}x_{11}x_{22}$ and $b_{13}b_{24} = b_{12}b_{34} = y_{11}y_{32}y_{12}y_{21}$, the double Pfaffian reduces to

$$\begin{aligned} \mathrm{Pf}(A,B) &= 1 + x_{11}x_{32}y_{31} + x_{21}x_{22}y_{22} + x_{12}y_{11}y_{12} + x_{31}y_{21}y_{32} \\ &\quad + x_{11}x_{22}y_{12}y_{21} + x_{21}x_{32}y_{11}y_{32} + x_{31}x_{12}y_{22}y_{31}. \end{aligned} \tag{4.63}$$

The $p-$th power of this double Pfaffian is expressed in terms of the unnormalized triangle monomials (4.31) (the triangle factors have been divided out in (4.64)) corresponding to the pair of labeled binary trees in (4.60) by

$$\begin{aligned} (G(X,Y))^p &= \sum_{\substack{p_1,p_2,\ldots,p_7 \\ p_1+p_2+\cdots+p_7=p}}^{\text{restricted}} \binom{p}{p_1,p_2,p_3,p_4,p_5,p_6,p_7} \\ &\quad \times \frac{\Phi_{T(((ab)c)k)_j}(X)\Phi_{T'(((a(bc))k')_j}(Y)}{\{a\,b\,k\})^{-1}(\{k\,c\,j\})^{-1}(\{b\,c\,k'\})^{-1}(\{a\,k'\,j\})^{-1}}. \end{aligned} \tag{4.64}$$

The restriction conditions (4.52) are

$$\begin{aligned} &p_1+p_5 = -a+b+k,\ p_2+p_6 = a-b+k,\ p_4+p_7 = a+b-k; \\ &p_3+p_7 = -k+c+j,\ p_2+p_5 = k-c+j,\ p_1+p_6 = k+c-j; \\ &p_3+p_6 = -b+c+k',\ p_4+p_5 = k-c+k',\ p_1+p_7 = b+c-k'; \\ &p_3+p_5 = -a+k'+j,\ p_2+p_7 = a-k'+j,\ p_4+p_6 = a+k'-j; \end{aligned} \tag{4.65}$$

These restriction conditions can be solved for the p_i as follows:

$$\begin{aligned} &p_1 = p-(a+k'+j), p_2 = p-(b+c+k'), \\ &p_3 = p-(a+b+k),\ p_4 = p-(k+c+j), \\ &p_5 = b+j+k+k'-p,\ p_6 = a+c+k+k'-p, \\ &p_7 = a+b+c+j-p. \end{aligned} \tag{4.66}$$

These relations identify the unknown multiplying factor in (4.53) as $I^{(p)}_{p_1,p_2,\ldots,p_7} = 1$. Thus, we obtain from (4.54) the result:

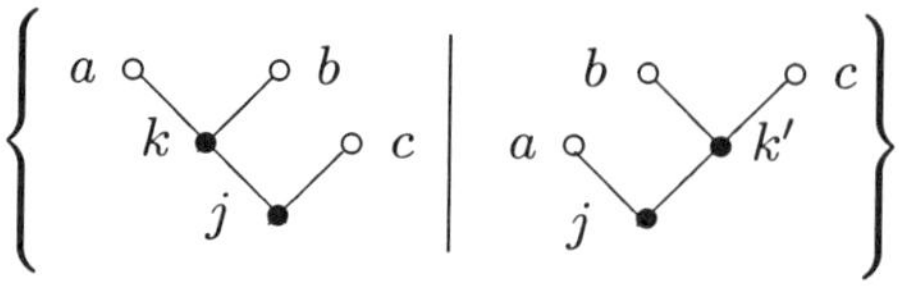

$$
\begin{aligned}
&= \left\{ \begin{array}{cc|cc} a & k & b & a \\ b & c & c & k' \\ k & j & k' & j \end{array} \right\} = \left(R^{(ab)c;\,a(bc)}_{T,T'} \right)_{k,j;\,k',j} \\
&= \langle T(((ab)c)k)_j \mid T'((a(bc))k')_j \rangle \\
&= (-1)^{a+b+c+j} \sqrt{(2k+1)(2k'+1)}\, \Delta(abk)\Delta(kcj))\Delta(bck')\Delta(ak'j) \\
&\quad \times \sum_{p}^{restricted} (-1)^p (p+1) \binom{p}{p_1, p_2, p_3, p_4, p_5, p_6, p_7} \\
&\qquad = \sqrt{(2k+1)(2k'+1)}\, W(abjc; kk'),
\end{aligned} \tag{4.67}
$$

where the p_i are defined by (4.66). Accounting for the introduction of the parameter p, this result is in complete agreement with (2.162). It is quite remarkable that the generating function method effects the summation over all WCG coefficients in (2.160).

$n + 1 = 4$: The generating function (4.46) for $n + 1 = 4$ (in an equivalent form) is often called the Bargmann-Wu generating function for the $9 - j$ coefficients (Bargmann [8], Wu [188]), but it is implicit in Schwinger's work. We give the rederivation in some detail, because it again confirms the details of relations (4.46) and (4.54) based on the double Pfaffian. We recall from (2.210) that we have obtained Wigner's $9 - j$ coefficient in the following form:

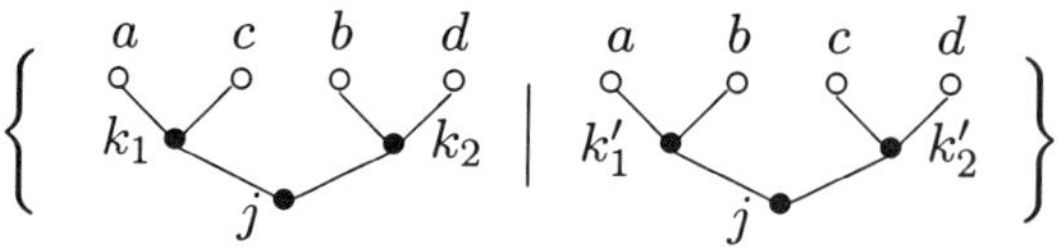

$$
\begin{aligned}
&= \left\{ \begin{array}{ccc|ccc} a & b & k_1 & a & c & k'_1 \\ c & d & k_2 & c & d & k'_2 \\ k_1 & k_2 & j & k'_1 & k'_2 & j \end{array} \right\} \\
&= \left(R^{(ac)(bd);(ab)(cd)}_{T,T} \right)_{k_1\,k_2\,j;\,k'_1\,k'_2\,j} \\
&= \langle T((ac)(bd)\,k_1 k_2)_j \mid T((ab)(bd)\,k'_1 k'_2)_j) \rangle
\end{aligned}
$$

$$= \sqrt{(2k_1+1)(2k_1'+1)(2k_2+1)(2k_2'+1)} \left\{ \begin{array}{ccc} a & b & k_1' \\ c & d & k_2' \\ k_1 & k_2 & j \end{array} \right\}$$

$$= \sqrt{(2k_1+1)(2k_1'+1)(2k_2+1)(2k_2'+1)} \sum_k (-1)^{a+k-k_1-k_1'}$$

$$\times (2k+1) W(dbjk_1; k_2k) W(bakc; k_1'k_1) W(k_1'cjd; kk_2'). \quad (4.68)$$

The two skew-symmetric matrices $A = A_{((13)(24))}(X), \pi' = (1324)$, and $B = A_{((12)(34))}(Y), e = (1234)$, that enter the double Pfaffian are

$$A = \begin{pmatrix} 0 & x_{21}x_{22}x_{33} & -x_{31} & x_{12}x_{21}x_{33} & x_{21}x_{23} \\ -x_{21}x_{22}x_{33} & 0 & -x_{11}x_{22}x_{33} & x_{32} & x_{13}x_{22} \\ x_{31} & x_{11}x_{22}x_{33} & 0 & x_{11}x_{12}x_{33} & x_{11}x_{23} \\ -x_{12}x_{21}x_{33} & -x_{32} & -x_{11}x_{12}x_{33} & 0 & x_{12}x_{13} \\ -x_{21}x_{23} & -x_{13}x_{22} & -x_{11}x_{23} & -x_{12}x_{13} & 0 \end{pmatrix}, \quad (4.69)$$

$$B = \begin{pmatrix} 0 & -y_{31} & y_{21}y_{22}y_{33} & y_{12}y_{21}y_{33} & y_{21}y_{23} \\ y_{31} & 0 & y_{11}y_{22}y_{33} & y_{11}y_{12}y_{33} & y_{11}y_{23} \\ -y_{21}y_{22}y_{33} & -y_{11}y_{22}y_{33} & 0 & y_{32} & y_{13}y_{22} \\ -y_{12}y_{21}y_{33} & -y_{11}y_{12}y_{33} & -y_{32} & 0 & y_{12}y_{13} \\ -y_{21}y_{23} & -y_{11}y_{23} & -y_{13}y_{22} & -y_{12}y_{13} & 0 \end{pmatrix}. \quad (4.70)$$

Since B corresponds to the identity permutation $e = (1234)$ (one-line notation) and $\pi = (3142)$, hence, $\pi_1^{-1} = 2, \pi_2^{-1} = 4, \pi_3^{-1} = 3, \pi_4^{-1} = 3$, application of (4.26) gives B from (4.25), after changing the x_{ij} to y_{ij}. Similarly, A is obtained from B in (4.70).

The double Pfaffian is obtained from relation (4.6) to be

$$\begin{aligned} \mathrm{Pf}(A,B) &= 1 + \sum_{1\le i<j\le 5} a_{ij}b_{ij} \quad (4.71) \\ &+(a_{12}a_{34} - a_{13}a_{24} + a_{23}a_{14})(b_{12}b_{34} - b_{13}b_{24} + b_{23}b_{14}) \\ &+(a_{12}a_{35} - a_{13}a_{25} + a_{23}a_{15})(b_{12}b_{35} - b_{13}b_{25} + b_{23}b_{15}) \\ &+(a_{12}a_{45} - a_{14}a_{25} + a_{24}a_{15})(b_{12}b_{45} - b_{14}b_{25} + b_{24}b_{15}) \\ &+(a_{13}a_{45} - a_{14}a_{35} + a_{34}a_{15})(b_{13}b_{45} - b_{14}b_{35} + b_{34}b_{15}) \\ &+(a_{23}a_{45} - a_{24}a_{35} + a_{34}a_{25})(b_{23}b_{45} - b_{24}b_{35} + b_{34}b_{25}). \end{aligned}$$

However, the family of relations as follows allows for a substantial simplification of the double Pfaffian:

$$a_{12}a_{34} = -a_{23}a_{14},\ b_{13}b_{24} = b_{23}b_{14},\ a_{12}a_{35} = -a_{23}a_{15},$$
$$b_{13}b_{25} = b_{23}b_{15},\ a_{12}a_{45} = a_{14}a_{25},\ b_{14}b_{25} = b_{24}b_{15}, \tag{4.72}$$
$$a_{14}a_{35} = a_{34}a_{15},\ b_{13}b_{45} = b_{14}b_{35},\ a_{23}a_{45} = -a_{34}a_{25},$$
$$b_{23}b_{45} = b_{24}b_{35}.$$

The double Pfaffian reduces to

$$\mathrm{Pf}(A,B) = 1 + \sum_{1\le i<j\le 5} a_{ij}b_{ij} - a_{13}a_{24}b_{12}b_{34} \tag{4.73}$$
$$-a_{13}a_{25}b_{12}b_{35} + a_{24}a_{15}b_{12}b_{45} + a_{13}a_{45}b_{34}b_{15} - a_{24}a_{35}b_{34}b_{25}.$$

This gives

$$G(X,Y) \tag{4.74}$$
$$= x_{12}x_{13}y_{12}y_{13} + x_{11}x_{23}y_{13}y_{22} + x_{11}x_{12}x_{33}y_{32}$$
$$+x_{13}x_{22}y_{11}y_{23} + x_{21}x_{23}y_{21}y_{23} + x_{21}x_{22}x_{33}y_{31}$$
$$+x_{32}y_{11}y_{12}y_{33} + x_{31}y_{21}y_{22}y_{33} - x_{31}x_{32}y_{31}y_{32}$$
$$-x_{11}x_{22}x_{33}y_{11}y_{22}y_{33} + x_{12}x_{13}x_{31}y_{21}y_{23}y_{32} + x_{21}x_{23}x_{32}y_{12}y_{13}y_{31}$$
$$-x_{13}x_{22}x_{31}y_{13}y_{22}y_{31} + x_{12}x_{21}x_{33}y_{12}y_{21}y_{33} - x_{11}x_{23}x_{32}y_{11}y_{23}y_{32}\,.$$

We denote the fifteen terms in this expression by $G_1(X,Y), G_2(X,Y), \ldots, G_{15}(X,Y)$ as read from left-to-right. While the ordering is arbitrary, this ordering of terms allows for easy comparison with the expression for the $9-j$ coefficient obtained below with the one presented in the Drake Handbook [53]. (Different orderings correspond to a redistribution of parameters.)

The generating function of the $9-j$ coefficients for the pair of labeled binary trees given by (4.68) is obtained from (4.44):

$$\frac{1}{(Pf(A,B))^2} = \frac{1}{\det(I-AB)} = \frac{1}{(1+G(X,Y))^2}$$
$$= 1 + \sum_{\substack{a,b,c,j\\ k_1,k_2,k_1',k_2'}} \frac{(-1)^{a+b+c+d+j}}{2j+1}$$
$$\times \Phi_{T((ac)(bd)\,k_1\,k_2)_j}(X)\Phi_{T((ab)(bd)\,k_1'\,k_2')_j}(Y)$$
$$\times \left(R_{T,T}^{(ac)(bd);(ab)(cd)}\right)_{k_1\,k_2\,j;\,k_1'\,k_2'\,j}. \tag{4.75}$$

Relation (4.54) now gives the following form of these recoupling coefficients:

$$\left(R_{T,T}^{(ac)(bd);(ab)(cd)}\right)_{k_1\,k_2\,j;\,k_1'\,k_2'\,j}$$

$$= \left\{ \begin{array}{ccc|ccc} a & b & k_1 & a & c & k_1' \\ c & d & k_2 & b & d & k_2' \\ k_1 & k_2 & j & k_1' & k_2' & j \end{array} \right\} \tag{4.76}$$

$$= \sqrt{(2k_1+1)(2k_1'+1)(2k_2+1)(2k_2'+1)} \left\{ \begin{array}{ccc} a & b & k_1' \\ c & d & k_2' \\ k_1 & k_2 & j \end{array} \right\};$$

$$\left\{ \begin{array}{ccc} a & b & k_1' \\ c & d & k_2' \\ k_1 & k_2 & j \end{array} \right\}$$

$$= (-1)^{a+b+c+d+j} \Delta(ack_1)\Delta(bdk_2)\Delta(k_1k_2j)\Delta(abk_1')\Delta(cdk_2')\Delta(k_1'k_2'j)$$

$$\times \sum_{p}^{\text{restricted}} \sum_{\substack{p_1,p_2,\ldots,p_{15} \\ p_1+\cdots+p_{15}=p}}^{\text{restricted}} (-1)^{p+q}(p+1) \binom{p}{p_1,p_2,\ldots,p_{15}}, \tag{4.77}$$

where $q = p_9 + p_{10} + p_{13} + p_{15}$. The restricted summation is over all $p_1, p_2, \ldots, p_{15}$ in the multinomial coefficient such that the following eighteen conditions are satisfied:

$$\begin{aligned} p_2+p_3+p_{10}+p_{15} &= -a+c+k_1, \\ p_5+p_6+p_{12}+p_{14} &= a-c+k_1, \\ p_8+p_9+p_{11}+p_{13} &= a+c-k_1; \\ p_1+p_3+p_{11}+p_{14} &= -b+d+k_2, \\ p_4+p_6+p_{10}+p_{13} &= b-d+k_2, \\ p_7+p_9+p_{12}+p_{15} &= b+d-k_2; \\ p_1+p_4+p_{11}+p_{13} &= -k_1+k_2+j, \\ p_2+p_5+p_{12}+p_{15} &= k_1-k_2+j, \\ p_3+p_6+p_{10}+p_{14} &= k_1+k_2-j; \end{aligned} \tag{4.78}$$

$$\begin{aligned} p_4+p_7+p_{10}+p_{15} &= -a+b+k_1', \\ p_5+p_8+p_{11}+p_{14} &= a-b+k_1', \\ p_6+p_9+p_{12}+p_{13} &= a+b-k_1'; \\ p_1+p_7+p_{12}+p_{14} &= -c+d+k_2', \end{aligned}$$

$$\begin{aligned}
p_2 + p_8 + p_{10} + p_{13} &= c - d + k_2',\\
p_3 + p_9 + p_{11} + p_{15} &= c + d - k_2';\\
p_1 + p_2 + p_{12} + p_{13} &= -k_1' + k_2' + j,\\
p_4 + p_5 + p_{11} + p_{15} &= k_1' - k_2' + j,\\
p_7 + p_8 + p_{10} + p_{14} &= k_1' + k_2' - j.
\end{aligned}$$

These relations may be reduced in a variety of ways; different reductions give different expressions of the $9-j$ coefficient. We effect the reduction in a manner that gives the final form (4.85) below (Wu's form), which also demonstrates their consistency. Adding the relations corresponding to a given triangle gives the six relations:

$$\begin{aligned}
p_1 + p_4 + p_7 &= p - (a + c + k_1),\\
p_2 + p_5 + p_8 &= p - (b + d + k_2),\\
p_7 + p_8 + p_9 &= p - (k_1 + k_2 + j),\\
p_1 + p_2 + p_3 &= p - (a + b + k_1'),\\
p_4 + p_5 + p_6 &= p - (c + d + k_2'),\\
p_3 + p_6 + p_9 &= p - (k_1' + k_2' + j).
\end{aligned} \tag{4.79}$$

These conditions may be written in the form of a 3×3 matrix array with fixed row and column sums:

$$\begin{pmatrix} p_1 & p_2 & p_3 \\ p_4 & p_5 & p_6 \\ p_7 & p_8 & p_9 \end{pmatrix} \begin{matrix} p - t_1 \\ p - t_2 \\ p - t_3 \end{matrix}, \tag{4.80}$$
$$p - t_4\, p - t_5\, p - t_6$$

where the t_i are the sums of quantum numbers of the six triangles $\langle abk_1'\rangle, \langle cdk_2'\rangle, \langle k_1'k_2'j\rangle, \langle ack_1\rangle, \langle bdk_2\rangle, \langle k_1k_2j\rangle$ associated with the two labeled binary trees in (4.68), and also are the row and column sums of the quantum numbers in the symbol in (4.68) for the $9-j$ coefficient:

$$\begin{aligned}
&t_1 = a + b + k_1',\ t_2 = c + d + k_2',\ t_3 = k_1 + k_2 + j,\\
&t_4 = a + c + k_1,\ t_5 = b + d + k_2,\ t_6 = k_1' + k_2' + j.
\end{aligned} \tag{4.81}$$

Each set of $p_1, p_2, \ldots, p_9$ that satisfies the row-column sum conditions in the matrix (4.80) determines a compatible system (4.78) for the set of $p_{10}, p_{11}, \ldots, p_{15}$ that satisfies the relations:

$$
\begin{aligned}
&p_{10} + p_{15} = -p + p_1 + s_1, p_{11} + p_{14} = -p + p_2 + s_2,\\
&p_{12} + p_{13} = -p + p_3 + s_3;\\
&p_{12} + p_{14} = -p + p_4 + s_4, p_{10} + p_{13} = -p + p_5 + s_5,\\
&p_{11} + p_{15} = -p + p_6 + s_6;\\
&p_{11} + p_{13} = -p + p_7 + s_7, p_{12} + p_{15} = -p + p_8 + s_8,\\
&p_{10} + p_{14} = -p + p_9 + s_9;
\end{aligned} \tag{4.82}
$$

$$
\begin{aligned}
&s_1 = b + c + k_1 + k_1', s_2 = a + d + k_1' + k_2, s_3 = a + b + k_2' + j,\\
&s_4 = a + d + k_1 + k_2', s_5 = b + c + k_2 + k_2', s_6 = c + d + k_1' + j,\\
&s_7 = a + c + k_2 + j, s_8 = b + d + k_1 + j, s_9 = k_1 + k_2 + k_1' + k_2'.
\end{aligned} \tag{4.83}
$$

These relations can be solved for any five of the $p_{10}, p_{11}, \ldots, p_{15}$ in terms of the remaining one. For example, if we choose $p_{15} = s$ to be the arbitrary parameter, then we have

$$
\begin{aligned}
&p_{10} = -s - p + p_1 + s_1,\ p_{11} = -s - p + p_6 + s_6,\\
&p_{12} = -s - p + p_8 + s_8,\\
&p_{13} = s + p_5 - p_1 - s_1 + s_5,\ p_{14} = s + p_2 - p_6 + s_2 - s_6,\\
&p_{15} = s.
\end{aligned} \tag{4.84}
$$

We thus arrive at the following expression for the Wigner $9-j$ coefficient:

$$
\begin{aligned}
&\left\{ \begin{matrix} a & b & k_1' \\ c & d & k_2' \\ k_1 & k_2 & j \end{matrix} \right\} \qquad (4.85)\\
&= \Delta(ack_1)\Delta(bdk_2)\Delta(k_1k_2j)\Delta(abk_1')\Delta(cdk_2')\Delta(k_1'k_2'j)\\
&\quad \times \sum_{p}^{\text{restricted}} \sum_{\mathbf{p}_1}^{\text{restricted}} \sum_{\mathbf{p}_2}^{\text{restricted}} (-1)^{p_{10}+p_{11}+p_{12}} (p+1) \left(\begin{matrix} p \\ \mathbf{p}_1, \mathbf{p}_2 \end{matrix} \right),
\end{aligned}
$$

where $\mathbf{p}_1 = (p_1, p_2, \ldots, p_9)$ and $\mathbf{p}_2 = (p_{10}, p_{11}, \ldots, p_{15})$. The summations in (4.85) are implemented as follows: For each set of quantum numbers $a, b, c, d, j, k_1, k_2, k_1', k_2'$ appearing in the $9-j$ coefficient symbol and satisfying the triangle conditions

$$
\langle a\, c\, k_1\rangle, \langle b\, d\, k_2\rangle, \langle k_1\, k_2\, j\rangle, \langle a\, b\, k_1'\rangle, \langle c\, d\, k_2'\rangle, \langle k_1'\, k_2'\, j\rangle, \tag{4.86}
$$

we select any nonnegative integer p such that the following two conditions are met:

$$\begin{aligned} p &\geq \max(t_1, t_2, t_3, t_4, t_5, t_6), \\ p &\leq \min(p_1 + s_1, p_2 + s_2, \ldots, p_9 + s_9), \end{aligned} \tag{4.87}$$

where $t_1, t_2, \ldots, t_6$ are defined by (4.81) and $s_1, s_2, \ldots, s_9$ are defined by (4.83). Then, the summations over $\mathbf{p}_1$ and $\mathbf{p}_2$ are over all nonnegative compositions of p such the row-column conditions in (4.80) are satisfied. The summation over p is then over the finite set of all p satisfying (4.87). The phase in (4.85) is obtained from that in (4.77), using relations (4.80) and (4.82), as follows: The identity

$$(-1)^{a+b+c+d+j}(-1)^{p+p_9+p_{10}+p_{13}+p_{15}} = (-1)^{p_{10}+p_{11}+p_{12}} \tag{4.88}$$

is a consequence of

$$\begin{aligned} &(-1)^{p+p_9+p_{10}+p_{13}+p_{15}}(-1)^{p_{10}+p_{11}+p_{12}} = (-1)^{p+p_9+(p_{12}+p_{13})+(p_{11}+p_{15})} \\ &= (-1)^{p+p_9+p_3+p_6+s_3+s_6} = (-1)^{s_3+s_6-t_6} = (-1)^{a+b+c+d+j}. \end{aligned}$$

Relation (4.85) is exactly that of Wu [188], as interpreted in the Drake Handbook [53] (see the discussion below).

Quite remarkably, just as the generating function for the $6-j$ coefficient gives the sum over four WCG coefficients in terms of the product of $\delta-$triangles (and associated dimension factors), multiplied by an integer that is a restricted sum of multinomial coefficients of order seven, so also does the generating function for the $9-j$ coefficient give the sum over three Racah coefficients in terms of the product of $\delta-$triangles (and associated dimension factors), multiplied by an integer that is a restricted sum of multinomial coefficients of order fifteen.

It is by no means a simple task to reduce the form (4.54) to a more manageable form, as illustrated by the intricacies of the special cases given above. What has been achieved, however, is a universal form (see Refs. [122, 123]) based on the combinatorial concepts of pairs of labeled binary trees, the uniform notation of triangle coefficients, the MacMahon master theorem, the double Pfaffian, and the skew-symmetric matrix of a binary tree, all of which exhibit the remarkable similarity of structure of the $6-j$ and $9-j$ coefficients, as expressed by a restricted summation of integral weighted multinomial coefficients.

The indeterminates that appear in the $3 \times n$ matrix X and the $3 \times n$ matrix $Y = X'$ in the generating relation (4.46), and in the matrix elements of the recoupling matrix in (4.54), are related to the triangles of the pair of binary trees through their occurrence in the triangle monomials in (4.53). This suggests a relationship with the cubic graph matrices $A_{3\times 2n}(x; y)$ introduced in Sect. 3.4.3 of Chapter 3. We illustrate this in the next section for the $6-j$ and $9-j$ coefficients.

4.6.1 Cubic graph geometry of the $6-j$ and $9-j$ coefficients

The double Pfaffian of the tetrahedron

We make the following identification of indeterminates x_{ij} and y_{ij} that occur in the skew-symmetric matrix A and B in relations (4.61) and (4.62) with the variables x_i and y_i in a cubic graph matrix (see (3.138)):

$$\begin{pmatrix} x_{11} & x_{12} & y_{11} & y_{12} \\ x_{21} & x_{22} & y_{21} & y_{22} \\ x_{31} & x_{32} & y_{31} & y_{32} \end{pmatrix} = \begin{pmatrix} y_1 & y_6 & x_5 & x_1 \\ y_5 & y_4 & x_4 & x_3 \\ x_6 & y_2 & y_3 & x_2 \end{pmatrix}. \tag{4.89}$$

This association is made from the pair of binary trees in (4.67), which corresponds to $x_1 = y_1 = a$, $x_2 = y_2 = j$, $x_3 = y_3 = k'$, $x_4 = y_4 = c$, $x_5 = y_5 = b.\, x_6 = y_6 = k$. The double Pfaffian (4.63) is given in terms of the x_i and y_i variables in the cubic graph matrix (4.89) by

$$1 + G(x, y) = 1 + V_3 + E_4, \tag{4.90}$$

where V_3 is called a *vertex function* and E_4 an *edge function*, as explained below, and given by

$$V_3 = y_1y_2y_3 + x_3y_4y_5 + x_1x_5y_6 + x_2x_4x_6, \tag{4.91}$$

$$E_4 = x_1y_1x_4y_4 + x_2y_2x_5y_5 + x_3y_3x_6y_6. \tag{4.92}$$

The tetrahedron corresponding to the double Pfaffian (4.90)-(4.92) is obtained from the columns of the cubic graph matrix (4.89) as follows, where the lines of a pair of adjacent points are labeled by the internal coordinates that share a common index:

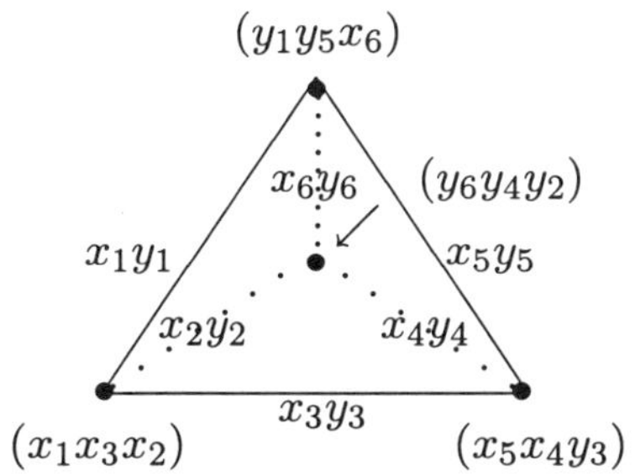

(4.93)

We give below a rule for reading the vertex function V_3 and the edge function E_4 directly off this labeled tetrahedron.

The expansion of $(1 + V_4 + E_3)^{-2}$ then gives the $6-j$ coefficient as the coefficient multiplying the triangle monomials, just as obtained in (4.67) from the double Pfaffian in terms of the X, Y variables.

The double Pfaffian of the $9-j$ cubic graph

We make the following identification of indeterminates x_{ij} and y_{ij} that occur in the skew-symmetric matrix A and B in relations (4.69) and (4.70) with the variables x_i and y_i in a cubic graph matrix (see p. 221)):

$$\begin{pmatrix} x_{11} & x_{12} & x_{13} & y_{11} & y_{12} & y_{13} \\ x_{21} & x_{22} & x_{23} & y_{21} & y_{22} & y_{23} \\ x_{31} & x_{32} & x_{33} & y_{31} & y_{32} & y_{33} \end{pmatrix} = \begin{pmatrix} x_2 & x_5 & y_3 & y_2 & y_1 & x_8 \\ x_1 & x_4 & y_6 & y_5 & y_4 & x_7 \\ x_3 & x_6 & y_9 & y_8 & y_7 & x_9 \end{pmatrix}. \tag{4.94}$$

This association is made from the pair of binary trees in (4.68), which corresponds to $x_1 = y_1 = c$, $x_2 = y_2 = a$, $x_3 = y_3 = k_1$, $x_4 = y_4 = d$, $x_5 = y_5 = b$, $x_6 = y_6 = k_2$, $x_7 = y_7 = k_2'$, $x_8 = y_8 = k_1'$, $x_9 = y_9 = j$. The double Pfaffian is now obtained from (4.74) in terms of the x_i and y_i variables in the cubic graph matrix (4.94) as

$$\begin{aligned} 1 + G(x,y) = 1 &+ y_1y_3x_5x_8 + y_4y_6x_2x_8 + y_7y_9x_2x_5 \\ &+ y_2y_3x_4x_7 + y_5y_6x_1x_7 + y_8y_9x_1x_4 \\ &+ y_1y_2x_6x_9 + y_4y_5x_3x_9 - y_7y_8x_3x_6 \\ &- x_2y_2x_4y_4x_9y_9 + x_3y_3x_5y_5x_7y_7 \\ &+ x_1y_1x_6y_6x_8y_8 - x_3y_3x_4y_4x_8y_8 \\ &+ x_1y_1x_5y_5x_9y_9 - x_2y_2x_6y_6x_7y_7. \end{aligned} \tag{4.95}$$

The cubic graph corresponding to the this double Pfaffian is obtained from the columns of the cubic graph matrix (4.94) as follows, where the lines of a pair of adjacent points are labeled by the internal coordinates that share a common index:

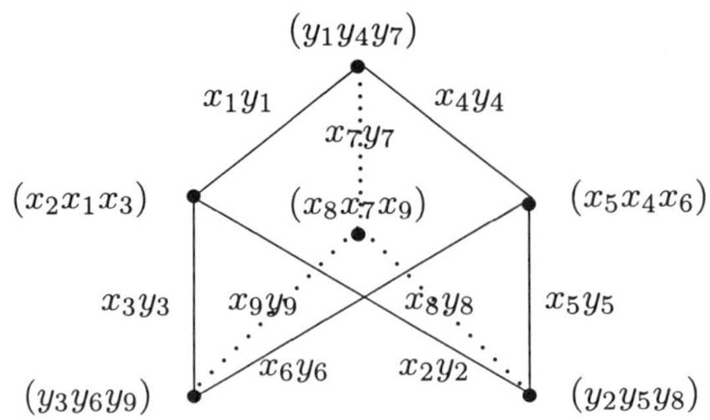

(4.96)

The expansion of $(1 + G(x,y))^{-2}$ then gives the $9-j$ coefficient as the coefficient multiplying the triangle monomials, just as obtained in (4.75)-(4.85) from the double Pfaffian in terms of the X, Y variables.

We have the following general result from the preceding sections on double Pfaffians and the skew-symmetric matrix of a binary tree:

A double Pfaffian can be associated with every cubic graph $\mathbb{G}(T,T')$ *of angular momentum type.*

The question arises: Can the double Pfaffian associated with a pair of labeled binary trees be read off directly from its associated cubic graph? Such rules for the $6-j$ coefficient and the $9-j$ coefficient are stated directly from the cubic graph by Biedenharn *et al.* [16] and in the Drake Handbook [53]. The rules below were found empirically from the results given by Bargmann [8] and Wu [188] prior to the discovery of the double Pfaffian, and, for the $9-j$ coefficients, the cited rule is given in a modified version of the double Pfaffian.

The rule for the tetrahedron (4.93) is very simple:

The tetrahedron rule: The vertex function V_3 and the edge function E_4 then have the following geometrical significance: First, we interchange the x_i and y_i coordinates labeling the points of the tetrahedron. Then, we have the following rules for obtaining the double Pfaffian:

V_3 : Multiply together the coordinates of each vertex and sum over all four vertices.

E_4 : Multiply together the coordinates of all pairs of disjoint edges and sum over all the three pairs of edges.

Disjoint edges of a cubic graph are defined to be edges having no common vertex. These rules give the forms (4.90)-(4.92).

The $9-j$ *cubic graph rule:* The Drake Handbook version replaces the double Pfaffian (4.95) by the following form:

$$1+H(x,y)=1-V_4+E_6, \tag{4.97}$$

where the vertex function V_4 and the edge function E_6 are defined by

$$\begin{aligned} V_4 = \quad & y_1y_2x_6x_9+y_1y_3x_5x_8+y_2y_3x_4x_7 \\ & +y_4y_5x_3x_9+y_4y_6x_2x_8+y_5y_6x_1x_7 \\ & +y_7y_8x_3x_6+y_7y_9x_2x_5+y_8y_9x_1x_4; \end{aligned} \tag{4.98}$$

$$E_6=\det\begin{pmatrix} x_1y_1 & x_2y_2 & x_3y_3 \\ x_4y_4 & x_5y_5 & x_6y_6 \\ x_7y_7 & x_8y_8 & x_9y_9 \end{pmatrix}. \tag{4.99}$$

These functions have the following interpretation in terms of the labeled cubic graph (4.96). Interchange the symbols x and y in the cubic graph (4.96), and then define:

V_4 vertex function: multiply together all the coordinates of each pair of adjacent vertices, divide out the coordinates with a common subscript, and sum over all nine such pairs.

E_6 edge function: multiply together the edge coordinates corresponding to any three edges having no common vertex, there being nine such combinations, and add the nine together with $\pm$ signs to give the determinant (4.99).

Except for signs, the expression (4.97) is the same as that for the double Pfaffian (4.95) (These latter signs are uniquely determined from the definition of the double Pfaffian and the skew-symmetric matrices A and B.) However, the signs in the different expressions compensate, and the expansion of $(1+H(x,y))^{-2}$ and $(1+G(x,y))^{-2}$ work out to the same final form (4.85) for the $9-j$ coefficient.

The answer to the question posed above is unsettled: We do not know the general rule for reading the double Pfaffian from the associated labeled cubic graph. Such a rule would be a unifying result of some importance.

The indeterminates $x_{3i}, y_{3i}, i = 1, 2, \ldots, 9$ in relation (4.94) can be realized in many ways in terms of a cubic graph matrix, the only requirement being that the subscript placement encodes the triangles of the associated cubic graph. In Biedenharn *et al.* [16], the following variables are used:

$$\begin{pmatrix} x_{11} & x_{12} & x_{13} & y_{11} & y_{12} & y_{13} \\ x_{21} & x_{22} & x_{23} & y_{21} & y_{22} & y_{23} \\ x_{31} & x_{32} & x_{33} & y_{31} & y_{32} & y_{33} \end{pmatrix} = \begin{pmatrix} x_2 & y_8 & y_3 & y_2 & y_1 & x_6 \\ x_1 & x_9 & x_4 & x_8 & y_9 & x_7 \\ x_3 & y_4 & x_5 & y_6 & y_7 & y_5 \end{pmatrix}. \tag{4.100}$$

Regrettably, several errors were made in Sect. 2.4 of Ref. [16]: Rather than giving the detailed corrections, we note only that relation (16) for the $9-j$ coefficients is corrected by interchanging the triangle sums t_3 and t_4, and taking into account the equality of the two notations as follows for the $9-j$ coefficient:

$$\begin{Bmatrix} j_1 & j_2 & j_3 \\ j_9 & j_8 & j_4 \\ j_7 & j_6 & j_5 \end{Bmatrix} = \begin{Bmatrix} a & b & k_1' \\ c & d & k_2' \\ k_1 & k_2 & j_5 \end{Bmatrix}. \tag{4.101}$$

We aso note that the number of free parameters in the summation (4.85) for the $9-j$ coefficient is four, there being two free parameters in (4.80). It is possible to reduce this number to three as shown by Yutsis and Bandzaitis [193], but caution must be exercised, since such reductions can obscure the combinatorial origins of the $3n-j$ coefficients.

We also take this opportunity to point out that the form for the $9-j$ coefficient given by relation (3.326) in Biedenharn and Louck [21] contains two errors: The repeated factor $(a+d+h+1-z)!$ should occur only once; and the factor $(b+f-a-j+x+z)!$ should be replaced by $(b+j-a-f+x+z)!$ Moreover, in Vol. 9, on p. 473, a line should be removed from one of the cubic graphs, and on p. 476, the second cubic graph diagram should be replaced by the second one in relations (3.126).

Symmetries

The symmetries of the WCG coefficients and the $6-j$ coefficients are discussed in Sect. 1.4, Chapter 1, and in a comprehensive manner in Ref. [21]. In particular, it is the rotation-inversions that transform the regular tetrahedron into itself (the group T_d in point-group terminology) that give rise to 24 permutations in S_4 of $1, 2, 3, 4, 5, 6$ that leave the $6-j$ coefficient invariant. In terms of the labeling of the tetrahedron in (4.93), these are the twenty-four permutations that leave the edge function E_4 and vertex function V_3 invariant upon setting $x_i = y_i, i = 1, 2. \ldots, 6$. This set consists of those permutations of the pairs $(14)(36)(25)$ among themselves, together with no permutations of the entries within each pair, or of the entries in any two of the pairs (four permutations, including the identity). This invariance group also has the second interpretation as the set of permutations induced on the subscripts $1, 2, \ldots, 6$ by the twenty-four permutations of the four 3−sets of unordered (all orders equal) integers $\Delta_1 = \{1, 2, 3\}$, $\Delta_2 = \{3, 4, 5\}, \Delta_3 = \{1, 5, 6\}, \Delta_4 = \{2, 4, 6\}$, where we standardize the notation by writing $1 < 2 < 3 < 4 < 5 < 6$. For example, the interchange of Δ_1 and Δ_2 induces the permutation $(1, 2, 3, 4, 5, 6) \mapsto (5, 4, 3, 2, 1, 6)$. The additional symmetries that involve linear combinations of angular momentum labels giving a total of 144 symmetries of the $6-j$ coefficient have no such geometrical interpretation. In general, a $3n-j$ coefficient $(n \geq 2)$ has the $(2n)!$ symmetries corresponding to the permutations of its $2n$ triangles in the form (4.54) (the second interpretation above), and further symmetries are unknown. The triangle symmetries are closely related to the choices of the p_i that solve relations (4.52); they are built into the structure of the cubic graph matrices (3.131).

4.7 Concluding Remarks on Angular Momentum Theory

We have attempted to place the quantum theory of angular momentum within the mathematical framework of combinatorial mathematics in a natural way. One of the most striking features is that the abstract entities of combinatorics become the quantities that label the mathematical functions that arise, as we have demonstrated numerous times, beginning with pairs of labeled binary trees without common forks, and following the natural evolving pathways. The unifying concepts of a recoupling matrix and of a fundamental triangle coefficient are principal examples.

Complementary approaches to the methods presented here can be found in the publications of many others, for example, Judd [85, 86] and Judd and Lister [87].

The are many unsolved problems. While many results have been presented in an algorithmic form, these forms are often too unruly for the derivation of fully explicit algebraic formulas. Unsolved problems include the following:

1. Enumeration of a complete set of nonisomorphic trivalent trees of order n and an associated counting formula.
2. Enumeration of a complete set of nonisomorphic cubic graphs on $2n$ points and an associated counting formula.
3. Enumeration of a complete set of nonisomorphic cubic graphs on $2n$ points of angular momentum type and an associated counting formula.
4. Presentation of a more tractable method of obtaining the skew-symmetric matrix of a binary tree.
5. Determination of the multiplying integers $I^{(p)}_{p_1,p_2,\ldots,p_h}$ in the universal form of recoupling coefficients.
6. Exposition of a geometric basis for the relationship between cubic graphs and double Pfaffians.
7. Development of a general theory of triangles and graphs.
8. Development of a more comprehensive theory of paths associated with recoupling matrices.

We turn next to the development for the general unitary group $U(n)$ of some aspects of $U(2)$ presented in the first four chapters. But the goal is far more modest, since a theory of recoupling coefficients in the sense of $3n-j$ coefficients is mostly unknown. The theory of the unitary irreducible representations themselves, and of the analog of the $U(2)$ WCG coefficients (one Kronecker product of irreducible representations), is already rich in structure. The proliferation of multiplicities for more than a single Kronecker product is overwhelming. Combinatorial objects that make their appearance and have a very prominent role include: semistandard Young tableaux, Gelfand-Tsetlin patterns, Littlewood-Richardson numbers, Sylvester's identity and the complete homogeneous symmetric functions, shift-operators, and operator-valued polynomials as tensor operators.

Chapter 5

The $D^\lambda-$Polynomials: Form

5.1 Overview

Chapter 5 through Chapter 9 of this monograph develop the generalization of the $SU(2)$ D^j-polynomials that occur in angular momentum theory to the general unitary group $U(n)$, indeed, the general linear group $GL(n, \mathbf{C})$. These chapters draw freely from the subject matter presented in Compendiums A and B, where summaries of various mathematical concepts are given, so as not to interfere with the flow of the applications of these notions in the chapters. The notations and properties of partitions, semistandard Young-Weyl tableaux, and Gelfand-Tsetlin patterns presented in Sects. 11.1-11.3, Compendium B, should be consulted before reading this chapter. Section 10.6 in Compendium A on inner products is also quite important. The structure of the $U(2)$ solid harmonics in Sect. 1.7, Chapter 1 is the model for generalization.

One of the principal objectives of Chapters 5-8 is to develop the properties of a class of matrices, labeled by partitions $\lambda \in \mathbb{P}\mathrm{ar}_n$ and denoted by $D^\lambda(Z)$, whose dimension is equal to $\mathrm{Dim}\lambda$, the Weyl dimension formula. These matrices are functions of n^2 variables $(z_{ij})_{1\leq i,j\leq n}$, which, for the most part, may be considered to be indeterminates. The elements of the matrix are functions of the form $D^\lambda_{r,c}(Z)$, $r, c \in \mathbb{I}$, where $\mathbb{I}$ is an indexing set that enumerate rows r and columns c of the matrix. These may be chosen to be either ordered pairs of semistandard Young-Weyl (SSYW) tableaux of shape λ, or ordered pairs of triangular Gelfand-Tsetlin (GT) patterns of shape λ (or by other schemes). We choose to denote the elements of this matrix by double GT patterns of shape λ, rather that the equivalent sets of double $SSYW$ tableaux of shape λ. We use several different ways of displaying this pair of patterns, depending on the context. Since the partition $\lambda \in \mathbb{P}\mathrm{ar}_n$ is usually specified, it is useful to show this label separately from the remaining $n-1$ rows of the triangular GT patterns that accompany the partition λ. We use the following three notations to serve various needs:

$$D\begin{pmatrix} m' \\ \lambda \\ m \end{pmatrix}(Z) = D\left(\begin{array}{c|c} \lambda & \lambda \\ m & m' \end{array}\right)(Z) = D^{\lambda}_{m,m'}(Z). \tag{5.1}$$

In the first symbol, the upper pattern $\binom{\lambda}{m'}$ is inverted over the lower pattern $\binom{\lambda}{m}$ so that the common partition is exhibited only once, but this inversion carries no other informational content, and is often referred to in normal (uninverted) form (see (5.6)). The center symbol is useful, for clarity, in certain specializations of the patterns m and m' that are encountered. The final symbol is patterned after that used for the D^j–polynomials of angular momentum theory and exhibits most clearly the parallel structures. In each of these symbols, we employ the convention:

$$m \in \mathbb{G}_\lambda \ \text{ means } \begin{pmatrix} \lambda \\ m \end{pmatrix} \in \mathbb{G}_\lambda, \tag{5.2}$$

where $\mathbb{G}_\lambda$ denotes the set of all GT patterns corresponding to the choice of all subpatterns m (see Sect. 11.3, Compendium B). The third notation, which separates the $n-1$ rows of partitions in the symbols m and m' from the parent partition λ, is the simplest notation, and more in accord with the general indexing scheme $D^{\lambda}_{r,c}$, but can be deceptive. We use the lexicographic order relation on such patterns given in Compendium B to write out or visualize the elements of the matrix, with m denoting rows and m' denoting columns. It is the custom in the physics literature, which we shall follow, to enumerate the rows and columns in decreasing order from the greatest pattern to the least pattern as read from top-to-bottom down the rows and left-to-right across the columns. The polynomials denoted by (5.1) are called D^λ–polynomials, and the corresponding matrices are called D^λ–matrices.

In order to have some insight into the nature of the D^λ–polynomials and the corresponding matrices, we collect together here in this Overview many of their main properties, and either give the proofs or indicate where the proof of a given property can be found.

Let $Z = (z_{ij})_{1\leq i,j\leq n}$ be a set of n^2 variables or indeterminates, and $A = (a_{ij})_{1\leq i,j\leq n}$ a corresponding set of n^2 nonnegative integers. The Maclaurin polynomials defined by

$$\frac{Z^A}{A!} = \prod_{i,j=1}^{n} \frac{z_{ij}^{a_{ij}}}{a_{ij}!} \tag{5.3}$$

are basic to our presentation of the properties of D^λ–polynomials. The polynomials (5.3) are orthogonal in the inner product (,), this inner product being defined and discussed in Sect. 1.3.1, Chapter 1:

$$(Z^A, Z^B) = \delta(A, B)A!, \quad A! = \prod_{i,j=1}^{n} a_{ij}!, \tag{5.4}$$

where we sometimes write the Kronecker delta in the form $\delta(A, B)$ for complex entities such as the arrays A and B. The polynomials Z^A are also homogeneous of degree

$$\begin{aligned} \alpha_i &= \sum_{j=1}^{n} a_{ij} \text{ in the variables} \\ &\quad z_i = (z_{i1}, z_{i2}, \ldots, z_{in}) \text{ in row } i \text{ of } Z; \\ \alpha'_j &= \sum_{i=1}^{n} a_{ij} \text{ in the variables} \\ &\quad z^j = (z_{1j}, z_{2j}, \ldots, z_{nj}) \text{ in column } j \text{ of } Z. \end{aligned} \tag{5.5}$$

The D^λ–polynomials are invertible real linear transformations of the Z^A polynomials that preserve their homogeneity properties in the rows and columns of Z :

$$D\begin{pmatrix} m' \\ \lambda \\ m \end{pmatrix}(Z) = \sum_{A \in \mathbb{M}^p_{n\times n}(\alpha,\alpha')} C\begin{pmatrix} m' \\ \lambda \\ m \end{pmatrix}(A)\,\frac{Z^A}{A!}, \tag{5.6}$$

where, for given double GT patterns $\binom{\lambda}{m}$ and $\binom{\lambda}{m'}$ of the same shape (partition) $\lambda \vdash p$, the compositions $\alpha \vdash p$ and $\alpha' \vdash p$ are the weights of the lower and upper patterns:

$$\alpha = (\alpha_1, \ldots, \alpha_n) = W\binom{\lambda}{m}, \; \alpha' = (\alpha'_1, \ldots, \alpha'_n) = W\binom{\lambda}{m'}. \tag{5.7}$$

The summation in (5.6) is over all matrix arrays $A = (a_{ij})_{1\leq i,j\leq n}$ of nonnegative integer exponents in the set $\mathbb{M}^p_{n\times n}(\alpha, \alpha')$ defined by

$$\mathbb{M}^p_{n\times n}(\alpha, \alpha') \left\{ A \;\middle|\; \begin{array}{l} \sum_{j=1}^{n} a_{ij} = \alpha_i,\ i = 1, \ldots, n \\ \sum_{i=1}^{n} a_{ij} = \alpha'_j,\ j = 1, \ldots, n \end{array} \right\}. \tag{5.8}$$

These are just the conditions of homogeneity given by (5.5), the total degree being the nonnegative integer p. If the partition $\lambda \in \mathbb{P}ar_n$ is selected in (5.6), then $p = |\lambda| = \lambda_1 + \lambda_2 + \cdots + \lambda_n$, whereas if the nonnegative integer p is selected, then $\lambda \vdash p$. The partition λ and the nonnegative integer p stand in this relationship throughout this monograph, even when not noted.

The transformation (5.6) is invertible because we require the transformation coefficients $C\begin{pmatrix} m' \\ \lambda \\ m \end{pmatrix}(A)$ to be the elements of a real orthogonal matrix, up to multiplying factors as described below in (5.13)-(5.14). The inversion is given by

$$\frac{Z^A}{A!} = \sum_{\lambda \vdash p} \sum_{m,m' \in \mathbb{G}_\lambda(\alpha,\alpha')} \frac{1}{M(\lambda)A!} C\begin{pmatrix} m' \\ \lambda \\ m \end{pmatrix}(A) D\begin{pmatrix} m' \\ \lambda \\ m \end{pmatrix}(Z),$$

$$\text{each } A \in \mathbb{M}^p_{n\times n}(\alpha,\alpha'), \tag{5.9}$$

where $\mathbb{G}_\lambda(\alpha,\alpha')$ is the set of double GT patterns of weight (α,α') and shape λ, and the quantity $M(\lambda)$ is an invariant normalizing factor defined by

$$M(\lambda) = \prod_{i=1}^{n} (\lambda_i + n - i)! \Big/ 1!2!\cdots(n-1)!\mathrm{Dim}\lambda; \tag{5.10}$$

$$\mathrm{Dim}\lambda = \prod_{1\le i<j\le n} (\lambda_i - \lambda_j + j - i) \Big/ 1!2!\cdots(n-1)!. \tag{5.11}$$

The factor Dimλ is the Weyl dimension formula that gives the number of GT patterns (and SSYW tableau) in the set $\mathbb{G}_\lambda$ of all GT patterns of shape λ.

The orthogonality and normalization of the D^λ–polynomials is expressed for all pairs $\lambda, \lambda' \in \mathbb{P}ar_n(p)$ by

$$\left(D\begin{pmatrix} m' \\ \lambda \\ m \end{pmatrix}(Z), D\begin{pmatrix} m''' \\ \lambda' \\ m'' \end{pmatrix}(Z) \right) = \delta_{m,m''}\delta_{m',m'''}\delta_{\lambda,\lambda'}M(\lambda). \tag{5.12}$$

The normalization of the D^λ–polynomials to the factor $M(\lambda)$ is made so that the D^λ–matrices give the unit matrix when evaluated at $Z = I_n : D^\lambda(I_n) = I_{\mathrm{Dim}\lambda}$. From the orthogonality relations (5.4) for the Maclaurin monomials $Z^A/A!$, it follows that the orthogonality relations for the D^λ–polynomials are expressed in terms of the C–coefficients for all pairs $\lambda, \lambda' \in \mathbb{P}\mathrm{ar}_n(p)$ by

$$\sum_{A \in \mathbb{M}^p_{n\times n}(\alpha,\alpha')} R\begin{pmatrix} m' \\ \lambda \\ m \end{pmatrix}(A) R\begin{pmatrix} m''' \\ \lambda' \\ m'' \end{pmatrix}(A) = \delta_{m,m''}\delta_{m',m'''}\delta_{\lambda,\lambda'}, \tag{5.13}$$

where the new transformation coefficients are defined by

$$R\begin{pmatrix} m' \\ \lambda \\ m \end{pmatrix}(A) = \frac{1}{\sqrt{M(\lambda)A!}} C \begin{pmatrix} m' \\ \lambda \\ m \end{pmatrix}(A). \tag{5.14}$$

The orthogonality relation (5.4) applied to (5.9) then also gives

$$\sum_{\lambda \vdash p} \sum_{m,m' \in G_\lambda(\alpha,\alpha')} R\begin{pmatrix} m' \\ \lambda \\ m \end{pmatrix}(A) R\begin{pmatrix} m' \\ \lambda \\ m \end{pmatrix}(B) = \delta(A,B),$$

$$\text{for all pairs } A, B \in \mathbb{M}^p_{n\times n}(\alpha, \alpha'). \tag{5.15}$$

The orthogonality relations (5.13) and (5.15) depend on two distinct indexing sets: The set that appears in (5.13) is $\mathbb{M}^p_{n\times n}(\alpha, \alpha')$, which contains all $n \times n$ matrix arrays A of prescribed row and column sums, as given by the weights α and α' of the lower and upper GT patterns of shape λ; and the set that appears in (5.15) is the set of all GT patterns of shape λ, which is denoted by $\mathbb{G}_\lambda(\alpha, \alpha')$, having α as the prescribed weight of the lower pattern and α' as that of the upper pattern, these weights being held fixed as λ runs over all compositions of p. For the enumeration of the second set, the Kostka number $K(\lambda, \alpha)$ that gives the number of GT patterns $\binom{\lambda}{m}$ of weight α is required. Thus, the two indexing sets are $\mathbb{M}^p_{n\times n}(\alpha, \alpha')$ and $\cup_{\lambda \vdash p} \mathbb{G}_\lambda(\alpha, \alpha')$, which are, respectively, of cardinality $|\mathbb{M}^p_{n\times n}(\alpha, \alpha')|$ and $\sum_{\lambda \vdash p} |\mathbb{G}_\lambda(\alpha, \alpha')|$. If desired, the $R-$quantities defined by (5.14) can first be partitioned into subsets of all partitions λ of the nonnegative integer p, followed by the set of all patterns having prescribed weights α and α'. These patterns can then be assembled into a square matrix of order $|\mathbb{M}^p_{n\times n}(\alpha, \alpha')|$ with rows enumerated by the matrix arrays A in the set $\mathbb{M}^p_{n\times n}(\alpha, \alpha')$ and columns enumerated by the GT patterns in the set $\cup_{\lambda \vdash p} \mathbb{G}_\lambda(\alpha, \alpha')$. The "squareness" of this real orthogonal matrix then requires that

$$\sum_{\lambda \vdash p} K(\lambda, \alpha) K(\lambda, \alpha') = |\mathbb{M}^p_{n\times n}(\alpha, \alpha')|. \tag{5.16}$$

This is a famous relation known as the Robinson, Schensted, Knuth identity, and proved combinatorially by Knuth [91]. The above interpretation of this identity was pointed out in Ref. [118]. This is a classical example in combinatorics where closed formulas for neither the Kostka numbers $K(\lambda, \alpha)$ nor the number of matrix arrays $\mathbb{M}^p_{n\times n}(\alpha, \alpha')$ are known, but a relation between them exists.

It is, of course, possible to calculate all the numbers in (5.16) by the highly inefficient method of enumerating all GT patterns of shape λ (see Sect. 11.3 in Compendium B) and then determining the weight $\alpha = (\alpha_1, \alpha_2, \ldots, \alpha_n)$ from the formula

$$\alpha_j = \sum_{i=1}^{j} m_{ij} - \sum_{i=1}^{j-1} m_{i\,j-1}, j = 1, 2, \ldots, n, \tag{5.17}$$

where $\alpha_1 = m_{11}$ and $(\lambda_1, \lambda_2, \ldots, \lambda_n) = (m_{1n}, m_{2n}, \ldots, m_{1n})$. The full set of GT patterns, Dimλ in number, is then listed, and the multiplicity of a given weight α recorded. This is the Kosta number $K(\lambda, \alpha)$. The list is then expanded to include all $\lambda \vdash p$, and from the expanded list the number of occurrences of a given pair (α, α') is counted, which is the number $|\mathbb{M}^p_{n\times n}(\alpha, \alpha')|$.

There are many orthogonal transformations between the polynomials $Z^A/A!$ and some set of D^λ–polynomials that might be considered. Our interest in this monograph is in a very specific family of D^λ–polynomials for which the orthogonal transformation is completely specified, and which endows these polynomials with many special properties. Accordingly, we state at the outset a set of relations that fully defines these polynomials, although the verification of some of these properties is intricate. These details are deferred to Chapters 6-9. The description of the transformation C^λ–coefficients in the definition of the D^λ–polynomials is a major objective.

Relation (5.6) can also be formulated as a relationship between matrices of dimension Dimλ : The lower GT patterns $\binom{\lambda}{m} \in G_\lambda$ enumerate the rows, and the upper GT patterns $\binom{\lambda}{m'} \in G_\lambda$ enumerate the columns of a matrix $D^\lambda(Z)$, as well as those of a matrix $C^\lambda(A)$. For the explicit enumeration of the elements of such matrices, we need to order the set of GT patterns corresponding to a given shape λ, so that the D^λ–polynomials and the C^λ–coefficients can be entered into the matrix in a specified manner. For this, we first associate with each GT pattern $\binom{\lambda}{m}$ (see Sect. 11.3) the sequence of length $n(n-1)/2$ obtained by placing rows $n-1, n-2, \ldots, 2, 1$ in a single row sequence $R_\lambda(m)$ defined as follows:

$$R_\lambda(m) = (m_{1\,n-1}, m_{2\,n-1}, m_{n-1\,n-1}; \;\cdots\; ; m_{1\,2}, m_{2\,2}; m_{11}). \tag{5.18}$$

We then use the lexicographic order that

$$R_\lambda(m) > R_\lambda(m'), \tag{5.19}$$

if and only if the first nonzero entry in the difference sequence $R_\lambda(m) - R_\lambda(m')$ is positive. In one-to-one correspondence with the ordering

(5.19), we also write

$$\begin{pmatrix} \lambda \\ m \end{pmatrix} > \begin{pmatrix} \lambda \\ m' \end{pmatrix}, \text{ if and only if } R_\lambda(m) > R_\lambda(m'). \tag{5.20}$$

This order relation is a complete order relation; it is regarded simply as a convenient way for placing the D^λ–polynomials into a matrix and has no deeper significance here.

The rows and columns of the matrix $D^\lambda(Z)$ are indexed by the lower and upper GT patterns, respectively, of the polynomials (5.6), where the rows are ordered left-to-right as read across the columns from the greatest to the least pattern, and the columns are ordered top-to-bottom down the rows by the greatest to the least pattern. The elements of the matrix $C^\lambda(A)$ of coefficients in (5.6) are indexed in the same way. But now it is necessary to defined the C^λ–coefficients in (5.6) for all weights $\alpha, \alpha' \in \mathbb{W}_\lambda$, where $\mathbb{W}_\lambda$ denotes the set of all weights of the set of GT patterns $\mathbb{G}_\lambda$. The Maclaurin polynomials $Z^A/A!$ are defined for all matrices of exponents $A = (a_{ij})_{1\leq i,j\leq n}$ having nonnegative elements a_{ij}. In particular, they are defined for all weights $\alpha, \alpha' \in \mathbb{W}_\lambda$. We accordingly define the C^λ–coefficients in the linear transformation for all A corresponding to all $\alpha, \alpha' \in \mathbb{W}_\lambda$ by

$$\begin{aligned}
C\begin{pmatrix} m' \\ \lambda \\ m \end{pmatrix}(A) &= C\begin{pmatrix} m' \\ \lambda \\ m \end{pmatrix}(A), \text{ for } A \in \mathbb{M}^p_{n\times n}(\alpha, \alpha'), \\
&\qquad \alpha = W\begin{pmatrix} \lambda \\ m \end{pmatrix}, \ \alpha' = W\begin{pmatrix} \lambda \\ m' \end{pmatrix}; \\
C\begin{pmatrix} m' \\ \lambda \\ m \end{pmatrix}(A) &= 0, \text{ for } A \in \mathbb{M}^p_{n\times n}(\alpha, \alpha'),\ \alpha, \alpha' \in \mathbb{W}_\lambda, \\
&\qquad \alpha \neq W\begin{pmatrix} \lambda \\ m \end{pmatrix} \text{ or } \alpha' \neq W\begin{pmatrix} \lambda \\ m' \end{pmatrix}.
\end{aligned} \tag{5.21}$$

This definition of the C^λ–coefficients for all A having row and columns sums corresponding to all weights $\alpha, \alpha' \in \mathbb{W}_\lambda$ allows relation (5.6) to be written in the form

$$D\begin{pmatrix} m' \\ \lambda \\ m \end{pmatrix}(Z) = \sum_{A\in\mathbb{M}^p_{n\times n}} C\begin{pmatrix} m' \\ \lambda \\ m \end{pmatrix}(A)\,\frac{Z^A}{A!}, \tag{5.22}$$

where the set $\mathbb{M}^p_{n\times n}$ is defined by the union

$$\mathbb{M}^p_{n\times n} = \bigcup_{\alpha,\alpha'\in\mathbb{W}_\lambda} \mathbb{M}^p_{n\times n}(\alpha,\alpha'). \tag{5.23}$$

Relation (5.22) can now be written in the concise matrix form:

$$D^\lambda(Z) = \sum_{A\in\mathbb{M}^p_{n\times n}} \frac{Z^A}{A!}\, C^\lambda(A), \tag{5.24}$$

where the elements of the matrix $C^\lambda(A)$ are defined by

$$\Big(C^\lambda(A)\Big)_{m,m'} = C\begin{pmatrix} m' \\ \lambda \\ m \end{pmatrix}(A), \text{ each } A\in\mathbb{M}^p_{n\times n}. \tag{5.25}$$

Thus, in the matrix form (5.24), the monomial term $Z^A/A!$ appears as a scalar multiplier of the numerical-valued matrix $C^\lambda(A)$. By design, the matrix elements of relation (5.24) give back relation (5.6) in consequence of definition (5.21).

A combinatorial interpretation of the elements of the matrices $C^\lambda(A)$ would place the definition of the D^λ–matrix in a purely combinatorial context. Such an interpretation can be given in terms of digraphs as we show in Chapter 6. But, now, in the spirit of the present chapter, we continue with the basic definitions and form of results.

5.2 Defining Relations for the D^λ–Polynomials

The Kronecker product (see Sect. 10.5, Compendium A) $D^{\lambda'}(Z)\otimes D^\lambda(Z)$ of two D^λ–polynomials is a well-defined matrix polynomial (matrix with polynomial elements). In particular, for the partitions $\lambda = (0^n)$ and $\lambda = (1\,0^{n-1})$, we require that $D^{(0^n)}(Z) = I_{\mathrm{Dim}\lambda}$ and $D^{(\lambda\, 0^{n-1})}(Z) = Z$. The defining relation for the general D^λ–polynomials is taken to be the following:

$$\sum_{\tau=1}^{n}\oplus\; D^{\lambda+e_\tau}(Z) = R^\lambda\Big(Z\otimes D^\lambda(Z)\Big)\,(R^\lambda)^T, \tag{5.26}$$

where e_τ is the unit row vector of length n with 1 as part τ and 0 as all other parts, each $\tau = 1,2,\ldots,n$. Thus, the partition $\lambda+e_\tau$ has λ_τ shifted upward by 1, the other parts remaining unchanged. Should $\lambda+e_\tau$ violate the conditions of being a partition, the corresponding $D^{\lambda+e_\tau}$–polynomial is to be omitted from the left-hand side. The real orthogonal matrix R^λ (not to be confused with $C^\lambda(A)$) is a matrix that is fully defined in Sects. (6.1)-(6.4) of Chapter 6.

Relation (5.26) can be viewed as a recurrence relation: If we take the elements of the real orthogonal matrix R^λ as known for all $\lambda \in \mathbb{P}\mathrm{ar}_n$, and the matrix $D^\lambda(Z)$ as known for all partitions $\lambda \vdash p$, then this relation determines the elements of each matrix $D^{\lambda+e_r}(Z)$ (see (5.36) below). The procedure can then be repeated to generate the elements of each matrix $D^{\lambda+e_r+e_\sigma}(Z)$, etc. The elements of the D^λ–polynomials for $\lambda \vdash p+1$ are uniquely determined in terms of those for $\lambda \vdash p$. Since they are given by $D^{(0^n)}(Z) = I_{\mathrm{Dim}\lambda}$ for $p = 0$, the iteration procedure is fully defined.

It is useful to motivate definition (5.26). It turns out (Chapter 9) that the specialization of the matrix Z to a unitary matrix $Z = U \in U(n)$ gives an irreducible unitary matrix representation $D^\lambda(U)$ of $U(n)$:

$$\begin{aligned} D^\lambda(U')D^\lambda(U) &= D^\lambda(U'U), \text{ each pair } U', U \in U(n), \\ D^\lambda(U^\dagger) &= \left(D^\lambda(U)\right)^\dagger . \end{aligned} \tag{5.27}$$

It is also the case that the Kronecker product of two irreducible unitary representations of $U(n)$ is completely reducible into a direct sum of such irreducible unitary representations by a real orthogonal similarity transformation:

$$C^{(\mu,\nu)}(D^\mu(U) \otimes D^\nu(U))(C^{(\mu,\nu)})^T = \sum_\lambda \oplus\, c^\lambda_{\mu\nu}\, D^\lambda(U), \tag{5.28}$$

where $\mu, \nu, \lambda \in \mathbb{P}ar_n$, the nonnegative integers $c^\lambda_{\mu\nu}$ are the Littlewood-Richardson [104] numbers, and $C^{(\mu,\nu)}$ is a real orthogonal matrix of dimension

$$\mathrm{Dim}\, C^{(\mu,\nu)} = \mathrm{Dim}\, \mu\, \mathrm{Dim}\, \nu = \sum_\lambda c^\lambda_{\mu\nu}\, \mathrm{Dim}\lambda. \tag{5.29}$$

Relation (5.26) is just the specialization of (5.28) to the partition $\mu = (1\, 0^{n-1})$ with the replacement of ν by λ and U by Z. We have taken (5.26) as a definition, and the proof that it uniquely defines the D^λ–polynomials does not depend, of course, on this motivation.

We next give the form of elements of the matrix R^λ that enter into the defining relation (5.26), together with their orthogonality relations; they are numerical coefficients given by the bra-ket matrix elements of a fundamental tensor operator, which, in turn, are simple functions defined over diagraphs. A direct proof of the orthogonality relations follows from Sylvester's identity. These results are proved in Chapter 6. Using these results, we then list below a number of explicit relations for the D^λ–polynomials implied by the defining relation (5.26). The elements of the real orthogonal matrix R^λ are given by

$$R^\lambda_{(\tau,m')(i,m)} = \left\langle \begin{matrix} \lambda + e_\tau \\ m' \end{matrix} \middle| t_{i\tau} \middle| \begin{matrix} \lambda \\ m \end{matrix} \right\rangle. \tag{5.30}$$

In this relation, the matrix R^λ is of dimension $n\,\mathrm{Dim}\lambda$ with rows and columns enumerated as follows:

$$\begin{aligned} &\text{rows: } (\tau, m'), \tau = 1, 2, \ldots, n; m' \in \mathbb{G}_{\lambda+e_\tau}, \\ &\text{columns: } (i, m), i = 1, 2, \ldots, n; m \in \mathbb{G}_\lambda. \end{aligned} \tag{5.31}$$

That this real orthogonal matrix is square is assured by the identity:

$$n\,\mathrm{Dim}\lambda = \sum_{\tau=1}^{n} \mathrm{Dim}(\lambda + e_\tau). \tag{5.32}$$

As remarked above, the bra-ket coefficients in the right-hand side of (5.30) are fully defined in Chapter 6, but, here the important property is the orthogonality relations for the rows and columns of the R^λ–matrix given, respectively, by

$$\begin{aligned} \sum_{i=1}^{n} \sum_{m \in \mathbb{G}_\lambda} \left\langle \begin{matrix} \lambda + e_\rho \\ m' \end{matrix} \middle| t_{i\rho} \middle| \begin{matrix} \lambda \\ m \end{matrix} \right\rangle \left\langle \begin{matrix} \lambda + e_\tau \\ m'' \end{matrix} \middle| t_{i\tau} \middle| \begin{matrix} \lambda \\ m \end{matrix} \right\rangle &= \delta_{\rho\tau}\delta_{m',m''}, \\ \sum_{\tau=1}^{n} \sum_{m'' \in \mathbb{G}_{\lambda+e_\tau}} \left\langle \begin{matrix} \lambda + e_\tau \\ m'' \end{matrix} \middle| t_{i\tau} \middle| \begin{matrix} \lambda \\ m \end{matrix} \right\rangle \left\langle \begin{matrix} \lambda + e_\tau \\ m'' \end{matrix} \middle| t_{j\tau} \middle| \begin{matrix} \lambda \\ m' \end{matrix} \right\rangle &= \delta_{ij}\delta_{m,m'}. \end{aligned} \tag{5.33}$$

We now prove the following four identities, which become explicit when the numerical objects, expressed as the bra-ket matrix elements,

$$\left\langle \begin{matrix} \lambda + e_\tau \\ m' \end{matrix} \middle| t_{i\tau} \middle| \begin{matrix} \lambda \\ m \end{matrix} \right\rangle \tag{5.34}$$

appearing in the identities are specified. The first relation gives the action of all the z_{ij} on the D^λ–polynomials, and the second gives their recursive definition, which uniquely determines them:

1. Action of z_{ij} :

$$z_{ij}D\begin{pmatrix} m' \\ \lambda \\ m \end{pmatrix}(Z) = \sum_{\tau=1}^{n} \sum_{m'',m'''\in\mathbb{G}_{\lambda+e_\tau}} \left\langle \begin{matrix} \lambda+e_\tau \\ m'' \end{matrix} \middle| t_{i\tau} \middle| \begin{matrix} \lambda \\ m \end{matrix} \right\rangle$$

$$\times \left\langle \begin{matrix} \lambda+e_\tau \\ m''' \end{matrix} \middle| t_{j\tau} \middle| \begin{matrix} \lambda \\ m' \end{matrix} \right\rangle D\begin{pmatrix} m''' \\ \lambda+e_\tau \\ m'' \end{pmatrix}(Z). \tag{5.35}$$

2. Raising recurrence relation:

$$D\begin{pmatrix} m''' \\ \lambda+e_\tau \\ m'' \end{pmatrix}(Z) = \sum_{i,j=1}^{n} \sum_{m,m'\in\mathbb{G}_{\lambda}} \left\langle \begin{matrix} \lambda+e_\tau \\ m'' \end{matrix} \middle| t_{i\tau} \middle| \begin{matrix} \lambda \\ m \end{matrix} \right\rangle$$

$$\times \left\langle \begin{matrix} \lambda+e_\tau \\ m''' \end{matrix} \middle| t_{j\tau} \middle| \begin{matrix} \lambda \\ m' \end{matrix} \right\rangle z_{ij}D\begin{pmatrix} m' \\ \lambda \\ m \end{pmatrix}(Z). \tag{5.36}$$

3. Action of $\partial/\partial z_{ij}$:

$$\frac{\partial}{\partial z_{ij}}D\begin{pmatrix} m' \\ \lambda \\ m \end{pmatrix}(Z) = \sum_{\tau=1}^{n} \sum_{m'',m'''\in\mathbb{G}_{\lambda-e_\tau}} \frac{M(\lambda)}{M(\lambda-e_\tau)} \tag{5.37}$$

$$\times \left\langle \begin{matrix} \lambda-e_\tau \\ m'' \end{matrix} \middle| t^{\dagger}_{i\tau} \middle| \begin{matrix} \lambda \\ m \end{matrix} \right\rangle \left\langle \begin{matrix} \lambda-e_\tau \\ m''' \end{matrix} \middle| t^{\dagger}_{j\tau} \middle| \begin{matrix} \lambda \\ m' \end{matrix} \right\rangle D\begin{pmatrix} m''' \\ \lambda-e_\tau \\ m'' \end{pmatrix}(Z).$$

4. Lowering recurrence relation:

$$D\begin{pmatrix} m''' \\ \lambda-e_\tau \\ m'' \end{pmatrix}(Z) = \sum_{i,j=1}^{n} \sum_{m,m'\in\mathbb{G}_{\lambda}} \frac{M(\lambda-e_\tau)}{M(\lambda)} \tag{5.38}$$

$$\times \left\langle \begin{matrix} \lambda-e_\tau \\ m'' \end{matrix} \middle| t^{\dagger}_{i\tau} \middle| \begin{matrix} \lambda \\ m \end{matrix} \right\rangle \left\langle \begin{matrix} \lambda-e_\tau \\ m''' \end{matrix} \middle| t^{\dagger}_{j\tau} \middle| \begin{matrix} \lambda \\ m' \end{matrix} \right\rangle \frac{\partial}{\partial z_{ij}}D\begin{pmatrix} m' \\ \lambda \\ m \end{pmatrix}(Z).$$

Proof. Relation (5.35) is the matrix element version of (5.26); relation (5.36) is the result of transferring the orthogonal matrices R^{λ} and $(R^{\lambda})^{T}$ in (5.26) to the right-hand side and taking matrix elements; relation (5.37) is obtained from (5.35) by using the Hermitian conjugate relation $(z_{ij})^{\dagger} = \partial/\partial z_{ij}$ for the inner product (,); and (5.38) is obtained from (5.37) by using the orthogonality relations in their conjugated form, where the conjugate coefficients are given by

$$\left\langle \begin{matrix} \lambda - e_\tau \\ m' \end{matrix} \middle| t^\dagger_{i\tau} \middle| \begin{matrix} \lambda \\ m \end{matrix} \right\rangle = \left\langle \begin{matrix} \lambda \\ m \end{matrix} \middle| t_{i\tau} \middle| \begin{matrix} \lambda - e_\tau \\ m' \end{matrix} \right\rangle, \tag{5.39}$$

since the coefficients are real. □

The unit tensor operator coefficient for $\lambda = (0^n)$ is given by (see Chapter 6, especially the examples in Sect. 6.2)

$$\left\langle \begin{matrix} e_\tau \\ (0) \end{matrix} \middle| t_{i\tau} \middle| \begin{matrix} 0^n \\ (0) \end{matrix} \right\rangle = \delta_{i,n}, \text{ each } \tau = 1, 2, \ldots, n. \tag{5.40}$$

Thus, starting with

$$D\begin{pmatrix} (0) \\ 0^n \\ (0) \end{pmatrix}(Z) = 1, \tag{5.41}$$

we calculate from (5.36) that $D^{(1\,0^{n-1})}(Z) = Z$; that is,

$$D\begin{pmatrix} j \\ 1\,0^{n-1} \\ i \end{pmatrix}(Z) = z_{ij}, \tag{5.42}$$

where the GT patterns denoted by i and j are the unique patterns whose weights are the unit row vectors $W\left(\begin{smallmatrix} 1\;0^{n-1} \\ i \end{smallmatrix}\right) = e_i, W\left(\begin{smallmatrix} 1\;0^{n-1} \\ j \end{smallmatrix}\right) = e_j$. Using this result back in (5.36), and the assumed known values of the coefficients (5.34), we construct the polynomial for partitions $(2, 0^{n-1})$ and $(1, 1, 0^{n-2})$, then, from these, the polynomials for $(3, 0^{n-1}), (2, 1, 0^{n-2})$, $(1, 1, 1, 0^{n-3}), \ldots$. Thus, the raising recurrence relation (5.36) determines uniquely the general $D^\lambda(Z)$ and $C^\lambda(A)$ matrices in (5.24). For completeness, we have given the conjugate relations as well.

If the D^λ–polynomials are known by iteration of either (5.36), or their derivatives by iteration of (5.37), or by some other means, then the $C^\lambda(A)$–coefficients are also obtained from the inner product:

$$\begin{aligned} \left(Z^A, D\begin{pmatrix} m' \\ \lambda \\ m \end{pmatrix}(Z) \right) &= \left(\frac{\partial}{\partial Z}\right)^A D\begin{pmatrix} m' \\ \lambda\ (Z) \\ m \end{pmatrix}\Bigg|_{Z=0} \\ &= C\begin{pmatrix} m' \\ \lambda \\ m \end{pmatrix}(A). \end{aligned} \tag{5.43}$$

That relations (5.35)-(5.37) and (5.36)-(5.38)) are conjugate relations is put more in evidence by using the normalized D^λ–polynomials:

$$\widehat{D}\begin{pmatrix} m' \\ \lambda \\ m \end{pmatrix}(Z) = \frac{1}{\sqrt{M(\lambda)}} D\begin{pmatrix} m' \\ \lambda \\ m \end{pmatrix}(Z). \tag{5.44}$$

Thus, we obtain the following matrix element expressions:

$$\left(\widehat{D}\begin{pmatrix} m''' \\ \lambda+e_\tau \\ m'' \end{pmatrix}(Z), z_{ij}\widehat{D}\begin{pmatrix} m' \\ \lambda \\ m \end{pmatrix}(Z)\right) \tag{5.45}$$

$$= \sqrt{\frac{M(\lambda+e_\tau)}{M(\lambda)}} \left\langle \begin{matrix} \lambda+e_\tau \\ m'' \end{matrix} \middle| t_{i\tau} \middle| \begin{matrix} \lambda \\ m \end{matrix} \right\rangle \left\langle \begin{matrix} \lambda+e_\tau \\ m''' \end{matrix} \middle| t_{j\tau} \middle| \begin{matrix} \lambda \\ m' \end{matrix} \right\rangle;$$

$$\left(\widehat{D}\begin{pmatrix} m''' \\ \lambda-e_\tau \\ m'' \end{pmatrix}(Z), \frac{\partial}{\partial z_{ij}}\widehat{D}\begin{pmatrix} m' \\ \lambda \\ m \end{pmatrix}(Z)\right) \tag{5.46}$$

$$= \sqrt{\frac{M(\lambda)}{M(\lambda-e_\tau)}} \left\langle \begin{matrix} \lambda-e_\tau \\ m'' \end{matrix} \middle| t^\dagger_{i\tau} \middle| \begin{matrix} \lambda \\ m \end{matrix} \right\rangle \left\langle \begin{matrix} \lambda-e_\tau \\ m''' \end{matrix} \middle| t^\dagger_{j\tau} \middle| \begin{matrix} \lambda \\ m' \end{matrix} \right\rangle.$$

Re-expressing these two results in terms of the D^λ–polynomials then gives agreement with relations (5.35)-(5.38).

The procedure above places the entire burden for the determination of the D^λ–polynomials on knowledge of the family of R^λ–coefficients (5.30). Given these coefficients, it is, of course, not easy to effect the general iteration of any of relations (5.35)-(5.38). We develop in Chapter 7 more effective means for doing this. The $t_{i\tau}$ unit tensor operator appearing in the coefficient (5.30) may be interpreted as a *shift or transfer* of the GT pattern $\binom{\lambda}{m}$ to the new pattern $\binom{\lambda+e_\tau}{m'}$. The algebra of these *fundamental shift operators* is basic to our approach. This algebra is developed in Chapter 6, where they are defined combinatorially as discrete functions over arc digraphs.

The following list gives a number of properties of the D^λ–polynomials. Either the proof of each item is given or it is indicated where it is to be found. Ultimately, these properties are all consequences of the basic relations (5.26), (5.35)-(5.38), and the coefficients (5.30). These properties suggest that the matrices $D^\lambda(Z)$ be called *matrix Schur functions:*

1. Transpositional symmetry:

$$\begin{aligned}\left(D^\lambda(Z)\right)^T &= D^\lambda(Z^T),\\ \left(C^\lambda(A)\right)^T &= C^\lambda(A^T).\end{aligned} \tag{5.47}$$

   The first property follows from (5.26) by replacing Z by Z^T, followed by transposing the resulting relation. The second relation follows from the first and (5.24). A second proof is given by (7.12), Chapter 7.

2. Diagonal property: For each weight $\alpha \in \mathbb{W}_\lambda$, we have

$$\begin{aligned} D\begin{pmatrix} m' \\ \lambda \\ m \end{pmatrix}(\mathrm{diag}(x_1, x_2, \ldots, x_n)) &= \delta_{m,m'} x_1^{\alpha_1} x_2^{\alpha_2} \cdots x_n^{\alpha_n},\\ C\begin{pmatrix} m' \\ \lambda \\ m \end{pmatrix}(\mathrm{diag}(\alpha_1, \alpha_2, \cdots, \alpha_n)) &= \delta_{m,m'} \alpha_1! \alpha_2! \cdots \alpha_n!,\end{aligned} \tag{5.48}$$

   where Z and A are the diagonal matrices $Z = \mathrm{diag}(x_1, x_2, \ldots, x_n)$ and $A = \mathrm{diag}(\alpha_1, \alpha_2, \ldots, \alpha_n)$. These results are proved directly from (5.6) by specializaling to $Z = \mathrm{diag}(x_1, x_2, \ldots, x_n)$ and using the relation

$$D^\lambda(I_n) = I_{\mathrm{Dim}\lambda}, \tag{5.49}$$

   which is a consequence of the multiplication property given by the next Item.

3. Multiplication property:

$$D^\lambda(X) D^\lambda(Y) = D^\lambda(XY), \text{ for arbitrary } X, Y. \tag{5.50}$$

   This property is fundamental. It follows from the fact that the iteration of relation (5.26) is independent of properties of Z : It is the partitions that are generated by the iteration. The multiplication property must hold for arbitrary X, Y. An independent proof is given in Sect. 5.2.1.

4. Kronecker product: Let $\mu, \nu \in \mathbb{P}ar_n$. Then, the following identity holds:

$$C^{(\mu,\nu)}(D^\mu(Z) \otimes D^\nu(Z))(C^{(\mu,\nu)})^T = \sum_\lambda \oplus\, c_{\mu\nu}^\lambda\, D^\lambda(Z), \tag{5.51}$$

where $C^{(\mu,\nu)}$ is a real orthogonal matrix of dimension given by (5.29). We leave the summation over λ unspecified; it is determined by the values that λ can assume for given μ, ν in the Littlewood-Richardson numbers $c^{\lambda}_{\mu\nu}$, which are discussed in Sects. 11.3.6-11.3.8, Compendium B. The form (5.51) already follows from that of (5.28), since the properties of the Kronecker product $D^{\mu}(U) \otimes D^{\nu}(U)$ extend to those of $D^{\mu}(Z) \otimes D^{\nu}(Z)$ by the argument in Item 3. The determination of a real orthogonal matrix $C^{(\mu,\nu)}$ that effects the transformation of the Kronecker product of two $D-$polynomials to a direct sum of $D-$ polynomials is another matter; it is difficult. Various aspects of the problem are addressed in Sect. 6.7, Chapter 6, and its relation to tensor operators is presented in Chapter 9.

5. Group property: If $Z \in GL(n, \mathbb{C})$, the matrices $D^{\lambda}(Z)$ are irreducible representations of $GL(n, \mathbb{C})$. If specialized still further to $Z = U \in U(n)$, and multiplied by appropriate integer powers of $\det U$, all inequivalent, irreducible, unitary representations of the unitary group $U(n)$ are obtained. If Z is specialized to belong to any matrix group, continuous or finite, representations of that group are obtained. Indeed, this applies as well to any multiplicative matrix algebra. This representation property is a consequence of the multiplication property given in Item 3 above.

6. Trace functions: Define $T_{\lambda}(Z) = \operatorname{trace} D^{\lambda}(Z)$. Then, these functions satisfy, in consequence of the Kronecker product relation (5.51), the identity:
$$T_{\mu}(Z)T_{\nu}(Z) = \sum_{\lambda} c^{\lambda}_{\mu\nu}\, T_{\lambda}(Z). \tag{5.52}$$
This result is a consequence of the fact that the Kronecker product in Item 4 is brought to a direct sum by a similarity transformation, and the detailed form of the real orthogonal matrix $C^{(\mu,\nu)}$ is not needed.

7. Schur functions: The polynomials $T_{\lambda}(Z)$ are not symmetric functions in the variables z_{ij}, but if Z is specialized to the diagonal matrix, $Z = \operatorname{diag}(x_1, x_2, \ldots, x_n)$, then, by the diagonal property in Item 2, they become symmetric functions in these variables, in which case, we have the relation to Schur functions given by
$$s_{\lambda}(x) = T_{\lambda}(\operatorname{diag}(x_1, x_2, \ldots, x_n)). \tag{5.53}$$
The Kronecker product relation in Item 4 then gives
$$s_{\mu}(x)s_{\nu}(x) = \sum_{\lambda} c^{\lambda}_{\mu\nu}\, s_{\lambda}(x). \tag{5.54}$$

It is interesting that relation (5.29) for the Weyl dimension, relation (5.51) for the Kronecker product reduction, relation (5.52) for the trace functions, and relation (5.54) for the Schur functions all stand in the abstract relationship between partitions $\mu, \nu, \lambda \in \mathbb{P}ar_n$ that may be symbolized by

$$\mu \boxtimes \nu \quad R \quad \sum_{\lambda} \boxplus c^{\lambda}_{\mu\nu}\, \lambda, \tag{5.55}$$

where R denotes the "relationship" between the left- and right-hand sides of this expression. Of course, it is the Kronecker product relation that is the master relation which gives all the others by an appropriate operation. Nonetheless, the question arises as to why the other three cannot be iterated in the fashion of the Kronecker product in the manner of (5.26) to arrive at formulas giving $\mathrm{Dim}\,\lambda, T_\lambda(Z), e_\lambda(x)$. It is the properties of the relation R that determines the answer. In the cases at hand, it is only the similarity relation in the Kronecker product for which R is reversible: The operations that lead to the other relations destroy the reversibility; that is, it is not possible to invert relations (5.29) for the Weyl dimension, nor (5.52) for the trace functions, nor (5.54) for the Schur functions.

We continue our list of properties of the D^λ–polynomials with the reduction (subduction) relation that shows how the D^μ–polynomials, $\mu \in \mathbb{P}\mathrm{ar}_{n-1}$, are obtained from the D^λ–polynomials, $\lambda \in \mathbb{P}\mathrm{ar}_n$, and with a discussion of the form of an induction method for the construction of the D^λ–polynomials that uses the D^μ–polynomials:

(i). Reduction (subduction) relation. Let $Z_{n-1} = (z_{ij})_{1 \le i,j \le n-1}$ denote the submatrix of order $n-1$ obtained from the matrix $Z_n = (z_{ij})_{1\le i,j\le n}$ of order n by striking row n and column n. The matrix A_{n-1} is similarly defined in terms of $A_n = (a_{ij})_{1\le i,j\le n}$. Then, we have the direct sum relations:

$$\begin{aligned} D^\lambda\begin{pmatrix} Z_{n-1} & \mathbf{0} \\ \mathbf{0} & 1 \end{pmatrix} &= \sum_{\mu \prec \lambda} \oplus D^\mu(Z_{n-1}), \\ C^\lambda\begin{pmatrix} A_{n-1} & \mathbf{0} \\ \mathbf{0} & 1 \end{pmatrix} &= \sum_{\mu \prec \lambda} \oplus C^\mu(A_{n-1}). \end{aligned} \tag{5.56}$$

The notation $\mu \prec \lambda$ denotes any partition $\nu \in \mathbb{P}\mathrm{ar}_{n-1}$ whose parts falls between those of $\lambda \in \mathbb{P}\mathrm{ar}_n : \lambda_1 \ge \mu_1 \ge \lambda_2 \ge \mu_2 \ge \cdots \ge \lambda_{n-1} \ge \mu_{n-1} \ge \lambda_n$ (see Sect. 11.3, Compendium B). This is just the statement of the explicit way in which Weyl's group-subgroup rule is implemented into the definition of the D^λ–polynomials.

(ii). Induction relation. The $D^\lambda-$polynomials, $\lambda \in \mathbb{P}\mathrm{ar}_n$, can be obtained from a summation over products of $D^\mu-$polynomials, $\mu \in \mathbb{P}\mathrm{ar}_{n-1}$ at level $n-1$, and certain monomials in the n variables $\mathbf{z}_n = (z_{n1}, z_{n2}, \ldots, z_{n,n})$ from row n of Z_n and in the n variables $\mathbf{z}^n = (z_{1n}, z_{2n}, \ldots, z_{n-1,n})$ from column n. The terms in each product in the summations are multiplied by appropriate numerical coefficients and summed over the partitions μ and the power exponents of the variables $\mathbf{z}_n$ and $\mathbf{z}^n$ such that the homogeneity properties of the $D^\lambda-$polynomials in the rows and columns of Z_n are fulfilled. It is clear from (5.6) that such a recurrence relation can be obtained simply by factoring off from $Z^A/A!$ the factor $Z^{A'}/(A')!, A' = (z_{ij})_{1\leq i,j\leq n-1}$ and expanding this factor in terms of the $D^\mu-$polynomials at level $n-1$ by using (5.9). We defer this derivation to Sect. 7.3.2, Chapter 7, where some special $D^\lambda-$polynomials are also derived.

5.2.1 Another proof of the multiplication rule

A proof the multiplication rule (5.50) of the $D^\lambda-$matrices

$$D^\lambda(X)D^\lambda(Y) = D^\lambda(XY), \ X, Y \text{ arbitrary } n\times n \text{ matrices} \tag{5.57}$$

was given in Item 3, using the defining relation (5.26). Here we give a second proof. In Sect. 1.6.3 of Chapter 1, we gave a combinatorial proof of the multiplication for the special D^p-polynomials corresponding to partitions $(p\,0^{n-1})$ with one part nonzero. This result and existence of a real orthogonal matrix C that reduces repeated Kronecker products of this type is all that is needed for the general case. Thus, the following two relations hold, where the reduction is into a matrix direct sum of c^λ_q repeated identical block matrices along the diagonal for each $\lambda \vdash |q| = q_1 + q_2 + \cdots + q_m$, where q is the composition $q = (q_1, q_2, \ldots, q_m)$:

$$\begin{aligned} C\left(D^{q_1}(X)\otimes D^{q_2}(X)\otimes\cdots\otimes D^{q_m}(X)\right)C^T &= \sum_{\lambda\vdash|q|}\oplus\, c^\lambda_q\, D^\lambda(X),\\ C\left(D^{q_1}(Y)\otimes D^{q_2}(Y)\otimes\cdots\otimes D^{q_m}(Y)\right)C^T &= \sum_{\lambda\vdash|q|}\oplus\, c^\lambda_q\, D^\lambda(Y). \end{aligned} \tag{5.58}$$

Multiplying these two relations together and using the multiplication properties of Kronecker products and the direct sum gives

$$\begin{aligned} &C\left(D^{q_1}(XY)\otimes D^{q_2}(XY)\otimes\cdots\otimes D^{q_m}(XY)\right)C^T\\ &= \sum_{\lambda\vdash|q|}\oplus\, c^\lambda_q\, D^\lambda(X)D^\lambda(Y) = \sum_{\lambda\vdash|q|}\oplus\, c^\lambda_q\, D^\lambda(XY), \end{aligned} \tag{5.59}$$

which implies $D^\lambda(X)D^\lambda(Y) = D^\lambda(XY)$. Since relation (5.59) is true for every composition q of nonnegative integers into m parts for each $m \geq 2$, all partitions $\lambda \in \mathbb{P}ar_n$ can be obtained: The multiplication rule (1.245) for partition $(p\,0^{n-1})$ extends to every partition $\lambda \in \mathbb{P}ar_n$.

5.3 The D^λ–Polynomials Over $l \times n$ Variables

Let $\lambda \in \mathbb{P}\mathrm{ar}_l$ and $(\lambda\ 0^{n-l}) \in \mathbb{P}\mathrm{ar}_n$. If we set all variables in the matrix $Z_n = Z_{n\times n}$ in rows $l+1, l+2, \ldots, n$ to zero, we are left with a matrix $Z_{l\times n}$ of ln variables. Moreover, the D^λ–polynomials defined by relation (5.6) are zero unless the partition $\lambda \in \mathbb{P}\mathrm{ar}_n$ has the form $(\lambda\,0^{n-l}), \lambda \in \mathbb{P}\mathrm{ar}_l$. Conversely, if we choose the partition in the D^λ–polynomial to be of the form $(\lambda, 0^{n-l}), \lambda \in \mathbb{P}\mathrm{ar}_l$ and choose the partitions in rows $l, l+1, \ldots, n$ of the lower pattern to be maximal, that is, $m_k = (\lambda, 0^{k-l}), k = n-1, n-2, \ldots, l$, then the polynomial is independent of the variables in the matrix $Z_{n\times n}$ that appear in rows $l+1, l+2, \ldots, n$, hence, these variables can be set to 0. We denote (see 5.1 for alternative notation) the polynomials obtained by either of these operations by

$$D\left(\begin{array}{c|c} \lambda & \lambda\ 0^{n-l} \\ m & m' \end{array}\right)(Z) = D\begin{pmatrix} m' \\ \lambda\ 0^{n-l} \\ \max \\ \lambda \\ m \end{pmatrix}(Z), \tag{5.60}$$

where $\lambda \in \mathbb{P}\mathrm{ar}_l$, and it is to be understood from the notation that the matrix of indeterminates Z is $l \times n : Z = (z_{ij})_{1\leq i\leq l, 1\leq j\leq n}$. There are no constraints, other than betweenness, in the pair of GT patterns

$$\begin{pmatrix} \lambda \\ m \end{pmatrix} \text{and} \begin{pmatrix} \lambda\ 0^{n-l} \\ m' \end{pmatrix},\ \lambda \in \mathbb{P}ar_l. \tag{5.61}$$

The pair of SSYW tableaux corresponding to these patterns are both of shape $\lambda \in \mathbb{P}\mathrm{ar}_l$, the first one being filled-in by the standard rule with $1's, 2's, \ldots, l's$, the second with $1's, 2's, \ldots, n's$. The number of such double GT patterns and of the number of SSYW tableau for given $\lambda \in \mathbb{P}\mathrm{ar}_l$ is the product of Weyl dimensions given by

$$\mathrm{Dim}\lambda\ \mathrm{Dim}(\lambda\ 0^{n-l}). \tag{5.62}$$

Many of the properties of the D^λ–polynomials defined over n^2 variables given in the preceding sections carry over directly to the restricted

$D^\lambda-$polynomials defined over $l \times n$ commuting indeterminates, where now we write:

$$Z^A = \prod_{i=1}^{l}\prod_{j=1}^{n} z_{ij}^{a_{ij}},\ A! = \prod_{i=1}^{l}\prod_{j=1}^{n}(a_{ij})!. \tag{5.63}$$

We list some of these properties that are self-explanatory:

1. Relation to Maclaurin polynomials:

$$D\left(\begin{array}{c|c} \lambda & \lambda\ 0^{n-l} \\ m & m' \end{array}\right)(Z) \\ = \sum_{A\in\mathbb{M}^p_{l\times n}(\alpha,\alpha')} C\left(\begin{array}{c|c} \lambda & \lambda\ \ 0^{n-l} \\ m & m' \end{array}\right)(A)\frac{Z^A}{A!}. \tag{5.64}$$

2. Inversion:

$$\frac{Z^A}{A!} = \sum_{\lambda\vdash p}\ \sum_{m\in\mathbb{G}_\lambda(\alpha),\, m'\in\mathbb{G}_{(\lambda\ 0^{n-l})}(\alpha')} \frac{1}{M(\lambda\ 0^{n-l})A!} \\ \times C\left(\begin{array}{c|c} \lambda & \lambda\ 0^{n-l} \\ m & m' \end{array}\right)(A)\, D\left(\begin{array}{c|c} \lambda & \lambda\ 0^{n-l} \\ m & m' \end{array}\right)(Z), \\ \text{each } A\in\mathbb{M}^p_{l\times n}(\alpha,\alpha'). \tag{5.65}$$

3. Orthogonality relations for Maclaurin and $D-$polynomials:

$$(Z^A, Z^B) = A!\delta(A,B). \tag{5.66}$$

$$\left(D\left(\begin{array}{c|c} \lambda & \lambda\ \ 0^{n-l} \\ m & m' \end{array}\right)(Z), D\left(\begin{array}{c|c} \lambda & \lambda\ 0^{n-l} \\ m'' & m''' \end{array}\right)(Z)\right) \\ = \delta_{m,m''}\delta_{m',m'''}M(\lambda\ 0^{n-l}). \tag{5.67}$$

4. Orthogonality relations for transformation coefficients:

$$R\left(\begin{array}{c|c} \lambda & \lambda\ 0^{n-l} \\ m & m' \end{array}\right)(A) \\ = \frac{1}{\sqrt{M(\lambda\ 0^{n-l})A!}} C\left(\begin{array}{c|c} \lambda & \lambda\ 0^{n-l} \\ m & m' \end{array}\right)(A); \tag{5.68}$$

$$\sum_{A\in\mathbb{M}^p_{l\times n}(\alpha,\alpha')} R\left(\begin{array}{c|c}\lambda & \lambda\ 0^{n-l}\\ m & m'\end{array}\right)(A)$$
$$\times R\left(\begin{array}{c|c}\lambda & \lambda\ 0^{n-l}\\ m'' & m'''\end{array}\right)(A) = \delta_{m,m''}\delta_{m',m'''}; \tag{5.69}$$

$$\sum_{\lambda\vdash p}\ \sum_{m\in\mathbb{G}_\lambda(\alpha),\, m'\in\mathbb{G}_{(\lambda\ 0^{n-l})}(\alpha')} R\left(\begin{array}{c|c}\lambda & \lambda\ 0^{n-l}\\ m & m'\end{array}\right)(A)$$
$$\times R\left(\begin{array}{c|c}\lambda & \lambda\ 0^{n-l}\\ m & m'\end{array}\right)(B) = \delta(A,B). \tag{5.70}$$

5. Transposition property:

$$D\left(\begin{array}{c|c}\lambda & \lambda\ 0^{n-l}\\ m & m'\end{array}\right)(Z^T) = D\left(\begin{array}{c|c}\lambda\ 0^{n-l} & \lambda\\ m' & m\end{array}\right)(Z). \tag{5.71}$$

The identity $M(\lambda) = M(\lambda\ 0^{n-l})$ between measure factors also holds.

5.4 Vector Space Aspects

The D^λ–polynomials are an orthogonal basis of a vector space of polynomials. We next itemize the properties of the vector spaces that are relevant to the D^λ–polynomials defined over the $l \times n$ commuting indeterminates $(z_{ij})_{1\le i\le l, 1\le j\le n}$, which are arranged in an $l \times n$ matrix Z with row i and column j given, respectively, by

$$\begin{aligned} z_i &= (z_{i1}, z_{i2}, \ldots, z_{in}), \\ z^j &= (z_{1j}, z_{2j}, \ldots, z_{lj}). \end{aligned} \tag{5.72}$$

We introduce the following vector spaces:

1. Vector space of homogeneous polynomials of total degree p :

$$\mathcal{P}^p_{l\times n} = \left\{\begin{array}{l}\text{set of polynomials homogeneous of}\\ \text{degree } p \text{ in the } l\times n \text{ indeterminates}\\ Z = (z_{ij})_{1\le i\le l,1\le j\le n}\end{array}\right\}. \tag{5.73}$$

2. Vector space of homogeneous polynomials of fixed degree in the row vectors z_i and column vectors z^j of Z : Let $\alpha = (\alpha_1, \alpha_2, \ldots, \alpha_l)$ and $\alpha' = (\alpha'_1, \alpha'_2, \ldots, \alpha'_n)$ be nonnegative compositions of p into l and n parts, respectively: $\alpha \in \mathbb{C}_l(p), \alpha' \in \mathbb{C}_n(p)$. The subspace of polynomials $\mathcal{P}^p_{l\times n}(\alpha, \alpha')$ of the vector space $\mathcal{P}^p_{l\times n}$ is defined by

$$\mathcal{P}^p_{l\times n}(\alpha, \alpha') = \left\{ \begin{array}{l} \text{set of polynomials homogeneous of} \\ \text{degree } \alpha_i \text{ in the } z_i = (z_{i1}, z_{i2}, \ldots, z_{in}), \\ \text{each } i = 1, 2, \ldots, l; \text{ and of degree } \alpha'_j \\ \text{in the } z^j = (z_{1j}, z_{2j}, \ldots, z_{lj}), \\ \text{each } j = 1, 2, \ldots, n \end{array} \right\}. \tag{5.74}$$

A polynomial in the space $\mathcal{P}^p_{l\times n}(\alpha, \alpha')$ is denoted by $P^p_{l\times n}(\alpha, \alpha')$ with values $P^p_{l\times n}(\alpha, \alpha')(Z)$.

We have given two orthonormal bases of the space $\mathcal{P}^p_{l\times n}(\alpha, \alpha')$: the normalized Maclaurin monomials and the normalized $D^\lambda-$polynomials in $l \times n$ variables that are homogeneous of degree α and α' in the row and column vectors of Z. These bases are described by

$$\mathcal{B}^p_{l\times n}(\alpha, \alpha') = \left\{ \frac{Z^A}{\sqrt{A!}} \;\middle|\; A \in \mathbb{M}^p_{l\times n}(\alpha, \alpha') \right\}, \tag{5.75}$$

$$\mathbb{B}^p_{l\times n}(\alpha, \alpha') = \bigcup_{\lambda \in \mathbb{P}ar_l(p)} \mathbb{B}^\lambda_{l\times n}(\alpha, \alpha'), \tag{5.76}$$

where $\mathbb{P}ar_l(p)$ is the set of partitions $\lambda \vdash p$ into not more than l nonzero parts, and $\mathbb{B}^\lambda_{l\times n}(\alpha, \alpha')$ is the set of all orthonormal $D^\lambda-$polynomials in the $l \times n$ indeterminates Z :

$$\begin{aligned} &\mathbb{B}^\lambda_{l\times n}(\alpha, \alpha') \qquad (5.77) \\ &= \left\{ \frac{1}{\sqrt{M(\lambda\, 0^{n-l})}} D \begin{pmatrix} \lambda & \lambda\, 0^{n-l} \\ m & m' \end{pmatrix} (Z) \;\middle|\; \begin{array}{l} m \in \mathbb{G}_\lambda(\alpha), \\ m' \in \mathbb{G}_{(\lambda\, 0^{n-l})}(\alpha') \end{array} \right\}. \end{aligned}$$

The dimension of the space $\mathcal{P}^p_{l\times n}(\alpha, \alpha')$ is given by the cardinalities of various sets as follows:

$$\begin{aligned} \mathrm{Dim}\mathcal{P}^p_{l\times n}(\alpha, \alpha') &= |\mathbb{M}^p_{l\times n}(\alpha, \alpha')| = |\mathbb{B}^p_{l\times n}(\alpha, \alpha')| \\ &= \sum_{\lambda \vdash p} |\mathbb{B}^\lambda_{l\times n}(\alpha, \alpha') = \sum_{\lambda \vdash p} K(\lambda, \alpha) K((\lambda\, 0^{n-l}), \alpha'). \end{aligned} \tag{5.78}$$

A polynomial in the space $\mathcal{P}^p_{l\times n}$ is denoted $P^p_{l\times n}$ with values $P^p_{l\times n}(Z)$. This space is then given by the direct sum of perpendicular subspaces as follows:

$$\mathcal{P}^p_{l\times n} = \sum_{\alpha\in\mathbb{C}_l(p),\, \alpha'\in\mathbb{C}_n(p)} \oplus\, \mathcal{P}^p_{l\times n}(\alpha,\alpha'). \tag{5.79}$$

The dimension of the space $\mathcal{P}^p_{l\times n}$ is given by the cardinalities of various sets as follows:

$$\mathrm{Dim}\mathcal{P}^p_{l\times n} = \binom{ln+p-1}{p} \tag{5.80}$$

$$= \sum_{\lambda\vdash p} \sum_{\alpha\in\mathbb{C}_l(p),\alpha'\in\mathbb{C}_n(p)} |M^p_{l\times n}(\alpha,\alpha')| = \sum_{\lambda\vdash p} \mathrm{Dim}\,\lambda\,\mathrm{Dim}\,(\lambda\,0^{n-l}).$$

The vector space $\mathcal{P}^\lambda_{l\times n}$ of polynomials has the following general property: Let $P^\lambda_{l\times n}(Z)$ be an arbitrary polynomial in this space. Then, the following polynomials are also in the vector space $\mathcal{P}^\lambda_{l\times n}$:

$$P^\lambda_{l\times n}(XZ) \text{ and } P^\lambda_{l\times n}(ZY), \tag{5.81}$$

for each $l\times l$ matrix X and each $n\times n$ matrix Y. Moreover, for each $\lambda\in\mathbb{P}ar_l$, no subspace of $\mathcal{P}^\lambda_{l\times n}$ has the property (5.81); that is,

The space of polynomials $\mathcal{P}^\lambda_{l\times n}$ is invariant and irreducible with respect to the action of left and right matrix transformations of its variables.

This property is a standard application of Schur's lemma (Sect. 10.7.2, Compendium A)

The left and right transformation properties of the D–polynomials are as follows:

$$D\left(\begin{array}{c|c} \lambda & \lambda\,0^{n-l} \\ m & m' \end{array}\right)(X^T Z) \tag{5.82}$$

$$= \sum_{m''\in\mathbb{G}_\lambda} D\begin{pmatrix} m \\ \lambda \\ m'' \end{pmatrix}(X)\, D\left(\begin{array}{c|c} \lambda & \lambda\,0^{n-l} \\ m'' & m' \end{array}\right)(Z);$$

$$D\left(\begin{array}{c|c} \lambda & \lambda\,0^{n-l} \\ m & m' \end{array}\right)(ZY) \tag{5.83}$$

$$= \sum_{m'''\in\mathbb{G}_{(\lambda\ 0^{n-l})}} D\begin{pmatrix} m' \\ \lambda\,0^{n-l} \\ m''' \end{pmatrix}(Y)\, D\left(\begin{array}{c|c} \lambda & \lambda\,0^{n-l} \\ m & m''' \end{array}\right)(Z).$$

The results of this subsection are, of course, valid for $l = n$. The results for the $l \times n$ $D-$polynomials do not require independent verification: They follow from the $n \times n$ relations by the complementarity of zeros of the Z variables and those of the discrete A variables.

5.5 Fundamental Structural Relations and Principal Objectives

5.5.1 Structural relations

We summarize in this section the three most significant structural properties of the $D^\lambda-$matrices in n^2 variables z_{ij} and of the $C^\lambda-$matrices in n^2 integral variables a_{ij}, together with some remarks on remaining objectives. The objects of interest are the matrices:

$$D^\lambda(Z) = \sum_{A \in \mathbb{M}^p_{n\times n}} \frac{Z^A}{A!} C^\lambda(A), \text{ each } \lambda \in \mathbb{P}\mathrm{ar}_n, \tag{5.84}$$

$$C^\lambda(A) = \left(Z^A, D^\lambda(Z)\right) = (\partial/\partial Z)^A \, D^\lambda(Z)\Big|_{Z=0}, \tag{5.85}$$

where we have extended the inner product $(\,,)$ in the obvious way to a matrix. Then, we have the following properties:

1. Multiplication of $D^\lambda-$matrices:
$$D^\lambda(X)D^\lambda(Y) = D^\lambda(XY), \text{ for arbitrary } X \text{ and } Y. \tag{5.86}$$

2. Multiplication of $C^\lambda-$matrices:
$$C^\lambda(A)C^\lambda(B) = \sum_{C \in M^p_{n\times n}} \left\{ \begin{array}{c} C \\ A\,B \end{array} \right\} C^\lambda(C), \text{ for arbitrary } A, B \in \mathbb{M}^p_{n\times n}. \tag{5.87}$$

3. Kronecker product similarity equivalence:
$$D^\mu(Z) \otimes D^\nu(Z) \sim \sum_\lambda \oplus\, c^\lambda_{\mu\nu}\, D^\lambda(Z). \tag{5.88}$$

The proof of the multiplication property of the $D^\lambda-$matrices has already been given. The proof of the multiplication property of the $C^\lambda-$matrices is a direct consequence of this multiplication property of functions, and of the transformation and orthogonality property of the Maclaurin monomials. The structural form of the Kronecker product has also been proved, but it remains to determine the transformation coefficients that effect the explicit reduction in the general case.

5.5.2 Principal objectives and tasks

In this chapter, we have given the form and demonstrated the uniqueness of the D^λ−polynomials and the C^λ−coefficients based on the defining relation (5.26); it remains to give the fundamental tensor operator coefficients in relation (5.36). Also, we must give explicit methods for determining the C^λ−coefficients, hence, the D^λ−polynomials. We carry this out in several chapters, as follows:

1. Chapter 6: The coefficients (5.30) are derived, using diagraphs. It is proved that these coefficients effect the reduction of the Kronecker product $Z \otimes D^\lambda(Z)$. A combinatorial result, Sylvester's identity, is a basic result for this derivation. The reduction of the general Kronecker product of a D^μ-polynomial and a D^ν−polynomal into a direct sum of D^λ−polynomial is considered from several points of view, including operator-valued D^λ−polynomials.

2. Chapter 7: The structure of the D^λ−polynomials is examined with a focus on a complete derivation of the CG coefficients associated with the totally symmetric $D^{(p\ 0^{n-l})}$−polynomials, and on an associated recurrence formula for the general D^λ−polynomials and the general C^λ−coefficients. The case for partitions with two nonzero parts is also solved completely.

3. Chapter 8: The D^λ−polynomials are presented as the simultaneous eigenvectors of a complete set of commuting Hermitian differential operators. This is a Lie algebraic characterization, but realized here in terms of the more fundamentatl algebra of a set of shift operators $t_{i\tau}$ introduced in Chapter 6. These shift operators, which are related to digraphs, are the basic combinatorial objects in the theory of the D^λ−polynomials, and, indeed, in the theory of irreducible tensor operators.

4. Chapter 9: The theory of irreducible tensor operators is developed from the viewpoint of the role of the fundamental tensor (shift) operators, reduced matrix elements, and coupling theory. The role of the Littlewood-Richardson numbers is examined critically and from several viewpoints.

Chapter 6

Operator Actions in Abstract Hilbert Space

6.1 Introductory Remarks

The unifying structure underlying this work is the set of *fundamental shift operators* $\{t_{i\tau} \,|\, i, \tau = 1, 2, \ldots n\}$ out of which we shall build almost all other mathematical objects of interest in this monograph. These operators and the related sets of orthogonal coefficients

$$\left(R^{\lambda}\right)_{(\tau,m')(i,m)} = \left\langle \begin{matrix} \lambda + e_\tau \\ m' \end{matrix} \middle| t_{i\tau} \middle| \begin{matrix} \lambda \\ m \end{matrix} \right\rangle, \tag{6.1}$$

which effect the reduction of the direct product

$$R^{\lambda}\left(Z \otimes D^{\lambda}(Z)\right)(R^{\lambda})^{T} = \sum_{\tau=1}^{n} \oplus\, D^{\lambda+e_\tau}(Z). \tag{6.2}$$

are basic to our methods for constructing the polynomials

$$D\begin{pmatrix} m' \\ \lambda \\ m \end{pmatrix}(Z). \tag{6.3}$$

Following Gelfand and Tsetlin [62], we assume, as given, a separable model Hilbert space with the following structure:

$$H = \sum_{p=0}^{\infty} \sum_{\lambda \in \mathbb{P}\mathrm{ar}_n(p)} \oplus\, H_\lambda,\ H_\lambda \perp H_{\lambda'},\ \lambda \neq \lambda',\ \text{no repetitions}, \tag{6.4}$$

where each subspace $H_\lambda, \lambda \in \mathbb{P}\mathrm{ar}_n(p)$ possesses an orthonormal basis with vectors enumerated by GT patterns and denoted in the Dirac bra-ket notation by

$$\mathbf{B}_\lambda = \left\{ \left| \begin{array}{c} \lambda \\ m \end{array} \right\rangle \middle| m \text{ is a GT pattern} \right\}. \tag{6.5}$$

In what follows in this chapter, an explicit realization of such a Hilbert space is not required. The entire structure can be regarded as a formal set of rules for constructing a set of coefficients whose various properties can be explicitly verified.

A fundamental shift operator $t_{i\tau}$ is a mapping of the subspace $H_\lambda \subset H$ to the subspace $H_{\lambda+e_\tau} \subset H$:

$$t_{i\tau} : H_\lambda \to H_{\lambda+e_\tau}, \text{ each } i = 1, 2, \ldots, n, \tag{6.6}$$

where $H_{\lambda+e_\tau} = \mathbf{0}$, if $\lambda + e_\tau$ is not a partition. It is this shift property on an arbitrary vector subspace $H_\lambda, \lambda \in \mathbb{P}\mathrm{ar}_n$ from which we derive the term *fundamental shift operator.*

After giving examples for $n = 1, 2, 3$, the general mapping $t_{i\tau}$ and its relation to diagraphs is presented in Sect. 6.3 by giving the explicit operator action on the basis $\mathbf{B}_\lambda$. That these are the coefficients (6.1) that reduce the Kronecker product $Z \otimes D^\lambda(Z)$ to the direct sum form (6.2) is proved in Sect. 6.4.

6.2 Action of Fundamental Shift Operators

We uniformly express partitions as $\lambda = (m_{1,n}, m_{2,n}, \ldots, m_{n,n})$, and coefficients in the notations of the so-called partial hooks: $p_{ij} = m_{i,j} + j - i$. (It is also convenient not to clutter the following relations with numbers.)

Examples:

n = 1:

$$t_{11}|\lambda_1\rangle = |\lambda_1 + 1\rangle, \; t_{11}^\dagger|\lambda_1\rangle = |\lambda_1 - 1\rangle, \; t_{11}^\dagger|0\rangle = 0.$$

n = 2:

$$t_{11} \left| \begin{array}{cc} \lambda_1 & \lambda_2 \\ \multicolumn{2}{c}{m_{11}} \end{array} \right\rangle = \sqrt{\frac{p_{11} - p_{22} + 1}{p_{12} - p_{22}}} \left| \begin{array}{cc} \lambda_1 + 1 & \lambda_2 \\ \multicolumn{2}{c}{m_{11} + 1} \end{array} \right\rangle,$$

$$t_{21} \left| \begin{array}{cc} \lambda_1 & \lambda_2 \\ \multicolumn{2}{c}{m_{11}} \end{array} \right\rangle = \sqrt{\frac{p_{12} - p_{11}}{p_{12} - p_{22}}} \left| \begin{array}{cc} \lambda_1 + 1 & \lambda_2 \\ \multicolumn{2}{c}{m_{11}} \end{array} \right\rangle,$$

$$t_{12}\left|\begin{array}{cc}\lambda_1 & \lambda_2 \\ \multicolumn{2}{c}{m_{11}}\end{array}\right\rangle = -\sqrt{\frac{p_{11}-p_{12}+1}{p_{22}-p_{12}}}\left|\begin{array}{cc}\lambda_1 & \lambda_2+1 \\ \multicolumn{2}{c}{m_{11}+1}\end{array}\right\rangle,$$

$$t_{22}\left|\begin{array}{cc}\lambda_1 & \lambda_2 \\ \multicolumn{2}{c}{m_{11}}\end{array}\right\rangle = \sqrt{\frac{p_{22}-p_{11}}{p_{22}-p_{12}}}\left|\begin{array}{cc}\lambda_1 & \lambda_2+1 \\ \multicolumn{2}{c}{m_{11}}\end{array}\right\rangle;$$

$$t^{\dagger}_{11}\left|\begin{array}{cc}\lambda_1 & \lambda_2 \\ \multicolumn{2}{c}{m_{11}}\end{array}\right\rangle = \sqrt{\frac{p_{11}-p_{22}}{p_{12}-p_{22}-1}}\left|\begin{array}{cc}\lambda_1-1 & \lambda_2 \\ \multicolumn{2}{c}{m_{11}-1}\end{array}\right\rangle,$$

$$t^{\dagger}_{21}\left|\begin{array}{cc}\lambda_1 & \lambda_2 \\ \multicolumn{2}{c}{m_{11}}\end{array}\right\rangle = \sqrt{\frac{p_{12}-p_{11}-1}{p_{12}-p_{22}-1}}\left|\begin{array}{cc}\lambda_1-1 & \lambda_2 \\ \multicolumn{2}{c}{m_{11}}\end{array}\right\rangle,$$

$$t^{\dagger}_{12}\left|\begin{array}{cc}\lambda_1 & \lambda_2 \\ \multicolumn{2}{c}{m_{11}}\end{array}\right\rangle = -\sqrt{\frac{p_{11}-p_{12}}{p_{22}-p_{12}-1}}\left|\begin{array}{cc}\lambda_1 & \lambda_2-1 \\ \multicolumn{2}{c}{m_{11}-1}\end{array}\right\rangle,$$

$$t^{\dagger}_{22}\left|\begin{array}{cc}\lambda_1 & \lambda_2 \\ \multicolumn{2}{c}{m_{11}}\end{array}\right\rangle = \sqrt{\frac{p_{22}-p_{11}-1}{p_{22}-p_{12}-1}}\left|\begin{array}{cc}\lambda_1 & \lambda_2-1 \\ \multicolumn{2}{c}{m_{11}}\end{array}\right\rangle.$$

n=3:

$$t_{11}\left|\begin{array}{c}\lambda_1 \quad \lambda_2 \quad \lambda_3 \\ m_{12}\ m_{22} \\ m_{11}\end{array}\right\rangle$$

$$= \sqrt{\frac{(p_{12}-p_{23}+1)(p_{12}-p_{33}+1)(p_{13}-p_{22})}{(p_{12}-p_{22}+1)(p_{13}-p_{23})(p_{13}-p_{33})}}\sqrt{\frac{p_{11}-p_{22}+1}{p_{12}-p_{22}}}\left|\begin{array}{c}\lambda_1+1 \quad \lambda_2 \quad \lambda_3 \\ m_{12}+1\ m_{22} \\ m_{11}+1\end{array}\right\rangle$$

$$-\sqrt{\frac{(p_{22}-p_{23}+1)(p_{22}-p_{33}+1)(p_{13}-p_{12})}{(p_{22}-p_{12}+1)(p_{13}-p_{23})(p_{13}-p_{33})}}\sqrt{\frac{p_{11}-p_{12}+1}{p_{22}-p_{12}}}\left|\begin{array}{c}\lambda_1+1 \quad \lambda_2 \quad \lambda_3 \\ m_{12}\ m_{22}+1 \\ m_{11}+1\end{array}\right\rangle,$$

$$t_{21}\left|\begin{array}{c}\lambda_1 \quad \lambda_2 \quad \lambda_3 \\ m_{12}\ m_{22} \\ m_{11}\end{array}\right\rangle$$

$$= \sqrt{\frac{(p_{12}-p_{23}+1)(p_{12}-p_{33}+1)(p_{13}-p_{22})}{(p_{12}-p_{22}+1)(p_{13}-p_{23})(p_{13}-p_{33})}}\sqrt{\frac{p_{12}-p_{11}}{p_{12}-p_{22}}}\left|\begin{array}{c}\lambda_1+1 \quad \lambda_2 \quad \lambda_3 \\ m_{12}+1\ m_{22} \\ m_{11}\end{array}\right\rangle$$

$$+\sqrt{\frac{(p_{22}-p_{23}+1)(p_{22}-p_{33}+1)(p_{13}-p_{12})}{(p_{22}-p_{12}+1)(p_{13}-p_{23})(p_{13}-p_{33})}}\sqrt{\frac{p_{22}-p_{11}}{p_{22}-p_{12}}}\left|\begin{array}{c}\lambda_1+1 \quad \lambda_2 \quad \lambda_3 \\ m_{12}\ m_{22}+1 \\ m_{11}\end{array}\right\rangle,$$

$$t_{31}\left|\begin{array}{c}\lambda_1 \quad \lambda_2 \quad \lambda_3 \\ m_{12}\ m_{22} \\ m_{11}\end{array}\right\rangle = \sqrt{\frac{(p_{13}-p_{12})(p_{13}-p_{22})}{(p_{13}-p_{23})(p_{13}-p_{33})}}\left|\begin{array}{c}\lambda_1+1 \quad \lambda_2 \quad \lambda_3 \\ m_{12}\ m_{22} \\ m_{11}\end{array}\right\rangle;$$

$$
\begin{aligned}
&t_{12}\left|\begin{array}{ccc}\lambda_1 & \lambda_2 & \lambda_3\\ \multicolumn{3}{c}{m_{12}\ m_{22}}\\ \multicolumn{3}{c}{m_{11}}\end{array}\right\rangle\\
&= -\sqrt{\frac{(p_{12}-p_{13}+1)(p_{12}-p_{33}+1)(p_{23}-p_{22})}{(p_{12}-p_{22}+1)(p_{23}-p_{13})(p_{23}-p_{33})}}\sqrt{\frac{p_{11}-p_{22}+1}{p_{12}-p_{22}}}\left|\begin{array}{ccc}\lambda_1 & \lambda_2+1 & \lambda_3\\ \multicolumn{3}{c}{m_{12}+1\ m_{22}}\\ \multicolumn{3}{c}{m_{11}+1}\end{array}\right\rangle\\
&-\sqrt{\frac{(p_{22}-p_{13}+1)(p_{22}-p_{33}+1)(p_{23}-p_{12})}{(p_{22}-p_{12}+1)(p_{23}-p_{13})(p_{23}-p_{33})}}\sqrt{\frac{p_{11}-p_{12}+1}{p_{22}-p_{12}}}\left|\begin{array}{ccc}\lambda_1 & \lambda_2+1 & \lambda_3\\ \multicolumn{3}{c}{m_{12}\ m_{22}+1}\\ \multicolumn{3}{c}{m_{11}+1}\end{array}\right\rangle,
\end{aligned}
$$

$$
\begin{aligned}
&t_{22}\left|\begin{array}{ccc}\lambda_1 & \lambda_2 & \lambda_3\\ \multicolumn{3}{c}{m_{12}\ m_{22}}\\ \multicolumn{3}{c}{m_{11}}\end{array}\right\rangle\\
&= -\sqrt{\frac{(p_{12}-p_{13}+1)(p_{12}-p_{33}+1)(p_{23}-p_{22})}{(p_{12}-p_{22}+1)(p_{23}-p_{13})(p_{23}-p_{33})}}\sqrt{\frac{p_{12}-p_{11}}{p_{12}-p_{22}}}\left|\begin{array}{ccc}\lambda_1 & \lambda_2+1 & \lambda_3\\ \multicolumn{3}{c}{m_{12}+1\ m_{22}}\\ \multicolumn{3}{c}{m_{11}}\end{array}\right\rangle\\
&+\sqrt{\frac{(p_{22}-p_{13}+1)(p_{22}-p_{33}+1)(p_{23}-p_{12})}{(p_{22}-p_{12}+1)(p_{23}-p_{13})(p_{23}-p_{33})}}\sqrt{\frac{p_{22}-p_{11}}{p_{22}-p_{12}}}\left|\begin{array}{ccc}\lambda_1 & \lambda_2+1 & \lambda_3\\ \multicolumn{3}{c}{m_{12}\ m_{22}+1}\\ \multicolumn{3}{c}{m_{11}}\end{array}\right\rangle,
\end{aligned}
$$

$$
t_{32}\left|\begin{array}{ccc}\lambda_1 & \lambda_2 & \lambda_3\\ \multicolumn{3}{c}{m_{12}\ m_{22}}\\ \multicolumn{3}{c}{m_{11}}\end{array}\right\rangle = \sqrt{\frac{(p_{23}-p_{12})(p_{23}-p_{22})}{(p_{23}-p_{13})(p_{23}-p_{33})}}\left|\begin{array}{ccc}\lambda_1 & \lambda_2+1 & \lambda_3\\ \multicolumn{3}{c}{m_{12}\ m_{22}}\\ \multicolumn{3}{c}{m_{11}}\end{array}\right\rangle.
$$

$$
\begin{aligned}
&t_{13}\left|\begin{array}{ccc}\lambda_1 & \lambda_2 & \lambda_3\\ \multicolumn{3}{c}{m_{12}\ m_{22}}\\ \multicolumn{3}{c}{m_{11}}\end{array}\right\rangle\\
&= -\sqrt{\frac{(p_{12}-p_{13}+1)(p_{12}-p_{23}+1)(p_{33}-p_{22})}{(p_{12}-p_{22}+1)(p_{33}-p_{13})(p_{33}-p_{23})}}\sqrt{\frac{p_{11}-p_{22}+1}{p_{12}-p_{22}}}\left|\begin{array}{ccc}\lambda_1 & \lambda_2 & \lambda_3+1\\ \multicolumn{3}{c}{m_{12}+1\ m_{22}}\\ \multicolumn{3}{c}{m_{11}+1}\end{array}\right\rangle\\
&+\sqrt{\frac{(p_{22}-p_{13}+1)(p_{22}-p_{23}+1)(p_{33}-p_{12})}{(p_{22}-p_{12}+1)(p_{33}-p_{13})(p_{33}-p_{23})}}\sqrt{\frac{p_{11}-p_{12}+1}{p_{22}-p_{12}}}\left|\begin{array}{ccc}\lambda_1 & \lambda_2 & \lambda_3+1\\ \multicolumn{3}{c}{m_{12}\ m_{22}+1}\\ \multicolumn{3}{c}{m_{11}+1}\end{array}\right\rangle,
\end{aligned}
$$

$$
t_{23}\left|\begin{array}{c}\lambda_1 \quad \lambda_2 \quad \lambda_3\\ m_{12}\ m_{22}\\ m_{11}\end{array}\right\rangle
$$

$$
= -\sqrt{\tfrac{(p_{12}-p_{13}+1)(p_{12}-p_{23}+1)(p_{33}-p_{22})}{(p_{12}-p_{22}+1)(p_{33}-p_{13})(p_{33}-p_{23})}}\sqrt{\tfrac{p_{12}-p_{11}}{p_{12}-p_{22}}}\left|\begin{array}{c}\lambda_1 \quad \lambda_2 \quad \lambda_3+1\\ m_{12}+1\ m_{22}\\ m_{11}\end{array}\right\rangle
$$

$$
-\sqrt{\tfrac{(p_{22}-p_{13}+1)(p_{22}-p_{23}+1)(p_{33}-p_{12})}{(p_{22}-p_{12}+1)(p_{33}-p_{13})(p_{33}-p_{23})}}\sqrt{\tfrac{p_{22}-p_{11}}{p_{22}-p_{12}}}\left|\begin{array}{c}\lambda_1 \quad \lambda_2 \quad \lambda_3+1\\ m_{12}\ m_{22}+1\\ m_{11}\end{array}\right\rangle,
$$

$$
t_{33}\left|\begin{array}{c}\lambda_1 \quad \lambda_2 \quad \lambda_3\\ m_{12}\ m_{22}\\ m_{11}\end{array}\right\rangle = \sqrt{\tfrac{(p_{33}-p_{12})(p_{33}-p_{22})}{(p_{33}-p_{13})(p_{33}-p_{23})}}\left|\begin{array}{c}\lambda_1 \quad \lambda_2 \quad \lambda_3+1\\ m_{12}\ m_{22}\\ m_{11}\end{array}\right\rangle;
$$

Conjugate action:

$$
t_{11}^{\dagger}\left|\begin{array}{c}\lambda_1 \quad \lambda_2 \quad \lambda_3\\ m_{12}\ m_{22}\\ m_{11}\end{array}\right\rangle
$$

$$
= \sqrt{\tfrac{(p_{12}-p_{23})(p_{12}-p_{33})(p_{13}-p_{22}-1)}{(p_{12}-p_{22})(p_{13}-p_{23}-1)(p_{13}-p_{33}-1)}}\sqrt{\tfrac{p_{11}-p_{22}}{p_{12}-p_{22}-1}}\left|\begin{array}{c}\lambda_1-1 \quad \lambda_2 \quad \lambda_3\\ m_{12}-1\ m_{22}\\ m_{11}-1\end{array}\right\rangle
$$

$$
-\sqrt{\tfrac{(p_{22}-p_{23})(p_{22}-p_{33})(p_{13}-p_{12}-1)}{(p_{22}-p_{12})(p_{13}-p_{23}-1)(p_{13}-p_{33}-1)}}\sqrt{\tfrac{p_{11}-p_{12}}{p_{22}-p_{12}-1}}\left|\begin{array}{c}\lambda_1-1 \quad \lambda_2 \quad \lambda_3\\ m_{12}\ m_{22}-1\\ m_{11}-1\end{array}\right\rangle,
$$

$$
t_{21}^{\dagger}\left|\begin{array}{c}\lambda_1 \quad \lambda_2 \quad \lambda_3\\ m_{12}\ m_{22}\\ m_{11}\end{array}\right\rangle
$$

$$
= \sqrt{\tfrac{(p_{12}-p_{23})(p_{12}-p_{33})(p_{13}-p_{22}-1)}{(p_{12}-p_{22})(p_{13}-p_{23}-1)(p_{13}-p_{33}-1)}}\sqrt{\tfrac{p_{12}-p_{11}-1}{p_{12}-p_{22}-1}}\left|\begin{array}{c}\lambda_1-1 \quad \lambda_2 \quad \lambda_3\\ m_{12}-1\ m_{22}\\ m_{11}\end{array}\right\rangle
$$

$$
+\sqrt{\tfrac{(p_{22}-p_{23})(p_{22}-p_{33})(p_{13}-p_{12}-1)}{(p_{22}-p_{12})(p_{13}-p_{23}-1)(p_{13}-p_{33}-1)}}\sqrt{\tfrac{p_{22}-p_{11}-1}{p_{22}-p_{12}-1}}\left|\begin{array}{c}\lambda_1-1 \quad \lambda_2 \quad \lambda_3\\ m_{12}\ m_{22}-1\\ m_{11}\end{array}\right\rangle,
$$

$$
t_{31}^{\dagger}\left|\begin{array}{c}\lambda_1 \quad \lambda_2 \quad \lambda_3\\ m_{12}\ m_{22}\\ m_{11}\end{array}\right\rangle = \sqrt{\tfrac{(p_{13}-p_{12}-1)(p_{13}-p_{22}-1)}{(p_{13}-p_{23}-1)(p_{13}-p_{33}-1)}}\left|\begin{array}{c}\lambda_1-1 \quad \lambda_2 \quad \lambda_3\\ m_{12}\ m_{22}\\ m_{11}\end{array}\right\rangle;
$$

$$t_{12}^\dagger\left|\begin{array}{ccc}\lambda_1 & \lambda_2 & \lambda_3\\ & m_{12}\ m_{22} & \\ & m_{11} & \end{array}\right\rangle$$
$$= -\sqrt{\frac{(p_{12}-p_{13})(p_{12}-p_{33})(p_{23}-p_{22}-1)}{(p_{12}-p_{22})(p_{23}-p_{13}-1)(p_{23}-p_{33}-1)}}\sqrt{\frac{p_{11}-p_{22}}{p_{12}-p_{22}-1}}\left|\begin{array}{ccc}\lambda_1 & \lambda_2-1 & \lambda_3\\ & m_{12}-1\ m_{22} & \\ & m_{11}-1 & \end{array}\right\rangle$$
$$-\sqrt{\frac{(p_{22}-p_{13})(p_{22}-p_{33})(p_{23}-p_{12}-1)}{(p_{22}-p_{12})(p_{23}-p_{13}-1)(p_{23}-p_{33}-1)}}\sqrt{\frac{p_{11}-p_{12}}{p_{22}-p_{12}-1}}\left|\begin{array}{ccc}\lambda_1 & \lambda_2-1 & \lambda_3\\ & m_{12}\ m_{22}-1 & \\ & m_{11}-1 & \end{array}\right\rangle,$$

$$t_{22}^\dagger\left|\begin{array}{ccc}\lambda_1 & \lambda_2 & \lambda_3\\ & m_{12}\ m_{22} & \\ & m_{11} & \end{array}\right\rangle$$
$$= -\sqrt{\frac{(p_{12}-p_{13})(p_{12}-p_{33})(p_{23}-p_{22}-1)}{(p_{12}-p_{22})(p_{23}-p_{13}-1)(p_{23}-p_{33}-1)}}\sqrt{\frac{p_{12}-p_{11}-1}{p_{12}-p_{22}-1}}\left|\begin{array}{ccc}\lambda_1 & \lambda_2-1 & \lambda_3\\ & m_{12}-1\ m_{22} & \\ & m_{11} & \end{array}\right\rangle$$
$$+\sqrt{\frac{(p_{22}-p_{13})(p_{22}-p_{33})(p_{23}-p_{12}-1)}{(p_{22}-p_{12})(p_{23}-p_{13}-1)(p_{23}-p_{33}-1)}}\sqrt{\frac{p_{22}-p_{11}-1}{p_{22}-p_{12}-1}}\left|\begin{array}{ccc}\lambda_1 & \lambda_2-1 & \lambda_3\\ & m_{12}\ m_{22}-1 & \\ & m_{11} & \end{array}\right\rangle,$$

$$t_{32}^\dagger\left|\begin{array}{ccc}\lambda_1 & \lambda_2 & \lambda_3\\ & m_{12}\ m_{22} & \\ & m_{11} & \end{array}\right\rangle = \sqrt{\frac{(p_{23}-p_{12}-1)(p_{23}-p_{22}-1)}{(p_{23}-p_{13}-1)(p_{23}-p_{33}-1)}}\left|\begin{array}{ccc}\lambda_1 & \lambda_2-1 & \lambda_3\\ & m_{12}\ m_{22} & \\ & m_{11} & \end{array}\right\rangle;$$

$$t_{13}^\dagger\left|\begin{array}{ccc}\lambda_1 & \lambda_2 & \lambda_3\\ & m_{12}\ m_{22} & \\ & m_{11} & \end{array}\right\rangle$$
$$= -\sqrt{\frac{(p_{12}-p_{13})(p_{12}-p_{23})(p_{33}-p_{22}-1)}{(p_{12}-p_{22})(p_{33}-p_{13}-1)(p_{33}-p_{23}-1)}}\sqrt{\frac{p_{11}-p_{22}}{p_{12}-p_{22}-1}}\left|\begin{array}{ccc}\lambda_1 & \lambda_2 & \lambda_3-1\\ & m_{12}-1\ m_{22} & \\ & m_{11}-1 & \end{array}\right\rangle$$
$$+\sqrt{\frac{(p_{22}-p_{13})(p_{22}-p_{23})(p_{33}-p_{12}-1)}{(p_{22}-p_{12})(p_{33}-p_{13}-1)(p_{33}-p_{23}-1)}}\sqrt{\frac{p_{11}-p_{12}}{p_{22}-p_{12}-1}}\left|\begin{array}{ccc}\lambda_1 & \lambda_2 & \lambda_3-1\\ & m_{12}\ m_{22}-1 & \\ & m_{11}-1 & \end{array}\right\rangle,$$

$$t^{\dagger}_{23}\left|\begin{array}{c}\lambda_1 \quad \lambda_2 \quad \lambda_3\\ m_{12}\; m_{22}\\ m_{11}\end{array}\right\rangle$$

$$= -\sqrt{\frac{(p_{12}-p_{13})(p_{12}-p_{23})(p_{33}-p_{22}-1)}{(p_{12}-p_{22})(p_{33}-p_{13}-1)(p_{33}-p_{23}-1)}}\sqrt{\frac{p_{12}-p_{11}-1}{p_{12}-p_{22}-1}}\left|\begin{array}{c}\lambda_1 \quad \lambda_2 \quad \lambda_3-1\\ m_{12}-1\; m_{22}\\ m_{11}\end{array}\right\rangle$$

$$-\sqrt{\frac{(p_{22}-p_{13})(p_{22}-p_{23})(p_{33}-p_{12}-1)}{(p_{22}-p_{12})(p_{33}-p_{13}-1)(p_{33}-p_{23}-1)}}\sqrt{\frac{p_{22}-p_{11}-1}{p_{22}-p_{12}-1}}\left|\begin{array}{c}\lambda_1 \quad \lambda_2 \quad \lambda_3-1\\ m_{12}\; m_{22}-1\\ m_{11}\end{array}\right\rangle,$$

$$t^{\dagger}_{33}\left|\begin{array}{c}\lambda_1 \quad \lambda_2 \quad \lambda_3\\ m_{12}\; m_{22}\\ m_{11}\end{array}\right\rangle = \sqrt{\frac{(p_{33}-p_{12}-1)(p_{33}-p_{22}-1)}{(p_{33}-p_{13}-1)(p_{33}-p_{23}-1)}}\left|\begin{array}{c}\lambda_1 \quad \lambda_2 \quad \lambda_3-1\\ m_{12}\; m_{22}\\ m_{11}\end{array}\right\rangle.$$

□

Each fundamental shift operator $t_{i\tau}$ is defined by giving its explicit action on the basis $\mathbf{B}_\lambda$ of each subspace $H_\lambda \subset H$. The form of this action is

$$t_{i\tau}\left|\begin{array}{c}\lambda\\ m\end{array}\right\rangle = \sum_{m'\in\mathbb{G}_{\lambda+e_\tau}}\left\langle\begin{array}{c}\lambda+e_\tau\\ m'\end{array}\right| t_{i\tau}\left|\begin{array}{c}\lambda\\ m\end{array}\right\rangle\left|\begin{array}{c}\lambda+e_\tau\\ m'\end{array}\right\rangle. \tag{6.7}$$

In the bra-ket notation for the coefficient in this expression, $\binom{\lambda}{m}$ is called the initial GT pattern and $\binom{\lambda+e_\tau}{m'}$ the final GT pattern.

For a given shift operator $t_{i\tau}$, the collection of all possible final patterns can be identified more precisely. We first define a triangular pattern of n rows containing only $0's$ and $1's$ by

$$\Delta(\tau_i,\ \tau_{i+1},\dots,\tau_{n-1},\tau) = \left(\begin{array}{c} e_\tau\\ e_{\tau_{n-1}}\\ \vdots\\ e_{\tau_i}\\ (0)_{i-1}\end{array}\right), \tag{6.8}$$

where $1\le\tau_k\le k$ for $k=i,i+1,\dots,n-1$; the symbol e_{τ_k} denotes a unit row matrix of length k with 1 in position τ_k and 0 elsewhere; and $(0)_{i-1}$ denotes a triangular array of zeros with $i-1$ rows. The length of a row matrix e_{τ_k} in the pattern (6.8) can always be identified by the row of the GT pattern in which it occurs. We call a triangular pattern of $0's$ and $1's$

of the form (6.8) a *triangular shift pattern.* For example, such a triangular shift pattern is given for $n = 5, i = 2, \tau_2 = 2, \tau_3 = 1, \tau_4 = 3, \tau = 2$ by

$$\Delta(2,1,3,2) = \begin{pmatrix} 0 \;\; 1 \;\; 0 \;\; 0 \;\; 0 \\ 0 \;\; 0 \;\; 1 \;\; 0 \\ 1 \;\; 0 \;\; 0 \\ 0 \;\; 1 \\ 0 \end{pmatrix}. \tag{6.9}$$

We define $\Delta_i(\tau)$ to be the set of all triangular shift patterns in which rows 1 through $i-1$ contain all $0's$, a single 1 occurs at any position in row i; at any position in row $i+1$, ..., at any position in row $n-1$; at fixed position τ in row n; and at position 1 for $i = 1$:

$$\Delta_i(\tau) = \left\{ \Delta(\tau_i, \ \tau_{i+1}, \ldots, \tau_{n-1}, \tau) \,\middle|\, \begin{array}{l} \tau_j = 1, 2, \ldots, j; \text{ each} \\ j = i, i+1, \ldots, n-1 \end{array} \right\}, \tag{6.10}$$

for each $i = 1, 2, \ldots, n-1$, where also $\Delta_n(\tau)$ is defined by

$$\Delta_n(\tau) = \left\{ \begin{pmatrix} e_\tau \\ (0)_{n-1} \end{pmatrix} \right\}. \tag{6.11}$$

The cardinality of the set $\Delta_i(\tau)$ is given by

$$|\Delta_i(\tau)| = (n-1)!/(i-1)!, \ \ i = 1, 2, \ldots, n. \tag{6.12}$$

Example. For $n = 3$, the following sets of triangular shift patterns correspond to the shifts of the GT patterns for the fundamental shift operators t_{11}, t_{21}, t_{31} :

$$\begin{aligned} \Delta_1(1) &= \left\{ \begin{pmatrix} 1\,0\,0 \\ 1\,0 \\ 1 \end{pmatrix}, \begin{pmatrix} 1\,0\,0 \\ 0\,1 \\ 1 \end{pmatrix} \right\}; \\ \Delta_2(1) &= \left\{ \begin{pmatrix} 1\,0\,0 \\ 1\,0 \\ 0 \end{pmatrix}, \begin{pmatrix} 1\,0\,0 \\ 0\,1 \\ 0 \end{pmatrix} \right\}; \\ \Delta_3(1) &= \left\{ \begin{pmatrix} 1\,0\,0 \\ 0\,0 \\ 0 \end{pmatrix} \right\}. \qquad \Box \end{aligned} \tag{6.13}$$

Triangular shift patterns are the basic objects from which certain directed graphs, called *arc digraphs,* are constructed in Sect. 6.3.

Relation (6.7) can now be expressed more precisely as

$$t_{i\tau}\left|\begin{array}{c}\lambda\\ m\end{array}\right\rangle=\sum_{\delta_\tau\in\Delta_i(\tau)}\left\langle\left(\begin{array}{c}\lambda\\ m\end{array}\right)+\delta_\tau\left|t_{i\tau}\right|\begin{array}{c}\lambda\\ m\end{array}\right\rangle\left|\left(\begin{array}{c}\lambda\\ m\end{array}\right)+\delta_\tau\right\rangle. \quad (6.14)$$

It is also convenient to define the shift operator T_{δ_τ}, each $\delta_\tau \in \Delta_i(\tau)$ by

$$T_{\delta_\tau}\left|\begin{array}{c}\lambda\\ m\end{array}\right\rangle=\left\langle\left(\begin{array}{c}\lambda\\ m\end{array}\right)+\delta_\tau\left|t_{i\tau}\right|\begin{array}{c}\lambda\\ m\end{array}\right\rangle\left|\left(\begin{array}{c}\lambda\\ m\end{array}\right)+\delta_\tau\right\rangle. \quad (6.15)$$

Relation (6.14) is now written as

$$t_{i\tau}\left|\begin{array}{c}\lambda\\ m\end{array}\right\rangle=\sum_{\delta_\tau\in\Delta_i(\tau)}T_{\delta_\tau}\left|\begin{array}{c}\lambda\\ m\end{array}\right\rangle. \quad (6.16)$$

A characteristic of the coefficient

$$\left\langle\left(\begin{array}{c}\lambda\\ m\end{array}\right)+\delta_\tau\left|t_{i\tau}\right|\begin{array}{c}\lambda\\ m\end{array}\right\rangle, \delta_\tau\in\Delta_i(\tau), \quad (6.17)$$

in which the final pattern is expressed in terms of the shift $\delta_\tau \in \Delta_i(\tau)$ of the initial pattern, is the following: If the final GT pattern does not satisfy betweenness, that is, is nonlexical, then the coefficient is zero, by definition. But, in fact, the coefficient itself carries this zero because a linear factor under the radical becomes zero exactly at the "boundary" where the betweenness conditions are violated. For example, the coefficient

$$\left\langle\begin{array}{c}\lambda_1\quad\lambda_2\quad\lambda_3+1\\ m_{12}+1\ m_{22}\\ m_{11}+1\end{array}\left|t_{13}\right|\begin{array}{c}\lambda_1\quad\lambda_2\quad\lambda_3\\ m_{12}\ m_{22}\\ m_{11}\end{array}\right\rangle \quad (6.18)$$
$$=-\sqrt{\frac{(p_{12}-p_{13}+1)(p_{12}-p_{23}+1)(p_{33}-p_{22})}{(p_{12}-p_{22}+1)(p_{33}-p_{13})(p_{33}-p_{23})}}\sqrt{\frac{p_{11}-p_{22}+1}{p_{12}-p_{22}}}$$

is zero at $m_{12}=m_{13}$ $(p_{12}=p_{13}-1)$ and at $m_{22}=m_{33}$ $(p_{22}=p_{33})$, which, for given lexical initial GT pattern, are exactly the boundary values at which the final GT patterns becomes nonlexical, that is, betweenness is violated. This feature is general.

We are now in position to define the general coefficient in (6.14), where it is now convenient to set $\tau=\tau_n$. The coefficient is given by a product of factors that reflect its dependence on pairs of contiguous rows in the GT pattern $\binom{\lambda}{m}$:

$$\left\langle \left(\begin{matrix} \lambda \\ m \end{matrix} \right) + \Delta(\tau_i, \tau_{i+1}, \ldots, \tau_n) \,\middle|\, t_{i\tau_n} \,\middle|\, \begin{matrix} \lambda \\ m \end{matrix} \right\rangle = \prod_{l=i}^{n} A_{\tau_{l-1},\tau_l}(\mathbf{p}_{l-1};\mathbf{p}_l); \tag{6.19}$$

$$A_{\tau_{l-1},\tau_l}(\mathbf{p}_{l-1};\mathbf{p}_l) = S(\tau_{l-1} - \tau_l) \times \left[\prod_{\substack{j=1 \\ j\neq\tau_l}}^{l} \frac{p_{\tau_{l-1},l-1} - p_{j,l} + 1}{p_{\tau_l,l} - p_{j,l}} \prod_{\substack{k=1 \\ k\neq\tau_{l-1}}}^{l-1} \frac{p_{\tau_l,l} - p_{k,l-1}}{p_{\tau_{l-1},l-1} - p_{k,l-1} + 1} \right]^{1/2}, \quad \text{for } i+1 \leq l \leq n; \tag{6.20}$$

$$A_{i,\tau_i}(\mathbf{p}_{i-1};\mathbf{p}_i) = \left[\frac{\displaystyle\prod_{k=1}^{i-1} (p_{\tau_i,i} - p_{k,i-1})}{\displaystyle\prod_{\substack{j=1 \\ j\neq\tau_i}}^{i} (p_{\tau_i,i} - p_{j,i})} \right]^{1/2}, \quad \text{for } l = i, \text{ in which case } \tau_{i-1} = i. \tag{6.21}$$

In these relations, we have $\mathbf{p}_j = (p_{1,j}, p_{2,j}, \ldots, p_{j,j}), j = 1, 2, \ldots, n$, where the $p_{i,j}$ are the so-called *partial hooks* defined in terms of the entries of GT pattern $\binom{\lambda}{m}$ by $p_{i,j} = m_{i,j} + j - i$. The symbol $S(i - j)$ denotes the sign of $i - j$ with $S(0) = 1$, and, by definition, the factor (6.21) is 1 for $i = 1$.

The action of the Hermitian conjugate operators $t^{\dagger}_{i\tau}$ is similarly given:

$$t^{\dagger}_{i\tau} \left| \begin{matrix} \lambda \\ m \end{matrix} \right\rangle = \sum_{\delta_\tau \in \Delta_i(\tau)} T^{\dagger}_{\delta_\tau} \left| \begin{matrix} \lambda \\ m \end{matrix} \right\rangle \tag{6.22}$$

$$= \sum_{\delta_\tau \in \Delta_i(\tau)} \left\langle \left(\begin{matrix} \lambda \\ m \end{matrix} \right) - \delta_\tau \,\middle|\, t^{\dagger}_{i\tau} \,\middle|\, \begin{matrix} \lambda \\ m \end{matrix} \right\rangle \left| \left(\begin{matrix} \lambda \\ m \end{matrix} \right) - \delta_\tau \right\rangle,$$

where the coefficients are defined as follows:

$$\left\langle \left(\begin{array}{c} \lambda \\ m \end{array} \right) - \Delta(\tau_i, \tau_{i+1}, \ldots, \tau_n) \,\middle|\, t^\dagger_{i\tau_n} \,\middle|\, \begin{array}{c} \lambda \\ m \end{array} \right\rangle$$

$$= \prod_{l=i}^{n} A^\dagger_{\tau_{l-1},\tau_l}(\mathbf{p}_{l-1};\mathbf{p}_l); \tag{6.23}$$

$$A^\dagger_{\tau_{l-1},\tau_l}(\mathbf{p}_{l-1};\mathbf{p}_l) = S(\tau_{l-1} - \tau_l)$$

$$\times \left[\prod_{\substack{j=1 \\ j \neq \tau_l}}^{l} \frac{p_{\tau_{l-1},l-1} - p_{j,l}}{p_{\tau_l,l} - p_{j,l} - 1} \prod_{\substack{k=1 \\ k \neq \tau_{l-1}}}^{l-1} \frac{p_{\tau_l,l} - p_{k,l-1} - 1}{p_{\tau_{l-1},l-1} - p_{k,l-1}} \right]^{1/2},$$

$$\text{for } i+1 \leq l \leq n; \tag{6.24}$$

$$A^\dagger_{i,\tau_i}(\mathbf{p}_{i-1};\mathbf{p}_i) = \left[\frac{\prod_{k=1}^{i-1} (p_{\tau_i,i} - p_{k,i-1} - 1)}{\prod_{\substack{j=1 \\ j \neq \tau_i}}^{i} (p_{\tau_i,i} - p_{j,i} - 1)} \right]^{1/2},$$

$$\text{for } l = i, \text{ in which case } \tau_{i-1} = i. \tag{6.25}$$

For $i = 1$, by definition, the factor (6.25) is 1.

The above relations define completely the action of the fundamental shift operators $t_{i\tau}$ in the Hilbert space H. While presented here in an ad hoc fashion, we demonstrate below the fundamental origin of these operators in terms of diagraphs and prove their basic properties.

6.3 Arc Digraph Interpretation of the Fundamental Shift Operators

Each expression (6.20) and (6.21) for $A_{\rho,\sigma}(\mathbf{p}_{k-1}, \mathbf{p}_k)$ for each pair $\rho, \sigma = 1, 2, \ldots, k$ has a simple interpretation in terms of a labeled arc digraph $\vec{G}_k(\rho, \sigma), k \geq 2$. We begin with a graph $G_k(\rho, \sigma)$ having labeled points and lines as follows:

1. The $2k-1$ labeled points P of the graph are given by

$$P = \{x_1, x_2, \ldots, x_{k-1}, y_1, y_2, \ldots, y_k\}. \tag{6.26}$$

2. The $2(2k-3)$ edges $E_{\rho,\sigma}$ of the graph for selected $\rho, \sigma \in \{1, 2, \ldots, k\}$ and $\rho < k$ are the ordered pairs of points given by

$$E_{\rho,\sigma} = \left\{ \begin{array}{l|l} (x_\rho, x_i), (y_\sigma, x_i), & i = 1, 2, \ldots, k-1; i \neq \rho; \\ (x_\rho, y_j), (y_\sigma, y_j) & j = 1, 2, \ldots, k; j \neq \sigma \end{array} \right\}; \tag{6.27}$$

The $2(k-1)$ edges of the graph $E_{k,\sigma}$ of the graph for selected $\sigma \in \{1, 2, \ldots, k\}$ are the ordered pairs of points given by

$$E_{k,\sigma} = \left\{ \begin{array}{l|l} (y_\sigma, x_i), & i = 1, 2, \ldots, k-1; \\ (y_\sigma, y_j) & j = 1, 2, \ldots, k; j \neq \sigma \end{array} \right\}. \tag{6.28}$$

The arc digraph $\vec{G}_k(\rho, \sigma)$ is now obtained by assigning an arrow to each edge of the graph, where the direction of the arrow for each defined edge is given by

$$x_\rho \to x_i, x_\rho \to y_j; y_\sigma \to y_j, y_\sigma \to x_i. \tag{6.29}$$

Examples: The following diagrams illustrate these rules for three of the nine cases for $k = 3$. The points x_ρ and y_σ are represented by a $\bullet$ and points $x_i, i \neq \rho$, and $y_j, j \neq \sigma$, by a $\circ$:

$\vec{G}_3(1,1) =$ 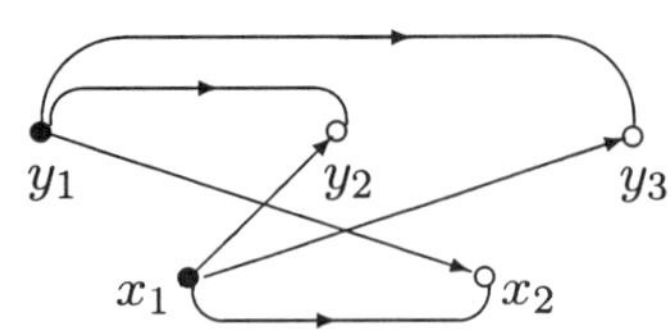

$\vec{G}_3(1,2) =$ 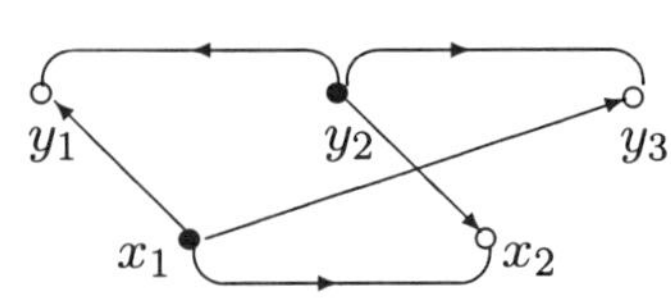

$\vec{G}_3(3,2) =$ 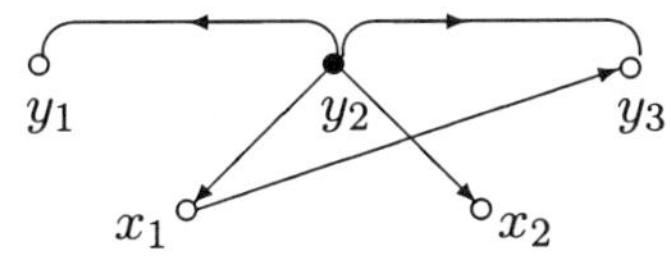

□

The relationship of the set $\left\{\vec{G}_k(\rho,\sigma) \mid \rho,\sigma = 1,2,\ldots,k\right\}$ of k^2 arc diagraphs to the properties of the set of triangular shift patterns $\Delta_i(\tau)$ originates from the contiguous rows $k-1$ and k in $\Delta_i(\tau)$, which has as shifts the pair of row matrices $e_\sigma = e_\sigma^{(k)}$ of length k and $e_\rho = e_\rho^{(k-1)}$ of length $k-1$ as depicted by

$$\begin{array}{lccccccccccc} e_\sigma^{(k)} = & 0 & 0 & \cdots & 0 & \cdots & 1 & \cdots & 0 & 0 \\ e_\rho^{(k-1)} = & & 0 & 0 & \cdots & 1 & \cdots & 0 & \cdots & 0 \end{array} \tag{6.30}$$

The 1 in the top row is in position σ, and the 1 in the bottom row is in position ρ, where for $\rho = k$ only 0 appears. In the digraph $G_k(\rho,\tau), k \geq 2$, each 0 is represented by a point $\circ$, each 1 by a point $\bullet$, and a directed arc goes from each $\bullet$ point to each $\circ$ point. The points of the arc digraph are then labeled as described above. Thus, the general labeled arc digraph $\vec{G}_k(\rho,\sigma)$ has the following picture:

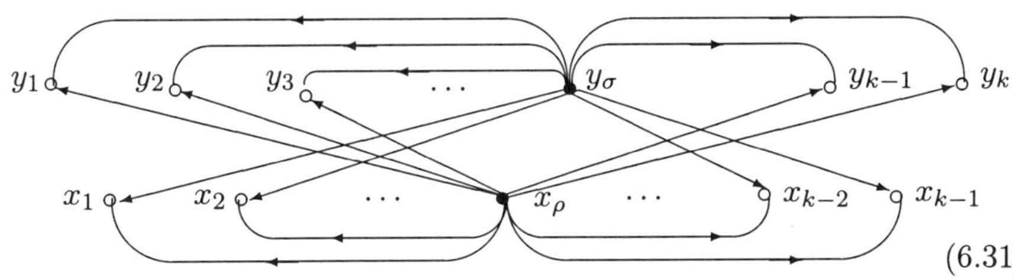

(6.31)

All lines are directed from $\bullet$ to $\circ$. For $\rho = k$, the $\bullet$ point labeled by x_ρ in the bottom is absent; all lines flow out of the single $\bullet$ point in the top row.

We now associate a function $A_{\rho,\sigma}(x;y), x = (x_1, x_2, \ldots, x_{k-1})$, $y = (y_1, y_2, \ldots, y_k)$ with the labeled arc digraph $\vec{G}_k(\rho,\sigma), k \geq 2$ by the following rules:

1. The edge (x_ρ, y_j) is assigned the linear factor $x_\rho - y_j + 1$,
2. The edge (x_ρ, x_i) is assigned the linear factor $x_\rho - x_i + 1$,

3. The edge (y_σ, x_i) is assigned the linear factor $y_\sigma - x_i$,
4. The edge (y_σ, y_j) is assigned the linear factor $y_\sigma - y_j$. (6.32)

These factors apply for each $i = 1, 2, \ldots, k-1$ with $i \neq \rho$, and each $j = 1, 2, \ldots, k$ with $j \neq \sigma$. The last rule assembles these factors, as follows: The factors associated with the edges going between rows are called *numerator factors* and are multiplied together to form the numerator $N_{\rho,\sigma}(x;y)$. The factors associated with the edges going within each row are called *denominator factors* and are multiplied together to form the denominator $D_{\rho,\sigma}(x;y)$. The function $A_{\rho,\sigma}(x;y)$ is then defined by

$$\begin{aligned} A_{\rho,\sigma}(x;y) &= S(\rho-\sigma)\sqrt{\frac{N_{\rho,\sigma}(x;y)}{D_{\rho,\sigma}(x;y)}} = \\ &= S(\rho-\sigma)\sqrt{\prod_{\substack{j=1\\ j\neq\sigma}}^{k} \frac{x_\rho - y_j + 1}{y_\sigma - y_j} \prod_{\substack{i=1\\ i\neq\rho}}^{k-1} \frac{y_\sigma - x_i}{x_\rho - x_i + 1}}, \\ &\qquad \text{for } \rho \leq k-1, \end{aligned} \tag{6.33}$$

$$\begin{aligned} A_{k,\sigma}(x;y) &= S(k-\sigma)\sqrt{\frac{N_{k,\sigma}(x;y)}{D_{k,\sigma}(x;y)}} \\ &= S(k-\sigma)\sqrt{\prod_{i=1}^{k-1}(y_\sigma - x_i) \Big/ \prod_{\substack{j=1\\ j\neq\sigma}}^{k}(y_\sigma - y_j)}, \\ &\qquad \text{for } \rho \leq k, \end{aligned} \tag{6.34}$$

where $S(\rho - \sigma) = \text{sign of}\,(\rho - \sigma)$ $(= 1 \text{ for } \rho = \sigma)$.

The shifts $x \to x - e_\rho, y \to y - e_\sigma$ in (6.33)-(6.34) give the conjugate functions:

$$\begin{aligned} A^\dagger_{\rho,\sigma}(x;y) &= S(\rho-\sigma)\sqrt{\prod_{\substack{j=1\\ j\neq\sigma}}^{k} \frac{x_\rho - y_j}{y_\sigma - y_j - 1} \prod_{\substack{i=1\\ i\neq\rho}}^{k-1} \frac{y_\sigma - x_i - 1}{x_\rho - x_i}}, \\ &\qquad \text{for } \rho \leq k-1; \end{aligned} \tag{6.35}$$

$$\begin{aligned} A^\dagger_{k,\sigma}(x;y) &= S(k-\sigma)\sqrt{\prod_{i=1}^{k-1}(y_\sigma - x_i - 1) \Big/ \prod_{\substack{j=1\\ j\neq\sigma}}^{k}(y_\sigma - y_j - 1)}, \\ &\qquad \text{for } \rho \leq k. \end{aligned} \tag{6.36}$$

The signs of the linear factors appearing under the radical in relations (6.33)-(6.36) are important. In the applications of these relations to be made, row k is always identified with the partial hooks associated with the partition occurring in row k of a GT pattern, and row $k-1$ with the partial hooks associated with the partition occurring in row $k-1$, so that $y_j = p_{j,k} = m_{j,k} + k - j, j = 1, 2, \ldots, k$; $x_i = p_{i,k-1} = m_{i,k-1} + k - 1 - i, i = 1, 2, \ldots, k-1$. The betweenness conditions on these two contiguous rows of a GT pattern regulate the signs of the linear factors in (6.33)-(6.36). As written, some factors are negative, some are positive, and some can be zero for certain GT patterns, but the product of factors is, however, always positive: The number of negative factors in the numerator always equals the number of negative factors in the denominator. Careful accounting of negative factors is necessary to avoid errors. The complex number $\sqrt{-1}$ is not to be introduced.

Examples: We list below the application of the rules (6.31)-(6.32) to the functions defined above for all arc digraphs for $n = 2, 3$:

$\vec{G}_2(1,1) =$ 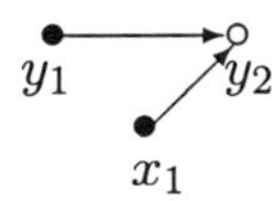

$$A_{1,1}(x;y) = \sqrt{\frac{x_1 - y_2 + 1}{y_1 - y_2}}$$

$\vec{G}_2(2,1) =$ 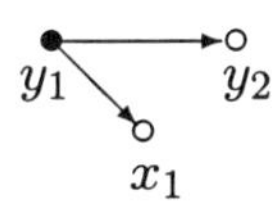

$$A_{2,1}(x;y) = \sqrt{\frac{y_1 - x_1}{y_1 - y_2}}$$

$\vec{G}_2(1,2) =$ 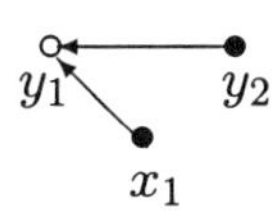

$$A_{1,2}(x;y) = -\sqrt{\frac{x_1 - y_1 + 1}{y_2 - y_1}}$$

$\vec{G}_2(2,2) =$ 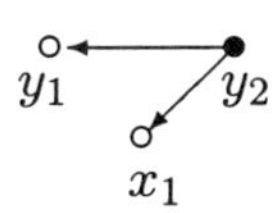

$$A_{2,2}(x;y) = \sqrt{\frac{y_2 - x_1}{y_2 - y_1}}$$

$\vec{G}_3(1,1) =$ 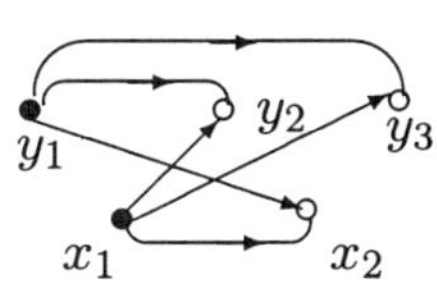

$$A_{1,1}(x;y) = \sqrt{\frac{(x_1 - y_2 + 1)(x_1 - y_3 + 1)(y_1 - x_2)}{(x_1 - x_2 + 1)(y_1 - y_2)(y_1 - y_3)}}$$

$\vec{G}_3(1,2) =$ 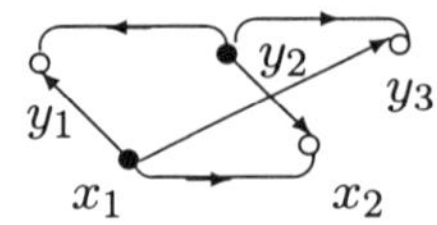 $A_{1,2}(x;y) = -\sqrt{\frac{(x_1-y_1+1)(x_1-y_3+1)(y_2-x_2)}{(x_1-x_2+1)(y_2-y_1)(y_2-y_3)}}$

$\vec{G}_3(1,3) =$ 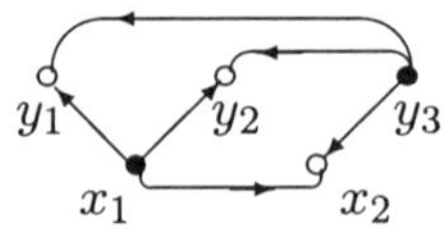 $A_{1,3}(x;y) = -\sqrt{\frac{(x_1-y_1+1)(x_1-y_2+1)(y_3-x_2)}{(x_1-x_2+1)(y_3-y_1)(y_3-y_2)}}$

$\vec{G}_3(2,1) =$ 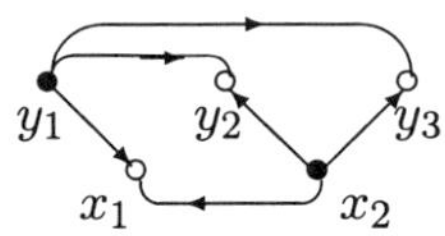 $A_{2,1}(x;y) = \sqrt{\frac{(x_2-y_2+1)(x_2-y_3+1)(y_1-x_1)}{(x_2-x_1+1)(y_1-y_2)(y_1-y_3)}}$

$\vec{G}_3(2,2) =$ 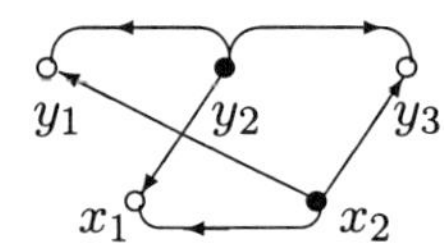 $A_{2,2}(x;y) = \sqrt{\frac{(x_2-y_1+1)(x_2-y_3+1)(y_2-x_1)}{(x_2-x_1+1)(y_2-y_1)(y_2-y_3)}}$

$\vec{G}_3(2,3) =$ 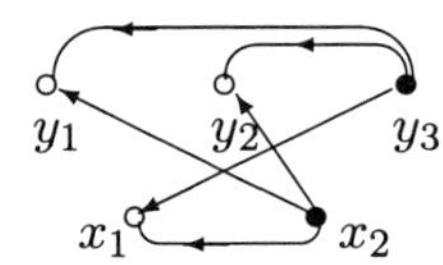 $A_{2,3}(x;y) = -\sqrt{\frac{(x_2-y_1+1)(x_2-y_2+1)(y_3-x_1)}{(x_2-x_1+1)(y_3-y_1)(y_3-y_2)}}$

$\vec{G}_3(3,1) =$ 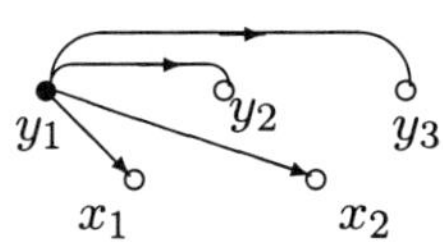 $A_{3,1}(x;y) = \sqrt{\frac{(y_1-x_1)(y_1-x_2)}{(y_1-y_2)(y_1-y_3)}}$

$\vec{G}_3(3,2) =$ 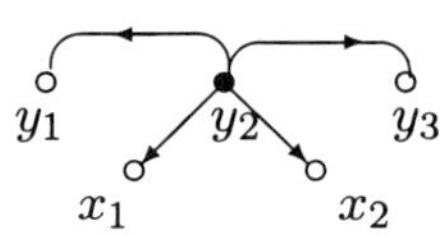 $A_{3,2}(x;y) = \sqrt{\frac{(y_2-x_1)(y_2-x_2)}{(y_2-y_1)(y_2-y_3)}}$

$\vec{G}_3(3,3) =$ 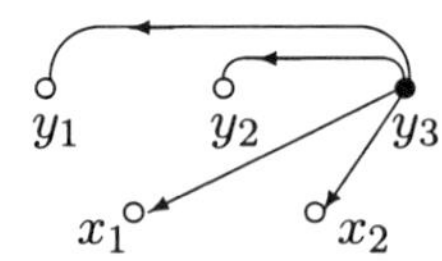 $A_{3,3}(x;y) = \sqrt{\frac{(y_3-x_1)y_3-x_2)}{(y_3-y_1)(y_3-y_2)}}$

□

6.4 Algebra of the Fundamental Shift Operators

In this section, we develop some preliminary properties of the algebra of the set of fundamental shift operators $\mathbb{T} = \{t_{i\tau} \mid i, \tau = 1, 2, \ldots, n\}$, but a more complete discussion must await developments in the theory of tensor operators in Chapter 9. For now, these operators are linear operators over the field of complex numbers on the model Hilbert space H, and their products are to be regarded as mappings $H \to H$ with action determined by repeated use of the defining relations (6.7) and (6.19)-(6.21)

One sees rather easily from the explicit results for $n = 1, 2, 3$ that the product $t_{i\tau}t_{j\rho}$ is not a fundamental shift operator in the set $\mathbb{T}$, but rather is a mapping $H_\lambda \to H_{\lambda+e_\rho+e_\tau}$. However, $t_{i\tau}t^\dagger_{j\tau}$ and $t^\dagger_{j\tau}t_{i\tau}$ are mappings of $H_\lambda \to H_\lambda$, each $\lambda \in \mathbb{P}ar_n$. Three significant properties of product actions of fundamental shift operators are the following (verified directly for $n = 2, 3$) :

1. Commutivity of shift operators with the same τ :
$$[t_{i\tau}, t_{j\tau}] = 0,\ [\mathrm{t}^\dagger_{\mathrm{i}\tau}, \mathrm{t}^\dagger_{\mathrm{j}\tau}] = 0. \tag{6.37}$$
This property is a consequence of the identity
$$T_{\delta_\tau}T_{\delta'_\tau}\left| \begin{matrix}\lambda\\ m\end{matrix} \right\rangle = T_{\delta'_\tau}T_{\delta_\tau}\left| \begin{matrix}\lambda\\ m\end{matrix} \right\rangle \tag{6.38}$$
$$= (factor)\left| \begin{pmatrix}\lambda\\ m\end{pmatrix} + \delta_\tau + \delta'_\tau \right\rangle,\ \text{each}\ \delta_\tau \in \Delta_i(\tau), \delta'_\tau \in \Delta_j(\tau).$$
The proof of this property is by direct verification. Summing this relation over all $\delta_\tau \in \Delta_i(\tau)$ and $\delta'_\tau \in \Delta_j(\tau)$ gives the desired result, using relation (6.16)

2. Column orthogonality: The basic relation is the *operator inner product* property expressed by
$$\sum_{\tau=1}^{n} t^\dagger_{i\tau}t_{j\tau} = \delta_{i,j}\mathbf{1}, \tag{6.39}$$
where $\mathbf{1}$ is the identity operator on H. Taking matrix elements of this relation, we obtain the column orthogonality relations:
$$\sum_{\tau=1}^{n}\sum_{m''\in\mathbb{G}_{\lambda+e_\tau}} \left\langle \begin{matrix}\lambda+e_\tau\\ m''\end{matrix} \right| t_{i\tau} \left| \begin{matrix}\lambda\\ m'\end{matrix} \right\rangle$$
$$\times \left\langle \begin{matrix}\lambda+e_\tau\\ m''\end{matrix} \right| t_{j\tau} \left| \begin{matrix}\lambda\\ m\end{matrix} \right\rangle = \delta_{i,j}\delta_{m',m}. \tag{6.40}$$

3. Row orthogonality: The basic relation is the *operator completeness* property expressed by

$$\sum_{i=1}^{n} t_{i\tau} t^{\dagger}_{i\rho} = \delta_{\tau\rho} \mathbf{N}_{\rho}, \tag{6.41}$$

where $\mathbf{N}_{\rho}$ is defined by its action on each subspace $H_{\lambda} \subset H$:

$$\mathbf{N}_{\rho} H_{\lambda} = \mathbf{0}, \text{ if and only if } t^{\dagger}_{i\rho} H_{\lambda} = \mathbf{0},\ i = 1, 2, ..., n. \tag{6.42}$$

This relation, in turn, implies the row orthogonality relations:

$$\sum_{i=1}^{n} \sum_{m \in \mathbb{G}_{\lambda}} \left\langle \begin{array}{c} \lambda + e_{\tau} \\ m' \end{array} \middle| t_{i\tau} \middle| \begin{array}{c} \lambda \\ m \end{array} \right\rangle \times \left\langle \begin{array}{c} \lambda + e_{\rho} \\ m'' \end{array} \middle| t_{i\rho} \middle| \begin{array}{c} \lambda \\ m \end{array} \right\rangle = \delta_{\tau,\rho} \delta_{m',m''}. \tag{6.43}$$

Proof. The proof of the basic relation (6.40) is by induction on n, using the matrix elements given by relations (6.19)-(6.25). For the purpose of this proof, we now write $t_{i\tau}$ as $t^{(n)}_{i\tau}$, etc. to make the role of n explicit. Relation (6.19) gives

$$\left\langle \begin{array}{c} \lambda + e_{\tau} \\ \mu + e_{\rho} \\ m' \end{array} \middle| t^{(n)}_{i\tau} \middle| \begin{array}{c} \lambda \\ \mu \\ m \end{array} \right\rangle = A_{\rho\tau}(x;y) \left\langle \begin{array}{c} \mu + e_{\rho} \\ m' \end{array} \middle| t^{(n-1)}_{i\rho} \middle| \begin{array}{c} \mu \\ m \end{array} \right\rangle, \tag{6.44}$$

in which now the indeterminates x and y are given by the partial hooks:

$$\begin{aligned} x &= (\mu_1 + n - 2, \mu_2 + n - 3, \ldots, \mu_{n-1}), \\ y &= (\lambda_1 + n - 1, \lambda_2 + n - 2, \ldots, \lambda_n). \end{aligned} \tag{6.45}$$

The factor $A_{\rho\tau}(x;y)$ is exactly the function (6.33)-(6.34) defined on the labeled arc digraph $\vec{G}_n(\rho, \tau)$. The *factorization property* (6.44) of the matrix elements of a fundamental shift operator at level n into the product $A_{\rho\tau}(x;y)$ times a fundamental shift operator at level $n-1$ is one of the basic properties of fundamental shift operators. The induction proof proceeds as follows. We assume that relation (6.40) is true at level $n-1$; that is, for n replaced by $n-1$. The relation is true for $n-1 = 1, 2$, as verified from the explicit relations given above. Then, the orthogonality relations (6.40) are true at level n, if and only if

$$\sum_{\tau=1}^{n} A_{\rho\tau}(x';y)A_{\sigma\tau}(x;y) = \delta_{\rho,\sigma}, \tag{6.46}$$

where $x' = x + e_\sigma^{(n-1)} - e_\rho^{(n-1)}, e_n^{(n-1)} = (0,\ldots,0)$.

The validity of (6.46) is a consequence of Sylvester's identity, which is discussed in Sect. 11.7, Compendium B; it is implemented in the present case as follows: We substitute from (6.33)-(6.34) the explicit expressions for $A_{\rho\tau}(x';y)$ and $A_{\sigma\tau}(x;y)$ into the left-hand side of relation (6.43). For $\sigma \neq \rho$, hence, $x'_\sigma = x_\sigma + 1, x'_\rho = x_\rho - 1, x'_i = x_i, i \neq \sigma, \rho$, relation (6.46) is the following:

$$\sqrt{factor}\sum_{\tau=1}^{n} \prod_{\substack{i=1\\ i\neq\sigma,\rho}}^{n} (y_\tau - x_i)\Big/ \prod_{\substack{j=1\\ j\neq\tau}}^{n}(y_\tau - y_j) = 0, \tag{6.47}$$

the summation to 0 being a direct consequence of Sylvester's identity. For $\sigma = \rho$, hence, $x'_i = x_i$, and $\rho < n$, we find

$$\sum_{\tau=1}^{n} (A_{\rho\tau}(x;y))^2 \tag{6.48}$$

$$= \frac{\prod_{j=1}^{n}(x_\rho - y_j + 1)}{\prod_{\substack{i=1\\ i\neq\rho}}^{n-1}(x_\rho - x_i + 1)} \sum_{\tau=1}^{n} \frac{\prod_{\substack{i=1\\ i\neq\rho}}^{n-1}(y_\tau - x_i)}{(x_\rho - y_\tau + 1)\prod_{\substack{j=1\\ j\neq\tau}}^{n}(y_\tau - y_j)} = 1.$$

To show that the sum is 1, we rewrite the summation term $\tau = 1$ to $\tau = n$ by setting $x_\rho + 1 = y_{n+1}$, extending the summation from $\tau = 1$ to $\tau = n+1$ and subtracting out the $\tau = n+1$ term. This gives for the summation term in (6.48) the result:

$$-\sum_{\tau=1}^{n+1} \frac{\prod_{\substack{i=1\\ i\neq\rho}}^{n-1}(y_\tau - x_i)}{\prod_{\substack{j=1\\ j\neq\tau}}^{n+1}(y_\tau - y_j)} + \frac{\prod_{\substack{i=1\\ i\neq\rho}}^{n-1}(x_\rho - x_i + 1)}{\prod_{j=1}^{n}(x_\rho - y_j + 1)}. \tag{6.49}$$

The first term is 0 by Sylvester's relation for $n+1$, and the second term is the reciprocal of the multiplying factor in front of the summation in (6.48), which proves the relation. For $\rho = n$, the above proof is modified to

$$\sum_{\tau=1}^{n} (A_{n\tau}(x;y))^2 = \sum_{\tau=1}^{n} \frac{\prod_{i=1}^{n-1}(y_\tau - x_i)}{\prod_{\substack{j=1 \\ j\neq\tau}}^{n}(y_\tau - y_j)} = 1. \tag{6.50}$$

The "trick" of extending the summation to $n+1$ in (6.48) because of the occurrence of an extra $\tau-$dependent factor arising in the denominator is used several times in subsequent proofs, sometimes without comment.

The proof of (6.43) is again by induction on n, following the proof of (6.40), where we need now to show that

$$\sum_{\rho=1}^{n} A_{\rho\tau}(x - e_\rho^{(n-1)}; y)\, A_{\rho\sigma}(x - e_\rho^{(n-1)}; y)$$
$$= \delta_{\tau,\sigma},\ \tau, \sigma = 1, 2, \ldots, n. \tag{6.51}$$

The proof of this relation, using Sylvester's identity, is similar to the proof of (6.46). □

The orthogonal coefficients

$$\left\langle \begin{matrix} \lambda + e_\tau \\ m' \end{matrix} \middle| t_{i\tau} \middle| \begin{matrix} \lambda \\ m \end{matrix} \right\rangle \tag{6.52}$$

are combinatorial objects defined fully as functions on labeled arc digraphs. It is these explicit coefficients that are the elements (5.30)-(5.31) of the real orthogonal matrix R^λ that is used to define the $D^\lambda-$matrices in (5.26).

6.5 Abstract Hilbert Space and $D^\lambda-$Polynomials

The purpose of this section is to relate the inner product (,) of polynomials, such as (5.12), to an inner product in the abstract Hilbert space H, defined by (6.4), in which the fundamental tensor operators $t_{i\tau}$ act. In physics, it is the custom to use orthonormalized bra-ket basis vectors in order to have some semblance of a uniform convention in presenting the action of operators acting in the spaces spanned by such bra-ket basis vectors. Because one of the important properties of the $D^\lambda-$polynomials is their relationship to representation theory of groups, they have not been normalized to unity, but rather to the invariant factor $M(\lambda)$ defined by (5.10)-(5.11), which gives $D^\lambda(I_n) = I_{\mathrm{Dim}\lambda}$. For the purposes of this section, it is convenient to normalize the D^λ–polynomials to unity in the inner product (5.12) by multiplying by $(M(\lambda))^{-1/2}$:

$$\widehat{D}^\lambda(Z) = (M(\lambda))^{-1/2} D^\lambda(Z). \tag{6.53}$$

The D^λ–polynomials carry two sets of labels, the lower GT pattern and the upper GT pattern $\binom{\lambda}{m}$ and $\binom{\lambda}{m'}$, each of which enumerates abstract basis vectors $\left| \begin{matrix} \lambda \\ m \end{matrix} \right\rangle$ and $\left| \begin{matrix} \lambda \\ m' \end{matrix} \right\rangle$ in the Hilbert space H defined by (6.4). What is needed to encode the same information in the abstract vector spaces as is encoded in the polynomial vector spaces spanned by the orthonormal vectors $\widehat{D}^\lambda_{m,m'}(Z)$ is a tensor product of spaces H such that each subspace H_λ in (6.4) shares the same partition $\lambda \in \mathbb{P}\mathrm{ar}_n$. For this analog, we introduce the concept of a *diagonal tensor product space:* The diagonal tensor product space of H is denoted $H_d = H \otimes_d H$; it is defined to be the direct sum of vector spaces given by

$$H_d = H \otimes_d H = \sum_{p=0}^{\infty} \sum_{\lambda \in \mathbb{P}\mathrm{ar}_n(p)} \oplus \left(H_\lambda \otimes_d H_\lambda \right), \tag{6.54}$$

where $H_\lambda \otimes_d H_\lambda$ is defined to be the finite-dimensional vector space of dimension $(\mathrm{Dim}\lambda)^2$ with orthonormal basis

$$\mathbf{B}_\lambda \otimes_d \mathbf{B}_\lambda = \left\{ \left| \begin{matrix} m' \\ \lambda \\ m \end{matrix} \right\rangle = \left| \begin{matrix} \lambda \\ m \end{matrix} \right\rangle \otimes_d \left| \begin{matrix} \lambda \\ m' \end{matrix} \right\rangle \middle| m, m' \in \mathbb{G}_\lambda \right\}. \tag{6.55}$$

This is a subspace of the tensor operator space $H \otimes H$ introduced in Sect. 10.5, Compendium A, which is obtained by restricting the basis $\mathbf{B}_\lambda \otimes \mathbf{B}_{\lambda'}$ to partitions $\lambda' = \lambda$, which is the meaning of the symbol $\otimes_d$. Such diagonal tensor operator spaces are linear vector spaces that have all the characteristics of the parent general tensor product space.

The correspondence between normalized D^λ–polynomials and abstract bra-ket basis vectors given by

$$\widehat{D} \begin{pmatrix} m' \\ \lambda \\ m \end{pmatrix} (Z) \rightarrow \left| \begin{matrix} m' \\ \lambda \\ m \end{matrix} \right\rangle \tag{6.56}$$

is to satisfy the inner product rule:

$$\left(\widehat{D}\begin{pmatrix} m''' \\ \lambda \\ m'' \end{pmatrix}(Z),\, \widehat{D}\begin{pmatrix} m' \\ \lambda \\ m \end{pmatrix}(Z)\right) = \left\langle \begin{matrix} m''' \\ \lambda \\ m'' \end{matrix} \middle| \begin{matrix} m' \\ \lambda \\ m \end{matrix} \right\rangle = \delta_{m'',m}\delta_{m''',m'}. \tag{6.57}$$

In this section, we extend this relation to general matrix elements of polynomials in the elements z_{ij} of Z in the left-hand side.

We first have the following transcription of relations (5.35) and (5.37) for the matrix elements of z_{ij} and $\partial/\partial z_{ij}$ on a normalized $\widehat{D}^{\lambda}$−polynomial:

$$\left(\widehat{D}\begin{pmatrix} m''' \\ \mu \\ m'' \end{pmatrix}(Z), z_{ij}\widehat{D}\begin{pmatrix} m' \\ \lambda \\ m \end{pmatrix}(Z)\right) = \left\langle \begin{matrix} m''' \\ \mu \\ m'' \end{matrix} \middle| \theta_{ij} \middle| \begin{matrix} m' \\ \lambda \\ m \end{matrix} \right\rangle, \tag{6.58}$$

$$\left(\widehat{D}\begin{pmatrix} m''' \\ \mu \\ m'' \end{pmatrix}(Z), \frac{\partial}{\partial z_{ij}}\widehat{D}\begin{pmatrix} m' \\ \lambda \\ m \end{pmatrix}(Z)\right) = \left\langle \begin{matrix} m''' \\ \mu \\ m'' \end{matrix} \middle| \theta^{\dagger}_{ij} \middle| \begin{matrix} m' \\ \lambda \\ m \end{matrix} \right\rangle, \tag{6.59}$$

where we have defined new " coordinates" and "derivatives" in the space $H \otimes_d H$ by

$$\theta_{ij} = M^{1/2}_{\text{op}}\left(\sum_{\tau=1}^{n}(t_{i\tau} \otimes_d t_{j\tau})\right) M^{-1/2}_{\text{op}}; \tag{6.60}$$

$$\theta^{\dagger}_{ij} = M^{-1/2}_{\text{op}}\left(\sum_{\tau=1}^{n}(t^{\dagger}_{i\tau} \otimes_d t^{\dagger}_{j\tau})\right) M^{1/2}_{\text{op}}. \tag{6.61}$$

In these relations, the quantity M_{op} is an operator in H_{λ} with eigenvalue given by the invariant factor $M(\lambda)$ defined by (5.10) for an arbitrary vector in H_{λ}; that is,

$$M_{\text{op}}\left| \begin{matrix} m' \\ \lambda \\ m \end{matrix} \right\rangle = M(\lambda)\left| \begin{matrix} m' \\ \lambda \\ m \end{matrix} \right\rangle, \tag{6.62}$$

with the obvious definition of the action of the diagonal operators $M^{1/2}_{\text{op}}$ and $M^{-1/2}_{\text{op}}$.

By definition, ordinary indeterminates (coordinates) z_{ij} and z_{kl} commute. Since the following relation holds for arbitrary matrix elements,

$$\left(\widehat{D}\left(\begin{array}{c} m''' \\ \mu \\ m'' \end{array}\right)(Z), z_{kl}z_{ij}\widehat{D}\left(\begin{array}{c} m' \\ \lambda \\ m \end{array}\right)(Z)\right) = \left\langle \begin{array}{c} m''' \\ \mu \\ m'' \end{array} \middle| \theta_{kl}\theta_{ij} \middle| \begin{array}{c} m' \\ \lambda \\ m \end{array} \right\rangle, \tag{6.63}$$

it follows also that the coordinates θ_{ij} and θ_{kl} commute:

$$[\theta_{ij}, \theta_{kl}] = 0, \text{ all } i,j,k,l = 1,2,\ldots,n. \tag{6.64}$$

Similarly, all the conjugates $\theta^\dagger_{ij}$ and $\theta^\dagger_{kl}$ commute.

The matrix elements in both sides of relation (6.58) are 0 unless μ has the form $\mu = \lambda + e_\tau, \tau = 1,2,\ldots,n$. We write this relation in the stated form to show the general equality over all matrix elements in the respective inner product spaces. Similarly, in relation (6.59), we have $\mu = \lambda - e_\tau$ for a nonzero relation.

To summarize: The diagonal tensor operator space (6.54) is the abstract analogy to the polynomial space:

$$\mathcal{P} = \sum_{p=0}^{\infty} \sum_{\lambda \in \mathbb{P}\mathrm{ar}_n(p)} \oplus \mathcal{P}_\lambda, \tag{6.65}$$

where each subspace $\mathcal{P}_\lambda, \lambda \in \mathbb{P}\mathrm{ar}_n(p)$ is the space of polynomials relative to the inner product (,) with the orthonormal basis given by

$$\mathcal{P}_\lambda = \left\{ \widehat{D}\left(\begin{array}{c} m' \\ \lambda \\ m \end{array}\right)(Z) \middle| m, m' \text{ are GT patterns} \right\}. \tag{6.66}$$

Relations (6.58)-(6.59) transfer matrix elements in the polynomial space $\mathcal{P}$ to matrix elements in the abstract Hilbert space $H \otimes_d H$ according to the mappings

$$z_{ij} \mapsto \theta_{ij}, \quad \frac{\partial}{\partial z_{ij}} \mapsto \theta^\dagger_{ij}\,. \tag{6.67}$$

Because the coordinates $\{z_{ij} \mid i,j = 1,2,\ldots,n\}$ are commuting, so are the $\{\theta_{ij} \mid i,j = 1,2,\ldots,n\}$, as well as the set of conjugates. The polynomial space $\mathcal{P}$ is an explicit model of the abstract Hilbert space H_d.

Relations (5.58)-(6.59) generalize to the following expression relating matrix elements in polynomial space $\mathcal{P}$ to matrix elements in the abstract Hilbert space $H_d = H \otimes_d H$:

$$\left(\widehat{D}\begin{pmatrix} m''' \\ \mu \\ m'' \end{pmatrix}(Z), P(Z, \frac{\partial}{\partial Z})\widehat{D}\begin{pmatrix} m' \\ \lambda \\ m \end{pmatrix}(Z)\right)$$
$$= \left\langle \begin{matrix} m''' \\ \mu \\ m'' \end{matrix} \middle| P(\Theta, \Theta^{\dagger}) \middle| \begin{matrix} m' \\ \lambda \\ m \end{matrix} \right\rangle, \tag{6.68}$$

where $P(Z, \frac{\partial}{\partial Z})$ is an arbitrary polynomial in the variables z_{ij} and $\partial/\partial z_{ij}$, and $P(\Theta, \Theta^{\dagger})$ is the **same** polynomial in the $\Theta = (\theta_{ij})_{1 \leq i,j \leq n}$ and $\Theta^{\dagger} = (\theta^{\dagger}_{ij})_{1 \leq i,j \leq n}$.

Toward our goal of giving as many properties of the complicated $C^{\lambda}(A)$–coefficients as possible, we give their expression as the following matrix elements in the abstract Hilbert space H, as follows from (5.22) and (6.68):

$$C\begin{pmatrix} m' \\ \lambda \\ m \end{pmatrix}(A) = \sqrt{M(\lambda)} \left\langle \begin{matrix} m' \\ \lambda \\ m \end{matrix} \middle| \Theta^{A} \middle| \begin{matrix} (0) \\ 0^{n} \\ (0) \end{matrix} \right\rangle. \tag{6.69}$$

Since the action of each fundamental unit tensor operator appearing in the definition (6.60) of θ_{ij} is fully defined on the Hilbert space H, relation (6.69) gives uniquely the matrix elements of $C^{\lambda}(A)$. This relation is used in Chapter 7 to give this evaluation.

6.6 Shift Operator Polynomials

We have not yet made any use of the commutation property $[t_{i\tau}, t_{j\tau}] = \mathbf{0}$, $i, j = 1, 2, \ldots, n$, of the fundamental shift operators. This commutation property is significant because powers of complex linear combinations such as

$$(z_1 t_{1\tau} + z_2 t_{2\tau} + \cdots + z_n t_{n\tau})^k, k = 0, 1, \ldots, \tag{6.70}$$

can be expanded in the usual way by the multinomial theorem. This property allows us to construct polynomials in the fundamental shift operators that have well-defined transformation properties under linear transformations. Such operator-valued polynomials are the basis of the tensor operator theory developed in Chapter 9. It is, however, useful to carry out some preliminary developments here to illustrate this aspect of the fundamental shift operators. Operator-valued polynomials in the fundamental shift operators $t_{i\tau}$ turn out to play a role similar to that of the commuting z_{ij} in the theory of the D^λ–polynomials.

In analogy with the matrix Z with elements z_{ij}, we arrange the elements of the set $\{t_{i\tau} \mid i,\tau = 1,2,\ldots,n\}$ into an $n \times n$ matrix T as follows:

$$T = \begin{pmatrix} t_{11} & t_{12} & \cdots & t_{1n} \\ t_{21} & t_{22} & \cdots & t_{2n} \\ \vdots & \vdots & \cdots & \vdots \\ t_{n1} & t_{n2} & \cdots & t_{nn} \end{pmatrix}. \tag{6.71}$$

For each matrix array of nonnegative integers $A = (a_{ij})_{1\le i,j\le n}$, the power T^A of T is defined by the *ordered product:*

$$T^A = \prod_{i=1}^{n} t_{in}^{a_{in}} \cdots \prod_{i=1}^{n} t_{i2}^{a_{i2}} \prod_{i=1}^{n} t_{i1}^{a_{i1}}, \tag{6.72}$$

in which the powers of the elements in column 1 appear to the right, followed by the powers of the elements in column 2, ..., followed by the powers of elements in column n. Because the components $t_{i\tau}, i = 1,2,\ldots,n$ appearing in each column T_i of T mutually commute, the shift operators in each product $\prod_{i=1}^{n} t_{i\tau}^{a_{i\tau}}, \tau \in \{1,2,\ldots,n\}$, can be permuted to any order without changing T^A. Thus, these operator-valued polynomials are the analogues of the Maclaurin polynomials Z^A. Under a left transformation of T by an arbitrary $n \times n$ matrix Z of indeterminates, we have that

$$T \to ZT = (ZT_1, ZT_2, \ldots, ZT_n), \tag{6.73}$$

so that only commuting components $t_{i\tau}$ occurring in the same column of T are mixed; hence, an arbitrary polynomial

$$P(T) = \sum_A C(A)\,\frac{T^A}{A!} \tag{6.74}$$

has an unambiguous meaning under arbitrary left transformations: $P(T) \to P(ZT)$.

The above relations can be applied to the D^λ–polynomials given by (5.6) to obtain operator-valued polynomials. We introduce a capital letter notation for operator-valued D–polynomials, as follows:

$$D\begin{pmatrix} \Gamma \\ \Lambda \\ M \end{pmatrix}(T) = \sum_{A\in \mathbb{M}^p_{n\times n}(W,\Delta)} C\begin{pmatrix} \Gamma \\ \Lambda \\ M \end{pmatrix}(A)\,\frac{T^A}{A!}, \tag{6.75}$$

where $\Lambda \in \mathbb{P}\mathrm{ar}_n, p = \Lambda_1 + \Lambda_2 + \cdots + \Lambda_n$, and M and Γ denote lower and upper GT patterns, and $W = W\binom{\Lambda}{M} \in \mathbb{W}_\Lambda$ and $\Delta = W\binom{\Lambda}{\Gamma} \in \mathbb{W}_\Lambda$

denote the weights of the lower and upper GT patterns. The notations Γ and Δ for the upper GT pattern and its weight anticipate a special role for these patterns in the theory of tensor operators in Chapter 9: The weight Δ of the GT pattern $\binom{\Lambda}{\Gamma}$ is denoted $\Delta = (\Delta_1, \Delta_2, \ldots, \Delta_n)$, where Δ_τ is the total shift $\Delta_\tau = \sum_{i=1}^{n} a_{i\tau}$ associated with the product term $t_{n\tau}^{a_{n\tau}} \cdots t_{2\tau}^{a_{2\tau}} t_{1\tau}^{a_{1\tau}}$ from column τ in the ordered product T^A. Thus, column τ in the matrix array $A^p_{n\times n}(W, \Delta)$ has sum Δ_τ, while row i still has sum given by the weight W_i of the lower GT pattern. From the point of view of GT patterns, both Δ and W are weights. The coefficients $C\begin{pmatrix}\Gamma\\ \Lambda\\ M\end{pmatrix}(A)$ are exactly those in (5.6).

The operator-valued polynomials (6.75) have the following matrix transformation property under left transformations by an arbitrary matrix Z :

$$D^\Lambda(ZT) = D^\Lambda(Z)D^\Lambda(T). \tag{6.76}$$

Such left transformations of $T \to ZT$ mix commuting elements of T occurring in a given column as shown in (6.73), and therefore satisfy the standard transformation rule (6.76). But right transformations $T \to TZ$ mix the noncommuting elements of T occurring in a given row: *Right transformations of the operator-valued polynomials (6.75) do not transform by right multiplication of* $D^\Lambda(T)$ *by* $D^\Lambda(Z)$. These are important properties in the theory of tensor operators given in Chapter 9.

The operator-valued polynomials (6.75) act in the Gelfand-Tsetlin model Hilbert space H defined by (6.4) through the action of the $t_{i\tau}$ on the basis $\mathbf{B}_\lambda$ of $H_\lambda \subset H$, as described in detail in Sects. 6.1-6.5. Thus, we have the action given by

$$D\begin{pmatrix}\Gamma\\ \Lambda\\ M\end{pmatrix}(T)\left|\begin{matrix}\lambda\\ m\end{matrix}\right\rangle \tag{6.77}$$

$$= \sum_{m'} \left\langle \begin{matrix}\lambda+\Delta\\ m'\end{matrix} \right| D\begin{pmatrix}\Gamma\\ \Lambda\\ M\end{pmatrix}(T)\left|\begin{matrix}\lambda\\ m\end{matrix}\right\rangle \left|\begin{matrix}\lambda+\Delta\\ m'\end{matrix}\right\rangle,$$

in which $\Gamma, M \in \mathbb{G}_\Lambda$, and $\Delta = W\binom{\Lambda}{\Gamma} \in \mathbb{W}_\Lambda$ is the shift-weight of the GT pattern Γ. The coefficients

$$\left\langle \begin{matrix}\lambda+\Delta\\ m'\end{matrix} \right| D\begin{pmatrix}\Gamma\\ \Lambda\\ M\end{pmatrix}(T)\left|\begin{matrix}\lambda\\ m\end{matrix}\right\rangle \tag{6.78}$$

in this relation are completely determined by the shift action of the fundamental shift operators $t_{i\tau}$ described in the sections above. These matrix elements are important for the theory of tensor operators considered in Chapter 9. They also enter into an alternative form of the coefficients in relation (6.79) below. But these relationships must await further developments in Chapter 7 and 8.

There is another important relation involving the operator-valued polynomials (6.75) that originates from relation (6.68). We choose the polynomial $P(Z, \partial/\partial Z)$ in (6.68) to be the $D^{\Lambda}-$polynomial and obtain:

$$\left(\widehat{D}\begin{pmatrix} m''' \\ \lambda+\Delta \\ m'' \end{pmatrix}(Z), D\begin{pmatrix} M' \\ \Lambda \\ M \end{pmatrix}(Z)\widehat{D}\begin{pmatrix} m' \\ \lambda \\ m \end{pmatrix}(Z)\right) \tag{6.79}$$

$$= \left\langle \begin{matrix} m''' \\ \lambda+\Delta \\ m'' \end{matrix} \middle| D\begin{pmatrix} M' \\ \Lambda \\ M \end{pmatrix}(\Theta) \middle| \begin{matrix} m' \\ \lambda \\ m \end{matrix} \right\rangle, \text{ each } \Delta \in \mathbb{W}_{\Lambda},$$

in which the pattern M' is any GT pattern of weight $\Delta \in \mathbb{W}_{\Lambda}$. We summarize the definitions of quantities in this result:

$$\left| \begin{matrix} m' \\ \lambda \\ m \end{matrix} \right\rangle = \left| \begin{matrix} \lambda \\ m \end{matrix} \right\rangle \otimes_d \left| \begin{matrix} \lambda \\ m' \end{matrix} \right\rangle, \tag{6.80}$$

$$\theta_{ij} = M_{\mathrm{op}}^{1/2}\left(\sum_{\tau=1}^{n}(t_{i\tau} \otimes_d t_{j\tau})\right) M_{\mathrm{op}}^{-1/2}. \tag{6.81}$$

The operator-valued $D^{\Lambda}-$polynomials in the right-hand side of (6.79) are defined by

$$D\begin{pmatrix} M' \\ \Lambda \\ M \end{pmatrix}(\Theta) = \sum_{A \in \mathbb{M}^{p}_{n\times n}(W,\Delta)} C\begin{pmatrix} M' \\ \Lambda \\ M \end{pmatrix} \frac{\Theta^{A}}{A!}, \tag{6.82}$$

where Θ is the $n \times n$ matrix of commuting operators (see (6.67))

$$\Theta = \begin{pmatrix} \theta_{11} & \theta_{12} & \cdots & \theta_{1n} \\ \theta_{21} & \theta_{22} & \cdots & \theta_{2n} \\ \vdots & & & \\ \theta_{n1} & \theta_{n2} & \cdots & \theta_{nn} \end{pmatrix}, \tag{6.83}$$

and Θ^{A} is the product (see (6.63)) given by

$$\Theta^{A} = \prod_{i,j=1}^{n} \theta_{ij}^{a_{ij}}. \tag{6.84}$$

Since the action of the fundamental tensor operators is completely defined in the right-hand side of (6.79), the matrix elements on the left are known. But this inner product is just that encountered in reducing the Kronecker product $D^{\Lambda}(Z) \otimes D^{\lambda}(Z)$ into a direct sum of irreducible representations. We consider this next.

6.7 Kronecker Product Reduction and Operator-valued Polynomials

The Kronecker product of a $D^{\Lambda}-$matrix and a $D^{\lambda}-$matrix is orthogonally equivalent to the direct sum of $D^{\lambda+\Delta}-$matrices, and a real orthogonal matrix $C^{(\Lambda\,\lambda)}$ of dimension $\mathrm{Dim}\Lambda\,\mathrm{Dim}\lambda$ effects the complete reduction:

$$C^{(\Lambda\,\lambda)}\left(D^{\Lambda}(Z)\otimes D^{\lambda}(Z)\right)\left(C^{(\Lambda\,\lambda)}\right)^{T} = \sum_{\Delta} \oplus\, c_{\Lambda\,\lambda}^{\lambda+\Delta} D^{\lambda+\Delta}(Z), \qquad (6.85)$$

where this result uses the fact that every Littlewood-Richardson number can be written in the form $c_{\Lambda\,\lambda}^{\lambda+\Delta}$ for some weight $\Delta \in \mathbb{W}_{\Lambda}$ (see relation 11.161, Compendium B). For this reason, we often refer to Δ as a *shift-weight.* The coefficient $c_{\Lambda\,\lambda}^{\lambda+\Delta}$ is defined to be zero if $D^{\lambda+\Delta}(Z)$ is not in the Kronecker product $D^{\Lambda}(Z) \otimes D^{\lambda}(Z)$. Relation (6.85) also implies

$$D^{\Lambda}(Z)\otimes D^{\lambda}(Z) = (C^{\Lambda\,\lambda})^{T}\left(\sum_{\Delta}\oplus\, c_{\Lambda\,\lambda}^{\lambda+\Delta} D^{\lambda+\Delta}(Z)\right) C^{(\Lambda\,\lambda)}. \qquad (6.86)$$

It is required that the orthogonal matrix $C^{(\Lambda\,\lambda)}$ in (6.85) should effect completely the reduction of the Kronecker product; that is, the direct sum on the right-hand side contains one and the same matrix $D^{\lambda+\Delta}(Z)$, repeated a number of times equal to the Littlewood-Richardson number, as expressed by the block diagonal form:

$$C^{(\Lambda\,\lambda)}\left(D^{\Lambda}(Z)\otimes D^{\lambda}(Z)\right)\left(C^{(\Lambda\,\lambda)}\right)^{T}$$

$$= \begin{pmatrix} \ddots & & \\ & \begin{pmatrix} D^{\lambda+\Delta}(Z) & & \\ & \ddots & \\ & & D^{\lambda+\Delta}(Z) \end{pmatrix} & \\ & & \ddots \end{pmatrix}. \qquad (6.87)$$

The repetition of the same irreducible $D-$matrix in the Kronecker product was already encountered in Sects. 2.1.1-2.1.3, Chapter 2 in the coupling theory of angular momenta, but did not occur until a three-fold Kronecker product was considered, whereas here the multiplicity of a given $D^{\Lambda+\Delta}-$polynomial occurs at the basic level of the Kronecker product itself. For the general coupling of n angular momenta, the CG numbers give the multiplicity of occurrence; here, it is the Littlewood-Richardson numbers that play this role at the very beginning.

Each parenthesis pair ( ) in (6.87) contains $c_{\Lambda\,\lambda}^{\lambda+\Delta}$ diagonal blocks, each block consisting of the same matrix $D^{\lambda+\Delta}(Z)$; and there is such a block, for given partitions Λ and λ for which the Littlewood-Richardson number is nonzero. For standardization, we also order the blocks in the matrix (6.87) from the least partition to the greatest partition in accordance with lexicographic order on the set of partitions in $\mathbb{P}\mathrm{ar}_n$; that is $\lambda > \lambda'$, if and only if the first nonzero entry in the difference $\lambda - \lambda'$ is positive. The consistency of matrix dimensions on each side of relation (6.87) is assured by the identity

$$\mathrm{Dim}\Lambda\,\mathrm{Dim}\lambda = \sum_{\Delta} c_{\Lambda,\lambda}^{\lambda+\Delta}\,\mathrm{Dim}(\lambda+\Delta). \tag{6.88}$$

We refer to the organization of the direct sum in (6.87) from the least to greatest value of $\Lambda + \Delta$ as the *standard Kronecker direct sum.*

It is important to fix precisely the notation in the reduction relation (6.87). We denote the elements of the real orthogonal matrix $C^{(\Lambda\,\lambda)}$ by the following notation:

$$\left(C^{(\Lambda\,\lambda)}\right)_{(k,\left(\begin{smallmatrix}\lambda+\Delta\\ m'\end{smallmatrix}\right));(M,m)} = C_k\left[\begin{pmatrix}\lambda+\Delta\\ m'\end{pmatrix}\begin{pmatrix}\Lambda\\ M\end{pmatrix}\begin{pmatrix}\lambda\\ m\end{pmatrix}\right], \tag{6.89}$$

where these labels have the following significance:

1. Rows of $C^{(\Lambda\,\lambda)}$: These are the pairs of labels $\left(k, \left(\begin{smallmatrix}\lambda+\Delta\\ m'\end{smallmatrix}\right)\right)$ that index the hierarchal order of the rows of the block diagonal matrix on the right in (6.87): the partitions in the set

$$\lambda + \Delta \in \{\mathbb{P}\mathrm{ar}_n \,|\, \Delta \in \mathbb{W}_\Lambda \text{ and } c_{\Lambda\,\lambda}^{\lambda+\Delta} \neq 0\} \tag{6.90}$$

index the block of repeated $D^{\lambda+\Delta}-$matrices enclosed by parentheses ( ) , ordered lexicographically; k indexes the blocks $k = 1, 2, \ldots, c_{\Lambda,\lambda}^{\lambda+\Delta}$ that contain the identical $D^{\lambda+\Delta}(Z)$ matrices; and the GT patterns $m' \in \mathbb{G}_{\lambda+\Delta}$ index the elements of each matrix $D^{\lambda+\Delta}(Z)$ in row m'. These GT patterns themselves are ordered lexicographically, as given by (5.18)-(5.20).

2. Columns of $C^{(\Lambda\,\lambda)}$: These are the pairs of GT pattern (M,m) that index the rows (M,m) of the Kronecker product:

$$\left(D^{\Lambda}(Z)\otimes D^{\lambda}(Z)\right)_{(M,m);(M',m')} = D\begin{pmatrix} M' \\ \Lambda \\ M \end{pmatrix}(Z)D\begin{pmatrix} m' \\ \Lambda \\ m \end{pmatrix}(Z). \tag{6.91}$$

We give explictly the orthogonality relations for the elements of the rows and columns, respectively, of the orthogonal matrix $C^{(\Lambda\,\lambda)}$:

$$\sum_{m,M} C_{k'}\left[\begin{pmatrix}\lambda+\Delta \\ m''\end{pmatrix}\begin{pmatrix}\Lambda \\ M\end{pmatrix}\begin{pmatrix}\lambda \\ m\end{pmatrix}\right] C_k\left[\begin{pmatrix}\lambda+\Delta \\ m'\end{pmatrix}\begin{pmatrix}\Lambda \\ M\end{pmatrix}\begin{pmatrix}\lambda \\ m\end{pmatrix}\right]$$
$$= \delta_{k',k}\,\delta_{m'',m'}, \tag{6.92}$$

$$\sum_{k=1}^{c^{\lambda+\Delta}_{\Lambda,\lambda}} \sum_{m''} C_k\left[\begin{pmatrix}\lambda+\Delta \\ m''\end{pmatrix}\begin{pmatrix}\Lambda \\ M'\end{pmatrix}\begin{pmatrix}\lambda \\ m'\end{pmatrix}\right] C_k\left[\begin{pmatrix}\lambda+\Delta \\ m''\end{pmatrix}\begin{pmatrix}\Lambda \\ M\end{pmatrix}\begin{pmatrix}\lambda \\ m\end{pmatrix}\right]$$
$$= \delta_{m',m}\,\delta_{M',M}. \tag{6.93}$$

The matrix elements of relation (6.87) are given in terms of the above conventions and notations by the following expression:

$$\sum_{M,m;M',m'} C_k\left[\begin{pmatrix}\lambda+\Delta \\ m''\end{pmatrix}\begin{pmatrix}\Lambda \\ M\end{pmatrix}\begin{pmatrix}\lambda \\ m\end{pmatrix}\right] D\begin{pmatrix} M' \\ \Lambda \\ M \end{pmatrix}(Z)D\begin{pmatrix} m' \\ \lambda \\ m \end{pmatrix}(Z)$$
$$\times C_{k'}\left[\begin{pmatrix}\lambda+\Delta' \\ m'''\end{pmatrix}\begin{pmatrix}\Lambda \\ M'\end{pmatrix}\begin{pmatrix}\lambda \\ m'\end{pmatrix}\right] = \delta_{k,k'}\,\delta_{\Delta,\Delta'}\,D\begin{pmatrix} m''' \\ \lambda+\Delta \\ m'' \end{pmatrix}(Z). \tag{6.94}$$

Similarly, if the orthogonal matrices in (6.87) are moved to the right-hand side, the following matrix element expression results:

$$D\begin{pmatrix} M' \\ \Lambda \\ M \end{pmatrix}(Z)D\begin{pmatrix} m' \\ \lambda \\ m \end{pmatrix}(Z)$$

$$= \sum_{k=1}^{c_{\Lambda\,\lambda}^{\lambda+\Delta}} \sum_{m'',m'''} C_k\left[\begin{pmatrix} \lambda+\Delta \\ m'' \end{pmatrix}\begin{pmatrix} \Lambda \\ M \end{pmatrix}\begin{pmatrix} \lambda \\ m \end{pmatrix}\right]$$

$$\times D\begin{pmatrix} m''' \\ \lambda+\Delta \\ m'' \end{pmatrix}(Z)C_k\left[\begin{pmatrix} \lambda+\Delta' \\ m''' \end{pmatrix}\begin{pmatrix} \Lambda \\ M' \end{pmatrix}\begin{pmatrix} \lambda \\ m' \end{pmatrix}\right]. \qquad (6.95)$$

This relation, in turn, gives the following result for the inner product (6.79) with initial and final normalized $D-$polynomials:

$$\left(\widehat{D}\begin{pmatrix} m''' \\ \lambda+\Delta \\ m'' \end{pmatrix}(Z),\ D\begin{pmatrix} M' \\ \Lambda \\ M \end{pmatrix}(Z)\widehat{D}\begin{pmatrix} m' \\ \lambda \\ m \end{pmatrix}(Z)\right)$$

$$= \left\langle \begin{matrix} m''' \\ \lambda+\Delta \\ m' \end{matrix} \middle| D\begin{pmatrix} M' \\ \Lambda \\ M \end{pmatrix}(\Theta) \middle| \begin{matrix} m'' \\ \lambda \\ m \end{matrix} \right\rangle$$

$$= \sqrt{\frac{M(\lambda+\Delta)}{M(\lambda)}} \sum_{k=1}^{c_{\Lambda\,\lambda}^{\lambda+\Delta}} C_k\left[\begin{pmatrix} \lambda+\Delta \\ m' \end{pmatrix}\begin{pmatrix} \Lambda \\ M \end{pmatrix}\begin{pmatrix} \lambda \\ m \end{pmatrix}\right]$$

$$\times C_k\left[\begin{pmatrix} \lambda+\Delta \\ m''' \end{pmatrix}\begin{pmatrix} \Lambda \\ M' \end{pmatrix}\begin{pmatrix} \lambda \\ m'' \end{pmatrix}\right], \text{ each } \Delta \in \mathbb{W}_\Lambda, \qquad (6.96)$$

where we have repeated the identity with (6.79). This is a key relation for determining the various sets of real orthogonal matrices that can effect the reduction of the Kronecker product to the standard Kronecker direct sum. It is called the *factorization lemma.*

The right-hand side of relation (6.87) is given by the Kronecker direct sum defined by

$$K_{\oplus}^{(\Lambda\,\lambda)} = \sum_{\Delta} \oplus \left(I_{c_{\Lambda\,\lambda}^{\lambda+\Delta}} \otimes D^{\lambda+\Delta}(Z)\right). \qquad (6.97)$$

It follows that every real orthogonal matrix $S^{(\Lambda\,\lambda)}$ of the form

$$S^{(\Lambda\,\lambda)} = \sum_{\Delta} \oplus \left(S_{(\Lambda\,\lambda)}^{\lambda+\Delta} \otimes I_{\mathrm{Dim}(\lambda+\Delta)}\right) \qquad (6.98)$$

commutes with the Kronecker direct sum; that is,

$$S^{(\Lambda\,\lambda)}K_{\oplus}^{(\Lambda\,\lambda)} = K_{\oplus}^{(\Lambda\,\lambda)}S^{(\Lambda\,\lambda)}. \tag{6.99}$$

The matrix $S_{(\Lambda\,\lambda)}^{\lambda+\Delta}$ is of order equal to the Littlewood-Richardson number $c_{\Lambda\,\lambda}^{\lambda+\Delta}$, and is itself a real orthogonal matrix. (It could be unitary, but we can choose all matrices to be real, as the defining relations for the $D-$ polynomials show.) Property (6.99) of Kronecker product reduction is completely analogous to that for the coupling of n angular momenta, as expressed by (2.44):

If $C^{(\Lambda\,\lambda)}$ is a real orthogonal matrix that effects the reduction of the Kronecker product $D^{\Lambda}(Z)\otimes D^{\lambda}(Z)$ to the Kronecker direct sum $K_{\oplus}^{(\Lambda\,\lambda)}$, then the real orthogonal matrix $R^{(\Lambda\,\lambda)}$ defined by

$$R^{(\Lambda\,\lambda)} = S^{(\Lambda\,\lambda)}C^{(\Lambda\,\lambda)} \tag{6.100}$$

effects exactly the same reduction.

We use the term *coupling scheme* to describe a given real orthogonal similarity transformation that effects the reduction of the Kronecker product $D^{\Lambda}(Z)\otimes D^{\lambda}(Z)$ to the standard Kronecker direct sum. Thus, given two coupling schemes $C^{(\Lambda\,\lambda)}$ and $R^{(\Lambda\,\lambda)}$, the matrix $S^{(\Lambda\,\lambda)}$ defined by

$$S^{(\Lambda\,\lambda)} = R^{(\Lambda\,\lambda)}\left(C^{(\Lambda\,\lambda)}\right)^{T} \tag{6.101}$$

is the *recoupling matrix* for these two coupling schemes.

The Clebsch-Gordan coefficients of the unitary group $U(n)$ $(GL(n,\mathbb{C}))$ are defined in this monograph to be the elements of a real orthogonal matrix that completely reduces the Kronecker product to Kronecker direct sum form, as shown in (6.85)-(6.87). Thus, each set of Clebsch-Gordan (CG) coefficients satisfies relation (6.96), where the left-hand side is known. But there are nondenumerably infinite many solutions to this relation, since, given any solution $C^{(\Lambda\,\lambda)}$, then $R^{(\Lambda\,\lambda)} = S^{(\Lambda\,\lambda)}C^{(\Lambda\,\lambda)}$ is also a solution, where $S_{(\Lambda\,\lambda)}^{\lambda+\Delta}$ in (6.98) is an arbitrary real orthogonal matrix of order $c_{\Lambda\,\lambda}^{\lambda+\Delta}$. A set of CG coefficients is determined only up to arbitrary real orthogonal transformations in *multiplicity space,* which is a real vector space of dimension equal to the Littlewood-Richardson number.

But this freedom in the definition of CG coefficients overlooks the possibility that there might exist a set of such coefficients whose properties are so "natural" that this "canonical" set is to be preferred over all others. For example, in the case of the binary addition of n angular momenta, the degeneracy of the multiplicity space is "broken" by the existence of the intermediate angular momenta. Perhaps there are properties of the Lie algebra of the unitary group such that when two copies are considered, as in the addition of angular momenta, there are "intermediate" sets of operators in the Lie algebra that give a complete set of commuting Hermitian operators that determine a preferred coupling scheme. Such coupling schemes have been considered by Racah [145] for SU(3), but have not proved very successful. But there may exist still other natural schemes.

A natural structure for selecting a preferred set of CG coefficients that suggests itself is the properties of the Littlewood-Richardson numbers, since the dimensions of the multiplicity spaces are determined by these numbers; perhaps, it is within this family of multiplicity spaces that multiplicity breaking principles reside. This is the idea developed extensively in numerous collaborations with Biedenharn. The viewpoint we adopt of the Littlewood-Richardson numbers is that they are to be considered as **functions** $I_{\Lambda,\Delta}$ defined over the set of all partitions $\lambda \in \mathbb{P}\mathrm{ar}_n$; hence, we write

$$c^{\lambda+\Delta}_{\Lambda,\lambda} = I_{\Lambda,\Delta}(\lambda). \tag{6.102}$$

It is known that the range of these functions is given by

$$I_{\Lambda,\Delta}(\lambda) \in \{0, 1, 2, \ldots, K(\Lambda, \Delta)\}, \tag{6.103}$$

where $K(\Lambda, \Delta)$ is the Kostka number. We can ask for the set of all partitions such that the Littlewood-Richardson function $I_{\Lambda,\Delta}$ takes on the constant value L :

$$\begin{gathered} \mathbb{P}_{\Lambda,\Delta}(L) = \{\lambda \in \mathbb{P}\mathrm{ar}_n \mid I_{\Lambda,\Delta}(\lambda) = L\}, \\ L \in \{0, 1, 2, \ldots, K(\Lambda, \Delta)\}. \end{gathered} \tag{6.104}$$

The sets of partitions $\mathbb{P}_{\Lambda,\Delta}(L)$ are called "level sets." For some such values of L, the level set $\mathbb{P}_{\Lambda,\Delta}(L)$ might be empty, but if the level sets satisfy the strong inclusion relations

$$\mathbb{P}_{\Lambda,\Delta}(0) \subset \mathbb{P}_{\Lambda,\Delta}(1) \subset \cdots \subset \mathbb{P}_{\Lambda,\Delta}(K(\Lambda, \Delta)), \tag{6.105}$$

there exists a natural orthogonal matrix that brings the Kronecker product $D^{\Lambda}(Z) \otimes D^{\lambda}(Z)$ to the standard Kronecker direct sum. Even if some set equalities occur in (6.105), there are still reasonable choices that can be made.

The *geometry of the level sets* $I_{\Lambda,\Delta}(L)$ is a key element in the problem of the reducing the Kronecker product $D^{\Lambda}(Z) \otimes D^{\lambda}(Z)$ to the standard Kronecker product direct sum. Since the Littlewood-Richardson numbers are well-studied objects (Macdonald [126], Stanley [163]), it may be possible to give a purely combinatorial formulation of the multiplicity problem. (We develop and cite numerous properties of Littlewood-Richardson numbers in Sects. 11.3.7-11.3.8, Compendium B and also in Sect. 9.6, Chapter 9.) A natural setting for considering Littlewood-Richardson is in the theory of tensor operators (initiated by Racah [144] for the group $SU(2)$), where they have a basic connection to the concept of null space. We develop this in Chapter 9 for $U(n)$. Tensor operators are important objects in many physical applications, and it is essential to know their relationship to CG coefficients.

We summarize the main results of this chapter:

A model separable Hilbert space H is introduced that has the minimal structure need to encode the full structure of the $D^{\lambda}-$polynomials. These polynomials are then unambiguously defined by the action of a set of fundamental shift operators. The action of each of these operators is defined combinatorially by arc digraphs, which give the rules for computing their matrix elements on an orthonormal basis of the Hilbert space H. This culminates naturally in an expression that gives the $C^{\lambda}-$coefficients, which express the $D^{\lambda}-$polynomials in terms of the Maclaurin polynomials, as the matrix elements of operator-valued Maclaurin polynomials in the fundamental shift operators.

Operator-valued $D^{\Lambda}-$polynomials now become the key objects for the study of the reduction of the Kronecker product $D^{\Lambda}(Z)\otimes D^{\lambda}(Z)$ into a fully reduced Kronecker direct sum $\sum_{\Delta}\oplus c^{\lambda+\Delta}_{\Lambda\,\lambda}\,D^{\lambda+\Delta}(Z)$. The result is the factorization lemma that relates the inner product of these three $D-$polynomials to the matrix elements of the operator-valued $D^{\Lambda}-$polynomials, and then, in turn, to the CG coefficients, which are the elements in an orthogonal matrix the effects the reduction of the Kronecker product itself.

The remaining chapters of this monograph fill in some of the rich (and complex) details of the above inter-related notions: Hilbert space—fundamental operator actions—operator-valued $D-$polynomials—Kronecker products—factorization lemma—Clebsch-Gordan coefficients—tensor operators.

6.8 More on Explicit Operator Actions

It is convenient for later use to introduce the *renormalized basic shift* operators $s_{i\tau}$ defined by

$$s_{i\tau}=M^{1/2}_{\mathrm{op}}t_{i\tau}M^{-1/2}_{\mathrm{op}}. \tag{6.106}$$

The action of these operators on the basis $\mathbf{B}_\lambda$ of H_λ defined by (6.5) is then given by

$$
\begin{aligned}
s_{i\tau}\left| \begin{matrix} \lambda \\ m \end{matrix} \right\rangle &= \sqrt{\frac{M(\lambda+e_\tau)}{M(\lambda)}}\, t_{i\tau}\left| \begin{matrix} \lambda \\ m \end{matrix} \right\rangle \\
&= \left[(p_{\tau n}+1)\prod_{\substack{k=1\\k\neq\tau}}^{n} \frac{p_{kn}-p_{\tau n}}{p_{kn}-p_{\tau n}-1} \right]^{1/2} t_{i\tau}\left| \begin{matrix} \lambda \\ m \end{matrix} \right\rangle .
\end{aligned} \tag{6.107}
$$

We then have also the following relations for the Hermitian conjugate:

$$
s_{i\tau}^\dagger = M_{\mathrm{op}}^{-1/2} t_{i\tau}^\dagger M_{\mathrm{op}}^{1/2}, \tag{6.108}
$$

with action given by

$$
\begin{aligned}
s_{i\tau}^\dagger\left| \begin{matrix} \lambda \\ m \end{matrix} \right\rangle &= \sqrt{\frac{M(\lambda)}{M(\lambda-e_\tau)}}\, t_{i\tau}^\dagger\left| \begin{matrix} \lambda \\ m \end{matrix} \right\rangle \\
&= \left[p_{\tau n}\prod_{\substack{k=1\\k\neq\tau}}^{n} \frac{p_{kn}-p_{\tau n}+1}{p_{kn}-p_{\tau n}} \right]^{1/2} t_{i\tau}^\dagger\left| \begin{matrix} \lambda \\ m \end{matrix} \right\rangle .
\end{aligned} \tag{6.109}
$$

We list the following results for the action of the shift operators $s_{i\tau}$ and their conjugates in the Hilbert space H_λ for $n = 2, 3$:

$n = 2.$

$$
s_{11}\left| \begin{matrix} \lambda_1 \quad \lambda_2 \\ m_{11} \end{matrix} \right\rangle = \sqrt{\frac{(p_{12}+1)(p_{11}-p_{22}+1)}{p_{12}-p_{22}+1}} \left| \begin{matrix} \lambda_1+1 \quad \lambda_2 \\ m_{11}+1 \end{matrix} \right\rangle,
$$

$$
s_{21}\left| \begin{matrix} \lambda_1 \quad \lambda_2 \\ m_{11} \end{matrix} \right\rangle = \sqrt{\frac{(p_{12}+1)(p_{12}-p_{11})}{p_{12}-p_{22}+1}} \left| \begin{matrix} \lambda_1+1 \quad \lambda_2 \\ m_{11} \end{matrix} \right\rangle,
$$

$$
s_{12}\left| \begin{matrix} \lambda_1 \quad \lambda_2 \\ m_{11} \end{matrix} \right\rangle = -\sqrt{\frac{(p_{22}+1)(p_{11}-p_{12}+1)}{p_{22}-p_{12}+1}} \left| \begin{matrix} \lambda_1 \quad \lambda_2+1 \\ m_{11}+1 \end{matrix} \right\rangle,
$$

$$
s_{22}\left| \begin{matrix} \lambda_1 \quad \lambda_2 \\ m_{11} \end{matrix} \right\rangle = \sqrt{\frac{(p_{22}+1)(p_{22}-p_{11})}{p_{22}-p_{12}+1}} \left| \begin{matrix} \lambda_1 \quad \lambda_2+1 \\ m_{11} \end{matrix} \right\rangle;
$$

$$s^\dagger_{11}\left|\begin{matrix}\lambda_1 & & \lambda_2\\ & m_{11} & \end{matrix}\right\rangle=\sqrt{\frac{p_{12}(p_{11}-p_{22})}{p_{12}-p_{22}}}\left|\begin{matrix}\lambda_1-1 & & \lambda_2\\ & m_{11}-1 & \end{matrix}\right\rangle,$$

$$s^\dagger_{21}\left|\begin{matrix}\lambda_1 & & \lambda_2\\ & m_{11} & \end{matrix}\right\rangle=\sqrt{\frac{p_{12}(p_{12}-p_{11}-1)}{p_{12}-p_{22}}}\left|\begin{matrix}\lambda_1-1 & & \lambda_2\\ & m_{11} & \end{matrix}\right\rangle,$$

$$s^\dagger_{12}\left|\begin{matrix}\lambda_1 & & \lambda_2\\ & m_{11} & \end{matrix}\right\rangle=-\sqrt{\frac{p_{22}(p_{11}-p_{12})}{p_{22}-p_{12}}}\left|\begin{matrix}\lambda_1 & & \lambda_2-1\\ & m_{11}-1 & \end{matrix}\right\rangle,$$

$$s^\dagger_{22}\left|\begin{matrix}\lambda_1 & & \lambda_2\\ & m_{11} & \end{matrix}\right\rangle=\sqrt{\frac{p_{22}(p_{22}-p_{11}-1)}{p_{22}-p_{12}}}\left|\begin{matrix}\lambda_1 & & \lambda_2-1\\ & m_{11} & \end{matrix}\right\rangle.$$

$n=3:$

$$\begin{aligned}
s_{11}\left|\begin{matrix}\lambda_1 & \lambda_2 & \lambda_3\\ m_{12} & m_{22} & \\ m_{11} & & \end{matrix}\right\rangle &= \sqrt{\frac{(p_{13}+1)(p_{12}-p_{23}+1)(p_{12}-p_{33}+1)(p_{13}-p_{22})}{(p_{12}-p_{22}+1)(p_{13}-p_{23}+1)(p_{13}-p_{33}+1)}}\\
&\quad\times\sqrt{\frac{p_{11}-p_{22}+1}{p_{12}-p_{22}}}\left|\begin{matrix}\lambda_1+1 & \lambda_2 & \lambda_3\\ m_{12}+1 & m_{22} & \\ m_{11}+1 & & \end{matrix}\right\rangle\\
&-\sqrt{\frac{(p_{13}+1)(p_{22}-p_{23}+1)(p_{22}-p_{33}+1)(p_{13}-p_{12})}{(p_{22}-p_{12}+1)(p_{13}-p_{23}+1)(p_{13}-p_{33}+1)}}\\
&\quad\times\sqrt{\frac{p_{11}-p_{12}+1}{p_{22}-p_{12}}}\left|\begin{matrix}\lambda_1+1 & \lambda_2 & \lambda_3\\ m_{12} & m_{22}+1 & \\ m_{11}+1 & & \end{matrix}\right\rangle,
\end{aligned}$$

$$\begin{aligned}
s_{21}\left|\begin{matrix}\lambda_1 & \lambda_2 & \lambda_3\\ m_{12} & m_{22} & \\ m_{11} & & \end{matrix}\right\rangle &= \sqrt{\frac{(p_{13}+1)(p_{12}-p_{23}+1)(p_{12}-p_{33}+1)(p_{13}-p_{22})}{(p_{12}-p_{22}+1)(p_{13}-p_{23}+1)(p_{13}-p_{33}+1)}}\\
&\quad\times\sqrt{\frac{p_{12}-p_{11}}{p_{12}-p_{22}}}\left|\begin{matrix}\lambda_1+1 & \lambda_2 & \lambda_3\\ m_{12}+1 & m_{22} & \\ m_{11}+1 & & \end{matrix}\right\rangle\\
&+\sqrt{\frac{(p_{13}+1)(p_{22}-p_{23}+1)(p_{22}-p_{33}+1)(p_{13}-p_{12})}{(p_{22}-p_{12}+1)(p_{13}-p_{23}+1)(p_{13}-p_{33}+1)}}\\
&\quad\times\sqrt{\frac{p_{22}-p_{11}}{p_{22}-p_{12}}}\left|\begin{matrix}\lambda_1+1 & \lambda_2 & \lambda_3\\ m_{12} & m_{22}+1 & \\ m_{11}+1 & & \end{matrix}\right\rangle,
\end{aligned}$$

$$s_{31}\left|\begin{matrix}\lambda_1 & \lambda_2 & \lambda_3\\ m_{12} & m_{22} & \\ m_{11} & & \end{matrix}\right\rangle=\sqrt{\frac{(p_{13}+1)(p_{13}-p_{12})(p_{13}-p_{22})}{(p_{13}-p_{23}+1)(p_{13}-p_{33}+1)}}\left|\begin{matrix}\lambda_1+1 & \lambda_2 & \lambda_3\\ m_{12} & m_{22} & \\ m_{11} & & \end{matrix}\right\rangle;$$

$$s_{12}\left|\begin{matrix}\lambda_1 \quad \lambda_2 \quad \lambda_3\\ m_{12}\ m_{22}\\ m_{11}\end{matrix}\right\rangle = -\sqrt{\frac{(p_{23}+1)(p_{12}-p_{13}+1)(p_{12}-p_{33}+1)(p_{23}-p_{22})}{(p_{12}-p_{22}+1)(p_{23}-p_{13}+1)(p_{23}-p_{33}+1)}}$$

$$\times\sqrt{\frac{p_{11}-p_{22}+1}{p_{12}-p_{22}}}\left|\begin{matrix}\lambda_1 \quad \lambda_2+1 \quad \lambda_3\\ m_{12}+1\ m_{22}\\ m_{11}+1\end{matrix}\right\rangle$$

$$-\sqrt{\frac{(p_{23}+1)(p_{22}-p_{13}+1)(p_{22}-p_{33}+1)(p_{23}-p_{12})}{(p_{22}-p_{12}+1)(p_{23}-p_{13}+1)(p_{23}-p_{33}+1)}}$$

$$\times\sqrt{\frac{p_{11}-p_{12}+1}{p_{22}-p_{12}}}\left|\begin{matrix}\lambda_1 \quad \lambda_2+1 \quad \lambda_3\\ m_{12}\ m_{22}+1\\ m_{11}+1\end{matrix}\right\rangle,$$

$$s_{22}\left|\begin{matrix}\lambda_1 \quad \lambda_2 \quad \lambda_3\\ m_{12}\ m_{22}\\ m_{11}\end{matrix}\right\rangle = -\sqrt{\frac{(p_{23}+1)(p_{12}-p_{13}+1)(p_{12}-p_{33}+1)(p_{23}-p_{22})}{(p_{12}-p_{22}+1)(p_{23}-p_{13}+1)(p_{23}-p_{33}+1)}}$$

$$\times\sqrt{\frac{p_{12}-p_{11}}{p_{12}-p_{22}}}\left|\begin{matrix}\lambda_1 \quad \lambda_2+1 \quad \lambda_3\\ m_{12}+1\ m_{22}\\ m_{11}+1\end{matrix}\right\rangle$$

$$+\sqrt{\frac{(p_{23}+1)(p_{22}-p_{13}+1)(p_{22}-p_{33}+1)(p_{23}-p_{12})}{(p_{22}-p_{12}+1)(p_{23}-p_{13}+1)(p_{23}-p_{33}+1)}}$$

$$\times\sqrt{\frac{p_{22}-p_{11}}{p_{22}-p_{12}}}\left|\begin{matrix}\lambda_1 \quad \lambda_2+1 \quad \lambda_3\\ m_{12}\ m_{22}+1\\ m_{11}+1\end{matrix}\right\rangle,$$

$$s_{32}\left|\begin{matrix}\lambda_1 \quad \lambda_2 \quad \lambda_3\\ m_{12}\ m_{22}\\ m_{11}\end{matrix}\right\rangle = \sqrt{\frac{(p_{23}+1)(p_{23}-p_{12})(p_{23}-p_{22})}{(p_{23}-p_{13}+1)(p_{23}-p_{33}+1)}}\left|\begin{matrix}\lambda_1 \quad \lambda_2+1 \quad \lambda_3\\ m_{12}\ m_{22}\\ m_{11}\end{matrix}\right\rangle;$$

$$s_{13}\left|\begin{matrix}\lambda_1 \quad \lambda_2 \quad \lambda_3\\ m_{12}\ m_{22}\\ m_{11}\end{matrix}\right\rangle = -\sqrt{\frac{(p_{33}+1)(p_{12}-p_{13}+1)(p_{12}-p_{23}+1)(p_{33}-p_{22})}{(p_{12}-p_{22}+1)(p_{33}-p_{13}+1)(p_{33}-p_{23}+1)}}$$

$$\times\sqrt{\frac{p_{11}-p_{22}+1}{p_{12}-p_{22}}}\left|\begin{matrix}\lambda_1 \quad \lambda_2 \quad \lambda_3+1\\ m_{12}+1\ m_{22}\\ m_{11}+1\end{matrix}\right\rangle$$

$$+\sqrt{\frac{(p_{33}+1)(p_{22}-p_{13}+1)(p_{22}-p_{23}+1)(p_{33}-p_{12})}{(p_{22}-p_{12}+1)(p_{33}-p_{13}+1)(p_{33}-p_{23}+1)}}$$

$$\sqrt{\frac{p_{11}-p_{12}+1}{p_{22}-p_{12}}}\left|\begin{matrix}\lambda_1 \quad \lambda_2 \quad \lambda_3+1\\ m_{12}\ m_{22}+1\\ m_{11}+1\end{matrix}\right\rangle,$$

$$s_{23}\left|\begin{matrix}\lambda_1 & \lambda_2 & \lambda_3\\ m_{12} & m_{22} & \\ m_{11} & & \end{matrix}\right\rangle = -\sqrt{\frac{(p_{33}+1)(p_{12}-p_{13}+1)(p_{12}-p_{23}+1)(p_{33}-p_{22})}{(p_{12}-p_{22}+1)(p_{33}-p_{13}+1)(p_{33}-p_{23}+1)}}$$
$$\times\sqrt{\frac{p_{12}-p_{11}}{p_{12}-p_{22}}}\left|\begin{matrix}\lambda_1 & \lambda_2 & \lambda_3+1\\ m_{12}+1 & m_{22} & \\ m_{11}+1 & & \end{matrix}\right\rangle$$
$$-\sqrt{\frac{(p_{33}+1)(p_{22}-p_{13}+1)(p_{22}-p_{23}+1)(p_{33}-p_{12})}{(p_{22}-p_{12}+1)(p_{33}-p_{13}+1)(p_{33}-p_{23}+1)}}$$
$$\times\sqrt{\frac{p_{22}-p_{11}}{p_{22}-p_{12}}}\left|\begin{matrix}\lambda_1 & \lambda_2 & \lambda_3+1\\ m_{12} & m_{22}+1 & \\ m_{11}+1 & & \end{matrix}\right\rangle,$$

$$s_{33}\left|\begin{matrix}\lambda_1 & \lambda_2 & \lambda_3\\ m_{12} & m_{22} & \\ m_{11} & & \end{matrix}\right\rangle = \sqrt{\frac{(p_{33}+1)(p_{33}-p_{12})(p_{33}-p_{22})}{(p_{33}-p_{13}+1)(p_{33}-p_{23}+1)}}\left|\begin{matrix}\lambda_1 & \lambda_2 & \lambda_3+1\\ m_{12} & m_{22} & \\ m_{11} & & \end{matrix}\right\rangle.$$

$$s_{11}^{\dagger}\left|\begin{matrix}\lambda_1 & \lambda_2 & \lambda_3\\ m_{12} & m_{22} & \\ m_{11} & & \end{matrix}\right\rangle = \sqrt{\frac{p_{13}(p_{12}-p_{23})(p_{12}-p_{33})(p_{13}-p_{22}-1)}{(p_{12}-p_{22})(p_{13}-p_{23})(p_{13}-p_{33})}}$$
$$\times\sqrt{\frac{p_{11}-p_{22}}{p_{12}-p_{22}-1}}\left|\begin{matrix}\lambda_1-1 & \lambda_2 & \lambda_3\\ m_{12}-1 & m_{22} & \\ m_{11}-1 & & \end{matrix}\right\rangle$$
$$-\sqrt{\frac{p_{13}(p_{22}-p_{23})(p_{22}-p_{33})(p_{13}-p_{12}-1)}{(p_{22}-p_{12})(p_{13}-p_{23})(p_{13}-p_{33})}}$$
$$\times\sqrt{\frac{p_{11}-p_{12}}{p_{22}-p_{12}-1}}\left|\begin{matrix}\lambda_1-1 & \lambda_2 & \lambda_3\\ m_{12} & m_{22}-1 & \\ m_{11}-1 & & \end{matrix}\right\rangle,$$

$$s_{21}^{\dagger}\left|\begin{matrix}\lambda_1 & \lambda_2 & \lambda_3\\ m_{12} & m_{22} & \\ m_{11} & & \end{matrix}\right\rangle = \sqrt{\frac{p_{13}(p_{12}-p_{23})(p_{12}-p_{33})(p_{13}-p_{22}-1)}{(p_{12}-p_{22})(p_{13}-p_{23})(p_{13}-p_{33})}}$$
$$\times\sqrt{\frac{p_{12}-p_{11}-1}{p_{12}-p_{22}-1}}\left|\begin{matrix}\lambda_1-1 & \lambda_2 & \lambda_3\\ m_{12}-1 & m_{22} & \\ m_{11} & & \end{matrix}\right\rangle$$
$$+\sqrt{\frac{p_{13}(p_{22}-p_{23})(p_{22}-p_{33})(p_{13}-p_{12}-1)}{(p_{22}-p_{12})(p_{13}-p_{23})(p_{13}-p_{33})}}$$
$$\times\sqrt{\frac{p_{22}-p_{11}-1}{p_{22}-p_{12}-1}}\left|\begin{matrix}\lambda_1-1 & \lambda_2 & \lambda_3\\ m_{12} & m_{22}-1 & \\ m_{11} & & \end{matrix}\right\rangle,$$

$$s_{31}^{\dagger}\left|\begin{matrix}\lambda_1 & \lambda_2 & \lambda_3\\ m_{12} & m_{22} & \\ m_{11} & & \end{matrix}\right\rangle = \sqrt{\frac{p_{13}(p_{13}-p_{12}-1)(p_{13}-p_{22}-1)}{(p_{13}-p_{23})(p_{13}-p_{33})}}\left|\begin{matrix}\lambda_1-1 & \lambda_2 & \lambda_3\\ m_{12} & m_{22} & \\ m_{11} & & \end{matrix}\right\rangle;$$

$$
\begin{aligned}
s_{12}^{\dagger}\left|\begin{matrix}\lambda_1 & & \lambda_2 & & \lambda_3\\ & m_{12} & & m_{22} & \\ & & m_{11} & & \end{matrix}\right\rangle = & -\sqrt{\frac{p_{23}(p_{12}-p_{13})(p_{12}-p_{33})(p_{23}-p_{22}-1)}{(p_{12}-p_{22})(p_{23}-p_{13})(p_{23}-p_{33})}} \\
& \times\sqrt{\frac{p_{11}-p_{22}}{p_{12}-p_{22}-1}}\left|\begin{matrix}\lambda_1 & & \lambda_2-1 & & \lambda_3\\ & m_{12}-1 & & m_{22} & \\ & & m_{11}-1 & & \end{matrix}\right\rangle \\
& -\sqrt{\frac{p_{23}(p_{22}-p_{13})(p_{22}-p_{33})(p_{23}-p_{12}-1)}{(p_{22}-p_{12})(p_{23}-p_{13})(p_{23}-p_{33})}} \\
& \times\sqrt{\frac{p_{11}-p_{12}}{p_{22}-p_{12}-1}}\left|\begin{matrix}\lambda_1 & & \lambda_2-1 & & \lambda_3\\ & m_{12} & & m_{22}-1 & \\ & & m_{11}-1 & & \end{matrix}\right\rangle,
\end{aligned}
$$

$$
\begin{aligned}
s_{22}^{\dagger}\left|\begin{matrix}\lambda_1 & & \lambda_2 & & \lambda_3\\ & m_{12} & & m_{22} & \\ & & m_{11} & & \end{matrix}\right\rangle - & \sqrt{\frac{p_{23}(p_{12}-p_{13})(p_{12}-p_{33})(p_{23}-p_{22}-1)}{(p_{12}-p_{22})(p_{23}-p_{13})(p_{23}-p_{33})}} \\
& \times\sqrt{\frac{p_{12}-p_{11}-1}{p_{12}-p_{22}-1}}\left|\begin{matrix}\lambda_1 & & \lambda_2-1 & & \lambda_3\\ & m_{12}-1 & & m_{22} & \\ & & m_{11} & & \end{matrix}\right\rangle \\
& +\sqrt{\frac{p_{23}(p_{22}-p_{13})(p_{22}-p_{33})(p_{23}-p_{12}-1)}{(p_{22}-p_{12})(p_{23}-p_{13})(p_{23}-p_{33})}} \\
& \times\sqrt{\frac{p_{22}-p_{11}-1}{p_{22}-p_{12}-1}}\left|\begin{matrix}\lambda_1 & & \lambda_2-1 & & \lambda_3\\ & m_{12} & & m_{22}-1 & \\ & & m_{11} & & \end{matrix}\right\rangle,
\end{aligned}
$$

$$
s_{32}^{\dagger}\left|\begin{matrix}\lambda_1 & & \lambda_2 & & \lambda_3\\ & m_{12} & & m_{22} & \\ & & m_{11} & & \end{matrix}\right\rangle = \sqrt{\frac{p_{23}(p_{23}-p_{12}-1)(p_{23}-p_{22}-1)}{(p_{23}-p_{13})(p_{23}-p_{33})}}\left|\begin{matrix}\lambda_1 & & \lambda_2-1 & & \lambda_3\\ & m_{12} & & m_{22} & \\ & & m_{11} & & \end{matrix}\right\rangle;
$$

$$
\begin{aligned}
s_{13}^{\dagger}\left|\begin{matrix}\lambda_1 & & \lambda_2 & & \lambda_3\\ & m_{12} & & m_{22} & \\ & & m_{11} & & \end{matrix}\right\rangle = & -\sqrt{\frac{p_{33}(p_{12}-p_{13})(p_{12}-p_{23})(p_{33}-p_{22}-1)}{(p_{12}-p_{22})(p_{33}-p_{13})(p_{33}-p_{23})}} \\
& \times\sqrt{\frac{p_{11}-p_{22}}{p_{12}-p_{22}-1}}\left|\begin{matrix}\lambda_1 & & \lambda_2 & & \lambda_3-1\\ & m_{12}-1 & & m_{22} & \\ & & m_{11}-1 & & \end{matrix}\right\rangle \\
& +\sqrt{\frac{p_{33}(p_{22}-p_{13})(p_{22}-p_{23})(p_{33}-p_{12}-1)}{(p_{22}-p_{12})(p_{33}-p_{13})(p_{33}-p_{23})}} \\
& \times\sqrt{\frac{p_{11}-p_{12}}{p_{22}-p_{12}-1}}\left|\begin{matrix}\lambda_1 & & \lambda_2 & & \lambda_3-1\\ & m_{12} & & m_{22}-1 & \\ & & m_{11}-1 & & \end{matrix}\right\rangle,
\end{aligned}
$$

$$s_{23}^{\dagger}\left|\begin{array}{c}\lambda_1\quad\lambda_2\quad\lambda_3\\ m_{12}\ m_{22}\\ m_{11}\end{array}\right\rangle = -\sqrt{\frac{p_{33}(p_{12}-p_{13})(p_{12}-p_{23})(p_{33}-p_{22}-1)}{(p_{12}-p_{22})(p_{33}-p_{13})(p_{33}-p_{23})}}$$
$$\times\sqrt{\frac{p_{12}-p_{11}-1}{p_{12}-p_{22}-1}}\left|\begin{array}{c}\lambda_1\quad\lambda_2\quad\lambda_3-1\\ m_{12}-1\ m_{22}\\ m_{11}\end{array}\right\rangle$$
$$-\sqrt{\frac{p_{23}(p_{22}-p_{13})(p_{22}-p_{23})(p_{33}-p_{12}-1)}{(p_{22}-p_{12})(p_{33}-p_{13})(p_{33}-p_{23})}}$$
$$\times\sqrt{\frac{p_{22}-p_{11}-1}{p_{22}-p_{12}-1}}\left|\begin{array}{c}\lambda_1\quad\lambda_2\quad\lambda_3-1\\ m_{12}\ m_{22}-1\\ m_{11}\end{array}\right\rangle,$$

$$s_{33}^{\dagger}\left|\begin{array}{c}\lambda_1\quad\lambda_2\quad\lambda_3\\ m_{12}\ m_{22}\\ m_{11}\end{array}\right\rangle = \sqrt{\frac{p_{33}(p_{33}-p_{12}-1)(p_{33}-p_{22}-1)}{(p_{33}-p_{13})(p_{33}-p_{23})}}\left|\begin{array}{c}\lambda_1\quad\lambda_2\quad\lambda_3-1\\ m_{12}\ m_{22}\\ m_{11}\end{array}\right\rangle.$$

Chapter 7

The $D^\lambda-$Polynomials: Structure

In this chapter, we derive explicit expressions for the $C^\lambda-$coefficients that occur in the expansion of the $D^\lambda-$polynomials in terms of Maclaurin polynomials, for the CG coefficients that reduce the Kronecker product of two $D^\lambda-$polynomials, and for their interrelations. The relationship of the $D^\lambda-$polynomials with the representations and Lie algebra of the general linear group is a major aspect of these polynomials, one that we present in Chapter 8. The operator methods of the last section have a major role, and most results have a combinatorial basis.

7.1 Combinatorial Formula for the $C^\lambda(A)$ Matrices

In the first method of determining the $D^\lambda-$polynomials, we use relation (6.69), Chapter 6:

$$\begin{aligned} C\begin{pmatrix} m' \\ \lambda \\ m \end{pmatrix}(A) &= \sqrt{M(\lambda)}\left\langle \begin{matrix} m' \\ \lambda \\ m \end{matrix} \,\middle|\, \Theta^A \,\middle|\, \begin{matrix} (0) \\ 0^n \\ (0) \end{matrix} \right\rangle \\ &= M(\lambda)\left\langle \begin{matrix} m' \\ \lambda \\ m \end{matrix} \,\middle|\, \prod_{i,j=1}^{n}\left(\sum_{\tau=1}^{n}(t_{i\tau}\otimes_d t_{j\tau})\right)^{a_{ij}} \,\middle|\, \begin{matrix} (0) \\ 0^n \\ (0) \end{matrix} \right\rangle, \end{aligned} \quad (7.1)$$

in which $\lambda \in \mathbb{P}\mathrm{ar}_n(p), A \in \mathbb{M}^p_{n\times n}(\alpha,\beta), \alpha = W\binom{\lambda}{m}, \beta = W\binom{\lambda}{m'}$.

We bring (7.1) to a more manageable form by defining the n^2 operators σ_{ij} as follows:

$$\sigma_{ij} = \sum_{\tau=1}^{n} (t_{i\tau} \otimes_d t_{j\tau}),\ i,j = 1,2,\ldots,n. \tag{7.2}$$

These n^2 operators mutually commute (see (6.64)), so that the factors in $\prod_{i,j=1}^{n} \sigma_{ij}^{a_{ij}}$ may be arranged in any of $p!/A!$ different ways. The fundamental shift operators in the summation in (7.2) corresponding to different τ do not commute; different orders of the σ_{ij} in the power product can lead to different expressions for the C^λ–coefficients. We consider them all by writing

$$\prod_{i,j=1}^{n} \sigma_{ij}^{a_{ij}} = \sigma_{i_p j_p} \cdots \sigma_{i_2 j_2} \sigma_{i_1 j_1}, \tag{7.3}$$

where $(i_1, i_2, \ldots, i_p)$ is a permutation of $(1^{\alpha_1}, 2^{\alpha_2}, \ldots, n^{\alpha_n})$, since $A \in \mathbb{M}^p_{n\times n}(\alpha,\beta)$ requires that the row index 1 appears α_1 times, 2 appears α_2 times, $\ldots$, n appears α_n times; similarly, $(j_1, j_2, \ldots, j_p)$ is a permutation of $(1^{\beta_1}, 2^{\beta_2}, \ldots, n^{\beta_n})$. We use these two permutation groups to classify the sequences (7.3), as we next describe.

Let $\mathbb{A}_\alpha$ and $\mathbb{A}_\beta$ denote the sets of sequences defined by

$$\begin{aligned} \mathbb{A}_\alpha &= \{\pi(1^{\alpha_1}, 2^{\alpha_2}, \ldots, n^{\alpha_n}) \mid \pi \in S_p\},\ \alpha \vdash p, \\ \mathbb{A}_\beta &= \{\pi(1^{\beta_1}, 2^{\beta_2}, \ldots, n^{\beta_n}) \mid \pi \in S_p\},\ \beta \vdash p, \end{aligned} \tag{7.4}$$

where the action of $\pi \in S_p$ is to permute the parts of the sequences; hence, there are $p!/\alpha!$ distinct sequences in the set $\mathbb{A}_\alpha$ and $p!/\beta!$ distinct sequences in the set $\mathbb{A}_\beta$. We refer to the sets of sequences $\mathbb{A}_\alpha$ and $\mathbb{A}_\beta$ as being of type α and of type β.

We also require the direct product of the symmetric groups $S_{\alpha_i}, i = 1,2,\ldots,n$:

$$S^\alpha = S_{\alpha_1} \times S_{\alpha_2} \times \cdots \times S_{\alpha_n}. \tag{7.5}$$

This subgroup of S_p then has the property

$$\rho(1^{\alpha_1}, 2^{\alpha_2}, \ldots, n^{\alpha_n}) = (1^{\alpha_1}, 2^{\alpha_2}, \ldots, n^{\alpha_n}), \tag{7.6}$$

for each $\rho \in S^\alpha$. Thus, $S^\alpha \subset S_p$ is the *isotropy group* (also called little group) of S_p that fixes the sequence $(1^{\alpha_1}, 2^{\alpha_2}, \ldots, n^{\alpha_n}) \in \mathbb{A}_\alpha$. More generally, if we choose the sequence $(i_1, i_2, \ldots, i_p)$ to be

$$(i_1, i_2, \ldots, i_p) = \pi(1^{\alpha_1}, 2^{\alpha_2}, \ldots, n^{\alpha_n}) \in \mathbb{A}_\alpha,\ \pi \in S_p, \tag{7.7}$$

then this sequence has the property

$$\left(\pi\,\rho\,\pi^{-1}\right)(i_1, i_2, \ldots, i_p) = (i_1, i_2, \ldots, i_p). \tag{7.8}$$

Thus, the $\pi-$conjugated group , $\pi S^\alpha \pi^{-1} \subset S_p$, is the isotropy group that fixes the sequence $(i_1, i_2, \ldots, i_p) = \pi(1^{\alpha_1}, 2^{\alpha_2}, \ldots, n^{\alpha_n})$.

Next, consider the action of S^α on the set $\mathbb{A}_\beta$. For this, we write each sequence in $\mathbb{A}_\beta$ in the form

$$\begin{aligned}&(b_1, b_2, \ldots, b_p) \\ &= (b_{11}, b_{12}, \ldots, b_{1\alpha_1}; \ldots; b_{i1}, b_{i2}, \ldots, b_{i\alpha_i}; \ldots; b_{n1}, b_{n2}, \ldots, b_{n\alpha_n}).\end{aligned} \tag{7.9}$$

The $i-$th symmetric group S_{α_i} in the direct product group S_α permutes the parts of the $i-$th subsequence $(b_{i1}, b_{i2}, \ldots, b_{i\alpha_i})$ in the sequence (7.9), leaving the other parts unchanged, this holding for each $i = 1, 2, \ldots, n$. Accordingly, the action of the group S^α on the set $\mathbb{A}_\beta$ gives another sequence in the set $\mathbb{A}_\beta$; this is also the case for each $\pi \in S_p$. Thus, the action of the isotropy group $\pi S^\alpha \pi^{-1}$ on the set $\mathbb{A}_\beta$ is to give a subset $\mathbb{A}_{\alpha,\beta}(\pi,\rho)$ of sequences in $\mathbb{A}_\beta$ defined by

$$\begin{aligned}&\mathbb{A}_{\alpha,\beta}(\pi,\rho) \\ &= \{[j_1, j_2, \ldots, j_p]_{\pi,\rho} = \left(\pi\,\rho\,\pi^{-1}\right)(b_1, b_2, \ldots, b_p) \in \mathbb{A}_\beta \,|\, \rho \in S^\alpha\}.\end{aligned} \tag{7.10}$$

In summary, we have the following result for the $C^\lambda-$coefficients:

$$\begin{aligned}&C\begin{pmatrix} m' \\ \lambda \\ m \end{pmatrix}(A) = M(\lambda) \\ &\times \sum_{\tau \in L(1^{\lambda_1}, 2^{\lambda_2}, \ldots, n^{\lambda_n})} \left\langle \begin{array}{c} \lambda \\ m \end{array} \middle| t_{i_p\tau_p} \cdots t_{i_2\tau_2} t_{i_1\tau_1} \middle| \begin{array}{c} 0^n \\ (0) \end{array} \right\rangle \\ &\qquad \times \left\langle \begin{array}{c} \lambda \\ m' \end{array} \middle| t_{j_p\tau_p} \cdots t_{j_2\tau_2} t_{j_1\tau_1} \middle| \begin{array}{c} 0^n \\ (0) \end{array} \right\rangle,\end{aligned} \tag{7.11}$$

where the sequences in this relation satisfy the following conditions:

1. *The sequence $(i_1, i_2, \ldots, i_p)$ is any sequence in the set $\mathbb{A}_\alpha$.*
2. *The sequence $(j_1, j_2, \ldots, j_p)$ is any sequence in the subset $\mathbb{A}_{\alpha,\beta}(\pi,\rho) \subset \mathbb{A}_\beta$.*
3. *The summation is over all $\tau = (\tau_1, \tau_2, \ldots, \tau_p)$ that are elements of the set $L(1^{\lambda_1}, 2^{\lambda_2}, \ldots, n^{\lambda_n})$ of lattice permutations of type λ.*

Proof. Items 1 and 2 summarize the results proved in (7.2)-(7.10), but we must still account for the property asserted in Item 3. The summation over τ in (7.11) was, before simplification to the asserted form, given by $\sum_{\tau_1=1}^{n}\sum_{\tau_2=1}^{n}\cdots\sum_{\tau_p=1}^{n}$. But the matrix elements between the initial vector $\left|\begin{matrix}0^n\\(0)\end{matrix}\right\rangle$ and final vectors $\left|\begin{matrix}\lambda\\m\end{matrix}\right\rangle$ and $\left|\begin{matrix}\lambda\\m'\end{matrix}\right\rangle$ are zero unless the shift $(0^n)\rightarrow e_{\tau_1}+e_{\tau_2}+\cdots+e_{\tau_k}$ effected by the action of the right subproduct $t_{i_k\tau_k}\cdots t_{i_2\tau_2}t_{i_1\tau_1}$ of $t_{i_p\tau_p}\cdots t_{i_2\tau_2}t_{i_1\tau_1}$ is a partition for each $k=1,2,\ldots,p$. But the sequence $e_{\tau_1}+e_{\tau_2}+\cdots+e_{\tau_k}$ is a partition if and only if the sequence $(\tau_1,\tau_2,\ldots,\tau_k)$ contains a number of $1's$ at least as great as the number of $2's$, a number of $2's$ at least as great as the number of $3's,\cdots$, a number $k-1's$ at least as great as the number of $k's$. (Since the entries in $e_{\tau_1}+e_{\tau_2}+\cdots+e_{\tau_k}$ sum to k, there is no $k+1$ in the sequence.) A sequence $\tau=(\tau_1,\tau_2,\ldots,\tau_p)$ such that each left subsequence $(\tau_1,\tau_2,\ldots,\tau_k), k=1,2,\ldots,p$, has these properties is called a *lattice permutation of type* λ (see Sect. 11.3.6, Compendium B). □

Relation (7.11) is quite a nice result, fully combinatorial in its structure. It is, of course, fully defined, since the action of each fundamental tensor operators on the initial Hilbert space $H_{0^{n-1}}$ is well-defined. We have considered the various forms of the C^{λ}–coefficients corresponding to all sequences $(i_1,i_2,\ldots,i_p)\in\mathbb{A}_{\alpha}$ to so as to include the rich variety of ways is which they can be expressed, since the various sequences give rise to different expressions of one and the same coefficient. It is, however, nontrivial to implement relation (7.11) to obtain fully explicit coefficients. We use this in the remaining sections to obtain various special results.

The interchange of the role of the sets of sequences $\mathbb{A}_{\alpha}$ and $\mathbb{A}_{\beta}$ in (7.4)-(7.11) leads to the transpositional symmetry expressed by

$$C\begin{pmatrix}m'\\\lambda\\m\end{pmatrix}(A)=C\begin{pmatrix}m\\\lambda\\m'\end{pmatrix}(A^T). \tag{7.12}$$

Even if the τ–sequence is a lattice permutation of type λ, a term in the summation in (7.11) can give 0 because a GT pattern that violates the betweenness conditions can result from the action of the product of fundamental shift operators, as illustrated by the following example:

Example: $n=4, p=7, \lambda=(4,2,1,0), \alpha=(2,2,1,2), (i_1.i_2,\ldots,i_7)=(1,1,2,2,3,4,4), \tau=(1,1,2,3,2,3,4)$: We have that

$$\left\langle\begin{matrix}4\;2\;1\;0\\4\;1\;0\\4\;0\\2\end{matrix}\right|t_{44}t_{43}t_{32}t_{23}t_{22}t_{11}t_{11}\left|\begin{matrix}0\;0\;0\;0\\0\;0\;0\\0\;0\\0\end{matrix}\right\rangle=0, \tag{7.13}$$

in consequence of the action of the operator $t_{23}t_{22}t_{11}t_{11}$ effecting the shift on $|(0)_n\rangle$ to the vector $\left|\begin{smallmatrix} 2\,1\,1\,0 \\ 2\,1\,1 \\ 2\,2 \\ 2 \end{smallmatrix}\right\rangle$ that violates the betweenness conditions. $\square$

The simplest of the family of D^λ–polynomials occurs for partitions having only one nonzero part. The C^λ–coefficients for $\lambda = (p\,0^{n-1})$ can be obtained by specializing (7.11). Thus, we prove next the following relation:

$$D^p_{\alpha,\beta}(Z) = \sqrt{\alpha!\beta!} \sum_{A \in \mathbb{M}^p_{n\times n}(\alpha,\beta)} \frac{Z^A}{A!}, \tag{7.14}$$

where $\alpha = W\left(\begin{smallmatrix} p & 0^{n-1} \\ & m \end{smallmatrix}\right)$ and $\beta = W\left(\begin{smallmatrix} p & 0^{n-1} \\ & m' \end{smallmatrix}\right)$. We see immediately from (7.11) that the only term in the summation that can "lift" the partition $(0, 0, \cdots, 0)$ to the partition $(p, 0, \ldots, 0)$ is the one corresponding to the single term $\tau_1 = \tau_2 = \cdots = \tau_p = 1$, all other terms between the initial vector and final vector giving zero. This property leads to the following result, where we use $M(p\ 0^{n-1}) = p!$:

$$\begin{aligned} C\begin{pmatrix} & m' & \\ p & 0^{n-1} & \\ & m & \end{pmatrix}(A) &= p!\left\langle \begin{matrix} p & 0^{n-1} \\ & m \end{matrix} \,\middle|\, \prod_{i=1}^{n} t_{i1}^{\alpha_i} \,\middle|\, \begin{matrix} 0^n \\ (0) \end{matrix} \right\rangle \\ &\times \left\langle \begin{matrix} p & 0^{n-1} \\ & m' \end{matrix} \,\middle|\, \prod_{j=1}^{n} t_{j1}^{\beta_j} \,\middle|\, \begin{matrix} 0^n \\ (0) \end{matrix} \right\rangle = \sqrt{\alpha!\beta!}, \end{aligned} \tag{7.15}$$

where we have used

$$\left\langle \begin{matrix} p & 0^{n-1} \\ & m \end{matrix} \,\middle|\, \prod_{i=1}^{n} t_{i1}^{\alpha_i} \,\middle|\, \begin{matrix} 0^n \\ (0) \end{matrix} \right\rangle = \sqrt{\frac{\alpha!}{p!}}, \tag{7.16}$$

and similarly for the second factor.

7.2 Reduction of the Kronecker Product $D^p(Z) \otimes D^\lambda(Z)$

In this section, we derive the CG coefficients that reduce the Kronecker product $D^p(Z) \otimes D^\lambda(Z)$ to standard Kronecker direct sum form. The Littlewood-Richardson numbers are either 0 or 1 for this case, and the CG coefficients are unique up to a sign. We use the factorization lemma (6.96), which reads as follows for the case at hand, where we can now drop the multiplicity index k :

$$D\begin{pmatrix} & (\gamma) & \\ p & 0^{n-1} & \\ & (\alpha) & \end{pmatrix}(\Theta) = \sqrt{\alpha!\gamma!}\sum_{A\in\mathbb{M}^{p}_{n\times n}(\alpha,\gamma)}\frac{\Theta^{A}}{A!}, \tag{7.17}$$

$$\left\langle \begin{matrix} m''' \\ \lambda+\delta \\ m'' \end{matrix} \middle| D\begin{pmatrix} & (\gamma) & \\ p & 0^{n-1} & \\ & (\alpha) & \end{pmatrix}(\Theta) \middle| \begin{matrix} m' \\ \lambda \\ m \end{matrix} \right\rangle$$

$$= \sqrt{\frac{M(\lambda+\delta)}{M(\lambda)}}\, C\left[\begin{pmatrix}\lambda+\delta \\ m''\end{pmatrix}\begin{pmatrix}p\ 0^{n-1} \\ (\alpha)\end{pmatrix}\begin{pmatrix}\lambda \\ m\end{pmatrix}\right] \tag{7.18}$$

$$\times C\left[\begin{pmatrix}\lambda+\delta \\ m'''\end{pmatrix}\begin{pmatrix}p\ 0^{n-1} \\ (\gamma)\end{pmatrix}\begin{pmatrix}\lambda \\ m'\end{pmatrix}\right], \text{ each } \delta\in\mathbb{W}_{(p\ 0^{n-1})}.$$

Thus, the rows and columns of the real orthogonal matrix $C^{(p\ 0^{n-1}),\lambda}$ that reduces the direct product

$$C^{(p\ 0^{n-1}),\lambda}\left(D^{p}(Z)\otimes D^{\lambda}(Z)\right)\left(C^{(p\ 0^{n-1}),\lambda}\right)^{T} = \sum_{\delta}\oplus D^{\lambda+\delta}(Z) \tag{7.19}$$

are given by

$$\left(C^{(p\ 0^{n-1}),\lambda}\right)_{\binom{\lambda+\delta}{m''};\binom{p\ 0^{n-1}}{(\alpha)}\binom{\lambda}{m}} = C\left[\begin{pmatrix}\lambda+\delta \\ m''\end{pmatrix}\begin{pmatrix}p\ 0^{n-1} \\ (\alpha)\end{pmatrix}\begin{pmatrix}\lambda \\ m\end{pmatrix}\right]. \tag{7.20}$$

The notation $\binom{p\ 0^{n-1}}{(\alpha)}$ in (7.18) designates the unique GT pattern with weight α, as illustrated by

$$\begin{pmatrix} p & & 0 & & 0 & & 0 \\ & a & & 0 & & 0 & \\ & & b & & 0 & & \\ & & & c & & & \end{pmatrix}, \quad \begin{aligned} p &= \alpha_1+\alpha_2+\alpha_3+\alpha_4, \\ a &= \alpha_1+\alpha_2+\alpha_3, \\ b &= \alpha_1+\alpha_2, \\ c &= \alpha_1, \end{aligned} \tag{7.21}$$

in which $\alpha = (\alpha_1, \alpha_2, \alpha_3, \alpha_4)$ can be an arbitrary composition of p. Moreover, the summation in the right-hand side of (7.19) is over all shift-weights $\delta \in \mathbb{W}_{(p\ 0^{n-1})}$ such that $\lambda+\delta$ is a partition. By definition, the CG coefficients (7.20) are also zero unless $\lambda+\delta$ is a partition.

Relation (7.18) is a fully explicit relation for determining the CG coefficients. The strategy is the following: Relation (7.18) is first specialized to the initial and final upper GT patterns m' and m''' being maximal, which means each $m'_{i,i} = m'_{i,i+1} = \cdots = m'_{i,n-1} = \lambda_i$, for each $i = 1, 2, n-1$, and similarly for m'''. Then, this relation is further specialized to the initial and final lower GT patterns m and m'' being maximal. The second expression determines, up to sign, the CG coefficient having maximal patterns, which is then used back in the first relation to determine the general CG coefficient.

For the implementation of this procedure, we evaluate the following maximal matrix elements of each fundamental shift operator $t_{i\tau}$ by the methods of the loop algebra described in Sects. 6.1-6.4, Chapter 6. These matrix elements have the very simple form given by

$$\left\langle \begin{matrix} \lambda + e_\tau \\ max \end{matrix} \middle| t_{j\tau} \middle| \begin{matrix} \lambda \\ max \end{matrix} \right\rangle = \delta_{j,\tau} \sqrt{\prod_{s=1}^{j-1} \frac{(p_{jn} - p_{sn} + 1)}{(p_{jn} - p_{sn})}}, 2 \leq j \leq n, \tag{7.22}$$

$$\left\langle \begin{matrix} \lambda + e_\tau \\ max \end{matrix} \middle| t_{1\tau} \middle| \begin{matrix} \lambda \\ max \end{matrix} \right\rangle = \delta_{1,\tau}.$$

Using relations (7.22), we are now able to implement the strategy above to obtain the following result:

The CG coefficients that reduce the Kronecker product $D^p(Z) \otimes D^\lambda(Z)$ are an invariant function times the matrix elements of the shift-operator polynomials $D^p_{\alpha,\delta}(T)$:

$$C\left[\begin{pmatrix} \lambda + \delta \\ m' \end{pmatrix} \begin{pmatrix} p\ 0^{n-1} \\ (\alpha) \end{pmatrix} \begin{pmatrix} \lambda \\ m \end{pmatrix}\right] \tag{7.23}$$

$$= \left[\prod_{1 \leq i < j \leq n} \frac{(p_{in} - p_{jn} - \delta_j + 1)_{\delta_j}}{(p_{in} - p_{jn} + \delta_i - \delta_j + 1)_{\delta_j}}\right]^{1/2} \left\langle \begin{matrix} \lambda + \delta \\ m' \end{matrix} \middle| D^p_{\alpha,\delta}(T) \middle| \begin{matrix} \lambda \\ m \end{matrix} \right\rangle,$$

$$D^p_{\alpha,\delta}(T) = \sqrt{\alpha!\delta!} \sum_{A \in \mathbb{M}^p_{n\times n}(\alpha,\delta)} \frac{T^A}{A!}, \text{ each pair } \alpha, \delta \vdash p, \tag{7.24}$$

in which $\lambda + \delta$ must be a partition for a nonzero result, $(z)_a$ is the rising factorial $(z)_a = z(z+1)\cdots(z+a-1)$, and T^A is the ordered product of fundamental shift operators given by

$$T^A = \prod_{i=1}^n t_{in}^{a_{in}} \cdots \prod_{i=1}^n t_{i2}^{a_{i2}} \prod_{i=1}^n t_{i1}^{a_{i1}}. \tag{7.25}$$

Proof. For maximal GT patterns, the second CG coefficient in the right-hand side of (7.18) is 0 unless $\gamma = \delta$; hence, we obtain:

$$\left\langle \begin{matrix} max \\ \lambda+\delta \\ m'' \end{matrix} \middle| D\begin{pmatrix} (\delta) \\ p\ \ 0^{n-1} \\ (\alpha) \end{pmatrix}(\Theta) \middle| \begin{matrix} max \\ \lambda \\ m \end{matrix} \right\rangle$$

$$= \sqrt{\frac{M(\lambda+\delta)}{M(\lambda)}}\, C\left[\begin{pmatrix} \lambda+\delta \\ m'' \end{pmatrix}\begin{pmatrix} p\ 0^{n-1} \\ (\alpha) \end{pmatrix}\begin{pmatrix} \lambda \\ m \end{pmatrix}\right] \tag{7.26}$$

$$\times C\left[\begin{pmatrix} \lambda+\delta \\ max \end{pmatrix}\begin{pmatrix} p\ 0^{n-1} \\ (\delta) \end{pmatrix}\begin{pmatrix} \lambda \\ max \end{pmatrix}\right],\ \text{each}\ \delta \in \mathbb{W}_{(p\ 0^{n-1})}.$$

In the evaluation of the $max - max$ matrix element in the left-hand side of this relation, only matrix elements of the max -max form (7.22) arise in the intermediate GT patterns from the repeated application of $\sum_{\tau=1}^{n}(t_{i\tau}\otimes_d t_{j\tau})$ to the initial vector $\left| \begin{matrix} max \\ \lambda \\ m \end{matrix} \right\rangle$; that is, the max-max GT patterns propagate to all intermediate GT patterns. Using this property reduces the left-hand side of (7.26) to

$$\left\langle \begin{matrix} max \\ \lambda+\delta \\ m'' \end{matrix} \middle| D\begin{pmatrix} (\delta) \\ p\ \ 0^{n-1} \\ (\alpha) \end{pmatrix}(\Theta) \middle| \begin{matrix} max \\ \lambda \\ m \end{matrix} \right\rangle$$

$$= \sqrt{\frac{M(\lambda+\delta)}{M(\lambda)}} \left\langle \begin{matrix} \lambda+\delta \\ max \end{matrix} \middle| t_{nn}^{\delta_n}\cdots t_{22}^{\delta_2}t_{11}^{\delta_1} \middle| \begin{matrix} \lambda \\ max \end{matrix} \right\rangle$$

$$\times \left\langle \begin{matrix} \lambda+\delta \\ m'' \end{matrix} \middle| D^{p}_{\alpha,\delta}(T) \middle| \begin{matrix} \lambda \\ m \end{matrix} \right\rangle. \tag{7.27}$$

Using again relation (7.22) to evaluate the first factor in this relation and combining the result with (7.26) gives the relation:

$$\left[\prod_{1\le i<j\le n} \frac{(p_{in}-p_{jn}+\delta_i-\delta_j)}{(p_{in}-p_{jn}+\delta_i)}\right]^{1/2} \left\langle \begin{matrix} \lambda+\delta \\ m'' \end{matrix} \middle| D^{p}_{\alpha,\delta}(T) \middle| \begin{matrix} \lambda \\ m \end{matrix} \right\rangle$$

$$= C\left[\begin{pmatrix} \lambda+\delta \\ m'' \end{pmatrix}\begin{pmatrix} p\ 0^{n-1} \\ (\alpha) \end{pmatrix}\begin{pmatrix} \lambda \\ m \end{pmatrix}\right] \tag{7.28}$$

$$\times C\left[\begin{pmatrix} \lambda+\delta \\ max \end{pmatrix}\begin{pmatrix} p\ 0^{n-1} \\ (\delta) \end{pmatrix}\begin{pmatrix} \lambda \\ max \end{pmatrix}\right],\ \text{each}\ \delta \in \mathbb{W}_{(p\ 0^{n-1})}.$$

Evaluating this relation for $m = max$ and $m'' = max$ gives 0 unless $\alpha = \delta$, in which case, we obtain:

$$\left[\prod_{1\le i<j\le n} \frac{(p_{in} - p_{jn} + \delta_i - \delta_j)}{(p_{in} - p_{jn} + \delta_i)}\right]^{1/2} \left\langle \begin{matrix} \lambda+\delta \\ max \end{matrix} \middle| D^p_{\delta,\delta}(T) \middle| \begin{matrix} \lambda \\ max \end{matrix} \right\rangle$$

$$= \left(C\left[\begin{pmatrix} \lambda+\delta \\ max \end{pmatrix} \begin{pmatrix} p\ 0^{n-1} \\ (\delta) \end{pmatrix} \begin{pmatrix} \lambda \\ max \end{pmatrix} \right] \right)^2 . \tag{7.29}$$

But again, using the $max - max$ propagation property as above, we obtain:

$$\left\langle \begin{matrix} \lambda+\delta \\ max \end{matrix} \middle| D^p_{\delta,\delta}(T) \middle| \begin{matrix} \lambda \\ max \end{matrix} \right\rangle = \left[\prod_{1\le i<j\le n} \frac{(p_{in} - p_{jn} + \delta_i - \delta_j)}{(p_{in} - p_{jn} + \delta_i)}\right]^{1/2}$$

$$\times \prod_{1\le i<j\le n} \frac{(p_{in} - p_{jn} + \delta_i - \delta_j + 1)_{\delta_j}}{(p_{in} - p_{jn} - \delta_j + 1)_{\delta_j}}. \tag{7.30}$$

Substituting (7.30) into (7.29) gives

$$C\left[\begin{pmatrix} \lambda+\delta \\ max \end{pmatrix} \begin{pmatrix} p\ 0^{n-1} \\ (\delta) \end{pmatrix} \begin{pmatrix} \lambda \\ max \end{pmatrix} \right] \tag{7.31}$$

$$= \left[\prod_{1\le i<j\le n} \frac{(p_{in} - p_{jn} + \delta_i - \delta_j)(p_{in} - p_{jn} + \delta_i - \delta_j + 1)_{\delta_j}}{(p_{in} - p_{jn} + \delta_i)(p_{in} - p_{jn} - \delta_j + 1)_{\delta_j}}\right]^{1/2},$$

where we have chosen the positive root. Finally, substituting (7.31) into (7.28) (changing m'' to m') gives the stated result (7.23)-(7.24). □

The entire calculation above can be repeated for the case in which the ordered product T^A defined by (7.25) is replaced by

$$(T_\pi)^A = \prod_{i=1}^{n} t_{i\tau_n}^{a_{i\tau_n}} \cdots \prod_{i=1}^{n} t_{i\tau_2}^{a_{i\tau_2}} \prod_{i=1}^{n} t_{i\tau_1}^{a_{i\tau_1}}, \tag{7.32}$$

in which $(\tau_1, \tau_2, \ldots, \tau_n) = \pi(1, 2, \ldots, n)$ is an arbitrary permutation $\pi \in S_n$. In this case, we define the operator-valued D^p−polynomials by

$$D^p_{\alpha,\delta}(T_\pi) = \sqrt{\alpha!\delta!} \sum_{A\in\mathbb{M}^p_{n\times n}(\alpha,\delta)} \frac{(T_\pi)^A}{A!}, \text{ each } \alpha, \delta \vdash p. \tag{7.33}$$

We then have that

$$C\left[\begin{pmatrix}\lambda+\delta\\ m'\end{pmatrix}\begin{pmatrix}p\ 0^{n-1}\\ (\alpha)\end{pmatrix}\begin{pmatrix}\lambda\\ m\end{pmatrix}\right] = \tag{7.34}$$

$$\left[\prod_{1\le i<j\le n}\frac{(p_{\tau_i n}-p_{\tau_j n}-\delta_{\tau_j}+1)_{\delta_{\tau_j}}}{(p_{\tau_i n}-p_{\tau_j n}+\delta_{\tau_i}-\delta_{\tau_j}+1)_{\delta_{\tau_j}}}\right]^{1/2}\left\langle\begin{matrix}\lambda+\delta\\ m'\end{matrix}\ \middle|\ D^p_{\alpha,\delta}(T_\pi)\ \middle|\ \begin{matrix}\lambda\\ m\end{matrix}\right\rangle.$$

This result must, of course, agree with (7.23), although its form is quite different for each $\pi\in S_n$. In general, we obtain $n!$ different forms of the same CG coefficients from (7.34) corresponding to the permutations $(\tau_1,\tau_2,\ldots,\tau_n)=\pi(1,2,\ldots,n), \pi\in S_n$. For $n=2$, the two forms for the WCG coefficients are those obtained by Wigner and Racah (see Ref. [21], Vol. 9), which, in turn, are equal in consequence of the Bailey transform of the ${}_3F_2$ hypergeometric functions of unit argument. There is a deeper theory, not explored here, underlying the relationships that exist between the different expressions for the CG coefficients corresponding to different permutations π. In the specialization, however, to the GT patterns $m = max$ and $m' = max$ there must be agreement, since the result is a single product of factors. We have

$$C\left[\begin{pmatrix}\lambda+\delta\\ max\end{pmatrix}\begin{pmatrix}p\ 0^{n-1}\\ (\delta)\end{pmatrix}\begin{pmatrix}\lambda\\ max\end{pmatrix}\right] \tag{7.35}$$

$$=\left[\prod_{1\le i<j\le n}\frac{(p_{\tau_i n}-p_{\tau_j n}+\delta_{\tau_i}-\delta_{\tau_j})(p_{\tau_i n}-p_{\tau_j n}+\delta_{\tau_i}-\delta_{\tau_j}+1)_{\delta_{\tau_j}}}{(p_{\tau_i n}-p_{\tau_j n}+\delta_{\tau_i})(p_{\tau_i n}-p_{\tau_j n}-\delta_{\tau_j}+1)_{\delta_{\tau_j}}}\right]^{1/2},$$

which, since the factors can be rearranged, is the same as (7.31).

Relation (7.23) and its general form (7.34), like (7.11) for the C^λ–coefficients, are quite nice expressions for the special CG coefficients in terms of the matrix elements of the operator-valued polynomials $D^p_{\alpha,\delta}(T_\pi)$.

As elements of a real orthogonal matrix, the CG coefficients (7.34) satisfy row and column orthogonality relations. Thus, from the notation for rows and columns given by (7.20), we have the orthogonality of rows and columns, respectively, given by

$$\sum_{\alpha\vdash p,\, m\in\mathbb{G}_\lambda} C\left[\begin{pmatrix}\lambda+\Delta\\ m'\end{pmatrix}\begin{pmatrix}p\ 0^{n-1}\\ (\alpha)\end{pmatrix}\begin{pmatrix}\lambda\\ m\end{pmatrix}\right]$$

$$\times C\left[\begin{pmatrix}\lambda+\Delta'\\ m''\end{pmatrix}\begin{pmatrix}p\ 0^{n-1}\\ (\alpha)\end{pmatrix}\begin{pmatrix}\lambda\\ m\end{pmatrix}\right]=\delta_{\Delta,\Delta'}\,\delta_{m',m''}, \tag{7.36}$$

$$\sum_{\Delta \vdash p,\, m'' \in \mathbb{G}_{\lambda+\Delta}} C\left[\begin{pmatrix} \lambda+\Delta \\ m'' \end{pmatrix}\begin{pmatrix} p\ 0^{n-1} \\ (\alpha) \end{pmatrix}\begin{pmatrix} \lambda \\ m \end{pmatrix}\right]$$
$$\times C\left[\begin{pmatrix} \lambda+\Delta \\ m'' \end{pmatrix}\begin{pmatrix} p\ 0^{n-1} \\ (\alpha') \end{pmatrix}\begin{pmatrix} \lambda \\ m' \end{pmatrix}\right] = \delta_{\alpha,\alpha'}\delta_{m,m'}, \qquad (7.37)$$

where $\Delta, \Delta' \vdash p$ and $\lambda+\Delta, \lambda+\Delta' \in \mathbb{P}\mathrm{ar}_n$. (We have set $\Delta = \delta \in \mathbb{W}(p\ 0^{n-1}), \Delta' = \delta' \in \mathbb{W}(p\ 0^{n-1})$ to avoid the awkward notation $\delta_{\delta,\delta'}$.)

Because the $D^{(p\ 0^{n-1})}(Z)$ polynomials (7.14) are invariant under all permutations of the rows and columns of Z, the CG coefficients (7.23) are called *totally symmetric.* They are the elements of the real orthogonal matrix that reduces the Kronecker product $D^p(Z) \otimes D^\lambda(Z) \sim \sum_{\lambda'} \oplus D^{\lambda'}(Z)$, where p abbreviates the partition $(p\ 0^{n-1}) \in \mathbb{P}\mathrm{ar}_n$ and λ is an arbitrary partition $\lambda \in \mathbb{P}\mathrm{ar}_n$. The explicit form of the reduction relation is given for $(p\ 0^{n-1}) \otimes \lambda \sim \lambda'$ by

$$\sum_{(\alpha),\, m,\, (\alpha'),\, m'} C\left[\begin{pmatrix} \lambda' \\ m'' \end{pmatrix}\begin{pmatrix} p\ \ 0^{n-1} \\ (\alpha) \end{pmatrix}\begin{pmatrix} \lambda \\ m \end{pmatrix}\right] D\begin{pmatrix} (\alpha') \\ p\ \ 0^{n-1} \\ (\alpha) \end{pmatrix}(Z) D\begin{pmatrix} m' \\ \lambda \\ m \end{pmatrix}(Z)$$
$$\times C\left[\begin{pmatrix} \lambda' \\ m''' \end{pmatrix}\begin{pmatrix} p\ \ 0^{n-1} \\ (\alpha') \end{pmatrix}\begin{pmatrix} \lambda \\ m' \end{pmatrix}\right] = D\begin{pmatrix} m''' \\ \lambda' \\ m'' \end{pmatrix}(Z). \qquad (7.38)$$

There are three other versions of this relation, obtained by using the orthogonality relations to move either or both of the CG coefficients on the left-side to the right-side.

7.3 Binary Tree Structure of the C^λ–Coefficients

This construction of the C^λ–coefficients is based on Pieri's rule (see relation (11.65), Compendium B), as encoded in the corresponding binary tree. In this section, we prove the following main result, where it is convenient to take the partitions in the form $(\lambda\ 0^{n-l}) \in \mathbb{P}\mathrm{ar}_n$, and also with $\lambda = (\lambda_1, \lambda_2, \ldots, \lambda_l) \in \mathbb{P}\mathrm{ar}_l$, so that we are considering partitions that have at most l nonzero parts, but in which Z is of order n :

The Main Result: *The $D^{(\lambda\ 0^{n-l})}$–polynomials for partitions with $\lambda = (\lambda_1, \lambda_2, \ldots, \lambda_l)$ having at most l nonzero parts are given by*

$$D\begin{pmatrix} m' \\ \lambda\ \ 0^{n-l} \\ m \end{pmatrix}(Z) = \sum_{A \in \mathbb{M}^p_{n\times n}(\alpha,\alpha')} C\begin{pmatrix} m' \\ \lambda\ \ 0^{n-l} \\ m \end{pmatrix}(A)\frac{Z^A}{A!}, \qquad (7.39)$$

$$
\begin{aligned}
&C\begin{pmatrix} & m' & \\ \lambda & 0^{n-l} & \\ & m & \end{pmatrix}(A) \\
&\quad = \frac{A!}{M(\lambda)} \sum_{B\in \mathbb{M}^p_{n\times l}(\alpha,\lambda)} \quad \sum_{C\in \mathbb{M}^p_{n\times l}(\alpha',\lambda)} \left\{ \begin{matrix} B \\ AC \end{matrix} \right\} \\
&\quad \times C\begin{pmatrix} & max & \\ \lambda & 0^{n-l} & \\ & m & \end{pmatrix}(B) C\begin{pmatrix} & max & \\ \lambda & 0^{n-l} & \\ & m' & \end{pmatrix}(C). \qquad (7.40)
\end{aligned}
$$

It is the occurrence of the maximal GT patterns in the expression for the $C-$coefficient that gives this result special significance, as exhibited by the CG coefficients at level l in relation (7.43) below.

The detailed notations and coefficients in the two relations (7.39)-(7.40) are as follows:

1. The matrix Z of indeterminates is $n \times n$, as is the matrix array $A \in \mathbb{M}^p_{n\times n}(\alpha, \alpha')$; the arrays in the summations are over all $B \in \mathbb{M}^p_{n\times l}(\alpha, \lambda)$ and $C \in \mathbb{M}^p_{n\times l} \in (\alpha', \lambda)$ because the maximal pattern in (7.40) forces columns $l+1, \ldots, n$ to be $0's$. The coefficient $\left\{ \begin{matrix} B \\ AC \end{matrix} \right\}$ is the combinatorial structure coefficient of the Maclaurin polynomials, as defined and discussed in Sect. 1.6.3-1.6.5, Chapter 1. We note explicitly that the GT pattern $\begin{pmatrix} \lambda & 0^{n-l} \\ & m \end{pmatrix}$ of shape $(\lambda\, 0^{n-l})$ is given by

$$
\begin{pmatrix}
\lambda_1 \quad \lambda_2 \quad \cdots \quad \lambda_l \quad 0 \quad \cdots \quad 0 \\
m_{1,n-1} \; m_{2,n-1} \cdots m_{l,n-1} \; 0 \quad \cdots \quad 0 \\
\vdots \\
m_{1,l+1} \quad m_{2,l+1} \quad \cdots \quad m_{l,l+1} \quad 0 \\
m_{1,l} \quad m_{2,l} \cdots \; m_{l,l} \\
\vdots \\
m_{1,2} \quad m_{2,2} \\
m_{1,1}
\end{pmatrix}, \qquad (7.41)
$$

where the upper right corner of the array is a triangle of $0's$ in which there are $j-l$ $0's$ in row j for $l+1 \le j \le n$. The number of GT patterns (7.41) is given by the Weyl dimension formula (5.11), specialized to the partitions $(\lambda\; 0^{n-l}), \lambda \in \mathbb{P}\mathrm{ar}_l$:

$$\mathrm{Dim}(\lambda\; 0^{n-l}) = \mathrm{Dim}\;\lambda \left[\prod_{i=1}^{l} \binom{\lambda_i + n - i}{n - l} \Big/ \prod_{i=1}^{l} \binom{n-i}{n-l}\right], \tag{7.42}$$

$$M(\lambda_1, \lambda_2, \ldots, \lambda_l, 0^{n-l}) = M(\lambda_1, \lambda_2, \ldots, \lambda_l),\;\text{all}\, n \geq l.$$

2. The C–coefficients in (7.40) having maximal upper GT patterns are given by

$$C\begin{pmatrix} & max & \\ \lambda & & 0^{n-l} \\ & m & \end{pmatrix}(B) = \sqrt{B!M(\lambda)} \tag{7.43}$$

$$\times \sum_{m^{(1)}, m^{(2)}, \ldots, m^{(n-1)}} \prod_{i=2}^{n} C\left[\begin{pmatrix}\lambda^{(i)}\\ m^{(i)}\end{pmatrix}\begin{pmatrix}\alpha_i\, 0^{l-1}\\ (\beta_i)\end{pmatrix}\begin{pmatrix}\lambda^{(i-1)}\\ m^{(i-1)}\end{pmatrix}\right].$$

The notations for the l–rowed totally symmetric CG coefficients in this relation have the following definitions:

(i). The $\lambda^{(i)}$ are the partitions read off the rows of the GT pattern (7.41), and adjoining $0's$, as needed, such that each partition $\lambda^{(i)} \in \mathbb{P}\mathrm{ar}_l$:

$$\lambda^{(i)} = (m_{1,i}, m_{2,i}, \ldots, m_{i,i}\; 0^{l-i}),\; i = 1, 2, \ldots, l-1;$$

$$\lambda^{(i)} = (m_{1,i}, m_{2,i}, \ldots, m_{l,i}),\; i = l, l+1, \ldots, n. \tag{7.44}$$

The final pattern $i = n$ in the product (7.43) of $n-1$ CG coefficients (each has l rows) has $\lambda^{(n)} = (m_{1,n}, m_{2,n}, \ldots, m_{l,n}) = (\lambda_1, \lambda_2, \ldots, \lambda_l)$, where the GT pattern $m^{(n)}$ of $l-1$ rows is maximal. The remaining $(l-1)$–rowed patterns $m^{(i)}, i = 2, 3, \ldots, n-1$ are general lexical GT patterns. It is very significant that the partitions $\lambda^{(i)}$ all belong to $\mathbb{P}\mathrm{ar}_l$.

(ii). The α_i, $i = 1, 2, \ldots, n$, are the components of the weight of the GT pattern (7.41):

$$\alpha = (\alpha_1, \alpha_2, \ldots, \alpha_n) = W\begin{pmatrix}\lambda & 0^{n-l}\\ & m\end{pmatrix}. \tag{7.45}$$

(iii). Each final pattern

$$\begin{pmatrix}\lambda^{(i)}\\ m^{(i)}\end{pmatrix} \in \mathbb{G}_{\lambda^{(i)}},\; \lambda^{(i)} \in \mathbb{P}\mathrm{ar}_l;\; i = 2, 3, \ldots, n, \tag{7.46}$$

as noted in Item (i), is a full triangular GT pattern of shape $\lambda^{(i)}$ containing $l(l+1)/2$ entries, except that the final pattern corresponding to $\lambda^{(n)} = \lambda \in \mathbb{P}\mathrm{ar}_l$ has maximal entries. This occurrence of the maximal labels in the last CG coefficient in the product is the result of enforcing the row-sum conditions $\sum_{j=1}^n a_{ij} = \lambda_i$ on the $n \times l$ matrix array B in (7.40).

(iv). The matrix array $B \in \mathbb{M}^p_{n\times l}(\alpha, \lambda)$ in the summation has n rows and l columns. The sequence $\beta_i = (b_{i1}, b_{i2}, \ldots, b_{il})$ is the i–th row of B, and is the weight of the GT pattern $G_{(\alpha_i\ 0^{n-l})}$. Thus, the center GT pattern in (7.43) is the unique pattern given by

$$\begin{pmatrix} \alpha_i\ 0^{l-1} \\ (\beta_i) \end{pmatrix} = \begin{pmatrix} \alpha_i & 0 & \cdots & 0 \\ & \vdots & & \\ & b_{i1}+b_{i2} & 0 & \\ & b_{i1} & & \end{pmatrix}, \tag{7.47}$$

where the top row is $\alpha_i = b_{i1} + b_{i2} + \cdots + b_{il}$.

(v). The right-most C–coefficients in (7.40) are defined by the same formula (7.43) by making the replacements $B \to C$, $m \to m'$ (the upper pattern in (7.39)-(7.40)) and reading off the partitions from the GT pattern (7.41) modified by replacing the $m_{i,j}$ by $m'_{i,j}$:

$$C\begin{pmatrix} max \\ \lambda \quad 0^{n-l} \\ m' \end{pmatrix}(C) = \sqrt{C!M(\lambda)} \tag{7.48}$$

$$\times \sum_{m'^{(1)},\ldots,m'^{(n-1)}} \prod_{i=2}^{n} C\left[\begin{pmatrix} \lambda'^{(i)} \\ m'^{(i)} \end{pmatrix}\begin{pmatrix} \alpha'_i\, 0^{l-1} \\ (\beta'_i) \end{pmatrix}\begin{pmatrix} \lambda'^{(i-1)} \\ m'^{(i-1)} \end{pmatrix}\right],$$

where $\lambda'^{(n)} = (\lambda_1, \lambda_2, \ldots, \lambda_l)$.

The CG coefficients occurring in relation (7.43) and given by

$$C\left[\begin{pmatrix} \lambda^{(i)} \\ m^{(i)} \end{pmatrix}\begin{pmatrix} \alpha_i\, 0^{l-1} \\ (\beta_i) \end{pmatrix}\begin{pmatrix} \lambda^{(i-1)} \\ m^{(i-1)} \end{pmatrix}\right] \tag{7.49}$$

are exactly those of the form (7.38) ($n = l$) that effect the coupling of the Kronecker product $D^{(\alpha_i\ 0^{l-1})}(Z) \otimes D^{\lambda^{(i-1)}}(Z)$ to $D^{\lambda^i}(Z)$, with a similar result for relation (7.48). This result implies:

The $C^{(\lambda\,0^{n-l})}$–coefficients in relation (7.40), hence, also the $D^{(\lambda\,0^{n-l})}$–polynomials in (7.39) are completely determined by the totally symmetric CG coefficients at level l.

Moreover, the "master" lower GT patterns (7.41) determines the partitions entering into the top row of each CG coefficient (7.49). A similar statement applies to (7.48) and the upper GT pattern.

Relation (7.40) is a quite complicated expression for the $C^{(\lambda\, 0^{n-l})}$–coefficients, containing as it does all the subsidiary special totally symmetric CG coefficients. Nonetheless, it is fully specific, since all these CG coefficients are known from the preceding section. While structurally inelegant, compared to relation to (7.11), it still has many combinatorial features, especially, the occurrence of the combinatorial structure coefficients originating from the transformation properties of Maclaurin polynomials. *It also shows that for $l = 2$, the general $C^{(\lambda_1,\lambda_2,0^{n-2})}$–coefficients depend entirely on the WCG coefficients of angular momentum theory, as carried out in detail in Sect. 7.3.1 below.* Of course, for $l = 1$, the above relations must give the $D^{p}_{\alpha,\alpha'}(Z)$ polynomials, as they do, accounting for the $l = 1$ CG coefficients given by $C\left[(m''_{11})(m'_{11})(m_{11})\right] = \delta_{m''_{11},m'_{11}+m_{11}}$.

We complete this subsection by giving the proof of the results expressed by(7.39)-(7.49). The method uses Pieri's rule (see Sect. 11.2.1, Compendium B), as encoded in a binary tree, as follows:

Proof. The mapping from the lower GT pattern in the coefficient (7.43) to the associated labeled binary tree is given by

$$\left(\begin{array}{ccccccccc}
\lambda_1 & & \lambda_2 & \cdots & \lambda_l & & 0 & \cdots & & 0 \\
& m_{1,n-1} & m_{2,n-1} & \cdots & m_{l,n-1} & 0 & & \cdots & 0 & \\
& & & & \vdots & & & & & \\
& & m_{1,l+1} & & m_{2,l+1} & \cdots & m_{l,l+1} & 0 & & \\
& & & m_{1,l} & & m_{2,l} \cdots & m_{l,l} & & & \\
& & & & \vdots & & & & & \\
& & & & m_{1,2} & m_{2,2} & & & & \\
& & & & m_{1,1} & & & & &
\end{array}\right)$$

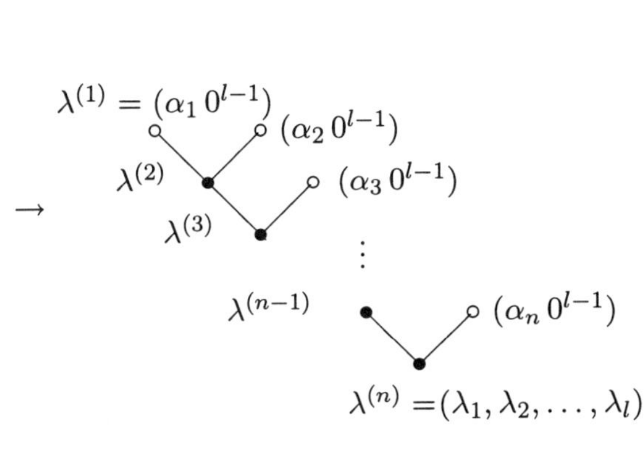

(7.50)

The partitions $(\alpha_i\, 0^{l-1}), i = 1, 2, \ldots, n$, each with l parts, that are assigned to the external $\circ$ points of the graph are the parts of the weight $\alpha = W\left(\begin{smallmatrix}\lambda & 0^{n-l}\\ m & \end{smallmatrix}\right)$ of the GT pattern on the left in (7.50), as given by

$$\begin{aligned}\alpha_i &= (m_{1,i} + m_{2,i} + \cdots + m_{i,i})\\ &\qquad -(m_{1,i-1} + m_{2,i-1} + \cdots + m_{i-1,i-1}),\ i = 1, 2, \ldots, l\\ \alpha_i &= (m_{1,i} + m_{2,i} + \cdots + m_{l,i})\\ &\qquad -(m_{1,i-1} + m_{2,i-1} + \cdots + m_{l,i-1}),\ i = l+1, l+2, \ldots, n,\end{aligned} \tag{7.51}$$

where $\alpha_1 = m_{1,1}$ and $\lambda_i = m_{i,n}, i = 1, 2, \ldots, l$. The partitions $\lambda^{(i)}, i = 2, 3, \ldots, n$, each with l parts, that are assigned to the internal $\bullet$ points of the binary graph are the rows of the GT pattern (7.43), adjusted with 0 parts, as given by (7.44).

The fork in the binary tree (7.50) with root $\lambda^{(i)}$ given by

$$\rightarrow\ C\left[\begin{pmatrix}\lambda^{(i)}\\ m^{(i)}\end{pmatrix}\begin{pmatrix}\alpha_i\, 0^{l-1}\\ (\beta_i)\end{pmatrix}\begin{pmatrix}\lambda^{(i-1)}\\ m^{(i-1)}\end{pmatrix}\right], \tag{7.52}$$

corresponds to the i–th CG coefficient in (7.43); it is this family of CG coefficients that consitute the elements of the real orthogonal matrix that effects the reduction of the Kronecker product

$$D^{(\alpha_i\, 0^{l-1})}(Z) \otimes D^{\lambda^{(i-1)}}(Z) \sim D^{\lambda^{(i)}}(Z). \tag{7.53}$$

But now by Pieri's rule (relation (11.65), Compendium B), the sequential coupling depicted in the binary tree (7.50) must effect exactly the coupling to the $D^{(\lambda\, 0^{n-l})}$–polynomial given by

$$D\begin{pmatrix} & max & \\ \lambda & 0^{n-l} & \\ & m & \end{pmatrix}(Z) = \sum_{A \in \mathbb{M}^p_{n\times l}(\alpha,\lambda)} C\begin{pmatrix} & max & \\ \lambda & 0^{n-l} & \\ & m & \end{pmatrix}(A)\,\frac{Z^A}{A!}, \tag{7.54}$$

where the corresponding $C^{(\lambda\ 0^{n-1})}$–coefficient is given by (7.43) (replace A in (7.54) by B). We now use (7.54) to obtain:

$$D\begin{pmatrix} & max & \\ \lambda & 0^{n-l} & \\ & m & \end{pmatrix}(ZY) = \sum_{A \in \mathbb{M}^p_{n\times l}(\alpha,\lambda)} C\begin{pmatrix} & max & \\ \lambda & 0^{n-l} & \\ & m & \end{pmatrix}(A)\,\frac{(ZY)^A}{A!}, \tag{7.55}$$

in which Z is $n \times n$ and Y is $n \times l$. We now use the multiplication property (5.50) to obtain

$$D\begin{pmatrix} & max & \\ \lambda & 0^{n-l} & \\ & m & \end{pmatrix}(ZY) = \sum_{m' \in \mathbb{G}_{(\lambda\, 0^{n-l})}} D\begin{pmatrix} & m' & \\ \lambda & 0^{n-l} & \\ & m & \end{pmatrix}(Z)$$
$$\times D\begin{pmatrix} & max & \\ \lambda & 0^{n-l} & \\ & m' & \end{pmatrix}(Y). \tag{7.56}$$

The orthogonality of the $D^{(\lambda\ 0^{n-l})}$–polynomials now gives

$$M(\lambda) D\begin{pmatrix} & m' & \\ \lambda & 0^{n-l} & \\ & m & \end{pmatrix}(Z) =$$
$$= \left(D\begin{pmatrix} & max & \\ \lambda & 0^{n-l} & \\ & m' & \end{pmatrix}(Y), D\begin{pmatrix} & max & \\ \lambda & 0^{n-l} & \\ & m & \end{pmatrix}(ZY) \right)$$
$$= \sum_{A \in \mathbb{M}^p_{n\times n}(\alpha,\alpha')} \sum_{B \in \mathbb{M}^p_{n\times l}(\alpha,\lambda)} \sum_{C \in \mathbb{M}^p_{n\times l}(\alpha',\lambda)} C\begin{pmatrix} max \\ \lambda\ 0^{n-l} \\ m \end{pmatrix}(B)$$
$$\times C\begin{pmatrix} max \\ \lambda\ 0^{n-l} \\ m' \end{pmatrix}(C) \left\{ \begin{matrix} B \\ AC \end{matrix} \right\} Z^A, \tag{7.57}$$

where the second relation (1.289) has been used to make the replacement

$$\left(\frac{Y^C}{C!}, \frac{(ZY)^B}{B!} \right) = \sum_{A \in \mathbb{M}^p_{n\times n}(\alpha,\alpha')} \left\{ \begin{matrix} B \\ AC \end{matrix} \right\} Z^A. \tag{7.58}$$

Relation (7.57) gives the coefficient as stated in (7.40) in the main result. □

The following result can be established by obvious modifications of the above:

The $D^{(\lambda\ 0^{n-l})}$–polynomials for partitions with $\lambda = (\lambda_1, \lambda_2, \ldots, \lambda_l)$ having at most l nonzero parts are given by

$$D\begin{pmatrix} & m' & \\ \lambda & 0^{n-l} & \\ & m & \end{pmatrix}(Z) = \sum_{A \in \mathbb{M}^p_{n\times n}(\alpha,\alpha')} C\begin{pmatrix} & m' & \\ \lambda & 0^{n-l} & \\ & m & \end{pmatrix}(A)\, \frac{Z^A}{A!}, \tag{7.59}$$

$$C\begin{pmatrix} & m' & \\ \lambda & & 0^{n-l} \\ & m & \end{pmatrix}(A)$$

$$= \frac{A!}{M(\lambda)} \sum_{B\in \mathbb{M}^p_{l\times n}(\lambda,\alpha)} \quad \sum_{C\in \mathbb{M}^p_{l\times n}(\lambda,\alpha')} \left\{ \begin{matrix} C \\ B\, A \end{matrix} \right\}$$

$$\times C\begin{pmatrix} & m & \\ \lambda & & 0^{n-l} \\ & max & \end{pmatrix}(B) C\begin{pmatrix} & m' & \\ \lambda & & 0^{n-l} \\ & max & \end{pmatrix}(C). \qquad (7.60)$$

The $l = 1$ form of relation (7.40) has B and C given by the $n \times 1$ matrices of weights $B = \mathrm{col}(\alpha_1, \alpha_2, \ldots, \alpha_n)$ and $C = \mathrm{col}(\alpha'_1, \alpha'_2, \ldots, \alpha'_n)$, in which case we have from (1.275) and (7.15) that

$$\left\{ \begin{matrix} B \\ A\, C \end{matrix} \right\} = \frac{1}{A!}, \quad C\begin{pmatrix} max \\ p\ 0^{n-1} \\ (\alpha) \end{pmatrix}(A) = \sqrt{p!\,\alpha!}\,. \qquad (7.61)$$

Using these values in (7.40) and $M(p\ 0^{n-1}) = p!$, then gives the special D^p–polynomial (7.14). Similarly, relation (7.60) is verified for the special case $l = 1$.

7.3.1 The D^λ–polynomials for partitions with two nonzero parts

In this section, we apply the general results of the last section to the determination of the D^λ–polynomials for partitions $\lambda = (\lambda_1, \lambda_2, 0^{n-2})$ with two nonzero parts. As noted in the paragraphs below relation (7.49), these CG coefficients are just the CG coefficients for $l = 2$; that is, the $U(2)$ CG coefficients, which, in fact, are the $SU(2)$ WCG coefficients discussed extensively in Chaper 1. It is the purpose of this subsection to make this relationship explicit.

For $l = 2$, we obtain from (7.39)-(7.40) the results as follows in which $\lambda = (\lambda_1, \lambda_2)$. We repeat the relations fully, since they also make the general notation clear and provide another check of the general relations:

$$D\begin{pmatrix} & m' & \\ \lambda & & 0^{n-2} \\ & m & \end{pmatrix}(Z) = \sum_{A\in \mathbb{M}^p_{n\times n}(\alpha,\alpha')} C\begin{pmatrix} & m' & \\ \lambda & & 0^{n-l} \\ & m & \end{pmatrix}(A)\, \frac{Z^A}{A!}, \qquad (7.62)$$

$$
\begin{aligned}
&C\begin{pmatrix} & m' & \\ \lambda & 0^{n-2} & \\ & m & \end{pmatrix}(A) \\
&\quad = \frac{A!}{M(\lambda)} \sum_{B\in \mathbb{M}^p_{n\times 2}(\alpha,\lambda)} \; \sum_{C\in \mathbb{M}^p_{n\times 2}(\alpha',\lambda)} \left\{ \begin{matrix} B \\ AC \end{matrix} \right\} \\
&\quad \times C\begin{pmatrix} & max & \\ \lambda & 0^{n-2} & \\ & m & \end{pmatrix}(B)\, C\begin{pmatrix} & max & \\ \lambda & 0^{n-2} & \\ & m' & \end{pmatrix}(C).
\end{aligned} \tag{7.63}
$$

The lower GT pattern is given by

$$
\begin{pmatrix}
\lambda_1 & \lambda_2 & 0 & \cdots & 0 \\
m_{1,n-1} & m_{2,n-1} & 0 & \cdots & 0 \\
 & & \vdots & & \\
 & m_{1,3} & m_{2,3} & 0 & \\
 & m_{1,2} & m_{2,2} & & \\
 & & m_{1,1} & &
\end{pmatrix}. \tag{7.64}
$$

The upper pattern is an inverted pattern of this form with entries $m'_{i,j}$.

Applied to $l = 2$, relation (7.43) gives

$$
\begin{aligned}
&C\begin{pmatrix} & max & \\ \lambda & 0^{n-2} & \\ & m & \end{pmatrix}(B) = \sqrt{B!\, M(\lambda)} \\
&\times \prod_{i=2}^{n} C\left[\begin{pmatrix} m_{1,i} & m_{2,i} \\ & h_i \end{pmatrix} \begin{pmatrix} \alpha_i & 0 \\ & b_{i1} \end{pmatrix} \begin{pmatrix} m_{1,i-1} & m_{2,i-1} \\ & h_{i-1} \end{pmatrix} \right],
\end{aligned} \tag{7.65}
$$

where there is no summation in this relation (see below). The matrix array B is the $n \times 2$ matrix array given by

$$
B = \begin{pmatrix} b_{11} & b_{12} \\ b_{21} & b_{22} \\ \vdots & \vdots \\ b_{n1} & b_{n2} \end{pmatrix}, \tag{7.66}
$$

with row and column sums given by

$$
\sum_{j=1}^{2} b_{ij} = \alpha_i,\ i = 1, 2, \ldots n;\ \sum_{i=1}^{n} b_{ij} = \lambda_j,\ j = 1, 2. \tag{7.67}
$$

The entries in the C–coefficient

$$C\left[\begin{pmatrix} m_{1,i}\ m_{2,i} \\ h_i \end{pmatrix}\begin{pmatrix} \alpha_i\ 0 \\ b_{i1} \end{pmatrix}\begin{pmatrix} m_{1,i-1}\ m_{2,i-1} \\ h_{i-1} \end{pmatrix}\right],\ 1 = 1, 2, \ldots, n, \tag{7.68}$$

are the following:

(i). Partitions read off the GT pattern (7.64):

$$\begin{aligned} \lambda^{(1)} &= (m_{1,1}, 0)\ \lambda^{(i)} = (m_{1,i}, m_{2,i}),\ i = 2, 3, \ldots, n-1, \\ \lambda^{(n)} &= (m_{1,n}\ m_{n-1}) = (\lambda_1,\ \lambda_2). \end{aligned} \tag{7.69}$$

(ii). Weights read off the GT pattern (7.64):

$$\begin{aligned} &\alpha_1 = m_{1,1} = b_{11} + b_{12}, \\ &\alpha_2 = m_{1,2} + m_{2,2} - m_{1,1} = b_{21} + b_{22}, \\ &\alpha_i = m_{1,i} + m_{2,i} - m_{1,i-1} - m_{2,i-1} = b_{i1} + b_{i2},\ i = 3, 4, \ldots, n. \end{aligned} \tag{7.70}$$

(iii). Initial and final GT patterns:

$$\begin{aligned} \begin{pmatrix} \lambda^{(1)} \\ m^{(1)} \end{pmatrix} &= \begin{pmatrix} m_{1,1}\ 0 \\ h_1 \end{pmatrix}, \\ \begin{pmatrix} \lambda^{(i)} \\ m^{(i)} \end{pmatrix} &= \begin{pmatrix} m_{1,i}\ m_{2,i} \\ h_i \end{pmatrix},\ i = 2, 3, \ldots, n-1, \\ \begin{pmatrix} \lambda^{(n)} \\ m^{(n)} \end{pmatrix} &= \begin{pmatrix} \lambda_1\ \lambda_2 \\ \lambda_1 \end{pmatrix}, \\ h_i &= b_{11} + b_{21} + \cdots + b_{i1},\ i = 1, 2, \ldots, n, \\ h_n &= \lambda_1. \end{aligned} \tag{7.71}$$

The conversion of the C–coefficient (7.63) into SU(2) angular momentum notation in now an exercise in correspondences between notations, as already used for $n = 2$ in relation (1.297), Chapter 1. The correspondence is made such that the GT pattern in (7.64) corresponds to a standard labeled binary tree:

$$\begin{pmatrix} J_{n-1}+j & & J_{n-1}-j & & 0 & & \cdots & & 0 \\ & J_{n-2}+k_{n-2} & & J_{n-2}-k_{n-2} & 0 & \cdots & & 0 & \\ & & \vdots & \vdots & & & & & \\ & & J_2+k_2 & & J_2-k_2 & & 0 & & \\ & & & J_1+k_1 & & J_1-k_1 & & & \\ & & & & 2j_1 & & & & \end{pmatrix}$$

$\longrightarrow$ 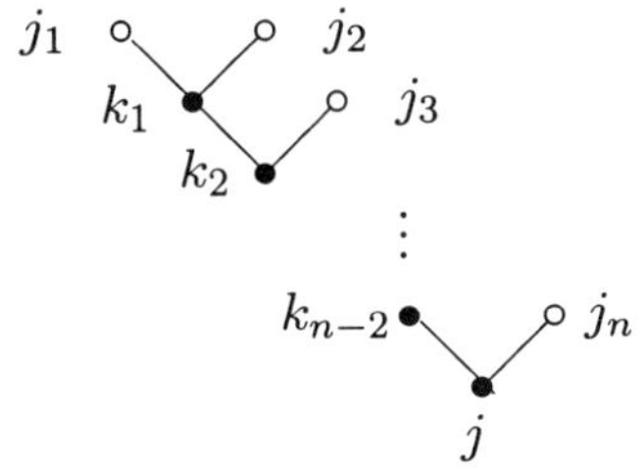

(7.72)

where $J_i = j_1 + j_2 + \cdots + j_{i+1},\ i = 1, 2, \ldots, n-1$. The weight of this pattern is $\alpha = (2j_1, 2j_2, \ldots, 2j_n)$. The invertible relations between labels is

$$\begin{aligned} &j_i = \alpha_i/2,\ k_i = (m_{1,i+1} - m_{2,i+1})/2, i = 1, 2, \ldots, n-1, \\ &k_{n-1} = (m_{1,n} - m_{2,n})/2 = (\lambda_1 - \lambda_2)/2 = j. \end{aligned} \tag{7.73}$$

The inversion of these relations is

$$m_{1,i+1} = J_i + k_i,\ m_{2,i+1} = J_i - k_i,\ i = 1, 2, \ldots, n-1. \tag{7.74}$$

The general rule is that the angular momentum quantum numbers (k, j, m) are associated with a GT pattern for $n = 2$ by the invertible rule

$$\begin{pmatrix} m_{12} & & m_{22} \\ & m_{11} & \end{pmatrix} = \begin{pmatrix} k+j & & k-j \\ & k+m & \end{pmatrix}, \tag{7.75}$$

$$\begin{aligned} &k = (m_{12} + m_{22})/2,\ j = (m_{12} - m_{22})/2, \\ &m = m_{11} - (m_{12} + m_{22})/2; \\ &m_{12} = k+, m_{22} = j - j,\ m_{1,1} = m + k. \end{aligned} \tag{7.76}$$

Thus, for k, j, m either all integers or all half-odd integers, the partition and betweenness conditions are all met, and conversely. The two notations encode the same information.

In the present application, we have the following equality between CG coeffients for $U(2)$ and those for $SU(2)$, the latter being expressed in angular momentum notation:

$$C\left[\begin{pmatrix} m'_{12} & & m'_{22} \\ & m'_{11} & \end{pmatrix}\begin{pmatrix} M_{12} & & M_{22} \\ & M_{11} & \end{pmatrix}\begin{pmatrix} m_{12} & & m_{22} \\ & m_{11} & \end{pmatrix}\right] = C^{j\ J\ j'}_{m\, M\, m'}, \tag{7.77}$$

$$\begin{aligned} j &= (m_{12} - m_{22})/2, & m &= m_{11} - (m_{12} + m_{22})/2, \\ J &= (M_{12} - M_{22})/2, & M &= M_{11} - (M_{12} + M_{22})/2, \\ j' &= (m'_{12} - m'_{22})/2, & m' &= m'_{11} - (m'_{12} + m_{22})/2. \end{aligned} \tag{7.78}$$

Countably infinitely many $U(2)$ CG coefficients equal the same $SU(2)$ CG coefficient: the k quantum label in the pattern (7.75) and similar patterns is irrelevant. This is because the k quantum number corresponds to the choice of a phase $e^{ik\phi}$ factor in the representations of the group $U(2)$ versus those of $SU(2)$, and the CG coefficients can be chosen to be real

It is also informative to note the following features of the two notations. The Clebsch-Gordan series rule is that

$$C\left[\begin{pmatrix} m'_{12} & & m'_{22} \\ & m'_{11} & \end{pmatrix}\begin{pmatrix} M_{12} & & M_{22} \\ & M_{11} & \end{pmatrix}\begin{pmatrix} m_{12} & & m_{22} \\ & m_{11} & \end{pmatrix}\right] = 0, \qquad (7.79)$$

unless the partition in the final pattern satisfies the relations

$$\begin{aligned}&(m'_{12}, m'_{22}) \in \\ &\{(m_{12}+M_{12}, m_{22}+M_{22}), (m_{12}+M_{12}-1, m_{22}+M_{22}+1), \\ &\quad \ldots, (m_{12}+M_{12}-r, m_{22}+M_{22}+r)\}, \end{aligned} \qquad (7.80)$$

where r is the nonnegative integer given by

$$r = \min(m_{12}-m_{22}, M_{12}-M_{22}). \qquad (7.81)$$

This rule is replaced by the addition of angular momentum (triangle rule) in the angular momentum WCG coefficients:

$$C^{j\;J\;j'}_{m\,M\,m'} = 0, \text{unless } j' \in \{j+J, j+J-1, \ldots, |j-J|\}. \qquad (7.82)$$

The $U(1)$ subgroup rule is that

$$C\left[\begin{pmatrix} m'_{12} & & m'_{22} \\ & m'_{11} & \end{pmatrix}\begin{pmatrix} M_{12} & & M_{22} \\ & M_{11} & \end{pmatrix}\begin{pmatrix} m_{12} & & m_{22} \\ & m_{11} & \end{pmatrix}\right] = 0, \qquad (7.83)$$

unless $m'_{11} = m_{11} + M_{11}$, and is replaced by the projection quantum number addition rule

$$C^{j\;J\;j'}_{m\,M\,m'} = 0, \text{ unless } m' = m + M. \qquad (7.84)$$

We next obtain the $C^{(\lambda_1\,\lambda_2\,0^{n-2})}(A)$–coefficients in angular momentum notation. For this, we introduce the abbeviated notation as follows for the GT patterns (7.72):

$$\begin{pmatrix} J_{n-1}+j & J_{n-1}-j & 0^{n-2} \\ & (\mathbf{j};\mathbf{k}) & \end{pmatrix}$$

$$= \begin{pmatrix} J_{n-1}+j & & J_{n-1}-j & & 0 & & \cdots & & 0 \\ & J_{n-2}+k_{n-2} & & J_{n-2}-k_{n-2} & 0 & \cdots & & 0 & \\ & & \vdots & & \vdots & & \vdots & & \\ & & J_2+k_2 & & J_2-k_2 & & 0 & & \\ & & & J_1+k_1 & & J_1-k_1 & & & \\ & & & & 2j_1 & & & & \end{pmatrix}, \tag{7.85}$$

where $\mathbf{j} = (j_1, j_2, \ldots, j_{n-1})$ and $\mathbf{k} = (k_1, k_2, \ldots, k_{n-2})$.

The fully explicit C−coefficient with two parts nonzero and with maximal upper pattern is given by a product of $n-1$ $SU(2)$ WCG coefficients, as follows:

$$\begin{aligned} C &\begin{pmatrix} \max \\ \\ J_{n-1}+j \;\; J_{n-1}-j \;\; 0^{n-2} \\ \\ (\mathbf{j};\mathbf{k}) \end{pmatrix} (A) \\ &= [M(J_{n-1}+j, J_{n-1}-j)A!]^{1/2} \\ &\times \prod_{i=1}^{n-1} C^{k_{i-1}\;\, j_{i+1}\;\, k_i}_{m_1+\cdots+m_i,\, m_{i+1},\, m_1+\cdots+m_{i+1}}, \end{aligned} \tag{7.86}$$

where we have the following relations and definitions of symbols:

(i). $$k_0 = j_1,\; k_{n-1} = j,\; m_1 + m_2 + \cdots + m_n = j,$$

$$M(J_{n-1}+j, J_{n-1}-j) = \frac{(J_{n-1}+j+1)!(J_{n-1}-j)!}{(2j+1)}. \tag{7.87}$$

(ii). $$p = 2J_{n-1},\; \alpha = (2j_1, 2j_2, \ldots, 2j_n),\; \lambda = (J_{n-1}+j,\; J_{n-1}-j),$$

$$A = \begin{pmatrix} j_1+m_1 & j_1-m_1 \\ j_2+m_2 & j_2-m_2 \\ \vdots & \vdots \\ j_n+m_n & j_n-m_n \end{pmatrix} \in \mathbb{M}_{n\times 2}(\alpha, \lambda). \tag{7.88}$$

The above results are now assembled with appropriate renaming of symbols and substituted into the main result (7.39)-(7.40) to obtain the following relation for the $D^{(\lambda\; 0^{n-2})}$−polynomials for partitions with at most two parts nonzero, expressed in terms of angular momentum quantum numbers and $SU(2)$ WCG coefficients:

$$D\begin{pmatrix} & (\mathbf{j}';\mathbf{k}') & \\ J_{n-1}+j & J_{n-1}-j & 0^{n-2} \\ & (\mathbf{j};\mathbf{k}) & \end{pmatrix}(Z) \tag{7.89}$$

$$= \sum_{A\in\mathbb{M}_{n\times n}(\alpha,\alpha')} C\begin{pmatrix} & (\mathbf{j}';\mathbf{k}') & \\ J_{n-1}+j & J_{n-1}-j & 0^{n-2} \\ & (\mathbf{j};\mathbf{k}) & \end{pmatrix}(A)\,\frac{Z^A}{A!},$$

where the C−coefficients have the following explicit form:

$$C\begin{pmatrix} & (\mathbf{j}';\mathbf{k}') & \\ J_{n-1}+j & J_{n-1}-j & 0^{n-2} \\ & (\mathbf{j};\mathbf{k}) & \end{pmatrix}(A) \tag{7.90}$$

$$= A! \sum_{B\in\mathbb{M}^p_{n\times 2}(\alpha,\lambda)} \sum_{C\in\mathbb{M}^p_{n\times 2}(\alpha',\lambda)} \sqrt{B!C!}\left\{ \begin{matrix} B \\ AC \end{matrix} \right\} \sum_{\substack{m_1+m_2+\cdots+m_n=j \\ m'_1+m'_2+\cdots+m'_n=j}}$$

$$\times \prod_{i=1}^{n-1} C^{k_{i-1}\; j_{i+1}\; k_i}_{m_1+\cdots+m_i,\, m_{i+1},\, m_1+\ldots+m_{i+1}} C^{k'_{i-1}\; j'_{i+1}\; k'_i}_{m'_1+\cdots+m'_i,\, m'_{i+1},\, m'_1+\ldots+m'_{i+1}}.$$

We recall, for convenience, the definitions of symbols:

1. Matrix array A :

$$A = \begin{pmatrix} a_{11} & a_{12} & \cdots & a_{1n} \\ a_{21} & a_{22} & \cdots & a_{2n} \\ \vdots & \vdots & \cdots & \vdots \\ a_{n1} & a_{n2} & \cdots & a_{nn} \end{pmatrix} \in \mathbb{M}^p_{n\times n}(\alpha,\alpha') \tag{7.91}$$

$$p = 2J_{n-1},\ \alpha=(2j_1,\ldots,2j_n),\ \alpha'=(2j'_1,\ldots,2j'_n),$$

with $J_{n-1}=j_1+j_2+\cdots+j_n$.

2. Matrix arrays B and C :

$$\begin{aligned} B &= \begin{pmatrix} b_{11} & b_{12} \\ b_{21} & b_{22} \\ \vdots & \vdots \\ b_{n1} & b_{n2} \end{pmatrix} \in \mathbb{M}^p_{n\times 2}(\alpha, \lambda), \\ C &= \begin{pmatrix} c_{11} & c_{12} \\ c_{21} & c_{22} \\ \vdots & \vdots \\ c_{n1} & c_{n2} \end{pmatrix} \in \mathbb{M}^p_{n\times 2}(\alpha', \lambda), \\ (\lambda_1, \lambda_2) &= (j_1 + j_2 + \cdots + j_n + j, j_1 + j_2 + \cdots + j_n - j). \end{aligned} \tag{7.92}$$

3. The structure coefficient:

$$\left\{ \begin{matrix} B \\ A\,C \end{matrix} \right\} = \sum_{H \in \mathbb{M}_{n\times n\times 2}(A,C,B)} \frac{1}{H!}. \tag{7.93}$$

The preceding results, (7.89)-(7.93), can also be applied to the case $j = j_1 + j_2 + \cdots + j_n$ to obtain the $D^p_{\alpha\,\beta}(Z)$ polynomials in relation (1.243) in terms of angular momentum notation. This provides a check on the above results, as also does the case for $n = 2$, which can be verified to reduce to (1.297). Thus, for $j = J_{n-1}$, the lower and upper GT patterns in (7.89) have $k_1 = k'_1 = J_1, k_2 = k'_2 = J_2, \ldots, k_{n-1} = k'_{n-1} = J_{n-2}$, which gives $p = 2J_{n-1}$ and

$$\begin{aligned} B &= \mathrm{col}(\alpha_1, \alpha_2, \ldots, \alpha_n) = (2j_1, 2j_2, \ldots, 2j_n), \\ C &= \mathrm{col}(\alpha'_1, \alpha'_2, \ldots, \alpha'_n) = (2j'_1, 2j'_2, \ldots, 2j'_n). \end{aligned} \tag{7.94}$$

Using the value of the structure constant $\left\{ \begin{matrix} B \\ A\,C \end{matrix} \right\} = \frac{1}{A!}$, and the fact that the sum over WCG coefficients gives 1, we reproduce relation (1.243) with parameters $p = 2(j_1 + j_2 + \cdots + j_n)$ and α and $\beta = \alpha'$ given by (7.94).

The result, recognizing that the $SU(3)$ representation functions for partitions with two nonzero parts could be expressed in terms of $SU(2)$ WCG coefficients, was found by Rowe and Repka [158] and in the more general context discussed above in Refs. [115, 122, 123].

Another interesting result comes from evaluating the polynomials (7.89) at the value $Z = P_\pi$, where P_π is the permutation matrix of order n with elements $(P_\pi)_{ij} = \delta_{i,\pi_j}$:

$$D\left(\begin{array}{c}(\mathbf{j}';\mathbf{k}')\\ \\ J_{n-1}+j\ \ J_{n-1}-j\ \ 0^{n-2}\\ \\ (\mathbf{j};\mathbf{k})\end{array}\right)(P_\pi) \tag{7.95}$$

$$= \frac{1}{(A_\pi)!}C\left(\begin{array}{c}(\mathbf{j}';\mathbf{k}')\\ \\ J_{n-1}+j\ \ J_{n-1}-j\ \ 0^{n-2}\\ \\ (\mathbf{j};\mathbf{k})\end{array}\right)(A_\pi), \tag{7.96}$$

where A_π is the matrix with elements given by

$$(A_\pi)_{k\,l} = \delta_{k,\pi_l}\,\delta_{j_k,j'_l}\,(2j_k). \tag{7.97}$$

The nonzero entries in A_π are in the same positions as the $1's$ in P_π. The entry in row k and column l is $2j_k = 2j'_l$, where l is determined by $\pi_l = k$, for $\pi = (\pi_1, \pi_2, \ldots, \pi_n)$. The structure constant in (7.90) is given by

$$\left\{\begin{array}{c} B \\ A_\pi\ C\end{array}\right\} = \delta_{B,C}\frac{1}{B!},\ \text{each}\ \pi \in S_n. \tag{7.98}$$

Thus, relation (7.89) and (7.90) reduce for $Z = P_\pi$ to the triangle coefficient of order $2(n-1)$ given by

$$D\left(\begin{array}{c}(\mathbf{j}';\mathbf{k}')\\ \\ J_{n-1}+j\ \ J_{n-1}-j\ \ 0^{n-2}\\ \\ (\mathbf{j};\mathbf{k})\end{array}\right)(P_\pi) \tag{7.99}$$

$$= \left\{\begin{array}{cccc|cccc} j_1 & k_1 & \cdots & k_{n-2} & j'_1 & k'_1 & \cdots & k'_{n-2}\\ j_2 & j_3 & \cdots & j_n & j'_2 & j'_3 & \cdots & j'_n\\ k_1 & k_2 & \cdots & k_{n-1} & k'_1 & k'_2 & \cdots & k'_{n-1}\end{array}\right\},$$

in which $k_{n-1} = k'_{n-1} = j$, and the angular momentum quantum numbers $\mathbf{j}' = (j'_1, j'_2, \cdots, j'_n)$ are the permutation $\pi = (\pi_1, \pi_2, \ldots, \pi_n)$ of $\mathbf{j} = (j_1, j_2, \cdots, j_n)$ given by

$$j'_l = j_k \text{ for the value of } l \text{ such that } \pi_l = k\ . \tag{7.100}$$

The triangle coefficient of order $2(n-1)$ in (7.99) may, of course, reduce to lower order for certain $\pi \in S_n$. Relation (7.99) correctly reduces to the unit matrix for $\pi = (1, 2, \ldots, n)$, the identity element in S_n, since then condition (7.100) gives the coefficient value $\delta_{\mathbf{j},\mathbf{j}'}$, and the triangle coefficient gives the factor $\delta_{\mathbf{k},\mathbf{k}'}$. Results relating the value of special D–polynomials at level n to $SU(2)$ $3n - j$ coefficients are largely unexplored. As an example of (7.99), we have the identity for $n = 3$ given by

$$D\left(\begin{array}{ccccc} & & 2j_3 & & \\ & j_2+j_3+k' & & j_2+j_3-k' & \\ j_1+j_2+j_3+j & & j_1+j_2+j_3-j & & 0 \\ & j_1+j_2+k & & j_1+j_2-k & \\ & & 2j_1 & & \end{array}\right)\left(\begin{array}{ccc} 0 & 0 & 1 \\ 0 & 1 & 0 \\ 1 & 0 & 0 \end{array}\right)$$

$$= \left\{\begin{array}{cc|cc} j_1 & k & j_3 & k' \\ j_2 & j_3 & j_2 & j_1 \\ k & j & k' & j \end{array}\right\}$$

$$= (-1)^{j_1+j_2+j_3-j}\sqrt{(2k+1)(2k'+1)}\; W(j_1 j_2 j j_3; kk'). \qquad (7.101)$$

7.3.2 Recurrence relation for the D^λ–polynomials

A recurrence relation giving the D^λ–polynomials at level n in terms of the variables $(z_{1n}, z_{2n} \ldots, z_{nn}), (z_{n1}, z_{n2} \ldots, z_{nn})$ and the D^μ–polynomials at level $n-1$ can be derived, as mentioned in Item (ii), p. 277. We begin with the general expression (5.6), Chapter 5, and write

$$\frac{Z^A}{A!} = \frac{z_{nn}^{a_{nn}}}{a_{nn}!} \times \prod_{i=1}^{n-1} \frac{z_{in}^{a_{in}}}{a_{in}!} \times \prod_{j=1}^{n-1} \frac{z_{nj}^{a_{nj}}}{a_{nj}!} \times \frac{(Z')^{A'}}{A'!}, \qquad (7.102)$$

where Z' and A' are the submatrices of Z and A given by $Z' = (z_{ij})_{1 \le i,j \le n-1}$ and $A' = (a_{ij})_{1 \le i,j \le n-1}$. We now use (5.9) at level $n-1$ to express $\frac{(Z')^{A'}}{A'!}$ in terms of the D^μ–polynomials, $\mu \in \mathbb{P}\mathrm{ar}_{n-1}$. This gives

$$D\left(\begin{array}{c} m' \\ \lambda \\ m \end{array}\right)(Z) = \sum_\mu \sum_{m'', m''' \in \mathbb{G}_\mu} \left[\left(\begin{array}{c} m' \\ \lambda \\ m \end{array}\right)\middle|\left(\begin{array}{c} m''' \\ \mu \\ m'' \end{array}\right)\right]$$

$$\times \frac{z_{nn}^{a_{nn}}}{a_{nn}!} \prod_{i=1}^{n-1} \frac{z_{in}^{\alpha_i-\beta_i}}{(\alpha_i-\beta_i)!} \prod_{j=1}^{n-1} \frac{z_{jn}^{\alpha'_i-\beta'_i}}{(\alpha'_i-\beta'_i)!} \times D\left(\begin{array}{c} m''' \\ \mu \\ m'' \end{array}\right)(Z'); \qquad (7.103)$$

$$\begin{aligned} \alpha &= W\binom{\lambda}{m},\ \alpha' = W\binom{\lambda}{m'}, \\ \beta &= W\binom{\mu}{m''},\ \beta' = W\binom{\mu}{m'''}, \qquad (7.104) \\ a_{nn} &= \alpha_n + \alpha'_n + |\mu| - |\lambda| \ge 0. \end{aligned}$$

The numerical coefficient $[\cdots]$ in this relation has the following definition:

$$\left[\begin{pmatrix} m' \\ \lambda \\ m \end{pmatrix} \middle| \begin{pmatrix} m''' \\ \mu \\ m'' \end{pmatrix}\right] = \sum_{A' \in \mathbb{M}^{|\mu|}_{(n-1)\times(n-1)}(\beta,\beta')} \frac{1}{M(\mu)A'!} C\begin{pmatrix} m' \\ \lambda \\ m \end{pmatrix}(A) C\begin{pmatrix} m''' \\ \mu \\ m'' \end{pmatrix}(A'), \tag{7.105}$$

in which $A \in \mathbb{M}^{|\lambda|}_{n\times n}(\alpha, \alpha')$ is given in terms of A' by

$$A = \left(\begin{array}{c|c} A' & \begin{matrix} \alpha_1 - \beta_1 \\ \vdots \\ \alpha_{n-1} - \beta_{n-1} \end{matrix} \\ \hline \alpha'_1 - \beta'_1 \cdots \alpha'_{n-1} - \beta'_{n-1} & a_{nn} \end{array}\right). \tag{7.106}$$

Since the C^λ–coefficients, $\lambda \in \mathbb{P}\mathrm{ar}_n$, and the C^μ–coefficients $\mu \in \mathbb{P}\mathrm{ar}_{n-1}$, are fully determined in terms of totally symmetric CG coefficients, relation (7.103) must likewise be expressible in terms of such CG coefficients: The summatation in (7.105) can be re-expressed to give the following result, which is in terms of totally symmetric CG coefficients:

$$\begin{aligned} \frac{1}{M(\lambda)}\left[\begin{pmatrix} m' \\ \lambda \\ m \end{pmatrix} \middle| \begin{pmatrix} m''' \\ \mu \\ m'' \end{pmatrix}\right] &= (a_{nn})!\sqrt{p!p'!\prod_{i=1}^{n-1}(a_{in})!\prod_{j=1}^{n-1}(a_{nj})! \Big/ a_n!a'_n!} \\ \times \sum_{\lambda'} C&\left[\begin{pmatrix} \lambda \\ m \end{pmatrix}\begin{pmatrix} p \;\; 0^{n-1} \\ (a) \end{pmatrix}\begin{pmatrix} \lambda' \\ \mu \\ m'' \end{pmatrix}\right] C\left[\begin{pmatrix} \lambda' \\ \mu \\ m'' \end{pmatrix}\begin{pmatrix} p' \;\; 0^{n-1} \\ (0) \end{pmatrix}\begin{pmatrix} \mu \;\; 0 \\ \mu \\ m'' \end{pmatrix}\right] \\ \times C&\left[\begin{pmatrix} \lambda \\ m' \end{pmatrix}\begin{pmatrix} p \;\; 0^{n-1} \\ (0) \end{pmatrix}\begin{pmatrix} \lambda' \\ m' \end{pmatrix}\right] C\left[\begin{pmatrix} \lambda' \\ m' \end{pmatrix}\begin{pmatrix} p' \;\; 0^{n-1} \\ (a') \end{pmatrix}\begin{pmatrix} \mu \;\; 0 \\ \mu \\ m''' \end{pmatrix}\right], \end{aligned} \tag{7.107}$$

where $|\lambda| = p + p' + |\mu|$. The GT patterns corresponding to the partitions $(p\ 0^{n-1})$ and $(p'\ 0^{n-1})$ are uniquely determined by the weights:

$$\begin{aligned} W\begin{pmatrix} p \;\; 0^{n-1} \\ (a) \end{pmatrix} &= (a_{1n}, a_{2n} \ldots, a_{n-1,n}, a_n) = W\begin{pmatrix} \lambda \\ m \end{pmatrix} - W\begin{pmatrix} \lambda' \\ \mu \\ m'' \end{pmatrix}, \\ W\begin{pmatrix} p' \;\; 0^{n-1} \\ (a') \end{pmatrix} &= (a_{n1}, a_{n2} \ldots, a_{n,n-1}, a'_n) = W\begin{pmatrix} \lambda \\ m' \end{pmatrix} - W\begin{pmatrix} \lambda' \\ \mu \\ m''' \end{pmatrix}, \\ p &= \sum_{i=1}^{n-1} a_{in} + a_n,\ p' = \sum_{j=1}^{n-1} a_{nj} + a'_n, a_n + a'_n = a_{nn}. \end{aligned} \tag{7.108}$$

Proof. We first give the special polynomials:

$$\begin{aligned} \frac{1}{\sqrt{p!}} D \begin{pmatrix} (0) \\ p \;\; 0^{n-1} \\ (a) \end{pmatrix} (Z) &= \prod_{i=1}^{n-1} \frac{z_{in}^{a_{in}}}{\sqrt{(a_{in})!}} \times \frac{z_{nn}^{a_n}}{\sqrt{a_n!}}, \\ \frac{1}{\sqrt{p'!}} D \begin{pmatrix} (a') \\ p' \;\; 0^{n-1} \\ (0) \end{pmatrix} (Z) &= \prod_{j=1}^{n-1} \frac{z_{nj}^{a_{nj}}}{\sqrt{(a_{nj})!}} \times \frac{z_{nn}^{a'_n}}{\sqrt{a'_n!}}. \end{aligned} \tag{7.109}$$

The following relation is obtained by using the orthogonality of the CG coefficients in (7.38) to move both to the right-hand side:

$$D \begin{pmatrix} (b') \\ p' \;\; 0^{n-1} \\ (b) \end{pmatrix} (Z)\, D \begin{pmatrix} m''' \\ \mu \;\; 0 \\ m'' \end{pmatrix} (Z) = \tag{7.110}$$

$$\sum_{\lambda', m, m'} C \left[\begin{pmatrix} \lambda' \\ m \end{pmatrix} \begin{pmatrix} p' \;\; 0^{n-1} \\ (b) \end{pmatrix} \begin{pmatrix} \mu \;\; 0 \\ m'' \end{pmatrix} \right] C \left[\begin{pmatrix} \lambda' \\ m' \end{pmatrix} \begin{pmatrix} p' \;\; 0^{n-1} \\ (b') \end{pmatrix} \begin{pmatrix} \mu \;\; 0 \\ m''' \end{pmatrix} \right] D \begin{pmatrix} m' \\ \lambda' \\ m \end{pmatrix} (Z),$$

where the sum is over all $\lambda' \in (p'\; 0^{n-1}) \otimes (\mu\; 0)$. We are free to specialize still further the GT patterns in the left-hand side of this relation. To obtain the form (7.103), we use a double application of (7.111) obtained by multiplying from the left by a $D^{(p\; 0^{n-1})}$–polynomial, followed by another expansion of the form (7.110). We then adapt the notation to the special products that appear in (7.103) to obtain:

$$\begin{aligned} &D \begin{pmatrix} (0) \\ p \;\; 0^{n-1} \\ (a) \end{pmatrix} (Z) D \begin{pmatrix} (a') \\ p' \;\; 0^{n-1} \\ (0) \end{pmatrix} (Z)\, D \begin{pmatrix} m''' \\ \mu \\ m'' \end{pmatrix} (Z') = \sum_{\lambda, m, m'} D \begin{pmatrix} m' \\ \lambda \\ m \end{pmatrix} (Z) \\ &\times \sum_{\lambda'} C \left[\begin{pmatrix} \lambda \\ m \end{pmatrix} \begin{pmatrix} p \;\; 0^{n-1} \\ (a) \end{pmatrix} \begin{pmatrix} \lambda' \\ \mu \\ m'' \end{pmatrix} \right] C \left[\begin{pmatrix} \lambda' \\ \mu \\ m'' \end{pmatrix} \begin{pmatrix} p' \;\; 0^{n-1} \\ (0) \end{pmatrix} \begin{pmatrix} \mu \;\; 0 \\ \mu \\ m'' \end{pmatrix} \right] \\ &\quad \times C \left[\begin{pmatrix} \lambda \\ m' \end{pmatrix} \begin{pmatrix} p \;\; 0^{n-1} \\ (0) \end{pmatrix} \begin{pmatrix} \lambda' \\ m' \end{pmatrix} \right] C \left[\begin{pmatrix} \lambda' \\ m' \end{pmatrix} \begin{pmatrix} p' \;\; 0^{n-1} \\ (a') \end{pmatrix} \begin{pmatrix} \mu \;\; 0 \\ \mu \\ m''' \end{pmatrix} \right]. \end{aligned} \tag{7.111}$$

It is very important that the GT patterns in the $D^{(\mu\; 0)}$–polynomials on the left-hand side of (7.110) have also been specialized by choosing the entries in row $n-1$ of the upper and lower pattern to be the maximal value $\mu \in \mathbb{P}\mathrm{ar}_{n-1}$. Then, the following identity holds, where now m'' and m''' denote GT patterns of $n-2$ rows:

$$D \begin{pmatrix} m''' \\ \mu \\ \mu \;\; 0 \\ \mu \\ m'' \end{pmatrix} (Z) = D \begin{pmatrix} m''' \\ \mu \\ m'' \end{pmatrix} (Z'). \tag{7.112}$$

The D^μ–polynomial in the left-hand side of (7.111) has the standard expansion in terms of $(Z')^{A'}/A'!$. This factor combines with those from (7.109) to give the factor $(mboxnumericalfactor Z^A/A!$ in the left-hand side of (7.111), where the numerical factor is the square-root factor in the right-hand side of (7.107). Thus, the coefficient (7.107) is obtained from the inversion of (7.111). □

Relation (7.103) can now be iterated upward by starting with $n = 2$ to obtain the D^λ–polynomials as the expansion (5.6) in terms of Maclaurin polynomials. The recurrence relation exhibits vividly that the $C^\lambda(A)$–coefficients at level n depend successively on the totally symmetric CG coefficients at levels $2, 3, \ldots, n$.

Relation (7.103) must give the $D^{(\lambda_1\,\lambda_2)}$–polynomials for $n = 2$. The GT patterns m'' and m''' are empty; $\mu_1 = a_{11} = \beta_1 = \beta_1'$; and $D(\mu_1)(z_{11}) = z_{11}^{a_{11}}$. All other symbols are such that $A \in \mathbb{M}^\lambda_{2\times 2}(\alpha, \alpha')$; the summation over $\mu_1 = a_{11}$ is therefore a summation over $A \in \mathbb{M}^{|\lambda|}_{2\times 2}(\alpha, \alpha')$. Also, the relation $p + p' = \lambda_1 + \lambda_2 - a_{11}$ must be satisfied. The C–coefficients in the right-hand side are standard $SU(2)$ WCG coefficients. Relations (7.103) and (7.107) give the following result in terms of the angular momentum notation (1.304)-(1.305):

$$
C\begin{pmatrix} \alpha_1' \\ \lambda_1\ \lambda_2 \\ \alpha_1 \end{pmatrix}(A) = (a_{11})! \left[\begin{pmatrix} \alpha_1' \\ \lambda_1\ \lambda_2 \\ \alpha_1 \end{pmatrix} \middle| (a_{11}) \right]
$$

$$
= M(\lambda)A! \sqrt{\binom{2k}{a_{12}}\binom{2k'}{a_{21}}} \sum_{j'} C^{j'\ \ k\ \ j}_{\kappa', \nu, m} C^{j''\ \ k'\ \ j'}_{j'', -k', \kappa'} C^{j'\ \ k\ \ j}_{\kappa'', -k, m'} C^{j''\ \ k'\ \ j'}_{j'', \nu', \kappa''}. \tag{7.113}
$$

$$
\begin{aligned}
j &= (\lambda_1 - \lambda_2)/2, m = \alpha_1 - (\lambda_1 + \lambda_2)/2,\ m' = \alpha_1' - (\lambda_1 + \lambda_2)/2, \\
j' &= (\lambda_1' - \lambda_2')/2, \kappa' = a_{11} - (\lambda_1' + \lambda_2')/2, \quad \kappa'' = \alpha_1' - (\lambda_1' + \lambda_2')/2, \\
j'' &= a_{11}/2,\ k = p/2,\ \nu = a_{12} - k,\ k' = p'/2,\ \nu' = a_{21} - k'.
\end{aligned}
$$

The summation in this relation is over all j' such that the triangles $\langle j'kj \rangle$ and $\langle j''k'j' \rangle$ are satisfied. The quantum number $j'' = a_{11}/2$ is specified from the left-hand side, as is the sum $k + k' = (\lambda_1 + \lambda_2 - a_{11})/2$, but otherwise k and k' can be chosen. There is no summation over projection quantum numbers—they are all fixed by the labels on the left-hand side and k and k'. Because of the difference in methods of assembly, relations (7.113) and (7.90) ($n = 2$) have quite different structures. An identity between WCG coefficients is implied.

Chapter 8

The General Linear and Unitary Groups

8.1 Background and Review

Exponential matrices are key objects underlying the relationship between the classical Lie groups and their algebras. Such matrices are of considerable interest on their own and some aspects of the subject are developed in Sect. 10.4.3, Compendium A. The formal relationship between a matrix $X \in M_{n\times n}(\mathbb{C})$, the set of complex matrices of order n, that appears as an exponent, and the exponential matrrix $Z(t)$ it defines is given by

$$Z(t) = e^{tX} = I_n + tX + \frac{1}{2!}(tX)^2 + \cdots, \;\; t \in \mathbb{R}. \tag{8.1}$$

By use of the Cayley-Hamilton theorem, the complex matrix e^{tX} can always be written as a sum over $I_n, X, \ldots, X^{n-1}$, with coefficients that depend on the parameter t and on the characteristic roots of X, but such functions are Maclaurin series expansions in t, where little can be said, in general, about the domain of convergence in t. The converse problem, where $Z(t)$ is given, and it is X that is to be determined, is still more difficult. The matrix $Z(t)$ is called a *matrix curve* through the identity $Z(0) = I_n$ (Sternberg [166]). It is a known result that the matrix tX does exist for each matrix curve $Z(t)$ through the origin if t belongs to a neighborhood of the origin. In what follows these properties are implicit, together with whatever differentiable properties are needed for the relations to make sense. It is not our intention to be rigorous, but only to make the connection between matrix Lie groups and their matrix algebras plausible. Two other basic relations are

$$\det Z(t) = e^{\mathrm{trace}(tX)}, \tag{8.2}$$

$$\left.\frac{dZ(t)}{dt}\right|_{t=0} = X. \tag{8.3}$$

The set of matrices defined by $\mathcal{Z} = \{Z(t) \mid t \in \mathbb{R}\}$ is an Abelian group under matrix multiplication, since matrix multiplication is associative, there is an identity element $Z(0) = I_n$ in $\mathcal{Z}$, the product $Z(t)Z(s) = A(t+s)$ belongs to $\mathcal{Z}$, and the inverse $A^{-1}(s) = A(-s)$ belongs to $\mathcal{Z}$, and satisfies $Z(-t)Z(t) = Z(t)Z(-t) = Z(0)$. Thus, we have that $\mathcal{Z} \subset GL(n,\mathbb{C})$. It is called a *one-parameter subgroup of* $GL(n,\mathbb{C})$. *The matrix X is called the generator* of the one-parameter subgroup $\mathcal{Z}$. The matrix X is also called the *infinitesimal element at the identity* of the subgroup $\mathcal{Z}$. For trace$X = 0$, the matrix curve satisfies $\det Z(t) = 1$. Conversely, if $\det Z(t) = 1$, we choose trace$X = 0$, and not trace$X = 2\pi mi, m \in \mathbb{Z}$. While our principal interest is in the matrix group $GL(n,\mathbb{C})$ and its unitary subgroup $U(n)$, it is useful to place these concepts in a more general setting, before returning below to what is relevant for this monograph.

The central idea (Weyl [178] (p.260), Sternberg [166] (p.234)) underlying the elements of a group $\mathcal{G}$ of matrices of order n and its infinitesimal matrix elements is the mapping of a group element to an infinitesimal element defined by

$$\left.\left(\frac{dG(t)}{dt}\right)\right|_{t=0} = A \in M_{n\times n}(\mathbb{C}), \tag{8.4}$$

where $G(t)$ is a differentiable curve of matrices, which means that each element $g_{ij}(t)$ of $G(t)$ is a differentiable function of t, where the matrix curve $G(t)$ also satisfies $G(t) \in \mathcal{G}$, all $t \in \mathbb{R}, G(0) = I_n$. A principal issue is the determination of the algebraic rules that the infinitesimal elements must obey in order that they are compatible with the properties of the group from which they are derived. These rules may be inferred systematically by examining the group laws (see Sect. 10.1, Compendium A). They fall into three classes:

1. Vector space properties: From $\exp t(\alpha A), \exp t(\beta B) \in \mathcal{G}, \exp t(\alpha A)$ $\exp t(\beta B) \in \mathcal{G}$, for $\alpha, \beta \in \mathbb{C}$, we infer that if A, B are infinitesimal elements of $\mathcal{G}$, then $\alpha A + \beta B$ is an infinitesimal element of $\mathcal{G}$ for all complex numbers α and β. Thus, the set of infinitesimal elements of a matrix group $\mathcal{G}$ is a linear vector space $V_{\mathcal{G}}$ of matrices of order n over the complex numbers.

2. Multiplication properties: The multiplication properties of the group $\mathcal{G}$ imply multiplication properties for the infinitesimal elements in the vector space $V_{\mathcal{G}}$. In anticipation of this, let $A * B$ denote the product of a pair of elements $(A, B) \in V_{\mathcal{G}} \times V_{\mathcal{G}}$, so that $A * B$ is

a mapping of the direct product vector space to the vector space itself: $V_{\mathcal{G}} \times V_{\mathcal{G}} \to V_{\mathcal{G}}$. An important group element for determining the product $A * B$ is the *commutator* of two differentiable curves $G(t) = \exp(tA)$ and $H(s) = \exp(sB)$ of matrices defined by

$$
\begin{aligned}
& G(t)H(s)G^{-1}(t)H^{-1}(s) \\
& = (I + tA + \frac{t^2}{2}A^2 + \cdots)(I + sB + \frac{t^2}{2}B^2 + \cdots) \\
& \times(I - tA + \frac{t^2}{2}A^2 + \cdots)(I - sB + \frac{s^2}{2}B^2 + \cdots) \\
& = I + ts(AB - BA) \qquad (8.5) \\
& \quad +(\text{terms in } s, t \text{ of total degree greater than } 2).
\end{aligned}
$$

Based on this product rule for the commutator of group elements, the product $A * B$ of two matrices in the vector space $V_{\mathcal{G}}$ is defined by the commutator of these matrices:

$$A * B = [A, B] = AB - BA. \qquad (8.6)$$

A closely related result uses the relation

$$G(t)H(s)G^{-1}(t) = \exp\left(sG(t)BG^{-1}(t)\right) \in \mathcal{G}. \qquad (8.7)$$

The matrix $G(t)BG^{-1}(t)$ is the element of the vector space $V_{\mathcal{G}}$ corresponding to the group element $G(t)H(s)G^{-1}(t)$. But we have

$$
\begin{aligned}
& A(t)B(s)G^{-1}(t) = (I + tA + \cdots)B(I - tA + \cdots) \\
& = I + t(AB - BA) + \text{ higher order terms in } t. \qquad (8.8)
\end{aligned}
$$

The commutator $AB - BA$ is the infinitesimal element (matrix) corresponding to the group matrix $G(t)H(s)G^{-1}(t)$. We could also use the Baker-Campell-Hausdorff identity (see relation (10.122), Compendium A):

$$e^{tA}Be^{-tA} = \sum_{k \geq 0} \left\langle A^k B \right\rangle \frac{t^k}{k!} \qquad (8.9)$$

to obtain this result: *Group similarity transformations of a matrix belonging to the vector space* $V_{\mathcal{G}}$ *are matrices in* $V_{\mathcal{G}}$.

3. Special algebraic properties: The product rule $A * B = AB - BA$ for two matrices of the vector space $V_{\mathcal{G}}$ satisfies the rules:

   (i). skew-symmetric: $B * A = -A * B$.

(ii). Jacobi identity: $A * (B * C) + B * (C * A) + C * (A * B) = \mathbf{0}$.

It may be verified by the above methods, that the skew-symmetry is a consequence of the group requirement of an inverse, and the Jacobi identity is a consequence of the group requirement of associativity.

A set of mathematical objects that constitute a linear vector space over a field for which the product is skew-symmetric and satisfies the Jacobi identity is called a *Lie algebra.* For matrix groups over the field of complex numbers, the product is the matrix commutator $[A, B] = AB - BA$, which trivially obeys the rules (i) and (ii), and constitutes a matrix Lie algebra. *The vector space $V_{\mathcal{G}}$ of skew-symmetric matrices with product of vectors given by the matrix commutator is the Lie algebra of the matrix group $\mathcal{G}$.*

An abstract Lie algebra can be defined by postulating that it elements satisfy the above rules without reference to a Lie group Thus, while a Lie group, by the structure of the differentiable manifold over which it is defined, possesses a Lie algebra, the inverse problem of determining the group, if it exists, from a given Lie algebra, is a difficult task that has been the subject of an enormous wealth of literature of the twentieth century and the last part of the ninetieth.

The discovery that quantum mechanics can be obtained from classical mechanics by mapping classical dynamical quantities into operators acting in Hilbert space, and, in particular, that classical angular momentum of a single particle in $\mathbb{R}^3$ becomes the infinitesimal elements of the Lie group $SO(3, \mathbb{R}^3)$, was recognized early on by Weyl [177], Wigner [181], and Eckart [54] as a fundamental expression of invariance principles in nonrelativistic quantum theory. Moreover, the natural extension of this to the covering group $SU(2)$ provided the mathematical description for the existence of intrinsic, internal attributes of particles called *spin.* Thus, quantum theory, applied to many-particle composite systems at all levels of the microscopic world, molecular, atomic, nuclear, and particle, became a natural and fertile arena for the development and applications of Lie groups and Lie algebras. Indeed, group symmetries and invariance principles are foundational blocks of modern physics.

8.2 $GL(n, \mathbb{C})$ and its Unitary Subgroup $U(n)$

We return now from the brief excursus into Lie groups and Lie algebras to our principal needs: The Lie algebra of the general linear group $GL(n, \mathbb{C})$ and of its unitary subgroup $U(n)$. For the general linear group, the matrix X occuring in the matrix curve $G(t) = \exp tX$ is an arbitrary complex matrix $X \in M_{n \times n}(\mathbb{C})$. Thus, *the Lie algebra of $GL(n, \mathbb{C})$ is the vector* *space of complex matrices $M_{n\times n}(\mathbb{C})$ over the field of complex numbers $\mathbb{C}$, equipped with the commutator product $X * Y = [X, Y]$. The standard notation for this Lie algebra is $gL(n, \mathbb{C}) = V_{GL(n,\mathbb{C})}$.*

Each matrix $X \in M_{n\times n}(\mathbb{C})$ can be expressed as the sum

$$X = \sum_{i,j=1}^{n} x_{ij} e_{ij}, \tag{8.10}$$

where $e_{ij} = e_i^T e_j$ is the *matrix unit* having a 1 in row i and column j and 0 for all other entries. The set of matrix units $E = \{e_{ij} \mid i, j = 1, 2, \ldots, n\}$ is thus a *basis* for all matrices in $M_{n\times n}(\mathbb{C})$. While the matrix units in E satisfy the multiplication rule

$$e_{ij} e_{kl} = \delta_{j,k} e_{il}, \tag{8.11}$$

it is the commutator product

$$[e_{ij}, e_{kl}] = \delta_{j,k} e_{il} - \delta_{l,i} e_{kj} \tag{8.12}$$

that is the key relation for the Lie algebra $gl(n, \mathbb{C})$ with elements (8.10).

For the unitary subgroup $U(n) \subset GL(n, \mathbb{C})$, a differentiable matrix curve through the origin is given by

$$U(t) = \exp(tH), \text{H } \textit{skew-Hermitian}; \left(\frac{dU(t)}{dt}\right)\Bigg|_{t=0} = H. \tag{8.13}$$

Thus, *the Lie algebra $u(n)$ of the unitary group $U(n)$ is given by the set of complex skew-Hermitian matrices over the field of complex numbers $\mathbb{C}$ with product $H * K = [H, K]$ given by the commutator.* Since

$$H = \sum_{i,j=1}^{n} h_{ij} e_{ij}, \tag{8.14}$$

with only the conditions $h_{ij} = -h_{ji}^*$, a basis of the Lie algebra is still the set E of matrix units.

If we restrict $GL(n, \mathbb{C})$ and $U(n)$ to the subgroups $SL(n, \mathbb{C})$ and $SU(n)$ having unit determinant, then the conditions trace$X = 0$ and trace$H = 0$ must be enforced on the elements $X \in gl(n, \mathbb{C})$ and $H \in su(n)$ of the Lie algebras.

A *matrix representation* Γ_k of the Lie algebra $gl(n, \mathbb{C})$ in the basis E is a set of n^2 complex matrices, each of dimension $k \in \mathbb{N}$:

$$\Gamma_k = \{M_{ij} \in M_{k\times k}(\mathbb{C}) \mid i, j = 1, 2, \ldots, n\}, \tag{8.15}$$

where the matrices in this set satisfy the same commutator relations as the defining matrix units (8.12):

$$[M_{ij}, M_{kl}] = \delta_{j,k} M_{il} - \delta_{l,i} M_{kj}. \tag{8.16}$$

An *operator representation,* or realization, Γ_k of the Lie algebra $gl(n,\mathbb{C})$ in the basis E is a set of linear operators with an invariant action defined in an inner product space V_k of dimension k :

$$\Gamma_k = \{T_{ij} : V_k \to V_k \mid i,j = 1,2,\ldots,n\}, \tag{8.17}$$

where the n^2 operators in this set satisfy the same commutator relations as the defining matrix units (8.12) and matrix representation (8.16):

$$[T_{ij}, T_{kl}] = \delta_{j,k} T_{il} - \delta_{l,i} T_{kj}. \tag{8.18}$$

This operator representation is equivalent to the matrix representation, since each such operator may be replaced by the $k \times k$ matrix representing the linear action of the operator in V_k.

We next give two differential operator representations of the algebra $gl(n,\mathbb{C})$ in the vector space $\mathcal{V}_N$ of homogeneous polynomials of degree N :

$$\mathcal{L}_{ij} = \sum_{h=1}^{n} z_{ih} \frac{\partial}{\partial z_{jh}}, \; i,j = 1,2,\ldots,n; \tag{8.19}$$

$$\mathcal{R}_{ij} = \sum_{h=1}^{n} z_{hi} \frac{\partial}{\partial z_{hj}}, \; i,j = 1,2,\ldots,n. \tag{8.20}$$

Since each of these operators is homogeneous of degree 0, the action of each maps the vector space $\mathcal{V}_N$ into itself: $\mathcal{L}_{ij} : \mathcal{V}_N \to \mathcal{V}_N, \;\; \mathcal{R}_{ij} : \mathcal{V}_N \to \mathcal{V}_N$. The $gl(n,\mathbb{C})$ operator commutator relations (8.18) are verified for each set $\{\mathcal{L}_{ij}\}$ and $\mathcal{R}_{ij}\}$ by direct computation:

$$\begin{aligned} [\mathcal{L}_{ij}, \mathcal{L}_{kl}] &= \delta_{j,k} \mathcal{L}_{il} - \delta_{l,i} \mathcal{L}_{kj}, \\ [\mathcal{R}_{ij}, \mathcal{R}_{kl}] &= \delta_{j,k} \mathcal{R}_{il} - \delta_{l,i} \mathcal{R}_{kj}. \end{aligned} \tag{8.21}$$

In addition, we have the property that each operator in the set $\{\mathcal{L}_{ij}\}$ commutes with each operator in the set $\{\mathcal{R}_{ij}\}$:

$$[\mathcal{L}_{ij}, \mathcal{R}_{kl}] = 0, \; i,j,k,l = 1,2,\ldots,n. \tag{8.22}$$

The left and right actions of matrix transformations of the vector space $\mathcal{V}_N$ are the source of the mutual commutation of the associated operator Lie algebras (8.22), as discussed in detail in Sect. 10.6, Compendium A.

The action of each of the sets of operators (8.19)-(8.20) on the D^λ–nomials is fully determined by the following matrix element relations, which are a special case of (6.67):

$$\left(\widehat{D}\begin{pmatrix} m''' \\ \lambda \\ m'' \end{pmatrix}(Z), \mathcal{L}_{ij}\widehat{D}\begin{pmatrix} m' \\ \lambda \\ m \end{pmatrix}(Z)\right)$$

$$= \delta_{m''',m'}\left\langle \begin{matrix} \lambda \\ m'' \end{matrix} \middle| L_{ij} \middle| \begin{matrix} \lambda \\ m \end{matrix} \right\rangle, \tag{8.23}$$

$$\left(\widehat{D}\begin{pmatrix} m''' \\ \lambda \\ m'' \end{pmatrix}(Z), \mathcal{R}_{ij}\widehat{D}\begin{pmatrix} m' \\ \lambda \\ m \end{pmatrix}(Z)\right)$$

$$= \delta_{m'',m}\left\langle \begin{matrix} \lambda \\ m''' \end{matrix} \middle| L_{ij} \middle| \begin{matrix} \lambda \\ m' \end{matrix} \right\rangle, \tag{8.24}$$

where the operators L_{ij} are defined on each abstract Hilbert space H_λ by

$$L_{ij} = \sum_{\tau=1}^{n} s_{i\tau}s^\dagger_{j\tau},\ i,j = 1,2,\ldots,n. \tag{8.25}$$

An important consequence of (8.23)-(8.24) is following: The abstract operators $\{L_{ij}\}$ defined $L_{ij} = \sum_{\tau=1}^{n} s_{i\tau}s^\dagger_{j\tau}$ satisfy the commutation relations

$$[L_{ij}, L_{kl}] = \delta_{j,k}L_{il} - \delta_{l,i}L_{kj} \tag{8.26}$$

on each space H_λ, hence, on the separable Hilbert space H. Moreover, we have the Hermitian conjugate relation $L^\dagger_{ij} = L_{ji}$. These properties are summarized as follows:

The action of the differential operator realizations $\{\mathcal{L}_{ij}\}$ and $\{\mathcal{R}_{ij}\}$ of the Lie algebra $gl(n,\mathbb{C})$ in polynomial space is effected by the action of the realization $\{L_{ij}\}$ of the Lie algebra in the abstract Hilbert space H_λ, which, in turn, is fully determined by the combinatorially defined fundamental shift operator action. The matrix representations of $\{\mathcal{L}_{ij}\}$ and $\{\mathcal{R}_{ij}\}$ so obtained are identical.

We point out that, despite appearances, *the realization (8.25) of the abstract Lie algebra $gl(n,\mathbb{C})$ is not the familiar harmonic oscillator realization, as realized through the boson operators.* The boson operator a_{ij} corresponds to x_{ij} in relation (6.67). The $\{s_{ij}\}$ and their conjugates do not satisfy the commutation relations of boson operators.

Examples: The action of the operators L_{ij} of the Lie algebra in the abstract Hilbert space H_λ for $n = 2, 3$ can be obtained directly from the explicit actions of the fundamental shift operator and their conjugates given at the end of Chapter 6:

$n = 2:$

$$E_{11}\left| \begin{matrix} \lambda_1 & & \lambda_2 \\ & m_{11} & \end{matrix} \right\rangle = p_{11}\left| \begin{matrix} \lambda_1 & & \lambda_2 \\ & m_{11} & \end{matrix} \right\rangle,$$

$$E_{22}\left| \begin{matrix} \lambda_1 & & \lambda_2 \\ & m_{11} & \end{matrix} \right\rangle = (p_{12} + p_{22} - p_{11} - 1)\left| \begin{matrix} \lambda_1 & & \lambda_2 \\ & m_{11} & \end{matrix} \right\rangle,$$

$$E_{12}\left| \begin{matrix} \lambda_1 & & \lambda_2 \\ & m_{11} & \end{matrix} \right\rangle = \sqrt{(p_{12} - p_{11} - 1)(p_{11} - p_{22} + 1)}\left| \begin{matrix} \lambda_1 & & \lambda_2, \\ & m_{11} + 1 & \end{matrix} \right\rangle,$$

$$E_{21}\left| \begin{matrix} \lambda_1 & & \lambda_2 \\ & m_{11} & \end{matrix} \right\rangle = \sqrt{(p_{12} - p_{11})(p_{11} - p_{22})}\left| \begin{matrix} \lambda_1 & & \lambda_2, \\ & m_{11} - 1 & \end{matrix} \right\rangle.$$

$n = 3:$

$$E_{11}\left| \begin{matrix} \lambda_1 & \lambda_2 & \lambda_3 \\ m_{12} & m_{22} & \\ & m_{11} & \end{matrix} \right\rangle = p_{11}\left| \begin{matrix} \lambda_1 & \lambda_2 & \lambda_3 \\ m_{12} & m_{22} & \\ & m_{11} & \end{matrix} \right\rangle,$$

$$E_{22}\left| \begin{matrix} \lambda_1 & \lambda_2 & \lambda_3 \\ m_{12} & m_{22} & \\ & m_{11} & \end{matrix} \right\rangle = (p_{12} + p_{22} - p_{11} - 1)\left| \begin{matrix} \lambda_1 & \lambda_2 & \lambda_3 \\ m_{12} & m_{22} & \\ & m_{11} & \end{matrix} \right\rangle,$$

$$E_{12}\left| \begin{matrix} \lambda_1 & \lambda_2 & \lambda_3 \\ m_{12} & m_{22} & \\ & m_{11} & \end{matrix} \right\rangle = \sqrt{(p_{12} - p_{11} - 1)(p_{11} - p_{22} + 1)}\left| \begin{matrix} \lambda_1 & \lambda_2 & \lambda_3 \\ m_{12} & m_{22} & \\ & m_{11} + 1 & \end{matrix} \right\rangle,$$

$$E_{21}\left| \begin{matrix} \lambda_1 & \lambda_2 & \lambda_3 \\ m_{12} & m_{22} & \\ & m_{11} & \end{matrix} \right\rangle = \sqrt{(p_{12} - p_{11})(p_{11} - p_{22})}\left| \begin{matrix} \lambda_1 & \lambda_2 & \lambda_3 \\ m_{12} & m_{22} & \\ & m_{11} - 1 & \end{matrix} \right\rangle,$$

$$E_{33}\left| \begin{matrix} \lambda_1 & \lambda_2 & \lambda_3 \\ m_{12} & m_{22} & \\ & m_{11} & \end{matrix} \right\rangle = (p_{13} + p_{23} + p_{33} - p_{12} - p_{22})\left| \begin{matrix} \lambda_1 & \lambda_2 & \lambda_3 \\ m_{12} & m_{22} & \\ & m_{11} & \end{matrix} \right\rangle,$$

$$\begin{aligned} E_{13}&\left| \begin{matrix} \lambda_1 & \lambda_2 & \lambda_3 \\ m_{12} & m_{22} & \\ & m_{11} & \end{matrix} \right\rangle \\ &= \sqrt{\frac{(p_{12} - p_{23} + 1)(p_{13} - p_{12} - 1)(p_{12} - p_{33} + 1)(p_{11} - p_{22} + 1)}{(p_{12} - p_{22})(p_{12} - p_{22} + 1)}}\left| \begin{matrix} \lambda_1 & \lambda_2 & \lambda_3 \\ m_{12} + 1 & m_{22} & \\ & m_{11} + 1 & \end{matrix} \right\rangle \\ &- \sqrt{\frac{(p_{23} - p_{22} - 1)(p_{22} - p_{33} + 1)(p_{13} - p_{22} - 1)(p_{12} - p_{11} - 1)}{(p_{12} - p_{22} - 1)(p_{12} - p_{22})}}\left| \begin{matrix} \lambda_1 & \lambda_2 & \lambda_3 \\ m_{12} & m_{22} + 1 & \\ & m_{11} + 1 & \end{matrix} \right\rangle, \end{aligned}$$

$$E_{23}\left|\begin{matrix}\lambda_1 & & \lambda_2 & & \lambda_3\\ & m_{12} & & m_{22} & \\ & & m_{11} & & \end{matrix}\right\rangle$$

$$=\sqrt{\frac{(p_{13}-p_{12}-1)(p_{12}-p_{23}+1)(p_{12}-p_{33}+1)(p_{12}-p_{11})}{(p_{12}-p_{22})(p_{12}-p_{22}+1)}}\left|\begin{matrix}\lambda_1 & & \lambda_2 & & \lambda_3\\ & m_{12}+1 & & m_{22} & \\ & & m_{11} & & \end{matrix}\right\rangle$$

$$+\sqrt{\frac{(p_{13}-p_{22}-1)(p_{23}-p_{22}-1)(p_{22}-p_{33}+1)(p_{11}-p_{22})}{(p_{12}-p_{22}-1)(p_{12}-p_{22})}}\left|\begin{matrix}\lambda_1 & & \lambda_2 & & \lambda_3\\ & m_{12} & & m_{22}+1 & \\ & & m_{11} & & \end{matrix}\right\rangle,$$

$$E_{31}\left|\begin{matrix}\lambda_1 & & \lambda_2 & & \lambda_3\\ & m_{12} & & m_{22} & \\ & & m_{11} & & \end{matrix}\right\rangle$$

$$=\sqrt{\frac{(p_{12}-p_{23})(p_{13}-p_{12})(p_{12}-p_{33})(p_{11}-p_{22})}{(p_{12}-p_{22}-1)(p_{12}-p_{22})}}\left|\begin{matrix}\lambda_1 & & \lambda_2 & & \lambda_3\\ & m_{12}-1 & & m_{22} & \\ & & m_{11}-1 & & \end{matrix}\right\rangle$$

$$-\sqrt{\frac{(p_{23}-p_{22})(p_{22}-p_{33})(p_{13}-p_{22})(p_{12}-p_{11})}{(p_{12}-p_{22})(p_{12}-p_{22}+1)}}\left|\begin{matrix}\lambda_1 & & \lambda_2 & & \lambda_3\\ & m_{12} & & m_{22}-1 & \\ & & m_{11}-1 & & \end{matrix}\right\rangle,$$

$$E_{32}\left|\begin{matrix}\lambda_1 & & \lambda_2 & & \lambda_3\\ & m_{12} & & m_{22} & \\ & & m_{11} & & \end{matrix}\right\rangle$$

$$=\sqrt{\frac{(p_{13}-p_{12})(p_{12}-p_{23})(p_{12}-p_{33})(p_{12}-p_{11}-1)}{(p_{12}-p_{22}-1)(p_{12}-p_{22})}}\left|\begin{matrix}\lambda_1 & & \lambda_2 & & \lambda_3\\ & m_{12}-1 & & m_{22} & \\ & & m_{11} & & \end{matrix}\right\rangle$$

$$+\sqrt{\frac{(p_{13}-p_{22})(p_{23}-p_{22})(p_{22}-p_{33})(p_{11}-p_{22}+1)}{(p_{12}-p_{22})(p_{12}-p_{22}+1)}}\left|\begin{matrix}\lambda_1 & & \lambda_2 & & \lambda_3\\ & m_{12} & & m_{22}-1 & \\ & & m_{11} & & \end{matrix}\right\rangle.$$

We next derive the general action of the Lie algebra in the Hilbert space H_λ, results first given by Gelfand and Tsetlin [62] (see also Gelfand and Graev [60]), and subsequently derived by a number of other authors (see, for example, Moshinsky [138] and Baird and Biedenharn [7]). It will be noticed in the examples above that the actions of the generators $\{E_{ij}|i,j=1,2,\ldots,n-1\}$ defining the Lie subalgebra $gl(n-1,\mathbb{C})$ are the same in both $gl(n-1,\mathbb{C})$ and $gl(n,\mathbb{C})$. This is, of course, a general feature that reflects the classification of the representations in accordance with the subalgebra embeddings

$$gl(n,\mathbb{C})\supset gl(n-1,\mathbb{C})\supset\cdots\supset gl(1,\mathbb{C}). \tag{8.27}$$

The general relationship is given by

$$\left\langle\begin{matrix}\lambda'\\ \mu'\\ m'\end{matrix}\right|E_{ij}^{(n)}\left|\begin{matrix}\lambda\\ \mu\\ m\end{matrix}\right\rangle$$

$$=\delta_{\lambda',\lambda}\delta_{\mu',\mu}\left\langle\begin{matrix}\mu\\ m'\end{matrix}\right|E_{ij}^{(n-1)}\left|\begin{matrix}\mu\\ m\end{matrix}\right\rangle,\ 1\le i\le j\le n-1, \tag{8.28}$$

where the partitions λ, $\lambda' \in \mathbb{P}ar_n$ and μ, $\mu' \in \mathbb{P}ar_{n-1}$ satisfy the betweenness conditions $\mu \prec \lambda$ and $\mu' \prec \lambda'$. The generators at level n and level $n-1$ are expressed in terms of fundamental shift operators by

$$E_{ij}^{(n)} = \sum_{\tau=1}^{n} s_{i\tau} s_{j\tau}^{\dagger},\; E_{ij}^{(n-1)} \sum_{\rho=1}^{n-1} s_{i\rho}^{(n-1)} s_{j\rho}^{(n-1)\dagger}. \tag{8.29}$$

The proof of relation (8.28) is an application of Sylvester's identity (Sect. 11.7, Compendium B): Using the reduction rule (6.19), we find that relation (8.28) is valid if and only if the following relation is true:

$$\sum_{\tau=1}^{n} y_\tau \prod_{\substack{k=1\\k\neq\tau}}^{n} \frac{y_k - y_\tau + 1}{y_k - y_\tau} \left(A_{\rho\tau}^{\dagger}(x,y)\right)^2 = x_\rho \prod_{\substack{l=1\\l\neq\rho}}^{n-1} \frac{x_l - x_\rho + 1}{x_l - x_\rho}. \tag{8.30}$$

The variables x and y variables are given in terms of the partitions λ and μ by

$$\begin{aligned} x &= (\mu_1 + n - 2, \mu_2 + n - 3, \ldots, \mu_{n-1}), \\ y &= (\lambda_1 + n - 1, \lambda_2 + n - 2, \ldots, \lambda_n). \end{aligned} \tag{8.31}$$

This relation is proved by substituting $A_{\rho\tau}^{\dagger}(x,y)$ from (6.35)-(6.36) into this relation. The simplification to the right-hand side is effected by using Sylvester's identity to carry out the summation over τ in much the same manner as done in proving relations (6.46)-(6.48). We omit the details. □

We are now in position to derive the Gelfand-Tsetlin result for action of the operator $E_{ij}^{(n)}$ on the basis $\mathcal{B}_\lambda$ of the Hilbert space H_λ. From relation (8.28), we have

$$E_{ij}^{(n)} \left| \begin{matrix} \lambda \\ \mu \\ m \end{matrix} \right\rangle = \sum_{m' \in \mathbb{G}_\mu} \left\langle \begin{matrix} \mu \\ m' \end{matrix} \right| E_{ij}^{(n-1)} \left| \begin{matrix} \mu \\ m \end{matrix} \right\rangle \left| \begin{matrix} \lambda \\ \mu \\ m' \end{matrix} \right\rangle, \tag{8.32}$$

for all $1 \le i \le j \le k \le n-1$. Thus, recursively, to obtain the matrix elements of the family of all generators, it is necessary and sufficient to determine the action

$$E_{in}^{(n)} \left| \begin{matrix} \lambda \\ m \end{matrix} \right\rangle = \sum_{m' \in \mathbb{G}_\lambda} \left\langle \begin{matrix} \lambda \\ m' \end{matrix} \right| E_{in}^{(n)} \left| \begin{matrix} \lambda \\ m \end{matrix} \right\rangle \left| \begin{matrix} \lambda \\ m' \end{matrix} \right\rangle, \tag{8.33}$$

for $i = 1, 2, \ldots, n$, since, from these, we can also obtain the action of the Hermitian conjugate generators $E_{ni}^{(n)}$.

The matrix element in (8.33) is given by

$$\left\langle \begin{matrix} \lambda \\ m' \end{matrix} \middle| E_{in}^{(n)} \middle| \begin{matrix} \lambda \\ m \end{matrix} \right\rangle = \sum_{\tau=1}^{n} \left\langle \begin{matrix} \lambda \\ m' \end{matrix} \middle| s_{i\tau}^{(n)} s_{n\tau}^{(n)\dagger} \middle| \begin{matrix} \lambda \\ m \end{matrix} \right\rangle \tag{8.34}$$

$$= \sum_{\tau=1}^{n} \left\langle \begin{matrix} \lambda \\ m' \end{matrix} \middle| s_{i\tau}^{(n)} \middle| \begin{matrix} \lambda - e_\tau \\ m \end{matrix} \right\rangle \left\langle \begin{matrix} \lambda - e_\tau \\ m \end{matrix} \middle| s_{n\tau}^{(n)\dagger} \middle| \begin{matrix} \lambda \\ m \end{matrix} \right\rangle,$$

since $s_{n\tau}^{(n)\dagger}$ effects the shift $\left| \begin{matrix} \lambda \\ m \end{matrix} \right\rangle \rightarrow \left| \begin{matrix} \lambda - e_\tau \\ m \end{matrix} \right\rangle$. From (6.7)-(6.8), the action of the shift operator $s_{i\tau}^{(n)}$ in the first matrix element is to effect the transformation of the initial vector by

$$\left| \begin{matrix} \lambda - e_\tau \\ m \end{matrix} \right\rangle \rightarrow \left| \begin{matrix} \lambda \\ m + \Delta(\tau_i, \ldots, \tau_{n-1}) \end{matrix} \right\rangle, \tag{8.35}$$

in which each $\tau_l, i \leq \tau_l \leq n-1$, can be any of the values $1 \leq \tau_l \leq l$. Thus, the matrix elements of the generators $E_{in}^{(n)}$ in (8.34) are given by

$$\left\langle \begin{matrix} \lambda \\ m + \Delta(\tau_i, \ldots, \tau_{n-1}) \end{matrix} \middle| E_{in}^{(n)} \middle| \begin{matrix} \lambda \\ m \end{matrix} \right\rangle$$

$$= \sum_{\tau=1}^{n} \left\langle \begin{matrix} \lambda \\ m + \Delta(\tau_i, \ldots, \tau_{n-1}) \end{matrix} \middle| s_{i\tau}^{(n)} \middle| \begin{matrix} \lambda - e_\tau \\ m \end{matrix} \right\rangle$$

$$\times \left\langle \begin{matrix} \lambda - e_\tau \\ m \end{matrix} \middle| s_{n\tau}^{(n)\dagger} \middle| \begin{matrix} \lambda \\ m \end{matrix} \right\rangle. \tag{8.36}$$

We now use relations (6.44), (6.107)-(6.109), and (6.19)-(6.25) to obtain the factors in (8.36):

$$\left\langle \begin{matrix} \lambda \\ m + \Delta(\tau_i, \ldots, \tau_{n-1}) \end{matrix} \middle| s_{i\tau}^{(n)} \middle| \begin{matrix} \lambda - e_\tau \\ m \end{matrix} \right\rangle = \sqrt{\frac{M(\lambda)}{M(\lambda - e_\tau)}}$$

$$\times A_{\tau_{n-1},\tau}(x; y') \, \langle m + \Delta(\tau_i, \ldots, \tau_{n-1}) \,|\, t_{i,\tau_{n-1}}^{(n-1)} \,|\, m \rangle, \tag{8.37}$$

$$\left\langle \begin{matrix} \lambda - e_\tau \\ m \end{matrix} \middle| s_{n\tau}^{(n)\dagger} \middle| \begin{matrix} \lambda \\ m \end{matrix} \right\rangle = \sqrt{\frac{M(\lambda)}{M(\lambda - e_\tau)}} \, A_{\tau_{n-1},\tau}^{\dagger}(x; y), \tag{8.38}$$

where we have defined x, y, and y' in terms of partial hooks by

$$\begin{aligned} x &= (p_{1\,n-1}, p_{2\,n-1}, \ldots, p_{n-1\,n-1}), \\ y &= (p_{1n}, p_{2n}, \ldots, p_{nn}), \\ y' &= (p_{1n}, p_{2n}, \ldots, p_{nn}) - e_\tau. \end{aligned} \tag{8.39}$$

Using the value of the $M(\lambda)/M(\lambda - e_\tau)$ from relation (6.109) and substituting relations (8.37)-(8.38) into (8.36), we obtain the following result for the matrix elements of the generators $E_{in} = E_{in}^{(n)}, i = 1, 2, \ldots, n-1$:

$$\left\langle \begin{matrix} \lambda \\ m + \Delta(\tau_i, \ldots, \tau_{n-1}) \end{matrix} \middle| E_{in} \middle| \begin{matrix} \lambda \\ m \end{matrix} \right\rangle = \sqrt{\frac{\prod_{k=1}^{n}(p_{kn} - p_{\tau_{n-1},n-1} - 1)}{\prod_{\substack{l=1 \\ l \neq \tau_{n-1}}}^{n-1} (p_{l,n-1} - p_{\tau_{n-1},n-1} - 1)}} \times \left\langle m + \Delta(\tau_i, \ldots, \tau_{n-1}) \middle| t_{i,\tau_{n-1}}^{(n-1)} \middle| m \right\rangle . \tag{8.40}$$

As remarked above, each τ_l, $l = 1, 2, \ldots, n-1$, in this relation can be $1 \leq \tau_l \leq l$. In obtaining relation (8.40) from (8.36), we have again used Sylvester's identity to effect the summation over τ that occurs (see (6.33)-(6.36)):

$$\sum_{\tau=1}^{n} A_{\tau_{n-1},\tau}(x; y') A_{n,\tau}^{\dagger}(x; y) \left(y_\tau \prod_{\substack{k=1 \\ k \neq \tau}}^{n} \frac{(y_k - y_\tau + 1)}{(y_k - y_\tau)} \right) = \sqrt{\frac{\prod_{k=1}^{n}(y_k - x_{\tau_{n-1}} - 1)}{\prod_{\substack{l=1 \\ l \neq \tau_{n-1}}}^{n-1} (x_l - x_{\tau_{n-1}} - 1)}}. \tag{8.41}$$

The matrix elements of the fundamental shift operators in (8.40) are, of course, given explicitly by the arc digraph rules given in detail in Sect. 6.3, Chapter 6.

From the Hermitian conjugate relation to (8.40), we also obtain the matrix elements of $E_{nj}, j = 1, 2, \ldots, n-1$:

$$\left\langle \begin{array}{c}\lambda\\ m-\Delta(\tau_j,\ldots,\tau_{n-1})\end{array} \middle| E_{nj} \middle| \begin{array}{c}\lambda\\ m\end{array}\right\rangle = \sqrt{\frac{\prod_{k=1}^{n}(p_{kn}-p_{\tau_{n-1},n-1})}{\prod_{\substack{l=1\\ l\neq\tau_{n-1}}}^{n-1}(p_{l,n-1}-p_{\tau_{n-1},n-1})}} \times \left\langle m-\Delta(\tau_j,\ldots,\tau_{n-1}) \middle| s_{j\tau}^{(n-1)\dagger} \middle| m\right\rangle. \tag{8.42}$$

Again, each τ_l, $l = 1, 2, \ldots, n-1$, in this relation can be $1 \leq \tau_l \leq l$.

For $i = n$ in (8.36), the above procedure is modified as follows to obtain the matrix elements of E_{nn} :

$$\left\langle \begin{array}{c}\lambda\\ m\end{array} \middle| E_{nn} \middle| \begin{array}{c}\lambda\\ m\end{array}\right\rangle = \sum_{\tau=1}^{n}\left(A^{\dagger}_{n,\tau}(x;y)\right)^2 \left(y_\tau \prod_{\substack{k=1\\ k\neq\tau}}^{n}\frac{(y_k-y_\tau+1)}{(y_k-y_\tau)}\right)$$

$$= \sum_{\rho=1}^{n}\frac{y_\tau\prod_{l=1}^{n-1}(y_\tau-x_l-1)}{\prod_{\substack{k=1\\ k\neq\tau}}^{n}(y_\tau-y_k)} = e_1(y_1,\ldots,y_n) - e_1(x_1+1,\ldots,x_{n-1}+1)$$

$$= (y_1+\cdots+y_n) - (x_1+\cdots+x_{n-1}) - (n-1)$$

$$= (m_{1,n}+\cdots+m_{n,n}) - (m_{1,n-1}+\cdots+m_{n-1,n-1}). \tag{8.43}$$

The derivation given above of the matrix elements of the generators $\{E_{ij}\}$ of $GL(n, \mathbb{C})$ is purely combinatorial, since the complete results have been derived from the combinatorially defined fundamental shift operators and Sylvester's identity, which also may be given a combinatorial proof, as shown by Good [66] (see also the review by Bhatnagar [13]).

8.3 Complete Set of Commuting Hermitian Operators

The polynomials

$$D\begin{pmatrix} m' \\ \lambda \\ m \end{pmatrix}(Z) \tag{8.44}$$

may be uniquely characterized, up to a normalization factor, as the solutions of a set of partial differential equations. The D^λ−polynomials are the simultaneous eigenfunctions of a complete set of Hermitian operators (Ref. [107]), using the language of quantum mechanical observables. This characterization is, in turn, a consequence of the Lie algebraic properties of the D^λ−polynomials, as we develop briefly in this section.

8.3.1 The Gelfand invariants

The Gelfand invariants of $GL(n, \mathbf{C})$ are a set of polynomial operators in the generators, each of which commutes with all of the generators. Let $E_{ij}, i, j = 1, 2, \ldots, n$, denote the Weyl basis of the Lie algebra $gl(n, \mathbf{C})$:

$$[E_{ij}, E_{kl}] = \delta_{j,k}E_{il} - \delta_{i,l}E_{kj}. \tag{8.45}$$

We define the $n \times n$ matrix E_n of generators by

$$E_n = \begin{pmatrix} E_{11} & E_{12} & \cdots & E_{1n} \\ E_{21} & E_{22} & \cdots & E_{2n} \\ \vdots & \vdots & \cdots & \vdots \\ E_{n1} & E_{n2} & \cdots & E_{nn} \end{pmatrix}. \tag{8.46}$$

We define the trace of the power q of this matrix of generators to be the polynomial in which the generators are ordered, as given by

$$\mathbf{Tr}(E_n)^q = \sum_{i_1, i_2, \ldots, i_q = 1}^{n} E_{i_1 i_2} E_{i_2 i_3} \cdots E_{i_q i_1}. \tag{8.47}$$

It is nontrivial, but straightforward, to prove from this definition (Gelfand [58], Ref. [107]) that

$$[\mathbf{Tr}(E_n)^q, E_{ij}] = 0, i, j = 1, 2, \ldots, n; \text{ all } q \geq 1. \tag{8.48}$$

It then follows that

$$[\mathbf{Tr}(E_n)^q, \mathbf{Tr}(E_n)^{q'}] = 0, \text{ all } q, q' \geq 1. \tag{8.49}$$

The collection of n ordered homogeneous polynomial forms $\mathbf{Tr}(E_n)^q$ of degree $q \geq 1$ in the generators E_{ij} constitute the Gelfand invariants, or *center of the Lie algebra* $gl(n, \mathbb{C})$.

The Weyl generators E_{ij} and the Gelfand invariants $\mathbf{Tr}(E_n)^q$ act in the separable Hilbert space H introduced in Sect. 6.1, Chapter 6. The operator E_{ji} is the Hermitian conjugate of E_{ij}, from which it follows that the Gelfand invariants are Hermitian operators on H. Thus, the set of Gelfand invariants can be simultaneously diagonalized on the space H.

We have given in Sect. 8.2 the action of the Lie algebra in the space H, including the subspaces that are invariant and irreducible. These are the subspaces H_λ with the GT orthonormal basis vectors $\mathbf{B}_\lambda$ given by relation (6.5). The explicit action on $\mathbf{B}_\lambda$ of the generators E_{ij} uniquely determines, of course, the eigenvalues $\Lambda_n^q(\lambda)$ in the eigenvector-eigenvalue relation

$$\mathbf{Tr}(E_n)^q \left| \begin{matrix} \lambda \\ m \end{matrix} \right\rangle = \Lambda_n^q(\lambda) \left| \begin{matrix} \lambda \\ m \end{matrix} \right\rangle. \tag{8.50}$$

These eigenvalues are the symmetric functions in the partial hooks $p_{in} = x_i = \lambda_i + n - i$ given by

$$\Lambda_n^q(\lambda) = \sum_{s=0}^{n} (-1)^{s+1} e_s(x+1) h_{q-s+1}(x), \tag{8.51}$$

where the $e_s(x+1) = e_s(x_1+1, x_2+1, \ldots, x_n+1), e_0(x+1) = 1$ are the elementary symmetric functions in $x+1$, and the $h_k(x), k \geq 0$ are the complete homogeneous symmetric functions defined in Sect. 11.6, Compendium B. We prove this result for the eigenvalues of $\mathbf{Tr}(E_n)^q$ in Sect. 8.5 below. The polynomial operators $\mathbf{Tr}(E_n)^q$ corresponding to $q = 1, 2, \ldots, n$ are algebraically independent, since arbitrary orderings of the E_{ij} always gives a form of degree q plus lower order terms, as pointed out by Gelfand [58]; equivalently, the symmetric functions (8.51) are algebraically independent for $q = 1, 2, \ldots, n$, since the term of highest total degree in Λ_n^q is the power sum symmetric function $(x_1)^q + (x_2)^q + \cdots + (x_n)^q$.

The Gelfand invariants (8.47) can be introduced at every level $n \geq 1$; that is, we can define the Gelfand operators forms $\mathbf{Tr}(E_k)^q$ for each $k = 1, 2, \ldots, n$, to be the ordered polynomial of degree q in the Weyl generators $E_{ij}, i, j = 1, 2, \ldots, k$, of the Lie algebra $gl(k, \mathbb{C})$:

$$\mathbf{Tr}(E_k)^q = \sum_{i_1, i_2, \ldots, i_q = 1}^{k} E_{i_1 i_2} E_{i_2 i_3} \cdots E_{i_q i_1}. \tag{8.52}$$

Then, because of the embedding of the Lie algebras (8.27) and the associated action of the generators, the collection of $n(n+1)/2$ Hermitian operators

$$\mathbf{Tr}(E_k)^q,\ q = 1, 2, \ldots, k;\ k = 1, 2, \ldots, n, \tag{8.53}$$

mutually commute. The simultaneous eigenvectors of these $n(n+1)/2$ commuting Hermitian operators determine, up to choices in the normalization, the orthonormal basis $\mathbf{B}_\lambda$ of the space H_λ :

$$\mathbf{Tr}(E_k)^q \left| \begin{matrix} \lambda \\ m \end{matrix} \right\rangle = \Lambda_k^q(m_k) \left| \begin{matrix} \lambda \\ m \end{matrix} \right\rangle, \tag{8.54}$$

where $m_k = (m_{1,k}, m_{2,k}, \ldots, m_{k,k})$ denotes row k of the GT pattern. It is this complete set of $n(n+1)/2$ algebraically independent mutually commuting Hermitian operators that fixes the number of entries $m_{i,j}$ in the GT pattern $\binom{\lambda}{m}$ to be $n(n+1)/2$.

8.4 Differential Operator Actions

The abstract results for the Lie algebra of $gl(n, \mathbb{C})$ acting in the separable abstract Hilbert space H are transcribed into properties of differential equations by relations (8.19)-(8.20) in Sect. 8.2. It follows from these relations that the $D^\lambda-$polynomials satisfy the following differential equation eigenfunction-eigenvalue relations:

$$\mathbf{Tr}(\mathcal{L}_k)^q \widehat{D} \begin{pmatrix} m' \\ \lambda \\ m \end{pmatrix} (Z) = \Lambda_k^q(m_k) \widehat{D} \begin{pmatrix} m' \\ \lambda \\ m \end{pmatrix} (Z), \tag{8.55}$$

$$\mathbf{Tr}(\mathcal{R}_k)^q \widehat{D} \begin{pmatrix} m' \\ \lambda \\ m \end{pmatrix} (Z) = \Lambda_k^q(m'_k) \widehat{D} \begin{pmatrix} m' \\ \lambda \\ m \end{pmatrix} (Z), \tag{8.56}$$

for $q = 1, 2, \ldots, k;\ k = 1, 2, \ldots, n$, where

$$\mathbf{Tr}(\mathcal{L}_k)^q = \sum_{i_1, i_2, \ldots, i_q = 1}^{k} \mathcal{L}_{i_1 i_2} \mathcal{L}_{i_2 i_3} \cdots \mathcal{L}_{i_q i_1}, \tag{8.57}$$

$$\mathbf{Tr}(\mathcal{R}_k)^q = \sum_{i_1, i_2, \ldots, i_q = 1}^{k} \mathcal{R}_{i_1 i_2} \mathcal{R}_{i_2 i_3} \cdots \mathcal{R}_{i_q i_1}. \tag{8.58}$$

There are $n(n+1)$ mutually commuting Hermitian operators in each of the two sets (8.57) and (8.58) corresponding to $1 \leq q \leq k; 1 \leq k \leq n$. But the $\mathcal{L}$ and $\mathcal{R}$ invariants of the same degree at level n are equal:

$$\mathbf{Tr}(\mathcal{L}_n)^q = \mathbf{Tr}(\mathcal{R}_n)^q,\ q = 1, 2, \ldots, n. \tag{8.59}$$

Thus, there are n^2 algebraically independent mutually commuting Hermitian operators in the two sets (8.57)-(8.58), and these operators constitute a complete set: The $D^{\lambda}-$polynomials are fully determined, up to normalization, by the set of differential equations (8.55)-(8.56).

8.5 Eigenvalues of the Gelfand Invariants

As pointed out above, the known action of the Lie algebra generators E_{ij} on the orthonormal basis vectors $\mathbf{B}_{\lambda}$ uniquely determines the eigenvalues in the eigenvector-eigenvalue relation

$$\mathbf{Tr}(E_n)^q \left| \begin{matrix} \lambda \\ m \end{matrix} \right\rangle = \Lambda_n^q(\lambda) \left| \begin{matrix} \lambda \\ m \end{matrix} \right\rangle . \tag{8.60}$$

Since the eigenvalues depend only on the partition λ, we can set $m = max$, so that

$$\mathbf{Tr}(E_n)^q \left| \begin{matrix} \lambda \\ max \end{matrix} \right\rangle = \Lambda_n^q(\lambda) \left| \begin{matrix} \lambda \\ max \end{matrix} \right\rangle , \tag{8.61}$$

which may also be written in matrix element form as

$$\left\langle \begin{matrix} \lambda \\ max \end{matrix} \right| \mathbf{Tr}(E_n)^q \left| \begin{matrix} \lambda \\ max \end{matrix} \right\rangle = \Lambda_n^q(\lambda). \tag{8.62}$$

The evaluation of the eigenvalue is a purely combinatorial problem: We pick a given generator $E_{ij}, i \neq j$, in a selected single term in the trace expression (8.47) for the polynomial invariant $\mathbf{Tr}(E_n)^q$ and move it to the extreme right for $i < j$, and to the extreme left for $i > j$, by using repeatedly the commutation relations

$$\begin{aligned} E_{ij}E_{kl} &= E_{kl}E_{ij} + \delta_{j,k}E_{il} - \delta_{i,l}E_{kj}, i < j, \\ E_{kl}E_{ij} &= E_{ij}E_{kl} + \delta_{i,l}E_{kj} - \delta_{j,k}E_{il}, i > j. \end{aligned} \tag{8.63}$$

The properties

$$\begin{aligned} \left\langle \begin{matrix} \lambda \\ max \end{matrix} \right| \cdots E_{ij} \left| \begin{matrix} \lambda \\ max \end{matrix} \right\rangle &= 0, i < j, \\ \left\langle \begin{matrix} \lambda \\ max \end{matrix} \right| E_{ij} \cdots \left| \begin{matrix} \lambda \\ max \end{matrix} \right\rangle &= 0, i > j, \end{aligned} \tag{8.64}$$

are then used to reduce the polynomial $\mathbf{Tr}(E_n)^q$ in the matrix element (8.62) to a polynomial of lower degree in the generators. One then selects another such generator $E_{ij}, i \neq j$, in the same term and repeats the entire process. This process is continued until the selected term is reduced to a polynomial in the diagonal generators $E_{ii}, i = 1, 2, \ldots, n$. Carrying this process out for each term in $\mathbf{Tr}(E_n)^q$ reduces the matrix element (8.62) uniquely to the form

$$\left\langle \begin{matrix} \lambda \\ max \end{matrix} \middle| \Lambda_n^q(E_{11}, E_{22}, \cdots, E_{nn}) \middle| \begin{matrix} \lambda \\ max \end{matrix} \right\rangle = \Lambda_n^q(\lambda), \tag{8.65}$$

since

$$E_{ii} \left| \begin{matrix} \lambda \\ max \end{matrix} \right\rangle = \lambda_i \left| \begin{matrix} \lambda \\ max \end{matrix} \right\rangle, i = 1, 2, \ldots, n. \tag{8.66}$$

Examples: Applied to $\mathbf{Tr}(E_n)^1$ and $\mathbf{Tr}(E_n)^2$, this method gives:

$$\mathbf{Tr}(E_n)^1 = \sum_{i=1}^{n} E_{ii} \to \Lambda_n^1(\lambda) = \lambda_1 + \lambda_2 + \cdots + \lambda_n, \tag{8.67}$$

$$\begin{aligned} \mathbf{Tr}(E_n)^2 &= \sum_{i,j=1}^{n} E_{ij}E_{ji} \to \Lambda_n^2(E_{11}, E_{22}, \ldots, E_{nn}) \\ &= \sum_{i=1}^{n} E_{ii}^2 + \sum_{1 \le i < j \le n} (E_{ii} - E_{jj}) \\ &\to \Lambda_n^2(\lambda) = \sum_{i=1}^{n} \lambda_i^2 + \sum_{1 \le i < j \le n} (\lambda_i - \lambda_j). \end{aligned} \tag{8.68}$$

These two eigenvalues may be rewritten in terms of the partial hook variables $x_i = \lambda_i + n - i$ as the first two symmetric functions given by

$$\begin{aligned} \Lambda_n^1(\lambda) &= p_1(x) - \binom{n}{2}, \\ \Lambda_n^2(\lambda) &= p_2(x) - (n-1)p_1(x) + \binom{n}{3}, \\ \Lambda_n^3(\lambda) &= p_3(x) - (n - \frac{3}{2})p_2(x) + \binom{n-1}{2} p_1 \\ &\quad - \frac{1}{2}p_1^2(x) - \binom{n}{4}, \end{aligned} \tag{8.69}$$

where the $p_i(x)$ are the power sum symmetric functions in the n variables $x = (x_1, x_2, \ldots, x_n)$. We have also listed the result of this procedure for $\Lambda_n^3(\lambda)$. This commutation method can be continued to calculate the eigenvalues $\Lambda_n^q(\lambda)$.

The eigenvalues $\Lambda_n^q(\lambda)$ of the Gelfand invariants $\mathbf{Tr}(E_n)^q$ determine a new class of symmetric functions, which we next describe, motivated by the above considerations, but now defined over arbitrary variables $x = (x_1, x_2, \ldots, x_n), 0 \le q \le n$. After describing briefly these new symmetric functions, denoted $b_k(x)$, we will prove that

$$\Lambda_n^q(\lambda) = b_q(x), \text{ for } x_i = \lambda_i + n - i. \tag{8.70}$$

8.5.1 A new class of symmetric functions

In this section, we first define a new class $b_q(x), q = 0, 1, 2, \ldots,$ of symmetric functions of n variables $x = (x_1, x_2, \ldots, x_n)$, and then prove relation (8.70). The symmetric functions $b_q(x)$ in n variables $x = (x_1, x_2, \ldots, x_n)$ are defined by

$$V_n(x) b_q(x) = \sum_{j=1}^{n} x_j^q V_n(x - e_j), \tag{8.71}$$

where $V_n(x)$ is the Vandermonde determinant. We then obtain the following expression for the symmetric functions $b_q(x)$:

$$b_q(x) = -\sum_{j=1}^{n} x_j^q \left[\prod_{i=1}^{n} (x_j - x_i - 1) / \prod_{\substack{i=1 \\ i \neq j}}^{n} (x_j - x_i) \right]. \tag{8.72}$$

We have included the term $i = j$ in the product $\prod_{i=1}^{n}(x_j - x_i - 1)$ so that its reduction to the expression given by (8.73) below is clearly a symmetric polynomial. Relation (8.72) can be brought to the following explicit form in terms of the elementary symmetric functions and the complete homogeneous symmetric functions by using Sylvester's formula (see Sect. 11.7, Compendium B):

$$\begin{aligned} b_q(x) &= \sum_{s=0}^{n} (-1)^{s+1} e_s(x+1) \sum_{j=1}^{n} \left[x_j^{n-s+q} / \prod_{\substack{i=1 \\ i \neq j}}^{n} (x_j - x_i) \right] \\ &= \sum_{s=0}^{q+1} (-1)^{s+1} e_s(x+1) h_{q-s+1}(x). \end{aligned} \tag{8.73}$$

In this relation, the elementary symmetric functions have the properties $e_0(x) = 1, e_s(x) = 0, s > n$, and the shifted function is given by

$$e_s(x+1) = e_s(x_1+1, x_2+1, \ldots, x_n+1). \tag{8.74}$$

The second summation expression in (8.73) terminates at $s = q+1$ in consequence of relation (11.317), Compendium B; for $q > n$, it terminates at $q = n$ in consequence of the elementary symmetric function being 0 for $s > n$. Also, we have that $h_0(x) = e_0(x) = 1$ and $h_1(x) = e_1(x)$, so that relation (8.73) gives $b_0(x) = n$.

The symmetric functions $b_q(x)$ defined by (8.71) satisfy the following three identities:

1. The first identity is obtained from (8.71) by replacing x by $x+y = (x_1+a, x_2+a, \ldots, x_n+a)$ for $y = (a, a, \ldots, a)$ and expanding $(x_j+a)^q$. Effecting these steps gives

$$b_q(x+y) = \sum_{k=0}^{q} \binom{q}{k} b_k(x) a^{q-k}. \tag{8.75}$$

2. The second identity is obtained from (8.75) by first setting $x_n = 0$, then replacing x_i by $x_i - x_n, i = 1, 2, \ldots, n-1$, followed by setting $a = x_n$, which gives

$$b_q(x) = \sum_{k=0}^{q} \binom{q}{k} b_k(x_1 - x_n, x_2 - x_n, \ldots, x_{n-1} - x_n, 0) x_n^{q-k}. \tag{8.76}$$

3. The third identity is obtained from (8.71) by setting $x_n = 0$, which gives

$$\begin{aligned} b_q(x_1, x_2, \ldots, x_{n-1}, 0) = b_q(x_1, x_2, \ldots, x_{n-1}) \\ -b_{q-1}(x_1, x_2, \ldots, x_{n-1}), \end{aligned} \tag{8.77}$$

where we have used $(x_j - 1)/x_j = 1 - (1/x_j)$ in obtaining this result. This relation is valid for all $q \geq 1$, where $b_0(x_1, \ldots, x_{n-1}) = n-1$.

Relations (8.75)-(8.77) are important because they can be iterated to construct the general symmetric function $b_q(x_1, \cdots, x_n)$. Thus, one starts with the initial condition $b_k(x_1) = x_1^k$ for $n = 1$ and all $k \geq 0$ in (8.76) and obtains $b_k(x_1, 0) = x_1^{k-1}(x_1 - 1)$, hence,

$$b_k(x_1 - x_2, 0) = (x_1 - x_2)^{k-1}(x_1 - x_2 - 1),\ k \geq 1. \tag{8.78}$$

This result is then substituted into (8.76) to obtain

$$\begin{aligned} b_q(x_1, x_2) &= 2x^q + \sum_{k=1}^{q} \binom{q}{k} [(x_1 - x_2)^{k-1}(x_1 - x_2 - 1)] x_2^{q-k} \\ &= x_1^q + x_2^q - \sum_{k=1}^{q} \binom{q}{k} (x_1 - x_2)^{k-1} x_2^{q-k}, \end{aligned} \tag{8.79}$$

since $b_0(x_1 - x_2, 0) = 2$. The polynomial $b_q(x_1, x_2)$ does not appear to be symmetric, but by expanding $(x_1 - x_2)^{k-1}$, using the binomial theorem, and performing the internal summation,

$$\sum_{k=s+1}^{q} (-1)^{k-s-1} \binom{q}{k} \binom{k-1}{s} = 1, \tag{8.80}$$

it can be brought to the symmetric form:

$$b_q(x_1, x_2) = x_1^q + x_2^q - \sum_{s=0}^{q-1} x_1^s x_2^{q-1-s}. \tag{8.81}$$

This process can be continued for $n = 3$ to obtain $b_q(x_1, x_2, x_3)$ from (8.75)-(8.76), etc. Polynomials satisfying (8.75)-(8.76) and the condition $b_0(x_1, x_2, \ldots, x_n) = n$ and $b_k(x_1) = x_1^k, k \geq 0$ uniquely determine the polynomials $b_q(x_1, x_2, \ldots, x_n)$, hence, those satisfying (8.71), as given by (8.73).

The form of relation (8.75) suggests that the symmetric functions $b_q(x)$ be called *symmetric functions of binomial type,* since they provide a generalization to symmetric functions in any number of variables n to Rota's functions of binomial type in two single variables x and y (see Ref. [98, p.125]), which satisfy

$$p_q(x + y) = \sum_{k=0}^{q} \binom{q}{k} p_k(x) p_{q-k}(y). \tag{8.82}$$

This is just the form of (8.75) for $n = 1$.

8.5.2 The general eigenvalues of the Gelfand invariants

In this section, we sketch the proof that

$$\Lambda_n^q(\lambda) = b_q(x), \text{ for } x_i = \lambda_i + n - i, \tag{8.83}$$

by showing that these eigenvalues satisfy the three relations (8.75)-(8.77).

The relation

$$\sum_{i_1,i_2,\ldots,i_q=1}^{n} (E_{i_1i_2}+a\delta_{i_1,i_2})(E_{i_2i_3}+a\delta_{i_2,i_3})\cdots(E_{i_qi_1}+a\delta_{i_q,i_1})$$
$$=\sum_{k=0}^{q}\binom{q}{k}\mathbf{Tr}(E_n)^k a^{q-k}, \tag{8.84}$$

in which $\mathbf{Tr}E_n^0=n$, is proved by direct expansion of the left-hand side. We now apply exactly the same steps to the left-hand side of this relation as were used in reducing (8.62) to (8.65) to obtain

$$\Lambda_n^q(\lambda+y)=\sum_{k=0}^{q}\binom{q}{k}\Lambda_n^k(\lambda)a^{q-k}, \tag{8.85}$$

where $y=(a,a,\ldots,a)$. In this relation, we first set $\lambda_n=0$, then replace λ_i by $\lambda_i-\lambda_n, i=1,2,\ldots,n-1$, and then choose $a=\lambda_n$, which gives the relation

$$\Lambda_n^q(\lambda)=\sum_{k=0}^{q}\binom{q}{k}\Lambda_n^k(\lambda_1-\lambda_n,\ldots,\lambda_{n-1}-\lambda_n,0)\lambda_n^{q-k}. \tag{8.86}$$

Thus, the first two relations (8.75)-(8.76) are satisfied by the eigenvalues $\Lambda_n^q(\lambda)$.

It is somewhat more difficult to prove that relation (8.77) is also satisfied. We first prove:

$$\left\langle \begin{matrix}\lambda & 0\\ max & \end{matrix} \left| \mathbf{Tr}(E_n)^q \right| \begin{matrix}\lambda & 0\\ max & \end{matrix} \right\rangle$$
$$=\sum_{k=1}^{q}\binom{q-1}{k-1}\left\langle \begin{matrix}\lambda\\ max\end{matrix} \left| \mathbf{Tr}(E_{n-1})^k \right| \begin{matrix}\lambda\\ max\end{matrix} \right\rangle, \tag{8.87}$$

in which $\lambda=(\lambda_1,\lambda_2,\ldots,\lambda_{n-1})$; that is, $\lambda_n=0$ in the right-hand side of this relation. The method of proof follows that of Sect. 8.5: The action of each $E_{in}, i=1,2,\ldots,n$ on a maximal vector with $\lambda_n=0$ gives the zero vector. Thus, the commutation relations for the generators E_{ij} are used to move every E_{in} in $\mathbf{Tr}(E_n)^q$ to the right-most position. We illustrate this for $q=2,3$:

$$\sum_{i,j=1}^{n} E_{ij}E_{ji} \to \sum_{i=1}^{n-1}\sum_{j=1}^{n} E_{ij}E_{ji}$$
$$= \sum_{i,j=1}^{n-1} E_{ij}E_{ji} + \sum_{i=1}^{n-1} E_{in}E_{ni} \to \sum_{i,j=1}^{n-1} E_{ij}E_{ji} + \sum_{i=1}^{n-1} E_{ii}. \quad (8.88)$$

Similarly, for $n = 3$, we have the sequence of steps given by

$$\sum_{i,j,k=1}^{n} E_{ij}E_{jk}E_{ki} \to \sum_{i=1}^{n-1}\sum_{j,k=1}^{n} E_{ij}E_{jk}E_{ki} = \sum_{i,j,k=1}^{n-1} E_{ij}E_{jk}E_{ki}$$
$$+ \sum_{i,k=1}^{n-1} E_{in}E_{nk}E_{ki} + \sum_{i,j=1}^{n-1} E_{ij}E_{jn}E_{ni} + \sum_{i=1}^{n-1} E_{in}E_{nn}E_{ni}$$
$$\to \sum_{i,j,k=1}^{n-1} E_{ij}E_{jk}E_{ki} + \sum_{i,k=1}^{n-1} E_{nk}E_{in}E_{ki} + \sum_{i,k=1}^{n-1} E_{ik}E_{ki}$$
$$+ \sum_{i,j=1}^{n-1} E_{ij}E_{ji} + \sum_{i=1}^{n-1} E_{in}E_{in}$$
$$\to \sum_{i,j,k=1}^{n-1} E_{ij}E_{jk}E_{ki} + 2\sum_{i,j=1}^{n-1} E_{ij}E_{ji} + \sum_{i=1}^{n-1} E_{ii}. \quad (8.89)$$

These two results validate relation (8.87) for $q = 2, 3$. The general proof is considerable more intricate, but can be carried out to validate the general relation (see Ref. [117]).

In terms of eigenvalues, relation (8.87) is expressed as

$$\Lambda_n^q(\lambda, 0) = \sum_{k=1}^{q} \binom{q-1}{k-1} \Lambda_{n-1}^k(\lambda), \;\; \lambda = (\lambda_1, \lambda_2, \ldots, \lambda_{n-1}). \quad (8.90)$$

Using the binomial coefficient identity

$$\binom{q-1}{k-1} = \binom{q}{k} - \binom{q-1}{k} \quad (8.91)$$

in (8.90), and, in turn, relation (8.84) now gives

$$\Lambda_n^q(\lambda, 0) = \Lambda_{n-1}^q(\lambda + 1) - \Lambda_{n-1}^{q-1}(\lambda + 1). \quad (8.92)$$

We have now shown that the eigenvalues $\Lambda_n^q(\lambda)$ satisfy relations (8.85), (8.86) and (8.92). But setting $\Lambda_n^q(\lambda) = b_q(x)$, $x_i = \lambda_i + n - 1$, we find that these relations are just those given by (8.75)-(8.77), which uniquely determine the $b_q(x)$. Since relation (8.73) is correct for $q = 1, 2, 3$, as given by (8.69), it is true in general:

The eigenvalues of the Gelfand invariants for the general linear group are symmetric functions of binomial type.

Remarks: It is natural from the point of view of Lie algebras to consider next two identical copies of the generators of $U(n)$, each acting in its own Hilbert space, with the sum of the two sets of generators now acting in the tensor product space. This is the approach used in the addition of two angular momenta, which led directly to the WCG coefficients as the transformation coefficients relating two orthonormal bases corresponding to the eigenvectors of two different sets of four Hermitian commuting operators, a set for the coupled and uncoupled orthonormal bases. This counting does not balance for the Lie algebra of the direct product group $U(n) \times U(n)$ $(n \geq 3)$, where the counting goes as follows:

(i). Two uncoupled systems: $\mathbf{Tr}(E_m^{(1)})^q$ and $\mathbf{Tr}(E_m^{(2)})^q$, $1 \leq m \leq q \leq n$. This gives $n(n+1)$ mutually commuting Hermitian operators, which determine the uncoupled orthonormal basis.

(ii). Coupled system: $2n$ $U(n)$ invariants $\mathbf{Tr}(E_n^{(1)})^q, q = 1, 2, \ldots, n$, and $\mathbf{Tr}(E_n^{(2)})^q$, $q = 1, 2, \ldots, n$ from the uncoupled system 1 and system 2, together with the $\mathbf{Tr}(E_m)^q; 1 \leq m \leq q \leq n$, mutually commuting Hermitian operators associated with the coupled system. This gives $2n + n(n+1)/2$ mutually commuting Hermitian operators. But the degree one $U(1)$ invariant for the coupled system is the sum of the two degree one $U(1)$ invariants for the uncoupled system; hence, the number of mutually commuting Hermitian operators is $(n^2 + 5n - 2)/2$.

(iii). The deficit of operators between the uncoupled system in Item 1 and the coupled system in Item 2 is $(n-1)(n-2)/2$.

For $U(2)$, the counting still balances for the uncoupled and coupled bases. For $n \geq 3$, attempts to determine and utilize the extra $1, 3, 6, \ldots$ operators have not been very successful (see Racah [145] for an interesting paper along these lines). We approach this problem through the theory of $U(n)$ irreducible tensor operators, whose matrix elements provide coefficients that reduce the Kronecker product, and also give the sought after CG coefficients.

Chapter 9

Tensor Operator Theory of the Unitary Group

9.1 Introduction

Tensor operators arise naturally in physical theory because of their role in formulating interactions in physical systems modeled by unitary symmetry. They arise in the study of the properties of the unitary group because their matrix elements provide the elements (see (6.85)) of a real orthogonal matrix $C^{(\Lambda\,\lambda)}$ that brings the Kronecker product $D^{\Lambda}(U)\otimes D^{\lambda}(U)$ to the Kronecker direct sum form (6.87) in which all repeated irreducible representations have been completely separated into identical blocks. Thus, tensor operators have several roles. It is our purpose in this chapter to define such irreducible tensor operators and to show how their matrix elements are used to effect this reduction of the Kronecker product.

The space in which tensor operators act is the model Hilbert space discussed in Sect. 6.1, Chapter 6 . Each subspace $H_\lambda \subset H$ in the direct sum space

$$H = \sum_{p=0}^{\infty} \sum_{\lambda \in \mathbb{P}\mathrm{ar}_n(p)} \oplus H_\lambda, \quad H_\lambda \perp H_{\lambda'},\ \lambda \neq \lambda', \quad \text{no repetitions}, \tag{9.1}$$

possesses the orthonormal basis

$$\mathbf{B}_\lambda = \left\{ \left| \begin{array}{c} \lambda \\ m \end{array} \right\rangle \middle| m \text{ is a GT pattern} \right\}. \tag{9.2}$$

The model Hilbert space H then has the additional properties as follows:

1. A group action T_U is defined on all of H such that each orthonormal vector $\left|\begin{smallmatrix}\lambda\\ m'\end{smallmatrix}\right\rangle \in \mathbf{B}_\lambda$ undergoes the linear transformation

$$T_U\left|\begin{matrix}\lambda\\ m'\end{matrix}\right\rangle = \sum_m D\begin{pmatrix}m'\\ \lambda\\ m\end{pmatrix}(U)\left|\begin{matrix}\lambda\\ m\end{matrix}\right\rangle, \text{ each } U \in U(n). \quad (9.3)$$

2. The infinitesimal operators of the transformations $\{T_U \mid U \in U(n)\}$ define the Lie algebra of the group with basis $\{E_{ij}|i, j = 1, 2, \ldots, n\}$, where the elements of this basis satisfy the commutation relations

$$[E_{ij}, E_{kl}] = \delta_{j,k}E_{il} - \delta_{l,i}E_{kj}. \quad (9.4)$$

We now define a general tensor operator in $U(n)$, and what is known as an irreducible tensor operator in $U(n)$, where the action of these operators is defined in the model Hilbert space H :

(i). *A tensor operator* $\mathbf{T}$ *in* $U(n)$ *of rank* $k \in \mathbb{P}$ *is a set of operators*

$$\mathbf{T} = \{T_h \mid h = 1, 2, \ldots, k\}, \quad (9.5)$$

where the operators $T_h : H \to H$ *in the set are called the components of* $\mathbf{T}$*, and which themselves, under the group action* $T_U : H \to H$ *corresponding to the bases transformations* $\mathbf{B}_\lambda$ *of* $H_\lambda, \lambda \in \mathbb{P}ar_n$*, given by (9.3), undergo the following linear similarity transformation for each* $U \in U(n)$:

$$T_U T_{h'} T_{U^{-1}} = \sum_{h=1}^{k} D_{h\,h'}(U)\, T_h, \text{ each } h' = 1, 2, \ldots, k, \quad (9.6)$$

where the matrix $D(U) = (D_{h\,h'}(U))_{1\le h,h'\le k}$ *is a matrix representation of order* k *of* $U(n)$*; that is,* $D(U)D(U') = D(UU')$*, for all* $U, U' \in U(n)$.

(ii). *An irreducible tensor operator* $\mathbf{T}_\Lambda, \Lambda \in \mathbb{P}ar_n$*, in* $U(n)$ *of rank* $Dim\,\Lambda$ *is a set of operators*

$$\mathbf{T}\begin{pmatrix}\Lambda\\ \bullet\end{pmatrix} = \left\{T\begin{pmatrix}\Lambda\\ M\end{pmatrix} \,\middle|\, M \in G_\Lambda\right\}, \quad (9.7)$$

where the operators $T\binom{\Lambda}{M} : H \to H$ *in the set are called the components of* $\mathbf{T}\binom{\Lambda}{\bullet}$ *, and which themselves, under the group action*

$T_U : H \rightarrow H$ of the space H corresponding to the basis transformations $\mathbf{B}_\lambda$ of $H_\lambda, \lambda \in \mathbb{P}ar_n$, given by (9.3), undergo the following linear similarity transformation for each $U \in U(n)$:

$$T_U T\begin{pmatrix} \Lambda \\ M' \end{pmatrix} T_{U^{-1}} = \sum_{M \in \mathbb{G}_\Lambda} D \begin{pmatrix} M' \\ \Lambda \\ M \end{pmatrix} (U) T\begin{pmatrix} \Lambda \\ M \end{pmatrix}, \quad \text{each}\, M' \in G_\Lambda, \tag{9.8}$$

where the matrix $D^\Lambda(U)$ is exactly the unitary irreducible matrix representation of $U(n)$, with all the properties discussed in Chapters 5-8.

We have introduced the notation with the $\bullet$ in place of the lower GT pattern in (9.7) to denote the tensor operator itself, this notation being suggestive that the components of the tensor operator are obtained by replacing $\bullet$ by the GT pattern M.

The infinitesimal operator version of the similarity transformation (9.8) gives the Lie algebraic definition of an irreducible tensor operator in $U(n)$ in terms of the transformation properties of the components under commutation with the generators $E_{ij}, i, j = 1, 2, \ldots, n$:

$$\left[E_{ij}, T\begin{pmatrix} \Lambda \\ M' \end{pmatrix} \right] = \sum_{M \in \mathbb{G}_\Lambda} \left\langle \begin{matrix} \Lambda \\ M \end{matrix} \middle| E_{ij} \middle| \begin{matrix} \Lambda \\ M' \end{matrix} \right\rangle T\begin{pmatrix} \Lambda \\ M \end{pmatrix}. \tag{9.9}$$

We have already met the simplest tensor operators. These are the fundamental shift operators

$$t_\tau = (t_{1\tau}, t_{2\tau}, \cdots, t_{n\tau}) \tag{9.10}$$

that undergo the transformation

$$T_U t_{j\tau} T_{U^{-1}} = \sum_{i=1}^{n} u_{ij}\ t_{i\tau}, \quad j = 1, 2, \ldots, n, \tag{9.11}$$

corresponding to the transformation $T_U : H \rightarrow H$ of the space H. Thus, the fundamental shift operator is a tensor operator that transforms in accordance with the fundamental representation $D^{(10^{n-1})}(U) = U$ of $U(n)$. There are n such fundamental unit tensor operators, one for each $\tau = 1, 2 \ldots, n$. The action of each such tensor operator on the space H has already been given in some detail in Chapters 6-8. Indeed, we have adopted the viewpoint that these operators are the basic objects from which the general D^λ–polynomials over arbitrary indeterminate variables Z are built, a viewpoint that we now expand.

9.1.1 A basis of irreducible tensor operators

We have given the main results in Sect. 6.6, Chapter 6 under the guise of shift-operator polynomials, the properties of which we summarize here:

The power T^A of the matrix of fundamental tensor operators

$$T = \begin{pmatrix} t_{11} & t_{12} & \cdots & t_{1n} \\ t_{21} & t_{22} & \cdots & t_{2n} \\ \vdots & \vdots & & \vdots \\ t_{n1} & t_{n2} & \cdots & t_{nn} \end{pmatrix} \tag{9.12}$$

is defined to be the ordered product:

$$T^A = \prod_{i=1}^{n} t_{in}^{a_{in}} \cdots \prod_{i=1}^{n} t_{i2}^{a_{i2}} \prod_{i=1}^{n} t_{i1}^{a_{i1}}. \tag{9.13}$$

Since the components of each fundamental tensor operator t_τ mutually commute, the transformation (9.11) is among commuting components in column τ of (9.12). Thus, the transformation of T^A under the similarity transformation by T_U is well-defined and given by

$$T_U\, T^A\, T_{U^{-1}} = (UT)^A = \sum_{\gamma \vdash p} \sum_{B \in \mathbb{M}^p_{n\times n}(\gamma,\beta)} \left(Z^B, (UZ)^A\right) \frac{T^B}{B!}, \tag{9.14}$$

for each $A \in \mathbb{M}^p_{n\times n}(\alpha, \beta)$. The inner product coefficients $\left(Z^B, (UZ)^A\right)$ in this relation are exactly the same as those that occur in the expansion $(UZ)^A$ of the Maclaurin monomials Z^A :

$$T_U\, Z^A = (UZ)^A = \sum_{\gamma \vdash p} \sum_{B \in \mathbb{M}^p_n(\gamma,\beta)} \left(Z^B, (UZ)^A\right) \frac{Z^B}{B!}, \tag{9.15}$$

for each $A \in \mathbb{M}^p_{n\times n}(\alpha, \beta)$. The coefficients defined by the inner product are given by (1.258), Chapter 1:

$$\left(Z^B, (UZ)^A\right) = A!B! \sum_{C \in \mathbb{M}^p_{n\times n}(\alpha,\gamma)} \left\{ \begin{matrix} A \\ C\,B \end{matrix} \right\} U^C. \tag{9.16}$$

The shift-operator polynomials are defined by (see (6.75))

$$D\begin{pmatrix} \Gamma \\ \Lambda \\ M \end{pmatrix}(T) = \sum_{A \in \mathbb{M}^p_{n\times n}(W,\Delta)} C\begin{pmatrix} \Gamma \\ \Lambda \\ M \end{pmatrix}(A)\, \frac{T^A}{A!}, \tag{9.17}$$

where $W = (W_1, W_2, \ldots, W_n)$ and $\Delta = (\Delta_1, \Delta_2, \ldots, \Delta_n)$ are the weights of the lower and upper GT patterns:

$$W = W\begin{pmatrix} \Lambda \\ M \end{pmatrix}, \quad \Delta = W\begin{pmatrix} \Lambda \\ \Gamma \end{pmatrix}. \tag{9.18}$$

The C-coefficients in (9.17) are those occurring in the irreducible unitary representation D-polynomials given by (5.6). It follows at once from the transformation property (9.15) that the shift-operator polynomials defined by (9.17) undergo the similarity transformation T_U corresponding to the bases transformation (9.3) of $\mathbf{B}_\lambda$, each $\lambda \in \mathbb{P}\mathrm{ar}_n$, given by

$$T_U D\begin{pmatrix} \Gamma \\ \Lambda \\ M' \end{pmatrix}(T)\, T_{U^{-1}} = \sum_{M \in \mathbb{G}_\Lambda} D\begin{pmatrix} M' \\ \Lambda \\ M \end{pmatrix}(U) D\begin{pmatrix} \Gamma \\ \Lambda \\ M \end{pmatrix}(T). \tag{9.19}$$

Thus, for each shift-weight pattern $\Gamma \in \mathbb{G}_\Lambda$, the shift-operator-valued polynomials (9.17) define an irreducible tensor operator $D\begin{pmatrix} \Gamma \\ \Lambda \\ \bullet \end{pmatrix}(T)$ with components enumerated by the GT patterns $M \in \mathbb{G}_\Lambda$:

$$D\begin{pmatrix} \Gamma \\ \Lambda \\ \bullet \end{pmatrix}(T) = \left\{ D\begin{pmatrix} \Gamma \\ \Lambda \\ M \end{pmatrix}(T) \,\middle|\, M \in G_\Lambda \right\}, \text{ each } \Gamma \in \mathbb{G}_\Lambda. \tag{9.20}$$

The transformation (9.19) is expressed in the form (9.7) by

$$T_U \mathbf{D}\begin{pmatrix} \Gamma \\ \Lambda \\ \bullet \end{pmatrix}(T)\, T_{U^{-1})} = \mathbf{D}\begin{pmatrix} \Gamma \\ \Lambda \\ \bullet \end{pmatrix}(UT) = D^\Lambda(U)\mathbf{D}\begin{pmatrix} \Gamma \\ \Lambda \end{pmatrix}(T), \tag{9.21}$$

which holds for each $\Gamma \in \mathbb{G}_\Lambda$. It is very important to recognize that there is no such simple rule for transformations $T \to TU$ because right transformations mix noncommuting $(t_{i1}, t_{i2}, \ldots, t_{in})$ in row i of (9.12).

The operator-valued polynomials supply us with a set of Dim Λ irreducible tensor operators, one for each upper GT pattern Γ.

These irreducible tensor operators have a pivotal role in determining the CG coefficients of the unitary group $U(n)$, as shown below. We refer to these operator-valued polynomials as $\mathbf{D}^\Lambda$–tensor operators.

Relation (9.17) is invertible to the same form as (5.9) in consequence of the orthogonality (5.13) of the C^λ–coefficients:

$$\frac{T^A}{A!} = \sum_{\Lambda \vdash p} \sum_{M, \Gamma \in \mathbb{G}_\Lambda(W,\Delta)} \frac{1}{M(\Lambda) A!} C\begin{pmatrix} \Gamma \\ \Lambda \\ M \end{pmatrix}(A) D\begin{pmatrix} \Gamma \\ \Lambda \\ M \end{pmatrix}(T), \tag{9.22}$$

for each $A \in \mathbb{M}^p_{n\times n}(W, \Delta)$. That the number of operators in the two sets

$$\left\{ \frac{T^A}{A!} \,\middle|\, A \in \mathbb{M}^p_{n\times n}(W, \Delta) \right\}, \tag{9.23}$$

$$\left\{ D\begin{pmatrix} \Gamma \\ \Lambda \\ M \end{pmatrix}(T) \,\middle|\, \Lambda \vdash p;\ M, \Gamma \in \mathbb{G}_\Lambda(W, \Delta) \right\} \tag{9.24}$$

is equal is assured by the Robertson-Schensted-Knuth identity (5.16):

$$|\mathbb{M}^p_{n\times n}(W, \Delta)| = \sum_{\Lambda \vdash p} K(\Lambda, W) K(\Lambda, \Delta). \tag{9.25}$$

This is an important result because it assures that the set of tensor operators (9.24) is a basis for all tensor operator mappings from the Hilbert space H_λ to the Hilbert space $H_{\lambda+\Delta}$. It follows that:

The set of tensor operators

$$\left\{ D\begin{pmatrix} \Gamma \\ \Lambda \\ \bullet \end{pmatrix}(T) \,\middle|\, \Gamma \in \mathbb{G}_{\Lambda,\Delta} \right\}, \tag{9.26}$$

where $\mathbb{G}_{\Lambda,\Delta}$ denotes the subset of GT patterns $\Gamma \in \mathbb{G}_\Lambda$ defined by

$$\mathbb{G}_{\Lambda,\Delta} = \left\{ \begin{pmatrix} \Lambda \\ \Gamma \end{pmatrix} \,\middle|\, W\begin{pmatrix} \Lambda \\ \Gamma \end{pmatrix} = \Delta \right\}, \tag{9.27}$$

is a set of irreducible tensor operator mappings that spans the space of all irreducible $U(n)$ tensor operators of type Λ from the Hilbert space H_λ to the Hilbert space $H_{\lambda+\Delta}$.

More generally, we have the full set of irreducible tensor operators, DimΛ in number, as given by

$$\left\{ D\begin{pmatrix} \Gamma \\ \Lambda \\ \bullet \end{pmatrix}(T) \,\middle|\, \Gamma \in \mathbb{G}_\Lambda \right\}, \tag{9.28}$$

where the shift-weight Δ is the weight of the upper GT pattern:

$$\Delta = (\Delta_1, \Delta_2, \ldots, \Delta_n) = W\begin{pmatrix} \Lambda \\ \Gamma \end{pmatrix}. \tag{9.29}$$

Thus, there are DimΛ irreducible $D^\Lambda-$tensor operators, this set being partitioned into $K(\Lambda, \Delta)$ disjoint subsets of the form (9.26): This partitioning is exactly right for accommodating every degenerate subset $D^{\Lambda+\Delta}(Z)$ that occurs in the reduction of the Kronecker product $D^{\Lambda}(Z) \otimes D^{\lambda}(Z)$.

In a certain sense, this is a natural solution of the construction of the irreducible tensor operator of the unitary group, since it is, perhaps, the simplest possible extension of the irreducible representation D^{λ}–polynomials themselves to operator-valued D^{λ}–polynomials that have the same left transformation properties. It is a very natural result.

As elements of the vector space of operators acting in H, linear combinations of the $\mathbf{D}^{\Lambda}$–tensor operators can be formed with scalars I_{Γ} that are $U(n)$ invariants:

$$\mathbf{X}^{\Lambda}_{\bullet}(T) = \sum_{\Gamma \in \mathbb{G}_{\Lambda}} \mathbf{D}\begin{pmatrix} \Gamma \\ \Lambda \\ \bullet \end{pmatrix}(T) I_{\Gamma}. \tag{9.30}$$

Then, $\mathbf{X}^{\Lambda}_{\bullet}(T)$ is a $U(n)$ irreducible tensor operator of type Λ. We have written the $U(n)$ scalars to the right in (9.30), so that the matrix elements in H of the components $X\begin{pmatrix} \Lambda \\ M \end{pmatrix}(T)$ of such a tensor operator are given by

$$\left\langle \begin{matrix} \lambda' \\ m' \end{matrix} \middle| X\begin{pmatrix} \Lambda \\ M \end{pmatrix}(T) \middle| \begin{matrix} \lambda \\ m \end{matrix} \right\rangle$$
$$= \sum_{\Gamma \in \mathbb{G}_{\Lambda}} \left\langle \begin{matrix} \lambda' \\ m' \end{matrix} \middle| D\begin{pmatrix} \Gamma \\ \Lambda \\ M \end{pmatrix}(T) \middle| \begin{matrix} \lambda \\ m \end{matrix} \right\rangle I_{\Gamma}(\lambda), \tag{9.31}$$

$$I_{\Gamma}\left| \begin{matrix} \lambda \\ m \end{matrix} \right\rangle = I_{\Gamma}(\lambda)\left| \begin{matrix} \lambda \\ m \end{matrix} \right\rangle. \tag{9.32}$$

The preceding results appear to present a quite satisfactory answer to the problem of spanning the multiplicity space of identical irreducible representations $D^{\lambda+\Delta}(Z)$, but there are still unresolved issues: The tensor operators in the set (9.26) are not orthogonal; hence, the matrix elements in (9.31) do not provide the elements of a real orthogonal matrix $C^{(\Lambda\,\lambda)}$ in relations (6.85)-(6.87) that bring the Kronecker product to standard Kronecker direct sum form. The construction of such a set of tensor operators requires further considerations to which we next turn.

9.2 Unit Tensor Operators

We remarked at the beginning of this section that the matrix elements of irreducible tensor operators provide CG coefficients in the relation for the reduction of the Kronecker product to Kronecker direct sum form. This relation is described in detail in Sect. 6.6, Chapter 6. We now analyze the relation

$$\begin{aligned} &C^{(\Lambda\,\lambda)}\left(D^{\Lambda}(Z)\otimes D^{\lambda}(Z)\right)\left(C^{(\Lambda\,\lambda)}\right)^{T} \\ &= \sum_{\substack{\Delta\,\in\,\mathbb{W}_{\Lambda}\\ \lambda+\Delta\in\Lambda\otimes\lambda}} \oplus\left(I_{I_{\Lambda,\Delta}(\lambda)}\otimes D^{\lambda+\Delta}(Z)\right) \end{aligned} \tag{9.33}$$

from the viewpoint of matrix elements of irreducible tensor operators acting in the model Hilbert space H. We define a *unit* tensor operator, denoted

$$\widehat{T}\begin{pmatrix}\Gamma\\ \Lambda\\ \bullet\end{pmatrix},\ \Gamma\in\mathbb{G}_{\Lambda}, \tag{9.34}$$

to be an irreducible tensor operator in $U(n)$ with the additional property that the matrix elements of its action in H give orthogonal coefficients:

1. The shift action on the basis $\mathbf{B}_{\lambda}$ of H_{λ}, each $\Gamma\in\mathbb{G}_{\Lambda,\Delta}$, is given by

$$\begin{aligned} &\widehat{T}\begin{pmatrix}\Gamma\\ \Lambda\\ M\end{pmatrix}\left|\begin{matrix}\lambda\\ m\end{matrix}\right\rangle \\ &= \sum_{m'\in\mathbb{G}_{\lambda+\Delta}}\left\langle\begin{matrix}\lambda+\Delta\\ m'\end{matrix}\right|\widehat{T}\begin{pmatrix}\Gamma\\ \Lambda\\ M\end{pmatrix}\left|\begin{matrix}\lambda\\ m\end{matrix}\right\rangle\left|\begin{matrix}\lambda+\Delta\\ m'\end{matrix}\right\rangle, \end{aligned} \tag{9.35}$$

where the matrix elements in this expression are required to satisfy the orthogonality relations

$$\begin{aligned} &\sum_{M,m}\left\langle\begin{matrix}\lambda+\Delta\\ m'\end{matrix}\right|\widehat{T}\begin{pmatrix}\Gamma\\ \Lambda\\ M\end{pmatrix}\left|\begin{matrix}\lambda\\ m\end{matrix}\right\rangle \\ &\times\left\langle\begin{matrix}\lambda+\Delta\\ m''\end{matrix}\right|\widehat{T}\begin{pmatrix}\Gamma'\\ \Lambda\\ M\end{pmatrix}\left|\begin{matrix}\lambda\\ m\end{matrix}\right\rangle=\delta_{m',m''}\,\delta_{\Gamma,\Gamma'}\,, \\ &\qquad\text{each pair }\Gamma,\Gamma'\in\mathbb{G}_{\Lambda,\Delta}. \end{aligned} \tag{9.36}$$

The matrix elements of a unit tensor operator must have the following properties: The set $\mathbb{G}_{\Lambda,\Delta}$ of shift-weights must split into two disjoint subsets, $\mathbb{G}^{(1)}_{\Lambda,\Delta}$ and $\mathbb{G}^{(0)}_{\Lambda,\Delta}$, such that there are exactly $I_{\Lambda,\Delta}(\lambda)$ shift-weight patterns $\Gamma \in \mathbb{G}^{(1)}_{\Lambda,\Delta}$ such that the matrix elements in (9.35)-(9.36) are not all zero for Γ, $\Gamma' \in \mathbb{G}^{(1)}_{\Lambda,\Delta}$, and there must be exactly $K(\Lambda,\Delta) - I_{\Lambda,\Delta}(\lambda)$ shift-weight patterns $\Gamma'' \in \mathbb{G}^{(0)}_{\Lambda,\Delta}$ such that the matrix elements are zero:

$$\left\langle \begin{matrix} \lambda+\Delta \\ m' \end{matrix} \middle| \widehat{T}\begin{pmatrix} \Gamma'' \\ \Lambda \\ M \end{pmatrix} \middle| \begin{matrix} \lambda \\ m \end{matrix} \right\rangle = 0,$$

$$\text{for } \lambda, \Lambda, \lambda+\Delta \in \mathbb{P}\mathrm{ar}_n, \lambda+\Delta \in \Lambda \otimes \lambda; \qquad (9.37)$$

$$\text{all patterns } m \in \mathbb{G}_\lambda, M \in \mathbb{G}_\Lambda, \Gamma'' \in \mathbb{G}^{(0)}_{\Lambda,\Delta}, m' \in \mathbb{G}_{\lambda+\Delta}.$$

This zero matrix element property is an essential characteristic in this definition of a unit tensor operator. It is important to observe that requiring **certain** matrix elements of an operator to be zero does not imply that the operator itself is the zero operator.

2. A unit tensor operator (9.34) is an irreducible tensor operator; that is, satisfies the transformation rule

$$T_U \widehat{T}\begin{pmatrix} \Gamma \\ \Lambda \\ M' \end{pmatrix} T_{U^{-1}} = \sum_{M \in \mathbb{G}_\Lambda} D\begin{pmatrix} M' \\ \Lambda \\ M \end{pmatrix}(U)\widehat{T}\begin{pmatrix} \Gamma \\ \Lambda \\ M \end{pmatrix},$$

$$\text{each } \Gamma \in \mathbb{G}_\Lambda. \qquad (9.38)$$

We now use the two properties (9.37)-(9.38) of a unit tensor operator to deduce the relationship between the matrix elements of a unit tensor operator and the elements of the real orthogonal matrix $C^{(\Lambda\,\lambda)}$ in (9.33) that effects the full reduction to Kronecker direct sum form. We perform six lengthy, but straightforward steps: (i) Relation (9.38) is multiplied from the right by T_U; (ii) matrix elements of the modified transformation relation resulting from the first step between initial and final basis vectors indicated by

$$\left\langle \begin{matrix} \lambda+\Delta \\ m' \end{matrix} \middle| \cdots \middle| \begin{matrix} \lambda \\ m \end{matrix} \right\rangle \qquad (9.39)$$

are taken, and intermediate states supplied and summed over; (iii) the matrix elements of T_U in its action on a basis vector as given by the transformation rule (9.3) are inserted; (iv) the matrix elements of the Kronecker product are identified from relation (6.89); (v) the orthogonality relations (9.36) are used to move the unit tensor operator matrix elements to the left-hand side of the relation resulting from the first four steps; and (vi) various labels are renamed. This gives the following relation, which is of exactly the form that effects the desired reduction of the Kronecker product:

$$\sum_{(M,m),(M',m')} \left\langle \begin{matrix} \lambda+\Delta \\ m'' \end{matrix} \middle| \widehat{T}\begin{pmatrix} \Gamma \\ \Lambda \\ M \end{pmatrix} \middle| \begin{matrix} \lambda \\ m \end{matrix} \right\rangle$$
$$\times \left\langle \begin{matrix} \lambda+\Delta \\ m''' \end{matrix} \middle| \widehat{T}\begin{pmatrix} \Gamma \\ \Lambda \\ M' \end{pmatrix} \middle| \begin{matrix} \lambda \\ m' \end{matrix} \right\rangle \left(D^{\Lambda}(U) \otimes D^{\lambda}(U)\right)_{(M,m),(M',m')}$$
$$= D\begin{pmatrix} m''' \\ \lambda+\Delta \\ m'' \end{pmatrix}(U), \text{ each } \Gamma \in \mathbb{G}^{(1)}_{\Lambda,\Delta}. \tag{9.40}$$

Relation (9.40) is the result needed to identify the elements of a matrix $C^{(\Lambda\,\lambda)}$ that effects the full reduction given by (9.33). The rows $\left(\Gamma, \binom{\lambda+\Delta}{m''}\right)$ and columns (M, m) of a real orthogonal matrix $C^{(\Lambda\,\lambda)}$ that brings the Kronecker product in (9.33) to full Kronecker direct sum form are given by

$$\left(C^{(\Lambda\,\lambda)}\right)_{(\Gamma,\binom{\lambda+\Delta}{m''});(M,m)} = \left\langle \begin{matrix} \lambda+\Delta \\ m'' \end{matrix} \middle| \widehat{T}\begin{pmatrix} \Gamma \\ \Lambda \\ M \end{pmatrix} \middle| \begin{matrix} \lambda \\ m \end{matrix} \right\rangle,$$
$$\lambda, \Lambda, \lambda+\Delta \in \mathbb{P}\mathrm{ar}_n, \lambda+\Delta \in \Lambda \otimes \lambda, \tag{9.41}$$
$$m \in \mathbb{G}_{\lambda}, M \in \mathbb{G}_{\Lambda}, \Gamma \in \mathbb{G}^{(1)}_{\Lambda,\Delta}, m'' \in \mathbb{G}_{\lambda+\Delta}.$$

The elements of the full matrix $C^{(\Lambda\,\lambda)}$ of order $\mathrm{Dim}\Lambda\,\mathrm{Dim}\lambda$ are obtained by letting the shift-weight Δ take on all weights $\Delta \in \mathbb{W}_{\Lambda}$. These matrix elements of $C^{(\Lambda\,\lambda)}$ can then be arranged in an order that effects exactly the reduction (9.33).

The main result for tensor operator theory can be summarized as follows:

The shift action and transformation properties under unitary similarity transformations of a unit tensor operator imply that its matrix elements are the elements of an orthogonal matrix that reduces the Kronecker product of two irreducible representations of $U(n)$ to full Kronecker direct sum form.

This is a very nice result. It places the burden of resolving the multiplicity problem for the Kronecker product of two irreducible representations of $U(n)$ on that of defining unit tensor operators of general type Λ. But we already have a basis for all tensor operators of type Λ—the D^{Λ}–operator-valued polynomials. Thus, there is a pathway for defining sets of unit tensor operators. Indeed, there are several such pathways, as we discuss, after summarizing in a slightly different notation the properties of general unit tensor operators given above.

9.2.1 Summary of properties of unit tensor operators

The properties of an orthogonal set of unit tensor operators can be summarized by statements about the action of the operators and their conjugates on the basis $\mathbf{B}_{\lambda}$ of $H_{\lambda} \subset H$. We enumerate a set of orthogonal unit tensor operators having prescribed shift-weight Δ by a set of *operator patterns,* as we now call the upper GT patterns $\Gamma \in \mathbb{G}_{\Lambda}$:

$$\Gamma^{t}_{\Delta} : t = 1, 2, \ldots, I_{\Lambda,\Delta}(\lambda), \text{ each } \Gamma^{t}_{\Delta} \in \mathbb{G}_{\Lambda,\Delta}. \tag{9.42}$$

The set of operator patterns $\Gamma^{t}_{\Delta} \in \mathbb{G}_{\Lambda,\Delta}$ that have the same shift Δ contains $K(\Lambda, \Delta)$ patterns. We do not yet specify which of patterns (9.42) are the $I_{\Lambda,\Delta}(\lambda) \leq K(\Lambda, \Delta)$ enumerated by Γ^{t}_{Δ}, $t = 1, 2, \ldots, I_{\Lambda,\Delta}(\lambda)$.

The concept of null space is an essential feature of the definition of unit tensor operators. Relation (9.37) expresses the property that the entire vector space H_{λ} is annihilated by certain of the unit tensor operators even though $\lambda + \Delta \in \Lambda \otimes \lambda$. We refer to this annihilation of an entire vector space as a *characteristic null space.* The existence of characteristic null spaces of unit tensor operators is unavoidable—the unit tensor operators must, for all values of the Littlewood-Richardson numbers, provide orthogonal matrix elements that bring the Kronecker product $D^{\Lambda}(U) \otimes D^{\lambda}(U)$ to Kronecker direct sum form. Since the greatest value of the Littlewood-Richardson number is given by the Kostka numbers; that is, $I^{\max}_{\Lambda,\Delta}(\lambda) = K(\Lambda, \Delta)$, and there is a denumerable infinity of partitions $\lambda \in \mathbb{P}\mathrm{ar}_{n}$ for which this is true, all $K(\lambda, \Delta)$ unit tensor operators for $t = 1, 2, \ldots, K(\lambda, \Delta)$ are required, but, for those Littlewood-Richardson numbers less than the maximum $K(\Lambda, \Delta)$, *null operators* and associated characteristic null spaces must occur; such null space properties are controlled by those of the Littlewood-Richardon and Kostka numbers. For the development of this viewpoint, it is essential to consider the Littlewood-Richardson numbers as functions $I_{\Lambda,\Delta}$ of the partitions $\lambda \in \mathbb{P}\mathrm{ar}_{n}$ with values $c^{\lambda+\Delta}_{\Lambda\,\lambda} = I_{\Lambda,\Delta}(\lambda)$.

It is convenient to adopt the following R_{t} notation in place of the matrix element notation in (9.35):

$$\widehat{T}\begin{pmatrix}\Gamma^t_\Delta\\ \Lambda\\ M\end{pmatrix}\left|\begin{matrix}\lambda\\ m\end{matrix}\right\rangle = \sum_{m'\in\mathbb{G}_{\lambda+\Delta}} R_t\left[\begin{pmatrix}\lambda+\Delta\\ m'\end{pmatrix}\begin{pmatrix}\Lambda\\ M\end{pmatrix}\begin{pmatrix}\lambda\\ m\end{pmatrix}\right]\left|\begin{matrix}\lambda+\Delta\\ m'\end{matrix}\right\rangle,$$
$$t=1,2,\ldots,I_{\Lambda,\Delta}(\lambda); \tag{9.43}$$

$$\widehat{T}\begin{pmatrix}\Gamma^r_\Delta\\ \Lambda\\ M\end{pmatrix}\left|\begin{matrix}\lambda\\ m\end{matrix}\right\rangle = \mathbf{0},\ r=I_{\Lambda,\Delta}(\lambda)+1,\ldots,K(\Lambda,\Delta). \tag{9.44}$$

It follows from these results that the action of the conjugate unit tensor operator on the basis $\mathbf{B}_{\lambda+\Delta}$ of $H_{\lambda+\Delta}\subset H$ is given by

$$\widehat{T}^\dagger\begin{pmatrix}\Gamma^t_\Delta\\ \Lambda\\ M\end{pmatrix}\left|\begin{matrix}\lambda+\Delta\\ m'\end{matrix}\right\rangle = \sum_{m\in\mathbb{G}_{\lambda}} R_t\left[\begin{pmatrix}\lambda\\ m\end{pmatrix}\begin{pmatrix}\Lambda\\ M\end{pmatrix}\begin{pmatrix}\lambda+\Delta\\ m'\end{pmatrix}\right]\left|\begin{matrix}\lambda\\ m\end{matrix}\right\rangle,$$
$$t=1,2,\ldots,I_{\Lambda,\Delta}(\lambda); \tag{9.45}$$

$$\widehat{T}^\dagger\begin{pmatrix}\Gamma^r_\Delta\\ \Lambda\\ M\end{pmatrix}\left|\begin{matrix}\lambda+\Delta\\ m'\end{matrix}\right\rangle = \mathbf{0},\ r=I_{\Lambda,\Delta}(\lambda)+1,\ldots,K(\Lambda,\Delta). \tag{9.46}$$

The matrix elements of a unit tensor operator and its conjugate are related by

$$\left\langle\begin{matrix}\lambda\\ m\end{matrix}\right|\widehat{T}^\dagger\begin{pmatrix}\Gamma^t_\Delta\\ \Lambda\\ M\end{pmatrix}\left|\begin{matrix}\lambda+\Delta\\ m'\end{matrix}\right\rangle = \left\langle\begin{matrix}\lambda+\Delta\\ m'\end{matrix}\right|\widehat{T}\begin{pmatrix}\Gamma^t_\Delta\\ \Lambda\\ M\end{pmatrix}\left|\begin{matrix}\lambda\\ m\end{matrix}\right\rangle. \tag{9.47}$$

The action of a unit tensor operators is to effect an upward shift from a basis vector in H_λ to a basis vector in $H_{\lambda+\Delta}$, while that of the conjugate is a downward shift from $H_{\lambda+\Delta}$ to H_λ, this shift property holding when the respective vector spaces H_λ and $H_{\lambda+\Delta}$ are not in the characteristic null space. The null unit tensor operators and characteristic null space in (9.44) are essential features of the definition of unit tensor operators, and similarly for (9.46).

To emphasize that the relations in this subsection apply to unit tensor operators, we now denote the real orthogonal matrix $C^{(\Lambda\,\lambda)}$ in relation (9.33) by $R^{(\Lambda\,\lambda)}$. The rows and columns of the elements of this matrix are enumerated as follows:

$$\left(R^{(\Lambda\,\lambda)}\right)_{(t,\binom{\lambda+\Delta}{m'});(M,m)} = R_t\left[\begin{pmatrix}\lambda+\Delta\\ m'\end{pmatrix}\begin{pmatrix}\Lambda\\ M\end{pmatrix}\begin{pmatrix}\lambda\\ m\end{pmatrix}\right]. \tag{9.48}$$

We take careful note of the ranges of the various parameters that appear in this result, which are

$$\begin{gathered}\lambda, \Lambda, \lambda+\Delta \in \mathbb{P}\mathrm{ar}_n,\ \lambda+\Delta \in \Lambda \otimes \lambda, \\ m \in \mathbb{G}_\lambda,\ M \in \mathbb{G}_\Lambda,\ m' \in \mathbb{G}_{\lambda+\Delta},\ t = 1,2,\ldots, I_{\Lambda,\Delta}(\lambda).\end{gathered} \tag{9.49}$$

The shift-weight Δ then assumes all values $\Delta \in \mathbb{W}_\Lambda$ such that $\lambda+\Delta \in \Lambda \otimes \lambda$ to obtain the matrix $R^{(\Lambda\,\lambda)}$, a result confirmed by the dimensional identity:

$$\sum_{\Delta \in \mathbb{W}_\Lambda,\, \lambda+\Delta \in \Lambda\otimes\lambda} I_{\Lambda,\Delta}(\lambda)\mathrm{Dim}(\lambda+\Delta) = \mathrm{Dim}\Lambda\,\mathrm{Dim}\lambda. \tag{9.50}$$

The unit tensor operators of type Λ having operator patterns in the set (9.43), which have nonzero matrix elements, satisfy relations that can be written as operator identities on the space H, and also as orthogonality relations. We write the basis vector to the right of relations (9.51) and (9.53) below to avoid any ambiguity as to the vector space on which these operator relations hold:

Operator identity:

$$\sum_{t=1}^{I_{\Lambda,\Delta}(\lambda)} \widehat{T}^\dagger \begin{pmatrix} \Gamma^t_\Delta \\ \Lambda \\ M' \end{pmatrix} \widehat{T} \begin{pmatrix} \Gamma^t_\Delta \\ \Lambda \\ M \end{pmatrix} \left| \begin{matrix} \lambda \\ m \end{matrix} \right\rangle = \delta_{M',M} \left| \begin{matrix} \lambda \\ m \end{matrix} \right\rangle . \tag{9.51}$$

Orthogonality of the rows:

$$\begin{aligned}\sum_{t=1}^{I_{\Lambda,\Delta}(\lambda)} \sum_{m''} R_t \left[\begin{pmatrix} \lambda+\Delta \\ m'' \end{pmatrix} \begin{pmatrix} \Lambda \\ M' \end{pmatrix} \begin{pmatrix} \lambda \\ m' \end{pmatrix} \right] R_t \left[\begin{pmatrix} \lambda+\Delta \\ m'' \end{pmatrix} \begin{pmatrix} \Lambda \\ M \end{pmatrix} \begin{pmatrix} \lambda \\ m \end{pmatrix} \right] \\ = \delta_{M',M}\, \delta_{m',m} \, .\end{aligned} \tag{9.52}$$

Conjugate operator identity:

$$\sum_{M \in \mathbb{G}_\Lambda} \widehat{T} \begin{pmatrix} \Gamma^{t'}_\Delta \\ \Lambda \\ M \end{pmatrix} \widehat{T}^\dagger \begin{pmatrix} \Gamma^t_\Delta \\ \Lambda \\ M \end{pmatrix} \left| \begin{matrix} \lambda+\Delta \\ m' \end{matrix} \right\rangle = \delta_{t',t} \left| \begin{matrix} \lambda+\Delta \\ m' \end{matrix} \right\rangle . \tag{9.53}$$

Orthogonality of the columns:

$$\begin{aligned}\sum_{m \in \mathbb{G}_\lambda,\, M \in \mathbb{G}_\Lambda} R_{t'} \left[\begin{pmatrix} \lambda+\Delta \\ m'' \end{pmatrix} \begin{pmatrix} \Lambda \\ M \end{pmatrix} \begin{pmatrix} \lambda \\ m \end{pmatrix} \right] R_t \left[\begin{pmatrix} \lambda+\Delta \\ m' \end{pmatrix} \begin{pmatrix} \Lambda \\ M \end{pmatrix} \begin{pmatrix} \lambda \\ m \end{pmatrix} \right] \\ = \delta_{t',t}\, \delta_{m'',m'} \, .\end{aligned} \tag{9.54}$$

We remark again that we have not identified the operator patterns in (9.43) and (9.44) associated with the split of the set of $K(\Lambda, \Delta)$ unit tensor operators into two disjoint subsets, non-null and null tensor operators. We make such an identification subsequently. Nor have we yet indicated how the matrix elements of a set of unit tensor operator are to be calculated. We turn next to these issues.

9.2.2 Explicit unit tensor operators

We have already determined a basis set of all irreducible tensor operators of type Λ having operator patterns $\Gamma \in \mathbb{G}_{\Lambda,\Delta}$ with shift-weight Δ. These are the operator-valued $D^{\Lambda}-$polynomials, enumerated by operator patterns, as follows:

$$D\begin{pmatrix} \Gamma^1_\Delta \\ \Lambda \\ \bullet \end{pmatrix}(T),\ D\begin{pmatrix} \Gamma^2_\Delta \\ \Lambda \\ \bullet \end{pmatrix}(T),\ \ldots,\ D\begin{pmatrix} \Gamma^{K(\Lambda,\Delta)}_\Delta \\ \Lambda \\ \bullet \end{pmatrix}(T), \tag{9.55}$$

where we still do not identify explicitly the operator patterns $\Gamma^s_\Delta \in \mathbb{G}_{\Lambda,\Delta}, s = 1, 2, \ldots, K(\Lambda, \Delta)$. But, by (9.26)-(9.27), the set of irreducible tensor operators spans the space of irreducible tensor operator mappings from H_λ to $H_{\lambda+\Delta}$. It must, therefore, be the case that each unit tensor operator in the set

$$\hat{T}\begin{pmatrix} \Gamma^1_\Delta \\ \Lambda \\ \bullet \end{pmatrix}, \hat{T}\begin{pmatrix} \Gamma^2_\Delta \\ \Lambda \\ \bullet \end{pmatrix}, \ldots, \hat{T}\begin{pmatrix} \Gamma^{K(\Lambda,\Delta)}_\Delta \\ \Lambda \\ \bullet \end{pmatrix} \tag{9.56}$$

is a linear combination of those in the set (9.55), and conversely:

$$\hat{T}\begin{pmatrix} \Gamma^s_\Delta \\ \Lambda \\ \bullet \end{pmatrix} = \sum_{s'=1}^{K(\Lambda,\Delta)} D\begin{pmatrix} \Gamma^{s'}_\Delta \\ \Lambda \\ \bullet \end{pmatrix}(T) A_{s,s'}, \tag{9.57}$$

$$D\begin{pmatrix} \Gamma^s_\Delta \\ \Lambda \\ \bullet \end{pmatrix}(T) = \sum_{s'=1}^{K(\Lambda,\Delta)} \hat{T}\begin{pmatrix} \Gamma^{s'}_\Delta \\ \Lambda \\ \bullet \end{pmatrix} B_{s,s'}, \tag{9.58}$$

where $s = 1, 2, \ldots, K(\Lambda, \Delta)$ in each of these relations, and $A_{s,s'}$ and $B_{s,s'}$ are $U(n)$ invariants (scalars in the vector space of operators).

Consider now the set of vectors in $H_{\lambda+\Delta}$ enumerated by

$$D\begin{pmatrix}\Gamma^1_\Delta\\ \Lambda\\ M\end{pmatrix}(T)\left|\begin{matrix}\lambda\\ m\end{matrix}\right\rangle,\ D\begin{pmatrix}\Gamma^2_\Delta\\ \Lambda\\ M\end{pmatrix}(T)\left|\begin{matrix}\lambda\\ m\end{matrix}\right\rangle,\ \ldots,\tag{9.59}$$

$$D\begin{pmatrix}\Gamma^{I_{\Lambda,\Delta}(\lambda)}_\Delta\\ \Lambda\\ M\end{pmatrix}(T)\left|\begin{matrix}\lambda\\ m\end{matrix}\right\rangle,\ \ldots,\ D\begin{pmatrix}\Gamma^{K(\Lambda,\Delta)}_\Delta\\ \Lambda\\ M\end{pmatrix}(T)\left|\begin{matrix}\lambda\\ m\end{matrix}\right\rangle.$$

There are $K(\Lambda,\Delta)$ vectors in this set, but exactly $I_{\Lambda,\Delta}(\lambda)$ of them are linearly independent, since this is the number of irreducible tensor operator mappings from H_λ to $H_{\lambda+\Delta}$. Indeed, each subset of these $K(\Lambda,\Delta)$ vectors that contains $I_{\Lambda,\Delta}(\lambda)$ distinct vectors is linearly independent, and we take these to be the ones with operator patterns $\Gamma^1_\Delta, \Gamma^2_\Delta, \ldots, \Gamma^{I_{\Lambda,\Delta}(\lambda)}_\Delta$, as yet not identified explicitly. This freedom of choice of operator patterns in the operator-valued D^Λ–polynomials is proved below (see (9.79)).

We now apply the Gram-Schmidt procedure to the full set of $K(\Lambda,\Delta)$ vectors (9.59) in the order left-to-right. The application of the Gram-Schmidt procedure to the first $I_{\Lambda,\Delta}(\lambda)$ linearly independent vectors produces an orthornormal set of vectors; equally, if not more significantly, the continued application of the procedure to the remaining $K(\Lambda,\Delta) - I_{\Lambda,\Delta}(\lambda)$ vectors produces the **0** vector. (It is not generally pointed out that the Gram-Schmidt procedure can be applied to any set of vectors, linearly dependent or not, and always produces a set of orthogonal vectors equal to the number of linearly independent vectors in the set and a number of **0** vectors equal to the number of linearly dependent vectors in the set (for details, see Sect. 10.3.4, Compendium A). Thus, the application of the Gram-Schmidt procedure to the full set of vectors (9.59) generates a set of unit tensor operators with the following properties:

$$\begin{aligned}\widehat{D}\begin{pmatrix}\Gamma^t_\Delta\\ \Lambda\\ M\end{pmatrix}(T)\left|\begin{matrix}\lambda\\ m\end{matrix}\right\rangle &= \sum_{m'\in\mathbb{G}_{\lambda+\Delta}} R_t\left[\begin{pmatrix}\lambda+\Delta\\ m'\end{pmatrix}\begin{pmatrix}\Lambda\\ M\end{pmatrix}\begin{pmatrix}\lambda\\ m\end{pmatrix}\right]\left|\begin{matrix}\lambda\\ m\end{matrix}\right\rangle,\\ &\qquad t = 1,2,\ldots,I_{\Lambda,\Delta}(\lambda); \qquad (9.60)\\ \widehat{D}\begin{pmatrix}\Gamma^r_\Delta\\ \Lambda\\ M\end{pmatrix}(T)\left|\begin{matrix}\lambda\\ m\end{matrix}\right\rangle &= \mathbf{0},\ r = I_{\Lambda,\Delta}(\lambda)+1,\ldots,K(\Lambda,\Delta).\end{aligned}$$

The Gram-Schmidt procedure leading from relations (9.59) to relations (9.60) is to be carried out for each shift-weight $\Delta \in \mathbb{W}_\Lambda$. Thus, each of the irreducible tensor operators

$$\widehat{T}\begin{pmatrix} \Gamma^s_\Delta \\ \Lambda \\ \bullet \end{pmatrix} = \widehat{D}\begin{pmatrix} \Gamma^s_\Delta \\ \Lambda \\ \bullet \end{pmatrix}, \text{ each } \Delta \in \mathbb{W}_\Lambda, \tag{9.61}$$
$$\text{each } s = 1, 2, \ldots, K(\Lambda, \Delta),$$

is a unit tensor operator for which the R_t coefficients in (9.60) satisfy the orthogonality relations (9.51)-(9.54); that is, the real orthogonal matrix $R^{\Lambda\,\lambda)}$ effects the transformation (9.33) (replace $C^{(\Lambda\,\lambda)}$ by $R^{(\Lambda\,\lambda)}$) to the Kronecker direct sum. This property is true for every choice of operator patterns in the sequence (9.55), such that the first $I_{\Lambda,\Delta}(\lambda)$ of them are linearly independent: The Gram-Schmidt procedure automatically effects the split into non-null and null unit tensor operators, as exhibited in (9.60). We conclude:

There are many unit tensor operators whose matrix elements provide a set of elements of a real orthogonal matrix that effects the reduction of the Kronecker product of two unitary irreducible representations of $U(n)$ to the Kronecker direct sum of unitary irreducible representations.

If there is a natural choice for ordering the set of operator patterns $\{\Gamma\}$ that enumerate irreducible tensor operators, it must originate from additional properties of these patterns that in some manner assign distinguishing features, as yet unidentified.

9.3 Canonical Tensor Operators

The concept of characteristic null space of an irreducible tensor operator is unavoidable if the matrix elements of such irreducible tensor operators are to provide the elements of a real orthogonal matrix that reduces the Kronecker product of two unitary irreducible representations of $U(n)$ to the Kronecker direct sum of unitary irreducible representations. It was Biedenharn [7, 20, 117] who recognized the role of operator patterns Γ in effecting the shift-weight properties of irreducible tensor operators required to implement their properties into a comprehensible theory of Kronecker product reduction. While such $\Gamma-$patterns are GT patterns in their enumerative aspects, they possess no group-subgroup properties, which are intrinsic to ordinary GT patterns. In this respect, they may possess properties still to be discovered. This idea that characteristic null space had an important, if not definitive, role to play was made explicit in Ref. [107]. Optimistically, the irreducible tensor operators were called *canonical* tensor operators and assigned the symbols

$$\left\langle \begin{matrix} \Gamma \\ \Lambda \\ \bullet \end{matrix} \right\rangle = \left\{ \left\langle \begin{matrix} \Gamma \\ \Lambda \\ M \end{matrix} \right\rangle \middle| M \in \mathbb{G}_\Lambda \right\}, \text{ each } \Gamma \in \mathbb{G}_\Lambda. \tag{9.62}$$

The characteristic null space classification works very nicely for all unitary irreducible tensor operators in the so-called class of adjoint tensor operators, those with partition $\Lambda = (2, 1^{n-2}, 0)$ (Ref. [117]).

The properties of the Littlewood-Richardson numbers are a key ingredient in the characteristic null space approach to the theory of tensor operators. As noted earlier in relations (6.102)-(6.105), Chapter 6, we are led naturally to the viewpoint that the partitions $\lambda \in \mathbb{P}\mathrm{ar}_n$ in the Kronecker product $D^\Lambda(U) \otimes D^\lambda(U)$ are parameters that are to be varied over all $\lambda \in \mathbb{P}\mathrm{ar}_n$, while Λ and the weight Δ of the GT pattern $\binom{\Lambda}{\Gamma_s}$ are taken as fixed. This viewpoint then leads us to regard the Littlewood-Richardson numbers $c_{\Lambda\,\lambda}^{\lambda+\Delta} = I_{\Lambda,\Delta}(\lambda)$ as the values of a function $I_{\Lambda,\Delta}$ defined over the set of all partitions $\lambda \in \mathbb{P}\mathrm{ar}_n$.

The level sets of partitions defined by (6.104)-(6.105) and given by

$$\begin{aligned} &\mathbb{P}_{\Lambda,\Delta}(L) = \{\lambda \in \mathbb{P}\mathrm{ar}_n \mid I_{\Lambda\,\Delta}(\lambda) = L\}, \\ &L \in \{0, 1, \ldots, K(\Lambda, \Delta)\} \end{aligned} \tag{9.63}$$

are basic to the notion of canonical tensor operators. We do not preclude the possibility that, for certain L, the set $\mathbb{P}_{\Lambda,\Delta}(L)$ can be empty. We also require the unions of these level sets defined by

$$\mathbb{P}^{(s)}_{\Lambda,\Delta} = \bigcup_{L=0}^{K(\Lambda,\Delta)-s} \mathbb{P}_{\Lambda,\Delta}(L),\ s = 0, 1, \ldots, K(\Lambda, \Delta). \tag{9.64}$$

We now define a canonical tensor operator by the following three rules:

1. Each canonical tensor operator is a unit tensor operator; that is, it satisfies all the criteria given in Sect. 9.2 for a unit tensor operator.

2. The full set of canonical tensor operators of type Λ is totally ordered by the rule

$$\left\langle \begin{matrix} \Gamma \\ \Lambda \\ \bullet \end{matrix} \right\rangle > \left\langle \begin{matrix} \Gamma' \\ \Lambda \\ \bullet \end{matrix} \right\rangle, \tag{9.65}$$

if and only if the operator patterns Γ and Γ' satisfy

$$\begin{pmatrix} \Lambda \\ \Gamma \end{pmatrix} > \begin{pmatrix} \Lambda \\ \Gamma' \end{pmatrix}, \tag{9.66}$$

where these GT patterns are ordered in exactly the way described for ordinary GT patterns by relations (5.18)-(5.20), Chapter 5, which we repeat: The operator pattern $\binom{\Lambda}{\Gamma}$ is greater then the operator pattern $\binom{\Lambda}{\Gamma'}$ in (9.66), if and only if

$$(\Gamma_{n-1}, \Gamma_{n-2}, \ldots, \Gamma_1) > (\Gamma'_{n-1}, \Gamma'_{n-2}, \ldots, \Gamma'_1). \tag{9.67}$$

Each of these two sequences is of length $n(n-1)/2$, where the respective sequences Γ_i and Γ'_i, each of length i, is read off the row i of the GT pattern:

$$\begin{aligned} \Gamma_i &= (\Gamma_{1,i}, \Gamma_{2,i}, \ldots, \Gamma_{i,i}),\ i = 1, 2, \ldots, n-1, \\ \Gamma'_i &= (\Gamma'_{1,i}, \Gamma'_{2,i}, \ldots, \Gamma'_{i,i}),\ i = 1, 2, \ldots, n-1. \end{aligned} \tag{9.68}$$

The greater than symbol $>$ holds in (9.67), if and only if the first nonzero difference of the two corresponding sequences is positive.

3. Canonical tensor operators are here **defined** to be identical to the orthonormalized unit tensor operators constructed by the Gram-Schmidt procedure in Sect. 9.2.2 on the set of operator-valued $D^{\Lambda}-$ polynomials, where the operator-valued polynomials are systematically organized by the ordering of their operator patterns (read left-to-right in (9.55):

$$\begin{pmatrix} \Lambda \\ \Gamma^1_\Delta \end{pmatrix} > \begin{pmatrix} \Lambda \\ \Gamma^2_\Delta \end{pmatrix} > \cdots > \begin{pmatrix} \Lambda \\ \Gamma^{K(\Lambda,\Delta)}_\Delta \end{pmatrix}. \tag{9.69}$$

Thus, we have that

$$\begin{aligned} \left\langle \begin{matrix} \Gamma^t_\Delta \\ \Lambda \\ M \end{matrix} \right\rangle \left| \begin{matrix} \lambda \\ m \end{matrix} \right\rangle &= \widehat{D}\begin{pmatrix} \Gamma^t_\Delta \\ \Lambda \\ M \end{pmatrix}(T) \left| \begin{matrix} \lambda \\ m \end{matrix} \right\rangle,\ t = 1, 2, \ldots, I_{\Lambda,\Delta}(\lambda); \\ \left\langle \begin{matrix} \Gamma^r_\Delta \\ \Lambda \\ M \end{matrix} \right\rangle \left| \begin{matrix} \lambda \\ m \end{matrix} \right\rangle &= \mathbf{0},\ r = I_{\Lambda,\Delta}(\lambda) + 1, \ldots, K(\Lambda, \Delta). \end{aligned} \tag{9.70}$$

Examples: It useful to given examples of the GT pattern ordering (9.66):

1. $\Lambda = (2,1,0), \Delta = (1,1,1):$

$$\begin{pmatrix} 2 & & 1 & & 0 \\ & 2 & & 0 & \\ & & 1 & & \end{pmatrix} > \begin{pmatrix} 2 & & 1 & & 0 \\ & 1 & & 1 & \\ & & 1 & & \end{pmatrix}, \tag{9.71}$$

in consequence of $(2,0,1) > (1,1,1)$.

2. $\Lambda = (4,2,0), \Delta = (2,2,2):$

$$\begin{pmatrix} 4 & & 2 & & 0 \\ & 4 & & 0 & \\ & & 2 & & \end{pmatrix} > \begin{pmatrix} 4 & & 2 & & 0 \\ & 3 & & 1 & \\ & & 2 & & \end{pmatrix} > \begin{pmatrix} 4 & & 2 & & 0 \\ & 2 & & 2 & \\ & & 2 & & \end{pmatrix}, \tag{9.72}$$

in consequence of $(4,0,2) > (3,1,2) > (2,2,2)$. □.

We restate the factorization lemma (6.96), Chapter 6, in the notation of canonical tensor operators for the purpose of continuity with earlier published work:

$$\begin{aligned}
&\left(\widehat{D}\begin{pmatrix} m''' \\ \lambda+\Delta \\ m' \end{pmatrix}(Z), D\begin{pmatrix} M' \\ \Lambda \\ M \end{pmatrix}(Z)\widehat{D}\begin{pmatrix} m'' \\ \lambda \\ m \end{pmatrix}(Z)\right) \\
&= \left\langle \begin{matrix} m''' \\ \lambda+\Delta \\ m' \end{matrix} \middle| D\begin{pmatrix} M' \\ \Lambda \\ M \end{pmatrix}(\Theta) \middle| \begin{matrix} m'' \\ \lambda \\ m \end{matrix} \right\rangle \\
&= \sqrt{\frac{M(\lambda+\Delta)}{M(\lambda)}} \sum_{t=1}^{I_{\Lambda,\Delta}(\lambda)} \left\langle \begin{matrix} \lambda+\Delta \\ m' \end{matrix} \middle| \left\langle \begin{matrix} \Gamma^t_\Delta \\ \Lambda \\ M \end{matrix} \right\rangle \middle| \begin{matrix} \lambda \\ m \end{matrix} \right\rangle \\
&\qquad\qquad \times \left\langle \begin{matrix} \lambda+\Delta \\ m''' \end{matrix} \middle| \left\langle \begin{matrix} \Gamma^t_\Delta \\ \Lambda \\ M' \end{matrix} \right\rangle \middle| \begin{matrix} \lambda \\ m'' \end{matrix} \right\rangle;
\end{aligned} \tag{9.73}$$

$$\left\langle \begin{matrix} \lambda+\Delta \\ m' \end{matrix} \middle| \left\langle \begin{matrix} \Gamma^t_\Delta \\ \Lambda \\ M \end{matrix} \right\rangle \middle| \begin{matrix} \lambda \\ m \end{matrix} \right\rangle = R_t\left[\begin{pmatrix} \lambda+\Delta \\ m' \end{pmatrix}\begin{pmatrix} \Lambda \\ M \end{pmatrix}\begin{pmatrix} \lambda \\ m \end{pmatrix}\right],$$

$$t = 1, 2, \ldots, I_{\Lambda,\Delta}(\lambda); \tag{9.74}$$

$$\left\langle \begin{matrix} \lambda+\Delta \\ m' \end{matrix} \middle| \left\langle \begin{matrix} \Gamma^r_\Delta \\ \Lambda \\ M \end{matrix} \right\rangle \middle| \begin{matrix} \lambda \\ m \end{matrix} \right\rangle = 0,\ r = I_{\Lambda,\Delta}(\lambda)+1, \ldots, K(\Lambda,\Delta).$$

These last two matrix element relations are expressions of relations (9.60) and (9.70). The canonical tensor operators appearing in relations (9.73)-(9.74) are the ones with the ordered operator patterns (9.69).

The choice of the identification of patterns given by (9.69), as illustrated in (9.71)-(9.72), is arbitrary from the point of view of unit tensor operators, except that it must be verified that the set of operator-valued tensor operators (9.59) corresponding to this choice are linearly independent (We do this below). The persuasion for this being a natural choice based on properties of the operator patterns must be made separately. We will show in subsequent sections that this choice coincides with that made in many papers by Biedenharn and collaborators, including the author. Our view is, however, that it is a reasonable choice based on natural schema, and not a consequence of there being "no free choices," except trivial phase conventions, as often advanced by Biedenharn [14]. *Nonetheless, for reference purposes to the published literature, we will refer to this choice as canonical.*

The structure of the factorization lemma (9.73)-(9.74) can be investigated along the same lines as used in Sect. 7.2, Chapter 7. In particular, if we choose $m = m'' = max$ and $m' = m''' = max$, the relation reduces to one in which only the following maximal matrix elements appear:

$$\left\langle \begin{matrix} \lambda + \Delta \\ max \end{matrix} \middle| \left\langle \begin{matrix} \Gamma^t_\Delta \\ \Lambda \\ \Gamma^{t'}_\Delta \end{matrix} \right\rangle \middle| \begin{matrix} \lambda \\ max \end{matrix} \right\rangle, \; t, t' = 1, 2, \ldots, I_{\Lambda,\Delta}(\lambda), \tag{9.75}$$

since the selection of initial and final maximal patterns forces the lower pattern M to belong to the same shift-weight set $\mathbb{G}_{\Lambda,\Delta}$ as the upper pattern. Thus, the coefficients (9.75) can be arranged into a square matrix $L^{\lambda+\Delta}_{\Lambda,\lambda}$ of order equal to the Littlewood-Richardson number $I_{\Lambda,\Delta}(\lambda)$:

$$\left(L^{\lambda+\Delta}_{\Lambda,\lambda}\right)_{t',t} = \left\langle \begin{matrix} \lambda + \Delta \\ max \end{matrix} \middle| \left\langle \begin{matrix} \Gamma^t_\Delta \\ \Lambda \\ \Gamma^{t'}_\Delta \end{matrix} \right\rangle \middle| \begin{matrix} \lambda \\ max \end{matrix} \right\rangle, \tag{9.76}$$
$$t', t = 1, 2, \ldots, I_{\Lambda,\Delta}(\lambda).$$

The matrix $L^{\lambda+\Delta}_{\Lambda,\lambda}$ is of order 1 with entry 1 for partitions Λ having one part nonzero (see Sect. 7.2, Chapter 7). It is also nonsingular for partitions Λ with two parts nonzero, as can be demonstrated by applying the factorization lemma to the $D-$polynomials constructed in Sect. 7.3.1, Chapter 7. We next outline the application of the factorizations lemma to the general case.

9.3.1 Application of the factorization lemma to canonical tensor operators

Canonical tensor operators have been defined by (9.70) in terms of orthonormalized operator-valued $D^{\Lambda}-$polynomials. Nonetheless, it is useful to show the consistency with the factorization lemma.

The strategy for solving relations (9.73) for the canonical CG coefficients

$$\left\langle \begin{matrix} \lambda+\Delta \\ m' \end{matrix} \middle| \left\langle \begin{matrix} \Gamma_t \\ \Lambda \\ M \end{matrix} \right\rangle \middle| \begin{matrix} \lambda \\ m \end{matrix} \right\rangle \tag{9.77}$$

is to evaluate relation (9.73) first on the maximal GT patterns $m'' = max$ and $m''' = max$, followed also by $m' = max$ and $m = max$, just as we did in Sect. 7.2 for the case of $\Lambda = (p\ 0^{n-1})$. Thus, effecting the first step gives the following relation for each $t = 1, 2, \ldots, I_{\Lambda,\Delta}(\lambda)$:

$$\left\langle \begin{matrix} max \\ \lambda+\Delta \\ m' \end{matrix} \middle| D\begin{pmatrix} \Gamma_{t'} \\ \Lambda \\ M \end{pmatrix}(\Theta) \middle| \begin{matrix} max \\ \lambda \\ m \end{matrix} \right\rangle \tag{9.78}$$

$$= \sqrt{\frac{M(\lambda+\Delta)}{M(\lambda)}} \sum_{t=1}^{I_{\Lambda,\Delta}(\lambda)} \left\langle \begin{matrix} \lambda+\Delta \\ m' \end{matrix} \middle| \left\langle \begin{matrix} \Gamma_t \\ \Lambda \\ M \end{matrix} \right\rangle \middle| \begin{matrix} \lambda \\ m \end{matrix} \right\rangle \left(L^{\lambda+\Delta}_{\Lambda,\lambda}\right)_{t',t},$$

where the $L-$coefficients in the right-most position are the elements of the Littlewood-Richardson matrix defined by (9.76). By considering carefully the action of the fundamental shift operators t_{j_T} that effect the transformation between maximal upper states in the left-hand side of (9.78) (see Ref. [118]), the left-hand side of this relation can be shown to reduce to

$$\left\langle \begin{matrix} max \\ \lambda+\Delta \\ m' \end{matrix} \middle| D\begin{pmatrix} \Gamma_{t'} \\ \Lambda \\ M \end{pmatrix}(\Theta) \middle| \begin{matrix} max \\ \lambda \\ m \end{matrix} \right\rangle$$

$$= I_{\Delta}(\lambda)\sqrt{\frac{M(\lambda+\Delta)}{M(\lambda)}} \left\langle \begin{matrix} \lambda+\Delta \\ m' \end{matrix} \middle| D\begin{pmatrix} \Gamma_{t'} \\ \Lambda \\ M \end{pmatrix}(T) \middle| \begin{matrix} \lambda \\ m \end{matrix} \right\rangle, \tag{9.79}$$

$$I_{\Delta}(\lambda) = \left[\prod_{1 \le i < j \le n} \frac{(p_{i,n} - p_{j,n} + \Delta_i - \Delta_j)}{(p_{i,n} - p_{j,n} + \Delta_i)} \right]^{1/2}. \tag{9.80}$$

Taking this relation into account, we obtain the following specialized version of the factorization lemma:

$$I_\Delta(\lambda)\left\langle \begin{array}{c} \lambda+\Delta \\ m' \end{array} \middle| D\begin{pmatrix} \Gamma_{t'} \\ \Lambda \\ M \end{pmatrix}(T)\middle| \begin{array}{c} \lambda \\ m \end{array}\right\rangle$$

$$= \sum_{t=1}^{I_{\Lambda,\Delta}(\lambda)} \left\langle \begin{array}{c} \lambda+\Delta \\ m' \end{array} \middle| \left\langle \begin{array}{c} \Gamma_t \\ \Lambda \\ M \end{array}\right\rangle \middle| \begin{array}{c} \lambda \\ m \end{array}\right\rangle \left(L^{\lambda+\Delta}_{\Lambda,\lambda}\right)_{t',t}. \tag{9.81}$$

The choice of *max* in both the initial and final vectors in the left-hand side of (9.81) forces the GT pattern M of the operator-valued polynomial to be a pattern $\Gamma_s, s = 1,2,\ldots,K(\Lambda,\Delta)$ in the multiplicity set corresponding to the chosen $\Delta \in \mathbb{W}_\Lambda$. But in effecting the Gram-Schmitt procedure on the operator-valued polynomials, we have further restricted these GT patterns to be $\Gamma_{t''}, t'' = 1,2,\ldots,I_{\Lambda,\Delta}(\lambda)$. Thus, when restricted in this manner, (9.81) gives the following relation:

$$I_\Delta(\lambda)\left\langle \begin{array}{c} \lambda+\Delta \\ max \end{array} \middle| D\begin{pmatrix} \Gamma_{t'} \\ \Lambda \\ \Gamma_{t''} \end{pmatrix}(T)\middle| \begin{array}{c} \lambda \\ max \end{array}\right\rangle$$

$$= \left(D^{\lambda+\Delta}_{\Lambda,\lambda}\right)_{t'',t'} = \sum_{t=1}^{I_{\Lambda,\Delta}(\lambda)} \left(L^{\lambda+\Delta}_{\Lambda,\lambda}\right)_{t'',t} \left(L^{\lambda+\Delta}_{\Lambda,\lambda}\right)_{t',t}. \tag{9.82}$$

In matrix form, this relation becomes

$$D^{\lambda+\Delta}_{\Lambda,\lambda} = L^{\lambda+\Delta}_{\Lambda,\lambda}\left(L^{\lambda+\Delta}_{\Lambda,\lambda}\right)^T. \tag{9.83}$$

The matrix $D^{\lambda+\Delta}_{\Lambda,\lambda}$ is known from the action of the fundamental shift operators in the left-hand side of (9.82).

If $L^{\lambda+\Delta}_{\Lambda,\lambda}$ is any solution of (9.83), so is $L^{\lambda+\Delta}_{\Lambda,\lambda}\, S$, where S is an arbitrary real orthogonal matrix of order $I_{\Lambda,\Delta}(\lambda)$. *This is an intrinsic feature of any comprehensive theory that deals with the multiplicity space associated with reducing the Kronecker product to standard Kronecker direct sum form, which is linear space of order equal to the Littlewood-Richardson* $c^{\lambda+\Delta}_{\Lambda\,\lambda}$ *number.*The factorization lemma possesses this basic property.

We now make a special choice such that canonical tensor operators become the orthonormalized operator-valued polynomials. We choose the matrix $L^{\lambda+\Delta}_{\Lambda,\lambda}$ to be lower diagonal, as given by the following form:

$$L^{\lambda+\Delta}_{\Lambda,\lambda} = \begin{pmatrix} L_{1,1} & 0 & 0 & 0 & \cdots & 0 \\ L_{2,1} & L_{2,2} & 0 & 0 & \cdots & 0 \\ L_{3,1} & L_{3,2} & L_{3,3} & 0 & \cdots & 0 \\ \vdots & \vdots & \vdots & \vdots & \cdots & \vdots \\ L_{I,1} & L_{I,2} & L_{I,3} & L_{I,4} & \cdots & L_{I,I} \end{pmatrix}, \tag{9.84}$$

in which we have abbreviated $I = I_{\Lambda,\Delta}(\lambda)$ and also $L_{t',t} = \left(L^{\lambda+\Delta}_{\Lambda,\lambda}\right)_{t',t}$. Thus, relation (9.81) becomes

$$I_\Delta(\lambda)\left\langle \begin{matrix} \lambda+\Delta \\ m' \end{matrix} \middle| D\begin{pmatrix} \Gamma_{t'} \\ \Lambda \\ M \end{pmatrix}(T)\middle| \begin{matrix} \lambda \\ m \end{matrix} \right\rangle$$

$$= \sum_{t=1}^{t'} \left\langle \begin{matrix} \lambda+\Delta \\ m' \end{matrix} \middle| \left\langle \begin{matrix} \Gamma_t \\ \Lambda \\ M \end{matrix} \right\rangle \middle| \begin{matrix} \lambda \\ m \end{matrix} \right\rangle L_{t',t}(\lambda). \tag{9.85}$$

This triangular relation is invertible to the form:

$$\left\langle \begin{matrix} \lambda+\Delta \\ m' \end{matrix} \middle| \left\langle \begin{matrix} \Gamma_t \\ \Lambda \\ M \end{matrix} \right\rangle \middle| \begin{matrix} \lambda \\ m \end{matrix} \right\rangle$$

$$= I_\Delta(\lambda)\sum_{t'=1}^{t} \left\langle \begin{matrix} \lambda+\Delta \\ m' \end{matrix} \middle| D\begin{pmatrix} \Gamma_{t'} \\ \Lambda \\ M \end{pmatrix}(T)\middle| \begin{matrix} \lambda \\ m \end{matrix} \right\rangle L^{-1}_{t,t'}.$$

$$= \left\langle \begin{matrix} \lambda+\Delta \\ m' \end{matrix} \middle| \widehat{D}\begin{pmatrix} \Gamma_t \\ \Lambda \\ M \end{pmatrix}(T)\middle| \begin{matrix} \lambda \\ m \end{matrix} \right\rangle, \tag{9.86}$$

where $t = 1, 2, \ldots, I_{\Lambda,\Delta}(\lambda)$. For $t = 1$, the factor $I_\Delta(\lambda)L_{1,1}$ is the normalization factor of the operator-valued polynomial with operator pattern Γ_1, etc. :

The Gram-Schmidt procedure applied to the operator-valued D^Λ−polynomials to obtain the unit tensor operators (9.60) is a triangular procedure, as is the triangular method effected above that solves the factorization lemma. These two methods yield the same set of unit tensor operators, where, for consistency of enumeration, we label the successive unit tensor operators in the two constructions by the same set of ordered operator patterns.

We summarize the above results:

An explicit set of ordered canonical tensor operators given by

$$\left\langle \begin{matrix} \Gamma_1 \\ \Lambda \\ \bullet \end{matrix} \right\rangle > \left\langle \begin{matrix} \Gamma_2 \\ \Lambda \\ \bullet \end{matrix} \right\rangle > \cdots > \left\langle \begin{matrix} \Gamma_{I_{\Lambda,\Delta}(\lambda)} \\ \Lambda \\ \bullet \end{matrix} \right\rangle > \cdots > \left\langle \begin{matrix} \Gamma_{K(\Lambda,\Delta)} \\ \Lambda \\ \bullet \end{matrix} \right\rangle, \tag{9.87}$$

where the operator patterns are ordered by the rules (9.65)-(9.68), can be constructed from the factorization lemma such that they agree with the Gram-Schmidt orthonormalized operator-values polynomials,

$$\left\langle \begin{matrix} \Gamma_t \\ \Lambda \\ M \end{matrix} \right\rangle = \widehat{D} \begin{pmatrix} \Gamma_t \\ \Lambda \\ M \end{pmatrix} (T),\ t = 1, 2, \ldots, I_{\Lambda,\Delta}(\lambda), \tag{9.88}$$

and such that the following properties hold for each operator pattern $\Gamma^t_\Delta \in G_{\Lambda,\Delta}$, *each* $\Delta \in \mathbb{W}_\Lambda$:

$$\left\langle \begin{matrix} \Gamma_t \\ \Lambda \\ M \end{matrix} \right\rangle \left| \begin{matrix} \lambda \\ m \end{matrix} \right\rangle = \sum_{m' \in \mathbb{G}_{\lambda+\Delta}} R_t \left[\begin{pmatrix} \lambda+\Delta \\ m' \end{pmatrix} \begin{pmatrix} \Lambda \\ M \end{pmatrix} \begin{pmatrix} \lambda \\ m \end{pmatrix} \right] \left| \begin{matrix} \lambda+\Delta \\ m' \end{matrix} \right\rangle,$$

$$t = 1, 2, \ldots, I_{\Lambda,\Delta}(\lambda); \tag{9.89}$$

$$\left\langle \begin{matrix} \Gamma_r \\ \Lambda \\ \bullet \end{matrix} \right\rangle : H_\lambda \to \mathbf{0},\ H_\lambda \in \mathbf{N}_{\binom{\Lambda}{\Gamma_r}},$$

$$r = I_{\Lambda,\Delta}(\lambda) + 1, \ldots, K(\Lambda, \Delta). \tag{9.90}$$

The matrix elements of the canonical tensor operators in relation (9.89) then provide a set of CG coefficients that bring the Kronecker product $D^\Lambda \otimes D^\lambda$ to completely reduced form; that is, to the standard Kronecker direct sum.

We need to identify more precisely the full characteristic null space $\mathbf{N}\binom{\Lambda}{\Gamma^s_\Delta}$ of a given null canonical tensor operator implied by property (9.90); that is, the set of all $\lambda \in \mathbb{P}\mathrm{ar}_n$ such that

$$\left\langle \begin{matrix} \Gamma^s_\Delta \\ \Lambda \\ \bullet \end{matrix} \right\rangle \left| \begin{matrix} \lambda \\ m \end{matrix} \right\rangle = \mathbf{0},\ \text{all}\, \lambda \in \mathbf{N}\begin{pmatrix} \Lambda \\ \Gamma^s_\Delta \end{pmatrix}, \tag{9.91}$$

where Γ^s_Δ is any operator pattern $\Gamma^s_\Delta \in \mathbb{G}_{\Lambda,\Delta}$, $s = 1, 2, \ldots, K(\Lambda, \Delta)$. It follows from (9.90), relations (9.65)-(9.68), and definition (9.64) that the full characteristic null space is given by

$$\mathbf{N}\begin{pmatrix} \Lambda \\ \Gamma^s_\Delta \end{pmatrix} = \sum_{\lambda \in \mathbb{P}^s_{\Lambda,\Delta}} \oplus H_\lambda. \tag{9.92}$$

The characteristic null spaces of the respective canonical tensor operators (9.87) as read from right-to-left then satisfy the set inclusion relations

$$\mathbf{N}\begin{pmatrix} \Lambda \\ \Gamma^{K(\Lambda,\Delta)}_\Delta \end{pmatrix} \subseteq \cdots \subseteq \mathbf{N}\begin{pmatrix} \Lambda \\ \Gamma^2_\Delta \end{pmatrix} \subseteq \mathbf{N}\begin{pmatrix} \Lambda \\ \Gamma^1_\Delta \end{pmatrix}. \tag{9.93}$$

Two of the characteristic null spaces in (9.93) can be equal, as shown by Baclawski [6]. This can happen if and only if for certain values of $L \in \{0, 1, \ldots, K(\Lambda, \Delta)\}$ the corresponding set of partitions $\mathbb{P}_{\Lambda,\Delta}(L)$ defined by (9.64) is empty. In this case, a contiguous pair of characteristic null spaces will be equal. If the level sets $\mathbb{P}_{\Lambda,\Delta}(L)$ are all distinct, then the subset relations (9.93) are all proper, and the characteristic null space of a canonical tensor operator is a unique signature of the tensor operator. What actually occurs is, of course, completely determined by the properties of the Littlewood-Richardson numbers $I_{\Lambda,\Delta}(\lambda)$, viewed as functions $I_{\Lambda,\Delta}$ of $\lambda \in \mathbb{P}\mathrm{ar}_n$ that take on values in $\{0, 1, \ldots, K(\Lambda, \Delta)\}$. We defer this issue for now, taking it up again in detail in Sect. 9.5 for $U(3)$.

The results obtained in this section for canonical tensor operators hold independently of whether or not the characteristic null spaces are nested in the sense of strict inclusion in relations (9.93). (For $n = 3$, strict inclusion prevails.) It is the case, however, that the entire construction of canonical tensor operators and orthonormalized operator-valued $\widehat{D}^\Lambda$–polynomials can be carried out for any assignment of operator patterns, not just those in the ordered set (9.87)—every subset of the operator-valued polynomials containing $I_{\Lambda,\Delta}(\lambda)$ such operators is linearly independent on the space H_λ. *There is no implication in this construction that the ordering of the patterns (9.87) is special, nor that canonical tensor operators are "canonical" in the strong sense of there being essentialy "no free choices."*

9.3.2 Subgroup conditions and reduced matrix elements

The relation of irreducible tensor operators in $U(n)$ to those in $U(n-1)$ is important. Up until now, we have not paid sufficient attention to what values the final GT patterns m' in the matrix element

$$\left\langle \begin{matrix} \lambda+\Delta \\ m' \end{matrix} \middle| \left\langle \begin{matrix} \Gamma^t_\Delta \\ \Lambda \\ M \end{matrix} \right\rangle \middle| \begin{matrix} \lambda \\ m \end{matrix} \right\rangle \tag{9.94}$$

can assume for given $m \in \mathbb{G}_\lambda$ and $M \in \mathbb{G}_\Lambda$, other than requiring $m' \in \mathbb{G}_{\lambda+\Delta}$. (We have not paid careful attention to this in other matrix elements as well.) The values that m' can assume are, of course, dictated by the meaning of the rows of a GT pattern: Rows $M_k = (M_{1,k}, M_{2,k}, \ldots, M_{k,k}) \in \mathbb{P}\mathrm{ar}_k$ and $m_k = (m_{1,k}, m_{2,k}, \ldots, m_{k,k}) \in \mathbb{P}\mathrm{ar}_k$, for $k = 1, 2, \ldots, n-1$ are partitions of unitary irreducible representations of the unitary group $U(k)$, in addition to satisfying the betweenness relations of the GT patterns of shape Λ and λ. The meaning of row m'_k in the final pattern is then that it also must satisfy all betweenness relations for the GT patterns of shape $\lambda + \Delta$, and be a member of the Kronecker product $M_k \otimes m_k$; that is,

$$m'_k \in M_k \otimes m_k,\ k = 1, 2, \ldots, n-1. \tag{9.95}$$

We next incorporate these requirements into the matrix elements of a canonical tensor operator in this and the following sections.

The matrix elements of the set of canonical tensor operator

$$\left\langle \begin{matrix} \lambda+\Delta \\ m' \end{matrix} \middle| \left\langle \begin{matrix} \Gamma^t_\Delta \\ \Lambda \\ M \end{matrix} \right\rangle \middle| \begin{matrix} \lambda \\ m \end{matrix} \right\rangle = R_t\left[\begin{pmatrix} \lambda+\Delta \\ m' \end{pmatrix} \begin{pmatrix} \Lambda \\ M \end{pmatrix} \begin{pmatrix} \lambda \\ m \end{pmatrix}\right],$$

$$t = 1, 2, \ldots, I_{\Lambda,\Delta}(\lambda), \tag{9.96}$$

are defined for all values of the shift-weight $\Delta \in \mathbb{W}_\Lambda$ that provide exactly those CG coefficients that enter into the rows $\binom{t,\lambda+\Delta}{m'}$ and columns (M, m) of the real orthogonal matrix $R^{(\Lambda\,\lambda)}$ that reduces the Kronecker product $D^\Lambda(U) \otimes D^\lambda(U)$ to the Kronecker direct sum. The orthogonality relations for these CG coefficients are those given earlier by (9.52)-(9.54):

$$\sum_{t=1}^{I_{\Lambda,\Delta}(\lambda)} \sum_{m'' \in \mathbb{G}_{\lambda+\Delta}} R_t\left[\begin{pmatrix} \lambda+\Delta \\ m'' \end{pmatrix} \begin{pmatrix} \Lambda \\ M' \end{pmatrix} \begin{pmatrix} \lambda \\ m' \end{pmatrix}\right] R_t\left[\begin{pmatrix} \lambda+\Delta \\ m'' \end{pmatrix} \begin{pmatrix} \Lambda \\ M \end{pmatrix} \begin{pmatrix} \lambda \\ m \end{pmatrix}\right]$$
$$= \delta_{M',M}\, \delta_{m',m}, \tag{9.97}$$

$$\sum_{m \in \mathbb{G}_\lambda\, M \in \mathbb{G}_\Lambda} R_{t'}\left[\begin{pmatrix} \lambda+\Delta \\ m'' \end{pmatrix} \begin{pmatrix} \Lambda \\ M \end{pmatrix} \begin{pmatrix} \lambda \\ m \end{pmatrix}\right] R_t\left[\begin{pmatrix} \lambda+\Delta \\ m' \end{pmatrix} \begin{pmatrix} \Lambda \\ M \end{pmatrix} \begin{pmatrix} \lambda \\ m \end{pmatrix}\right]$$
$$= \delta_{t',t}\, \delta_{m'',m'},\ t', t = 1, 2, \ldots, I_{\Lambda,\Delta}(\lambda). \tag{9.98}$$

The $U(n-1)$ canonical tensor operators

$$\left\langle \begin{matrix} \lambda' + \Delta' \\ m'' \end{matrix} \middle| \left\langle \begin{matrix} \Gamma^{t'}_{\Delta'} \\ \Lambda' \\ M' \end{matrix} \right\rangle \middle| \begin{matrix} \lambda' \\ m' \end{matrix} \right\rangle = R_{t'} \left[\begin{pmatrix} \lambda' + \Delta' \\ m'' \end{pmatrix} \begin{pmatrix} \Lambda' \\ M' \end{pmatrix} \begin{pmatrix} \lambda' \\ m' \end{pmatrix} \right],$$

$$\begin{aligned} &\lambda',\ \Lambda',\ \lambda' + \Delta' \in \mathbb{P}\mathrm{ar}_{n-1}, \\ &m' \in \mathbb{G}_{\lambda'},\ M' \in \mathbb{G}_{\Lambda'},\ \lambda' + \Delta' \in \mathbb{G}_{\lambda'+\Delta'}, \\ &\Gamma^{t'}_{\Delta'} \in \mathbb{G}_{\Lambda',\Delta'},\ t' = 1, 2, \ldots, I_{\Lambda',\Delta'}(\lambda'),\ \text{each}\ \Delta' \in \mathbb{W}_{\Lambda'}, \end{aligned} \tag{9.99}$$

satisfy orthogonality relations of the same form as (9.97)-(9.98), now notated at level $n-1$.

For convenience of exposition, we assume that operator patterns are ordered at level n and at level $n-1$ in accordance with (9.87). Then, we also have the split of the set of $U(n-1)$ canonical tensor operators (9.99) into non-null and null tensor operators in accordance with the Littlewood-Richardson and Kostka numbers at level $n-1$. In particular, we have the set of null canonical tensor operators

$$\left\langle \begin{matrix} \Gamma^{r'}_{\Delta'} \\ \Lambda' \\ M' \end{matrix} \right\rangle \middle| \begin{matrix} \lambda' \\ m' \end{matrix} \right\rangle = \mathbf{0},\ r' = I_{\Lambda',\Delta'}(\lambda') + 1, \ldots, K(\Lambda', \Delta'). \tag{9.100}$$

We next give the relationship between the matrix elements of $U(n)$ and $U(n-1)$ canonical tensor operators and explain the notation for the transformation coefficients between the two sets of CG coefficients before giving the proof:

$$\begin{aligned} &\left\langle \begin{matrix} \lambda + \Delta \\ \lambda' + \Delta' \\ m'' \end{matrix} \middle| \left\langle \begin{matrix} \Gamma^{t}_{\Delta} \\ \Lambda \\ \Lambda' \\ M' \end{matrix} \right\rangle \middle| \begin{matrix} \lambda \\ \lambda' \\ m' \end{matrix} \right\rangle = \\ &\sum_{t'=1}^{I_{\Lambda',\Delta'}(\lambda')} \left\langle \begin{matrix} \lambda + \Delta \\ \lambda' + \Delta' \end{matrix} \middle| \begin{bmatrix} \Gamma^{t}_{\Delta} \\ \Lambda \\ \Lambda' \\ \Gamma^{t'}_{\Delta'} \end{bmatrix} \middle| \begin{matrix} \lambda \\ \lambda' \end{matrix} \right\rangle \left\langle \begin{matrix} \lambda' + \Delta' \\ m'' \end{matrix} \middle| \left\langle \begin{matrix} \Gamma^{t'}_{\Delta'} \\ \Lambda' \\ M' \end{matrix} \right\rangle \middle| \begin{matrix} \lambda' \\ m' \end{matrix} \right\rangle, \\ &\qquad\qquad \text{each}\ t = 1, \ldots, I_{\Lambda,\Delta}(\lambda). \end{aligned} \tag{9.101}$$

We summarize the meaning of the notations in this relation:

1. Rows $n-1$ of the $U(n)$ CG coefficient on the left in (9.101) have been displayed explicitly as $\lambda',\ \Lambda',\ \lambda' + \Delta' \in \mathbb{P}\mathrm{ar}_{n-1}$, where these partitions also satisfy the betweenness relations $\lambda' \prec \lambda$, $\Lambda' \prec \Lambda$, $\lambda' + \Delta' \prec$

$\lambda + \Delta$, and the GT patterns satisfy $m' \in \mathbb{G}_{\lambda'}, M' \in \mathbb{G}_{\Lambda'}, m'' \in \mathbb{G}_{\lambda'+\Delta'}$ and have $(n-2)$–rows. These partitions and GT patterns are those of the $U(n-1)$ CG coefficient on the right. For $U(n)$ canonical tensor operators, the shift-weight Δ belongs to $\mathbb{W}_\Lambda$ and has multiplicity $K(\Lambda, \Delta)$; and the index $t = 1, 2, \ldots, I_{\Lambda,\Delta}(\lambda)$ enumerates the $U(n)$ CG coefficients corresponding to the canonical tensor operators for which H_λ is not a characteristic null space. For $U(n-1)$ canonical tensor operators, the shift-weight Δ' belongs to $\mathbb{W}_{\Lambda'}$ has multiplicity $K(\Lambda', \Delta')$; and the index $t' = 1, 2, \ldots, I_{\Lambda',\Delta'}(\lambda')$ enumerates the $U(n-1)$ CG coefficients corresponding to the canonical tensor operators for which $H_{\lambda'}$ is not a characteristic null space.

2. The real coefficient

$$\left\langle \begin{matrix} \lambda+\Delta \\ \lambda'+\Delta' \end{matrix} \right| \left[\begin{matrix} \Gamma^t_\Delta \\ \Lambda \\ \Lambda' \\ \Gamma^{t'}_{\Delta'} \end{matrix} \right] \left| \begin{matrix} \lambda \\ \lambda' \end{matrix} \right\rangle \tag{9.102}$$

appearing in (9.101) inherits the two-rowed GT patterns $\binom{\lambda}{\lambda'}$, $\binom{\Lambda}{\Lambda'}$, and $\binom{\lambda+\Delta}{\lambda'+\Delta'}$ from the parent $U(n)$ CG coefficient on the left; the lower shift-weight pattern $\binom{\Lambda'}{\Gamma^{t'}_{\Delta'}}$ is inherited from the shift-weight pattern of the daughter $U(n-1)$ CG coefficient on the right. The matrix elements of the canonical tensor operator in $U(n-1)$ on the right inherits its patterns from the $U(n-1)$ subgroup labels of the parent $U(n)$ canonical tensor on the left, except for its upper shift-patterns, which are summed over. The coefficient (9.102) is called a $U(n) : U(n-1)$ *reduced matrix element.* Relation (9.101) is known as the $U(n) : U(n-1)$ *Wigner-Eckart theorem.*

Relation (9.101) can be inverted by using the orthogonality of the CG coefficients in $U(n-1)$ to obtain the $U(n) : U(n-1)$ reduced matrix elements in terms of the canonical $U(n)$ and $U(n-1)$ CG coefficients:

$$\left\langle \begin{matrix} \lambda+\Delta \\ \lambda'+\Delta' \end{matrix} \right| \left[\begin{matrix} \Gamma^t_\Delta \\ \Lambda \\ \Lambda' \\ \Gamma^{t'}_{\Delta'} \end{matrix} \right] \left| \begin{matrix} \lambda \\ \lambda' \end{matrix} \right\rangle \tag{9.103}$$

$$= \sum_{m',M'} \left\langle \begin{matrix} \lambda+\Delta \\ \lambda'+\Delta' \\ m'' \end{matrix} \right| \left\langle \begin{matrix} \Gamma^t_\Delta \\ \Lambda \\ \Lambda' \\ M' \end{matrix} \right\rangle \left| \begin{matrix} \lambda \\ \lambda' \\ m' \end{matrix} \right\rangle \left\langle \begin{matrix} \lambda'+\Delta' \\ m'' \end{matrix} \right| \left\langle \begin{matrix} \Gamma^{t'}_{\Delta'} \\ \Lambda' \\ M' \end{matrix} \right\rangle \left| \begin{matrix} \lambda' \\ m' \end{matrix} \right\rangle,$$

$$t = 1, \ldots, I_{\Lambda,\Delta}(\lambda);\ t' = 1, \ldots, I_{\Lambda',\Delta'}(\lambda').$$

Accompanying this relation are the zero matrix elements that are a consequence of the characteristic null space of the canonical tensor operators at level n and $n-1$:

$$\left\langle \begin{matrix} \lambda+\Delta \\ \lambda'+\Delta' \end{matrix} \right| \left[\begin{matrix} \Gamma^s_\Delta \\ \Lambda \\ \Lambda' \\ \Gamma^{s'}_{\Delta'} \end{matrix} \right] \left| \begin{matrix} \lambda \\ \lambda' \end{matrix} \right\rangle = 0, \tag{9.104}$$

where these relations hold for the pair of indices (s, s') given by the following two ranges—those not covered in (9.103):

$$\begin{aligned} s &= 1, \ldots, K(\Lambda, \Delta);\ s' = I_{\Lambda',\Delta'}(\lambda') + 1, \ldots, K(\Lambda', \Delta'), \\ s &= I_{\Lambda,\Delta}(\lambda) + 1, \ldots, K(\Lambda, \Delta);\ s' = 1, \ldots, K(\Lambda', \Delta'). \end{aligned} \tag{9.105}$$

The initial and final states in the reduced matrix element (9.103) are two-rowed GT patterns satisfying the betweenness relations. In evaluating these coefficients from (9.103), the GT pattern $m'' \in \mathbb{G}_{\lambda'+\Delta'}$ can be arbitrarily chosen, since the values of the $U(n) : U(n-1)$ reduced matrix elements are independent of m'' : they are $U(n) : U(n-1)$ invariants.

It cannot be emphasized too strongly that the following conditions must hold for the reduced matrix elements (9.102) to be defined (see Examples (9.110)-(9.113) below):

$$\begin{aligned} &\lambda, \lambda+\Delta \in \mathbb{P}\mathrm{ar}_n,\ \Gamma^s_\Delta \in \mathbb{G}_{\Lambda,\Delta},\ s = 1, \ldots, K(\Lambda, \Delta), \\ &\lambda', \lambda'+\Delta' \in \mathbb{P}\mathrm{ar}_{n-1},\ \Gamma^{s'}_{\Delta'} \in \mathbb{G}_{\Lambda',\Delta'},\ s' = 1, \ldots, K(\Lambda', \Delta'), \\ &\lambda' \prec \lambda,\ \Lambda' \prec \Lambda,\ \lambda'+\Delta' \prec \lambda+\Delta. \end{aligned} \tag{9.106}$$

The matrix elements of the canonical tensor operators are completely determined by the Gram-Schmidt orthogonalization of the operator-valued $D^\Lambda-$ and $D^{\Lambda'}-$polynomials; that is, by the relations:

$$\left\langle \begin{matrix} \Gamma^s_\Delta \\ \Lambda \\ \Lambda' \\ M \end{matrix} \right\rangle = \widehat{D} \begin{pmatrix} \Gamma^s_\Delta \\ \Lambda \\ \Lambda' \\ M \end{pmatrix} (T),\ \left\langle \begin{matrix} \Gamma^{s'}_{\Delta'} \\ \Lambda' \\ M \end{matrix} \right\rangle = \widehat{D} \begin{pmatrix} \Gamma^{s'}_{\Delta'} \\ \Lambda' \\ M \end{pmatrix} (T'),$$

$$s = 1, 2, \ldots, K(\Lambda, \Delta);\ s' = 1, 2, \ldots, K(\Lambda', \Delta'). \tag{9.107}$$

It is therefore also the case that all canonical $U(n) : U(n-1)$ reduced matrix elements (9.103) are fully determined: *there are no further free choices to be made.*

It is possible, however, to take a different viewpoint; namely, relation (9.101) may be regarded as giving a recursive construction of the canonical $U(n)$ tensor operators from the $U(n-1)$ tensor operators once the family of reduced matrix elements is known. In this viewpoint, the reduced matrix elements become the basic structural elements of the theory and, accordingly, the properties of reduced matrix elements should be developed on their own.

We note that the reduced $U(2) : U(1)$ matrix elements (9.103) for $n = 2$ are just the matrix elements of the $U(2)$ canonical tensor operator themselves in consequence of the $U(1)$ tensor operator being 0 unless $\lambda'_1 + \Delta'_1 = \lambda'_1 + \Lambda'_1$, in which case it is equal to 1, and all the GT patterns m', M', m'' are empty:

$$\left\langle \begin{matrix} \lambda_1+\Delta_1 \ \lambda_2+\Delta_2 \\ \lambda'_1+\Delta'_1 \end{matrix} \right\| \left[\begin{matrix} \Delta_1 \\ \Lambda_1 \quad \Lambda_2 \\ \Delta'_1 \end{matrix} \right] \left\| \begin{matrix} \lambda_1 \ \lambda_2 \\ \lambda'_1 \end{matrix} \right\rangle \tag{9.108}$$

$$= \left\langle \begin{matrix} \lambda_1+\Delta_1 \quad \lambda_2+\Delta_2 \\ \lambda'_1+\Delta'_1 \end{matrix} \right| \left\langle \begin{matrix} \Delta_1 \\ \Lambda_1 \quad \Lambda_2 \\ \Delta'_1 \end{matrix} \right\rangle \left| \begin{matrix} \lambda_1 \quad \lambda_2 \\ \lambda'_1 \end{matrix} \right\rangle .$$

We turn next to the development of some of the properties of reduced matrix elements. It is important to make the distinction between operator $\Gamma-$patterns and ordinary $M-$patterns. The entries in the various rows of a GT $M-$pattern always have the subgroup significance as emphasized in relation (9.95). *There are no such subgroup conditions on the various rows of an operator pattern: These patterns encode the shift-action of the tensor operator in making the transformations between vector spaces $H_\lambda \to H_{\lambda+\Delta}$ and $H'_{\lambda'} \to H'_{\lambda'+\Delta'}$.* It is, however, the case that the following two group properties must hold:

$$\lambda + \Delta \in \Lambda \otimes \lambda, \ \lambda' + \Delta' \in \Lambda' \otimes \lambda'. \tag{9.109}$$

The second property is inherited from the Wigner-Eckart theorem (9.101): it is responsible for a weak linkage between operator patterns and subgroups. It is this linkage that allows an interpretation of operator patterns given in Sect. 9.7.2 based on taking limits of a reduced matrix element as certain partition labels go to $-\infty$.

9.4 Properties of $U(n) : U(n-1)$ Reduced Matrix Elements

The view may be taken that the calculation of the $U(n) : U(n-1)$ reduced matrix elements in relation (9.101) is the primary problem. This would then allow the calculation of all canonical tensor operators to be completed at level n from those at level $n-1$, and then upward to level $n+1$, etc. The recursive construction of all $U(n)$ canonical tensor operators, starting at the $U(2)$ WCG level would then be complete. A procedure for determining the canonical reduced matrix elements at every level $n : n-1$ thus becomes a key objective.

We have remarked earlier that the reduced matrix elements of a set of canonical tensor operators are defined if and only if all patterns that appear in it are lexical. It is useful to illustrate this result.

Examples: Consider the following symbol for a special reduced matrix element (9.102):

$$\left\langle \begin{matrix} \lambda_1+3 & & \lambda_2+3 & & \lambda_3+3 \\ & \lambda'_1+\Delta'_1 & & \lambda'_2+\Delta'_3 & \end{matrix} \;\middle|\; \begin{bmatrix} \Gamma^s_{(3,3,3)} \\ 6 \quad 3 \quad 0 \\ \Gamma^{s'}_{(\Delta'_1,\Delta'_2)} \end{bmatrix} \;\middle|\; \begin{matrix} \lambda_1 & & \lambda_2 & & \lambda_3 \\ & \lambda'_1 & & \lambda'_2 & \end{matrix} \right\rangle, \tag{9.110}$$

where the operator pattern $\Gamma^s_{(3,3,3)}$ is any of the four having shift-weight $\Delta = (3,3,3)$:

$$\begin{aligned} &\Gamma^1_{(3,3,3)} = \begin{pmatrix} & 3 & \\ 6 & & 0 \end{pmatrix} > \Gamma^2_{(3,3,3)} = \begin{pmatrix} & 3 & \\ 5 & & 1 \end{pmatrix} \\ > \;&\Gamma^3_{(3,3,3)} = \begin{pmatrix} & 3 & \\ 4 & & 2 \end{pmatrix} > \Gamma^4_{(3,3,3)} = \begin{pmatrix} & 3 & \\ 3 & & 3 \end{pmatrix}. \end{aligned} \tag{9.111}$$

The lower operator pattern $\Gamma^{s'}_{(\Delta'_1,\Delta'_2)}$ can be any of the $\mathrm{Dim}(6\ 3\ 0) = 64$ lexical patterns of the form

$$\begin{pmatrix} \Lambda'_1 & & \Lambda'_2 \\ & \Delta'_1 & \end{pmatrix}, \; 6 \geq \Lambda'_1 \geq 3,\; 3 \geq \Lambda'_2 \geq 0,\; \Lambda'_1 \geq \Delta'_1 \geq \Lambda'_2. \tag{9.112}$$

The symbol (9.110) is defined if and only if, in addition to the upper and lower operator patterns being lexical, it is also the case that the initial and final two-rowed ket-symbols are lexical (satisfy betweenness).

An example of an undefined symbol (9.110) occurs for the lower operator pattern $\binom{3\ \ 0}{0}$ and the choice $\lambda_1 = \lambda_2 + 3$. In this case, the initial pattern is lexical for $\lambda'_1 = \lambda_2, \lambda_2+1, \lambda_2+2$ and for any λ'_2 satisfying $\lambda_3 \leq \lambda'_2 \leq \lambda_2$. But, for these initial values, the betweenness condition $\lambda_2 + 6 \geq \lambda'_1 \geq \lambda_2 + 3$ fails to be satisfied in the shifted final two-rowed pattern. (It is satisfied for the remaining choice of $\lambda'_1 = \lambda_2 + 3$; the coefficient is defined.)

Further examples of defined symbols (9.110) are

$$\left\langle \begin{matrix} \lambda_3+5 & & \lambda_3+4 & & \lambda_3+3 \\ & \lambda_3+5 & & \lambda_3+4 & \end{matrix} \right| \left[\begin{matrix} & \Gamma^{s}_{(3,3,3)} & \\ 6 & 3 & 0 \\ & \Gamma^{s'}_{(4,4)} & \end{matrix} \right] \left| \begin{matrix} \lambda_3+2 & & \lambda_3+1 & & \lambda_3 \\ & \lambda_3+1 & & \lambda_3 & \end{matrix} \right\rangle,$$

$$\Gamma^{1}_{(4,4))} = \begin{pmatrix} 6 & & 2 \\ & 4 & \end{pmatrix} > \Gamma^{2}_{(4,4)} = \begin{pmatrix} 5 & & 3 \\ & 4 & \end{pmatrix}. \tag{9.113}$$

All patterns in this symbol are lexical and the symbol is defined, although it might be 0. The reduced matrix elements (9.110) take on specific meanings in terms of the characteristic null space of the canonical $U(3)$ tensor operators, as discussed in Sect. 9.6 □

It is straightforward to derive from the betweenness rules the following conditions that the initial two-rowed pattern must satisfy for the final two-rowed pattern to be lexical:

Let the initial pattern $\binom{\lambda}{\lambda'}$ *in the reduced matrix element (9.102) satisfy* $\lambda \in \mathbb{P}ar_n$, $\lambda' \in \mathbb{P}ar_{n-1}$, $\lambda' \prec \lambda$; *that is, be lexical. Then, the shifted pattern* $\binom{\lambda+\Delta}{\lambda'+\Delta'}$ *is lexical; that is,* $\lambda+\Delta \in \mathbb{P}ar_n$, $\lambda'+\Delta' \in \mathbb{P}ar_{n-1}$, $\lambda'+\Delta' \prec \lambda+\Delta$, *if and only if the initial two-rowed pattern also satisfies the conditions:*

$$\lambda_i - \lambda'_i \geq \max(0, \Delta'_i - \Delta_i),\ \lambda'_i - \lambda_{i+1} \geq \max(0, \Delta_{i+1} - \Delta'_i),$$
$$i = 1, 2, \ldots, n-1. \tag{9.114}$$

9.4.1 Unit projective operators

The notation (9.102) for $U(n) : U(n-1)$ reduced matrix elements has been written in a form that suggests that the objects

$$\begin{bmatrix} \Gamma \\ \Lambda \\ \Gamma' \end{bmatrix},\ \Gamma, \Gamma' \in \mathbb{G}_{\Lambda}, \tag{9.115}$$

can be interpreted as operators acting in a Hilbert space. We call such operators *unit projective operators.* It is useful to develop this idea because it allows unit projective operators to be written as products of simpler unit projective operators, much in the same way that tensor operators can be written in terms of fundamental tensor operators.

A $U(n)$ tensor operator acts in the model Hilbert space

$$H = \sum_{p \geq 0} \sum_{\lambda \in \mathbb{P}\mathrm{ar}_n(p)} \oplus H_{\lambda}, \tag{9.116}$$

with orthonormal basis $\mathbf{B}_\lambda$ of the subspace $H_\lambda \subset H$. A $U(n-1)$ tensor operator acts in the model Hilbert space

$$H' = \sum_{p' \geq 0} \sum_{\lambda' \in \mathbb{P}\mathrm{ar}_{n-1}(p')} \oplus H_{\lambda'}, \tag{9.117}$$

with orthonormal basis $\mathbf{B}'_{\lambda'}$ of the subspace $H'_{\lambda'} \subset H'$.

In the same spirit of the model Hilbert spaces (9.116)-(9.117), we introduce the model separable Hilbert space

$$\mathcal{H} = \sum_{p \geq 0} \sum_{\lambda \in \mathbb{P}\mathrm{ar}_n(p)} \oplus \mathcal{H}_\lambda, \tag{9.118}$$

with an orthonormal basis of $\mathcal{H}_\lambda$ given by

$$\mathcal{B}_\lambda = \left\{ \left| \begin{array}{c} \lambda \\ \lambda' \end{array} \right\rangle \middle| \begin{array}{l} \lambda' \in \mathbb{P}\mathrm{ar}_{n-1}; \\ \lambda' \prec \lambda \end{array} \right\}. \tag{9.119}$$

We now introduce the tensor product space as follows:

$$\mathcal{H} \otimes H' = \sum_{p \geq 0} \sum_{\lambda \in \mathbb{P}\mathrm{ar}_n(p)} \sum_{\lambda' \prec \lambda} \oplus \left(\mathcal{H}_\lambda \otimes H'_{\lambda'} \right). \tag{9.120}$$

This is just another way of expressing the Hilbert space H itself, so that

$$H = \mathcal{H} \otimes H'. \tag{9.121}$$

Each orthonormal vector in the basis $H_\lambda \subset H$ is now expressed as

$$\left| \begin{array}{c} \lambda \\ \lambda' \\ m' \end{array} \right\rangle = \left| \begin{array}{c} \lambda \\ \lambda' \end{array} \right\rangle \otimes \left| \begin{array}{c} \lambda' \\ m' \end{array} \right\rangle, \ \text{all}\, \lambda' \prec \lambda, \ \text{all}\, m' \in \mathbb{G}'_{\lambda'}. \tag{9.122}$$

This form of the vector space exhibits clearly how $U(n)$ and $U(n-1)$ tensor operators and the associated $U(n) : U(n-1)$ projective operators are linked:

$$\left\langle \begin{array}{c} \Gamma \\ \Lambda \\ \Lambda' \\ M' \end{array} \right\rangle = \sum_{\Gamma'} \left[\begin{array}{c} \Gamma \\ \Lambda \\ \Lambda' \\ \Gamma' \end{array} \right] \otimes \left\langle \begin{array}{c} \Gamma' \\ \Lambda' \\ M' \end{array} \right\rangle, \tag{9.123}$$

with the Hermitian conjugate operator having a similar form:

$$\left\langle \begin{array}{c} \Gamma \\ \Lambda \\ \Lambda' \\ M' \end{array} \right\rangle^\dagger = \sum_{\Gamma'} \left[\begin{array}{c} \Gamma \\ \Lambda \\ \Lambda' \\ \Gamma' \end{array} \right]^\dagger \otimes \left\langle \begin{array}{c} \Gamma' \\ \Lambda' \\ M' \end{array} \right\rangle^\dagger. \tag{9.124}$$

Multiplication of relation (9.123) from the right by the operator

$$\mathcal{I} \otimes \left\langle \begin{matrix} \Gamma'' \\ \Lambda' \\ M' \end{matrix} \right\rangle^{\dagger}, \tag{9.125}$$

where $\mathcal{I}$ is the identity operator on $\mathcal{H}$, and summing over M', gives the expression for the unit projective operator in terms of $U(n)$ and $U(n-1)$ canonical tensor operators:

$$\begin{bmatrix} \Gamma \\ \Lambda \\ \Lambda' \\ \Gamma' \end{bmatrix} \otimes \mathbb{I}' = \sum_{M'} \left\langle \begin{matrix} \Gamma \\ \Lambda \\ \Lambda' \\ M' \end{matrix} \right\rangle \left(\mathcal{I} \otimes \left\langle \begin{matrix} \Gamma' \\ \Lambda' \\ M' \end{matrix} \right\rangle^{\dagger} \right). \tag{9.126}$$

Here $\mathbb{I}'$ is the identity operator on H'. From (9.122), the matrix elements of the operator identity (9.123) between bra-ket vectors

$$\left\langle \begin{matrix} \lambda+\Delta \\ \lambda'+\Delta' \\ m'' \end{matrix} \right| \cdots \left| \begin{matrix} \lambda \\ \lambda' \\ m' \end{matrix} \right\rangle \tag{9.127}$$

reproduces relation (9.101), when characteristic null spaces are taken into account. Similarly, the matrix elements of the operator identity (9.126) between bra-ket vectors

$$\left\langle \begin{matrix} \lambda+\Delta \\ \lambda'+\Delta' \end{matrix} \right| \quad \cdots \quad \left| \begin{matrix} \lambda \\ \lambda' \end{matrix} \right\rangle \mathcal{I} \otimes \mathbb{I}' \tag{9.128}$$

reproduces relation (9.103), when characteristic null spaces are taken into account.

The introduction in (9.118) of the separable Hilbert space $\mathcal{H}$ with properties (9.119)-(9.122) allows us to consider the action of unit projective operators in this space. The actions on the basis $\mathcal{B}_\lambda$ of each subspace $\mathcal{H}_\lambda \subset \mathcal{H}$ of the various unit projective operators are as follows:

$$\begin{bmatrix} \Gamma^t_\Delta \\ \Lambda \\ \Lambda' \\ \Gamma^{t'}_{\Delta'} \end{bmatrix} \left| \begin{matrix} \lambda \\ \lambda' \end{matrix} \right\rangle = \left\langle \begin{matrix} \lambda+\Delta \\ \lambda'+\Delta' \end{matrix} \right| \begin{bmatrix} \Gamma^t_\Delta \\ \Lambda \\ \Lambda' \\ \Gamma^{t'}_{\Delta'} \end{bmatrix} \left| \begin{matrix} \lambda \\ \lambda' \end{matrix} \right\rangle \left| \begin{matrix} \lambda+\Delta \\ \lambda'+\Delta' \end{matrix} \right\rangle, \tag{9.129}$$

$$t = 1, 2, \ldots, I_{\Lambda,\Delta}(\lambda);\; t' = 1, 2, \ldots, I_{\Lambda',\Delta'}(\lambda');$$

$$\begin{bmatrix} \Gamma^r_\Delta \\ \Lambda \\ \Lambda' \\ \Gamma^{s'}_{\Delta'} \end{bmatrix} \left| \begin{array}{c} \lambda \\ \lambda' \end{array} \right\rangle = \mathbf{0}, \tag{9.130}$$

$$r = I_{\Lambda,\Delta}(\lambda) + 1, \ldots, K(\Lambda, \Delta), \; s' = 1, 2, \ldots, K(\Lambda', \Delta');$$

$$\begin{bmatrix} \Gamma^s_\Delta \\ \Lambda \\ \Lambda' \\ \Gamma^{r'}_{\Delta'} \end{bmatrix} \left| \begin{array}{c} \lambda \\ \lambda' \end{array} \right\rangle = \mathbf{0}, \tag{9.131}$$

$$s = 1, 2, \ldots, K(\Lambda, \Delta), \; r' = I_{\Lambda',\Delta'}(\lambda') + 1, \ldots, K(\Lambda', \Delta').$$

A unit projective operator inherits it characteristic null space from its parent $U(n)$ canonical tensor operator and the daughter $U(n-1)$ canonical tensor operator. Thus, the characteristic null space of a unit projective operator is the vector space defined by

$$\mathbf{N}\begin{bmatrix} \Gamma^s_\Delta \\ \Lambda \\ \Lambda' \\ \Gamma^{s'}_{\Delta'} \end{bmatrix} = \sum_{\lambda \in \mathbb{P}^s_{\Lambda,\Delta} \cup \mathbb{P}^{s'}_{\Lambda',\Delta'}} \oplus \mathcal{H}_\lambda, \tag{9.132}$$

where the sets of partitions $\mathbb{P}^s_{\Lambda,\Delta}$ are defined by (9.64) at level n, with a similar definition of $\mathbb{P}^{s'}_{\Lambda',\Delta'}$ at level $n-1$.

Relations (9.53) and (9.51), respectively, for unit tensor operators, can be written in the following form for canonical tensor operators:

$$\sum_{M \in \mathbb{G}_\Lambda} \left\langle \begin{array}{c} \Gamma^{t'}_\Delta \\ \Lambda \\ M \end{array} \right\rangle \left\langle \begin{array}{c} \Gamma^t_\Delta \\ \Lambda \\ M \end{array} \right\rangle^\dagger = \delta_{t',t}, \text{ on } H_{\lambda+\Delta}. \tag{9.133}$$

$$\sum_{t=1}^{I_{\Lambda,\Delta}(\lambda)} \left\langle \begin{array}{c} \Gamma^t_\Delta \\ \Lambda \\ M' \end{array} \right\rangle^\dagger \left\langle \begin{array}{c} \Gamma^t_\Delta \\ \Lambda \\ M \end{array} \right\rangle = \delta_{M',M}, \text{ on } H_\lambda. \tag{9.134}$$

These operator orthogonality relations have been written in forms that account for the characteristic null space: They are valid on the indicated vector spaces and the values $t, t' = 1, 2, \ldots, I_{\Lambda,\Delta}(\lambda)$. There is a similar set of orthogonality relations for the $U(n-1)$ canonical tensor operators.

Operator relations similar to relations (9.133) and (9.134) for the unit projective operators can also be derived. Thus, we multiply together the two relations (9.123) and (9.124) with Γ replaced by Γ^t_Δ and $\Gamma^{t''}_\Delta$, respectively, sum over all patterns $\binom{\Lambda'}{M'}$, and use the orthogonality (9.133) for the respective $U(n)$ and $U(n-1)$ canonical tensor operators and obtain the following operator orthogonality relations:

$$\sum_{\Lambda' \prec \Lambda} \sum_{t'=1}^{I_{\Lambda',\Delta'}(\lambda')} \begin{bmatrix} \Gamma^t_\Delta \\ \Lambda \\ \Lambda' \\ \Gamma^{t'}_{\Delta'} \end{bmatrix} \begin{bmatrix} \Gamma^{t''}_\Delta \\ \Lambda \\ \Lambda' \\ \Gamma^{t'}_{\Delta'} \end{bmatrix}^\dagger = \delta_{t,t''}, \text{ on } \mathcal{H}_{\lambda+\Delta}$$

$$t, t'' = 1, 2, \ldots, I_{\Lambda,\Delta}(\lambda). \qquad (9.135)$$

We next replace Γ by $\Gamma^{t''}_\Delta$ in (9.124), multiply the resulting relation from the right by $\begin{bmatrix} \Gamma^{t''}_\Delta \\ \Lambda \\ \Lambda' \\ \Gamma^{t'''}_{\Delta'} \end{bmatrix} \otimes \mathbb{I}_{\lambda'}$, sum over t'', compare each side of the resulting relation, and obtain the following operator orthogonality relations:

$$\sum_{t=1}^{I_{\Lambda,\Delta}(\lambda)} \begin{bmatrix} \Gamma^t_\Delta \\ \Lambda \\ \Lambda'' \\ \Gamma^{t''}_{\Delta'} \end{bmatrix}^\dagger \begin{bmatrix} \Gamma^t_\Delta \\ \Lambda \\ \Lambda' \\ \Gamma^{t'}_{\Delta'} \end{bmatrix} = \delta_{t'',t'}\, \delta_{\Lambda'',\Lambda'}, \text{ on } \mathcal{H}_\lambda,$$

$$t'', t' = 1, 2, \ldots, I_{\Lambda',\Delta'}(\lambda'). \qquad (9.136)$$

Relations (9.135)-(9.136) now give the following orthogonality relations for the matrix elements of unit projective operators:

$$\sum_{\Lambda' \prec \Lambda} \sum_{t'=1}^{I_{\Lambda',\Delta'}(\lambda')} \left\langle \begin{matrix} \lambda+\Delta \\ \lambda'+\Delta' \end{matrix} \right| \begin{bmatrix} \Gamma^t_\Delta \\ \Lambda \\ \Lambda' \\ \Gamma^{t'}_{\Delta'} \end{bmatrix} \left| \begin{matrix} \lambda \\ \lambda' \end{matrix} \right\rangle \left\langle \begin{matrix} \lambda+\Delta \\ \lambda'+\Delta' \end{matrix} \right| \begin{bmatrix} \Gamma^{t''}_\Delta \\ \Lambda \\ \Lambda' \\ \Gamma^{t'}_{\Delta'} \end{bmatrix} \left| \begin{matrix} \lambda \\ \lambda' \end{matrix} \right\rangle = \delta_{t,t''},$$

$$t, t'' = 1, 2, \ldots, I_{\Lambda,\Delta}(\lambda); \qquad (9.137)$$

$$\sum_{t=1}^{I_{\Lambda,\Delta}(\lambda)} \left\langle \begin{matrix} \lambda+\Delta \\ \lambda'+\Delta' \end{matrix} \right| \begin{bmatrix} \Gamma^t_\Delta \\ \Lambda \\ \Lambda'' \\ \Gamma^{t''}_{\Delta'} \end{bmatrix} \left| \begin{matrix} \lambda \\ \lambda' \end{matrix} \right\rangle \left\langle \begin{matrix} \lambda+\Delta \\ \lambda'+\Delta' \end{matrix} \right| \begin{bmatrix} \Gamma^t_\Delta \\ \Lambda \\ \Lambda' \\ \Gamma^{t'}_{\Delta'} \end{bmatrix} \left| \begin{matrix} \lambda \\ \lambda' \end{matrix} \right\rangle = \delta_{t'',t'}\, \delta_{\Lambda'',\Lambda'},$$

$$t'', t' = 1, 2, \ldots, I_{\Lambda',\Delta'}(\lambda').$$

The two orthogonality relations (9.137) are valid independently of the operator interpretation of reduced matrix elements in terms of projective operators. However, such an operator interpretation places unit projective operators on a par with unit (canonical) tensor operators, but now with matrix elements taken between the more elemental orthonormal initial and final basis vectors having two-rowed GT patterns as their labels: $\left| \begin{matrix} \lambda \\ \lambda' \end{matrix} \right\rangle \in \mathcal{B}_\lambda$ and $\left| \begin{matrix} \lambda+\Delta \\ \lambda'+\Delta' \end{matrix} \right\rangle \in \mathcal{B}_{\lambda+\Delta}$. An approach based on unit projective operators approach determines the matrix elements of these operators first, and then defines canonical tensor operators by relation (9.123) with matrix elements (9.101).

We emphasize again that the $U(n) : U(n-1)$ reduced matrix elements are uniquely determined because of the equality of canonical tensor operators with the orthonormalized operator-valued $\widehat{D}^\Lambda$–polynomials. We restate the relations explicitly to emphasize this result:

$$\left\langle \begin{matrix} \lambda+\Delta \\ \lambda'+\Delta' \\ m'' \end{matrix} \right| \widehat{D}\begin{pmatrix} \Gamma^t_\Delta \\ \Lambda \\ \Lambda' \\ M' \end{pmatrix}(T) \left| \begin{matrix} \lambda \\ \lambda' \\ m' \end{matrix} \right\rangle =$$

$$\sum_{t'=1}^{I_{\Lambda',\Delta'}(\lambda')} \left\langle \begin{matrix} \lambda+\Delta \\ \lambda'+\Delta' \end{matrix} \right| \begin{bmatrix} \Gamma^t_\Delta \\ \Lambda \\ \Lambda' \\ \Gamma^{t'}_{\Delta'} \end{bmatrix} \left| \begin{matrix} \lambda \\ \lambda' \end{matrix} \right\rangle \left\langle \begin{matrix} \lambda'+\Delta' \\ m'' \end{matrix} \right| \widehat{D}\begin{pmatrix} \Gamma^{t'}_{\Delta'} \\ \Lambda' \\ M' \end{pmatrix}(T') \left| \begin{matrix} \lambda' \\ m' \end{matrix} \right\rangle,$$

$$\text{each } t = 1, \ldots, I_{\Lambda,\Delta}(\lambda);$$

$$\left\langle \begin{matrix} \lambda+\Delta \\ \lambda'+\Delta' \end{matrix} \right| \begin{bmatrix} \Gamma^t_\Delta \\ \Lambda \\ \Lambda' \\ \Gamma^{t'}_{\Delta'} \end{bmatrix} \left| \begin{matrix} \lambda \\ \lambda' \end{matrix} \right\rangle = \tag{9.138}$$

$$\sum_{m',\,M'} \left\langle \begin{matrix} \lambda+\Delta \\ \lambda'+\Delta' \\ m'' \end{matrix} \right| \widehat{D}\begin{pmatrix} \Gamma^t_\Delta \\ \Lambda \\ \Lambda' \\ M' \end{pmatrix}(T) \left| \begin{matrix} \lambda \\ \lambda' \\ m' \end{matrix} \right\rangle \left\langle \begin{matrix} \lambda'+\Delta' \\ m'' \end{matrix} \right| \widehat{D}\begin{pmatrix} \Gamma^{t'}_{\Delta'} \\ \Lambda' \\ M' \end{pmatrix}(T') \left| \begin{matrix} \lambda' \\ m' \end{matrix} \right\rangle,$$

$$\text{each } t = 1, \ldots, I_{\Lambda,\Delta}(\lambda),\ t' = 1, \ldots, I_{\Lambda',\Delta'}(\lambda').$$

9.4.2 The pattern calculus

The arc digraph technique discussed in Sect. 6.3 has a natural generalization to unit projective operators. These simple rules for writing out matrix elements, or factors thereof, are invaluable for calculations. They provide explicity defined $U(n) : U(n-1)$ invariant multiplicative factors common to every reduced matrix element having fixed shift-weights $\Delta \in \mathbb{W}_\lambda, \Delta' \in \mathbb{W}_{\lambda'}$.

We review these rules in this section. The rules consist of five steps:
1. The two-rowed shift-weight pattern. The first step replaces the unit vector shift-weights in (6.30) by general shift-weights as follows, where the Δ'_j are written between the Δ_i :

$$\begin{array}{lccccccccccc} \Delta = & \Delta_1 & & \Delta_2 & & \cdots & & \Delta_i & \cdots & \Delta_{n-1} & & \Delta_n \\ \Delta' = & & \Delta'_1 & & \Delta'_2 & \cdots & \Delta'_j & & \cdots & & \Delta'_{n-1} & \end{array} \tag{9.139}$$

2. The shift-weight labeled points. The configuration of shift-weights (9.139) is placed at the points of the two-rowed shift-weight pattern to obtain the labeled set of points:

$$\begin{array}{ccccccccccccc} \Delta_1 & & \Delta_2 & & \Delta_3 & & & \Delta_i & & & \Delta_{n-1} & & \Delta_n \\ \circ & & \circ & & \circ & & \cdots & \circ & & & \circ & & \circ \\ & \circ & & \circ & & \cdots & \circ & & \cdots & \circ & & \circ & \\ & \Delta'_1 & & \Delta'_2 & & & \Delta'_j & & & \Delta'_{n-2} & & \Delta'_{n-1} & \end{array} \tag{9.140}$$

3. The arrow diagram. A directed arc-line goes between each pair of labeled points in diagram (9.140) for which the shift-weights are unequal. This rule applies to all ordered pairs of points in the top row n, to all ordered pairs of points in the bottom row $n-1$, to all ordered pairs of points between the top row and bottom row, and to all ordered pairs of points between the bottom row and top row. Multiple directed arc-lines (arrows) go between pairs of points labeled by the shift-weights as given by the following rules for the respective four cases:

$$\begin{aligned} &\Delta_i \rightarrow \Delta_{i'}, \ \text{for}\, \Delta_i > \Delta_{i'}, \\ &\Delta'_j \rightarrow \Delta'_{j'}, \ \text{for}\, \Delta'_j > \Delta'_{j'}, \\ &\Delta_i \rightarrow \Delta'_j, \ \text{for}\, \Delta_i > \Delta'_j, \\ &\Delta'_j \rightarrow \Delta_i, \ \text{for}\, \Delta'_j > \Delta_i. \end{aligned} \tag{9.141}$$

The number of directed arcs or arrows that goes between the points is equal to the *difference of the respective shift-weights* in (9.141). Once the arrows have been assigned, the labels of the points by shift-weights are dropped, leaving behind an *arrow diagram,* which is just the two-rowed sequence of dots with arrows going between pairs of dots. Each unit projective operator of type Λ with upper shift-weight $\Delta \in G_{\Lambda,\Delta}$ and lower shift-weight $\Delta' \in G_{\Lambda',\Delta'}$ has a unique arrow diagram associated with it.

4. The labeled arc multigraph. The points of the arrow diagram of a unit projective operator of type Λ with shift-weights Δ and Δ' are now relabeled with the partial hooks associated with the partitions $\lambda \in \mathbb{P}\mathrm{ar}_n$ and $\lambda' \in \mathbb{P}\mathrm{ar}_{n-1}$ of the basis vector in $\mathcal{B}_\lambda$ on which the unit projective operator acts:

$$\begin{aligned} y_i &= \lambda_i + n - i,\ i = 1, 2, \ldots, n, \\ x_j &= \lambda'_j + n - 1 - j,\ j = 1, 2, \ldots, n-1. \end{aligned} \tag{9.142}$$

This gives the labeled arc multigraph with labeled points and multiple directed arrows going between points:

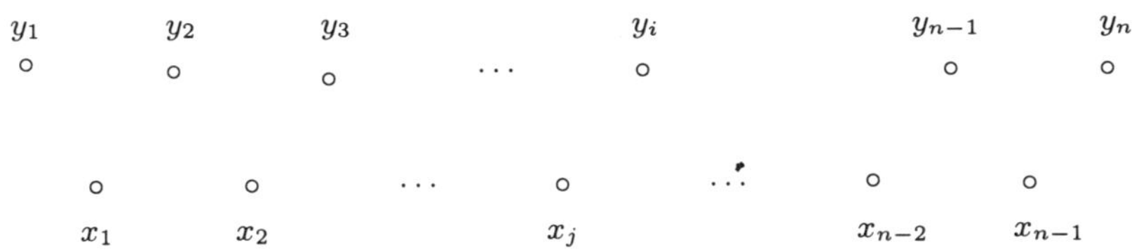

(9.143)

It is to be imagined that this diagram contains all the directed arc-lines (arrows) in accordance with the rules of Item 3. Without knowledge of the relative magnitudes of the differences of the shift-weights, a generic diagram cannot be drawn. But the entire process is clear from the example (9.152) below. We denote this labeled arc multigraph by $\vec{G}_{\Delta',\Delta}(x; y)$.

5. The unit projective operator shift-weight function. The arrow diagram of a unit projective operator is now used to assign a function, called the *shift-weight function* and denoted by $A_{\Delta',\Delta}(x; y)$, to each unit projective operator by the following rules, in which the indices i, i' and j, j' are pairs of generic points in rows n and $n-1$ between which there are directed arrows satisfying the rules of Item 3 and an associated product of linear factors:

(i). There are $\Delta_i - \Delta_{i'}$ arrows going from point i to point i' within row n. Linear factors:

$$\begin{aligned} &(y_i - y_{i'})_{\Delta_i - \Delta_{i'}} \\ &= (y_i - y_{i'})(y_i - y_{i'} + 1) \cdots (y_i - y_{i'} + \Delta_i - \Delta_{i'} - 1). \end{aligned} \tag{9.144}$$

(ii). There are $\Delta'_j - \Delta'_{j'}$ arrows going from point j to point j' within row $n-1$. Linear factors:

$$\begin{aligned}&(x_j - x_{j'} + 1)_{\Delta'_j - \Delta'_{j'}} \\ &= (x_j - x_{j'} + 1)(x_j - x_{j'} + 2) \cdots (x_j - x_{j'} + \Delta'_j - \Delta'_{j'}).\end{aligned} \tag{9.145}$$

(iii). There are $\Delta_i - \Delta'_j$ arrows going from point i in row n to point j within row $n-1$. Linear factors:

$$\begin{aligned}&(y_i - x_j)_{\Delta_i - \Delta'_j} \\ &= (y_i - x_j)(y_i - x_j + 1) \cdots (y_i - x_j + \Delta_i - \Delta'_j - 1).\end{aligned} \tag{9.146}$$

(iv). There are $\Delta'_j - \Delta_i$ arrows going from point j in row $n-1$ to point i within row n. Linear factors:

$$\begin{aligned}&(x_j - y_i + 1)_{\Delta'_j - \Delta_i} \\ &= (x_j - y_i + 1)(x_j - x_{j'} + 2) \cdots (x_j - x_{j'} + \Delta'_j - \Delta_i).\end{aligned} \tag{9.147}$$

We denote the product of all the factors in Items (i) and (ii) for arrows going within rows by $D_{\Delta',\Delta}(x;y)$, which is called the *denominator function;* the product of all the factors in Items (iii) and (iv) for arrows going between rows by $N_{\Delta',\Delta}(x;y)$, which is called the *numerator function.*

The *shift-weight function* of the unit projective operator in (9.115) is now defined by

$$A_{\Delta',\Delta}(x;y) = (\text{sign})\sqrt{\frac{N_{\Delta',\Delta}(x;y)}{D_{\Delta',\Delta}(x;y)}}\,. \tag{9.148}$$

This shift-weight function has several important features that we next discuss:

(i). The betweenness conditions $\lambda' \prec \lambda$ imply that $\lambda'_j \geq \lambda_i$, all $i \geq j+1$, and also $\lambda'_j \leq \lambda_i$, all $i \leq j$. In terms of the partial hooks (9.142), the betweenness conditions $\lambda' \prec \lambda$ imply that

$$\begin{aligned} y_i - x_j &\leq j + 1 - i, \text{ for all } i \geq j+1; \\ x_j - y_i + 1 &\leq i - j, \text{ for all } i \leq j. \end{aligned} \tag{9.149}$$

Similarly, the betweenness conditions $\lambda' + \Delta' \prec \lambda + \Delta$ imply that

$$\begin{aligned} y_i - x_j &\leq -(\Delta_i - \Delta'_j) + j + 1 - i, \text{ for all}\, i \geq j + 1; \\ x_j - y_i + 1 &\leq (\Delta_i - \Delta'_j) + i - j, \text{ for all}\, i \leq j. \end{aligned} \tag{9.150}$$

These conditions of betweenness imply that *the numerator function has no zeros for all partitions* $\lambda' \prec \lambda$ *and all partitions* $\lambda' + \Delta' \prec \lambda + \Delta$.

(ii). We next define an *inverted factor* in the numerator function based on the signs of the factors in (9.149)-(9.150). A single linear factor in the numerator function $N_{\Delta',\Delta}(x;y)$ of the form $y_i - x_j + s_{ij}$ is inverted if $i \geq j+1$, and a factor of the form $x_j - y_i + s_{ji}$ is inverted if $i \leq j$. This definition of an inverted factor is used to define the **sign** in the definition (9.148) of the shift-weight function:

$$(\text{sign}) = \text{number of inverted factors in the numerator function.} \tag{9.151}$$

(iii). A similar analysis can be applied to the denominator function $D_{\Delta',\Delta}(x;y)$ to conclude: (a) The denominator function contains the same number of factors as the numerator function; (b) the conditions of betweenness imply that the denominator function has no zeros for all partitions $\lambda' \prec \lambda$ and all partitions $\lambda'+\Delta' \prec \lambda+\Delta$; (c) the denominator function has inverted factors of the form $y_i - y_{i'} + s_{ii'}$ for $i \geq i' + 1$, and of the form $x_j - x_{j'} + s_{jj'}$ for $j \geq j' + 1$; (d) the number of inverted factors in the denominator function equals the number of inverted factors in the numerator function.

(iv). Only inverted factors can assume negative values, but the ratio of numerator and denominator functions is always positive. When considering products of shift-weight functions the square of a negative number under the radical can be encountered, and careful accounting of such factors is necessary to avoid the error of bringing a negative number outside the radical, since we always have $\alpha = \sqrt{(-\alpha)^2}$ for $\alpha \geq 0$. The complex number $\sqrt{-1}$ is never to be introduced.

(v). The shift-weight function is *common to all unit projective operators* with operator patterns $\Gamma^s_\Delta \in \mathbb{G}_{\Lambda,\Delta}$ and $\Gamma^{s'}_{\Delta'} \in \mathbb{G}_{\Lambda',\Delta'}$. It is not, in general, the full expression of the matrix elements of a unit projective operator; this occurs only for the special ones known as *extremal.* Extremal projective operators are those for which the shift-weights are permutations of the parts of the partition Λ, for which the multiplicity of each shift-weight is 1. This is the case, for example, for $\Lambda = (1^k\, 0^{n-k}), 1 \leq k \leq n$.

Example: The following arc multigraph for $\Lambda = (4,2,0), \Delta' = (1,3), \Delta = (3,2,1)$ illustrates the above rules for $n = 3$:

$$\vec{G}_{(1,3)(3,2,1)}(x_1, x_2; y_1, y_2, y_3)$$

$$= \tag{9.152}$$

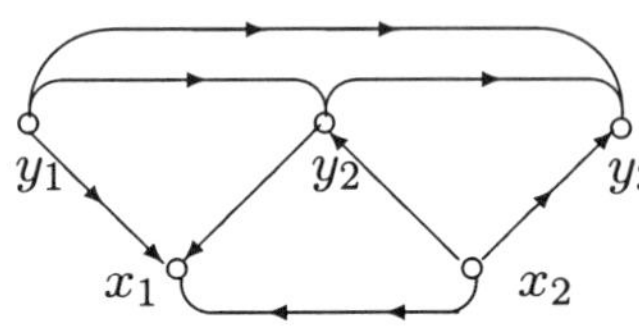

$$\begin{aligned} N_{(1,3)(3,2,1)}(x_1, x_2; y_1, y_2, y_3) &= (y_1 - x_1)(y_1 - x_1 + 1) \\ &\times (y_2 - x_1)(x_2 - y_2 + 1)(x_2 - y_3 + 1)(x_2 - y_3 + 2), \\ D_{(1,3)(3,2,1)}(x_1, x_2; y_1, y_2, y_3) &= (y_1 - y_2)(y_2 - y_3) \\ &\times (y_1 - y_3)(y_1 - y_3 + 1)(x_2 - x_1 + 1)(x_2 - x_1 + 2), \\ A_{(1,3)(3,2,1)}(x_1, x_2; y_1, y_2, y_3) & \\ &= \sqrt{\frac{N_{(1,3)(3,2,1)}(x_1, x_2; y_1, y_2, y_3)}{D_{(1,3)(3,2,1)}(x_1, x_2; y_1, y_2, y_3)}}. \end{aligned} \tag{9.153}$$

The sign is plus because the number of inverted factors in the numerator function (denominator function) is 2.

The fundamental shift operators $t_{i\tau}, \tau = 1, 2, \ldots, n$, introduced in Sects. 6.1 and 6.2, Chapter 6, are the simplest example of canonical tensor operators and the associated unit projective operators:

$$\begin{aligned} t_{i\tau} &= \left\langle \begin{matrix} & \tau & \\ 1 & 0^{n-1} & \\ & i & \end{matrix} \right\rangle = \sum_{\rho=1}^{n-1} \left[\begin{matrix} & \tau & \\ 1 & 0^{n-1} & \\ & \rho & \end{matrix} \right] \otimes \left\langle \begin{matrix} & \rho & \\ 1 & 0^{n-2} & \\ & i & \end{matrix} \right\rangle, \\ & \qquad\qquad i = 1, 2, \ldots, n-1, \\ t_{n\tau} &= \left\langle \begin{matrix} & \tau & \\ 1 & 0^{n-1} & \\ & (0) & \end{matrix} \right\rangle = \left[\begin{matrix} & \tau & \\ 1 & 0^{n-1} & \\ & (0) & \end{matrix} \right] \otimes \mathcal{I}'. \end{aligned} \tag{9.154}$$

The GT pattern notations τ, ρ, i denote the unique GT patterns having weight given, respectively, by the unit row vectors e_τ, e_ρ, e_i. The general pattern calculus rules applied to these fundamental canonical tensor operators gives the relations in Sects. 6.1-6.3, Chapter 6. We have, of course, completely defined the fundamental shift operators on their own in Chapter 6, where their role is primary.

We refer to the articles Ref. [19] and Ref. [117] for the proofs of properties of unit projective operators based on the pattern calculus.

9.4.3 Shift invariance of Kostka and Littlewood-Richardson numbers

Properties of the Kostka and Littlewood-Richardson numbers are very important for the splitting of canonical tensor operators on the space $H_\lambda \subset H$ into the disjoint subsets of non-null operators that provide the CG coefficients and the null operators with a characteristic null space, as described above. We anticipate several such general properties here so that they may be illustrated concretely in the special case of $U(3)$ canonical tensor operators presented in the next section. The step-function expressions for the Kostka numbers are proved and discussed in Refs. [18–20], and derived again in relation (11.144), Compendium B.

The Kostka numbers $K((\Lambda_1, \Lambda_2), (\Delta_1, \Delta_2))$ for $n = 2$ are given by the step function:

$$K(\Lambda, \Delta) = \begin{cases} 1, & \text{for } \Lambda_1 \leq \Delta_i \leq \Lambda_2. \\ 0, & \text{otherwise.} \end{cases} \tag{9.155}$$

The Kostka numbers $K((\Lambda_1, \Lambda_2, \Lambda_3), (\Delta_1, \Delta_2, \Delta_3))$ for $n = 3$ are given by the step function:

$$K(\Lambda, \Delta) = \Lambda_2 - \Lambda_3 + 1 - (\sigma_1 + \sigma_2 + \sigma_3), \tag{9.156}$$

where σ_i is the step function defined by

$$\sigma_i = \max(0, \Lambda_2 - \Delta_i), i = 1, 2, 3. \tag{9.157}$$

The result for the Kostka numbers for $n = 2$ is trivial, but the closed step-function formula for the Kostka numbers at level $n = 3$ is nontrivial; it is proved in Sect. 11.3.7, Compendium B. The following shift-invariance relations for the Kostka numbers and the Littlewood-Richardson numbers are also proved in Sects. 11.3.7-11.3.8, Compendium B:

$$\begin{aligned} &K(\Lambda + h, \Delta + h) = K(\Lambda, \Delta), \\ &I_{\Lambda,\Delta}(\lambda + h) = I_{\Lambda,\Delta}(\lambda), \\ &I_{\Lambda+h,\Delta+h}(\lambda) = I_{\Lambda,\Delta}(\lambda), \end{aligned} \tag{9.158}$$

where $h = (h, h, \ldots, h)(n$ parts) with h an arbitrary nonnegative integer. These relations are quite important for understanding level sets of partitions and for relating results in $U(n)$ and $SU(n)$.

A basic relation for canonical tensor operators is the following: Define the irreducible tensor operator $T(h)$ in $U(n)$ by

$$T(h) = \left\langle \begin{matrix} & (h) & \\ h & \cdots & h \\ & (h) & \end{matrix} \right\rangle, \tag{9.159}$$

in which all entries are equal to h. The action of this tensor operator on the basis $\mathbf{B}_\lambda$ of H_λ is to effect the shift h of all entries in the GT pattern:

$$T(h) \left| \begin{matrix} \lambda \\ m \end{matrix} \right\rangle = \left| \begin{matrix} \lambda + h \\ m(h) \end{matrix} \right\rangle, \tag{9.160}$$

where $m(h) \in \mathbb{G}_{\lambda+h}$ denotes the GT pattern $m \in \mathbb{G}_\lambda$ in which every entry is shifted upward by h. The tensor operator $T(h)$ commutes with every canonical tensor operator, indeed, with every irreducible tensor operator in $U(n)$:

$$\left\langle \begin{matrix} \Gamma \\ \Lambda \\ M \end{matrix} \right\rangle T(h) = T(h) \left\langle \begin{matrix} \Gamma \\ \Lambda \\ M \end{matrix} \right\rangle. \tag{9.161}$$

It follows from this property, applied to $U(n)$ and $U(n-1)$, that the canonical reduced matrix elements have the shift property expressed by

$$\left\langle \begin{matrix} \lambda + \Delta + h \\ \lambda' + \Delta' + h' \end{matrix} \right| \left[\begin{matrix} \Gamma_\Delta(h) \\ \Lambda + h \\ \Lambda' + h' \\ \Gamma_{\Delta'}(h) \end{matrix} \right] \left| \begin{matrix} \lambda \\ \lambda' \end{matrix} \right\rangle = \left\langle \begin{matrix} \lambda + \Delta \\ \lambda' + \Delta' \end{matrix} \right| \left[\begin{matrix} \Gamma_\Delta \\ \Lambda \\ \Lambda' \\ \Gamma_{\Delta'} \end{matrix} \right] \left| \begin{matrix} \lambda \\ \lambda' \end{matrix} \right\rangle, \tag{9.162}$$

where $h' = (h, \ldots, h)$ $(n-1$ parts). Thus, shifts of all entries in a unit projective operator by h, with a corresponding shift by h of all entries in the final vector, leaves the matrix elements of a unit projective operator invariant. By choosing $h = -\Lambda_n$, we see that the matrix elements of the general $U(n) : U(n-1)$ unit projective operator on the right can always be obtained from the $SU(n-1) : U(n-1)$ unit projective operator on the left having $\Lambda_n + h = 0$.

9.5 The Unitary Group $U(3)$

9.5.1 The $U(3)$ canonical tensor operators

The action of a $U(3)$ canonical tensor operator in the Hilbert space H is defined by its action on each basis vector of $H_\lambda \subset H$. We write this action, and some of the subsequent ones, out fully in terms of the tensor operator notations. While the GT pattern notation is space consuming, such relations must be exhibited in unequivocal form:

$$\left\langle \begin{matrix} \Gamma^s_\Delta \\ \Lambda_1\ \Lambda_2\ \Lambda_3 \\ \Lambda'_1\ \Lambda'_2 \\ M'_{11} \end{matrix} \right\rangle \left| \begin{matrix} \lambda_1\ \lambda_2\ \lambda_3 \\ \lambda'_1\ \lambda'_2 \\ m'_{11} \end{matrix} \right\rangle$$

$$= \sum_{\Delta'_1,\Delta'_2} \left\langle \begin{matrix} \lambda_1+\Delta_1\ \lambda_2+\Delta_2\ \lambda_3+\Delta_3 \\ \lambda'_1+\Delta'_1\ \lambda'_2+\Delta'_2 \\ m'_{11}+M'_{11} \end{matrix} \right| \left\langle \begin{matrix} \Gamma^s_\Delta \\ \Lambda_1\ \Lambda_2\ \Lambda_3 \\ \Lambda'_1\ \Lambda'_2 \\ M'_{11} \end{matrix} \right\rangle \left| \begin{matrix} \lambda_1\ \lambda_2\ \lambda_3 \\ \lambda'_1\ \lambda'_2 \\ m'_{11} \end{matrix} \right\rangle$$

$$\times \left| \begin{matrix} \lambda_1+\Delta_1\ \lambda_2+\Delta_2\ \lambda_3+\Delta_3 \\ \lambda'_1+\Delta'_1\ \lambda'_2+\Delta'_2 \\ m'_{11}+M'_{11} \end{matrix} \right\rangle, \tag{9.163}$$

where the summation is over all shift-weights $(\Delta'_1, \Delta'_2) \in \mathbb{W}_{(\Lambda'_1,\Lambda'_2)}$ such that the shifted pattern is lexical, and $(\lambda'_1 + \Delta'_1, \lambda'_2 + \Delta'_2) \in (\Lambda'_1, \Lambda'_2) \otimes (\lambda'_1, \lambda'_2)$. The matrix elements on the right-hand side of (9.163) are 0 for all $s = r = I_{\Lambda,\Delta}(\lambda) + 1, \ldots, K(\Lambda, \Delta)$, and, in general, not 0 for $s = t = 1, 2, \ldots, I_{\Lambda,\Delta}(\lambda)$. These conditions express the characteristic null space properties of the $U(3)$ canonical tensor operators.

The matrix elements of the $U(3)$ canonical tensor operator in relation (9.163) are given in terms of $U(3) : U(2)$ reduced matrix elements and the $U(2)$ WCG coefficients by the following relation:

$$\left\langle \begin{matrix} \lambda_1+\Delta_1\ \lambda_2+\Delta_2\ \lambda_3+\Delta_3 \\ \lambda'_1+\Delta'_1\ \lambda'_2+\Delta'_2 \\ m'_{11}+M'_{11} \end{matrix} \right| \left\langle \begin{matrix} \Gamma^t_\Delta \\ \Lambda_1\ \Lambda_2\ \Lambda_3 \\ \Lambda'_1\ \Lambda'_2 \\ M'_{11} \end{matrix} \right\rangle \left| \begin{matrix} \lambda_1\ \lambda_2\ \lambda_3 \\ \lambda'_1\ \lambda'_2 \\ m'_{11} \end{matrix} \right\rangle,$$

$$= \left\langle \begin{matrix} \lambda_1+\Delta_1\ \lambda_2+\Delta_2\ \lambda_3+\Delta_3 \\ \lambda'_1+\Delta'_1\ \lambda'_2+\Delta'_2 \end{matrix} \right| \left[\begin{matrix} \Gamma^t_\Delta \\ \Lambda_1\ \Lambda_2\ \Lambda_3 \\ \Lambda'_1\ \Lambda'_2 \\ \Delta'_1 \end{matrix} \right] \left| \begin{matrix} \lambda_1\ \lambda_2\ \lambda_3 \\ \lambda'_1\ \lambda'_2 \end{matrix} \right\rangle \tag{9.164}$$

$$\times \left\langle \begin{matrix} \lambda'_1+\Delta'_1\ \lambda'_2+\Delta'_2 \\ m'_{11}+M'_{11} \end{matrix} \right| \left\langle \begin{matrix} \Delta'_1 \\ \Lambda'_1\ \Lambda'_2 \\ M'_{11} \end{matrix} \right\rangle \left| \begin{matrix} \lambda'_1\ \lambda'_2 \\ m'_{11} \end{matrix} \right\rangle,$$

$$t = 1, 2, \ldots, I_{\Lambda,\Delta}(\lambda).$$

Also, the null operator and accompanying characteristic null space relations hold:

$$\left\langle \begin{matrix} \Gamma^r_\Delta \\ \Lambda_1\ \Lambda_2\ \Lambda_3 \\ \Lambda'_1\ \Lambda'_2 \\ M'_{11} \end{matrix} \right\rangle \left| \begin{matrix} \lambda_1\ \lambda_2\ \lambda_3 \\ \lambda'_1\ \lambda'_2 \\ m'_{11} \end{matrix} \right\rangle = \mathbf{0},\ r = I_{\Lambda,\Delta}(\lambda) + 1, \ldots, K(\Lambda, \Delta). \tag{9.165}$$

The reduced matrix element in (9.164) is independent of the choice of the $U(1)$ labels m'_{11}, M'_{11}, so that we can choose $m'_{11} = \lambda_1$ and $M'_{11} = \Delta'_1$ to obtain the relation:

$$\left\langle \begin{matrix} \lambda_1+\Delta_1 \; \lambda_2+\Delta_2 \; \lambda_3+\Delta_3 \\ \lambda'_1+\Delta'_1 \; \lambda'_2+\Delta'_2 \\ max \end{matrix} \right| \left\langle \begin{matrix} \Gamma^t_\Delta \\ \Lambda_1 \; \Lambda_2 \; \Lambda_3 \\ \Lambda'_1 \; \Lambda'_2 \\ \Delta'_1 \end{matrix} \right\rangle \left| \begin{matrix} \lambda_1 \; \lambda_2 \; \lambda_3 \\ \lambda'_1 \; \lambda'_2 \\ max \end{matrix} \right\rangle$$

$$= \left\langle \begin{matrix} \lambda_1+\Delta_1 \; \lambda_2+\Delta_2 \; \lambda_3+\Delta_3 \\ \lambda'_1+\Delta'_1 \; \lambda'_2+\Delta'_2 \end{matrix} \right\| \begin{bmatrix} \Gamma^t_\Delta \\ \Lambda_1 \; \Lambda_2 \; \Lambda_3 \\ \Lambda'_1 \; \Lambda'_2 \\ \Delta'_1 \end{bmatrix} \left\| \begin{matrix} \lambda_1 \; \lambda_2 \; \lambda_3 \\ \lambda'_1 \; \lambda'_2 \end{matrix} \right\rangle \tag{9.166}$$

$$\times \left\langle \begin{matrix} \lambda'_1+\Delta'_1 \; \lambda'_2+\Delta'_2 \\ max \end{matrix} \right| \left\langle \begin{matrix} \Delta'_1 \\ \Lambda'_1 \; \Lambda'_2 \\ \Delta'_1 \end{matrix} \right\rangle \left| \begin{matrix} \lambda'_1 \; \lambda'_2 \\ max \end{matrix} \right\rangle .$$

The $U(2)$ WCG coefficients are expressed in terms of angular momentum notation by

$$\left\langle \begin{matrix} \lambda'_1+\Delta'_1 \; \lambda'_2+\Delta'_2 \\ m'_{11}+M'_{11} \end{matrix} \right| \left\langle \begin{matrix} \Delta'_1 \\ \Lambda'_1 \; \Lambda'_2 \\ M'_{11} \end{matrix} \right\rangle \left| \begin{matrix} \lambda'_1 \; \lambda'_2 \\ m'_{11} \end{matrix} \right\rangle = C^{j'\;\;\; J'\;\;\; j''}_{m',\, M',\, m'+M'}, \tag{9.167}$$

$$\begin{aligned} &j' = (\lambda'_1 - \lambda'_2)/2,\ J' = (\Lambda'_1 - \Lambda'_2)/2,\ j'' = j' + (\Delta'_1 - \Delta'_2)/2, \\ &m' = m'_{11} - (\lambda'_1 + \lambda'_2)/2,\ M' = M'_{11} - (\Lambda'_1 + \Lambda'_2)/2, \\ &\Delta'_2 = \Lambda'_1 + \Lambda'_2 - \Delta'_1. \end{aligned} \tag{9.168}$$

Since $m' = j'$, $M' = j'' - j'$, $m' + M' = j''$ in the $U(2)$ WCG coefficient in (9.166), the explicit value of this coefficient is

$$\left\langle \begin{matrix} \lambda'_1+\Delta'_1 \; \lambda'_2+\Delta'_2 \\ \lambda'_1+\Delta'_1 \end{matrix} \right| \left\langle \begin{matrix} \Delta'_1 \\ \Lambda'_1 \; \Lambda'_2 \\ \Delta'_1 \end{matrix} \right\rangle \left| \begin{matrix} \lambda'_1 \; \lambda'_2 \\ \lambda'_1 \end{matrix} \right\rangle = C^{j'\;\; J'\;\;\;\;\; j''}_{j',j''-j',j''}$$

$$= \sqrt{\frac{(2j''+1)(2j')!}{(j'+j''-J')!(j'+j''+J'+1)!}}$$

$$= \sqrt{\frac{(\lambda'_1+\Delta'_1-\lambda'_2-\Delta'_2+1)(\lambda'_1-\lambda'_2)!}{(\lambda'_1-\lambda'_2+\Delta'_1-\Lambda'_1)!(\lambda'_1-\lambda'_2+\Delta'_1-\Lambda'_2+1)!}} . \tag{9.169}$$

Thus, relation (9.166) becomes

$$\left\langle \begin{matrix} \lambda_1+\Delta_1\ \lambda_2+\Delta_2\ \lambda_3+\Delta_3 \\ \lambda'_1+\Delta'_1\ \lambda'_2+\Delta'_2 \\ \lambda'_1+\Delta'_1 \end{matrix} \left| \left\langle \begin{matrix} \Gamma^t_\Delta \\ \Lambda_1\ \Lambda_2\ \Lambda_3 \\ \Lambda'_1\ \Lambda'_2 \\ \Delta'_1 \end{matrix} \right\rangle \right| \begin{matrix} \lambda_1\ \lambda_2\ \lambda_3 \\ \lambda'_1\ \lambda'_2 \\ \lambda'_1 \end{matrix} \right\rangle$$

$$= \left\langle \begin{matrix} \lambda_1+\Delta_1\ \lambda_2+\Delta_2\ \lambda_3+\Delta_3 \\ \lambda'_1+\Delta'_1\ \lambda'_2+\Delta'_2 \end{matrix} \left| \left[\begin{matrix} \Gamma^t_\Delta \\ \Lambda_1\ \Lambda_2\ \Lambda_3 \\ \Lambda'_1\ \Lambda'_2 \\ \Delta'_1 \end{matrix} \right] \right| \begin{matrix} \lambda_1\ \lambda_2\ \lambda_3 \\ \lambda'_1\ \lambda'_2 \end{matrix} \right\rangle \tag{9.170}$$

$$\times \sqrt{\frac{(\lambda'_1+\Delta'_1-\lambda'_2-\Delta'_2+1)(\lambda'_1-\lambda'_2)!}{(\lambda'_1-\lambda'_2+\Delta'_1-\Lambda'_1)!(\lambda'_1-\lambda'_2+\Delta'_1-\Lambda'_2+1)!}}.$$

$$t = 1, 2, \ldots, I_{\Lambda,\Delta}(\lambda).$$

Knowledge of the $U(1)$−maximal $U(3)$ CG coefficients gives the general $U(3) : U(2)$ reduced matrix elements, and conversely. These coefficients can then be used back in (9.164) to obtain the matrix elements of the general $U(3)$ canonical tensor operator, which then solves the problem of finding a set of CG coefficients that brings the Kronecker product to Kronecker direct sum form.

We turn next to the description of the characteristic null space; that is, the sets of partitions $\lambda \in \mathbb{P}\mathrm{ar}_3$ for which relations (9.165) hold.

9.6 Characteristic Null Spaces of $U(3)$ Canonical Tensor Operators

It is possible for $n = 3$ to enumerate the individual operator patterns having a prescribed shift-weight. The enumeration of all $U(3)$ operator patterns corresponding to the partition $(\Lambda_1, \Lambda_2, \Lambda_3)$ and shift-weight $(\Delta_1, \Delta_2, \Delta_3)$ can be obtained by direct enumeration of all eight cases from the Kostka number formula (9.156):

$$\begin{pmatrix} \Lambda \\ \Gamma^s_\Delta \end{pmatrix} \tag{9.171}$$

$$= \begin{pmatrix} \Lambda_1 & & \Lambda_2 & & \Lambda_3 \\ & \Delta_1+\Delta_2-\Lambda_3-\sigma_3-s+1 & & \Lambda_3+\sigma_3+s-1 & \\ & & \Delta_1 & & \end{pmatrix},$$

in which the shift weight $(\Delta_1, \Delta_2, \Delta_3) \in \mathbb{W}_{\Lambda_1,\Lambda_2,\Lambda_3}$, and the set of lexical patterns having this shift-weight, is enumerated by $s = 1, 2, \ldots, K(\Lambda, \Delta)$.

The characteristic null space for $n = 3$ is described in terms of the geometry of level sets in the Möbius plane, where the variables are adapted to the description of the shift-invariance properties (9.158) of the Kostka and the Littlewood-Richardson numbers. The Möbius plane is the Cartesian plane $\mathbb{R}^2$ in which the points are described in terms of the real numbers belonging to three axes with positive directions oriented at 120°. Each point in the plane is assigned three real coordinates (x_1, x_2, x_3) that are obtained by perpendicular projection onto the three axes, and, accordingly, the three numbers add to zero: $x_1 + x_2 + x_3 = 0$. The positive axes are also in the directions of the vertices of an equilateral triangle positioned at the origin $(0,0,0)$, the three axis being labeled counterclockwise. This arrangement of axes and perpendicular projections is shown in Fig. 9.1 below. The coordinates (x_1, x_2, x_3) are, of course, arbitrary real numbers adding to 0. For the description of the level sets of partitions, we use the *lattice points* of this space; that is, the set of points with integral coordinates. We require only the positive sector of lattice points $\mathbb{L}^+ \subset \mathbb{L}$, which is the set of lattice points $(x_1, x_2, x_3) \in \mathbb{L}$ that satisfy $x_1 \geq 1, x_2 \leq -2, x_3 \geq 1$. The set of lattice points $\mathbb{L}^+$ is now associated with partitions by using the **differences** of the partial hooks $p_{13} = \lambda_1 + 2, p_{23} = \lambda_2 + 1, p_{33} = \lambda_3$, as defined by $(x_1, x_2, x_3) = (p_{23} - p_{33}, p_{33} - p_{13}, p_{13} - p_{23})$; these differences take into account the shift-invariance properties (9.158). The use of partial hooks in place of the partitions themselves reflects the natural occurrence of partial hooks in all of unitary group theory. The partition associated with a point (x_1, x_2, x_3) is

$$\lambda = (\lambda_1, \lambda_2, \lambda_3) = (\lambda_3 + x_1 + x_3 - 2, \lambda_3 + x_1 - 1, \lambda_3). \tag{9.172}$$

In particular, the point $(x_1, x_2, x_3) = (1, -2, 1)$ corresponds to the partition $(\lambda_3, \lambda_3, \lambda_3)$. Each point in $\mathbb{L}^+$ gives the infinite number of partitions satisfying $\lambda_1 \geq \lambda_2 \geq \lambda_3$, all $\lambda_3 \geq 0$. Thus, the Möbius plane and its set of lattice points $\mathbb{L}^+$ is well-suited for the description of level sets, since all the points in $\mathbb{L}^+$ preserve the conditions on the partitions in the Hilbert space H_λ, and, moreover, cover all such partitions.

The level set $\mathbb{P}_{\Lambda,\Delta}(L)$ for given $L \in \{1, \ldots, K(\Lambda, \Delta)\}$ is depicted in Fig. 9.1 below. The marked points on the x_1−axis and x_3−axis are defined by

$$\begin{aligned} x_1^{(1)}(\Lambda,\Delta) &= \Delta_3 - \Lambda_3 - \sigma_1 + 1 + L - K, \\ x_1^{(2)}(\Lambda,\Delta) &= \Delta_3 - \Lambda_3 - \sigma_1 + 1; \\ \\ x_3^{(1)}(\Lambda,\Delta) &= \Delta_2 - \Lambda_3 - \sigma_3 + 1 + L - K, \\ x_3^{(2)}(\Lambda,\Delta) &= \Delta_2 - \Lambda_3 - \sigma_3 + 1, \end{aligned} \tag{9.173}$$

where we have abbreviated the Kostka number to $K = K(\Lambda, \Delta)$. The level sets of partitions $\mathbb{P}_{\Lambda,\Delta}(L) = \{\lambda \in \mathbb{P}\text{ar}_n \mid I_{\Lambda,\Delta}(\lambda) = L\}$ are subsets of lattice points in $\mathbb{L}^+$ in Fig. 9.1. It is always the case that

$$x_1^{(1)}(\Lambda, \Delta) \geq 0 \text{ and } x_3^{(1)}(\Lambda, \Delta) \geq 0, \tag{9.174}$$

as can be shown directly from the Kostka numbers:

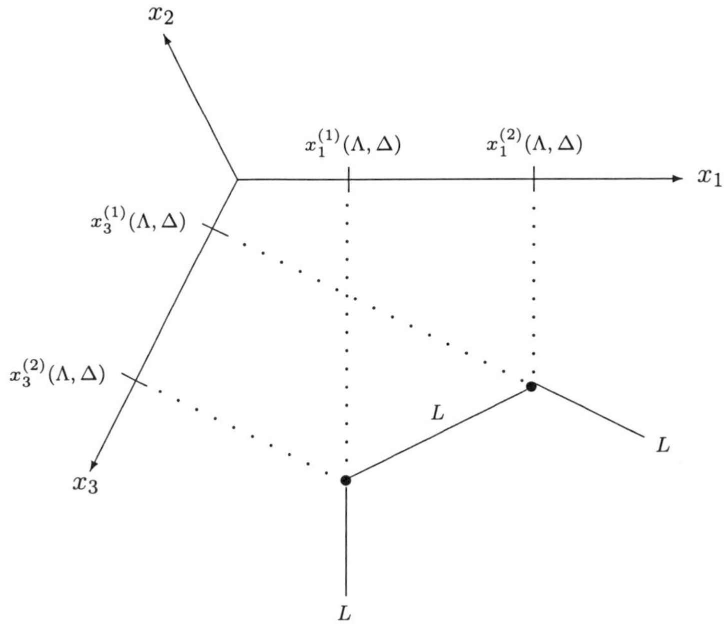

Figure 9.1. The level set $\mathbb{P}_{\Lambda,\Delta}(L)$ of partitions.

The level set $\mathbb{P}_{\Lambda,\Delta}(L)$ for given $L = \{1, \ldots, K\}$ shown in Fig. 9.1 is described in detail as follows (The case $L = 0$ is described separately in Fig. 9.2):

1. For $L = 1, 2, \ldots, K - 1$, the level set consists of the **single bent line of lattice points** marked by L.
2. For $L = K$, the two • points merge to a single point with coordinates $x_1^{(1)}(\Lambda, \Delta) = x_1^{(2)}(\Lambda, \Delta) = \Delta_3 - \Lambda_3 - \sigma_1 + 1$ and $x_3^{(1)}(\Lambda, \Delta) = x_3^{(2)}(\Lambda, \Delta) = \Delta_2 - \Lambda_3 - \sigma_3 + 1$. The level set $\mathbb{P}_{\Lambda,\Delta}(K)$ consists of all lattice points in the **two-dimensional set of lattice points** for which

$$x_1 \geq \Delta_3 - \Lambda_3 - \sigma_1 + 1 \text{ and } x_3 \geq \Delta_2 - \Lambda_3 - \sigma_3 + 1. \tag{9.175}$$

The level set of partitions $\mathbb{P}_{\Lambda,\Delta}(0)$ for which $I_{\Lambda,\Delta}(\lambda) = 0$ is the union of the regions I, II, III depicted in Fig. 9.2, as follows:

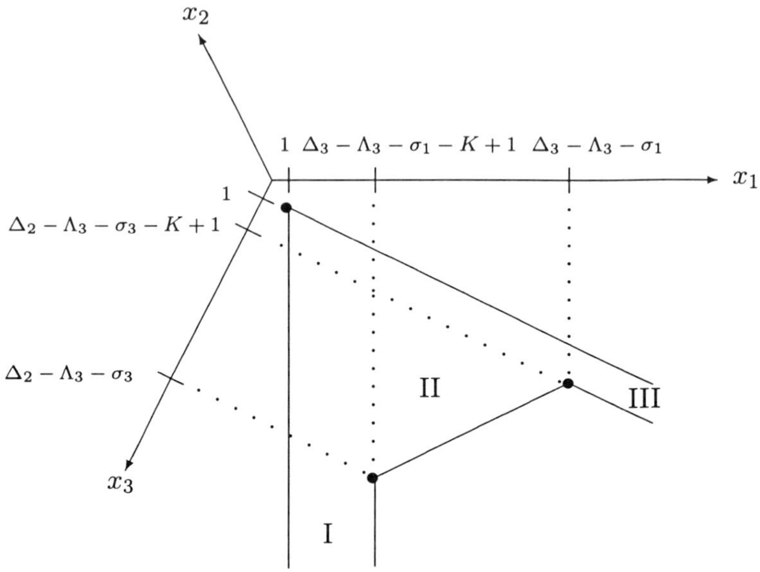

Figure 9.2. Level set of partitions for which $I_{\Lambda,\Delta}(\lambda) = 0$.

Regions I, II, and III are the subsets of lattice points with the following descriptions:

I: Strip $\mathbb{S}_K^{(I)}(\Lambda, \Delta)$:

$$1 \leq x_1 \leq \Delta_3 - \Lambda_3 - \sigma_1 - K + 1. \tag{9.176}$$

II: Strip $\mathbb{S}_K^{(III)}(\Lambda, \Delta)$:

$$1 \leq x_3 \leq \Delta_2 - \Lambda_3 - \sigma_3 - K + 1. \tag{9.177}$$

III: Equilateral triangle $T_K(\Lambda, \Delta)$:

$$\begin{aligned} &\Delta_3 - \Lambda_3 - \sigma_1 - K + 2 \leq x_1 \leq \Delta_3 - \Lambda_3 - \sigma_1, \\ &\Delta_2 - \Lambda_3 - \sigma_3 - K + 2 \leq x_3 \leq \Delta_2 - \Lambda_3 - \sigma_3, \\ &\Delta_2 - \sigma_3 + \Delta_3 - \sigma_1 - 2\Lambda_3 - 2K + 4 \leq -x_2 \\ &\qquad \leq \Delta_2 - \sigma_3 + \Delta_3 - \sigma_1 - 2\Lambda_3 - K + 2. \end{aligned} \tag{9.178}$$

The lattice points in Region I constitute a strip perpendicular to the x_1-axis of width $\Delta_2 - \Lambda_3 - \sigma_3 - K + 1$; those defined by Region III a

strip perpendicular to the x_3−axis of width $\Delta_2 - \Lambda_3 - \sigma_3 - K + 1$; and those defined by Region II an equilateral triangle $\mathbb{T}_K(\Lambda, \Delta)$ containing $K - 1$ lattice points on each edge. The triangle of lattice points is disjoint from the strips; it contains $|\,\mathbb{T}_K(\Lambda, \Delta)\,| = K(K-1)/2$ lattice points and is empty for $K = 1$. The following conventions are to hold in Fig. 9.2:

1. Strip $\mathbb{S}_K^{(I)}(\Lambda, \Delta)$ (Region I) is empty for $K = \Delta_3 - \Lambda_3 - \sigma_1 + 1$. In this case, the lower left $\bullet$ point of Region II falls on the line $x_1 = 1$ (see Fig. 9.4 below for an example); otherwise, Strip $\mathbb{S}_K^{(I)}(\Lambda, \Delta)$ contains at least one infinite line of lattice points.

2. Strip $\mathbb{S}_K^{(III)}(\Lambda, \Delta)$ (Region III) is empty for $K = \Delta_2 - \Lambda_3 - \sigma_3 + 1$. In this case, the upper right $\bullet$ point of Region II falls on the line $x_3 = 1$ (see Fig. 9.4 below for an example); otherwise, Strip $\mathbb{S}_K^{(III)}(\Lambda, \Delta)$ contains at least one infinite line of lattice points.

3. Equilateral triangle region $T_K(\Lambda, \Delta)$ (Region II) is empty for $K = 1$; otherwise, it contains $K(K-1)/2$ lattice points.

The partitions $(\lambda_1, \lambda_2, \lambda_3) \in \mathbb{P}_{\Lambda,\Delta}(0)$ are those for which there are no representations $D^{\lambda+\Delta}(U)$ in the Kronecker product $D^{\Lambda}(U) \otimes D^{\lambda}(U)$; these can be finite or infinite in number. The lattice points for which the Littlewood-Richardson number $I_{\Lambda,\Delta}(\lambda) = 0$ are a unique signature of the corresponding canonical tensor operator.

The diagrams above on level sets are obtained by direct calculation from formula (9.156) for the Kostka numbers, together with relation (11.178), Compendium B, which expresses the Littlewood-Richardson numbers in terms of Kostka numbers. The computations are tedious, but doable (see Ref. [18, 20]).

We recall from relations (9.96) and (9.101) that the index t with range $t = 1, 2, \ldots, I_{\Lambda,\Delta}(\lambda)$ enumerates the operator patterns $\Gamma^t_\Delta \in \mathbb{G}_{\Lambda,\Delta}$ that give canonical tensor operators whose matrix elements are CG coefficients, and that the index r with range $r = I_{\Lambda,\Delta}(\lambda) + 1, \ldots, K(\Lambda, \Delta)$ enumerates the operator patterns $\Gamma^r_\Delta \in \mathbb{G}_{\Lambda,\Delta}$ that give canonical tensor operators having the characteristic null space H_λ. From Fig. 9.1, we find that the characteristic null space of the canonical tensor operator $\left\langle \begin{matrix} \Gamma^r_\Delta \\ \Lambda \\ \bullet \end{matrix} \right\rangle$ consists of the partitions λ in the set

$$\mathbb{P}^{(r)}_{\Lambda,\Delta} = \bigcup_{L=0}^{K(\Lambda,\Delta)-r} \mathbb{P}_{\Lambda,\Delta}(L),\; r = I_{\Lambda,\Delta}(\lambda) + 1, \ldots, K(\Lambda, \Delta). \quad (9.179)$$

The characteristic null space is correspondingly given by

$$\mathbf{N}\begin{pmatrix} \Lambda \\ \Gamma^r_\Delta \end{pmatrix} = \left\{ H_\lambda \,\middle|\, \lambda \in \mathbb{P}^{(r)}_{\Lambda,\Delta} \right\}. \tag{9.180}$$

The sets of null space partitions $\mathbb{P}^{(r)}_{\Lambda,\Delta}$ exhibit the strict inclusion property

$$\mathbb{P}^{(K)}_{\Lambda,\Delta} \subset \mathbb{P}^{(K-1)}_{\Lambda,\Delta} \subset \cdots \subset \mathbb{P}^{(1)}_{\Lambda,\Delta}, \tag{9.181}$$

as shown diagrammatically in Fig. 9.3:

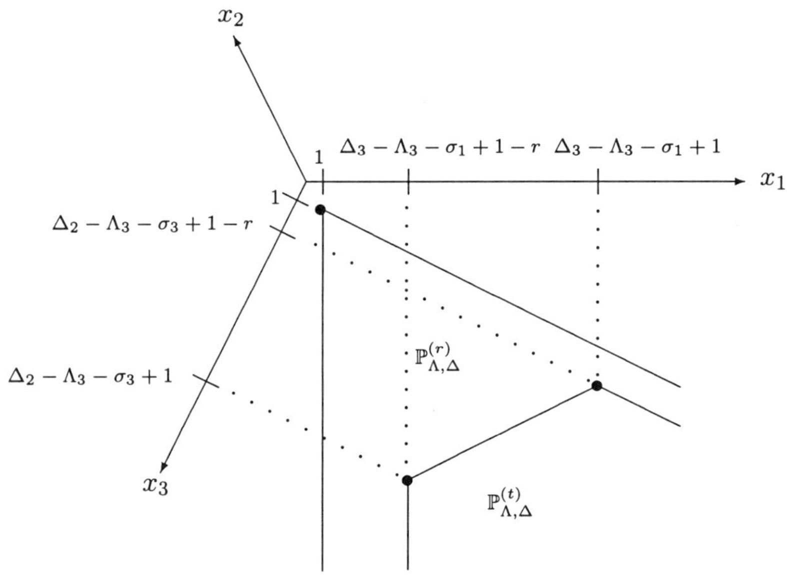

Figure 9.3. Characteristic null space partitions $\mathbb{P}^{(r)}_{\Lambda,\Delta}$.

The set of partitions $\mathbb{P}^{(r)}_{\Lambda,\Delta}$, each $r = I_{\Lambda,\Delta}(\lambda) + 1, \ldots, K(\Lambda, \Delta)$ consists of all lattice points interior to and on the boundary of the region enclosed by the two solid lines containing the common • point at $(1, -2, 1)$ and by the bent solid line containing the other two • points—these lattice points give the vector spaces in the characteristic null space of the canonical tensor operator $\left\langle \begin{smallmatrix} \Gamma^r_\Delta \\ \Lambda \\ \bullet \end{smallmatrix} \right\rangle$. The set of partitions $\mathbb{P}^{(t)}_{\Lambda,\Delta}$, each $t = 1, 2 \ldots, I_{\Lambda,\Delta}(\lambda)$, consists of all lattice points disjoint from the set $\mathbb{P}^{(r)}_{\Lambda,\Delta}$ —these lattice points give the vector spaces of the initial states of the canonical tensor operator $\left\langle \begin{smallmatrix} \Gamma^t_\Delta \\ \Lambda \\ \bullet \end{smallmatrix} \right\rangle$, whose matrix elements give the

CG coefficients that occur in the orthogonal matrix that brings $\Lambda \otimes \lambda$ to standard Kronecker direct sum form. *In all cases, the partitions themselves in these sets must be identified from the lattice coordinates by using the transformation (9.172).* The strict inclusion relations (9.181) for lattice points also imply strict inclusion relations for the characteristic null spaces of $U(3)$ canonical tensor operators:

$$\mathbf{N}\begin{pmatrix} \Lambda \\ \Gamma^1_\Delta \end{pmatrix} \supset \mathbf{N}\begin{pmatrix} \Lambda \\ \Gamma^2_\Delta \end{pmatrix} \supset \cdots \supset \mathbf{N}\begin{pmatrix} \Lambda \\ \Gamma^K_\Delta \end{pmatrix}. \tag{9.182}$$

These strict inclusion for characteristic null spaces hold for each $\Gamma_\Delta \in \mathbb{G}_{\Lambda,\Delta}$. We conclude: *Each $U(3)$ canonical tensor operator $\left\langle \begin{matrix} \Gamma \\ \Lambda \\ \bullet \end{matrix} \right\rangle, \Gamma \in \mathbb{G}_\Lambda$ has a unique characteristic null space $\mathbf{N}\binom{\Lambda}{\Gamma}$.*

Example: $\Lambda = (6,3,0), \Delta = (3,3,3)$, with $K = 4$:

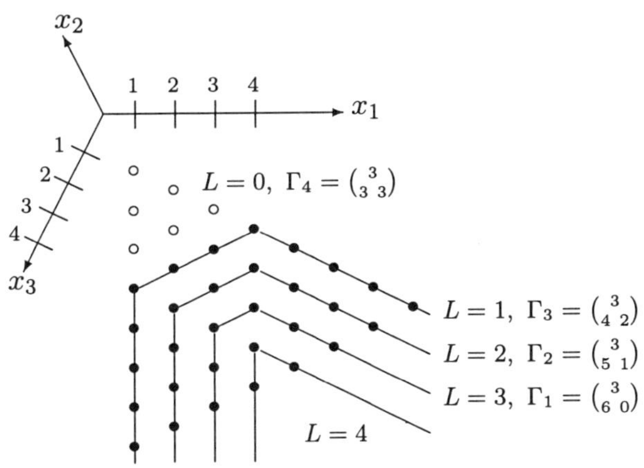

Figure 9.4. Level sets of partitions for $\Lambda = (6,3,0), \Delta = (3,3,3)$.

The values of λ for which $I_{(6,3,0),(3,3,3)}(\lambda) = 1,2,3$ are the lattice points belonging to the bent solid lines shown in Fig. 9.4; the values for $L = 0$ are the six lattice points $\circ$; and the values for $L = 4$ are all lattice points belonging to the two-dimensional pie-shaped region with the solid boundary lines containing the point $(4,-8,4)$, including the boundary. □

The null space diagrams given in Ref. [18] are valid only for $\Lambda_3 = 0$, a condition not always made explicit. This is actually not a restriction in consequence of the shift identities (9.158). Here we do not adjust $\Lambda_3 = 0$.

The methods presented thus far focus on the determination of the

$U(3)$ CG coefficients. But it is natural also to develop the rules for the determination of the $U(3):U(2)$ reduced matrix elements first, and then use (9.170) to obtain the $U(3)$ CG coefficients. For this purpose, we next develop additional properties of the reduced matrix elements.

9.7 The $U(3):U(2)$ Unit Projective Operators

The operator patterns (9.171) can also be used to enumerate the explicit lower operator patterns for the unit projective operators

$$\left\langle \begin{matrix} \lambda_1+\Delta_1\ \lambda_2+\Delta_2\ \lambda_3+\Delta_3 \\ \lambda'_1+\Delta'_1\ \lambda'_2+\Delta'_2 \end{matrix} \right| \left[\begin{matrix} \Gamma^s_\Delta \\ \Lambda_1\ \Lambda_2\ \Lambda_3 \\ \Gamma^{s'}_{\Delta'} \end{matrix} \right] \left| \begin{matrix} \lambda_1\ \lambda_2\ \lambda_3 \\ \lambda'_1\ \lambda'_2 \end{matrix} \right\rangle,$$

$$s=1,2,\ldots,K_{\Lambda,\Delta};\ s'=1,2,\ldots,K_{\Lambda,\Delta'}, \tag{9.183}$$

where we now use the notation Δ' to denote the three-part shift weight $\Delta'=(\Delta'_1,\Delta'_2,\Delta'_3)\in\mathbb{G}_{(\Lambda_1,\Lambda_2,\Lambda_3),(\Delta'_1,\Delta'_2,\Delta'_3)}$. The notational adjustment of the patterns (9.171) is

$$\begin{pmatrix} \Lambda \\ \Gamma^{s'}_{\Delta'} \end{pmatrix} = \begin{pmatrix} \Lambda_1 & & \Lambda_2 & & \Lambda_3 \\ & \Delta'_1+\Delta'_2-\Lambda_3-\sigma'_3-s'+1 & & \Lambda_3+\sigma'_3+s'-1 & \\ & & \Delta'_1 & & \end{pmatrix}, \tag{9.184}$$

where $s'=1,2,\ldots,K(\Lambda,\Delta')$, and the Kostka number is given by

$$\begin{aligned} K(\Lambda,\Delta') &= \Lambda_2-\Lambda_3+1-(\sigma'_1+\sigma'_2+\sigma'_3), \\ \sigma'_i &= \max(0,\Lambda_2-\Delta'_i),\ i=1,2,3. \end{aligned} \tag{9.185}$$

The action of a canonical unit projection operator on the basis $\mathcal{B}_\lambda$ of the Hilbert space $\mathcal{H}_\lambda\subset\mathcal{H}$ is given by

$$\begin{aligned} &\left[\begin{matrix} \Gamma^s_\Delta \\ \Lambda_1\ \Lambda_2\ \Lambda_3 \\ \Gamma^{s'}_{\Delta'} \end{matrix} \right] \left| \begin{matrix} \lambda_1\ \lambda_2\ \lambda_3 \\ \lambda'_1\ \lambda'_2 \end{matrix} \right\rangle \\ &= \left\langle \begin{matrix} \lambda_1+\Delta_1\ \lambda_2+\Delta_2\ \lambda_3+\Delta_3 \\ \lambda'_1+\Delta'_1\ \lambda'_2+\Delta'_2 \end{matrix} \right| \left[\begin{matrix} \Gamma^s_\Delta \\ \Lambda_1\ \Lambda_2\ \Lambda_3 \\ \Gamma^{s'}_{\Delta'} \end{matrix} \right] \left| \begin{matrix} \lambda_1\ \lambda_2\ \lambda_3 \\ \lambda'_1\ \lambda'_2 \end{matrix} \right\rangle \\ &\quad\times \left| \begin{matrix} \lambda_1+\Delta_1\ \lambda_2+\Delta_2\ \lambda_3+\Delta_3 \\ \lambda'_1+\Delta'_1\ \lambda'_2+\Delta'_2 \end{matrix} \right\rangle. \end{aligned} \tag{9.186}$$

Certain of the matrix elements of the unit projective operator in this expression are zero in consequence of null projective operators and their characteristic null space $\mathcal{H}_{(\lambda_1\,\lambda_2\,\lambda_3)}$, described as follows:

$$\left[\begin{array}{c} \Gamma^{s}_{\Delta} \\ \Lambda_1\ \Lambda_2\ \Lambda_3 \\ \Gamma^{s'}_{\Delta'} \end{array}\right] \left| \begin{array}{c} \lambda_1\ \lambda_2\ \lambda_3 \\ \lambda'_1\ \lambda'_2 \end{array} \right\rangle = \mathbf{0}, \tag{9.187}$$

for each value of s given by

$$s = I_{\Lambda,\Delta}(\lambda) + 1, \ldots, K(\Lambda, \Delta). \tag{9.188}$$

There is no characteristic null space associated with the values of s', since the multiplicity is 1; that is,

$$I_{(\Lambda'_1,\Lambda'_2),(\Delta'_1,\Delta'_2)}(\lambda'_1,\lambda'_2)) = K((\Lambda'_1,\Lambda'_2),(\Delta'_1,\Delta'_2)) = 1, \tag{9.189}$$

since the partition $(\Lambda'_1,\Lambda'_2) \in \mathbb{P}\mathrm{ar}_2$ and the shift-weight $(\Delta'_1,\Delta'_2) \in \mathbb{G}_{(\Lambda'_1,\Lambda'_2),(\Delta'_1,\Delta'_2)}$ uniquely determine s' in

$$(\Lambda'_1,\Lambda'_2) = (\Delta'_1 + \Delta'_2 - \Lambda_3 - \sigma'_3 - s' + 1,\ \Lambda_3 + \sigma'_3 + s' - 1). \tag{9.190}$$

Thus, the reduced matrix elements in (9.186) have value 0 for the values of s given by (9.188), and have values $\neq 0$ for $s = t = 1, 2, \ldots, I_{\Lambda,\Delta}(\lambda)$.

Remarks. The result given by Theorem 1.2 in Ref. [18] is **incorrect.** A unit projective operators should not be taken to be the zero operator on the entire Hilbert space $\mathcal{H}_\lambda$, all $\lambda \in \mathbb{P}\mathrm{ar}_3$, since this implies the corresponding canonical tensor operator (9.163) is the zero operator; this is inconsistent with the characteristic null space concept, which determines null, not zero, projective operators. Each non-null projective operator with a specified lower pattern in (9.186) effects the shift $(\lambda'_1,\lambda'_2) \to (\lambda'_1 + \Delta'_1, \lambda'_2 + \Delta'_2)$ of the initial partition $(\lambda'_1,\lambda'_2) \prec (\lambda_1,\lambda_2,\lambda_3)$ to the final partition $(\lambda'_1 + \Delta'_1, \lambda'_2 + \Delta'_2)$, and it is required that $(\lambda'_1+\Delta'_1, \lambda'_2+\Delta'_2) \prec (\lambda_1+\Delta_1, \lambda_2+\Delta_2, \lambda_3+\Delta_3)$; otherwise, the action is undefined (the coefficient in (9.186) is undefined). *Undefined coefficients are to be avoided in all relations.* Even if Theorem 1.2 is correctly formulated in the form above, the proof that the shifted final patterns are nonlexical for all $\lambda' \prec \lambda$ is incorrect as shown by the example (9.110). The computational aspects of $U(3):U(2)$ reduced matrix elements given in Ref. [18] are unaffected by the errors in Theorem 1.2. These computations lead to new classes of functions with exceptional properties (see Ref. [112]).

9.7.1 Coupling rules and Racah coefficients

There are many more important relations between canonical tensor operators and unit projective operators, the form of which depends on the so-called coupling rules whereby new irreducible tensors are constructed from the product of pairs of "simpler" irreducible tensor operators. Moreover, it is from this structure that the concept of Racah coefficients of the unitary group emerges. We deal briefly in this section with this algebra.

We consider that the following $U(n)$ CG coefficient is fully known and nonzero for a given pair of partitions $\Lambda, \Lambda' \in \mathbb{P}\mathrm{ar}_n$:

$$\left\langle \begin{matrix} \Lambda + \overline{\Delta} \\ M'' \end{matrix} \middle| \left\langle \begin{matrix} \overline{\Gamma} \\ \Lambda' \\ M' \end{matrix} \right\rangle \middle| \begin{matrix} \Lambda \\ M \end{matrix} \right\rangle, \tag{9.191}$$

where also $\overline{\Gamma} \in \mathbb{G}_{\Lambda', \overline{\Delta}}$ is a given operator pattern having the shift-weight $\overline{\Delta}$ for some $\overline{\Delta} \in \mathbb{W}_{\Lambda'}$. This CG coefficient supplies the numerical "coupling" coefficients needed to construct from two canonical tensor operators of type Λ and Λ' a new tensor operator of type $\Lambda + \overline{\Delta}$, as follows:

$$\begin{aligned} &\sum_{M', M} \left\langle \begin{matrix} \Lambda + \overline{\Delta} \\ M'' \end{matrix} \middle| \left\langle \begin{matrix} \overline{\Gamma} \\ \Lambda' \\ M' \end{matrix} \right\rangle \middle| \begin{matrix} \Lambda \\ M \end{matrix} \right\rangle \left\langle \begin{matrix} \Gamma' \\ \Lambda' \\ M' \end{matrix} \right\rangle \left\langle \begin{matrix} \Gamma \\ \Lambda \\ M \end{matrix} \right\rangle \\ &= \sum_{\Gamma'' \in \mathbb{G}_{\Lambda + \overline{\Delta}}} \left\{ \begin{pmatrix} \Lambda + \overline{\Delta} \\ \Gamma'' \end{pmatrix} \begin{pmatrix} \overline{\Gamma} \\ \Lambda' \\ \Gamma' \end{pmatrix} \begin{pmatrix} \Lambda \\ \Gamma \end{pmatrix} \right\} \left\langle \begin{matrix} \Gamma'' \\ \Lambda + \overline{\Delta} \\ M'' \end{matrix} \right\rangle . \end{aligned} \tag{9.192}$$

The expression on the left is called the "coupling of a pair of irreducible tensor operators." This coupling of a pair of irreducible tensor operators of type Λ and Λ' yields a new tensor operator of type $\Lambda + \overline{\Delta}$, since the left-hand side of expression (9.192) transforms under unitary similarity transformation $T_U(\cdots)T_{U^{-1}}$ in exactly the manner required of a tensor operator of type $\Lambda + \overline{\Delta}$. This result follows directly from the transformation properties of each of the irreducible tensor operators on the left and the properties of the coupling CG coefficients, and those that arise from the transformation of the product. This property is the analogue for the vector space of operators of the lower pattern coupling by a CG coefficient of the Kronecker product of two D-polynomials, $D^{\Lambda'}$ and D^{Λ}. Relation (9.192) is an operator identity on the model Hilbert space H with basis $\mathbf{B}_\lambda$ of $H_\lambda \subset H$; that is, the operator relation (9.192) is to be applied to a basis vector of $\mathbf{B}_\lambda$.

Relation (9.192) expresses the fact that the canonical tensor operators $\left\langle \begin{matrix} \Gamma'' \\ \Lambda + \overline{\Delta} \\ \bullet \end{matrix} \right\rangle$, $\Gamma'' \in \mathbb{G}_{\Lambda + \overline{\Delta}}$, are a basis for all tensors of type $\Lambda + \overline{\Delta}$. The "coefficients" in this linear combination of irreducible tensor operators on the right-hand side are $U(n)$ invariants. The notation

$$\left\{\begin{pmatrix}\Lambda+\overline{\Delta}\\ \Gamma''\end{pmatrix}\begin{pmatrix}\overline{\Gamma}\\ \Lambda'\\ \Gamma'\end{pmatrix}\begin{pmatrix}\Lambda\\ \Gamma\end{pmatrix}\right\} \tag{9.193}$$

denotes these $U(n)$ left-invariant operators, which we have placed to the left of the irreducible tensor operators in (9.192) (they could be placed to the right). These invariant operators are called *Racah invariants,* since, it turns out, their eigenvalues in the model Hilbert space H on which the operator relation (9.192) acts are a natural generalization of the Racah coefficients of angular momentum theory:

$$\left\{\begin{pmatrix}\Lambda+\overline{\Delta}\\ \Gamma''\end{pmatrix}\begin{pmatrix}\overline{\Gamma}\\ \Lambda'\\ \Gamma'\end{pmatrix}\begin{pmatrix}\Lambda\\ \Gamma\end{pmatrix}\right\}\left|\begin{matrix}\lambda+\Delta''\\ m''\end{matrix}\right\rangle$$
$$=\left\{\begin{pmatrix}\Lambda+\overline{\Delta}\\ \Gamma''\end{pmatrix}\begin{pmatrix}\overline{\Gamma}\\ \Lambda'\\ \Gamma'\end{pmatrix}\begin{pmatrix}\Lambda\\ \Gamma\end{pmatrix}\right\}(\lambda+\Delta'')\left|\begin{matrix}\lambda+\Delta''\\ m''\end{matrix}\right\rangle. \tag{9.194}$$

By convention, when relation (9.192) acts on a basis vector $\left|\begin{matrix}\lambda\\ m\end{matrix}\right\rangle\in\mathbf{B}_\lambda$, the left-invariant Racah operator has eigenvalue evaluated on the shifted vector space $H_{\lambda+\Delta''}, \Delta''=W\binom{\Lambda+\overline{\Delta}}{\Gamma''}$.

Great care must be exercised in coupling relations such as (9.192) in accounting for characteristic null space. Thus, to get a nonzero result from the left-hand side, we let the operator relation act on the basis vectors of a subspace $H_\lambda\subset H$ that is not in the null space of the product of the canonical tensor operators on the left. Putting these details aside for the moment, we use the orthogonality relation (9.53) for canonical tensor operators to obtain the following expression of the Racah left-invariant operator in terms of the canonical tensor operators (see also relation (9.199) below):

$$\left\{\begin{pmatrix}\Lambda+\overline{\Delta}\\ \Gamma''\end{pmatrix}\begin{pmatrix}\overline{\Gamma}\\ \Lambda'\\ \Gamma'\end{pmatrix}\begin{pmatrix}\Lambda\\ \Gamma\end{pmatrix}\right\}=\sum_{M'',M',M}\left\langle\begin{matrix}\Lambda+\overline{\Delta}\\ M''\end{matrix}\right|\left\langle\begin{matrix}\overline{\Gamma}\\ \Lambda'\\ M'\end{matrix}\right\rangle\left|\begin{matrix}\Lambda\\ M\end{matrix}\right\rangle$$
$$\times\left\langle\begin{matrix}\Gamma'\\ \Lambda'\\ M'\end{matrix}\right\rangle\left\langle\begin{matrix}\Gamma\\ \Lambda\\ M\end{matrix}\right\rangle\left\langle\begin{matrix}\Gamma''\\ \Lambda+\overline{\Delta}\\ M''\end{matrix}\right\rangle^{\dagger}. \tag{9.195}$$

Next, we take matrix elements of relations (9.192) and (9.195), account for characteristic null space, and obtain the following relations:

$$\sum_{m',M',M} \left\langle \begin{matrix} \Lambda+\overline{\Delta} \\ M'' \end{matrix} \middle| \left\langle \begin{matrix} \overline{\Gamma} \\ \Lambda' \\ M' \end{matrix} \right\rangle \middle| \begin{matrix} \Lambda \\ M \end{matrix} \right\rangle$$

$$\times \left\langle \begin{matrix} \lambda+\Delta+\Delta' \\ m'' \end{matrix} \middle| \left\langle \begin{matrix} \Gamma^{t'}_{\Delta'} \\ \Lambda' \\ M' \end{matrix} \right\rangle \middle| \begin{matrix} \lambda+\Delta \\ m' \end{matrix} \right\rangle \left\langle \begin{matrix} \lambda+\Delta \\ m' \end{matrix} \middle| \left\langle \begin{matrix} \Gamma^{t}_{\Delta} \\ \Lambda \\ M \end{matrix} \right\rangle \middle| \begin{matrix} \lambda \\ m \end{matrix} \right\rangle$$

$$= \delta_{\Delta+\Delta',\Delta''} \sum_{t''=1}^{I_{\Lambda+\overline{\Delta},\Delta''}(\lambda)} \left\{ \left(\begin{matrix} \Lambda+\overline{\Delta} \\ \Gamma^{t''}_{\Delta''} \end{matrix} \right) \left(\begin{matrix} \overline{\Gamma} \\ \Lambda' \\ \Gamma^{t'}_{\Delta'} \end{matrix} \right) \left(\begin{matrix} \Lambda \\ \Gamma^{t}_{\Delta} \end{matrix} \right) \right\} (\lambda+\Delta'')$$

$$\times \left\langle \begin{matrix} \lambda+\Delta'' \\ m'' \end{matrix} \middle| \left\langle \begin{matrix} \Gamma^{t''}_{\Delta''} \\ \Lambda+\overline{\Delta} \\ M'' \end{matrix} \right\rangle \middle| \begin{matrix} \lambda \\ m \end{matrix} \right\rangle; \tag{9.196}$$

$$\left\{ \left(\begin{matrix} \Lambda+\overline{\Delta} \\ \Gamma^{t''}_{\Delta''} \end{matrix} \right) \left(\begin{matrix} \overline{\Gamma} \\ \Lambda' \\ \Gamma^{t'}_{\Delta'} \end{matrix} \right) \left(\begin{matrix} \Lambda \\ \Gamma^{t}_{\Delta} \end{matrix} \right) \right\} (\lambda+\Delta'')$$

$$= \delta_{\Delta+\Delta',\Delta''} \sum_{M'',M',M,m',m} \left\langle \begin{matrix} \Lambda+\overline{\Delta} \\ M'' \end{matrix} \middle| \left\langle \begin{matrix} \overline{\Gamma} \\ \Lambda' \\ M' \end{matrix} \right\rangle \middle| \begin{matrix} \Lambda \\ M \end{matrix} \right\rangle$$

$$\times \left\langle \begin{matrix} \lambda+\Delta+\Delta' \\ m'' \end{matrix} \middle| \left\langle \begin{matrix} \Gamma^{t'}_{\Delta'} \\ \Lambda' \\ M' \end{matrix} \right\rangle \middle| \begin{matrix} \lambda+\Delta \\ m' \end{matrix} \right\rangle \left\langle \begin{matrix} \lambda+\Delta \\ m' \end{matrix} \middle| \left\langle \begin{matrix} \Gamma^{t}_{\Delta} \\ \Lambda \\ M \end{matrix} \right\rangle \middle| \begin{matrix} \lambda \\ m \end{matrix} \right\rangle$$

$$\times \left\langle \begin{matrix} \lambda+\Delta'' \\ m'' \end{matrix} \middle| \left\langle \begin{matrix} \Gamma^{t''}_{\Delta''} \\ \Lambda+\overline{\Delta} \\ M'' \end{matrix} \right\rangle \middle| \begin{matrix} \lambda \\ m \end{matrix} \right\rangle. \tag{9.197}$$

The GT pattern $m'' \in \mathbb{G}_{\Lambda+\Delta''}$ can be chosen arbitrarily in each of these relations. The ranges of the indices t, t', t'' for nonzero CG coefficients are the following:

$$\begin{aligned} &t = 1, 2, \ldots, I_{\Lambda,\Delta}(\lambda);\ t' = 1, 2, \ldots, I_{\Lambda',\Delta'}(\lambda+\Delta); \\ &t'' = 1, 2, \ldots, I_{\Lambda+\overline{\Delta},\Delta''}(\lambda). \end{aligned} \tag{9.198}$$

The operator pattern $\overline{\Gamma}$ is any pattern with shift-weight $\overline{\Delta} \in \mathbb{W}_{\Lambda'}$.

We recall that canonical tensor operators are identical to the orthonormalized operator-valued $\widehat{D}^{\Lambda}$–polynomials:

$$\left\langle \begin{matrix} \Gamma \\ \Lambda \\ M \end{matrix} \right\rangle = \widehat{D} \left(\begin{matrix} \Gamma \\ \Lambda \\ M \end{matrix} \right) (T). \tag{9.199}$$

This relation cannot be emphasized too strongly because it implies that the Racah invariant operators are uniquely defined—no freedom remains in specifying their properties, except possibly the choice of phase factors in the orthonormalization of the operator-valued $D-$polynomials themselves. The implementation of relation (9.197) to obtain an explicit formula in terms of the partial hooks p_{in} is, of course, a difficult problem.

Relation (9.197) is the $U(n)$ analogue of relation (2.160), which expresses an $SU(2)$ Racah coefficient as a sum over four $SU(2)$ WCG coefficients, but thus far in the theory of $U(n)$ Racah coefficients no simplification of the multiple summation in (9.197) is known.

For given partitions $\Lambda, \Lambda', \Lambda+\overline{\Delta} \in \mathbb{P}\mathrm{ar}_n$, the number of Racah coefficients given by relation (9.197) for each selected shift-weight pattern $\overline{\Gamma} \in G_{\Lambda',\overline{\Delta}}$ is the product of Littlewood-Richardson numbers given by

$$I_{\Lambda,\Delta}(\lambda) \times I_{\Lambda',\Delta'}(\lambda+\Delta) \times I_{\Lambda+\overline{\Delta},\Delta+\Delta'}(\lambda). \tag{9.200}$$

This counting relation holds, since there are no conditions on the coupling of Kronecker products of the subgroup row partitions as in the product of canonical tensor operators. For a nonzero Racah coefficient, it is only required that:

The final operator pattern $\Gamma^{t''}_{\Delta''}$ in the Racah coefficient (9.197) is only required to have the shift-weight $\Delta'' = \Delta + \Delta'$, which is the sum of the shift-weights associated with the patterns Γ^{t}_{Δ} and $\Gamma^{t'}_{\Delta'}$—there are no subgroup conditions.

$U(n)$ Racah coefficients are objects of great complexity.

Racah invariant operators satisfy orthogonality relations similar in form to those for the canonical CG coefficients:

$$\begin{gathered}
\sum_{t=1}^{I_{\Lambda,\Delta}(\lambda)} \sum_{t'=1}^{I_{\Lambda',\Delta'}(\lambda+\Delta)} \left\{ \begin{pmatrix} \Lambda+\overline{\Delta} \\ \Gamma^{t''}_{\Delta+\Delta'} \end{pmatrix} \begin{pmatrix} \overline{\Gamma} \\ \Lambda' \\ \Gamma^{t'}_{\Delta'} \end{pmatrix} \begin{pmatrix} \Lambda \\ \Gamma^{t}_{\Delta} \end{pmatrix} \right\} \\
\times \left\{ \begin{pmatrix} \Lambda+\overline{\Delta}' \\ \Gamma^{t'''}_{\Delta+\Delta'} \end{pmatrix} \begin{pmatrix} \overline{\Gamma}' \\ \Lambda' \\ \Gamma^{t'}_{\Delta'} \end{pmatrix} \begin{pmatrix} \Lambda \\ \Gamma^{t}_{\Delta} \end{pmatrix} \right\} = \delta_{t'',t'''}\,\delta_{\overline{\Gamma},\overline{\Gamma}'}; \\
\sum_{\overline{\Gamma}\in \mathrm{G}_{\Lambda'}} \sum_{t''=1}^{I_{\Lambda+\overline{\Delta},\Delta+\Delta'}(\lambda)} \left\{ \begin{pmatrix} \Lambda+\overline{\Delta} \\ \Gamma^{t''}_{\Delta+\Delta'} \end{pmatrix} \begin{pmatrix} \overline{\Gamma} \\ \Lambda' \\ \Gamma^{t'}_{\Delta'} \end{pmatrix} \begin{pmatrix} \Lambda \\ \Gamma^{t}_{\Delta} \end{pmatrix} \right\} \\
\times \left\{ \begin{pmatrix} \Lambda+\overline{\Delta} \\ \Gamma^{t''}_{\Delta+\Delta'} \end{pmatrix} \begin{pmatrix} \overline{\Gamma} \\ \Lambda' \\ \Gamma^{s'}_{\Delta'} \end{pmatrix} \begin{pmatrix} \Lambda \\ \Gamma^{s}_{\Delta} \end{pmatrix} \right\} = \delta_{t',s'}\,\delta_{t,s},
\end{gathered} \tag{9.201}$$

where these two left-invariant Racah operator relations are to be applied to the vector space $H_{\lambda+\Delta+\Delta'}$ to obtain the Racah coefficients (9.197). Relations (9.201) follow from the orthogonality of the $U(n)$ canonical tensor operators.

Only partitions and operator patterns appear in the definition of a $U(n)$ Racah coefficient. While exceedingly complex, the coherence brought to the structure of $U(n)$ Racah coefficients by Biedenharn's operator patterns is quite striking and elegant.

Examples:

1. $U(2)$ Racah invariants. For $n = 2$, we use the identification of $U(2)$ WCG coefficients and $SU(2)$ WCG coefficients given by (9.167)-(9.168) to write the right-hand side of relation (9.197) in the form

$$\begin{aligned}&\text{RHS of (9.197)}\\&=\sum_{\beta,\gamma} C^{b\ c\ k'}_{\beta,\gamma,\beta+\gamma} C^{k\ c\ j}_{\alpha+\beta,\gamma,\alpha+\beta+\gamma} C^{a\ b\ k}_{\alpha,\beta,\alpha+\beta} C^{a\ k'\ j}_{\alpha,\beta+\gamma,\alpha+\beta+\gamma},\end{aligned} \tag{9.202}$$

where the WCG coefficients in this result are listed in the same left to right order as in (9.197), and the angular momentum quantum numbers a, b, c, j, k, k' and associated projection quantum numbers $\alpha, \beta, \gamma, m, q, q'$ are identified from the four $U(2)$ WCG coefficients as follows:

$$\begin{aligned}&a = (\lambda_1 - \lambda_2)/2,\ \alpha = m'_{11} - M'_{11} - M_{11} - (\lambda_1 + \lambda_2)/2,\\&b = (\Lambda_1 - \Lambda_2)/2,\ \beta = M_{11} - (\Lambda_1 + \Lambda_2)/2,\\&c = (\Lambda'_1 - \Lambda'_2)/2,\ \gamma = M'_{11} - (\Lambda'_1 + \Lambda'_2)/2;\\&j = (\lambda_1 - \lambda_2 + \Delta_1 - \Delta_2 + \Delta'_1 - \Delta'_2)/2,\ m = \alpha + \beta + \gamma,\\&k = (\lambda_1 - \lambda_2 + \Delta_1 - \Delta_2)/2,\ q = \alpha + \beta,\\&k' = (\Lambda_1 - \Lambda_2 + \overline{\Delta}_1 - \overline{\Delta}_2)/2,\ q' = \beta + \gamma.\end{aligned} \tag{9.203}$$

From relation (2.160), the sum over WCG coefficients in (9.202) is $\sqrt{(2k+1)(2k'+1)}\ W(abjc; kk')$. Thus, we have the result:

$$\left\{ \begin{pmatrix} \Lambda_1 + \overline{\Delta}_1\ \Lambda_2 + \overline{\Delta}_2 \\ \Delta_1 + \Delta'_1 \end{pmatrix} \begin{pmatrix} \overline{\Delta}_1 \\ \Lambda'_1\ \Lambda'_2 \\ \Delta'_1 \end{pmatrix} \begin{pmatrix} \Lambda_1\ \Lambda_2 \\ \Delta_1 \end{pmatrix} \right\} (\lambda''_1, \lambda''_2)$$
$$= \sqrt{(2k+1)(2k'+1)}\ W(abjc; kk'), \tag{9.204}$$

where $(\lambda''_1, \lambda''_2) = (\lambda_1 + \Delta_1 + \Delta'_1, \lambda_2 + \Delta_2 + \Delta'_2)$.

2. The following Racah invariants are for the coupling of a pair of fundamental tensor operators (see (9.154) for notation):

$$\left\{\begin{pmatrix} 2 \;\; 0^{n-1} \\ \gamma \end{pmatrix}\begin{pmatrix} 1 \\ 1 \;\; 0^{n-1} \\ \tau \end{pmatrix}\begin{pmatrix} 1 \;\; 0^{n-1} \\ \rho \end{pmatrix}\right\}$$

$$= \sum_{m}\sum_{i,j=1}^{n}\left\langle \begin{matrix} 2 \;\; 0^{n-1} \\ m \end{matrix} \middle| \left\langle \begin{matrix} 1 \\ 1 \;\; 0^{n-1} \\ i \end{matrix} \right\rangle \middle| \begin{matrix} 1 \;\; 0^{n-1} \\ j \end{matrix} \right\rangle$$

$$\times \left\langle \begin{matrix} \tau \\ 1 \;\; 0^{n-1} \\ i \end{matrix} \right\rangle \left\langle \begin{matrix} \rho \\ 1 \;\; 0^{n-1} \\ j \end{matrix} \right\rangle \left\langle \begin{matrix} \gamma \\ 2 \;\; 0^{n-1} \\ m \end{matrix} \right\rangle^{\dagger}; \tag{9.205}$$

$$\left\{\begin{pmatrix} 1\,1\,0^{n-2} \\ \gamma \end{pmatrix}\begin{pmatrix} 2 \\ 1 \;\; 0^{n-1} \\ \tau \end{pmatrix}\begin{pmatrix} 1 \;\; 0^{n-1} \\ \rho \end{pmatrix}\right\}$$

$$= \sum_{m}\sum_{i,j=1}^{n}\left\langle \begin{matrix} 1\,1\,0^{n-2} \\ m \end{matrix} \middle| \left\langle \begin{matrix} 2 \\ 1 \;\; 0^{n-1} \\ i \end{matrix} \right\rangle \middle| \begin{matrix} 1 \;\; 0^{n-1} \\ j \end{matrix} \right\rangle$$

$$\times \left\langle \begin{matrix} \tau \\ 1 \;\; 0^{n-1} \\ i \end{matrix} \right\rangle \left\langle \begin{matrix} \rho \\ 1 \;\; 0^{n-1} \\ j \end{matrix} \right\rangle \left\langle \begin{matrix} \gamma \\ 1\,1\,0^{n-2} \\ m \end{matrix} \right\rangle^{\dagger}.$$

These Racah coefficient can be calculated from the known matrix elements of all fundamental tensor operators from the pattern calculus, and the known matrix elements of all $(2\,0^{n-1})$ tensor operators and $(1\,1\,0^{n-2})$ tensor operators, the latter being given completely by the pattern calculus, since all shift-weights are extremal. The results are nontrivial to display and are omitted. □

The relations for the coupling of irreducible tensor operators are formidable. But they simply supply the details for structurally elegant results: Ignoring the shift-weight patterns of the unit tensor operators in relation (9.196), it expresses the algebraic property that two irreducible tensors of type Λ and type Λ' can be coupled to one of type Λ'' by using a suitable CG coefficient. The additional feature coming from the coupling of unit tensor operator is that the basis property allows the coupled tensor operators to be again expressed as a sum of unit tensor operators.

The invariant scalars in the linear combination are then identified as Racah operators. Moreover, the relationships are more than just formal in content, since the canonical tensor operators are given explicitly in terms of the known operator-valued tensor operators in (9.199), and in some cases directly from the pattern calculus, as illustrated above.

9.7.2 Limit relations

It is a known rsult (see Ref. [21]) that $U(2)$ WCG coefficients have the following limit property:

$$\lim_{\lambda_2 \to -\infty} \left\langle \begin{matrix} \lambda_1+\Delta_1 \quad \lambda_2+\Delta_2 \\ m_{11}+M_{11} \end{matrix} \middle| \left\langle \begin{matrix} \Gamma_{11} \\ \Lambda_1 \ \Lambda_2 \\ M_{11} \end{matrix} \right\rangle \middle| \begin{matrix} \lambda_1 \ \lambda_2 \\ m \end{matrix} \right\rangle = \delta_{M_{11},\Gamma_{11}}, \tag{9.206}$$

where $(\Delta_1, \Delta_2) = (\Gamma_{11}, \Lambda_1 + \Lambda_2 - \Gamma_{11}$. Relation (9.206) is an interesting result because it gives the information that not only does the limit exist, but also that in this limit the shift-weight operator pattern Γ_{11} becomes the $U(1)$ group theoretical GT pattern M_{11}. This is the weak link between upper operator patterns and lower operator patterns mentioned in (9.109), since for $n = 2$ WCG coefficients and reduced matrix elements coincide. (The phase choice for $U(2)$ WCG coefficients is such that there is no sign factor in the right-hand side of (9.206).)

Let us assume for purposes of discussion that the following pair of limit relations holds:

$$\lim_{\lambda_n \to -\infty} \left\langle \begin{matrix} [\lambda+\Delta]_n \\ [\lambda'+\Delta']_{n-1} \end{matrix} \middle\| \begin{bmatrix} (\Gamma)_{n-2} \\ [\Gamma]_{n-1} \\ [\Lambda]_n \\ [\Lambda']_{n-1} \\ (\Gamma')_{n-2} \end{bmatrix} \middle\| \begin{matrix} [\lambda]_n \\ [\lambda']_{n-1} \end{matrix} \right\rangle \tag{9.207}$$

$$= \delta_{[\Gamma]_{n-1},[\Lambda']_{n-1}} \left\langle \begin{matrix} [\lambda+\Delta]_{n-1} \\ [\lambda'+\Delta']_{n-1} \end{matrix} \middle\| \begin{bmatrix} (\Gamma)_{n-2} \\ [\Lambda']_{n-1} \\ (\Gamma')_{n-2} \end{bmatrix}_{ext} \middle| \begin{matrix} [\lambda]_{n-1} \\ [\lambda']_{n-1} \end{matrix} \right\rangle;$$

$$\lim_{\lambda'_{n-1} \to -\infty} \left\langle \begin{matrix} [\lambda+\Delta]_{n-1} \\ [\lambda'+\Delta']_{n-1} \end{matrix} \middle\| \begin{bmatrix} (\Gamma)_{n-2} \\ [\Lambda']_{n-1} \\ (\Lambda')_{n-2} \end{bmatrix}_{ext} \middle| \begin{matrix} [\lambda]_{n-1} \\ [\lambda']_{n-1} \end{matrix} \right\rangle$$

$$= \left\langle \begin{matrix} [\lambda+\Delta]_{n-1} \\ [\lambda'+\Delta']_{n-2} \end{matrix} \middle\| \begin{bmatrix} (\Gamma)_{n-2} \\ [\Lambda']_{n-1} \\ (\Lambda')_{n-2} \end{bmatrix} \middle\| \begin{matrix} [\lambda]_{n-1} \\ [\lambda']_{n-2} \end{matrix} \right\rangle. \tag{9.208}$$

To make clear the role of various triangular patterns and partitions in these relations it has been necessary to make the notations more specific. An arbitrary partition $\lambda \in \mathbb{P}\mathrm{ar}_k$ is now denoted by $[\lambda]_k$; any triangular GT pattern m containing k rows is denoted by $(m)_k$; and any operator pattern Γ containing k rows by $(\Gamma)_k$; and any shift-weight Δ containing k components by $[\Delta]_k$. Thus, a unit projective operator of type $[\Lambda]_k$ is now denoted by

$$\begin{bmatrix} (\Gamma)_{k-1} \\ [\Lambda]_k \\ (\Gamma')_{k-1} \end{bmatrix}, \; k = 1, 2, \ldots, \tag{9.209}$$

where for $k = 1$ the pattern consists of the single entry $[\Lambda]_1 = \Lambda_1$.

The pair of limit relations (9.207)-(9.208) is hierarchical; that is, applies for each $n = 2, 3, \ldots$. The symbol on right-hand side of relation (9.207), which continues to the left-hand side of (9.208), is a new object, as indicated by the subscript ext = extension. The "matrix elements" of this object are between partitions of length n in both the initial and final position. It is called an *extended reduced matrix element;* it is a natural symbol for the indicated limit. The limit of an extended reduced matrix element is then taken again in relation (9.208) to complete the loop back to a reduced matrix element at level $n-1$. The process is then continued all the way down to the $U(2):U(1)$ level, for which

$$\langle \lambda_1 + \Delta_1 |\, [\Lambda_1] \,| \lambda_1 \rangle = \delta_{\Delta_1, \Lambda_1}. \tag{9.210}$$

To have relations of nontrivial content in (9.207)-(9.208), they are to be applied only to nonzero reduced matrix elements; that is, $[\lambda]_n$ and $[\lambda']_{n-1}$ should not belong to the characteristic null space.

The concept of an extended reduced matrix element can be motivated beyond its occurrence as a limit. The pattern calculus and the associated shift-weight function (9.148) for an extended unit projective operator are defined simply by adjoining a point n to the right-most position in row $n-1$ in the set of rules (9.139)-(9.150), where now $[\Delta']_n$ is assigned to the extended row $n-1$ and the coordinate $x_n = p_{n,n-1} - 1$ is the new variable associated to point n in row $n-1$. Two-rowed partitions of this sort of the same length are considered in Sect. 11.3.3, Compendium B. The extended shift-weight function $A^{\mathrm{ext}}_{[\Delta']_n, [\Delta]_n}([x]_n; [y]_n)$ function, where now $x = (x_1, x_2, \ldots, x_n)$ and $y = (y_1, y_2, \ldots, y_n)$, is well-defined.

The shift-weight function (9.148) and its extended version have finite limits of the form (9.207)-(9.208), subject to certain conditions: Define the integers $N'_{\Delta_n}, D_{\Delta_n}$ for the weight function (9.148) and the integers $N_{\Delta'_n}, D'_{\Delta'_n}$ for its extended version, as follows:

$$N'_{\Delta_n} = \text{number of loops incident on point } n \text{ in row } n \text{ from points within row } n-1 = \sum_{i=1}^{n-1} |\Delta'_i - \Delta_n|. \quad (9.211)$$

$$D_{\Delta_n} = \text{number of loops incident on point } n \text{ in row } n \text{ from points within row } n = \sum_{i=1}^{n-1} |\Delta_i - \Delta_n|. \quad (9.212)$$

$$N_{\Delta'_n} = \text{number of loops incident on point } n \text{ in row } n-1 \text{ from points within row } n = \sum_{i=1}^{n} |\Delta_i - \Delta'_n|. \quad (9.213)$$

$$D'_{\Delta'_n} = \text{number of loops incident on point } n \text{ in row } n-1 \text{ from points within row } n-1 = \sum_{i=1}^{n-1} |\Delta'_i - \Delta'_n|. \quad (9.214)$$

Then, the following limit relations of the shift-weight functions hold:

$$\begin{aligned} &\lim_{y_n \to -\infty} A_{[\Delta']_{n-1},[\Delta]_n}([x]_{n-1};[y]_n) = 0, \\ &\qquad\qquad \text{if and only if } N'_{\Delta_n} < D_{\Delta_n}, \\ &\lim_{y_n \to -\infty} A_{[\Delta']_{n-1},[\Delta]_n}([x]_{n-1};[y]_n) = A^{\text{ext}}_{[\Delta']_{n-1},[\Delta]_{n-1}}([x]_{n-1};[y]_{n-1}), \\ &\qquad\qquad \text{if and only if } N'_{\Delta_n} = D_{\Delta_n}; \end{aligned} \quad (9.215)$$

$$\begin{aligned} &\lim_{x_n \to -\infty} A^{\text{ext}}_{[\Delta']_n,[\Delta]_n}([x]_n;[y]_n) = 0, \\ &\qquad\qquad \text{if and only if } N_{\Delta'_n} < D'_{\Delta'_n}, \\ &\lim_{x_n \to -\infty} A^{\text{ext}}_{[\Delta']_n,[\Delta]_n}([x]_n;[y]_n) = A_{[\Delta']_{n-1},[\Delta]_n}([x]_{n-1};[y]_n), \\ &\qquad\qquad \text{if and only if } N_{\Delta'_n} = D'_{\Delta'_n}. \end{aligned} \quad (9.216)$$

The conditions in the limit relations (9.215)-(9.216) are just the trivial requirements that the number of linear numerator factors is less than or equal to the number of linear denominator factors associated with the point labeled y_n or x_n. The reason for giving these results is: *The limit relations (9.215)-(9.216) are valid for all cases of extremal unit projective operators.* These are the unit projective operators for which the only shift-weight patterns that can occur are permutations of the partition $[\Lambda]_n$; for extremal unit projective operators and their extended counterparts, the pattern calculus gives the complete reduced matrix element. This validates the limit relations (9.207)-(9.208) for all such extremal cases.

The violation of the conditions in (9.215)-(9.216) for the shift-weight functions, in which case these functions would be infinite in the indicated limit, does not invalidate relations (9.207)-(9.208) for the general case, since then there are other factors dependent on y_n and x_n in the denominator and numerator of the full expression for the reduced matrix element. Indeed, the validity of the limit relations can be demonstrated from the explicit calculation (Ref. [117]) of the reduced matrix elements for the system of all adjoint reduced matrix elements, which are those for partitions of the form $\Lambda = (2, 1^{n-2}, 0), n \geq 3$. We conjecture:

Conjecture: *The limit relations (9.207)-(9.208) are valid for all* $n = 2, 3, \ldots$.

Using this conjecture, we can now understand the origin of the Biedenharn operator patterns. Recall the crucial fact that row $n-1$ in the lower operator pattern, namely, the partition $[\Lambda']_{n-1}$, of the reduced matrix element expression (9.102) is a group theoretical label, since row $n-1$ expresses the group theoretical property that $[\lambda' + \Delta']_{n-1} \in [\Lambda']_{n-1} \otimes [\lambda']_n$. But now from the double limit (9.207)-(9.208), we have

$$\lim_{\lambda'_{n-1} \to -\infty} \lim_{\lambda_n \to -\infty} \left\langle \begin{matrix} [\lambda + \Delta]_n \\ [\lambda' + \Delta']_{n-1} \end{matrix} \right| \left| \begin{bmatrix} (\Gamma)_{n-2} \\ [\Gamma]_{n-1} \\ [\Lambda]_n \\ [\Gamma']_{n-1} \\ (\Gamma')_{n-2} \end{bmatrix} \right| \left| \begin{matrix} [\lambda]_n \\ [\lambda']_{n-1} \end{matrix} \right\rangle$$

$$= \delta_{[\Gamma]_{n-1}, [\Gamma']_{n-1}} \left\langle \begin{matrix} [\lambda + \Delta]_{n-1} \\ [\lambda' + \Delta']_{n-2} \end{matrix} \right| \left| \begin{bmatrix} (\Gamma)_{n-2} \\ [\Lambda']_{n-1} \\ (\Gamma')_{n-2} \end{bmatrix} \right| \left| \begin{matrix} [\lambda]_{n-1} \\ [\lambda']_{n-2} \end{matrix} \right\rangle . \quad (9.217)$$

This double limit expresses the property that the reduced matrix element is zero, unless row $[\Gamma]_{n-1}$ of the upper operator is equal to row $[\Gamma']_{n-1}$ of the lower pattern in which case $[\lambda' + \Delta']_{n-2} \in [\Gamma']_{n-2} \otimes [\lambda']_{n-2}$. We call property (9.217) *limit reduction of a reduced matrix element.*

The limit reduction process (9.217) can be continued downward to

$n = 2$, where relation (9.206) holds. We conclude:

Under full limit reduction of the reduced matrix element (9.102) from n down to 2, a reduced matrix element is 0, unless the upper operator pattern agrees with the lower operator pattern, in which case the group theoretical Kronecker product relation $[\lambda']_k + [\Delta]_k \in [\Gamma]_k \otimes [\lambda']_k, k = n-1, \ldots, 1$, holds, where the partition $[\lambda']_k$ is the first k parts of $[\lambda']_n$.

Relation (9.217) can be put in the context of the general $U(n)$ canonical tensor operator by using the general form of the reduction rule (9.101) from $U(n)$ to $U(n-1)$, which we now write in the detailed notation above:

$$\begin{aligned}
&\left\langle \begin{matrix} [\lambda+\Delta]_n \\ [\lambda'+\Delta']_{n-1} \\ (m')_{n-2} \end{matrix} \middle| \left\langle \begin{matrix} (\Gamma)_{n-1} \\ [\Lambda]_n \\ [\Lambda']_{n-1} \\ (M)_{n-2} \end{matrix} \right\rangle \middle| \begin{matrix} [\lambda]_n \\ [\lambda']_{n-1} \\ (m)_{n-2} \end{matrix} \right\rangle \\
&= \sum_{(\Gamma')_{n-2}} \left\langle \begin{matrix} [\lambda+\Delta]_n \\ [\lambda'+\Delta']_{n-1} \end{matrix} \middle\| \begin{bmatrix} (\Gamma)_{n-1} \\ [\Gamma]_{n-1} \\ [\Lambda]_n \\ [\Lambda']_{n-1} \\ (\Gamma')_{n-2} \end{bmatrix} \middle\| \begin{matrix} [\lambda]_n \\ [\lambda']_{n-1} \end{matrix} \right\rangle \\
&\times \left\langle \begin{matrix} [\lambda'+\Delta']_{n-1} \\ (m')_{n-2} \end{matrix} \middle| \left\langle \begin{matrix} (\Gamma')_{n-2} \\ [\Lambda']_{n-1} \\ (M)_{n-2} \end{matrix} \right\rangle \middle| \begin{matrix} [\lambda']_{n-1} \\ (m)_{n-2} \end{matrix} \right\rangle, \\
&[\Delta]_n = W\begin{pmatrix} [\Lambda]_n \\ (\Gamma)_{n-1} \end{pmatrix}, \ [\Delta']_{n-1} = W\begin{pmatrix} [\Lambda']_{n-1} \\ (\Gamma')_{n-2} \end{pmatrix}.
\end{aligned} \tag{9.218}$$

It is not necessary to take into account characteristic null space, since zero CG coefficients trivially satisfy limit reduction.

We now use this relation and the limit relation (9.217) to derive, by induction, the general result as follows, where we now revert back to the less-encumbered notation:

The limit relation (9.217), applied to every level $2, 3, \ldots, n$, *implies that*

$$LIM \left\langle \begin{matrix} \lambda+\Delta \\ m' \end{matrix} \middle| \left\langle \begin{matrix} \Gamma \\ \Lambda \\ M \end{matrix} \right\rangle \middle| \begin{matrix} \lambda \\ m \end{matrix} \right\rangle = \delta_{M,\Gamma}. \tag{9.219}$$

In this relation, LIM denotes the sequence of limits described as follows. Consider the GT subpattern in which the first down-diagonal is omitted:

$$\begin{pmatrix} \lambda_2 & & \lambda_3 & & \cdots & & \lambda_n \\ & m_{2,n-1} & & m_{3,n-1} & \cdots & m_{n-1,n-1} & \\ & & & \vdots & & & \\ & & m_{2,3} & & m_{3,3} & & \\ & & & m_{2,2} & & & \end{pmatrix}. \tag{9.220}$$

The operation LIM is described in terms of this modified GT pattern by the following sequence of limits:

$$\mathrm{LIM} = \begin{array}{c} (\lim_{\lambda_2\to-\infty}) \quad (\lim_{m_{2,n-1}\to-\infty} \quad \lim_{\lambda_3\to-\infty}) \quad \cdots \\ \cdots(\lim_{m_{2,2}\to-\infty} \quad \lim_{m_{3,3}\to-\infty}\cdots\lim_{\lambda_n\to-\infty}), \end{array} \tag{9.221}$$

where this sequence of limits is to be applied to the $U(n)$ canonical tensor operator in (9.219) from right-to-left. Thus, $\lim_{\lambda_n\to-\infty}$ is applied first, followed by $\lim_{m_{n-1,n-1}\to-\infty}, \ldots,$ followed by $\lim_{m_{2,2}\to-\infty}$, as read down the right-most up-diagonal. The limit process continues by next applying, first $\lim_{\lambda_{n-1}\to-\infty}, \ldots,$ followed by $\lim_{m_{2,3}\to-\infty}$, as read down the up-diagonal having λ_{n-1} in row n. This limit process continues all the way back to $\lim_{\lambda_3\to-\infty}$, followed by $\lim_{m_{2,n-1}\to-\infty}$, and at the last step $\lim_{\lambda_2\to-\infty}$. The parenthesis pairs in (9.221) are intended to show the sequence of limits in the various up-diagonals having $\lambda_n, \lambda_{n-1}, \ldots, \lambda_2$ in the top row n. Relation (9.219) applies to every nonzero matrix element of the general $U(n)$ canonical tensor operator. It has long known to be true for $n = 2$, and it is here conjectured to be a general result, based on the validity of the limit relation (9.217), which remains to be proved for the general case.(One method of proof might be from properties of the Gram-Schmidt orthonormalized operator-valued $D^{\Lambda}-$polynomials.)

We remark that different choices of phase factors in arriving at the canonical tensor operators and their reduced matrix elements can introduce a phase factor in (9.219). *Phases are to be chosen such that (9.217) and (9.219) hold for every n.*

We conclude this chapter on tensor operators with several remarks:

Remarks:

1. The origin of operator patterns is settled by (conjectured) limit reduction: They are uniquely assigned by the limit reduction process. For example, for $\Lambda = (4, 2, 0)$ and $\Delta = (2, 2, 2)$, the upper operator patterns obtained from the lower operator patterns by limit reduction are

$$\begin{pmatrix} & & 2 & & \\ & 2 & & 2 & \\ 4 & & 2 & & 0 \end{pmatrix}, \begin{pmatrix} & & 2 & & \\ & 3 & & 1 & \\ 4 & & 2 & & 0 \end{pmatrix}, \begin{pmatrix} & & 2 & & \\ & 4 & & 0 & \\ 4 & & 2 & & 0 \end{pmatrix}. \tag{9.222}$$

But the ordering of these patterns, respectively, by $\Gamma^3_{3,3,3} < \Gamma^2_{3,3,3} < \Gamma^1_{3,3,3}$ (see (9.65)-(9.68)and (9.71)) is still optional: There would appear to be no truly "canonical" tensor operators in the sense of no freedom of choice. But the ordering of objects is such a natural choice that the term canonical is, perhaps, not inappropriate.

2. Many more results for unit projective operators and Racah coefficients have been obtained in Refs. [16, 18, 19, 20, 21, 107, 108, 109, 111, 112, 115, 117, 118, 119, 120, 121, 122, 123] than can be presented here.

3. We have dealt with the reduction of a single Kronecker product. The consideration of multiple Kronecker products and the implied theory of multiple copies of $U(n)$ and the associated invariants going beyond Racah invariants, as was possible for $U(2)$ and the addition of several angular momenta, is, at the moment, beyond our reach. The potential relations going beyond the geometry of cubic graphs is beyond calculation, but perhaps within imagination.

4. Biedenharn's operator patterns are basic to the enumeration of invariants associated with the reduction of multiple Kronecker products of irreducible unitary representations of $U(n)$ to standard Kronecker direct sum form. An understanding of their structure, such as a proof of the conjectured limit reduction property outlined above, are essential to future progress in this subject.

5. It is difficult to obtain fully explicit formulas for the CG coefficients (and Racah coefficients) that effect the reduction of even a single Kronecker product. We have attempted to formulate most results in an algorithmic form that allows the generation of such formulas, leaving explicit formulas for the future of advanced symbolic computation.

6. Many of results presented here on operator patterns and tensor operators can be extended to $q-$tensor operator algebras as carried out by Biedenharn and Lohe [17].

Chapter 10

Compendium A. Basic Algebraic Objects and Structures

The material presented in this Chapter 11 and Chapter 12 as Compendiums A and B is intended to introduce vocabulary, notations, and concepts used in the monograph. We do not require the depth and detail of entire textbooks on the various topics, and the material is presented here to avoid digressions in the main text. Familiar topics are presented in a synoptic format, while less familiar subjects are developed in the detail required. Proofs, when indicated, are brief, since there are many excellent expositions on most topics, and it would distract from the main objective. With these qualifications in mind, we proceed with our collection of topics, often giving an orientation to it that suits our needs. We also reference textbooks and articles by topics at the end of each compendium, where in-depth discussions can be found.

10.1 Groups

A group G is a nonempty set of elements in which there is defined a binary operation $\star$, called product, such that the set is closed under the operation $\star$, is associative, contains an identity element, and each element has an inverse. In detail, the following rules hold:

1. Closure: For each pair of elements $g, g' \in G$, the product $g \star g' \in G$.

2. Associativity: For each triplet of elements $g, g', g'' \in G$, the product rule $(g \star g') \star g'' = g \star (g' \star g'')$ holds.

3. Identity: For each $g \in G$, the set G contains a distinguished element e such that $e \star g = g \star e$.

4. Inverse: For each element $g \in G$, there corresponds an element of G, denoted g^{-1} and called the inverse of g, such that $g \star g^{-1} = g^{-1} \star g = e$.

It may be proved from these axioms for an abstract group that the identity element e is unique, and then, in turn, that the inverse of a given element is unique. The number of elements in a group may be finite, countably infinite (denumerable), or nondenumerable. The symbol $\star$ denoting the binary operation in G has no intrinsic significance, nor does the term "product." Any intelligible mark will do, and often group multiplication is denoted by simple juxtaposition of group elements when no ambiguity can arise. It is important to observe, however, that in the definition of a group both the binary operation $\star$ must be specified together with the set G, the pair sometimes being written as $(\star, G)$ to emphasize that the two objects are needed.

A group is called *abelian* or *commutative* if each pair of elements in G commute, that is, $g \star g' = g' \star g$, for all $g, g' \in G$.

The set of all integers $\mathbf{Z} = \cdots, -2, -1, 0, 1, 2, \cdots$ under the binary operation of addition, denoted $+$, with identity 0 is an example of an abelian group, as is also the set of complex numbers $\{e^{i\phi} \,|\, 0 \le \phi < 2\pi\}$ of modulus 1 under the binary operation of ordinary multiplication with identity 1. An example of a nonabelian group is the set $GL(n, \mathbf{C})$ of all $n \times n$ nonsingular matrices with complex numbers as entries, where group multiplication is ordinary matrix multiplication, and the identity is the unit matrix $I_n = \mathrm{diag}(1, 1, \cdots, 1)$ with 1 along the diagonal. An important subgroup of $GL(n, \mathbf{C})$ is the set of unitary matrices, denoted $U(n)$, where each $U \in U(n)$ also satisfies $U^{-1} = U^\dagger = \left(U^T\right)^*$, where the operation T denotes matrix transposition and $*$ denotes complex conjugation.

The *symmetric* group S_n not only provided the historical context from which evolved the concept of an abstract group, it is, perhaps, the most important group in mathematics, physics, chemistry, and biology (see Lascoux and Schültzenberger [101], Lulek [125], and Wybourne [190]). It consists of elements, called permutations, which we write in the two-line form

$$\begin{pmatrix} 1 & 2 & \cdots & n \\ i_1 & i_2 & \cdots & i_n \end{pmatrix}, \tag{10.1}$$

where the sequence $(i_1, i_2, \ldots, i_n)$ is a rearrangement of the integers $(1, 2, \ldots, n)$. The symmetric group S_n contains $n!$ such elements corresponding to all possible $n!$ assignments of the sequence $(i_1, i_2, \ldots, i_n)$. Moreover, by definition, all permutation symbols corresponding to the

$n!$ rearrangements of columns are equal; for example,

$$\begin{pmatrix} 1 & 2 & 3 \\ 2 & 3 & 1 \end{pmatrix} = \begin{pmatrix} 2 & 3 & 1 \\ 3 & 1 & 2 \end{pmatrix}. \tag{10.2}$$

Multiplication of two permutations is defined by

$$\begin{pmatrix} 1 & 2 & \cdots & n \\ j_1 & j_2 & \cdots & j_n \end{pmatrix} \begin{pmatrix} 1 & 2 & \cdots & n \\ i_1 & i_2 & \cdots & i_n \end{pmatrix} \tag{10.3}$$

$$= \begin{pmatrix} i_1 & i_2 & \cdots & i_n \\ k_1 & k_2 & \cdots & k_n \end{pmatrix} \begin{pmatrix} 1 & 2 & \cdots & n \\ i_1 & i_2 & \cdots & i_n \end{pmatrix} = \begin{pmatrix} 1 & 2 & \cdots & n \\ k_1 & k_2 & \cdots & k_n \end{pmatrix}.$$

Thus, the columns of the permutation on the left are rearranged such that the top row agrees with the bottom row of the permutation on the right, and then the product permutation is read off by canceling the two common rows, as shown. This multiplication rule is closed over the collection of all $n!$ permutations; this rule may be verified to be associative. The identity element is

$$e = \begin{pmatrix} 1 & 2 & \cdots & n \\ 1 & 2 & \cdots & n \end{pmatrix}, \tag{10.4}$$

and the inverse of a given permutation is obtained by interchanging top and bottom rows in the permutation symbol:

$$\begin{pmatrix} 1 & 2 & \cdots & n \\ i_1 & i_2 & \cdots & i_n \end{pmatrix}^{-1} = \begin{pmatrix} i_1 & i_2 & \cdots & i_n \\ 1 & 2 & \cdots & n \end{pmatrix}. \tag{10.5}$$

For example,

$$\begin{pmatrix} 1 & 2 & 3 \\ 2 & 3 & 1 \end{pmatrix}^{-1} = \begin{pmatrix} 2 & 3 & 1 \\ 1 & 2 & 3 \end{pmatrix} = \begin{pmatrix} 1 & 2 & 3 \\ 3 & 1 & 2 \end{pmatrix}. \tag{10.6}$$

We often use the one-line notation

$$\pi = \begin{pmatrix} 1 & 2 & \cdots & n \\ \pi_1 & \pi_2 & \cdots & \pi_n \end{pmatrix} = (\pi_1, \pi_2, \ldots, \pi_n) \tag{10.7}$$

for a permutation $\pi \in S_n$ when no ambiguity can arise.

10.1.1 Group actions

Let G be a group and X a nonempty set. The group G is said to act on the set X if there is defined a mapping

$$A_g : x \mapsto x' = A_g(x) \in X, \text{ each } g \in G, \text{ each } x \in X, \tag{10.8}$$

such that

$$A_{g'}(A_g(x)) = A_{g'g}(x), \text{each pair } g', g \in G; \text{ each } x \in X, \tag{10.9}$$

$$A_e(A_g(x)) = A_g(x), \text{each } g \in G; \text{ each } x \in X, A_e = 1. \tag{10.10}$$

If we define $(A_{g'}A_g)(x) = A_{g'}(A_g(x))$, then we may write the sequential action (10.9) as the product of actions $A_{g'}A_g = A_{g'g}$ on X, each pair $g', g \in G$. Closure and associativity of group action then follow from the composition rule (10.9). The mapping A_g is sometimes called a $G-$operator and the set on which G acts, a $G-$space.

It is sometimes natural to define a *left* multiplication $g \cdot x$ of G on X. In this case, we write $L_g(x) = g \cdot x$, and refer to the left action of G on X, which is required to satisfy (10.9)-(10.10)). Similarly, it is sometimes natural to define a *right* multiplication $x * h$ of a group H on X, and write $R_h(x) = x * h$, which is required to satisfy (10.9)-(10.10).

Matrix groups, where group multiplication is ordinary matrix multiplication, illustrate these concepts nicely. Let G be the group of nonsingular complex matrices of order n, H the group of nonsingular complex matrices of order m, and X the set of $n \times m$ complex matrices. Define left multiplication by $g \cdot x = gx$ (matrix multiplication of $x \in X$ from the left by g), and right multiplication by $h * x = xh^{-1}$ (matrix multiplication of $x \in X$ from the right by the inverse of h.) Then, $L_g(x) = gx$ and $R_h(x) = xh^{-1}$ may be verified to satisfy the rules of group action. But, in this case, we also have

$$\begin{aligned} L_g(R_h(x)) &= L_g(xh^{-1}) = g(xh^{-1}) = gxh^{-1}, \\ R_h(L_g(x)) &= R_h(gx) = (gx)h^{-1} = gxh^{-1}, \end{aligned} \tag{10.11}$$

so that $L_gR_h = R_hL_g$ on X for all $g \in G$, all $h \in H$. *The two group actions mutually commute.* This is an important property for many applications.

Certain subsets of G and of X arise naturally from the concept of a group action A_g. Among these are the following:

1. The *orbit* $[x]$ of G in X containing $x \in X$ is the subset of X defined by
$$[x] = \{A_g(x) \mid g \in G\}. \tag{10.12}$$
A point $x \in X$ is said to be equivalent to point $x' \in X$ under the action of G on X, if there exists a $g \in G$ such that $A_g(x) = x'$. This equivalence is written $x \sim x'$. It then follows from the group property that $x \sim x$(reflexive), $x' \sim x$ (symmetric), and

$x \sim x', x' \sim x''$ imply $x \sim x''$ (transitive). Since each pair of points belonging to the same orbit are equivalent under this equivalency relation, and points belonging to distinct orbits are inequivalent, the set X may be partitioned into a union of disjoint orbits with respect to the action of G on X.

2. The *stabilizer, little group, or isotropy group* $S(x)$ of a point $x \in X$ is the subset of G defined by

$$S(x) = \{g \in G \mid A_g(x) = x\}. \tag{10.13}$$

The isotropy group $S(x)$ is the subset of all elements of G that is *fixed* under the action of G on X, where the group property of this subset is easily proved. Isotropy groups $S(x)$ and $S(x')$ of distinct points x and x' belonging to the same orbit $[x]$ of the action of G on X are conjugate to one another, specifically, if $x' = A_g(x)$, then $S(x') = gS(x)g^{-1}$.

3. An *invariant subset* I_X of X is the set of points in X defined by

$$I(X) = \{x \in X \mid A_g(x) \in I(X), \text{ each } g \in G\}. \tag{10.14}$$

An invariant subset includes itself in its definition; the full set X is an invariant subset of X. Invariant subsets are unions of orbits of G on X.

10.2 Rings

A ring $\mathcal{R}$ with an identity is a nonempty set of elements on which there is defined two binary operations, denoted $+$ and $\times$, and called addition and multiplication, that satisfy the following rules:

1. The set $\mathcal{R}$ is an abelian group with identity 0 under the binary operation $+$ of addition.

2. Multiplication of elements in the set $\{\mathcal{R} - \{0\}\}$ is associative: $(a \times b) \times c = a \times (b \times c)$.

3. Group multiplication $\times$ is distributive over addition, that is, for all triplets of elements a, b, c in $\mathcal{R}$, the following two rules are satisfied: $(a + b) \times c = a \times c + b \times c,\ c \times (a + b) = c \times a + c \times b$.

A ring is *commutative* if the multiplicative group $\{\mathcal{R} - \{0\}\}$ is abelian. We only consider rings with an identity, an refer to them simply as rings. Examples of commutative rings are the set $\mathbb{Z}$ of integers, the set $\mathbb{Q}$ of rational numbers, the set $\mathbb{R}$ of real numbers, and the set $\mathbb{C}$ of complex numbers. The set $GL(n, \mathbb{Q})$ of complex nonsingular matrices is a noncommutative ring under the binary operations of matrix addition and matrix multiplication.

For a commutative ring, the usual summation, product, and binomial expansion, etc. rules hold:

$$\sum_{i=1}^{n} a_i = a_1 + a_2 + \cdots + a_n,$$

$$\prod_{i=1}^{n} a_i = a_1 \times a_2 \times \cdots \times a_n = a_1 a_2 \cdots a_n, \tag{10.15}$$

$$(a+b)^n = \sum_{i=0}^{n} \binom{n}{i} a^i b^{n-i}.$$

where $\binom{n}{i}$ denotes the binomial coefficient $\binom{n}{i} = \frac{n!}{i!(n-i)!}$, and where the multiplicative symbol $\times$ is replaced by juxtaposition of ring elements.

The axioms for a *field* are obtained from those for a ring by replacing ring axiom (2) by

2a. The set $\{\mathcal{R} - \{0\}\}$ is an abelian group with identity 1 with respect to the binary operation of multiplication.

The set $\mathbb{R}$ of real numbers is a field, as is the set of complex numbers $\mathbb{C}$.

10.2.1 Rings of polynomials

For the definition of a polynomial, we require a *sequence* a of elements from a ring, as defined by

$$a = (a_0, a_1, a_2, \ldots, a_k, \ldots), \tag{10.16}$$

where only a finite number of the ring elements in the sequence are nonzero. In addition, we require a letter x from an alphabet. The letter x is called an indeterminate. It has no particular meaning, but may be individualized in a given context; it must, however, satisfy the multiplicative rule $x^j\, x^k = x^{j+k}$, for all integers j, k. From the sequence a of ring elements and the indeterminate x, we form the expression

$$p(x) = a_0 + a_1 x + a_2 x^2 + \cdots = \sum_{i \geq 0} a_i x^i, \tag{10.17}$$

where, by definition, $a_0x^0 = a_0$, and $\sum_{i\geq 0}$ denotes that only terms corresponding to nonzero ring elements are included in the summation. The ring element a_0 is called the *constant term* of the polynomial $p(x)$. If a_n is the last nonzero ring element in the sequence a, the expression $p(x)$ is called a polynomial of degree n in the indeterminate x with coefficients in the ring $\mathcal{R}$. In practice, we often write $p(x) = \sum_{i=0}^{n} a_i x^i$, and omit all terms of the form $0x^i$ corresponding to a ring element $a_i = 0 =$ additive ring identity.

The set of all polynomials in the indeterminate x with coefficients in the ring $\mathcal{R}$ is denoted $\mathcal{R}[x]$. The set $\mathcal{R}[x]$ is endowed with a ring structure by postulating the following two rules for adding and multiplying polynomials:

(i). Addition: $p(x) + p'(x) = \sum_{i\geq 0}(a_i + a'_i)x^i$. (10.18)

(ii). Multiplication: $p(x)p'(x) = \sum_{i\geq 0}\sum_{i'\geq 0} a_i a'_{i'} x^{i+i'}$

$$= \sum_{i''\geq 0}\Big(\sum_{i+i'=i''} a_i a'_{i'}\Big)x^{i''} = \sum_{i''\geq 0} a''_{i''} x^{i''}, \qquad (10.19)$$

where the indeterminate x commutes with all ring elements. That $\mathcal{R}[x]$ is a ring may be verified directly from these properties. The transition from $\mathcal{R}$ to $\mathcal{R}[x]$ is known as the *adjunction* of an indeterminate x (van der Waerden [173]). This use of a general indeterminate x is important in our view of polynomials, since we need the freedom of choosing it to have various meanings.

The generalization of the polynomial ring $\mathcal{R}[x]$ over a single indeterminate x to an alphabet of n commuting indeterminates $x_1, x_2, \ldots, x_n$ is effected in the obvious way by successive adjunction of the indeterminates, thus obtaining the expression

$$p(x_1, x_2, \ldots, x_n) = \sum_{i_1\geq 0, i_2\geq 0, \ldots, i_n\geq 0} a_{i_1,i_2,\ldots,i_n} x_1^{i_1} i_2^{i_2} \cdots x_n^{i_n},$$

$$\text{each } a_{i_1,i_2,\ldots,i_n} \in \mathcal{R}, \qquad (10.20)$$

where all but a finite number of the ring elements $a_{i_1,i_2,\ldots,i_n}$ are 0. The expression $p(x_1, x_2, \ldots, x_n)$ is called a polynomial in the indeterminates $x_1, x_2, \ldots, x_n$ with coefficients in the ring $\mathcal{R}$. The ring element $a_{0,0,\ldots,0}$ is called the *constant term* of the polynomial. Since the x_i can be arranged in any order whatsoever, the set of all polynomials in the indeterminates $x_1, x_2, \ldots, x_n$ with coefficients in the ring $\mathcal{R}$ may be denoted unambiguously by $\mathcal{R}_n[x] = \mathcal{R}[x_1, x_2, \ldots, x_n]$.

The rules of addition and multiplication that endow the set $\mathcal{R}_n[x]$ of polynomials with a ring structure may be put in the same form as those given by (10.18)-(10.19) for a single indeterminate x. This is accomplished by the following replacements: x is replaced by the sequence $x = (x_1, x_2, \ldots, x_n)$ of indeterminates, i by the sequence $i = (i_1, i_2, \ldots, i_n)$ of nonnegative integers, i' by the sequence $i' = (i'_1, i'_2, \ldots, i'_n)$ of nonnegative integers, and i'' by the sequence $i'' = (i''_1, i''_2, \ldots, i''_n)$ of nonnegative integers, where $i + i' == (i_1 + i'_1, i_2 + i'_2, \ldots, i_n + i'_n)$. Finally, for such sequences x and i, we define

$$x^i = x_1^{i_1} x_2^{i_2} \cdots x_n^{i_n}. \tag{10.21}$$

With these replacements, addition and multipliction of polynomials in the set $\mathcal{R}_n[x]$ take the same form (10.18)-10.19) as that for a single indeterminate. We call this algebraic structure the ring of polynomials $\mathcal{R}_n[x]$ in n indeterminates over $\mathcal{R}$.

In many applications of polynomial rings, it is necessary to choose the sequence of indeterminates $(x_1, x_2, \ldots, x_n)$ to be elements of ring $\mathcal{R}$. In this case, we still obtain a ring of polynomials if the x_i are replaced by any elements of the ring that mutually commute among themselves and with every element of $\mathcal{R}$.

Homogeneous polynomials have a central role in that an arbitrary polynomial in the ring $\mathcal{R}_n[x]$ may be expressed as a linear combination of homogeneous polynomials. A homogeneous polynomial $h_k(x) \in \mathcal{R}_n[x]$ is defined by the expression

$$h_k(x) = \sum_{\substack{i_1 \geq 0, i_2 \geq 0, \ldots, i_n \geq 0 \\ i_1 + i_2 + \cdots + i_n = k}} a_{i_1, i_2, \ldots, i_k} x_1^{i_1} x_2^{i_2} \cdots x_n^{i_n}, \tag{10.22}$$

and has the property

$$h_k(\lambda x_1, \lambda x_2, \ldots, \lambda x_n) = \lambda^k h_k(x_1, x_2, \ldots, x_n) \tag{10.23}$$

under the substitution $x_i \mapsto \lambda x_i, i = 1, 2, \ldots, n$, for each indeterminate, where λ is an indeterminate commuting with all elements of the ring and all other indeterminates. The polynomial $h_k(x)$ is said to be *homogeneous of total degree* k. It is evident that each polynomial $p(x) \in \mathcal{R}_n[x]$ can be expressed as a linear combination of the $h_k(x)$, $k \geq 0$ with coefficients $A_k \in \mathcal{R}$:

$$p(x) = \sum_{k \geq 0} A_k h_k(x). \tag{10.24}$$

10.2.2 Vector spaces of polynomials

A vector space is a mathematical object obtained from two sets: a field $\mathcal{K}$, which is a commutative ring with an identity and inverse with elements called *scalars,* and a second set $\mathcal{V}$ with elements called *vectors,* which is an abelian group with respect to a binary operation called addition and denoted $+$. By definition the product of any scalar $\alpha \in \mathcal{K}$ and any vector $v \in \mathcal{V}$ is a vector. It is denoted αv, and, in particular, we require $1v = v$, where 1 is the unit scalar in $\mathcal{K}$. The definition is completed by giving the distributive rules: For all scalars $\alpha, \beta \in \mathcal{K}$ and all vectors $v, v' \in \mathcal{V}$, we have

$$\alpha(v + v') = \alpha v + \alpha v'; \ (\alpha + \beta)v = \alpha v + \beta v. \tag{10.25}$$

Observe that we use the same symbol + for addition of scalars (field elements) and for addition of vectors, since no ambiguity will arise.

Examples of vector spaces abound. We illustrate first the example of polynomials, in n indeterminates $x = (x_1, x_2, \ldots, x_n)$ over the field $\mathbb{R}$ of real numbers or the field $\mathbb{C}$ of complex numbers. Since polynomials in set $\mathbb{R}_n[x]$ or $\mathbb{C}_n[x]$ constitute an abelian group with respect to addition, it is only necessary to observe that multiplication of a polynomial $p(x) = p(x_1, x_2, \ldots, x_n)$ in n indeterminates by a real number or complex number α is defined to be the polynomial obtained from $p(x)$ by multiplying all the coefficients of $p(x)$ by the scalar α :

$$\alpha p(x) = \sum_{i_1 \geq 0, i_2 \geq 0, \ldots, i_n \geq 0} (\alpha a_{i_1, i_2, \ldots, i_n}) x_1^{i_1} x_2^{i_2} \cdots x_n^{i_n}. \tag{10.26}$$

It now follows from the properties of the underlying field of real or complex numbers that the required distributive rules between field elements and polynomials (vectors) are satisfied. Thus, the polynomials in the set $\mathbb{R}_n[x]$ or $\mathbb{C}_n[x]$ constitute a linear vector space. The vector space property of polynomials makes no use of the fact that polynomials can also be multiplied. This additional property is, of course, essential for the development of the properties of polynomials. The multiplication rule for two polynomials,

$$q(y) = \sum_{j_1 \geq 0, j_2 \geq 0, \ldots, j_m \geq 0} b_j \, y^j \text{ and } p(x) = \sum_{i_1 \geq 0, i_2 \geq 0, \ldots, i_n \geq 0} a_i x^i, \tag{10.27}$$

over the field of real or complex numbers is the obvious extension of (10.19):

$$q(y)p(x) = \sum_{j \geq 0} \sum_{i \geq 0} b_j a_i y^j x^i. \tag{10.28}$$

The generalization of the concept of a vector space of polynomials over the matrix ring of real or complex matrices is an obvious extension of the above results. A matrix ring of polynomials in n indeterminates, denoted $\mathcal{M}_n[x]$, is obtained by selecting the general ring $\mathcal{R}_n$ to be the ring $\mathcal{M}_n$ of complex matrices of order n with unit matrix I_n, the identity of the ring. Ring addition and multiplication are matrix addition and multiplication. A general polynomial $P(x) \in \mathcal{M}_n[x]$ is given by the expression

$$P(x) = \sum_{i\geq 0} A_i x^i, \text{ each } A_i = A_{i_1,i_2,\ldots,i_n} \in \mathcal{M}_n. \tag{10.29}$$

The matrix $P(x)$ of order n is a linear combination of ring elements (matrices) with coefficients that are monomials in the indeterminates $x = (x_1, x_2, \ldots, x_n)$.

10.3 Abstract Hilbert Spaces

10.3.1 Inner product spaces

An *inner product space* $\mathbf{V}$ is a complex vector space $\mathcal{V}$ together with a mapping rule

$$(\ ,\ ) : \mathcal{V} \times \mathcal{V} \to \mathbb{C}, \tag{10.30}$$

such that, for each pair of vectors $u, v \in \mathcal{V}$, the mapping $(\ ,\ )$ is $(u, v) \mapsto \mathbb{C}$. The mapping $(\ ,\ )$ is called a *scalar or inner product* on $\mathcal{V}$. This mapping is required, for all vectors $u, v, w \in \mathcal{V}$ and all scalars $\alpha \in \mathbb{C}$, to possess the following properties:

1. Conjugation symmetric:$(u, v)^* = (v, u)$.
2. Additive: $(u + v, w) = (u, w) + (v, w)$.
3. Linear in second factor: $(u, \alpha v) = \alpha(u, v)$.
4. Antilinear in first factor: $(\alpha u, v) = \alpha^*(u, v)$.
5. Positivity: $(u, u) > 0, u > \mathbf{0}, (u, u) = 0,$ if and only if $u = \mathbf{0}$, where $\mathbf{0}$ is the zero vector.

Thus, the inner product space $\mathbf{V}$ is the pair, the vector space $\mathcal{V}$ and the relation R:

$$\mathbf{V} = (\mathcal{V}, R), R = (\ ,\ ). \tag{10.31}$$

One and the same vector space $\mathcal{V}$ may have several inner products defined on it: The inner product spaces $\mathbf{V} = (\mathcal{V}, R), R = (\ ,\ )$, and

$\mathbf{V}' = (\mathcal{V}, R'), R' = (\ ,\ )'$, are distinct mathematical objects. It is the custom, however, to speak of the inner product space $\mathcal{V}$ and to leave the inner product rule implicit, a practice we shall follow. Sometimes for clarity we speak of "an inner product on $\mathcal{V}$."

One of the simplest of inner products is often assigned to sequences of complex numbers

$$u = (u_1, u_2, \ldots),\ \ v = (v_1, v_2, \ldots), \tag{10.32}$$

in which only a finite number of the parts are nonzero:

$$(u, v) = \sum_{i \geq 1} u_i^*\, v_i. \tag{10.33}$$

This definition of an inner product also applies to real sequences for which $u^* = u, v^* = v, \ldots$.

A subset of vectors $\{v_1, v_2, \ldots, v_j\}$, each $v_i \in \mathcal{V}$, is said to be linearly independent if the linear relation

$$\alpha_1 v_1 + \alpha_2 v_2 + \cdots + \alpha_j v_j = \mathbf{0} \tag{10.34}$$

implies that all the scalars $\alpha_i, i = 1, 2, \ldots, j$ are equal to 0. In other words, a set of vectors is linearly dependent if at least one vector in the set is a linear combination of the remaining ones.

A subset of vectors $\mathcal{W} \subseteq \mathcal{V}$ is a subspace of $\mathcal{V}$ if $\mathcal{W}$ is a complex vector space on its own such that each $w \in \mathcal{W}$ is also a vector $w \in \mathcal{V}$. The vector space $\mathcal{W}$ is usually considered to inherit its inner product from $\mathcal{V}$; it is therefore also an inner product space. If there exists a nonzero vector $v \in \mathcal{V}$ such that $v \notin \mathcal{W}$, then $\mathcal{W}$ is a proper subspace of $\mathcal{V}$, which is denoted $\mathcal{W} \subset \mathcal{V}$. Any set of linearly independent vectors in $\mathcal{V}$ constitute a *basis* for some subspace $\mathcal{W} \subset \mathcal{V}$.

Definitions and concepts associated with the definition of an inner product $(\ ,\ )$ on $\mathcal{V}$ include the following:

1. The *norm* $\|u\|$ of a vector $u \in \mathcal{V}$ is defined by $\|u\| = \sqrt{(u, u)}$.
2. A vector $u \in \mathcal{V}$ is said to be *normalized* if $(u, u) = 1$.
3. Two vectors $u, v \in \mathcal{V}$ are said to be *orthogonal* or *perpendicular* if $(u, v) = 0$, this property being denoted by $u \perp v$.
4. An inner product space $\mathcal{V}$ is said to be finite-dimensional if there exists a set of vectors

$$\mathcal{B} = \{v_1, v_2, \ldots, v_n \mid v_i \perp v_j, \text{each pair } i, j = 1, 2, \ldots, n\},$$

such that each vector $v \in \mathcal{V}$ is a linear combination of the vectors in $\mathcal{B}$:

$$v = \alpha_1 v_1 + \alpha_2 v_2 + \cdots + \alpha_n v_n, \text{ each } \alpha_i \in \mathbb{C}. \tag{10.35}$$

The dimension of the vector space $\mathcal{V}$ is said to be n and is denoted by $\text{Dim}\mathcal{V} = n$. The set of vectors $\mathcal{B}$ is called a *basis* of $\mathcal{V}$. It is called an *orthogonal basis* if all pairs of vectors in the basis are perpendicular, and an *orthonormal basis* if all the perpendicular vectors are also normalized.

5. If the inner product vector space $\mathcal{W}$ is a subset of another inner product space $\mathcal{V}$, these two spaces having the same inner product, then a vector $v \in \mathcal{V}$ is said to be perpendicular to the space $\mathcal{W}$ if v is perpendicular to every vector in $\mathcal{W}$.

6. A *separable Hilbert space* $\mathcal{H}$ is an inner product space that possesses a set $\mathcal{B}$ of denumerably (countable) infinite orthonormal basis vectors

$$\mathcal{B} = \{(v_1, v_2, \ldots, v_i, \ldots) \mid (v_i, v_j) = \delta_{ij}, \text{ each pair } i, j = 1, 2, \ldots\} \tag{10.36}$$

such that: *The only vector* $v \in \mathcal{B}$ *perpendicular to* $\mathcal{B}$ *is the zero vector* $v = \mathbf{0}$.

We shall refer to a finite-dimensional inner product space $\mathcal{V}$ simply as a *finite-dimensional Hilbert space,* and use the term *separable Hilbert space* in the context of Item 6 above. The unqualified term *Hilbert space* will mean one or the other of these objects, finite-dimensional Hilbert space or separable Hilbert space, as implied by the context of usage. The term *abstract Hilbert space* is used in situations where the space $\mathcal{H}$ and its inner product are not explicitly specified.

10.3.2 Linear operators

Once a set of mathematical objects has been defined, the next step is to consider the kinds of rules or operations that can be effected on the set that leave the set *invariant;* that is, mappings of the set into itself. Our interest here is in linear mappings that leave a separable Hilbert space $\mathcal{H}$ invariant.

A mapping $T : \mathcal{H} \to \mathcal{H}$ of a separable Hilbert space $\mathcal{H}$ into itself is said to be *linear* if the following properties hold:

1. Additive: $T(u + v) = Tu + Tv$, all $u, v \in \mathcal{H}$.
2. Homogeneous: $T(\alpha v) = \alpha Tu$, all $u \in \mathcal{H}$, all $\alpha \in \mathbb{C}$.

The linear mapping T is also called a *linear operator* on $\mathcal{H}$.

One of the most often used properties of a linear operator that follows by induction on n from the definition is

$$T(\sum_{i\geq 1}\alpha_i u_i) = \sum_{i\geq 1}\alpha_i\ (Tu_i), \quad \text{all } u_i \in \mathcal{H}, \text{ all } \alpha_i \in \mathbb{C}. \tag{10.37}$$

This property subsumes such simpler ones as $T\,\mathbf{0} = \mathbf{0}, T(-u) = -Tu, T(u-v) = Tu - Tv$. We will encounter only linear operators, and therefore often drop "linear." We also describe the relation $v = Tu$ by the phrase "the action of the operator T on the vector u is the vector v."

Definitions and concepts associated with linear operators include the following:

1. Equal operators: If $Su = Tu$, all $u \in \mathcal{H}$, then, $S = T$.
2. Identity operator: If $\mathbb{I}\,u = u$, all $u \in \mathcal{H}$, then $\mathbb{I}$ is the identity operator on $\mathcal{H}$.
3. Null operator: If $\mathbf{N}_0\,u = \mathbf{0}$, all $u \in \mathcal{H}$ then $\mathbf{N}_0$ is the null operator on $\mathcal{H}$.
4. Invariant Hilbert space: A Hilbert space $\mathcal{H}$ is said to be invariant under the action of a linear operator T defined on $\mathcal{H}$ if and only if the set of vectors $\{Tu | u \in \mathcal{H}\}$ is a subset of $\mathcal{H}$.
5. Isomorphic Hilbert spaces: Let $\mathcal{H}$ and $\mathcal{H}'$ be separable Hilbert spaces with inner products $(\ ,\ )$ and $(\ ,\ )'$. Then, $\mathcal{H}$ is said to be *isomorphic* with $\mathcal{H}'$ if and only if there exists a one-to-one linear mapping L of $\mathcal{H}$ onto $\mathcal{H}'$ such that

   $$(u, v) = (Lu, Lv)', \text{ for all } u, v \in \mathcal{H}. \tag{10.38}$$

   The mapping L is called a *Hilbert space isomorphism* of $\mathcal{H}$ onto $\mathcal{H}'$. Hilbert space isomorphisms are very important for this monograph because for the calculation of many quantities such as representation functions of the unitary group $U(n)$, Clebsch-Gordan coefficients, etc., we choose an abstract Hilbert space that is sufficiently general such as to yield the most general possible structures for these mathematical entities. Applications to special situations must fall under the purview of the abstract theory.

The *product of two operators* S and T is denoted by ST, and is defined to be the operator such that $(ST)u = S(Tu)$, all $u \in \mathcal{H}$. Thus, if $v = Tu \in \mathcal{H}$ is the vector obtained from u by the action of the operator T, then $w = Sv \in \mathcal{H}$ is the vector obtained from v by the action of the operator S. Two operators S and T commute on $\mathcal{H}$ if $(ST)u = (TS)u$, all $u \in \mathcal{H}$.

We have the following definitions associated with the product of two linear operators acting on $\mathcal{H}$:

1. Inverse operator: If $(ST)u = (TS)u = u$, all $u \in \mathcal{H}$, then, the *inverse* of T is $S = T^{-1}$.
2. Commutator of two operators: If $(ST - TS)u = Cu$, all $u \in \mathcal{H}$, then $C = [S, T] = ST - TS$ is the *commutator* of S and T.

10.3.3 Inner products and linear operators

Definition of operators associated with the inner product of vectors include the following: Let T be a linear operator $T : \mathcal{H} \to \mathcal{H}$ on a separable Hilbert space $\mathcal{H}$ with the inner product $(\ ,\ )$. Then, we have the following list of definitions and properties, which must hold for all $u, v \in \mathcal{H}$:

1. If $(u, T^{\dagger}v) = (Tu, v)$, then $T^{\dagger}$ is the *adjoint* operator to T.
2. If $(u, Tv) = (Tu, v)$, then T is a *a self-adjoint operator, and* $T = T^{\dagger}$.
3. *If* $(Tu, Tv) = (u, v)$, *then* T *is a unitary* operator,and $T^{\dagger}T = TT^{\dagger} = \mathbb{I}$.
4. If $(Tu, Tv) = (T^{\dagger}u, T^{\dagger}v)$, then T is a *normal* operator, and $TT^{\dagger} = T^{\dagger}T$.
5. If $(u, Tv) = (Tu, v) = (T^{\dagger}u, Tv)$, then T is a *projection* operator, and $T = T^{\dagger} = TT$.
6. Invertible operator: If $(u, Tu) = (T^{-1}v, v)$, then T is said to be an *invertible operator,* and T^{-1} is called the *inverse* of T.

Self-adjoint operators on a separable Hilbert space are also called *Hermitian* operators. Hermitian, unitary, projection, and invertible operators are all normal operators.

The concept of operators acting in a Hilbert space is further enriched by the notions of eigenvalues and eigenvectors. Let T be a linear operator $T : \mathcal{H} \to \mathcal{H}$ on a separable Hilbert space $\mathcal{H}$. If there exists a vector $u \in \mathcal{H}$ and a scalar $t \in \mathbb{C}$ such that $Tu = tu$, the u is called an *eigenvector* and t an eigenvalue of T. Definitions and concepts associated with eigenvectors and eigenvalues are the following:

1. Diagonal operator on a basis $\mathcal{B}$: An operator D whose action on the basis $\mathcal{B} = \{u_1, u_2, \ldots\}$ of $\mathcal{H}$ is given by $Du_i = d_i u_i, i = 1, 2, \ldots\}$.

2. Diagonable operator on $\mathcal{H}$: An operator T such that there exists a basis $\mathcal{B}$ of $\mathcal{H}$ and an invertible operator S such that $T = S^{-1}DS$, where D is a diagonal operator on $\mathcal{B}$. An operator whose eigenvalues are distinct on a finite-dimensional subspace of $\mathcal{H}$ is diagonable on that subspace.

3. Set of simultaneously diagonal operators: Any set of mutually commuting normal operators can be diagonalized by a single unitary operator.

4. Complete set of commuting normal operators: A set of mutually commuting normal operators $N_k, k = 1, 2, \ldots$ such that their set of simultaneous eigenvectors $\{u_1, u_2, \ldots\}$ is an orthonormal basis of $\mathcal{H}$.

10.3.4 Orthonormalization methods

There are two important procedures for constructing an orthonormal basis of each finite-dimensional subspace of an inner product space, the Gram-Schmidt orthonormalization construction and the symmetric orthonormalization construction. Both make use of the *Gram* matrix.

Let $\mathbf{W}_n$ denote a set of n linearly independent vectors in an inner product space $\mathcal{V}$:

$$\mathbf{W}_n = \{w_1, w_2, \ldots, w_n \mid \text{each } w_i \in \mathcal{V}\}. \tag{10.39}$$

We arbitrarily arrange the vectors into an **ordered** sequence $w_1, w_2, \ldots, w_n$. The Gram matrix of order h of this ordered sequence of vectors is defined by

$$G_h(w) = \begin{pmatrix} (w_1, w_1) & (w_1, w_2) \cdots (w_1, w_h) \\ (w_2, w_1) & (w_2, w_2) \cdots (w_2, w_h) \\ & \vdots \\ (w_h, w_1) & (w_h, w_2) \cdots (w_h, w_h) \end{pmatrix}, \tag{10.40}$$

$$\text{each } h = 1, 2, \ldots, n.$$

Each Gram matrix $G_h(w)$ is a Hermitian positive definite matrix. The Hermitian property $G_h^\dagger(w) = G_h(w)$ is evident, since $(w_i, w_j) = (w_j, w_i)^*$. *Positive definite* means its eigenvalues are all real and positive. This result if proved by showing that if U is the unitary matrix that diagonalizes $G_h(w)$; that is, $U^\dagger G_h(w) U = D_h = \text{diag}(d_1^{(h)}, d_2^{(h)}, \ldots, d_h^{(h)})$, then

$d_j^{(h)} = (v_j, v_j)$, where $v_j = \sum_{i=1}^h u_{ij}^* w_i$. But then by the property of the inner product and the linear independence of the vectors $w_1, w_2, \ldots, w_h$, we have that each eigenvalue of $G_h(w)$ is real and positive.

The Gram-Schmidt construction gives the orthonormal set of vectors corresponding to the linearly independent sequence of vectors w_1, $w_2, \ldots, w_n$ to be the following:

$$u_j = \frac{1}{\|w_1\| \, \|w_2\| \cdots \|w_j\|} \times \det \begin{pmatrix} (w_1, w_1) & (w_1, w_2) & \cdots & (w_1, w_j) \\ (w_2, w_1) & (w_2, w_2) & \cdots & (w_2, w_j) \\ & & \vdots & \\ (w_{j-1}, w_1) & (w_{j-1}, w_2) & \cdots & (w_{j-1}, w_j) \\ w_1 & w_2 & \cdots & w_j \end{pmatrix}, \quad (10.41)$$

$$j = 1, 2, \ldots, n,$$

where $u_1 = w_1/\|w_1\|$. The determinant is defined by its expansion by row j in terms of cofactors, so that we have the relation:

$$u_j = \frac{\sum_{i=1}^j (-1)^{j-i} \det G_j^{(i)}(w))\, w_i}{\|w_1\| \, \|w_2\| \cdots \|w_j\|}, \quad (10.42)$$

where $G_j^{(i)}(w)$ is the submatrix of order $j-1$ of $G_j(w)$ obtained by striking row j and column i. Since (w_i, u_j) is given by a determinant of order j in which two rows are equal for $i = 1, 2, \cdots, j-1$, the vector $u_j \perp w_i$, each $i = 2, 3, \ldots, j-1$, hence, $u_1, u_2, \ldots, u_j$ are mutually perpendicular, this being true for each $j = 2, 3, \ldots, n$. The normalization may also be verified to be unity. Thus, the set of vectors $u_1, u_2, \ldots, u_n$ is an orthonormal set.

We point out that the Gram-Schmidt procedure works for any set of vectors $w_1, w_2, \ldots, w_n$, linearly independent or not, if we proceed by calculating, in turn, the unnormalized vectors $w'_1, w'_2, \ldots, w'_i$, corresponding to the numerators of relation (10.42), reject the first vector w'_{i+1} that gives a zero numerator, and then continue the process with $w'_1, w'_2, \ldots, w'_i, w'_{i+2}$, etc., always rejecting a vector with zero numerator. This process is continued until we have moved across the ordered set $w_1, w_2, \ldots, w_n$ of vectors. We normalize at each step for which a rejection does not occur. Thus, when the Gram-Schmidt process is implemented with this *rejection rule,* we arrive at a set of $n-r$ orthonormal vectors, where r is the number of rejections, which is also the number of linearly dependent vectors in the original set, each one, in fact, being identified explicitly. Thus, this version of the Gram-Schmidt procedure can also be used to split an arbitrary set of given vectors into a set of orthonormal vectors and a set that is linearly dependent on those in the orthonormal set.

Symmetric orthonormalization uses the fact that there is a unique square-root matrix $G_n^{1/2}(w)$ of a Hermitian positive definite matrix $G_n(w)$, which is defined to be the Hermitian matrix $G_n^{1/2}(w) = U^\dagger D_n^{1/2} U$, where U is the same unitary matrix that diagonalizes $G_n(w)$; that is,

$$G_n(w) = U^\dagger D_n U,\ D_n^{1/2} = \mathrm{diag}\left(\sqrt{d_1^{(n)}}, \sqrt{d_2^{(n)}}, \ldots, \sqrt{d_n^{(n)}}\right). \qquad (10.43)$$

There are, of course, arbitrary phase factors $exp(i\phi_j)$ that may multiply each column u_j of $U = (u_1\ u_2\ \cdots\ u_n)$ in diagonalizing $G_n(w)$, but whatever these choices are, we use the same U in defining $G_n^{1/2}(w)$. The relations

$$\begin{aligned} &G_n^{1/2}(w) G_n^{1/2}(w) = G_n(w),\ G_n^{-1/2}(w) = (G_n^{1/2}(w))^{-1}, \\ &G_n^{-1/2}(w) G_n(w) = G_n^{1/2}(w) \end{aligned} \qquad (10.44)$$

are easily verified. The orthonormal set of vectors corresponding to the linearly independent sequence of vectors $w_1, w_2, \ldots, w_n$ is now defined to be the following:

$$u_j = \sum_{i=1}^{n} (G_n^{-1/2}(w))_{i\,j}\, w_i,\ j = 1, 2, \ldots, n. \qquad (10.45)$$

It is readily verified that

$$\begin{aligned} (u_j, u_k) &= \sum_{i,l=1}^{n} (G_n^{-1/2}(w))^*_{i\,j} (G_n^{-1/2}(w))_{l\,k}\, (w_i, w_l) \\ &= \sum_{l=1}^{n} (G_n^{-1/2}(w))_{l\,k} \sum_{i=1}^{n} (G_n^{-1/2}(w))_{j\,i} (G_n(w))_{i\,l} \\ &= \sum_{l=1}^{n} (G_n^{1/2}(w))_{jl} (G_n^{-1/2}(w))_{l\,k} = \delta_{j,k}\,. \end{aligned} \qquad (10.46)$$

If the n vectors w_i are taken as the columns of an $n \times n$ nonsingular matrix $W = (w_1\ w_2\ \cdots\ w_n)$, then the Gram-Schmidt and symmetric orthonormalization procedures are mappings of W to a unitary matrix:

1. Gram-Schmidt: $W \mapsto U_W = W\Delta_W$, where Δ_W is upper triangular. This relation is invertible, hence, also $W = U_W\Delta_W^{-1}$.

2. Symmetric: $W \mapsto U_W = WG_W^{-1/2}$. This relation is invertible, hence, also $W = U_W G_W^{1/2}$. The matrices $G_W, G_W^{1/2}, G_W^{-1/2}$ are Hermitian, positive definite.

The unitary matrices are, of course, different in these two mappings. These results assert that every complex nonsingular diagonable matrix W may be written as the product of a unitary matrix and an upper diagonal matrix, as well as the product of a unitary and a Hermitian matrix. This second decomposition is known as the polar decomposition of W, since every unitary matrix can also be written in the form $U = \exp(i\Phi)$, where Φ is a Hermitian matrix.

10.3.5 Matrix representations of linear operators

Let $\mathcal{H}_n$ be a finite-dimensional Hilbert space with inner product $(\ ,\ )$ and orthonormal basis $\mathcal{B}_n = \{v_1, v_2, \ldots, v_n\}$. Let T be a linear operator with action on the basis $\mathcal{B}_n$ given by

$$Tv_j = \sum_{i=1}^{n} t_{ij}v_i,\ t_{ij} = (v_i, Tv_j),\ j = 1, 2, \ldots, n. \tag{10.47}$$

The matrix

$$M_T = (t_{ij})_{1\leq i,j\leq n} = \begin{pmatrix} t_{11}\ t_{12}\ \ldots\ t_{1n} \\ t_{21}\ t_{22}\ \ldots\ t_{2n} \\ \vdots \\ t_{n1}\ t_{n2}\ \ldots\ t_{nn} \end{pmatrix} \tag{10.48}$$

is said to be a *matrix representation* of the operator T on the basis $\mathcal{B}_n$ of $\mathcal{H}_n$. Observe that the indexing of rows and columns is such that the product rule in Item 2 below is obeyed.

Important properties of matrix representations of operators on $\mathcal{H}_n$ are the following:

1. On any given orthonormal basis of $\mathcal{H}_n$, the linear properties of operators are inherited by their matrix representations.

2. If S and T are represented by the matrix representations M_S and M_T on the same orthonormal basis $\mathcal{B}_n$ of $\mathcal{H}_n$, then the product operator ST has the matrix representation $M_{ST} = M_S M_T$ on the basis $\mathcal{B}_n$ of $\mathcal{H}_n$. In particular, if S and T are unitarily similar operators, $S = U^\dagger T U$, then the matrix representations M_S and M_T on the basis $\mathcal{B}_n$ of $\mathcal{H}_n$ are related by the unitarily similar matrices $M_S = M_U^\dagger M_T M_U$, where M_U is the matrix representation of the linear operator U on the basis $\mathcal{B}_n$ of $\mathcal{H}_n$.

3. If $M_U = (u_{ij})_{1\leq i,j\leq n}$ is the matrix representation of a unitary operator U on the basis $\mathcal{B}_n$ of $\mathcal{H}_n$, and M_T is the matrix representation of T on the basis $\mathcal{B}_n$ of $\mathcal{H}_n$, then the matrix representation of T on the new orthonormal basis

$$\mathcal{B}_n^{'} = \{v_1', v_2', \ldots, v_n'\},\ v_j' = \sum_{i=1}^{n} u_{ij} v_i, \qquad (10.49)$$
$$j = 1, 2, \ldots, n,$$

of $\mathcal{H}_n$ is given by the unitarily similar matrix $M_T' = M_U^\dagger M_T M_U$.

10.4 Properties of Matrices

Some of the more important properties of finite complex matrices are summarized in this section. One of the unifying concepts in the theory of matrices is the general result given below on determining the eigenvectors of a class of matrices known as normal.

The matrix A is *diagonable* if it is similar to a diagonal matrix D, otherwise, it is *nondiagonable.*The class of diagonable matrices includes the class of all normal matrices, where a normal matrix N is any matrix that commutes with its Hermitian conjugate: $NN^\dagger = N^\dagger N$. The class of normal matrices includes the following types (Perlis [142, p.195]): Hermitian, skew-Hermitian, real symmetric, real skew-symmetric, unitary, orthogonal, diagonal, and all matrices unitarily similar to normal matrices. Indeed, a complex matrix is *unitarily similar* to a diagonal matrix if and only if it is normal. By definition, a diagonable matrix A can be brought to diagonal form $A = S^{-1}DS$, but it need not be normal. It is normal if and only if its eigenvectors are linearly independent, which is always the case if its eigenvectors are distinct.

Let Z denote an $n \times n$ complex matrix. The determinant $\det(\lambda I - Z)$ is a polynomial of degree n in the complex parameter λ in which the coefficient of $\lambda^{n-k}, 0 \leq k \leq n$, is a homogeneous polynomial of degree k

in the n^2 elements z_{ij} of Z :

$$\det(\lambda I - Z) = \sum_{k=0}^{n} (-1)^k \lambda^{n-k} T_k(Z), \tag{10.50}$$

where the $T_k(Z)$ are the homogeneous polynomials of degree k in the z_{ij} defined as follows (Waring's formulas):

$$T_k(Z) = \sum_{a_1+2a_2+\cdots+ka_k=k} A_k(a)(\mathrm{tr}Z)^{a_1}(\mathrm{tr}Z^2)^{a_2}\cdots \mathrm{tr}(Z^k)^{a_k}, \tag{10.51}$$

$$A_k(a) = \frac{(-1)^{\sum a_{even}}}{1^{a_1}a_1!2^{a_2}a_2!\cdots k^{a_k}a_k!}, \quad k = 0, 1, \ldots , \tag{10.52}$$

$$T_0(Z) = 1, T_1(Z) = \mathrm{tr}Z,$$

$$T_2(Z) = \left((\mathrm{tr}Z)^2 - \mathrm{tr}(Z)^2\right)/2, \ \ldots, \ T_n(Z) = \det Z. \tag{10.53}$$

The algebraic equation of degree n given by

$$\det(\lambda I - Z) = \sum_{k=0}^{n} (-1)^k \lambda^{n-k} T_k(Z) = 0 \tag{10.54}$$

is called the *characteristic equation* of Z. The roots of the characteristic equation $\lambda_1, \lambda_2, \ldots, \lambda_n$; that is, the complex numbers satisfying

$$\prod_{k=1}^{n} (\lambda - \lambda_k) = 0, \tag{10.55}$$

are the eigenvalues of Z, some of which may be equal.

The Cayley-Hamilton theorem asserts that the matrix Z satisfies its own characteristic equation; that is,

$$\prod_{k=1}^{n} (\lambda_k I_n - Z) = \sum_{k=0}^{n} (-1)^k Z^{n-k} e_k(\lambda) = 0 \text{ (0 matrix)}, \tag{10.56}$$

where the $e_k(\lambda) = e_k(\lambda_1, \lambda_2, \ldots, \lambda_n)$ are the elementary symmetric functions in the roots (see Sect. 11.6, Compendium B).

The Cayley-Hamilton theorem for normal matrices Z depends only the distinct roots, say, $\lambda_1, \lambda_2, \ldots, \lambda_m$, of Z. Relation (10.56) is replaced by

$$\prod_{k=1}^{m}(\lambda_k I_n - Z) = \sum_{k=0}^{m}(-1)^k Z^{m-k} e_k(\lambda_1, \ldots, \lambda_m) = 0, 1 \leq m \leq n. \quad (10.57)$$

Every complex matrix with distinct eigenvalues is diagonable, but need not be normal. For example, the upper triangular matrix

$$\begin{pmatrix} a & b \\ 0 & b \end{pmatrix}, \; a \neq b, \quad (10.58)$$

is diagonable and nonnormal. The upper triangular Jordan matrix

$$\begin{pmatrix} a & b \\ 0 & a \end{pmatrix}, \; b \neq 0 \quad (10.59)$$

is nondiagonable and nonnormal. For general n, every nondiagonable matrix is similar to a Jordan matrix (see Perlis [142]).

10.4.1 Properties of normal matrices

Let A be a normal matrix of order n with distinct eigenvalues $\lambda_1, \lambda_2, \ldots, \lambda_k$ with eigenvalue λ_i repeated m_i times, $i = 1, 2, \ldots, k$, with $m_1 + m_2 + \cdots + m_k = n$. The principal idempotents E_i of A are defined by

$$E_i = \prod_{j \neq i}^{k} \frac{A - \lambda_j I_n}{\lambda_i - \lambda_j}, \; i = 1, 2, \ldots, k. \quad (10.60)$$

These principal idempotents of A have the following properties:

1. Hermitian: $E_i = E_i^{\dagger}$, $i = 1, 2, \ldots, k$.
2. Orthogonal projections: $E_i E_j = \delta_{i,j} E_i$.
3. Trace: $\mathrm{tr} E_i = m_i =$ rank of $E_i =$ the number of linearly independent columns (rows) of E_i, $i = 1, 2, \ldots, k$.
4. Resolution of the identity: $I_n = E_1 + E_2 + \cdots + E_k$.
5. Complete: $A = \lambda_1 E_1 + \lambda_2 E_2 + \cdots + \lambda_k E_k$.

The following relations are consequences of these basic properties:

(i). Eigenvectors of A :

$$
\begin{aligned}
&AE_i = E_iA = \lambda_i E_i, \ i = 1, 2, \ldots, k, \\
&Au_j(E_i) = \lambda_j u_j(E_i), \ \text{each column } u_j(E_i) \text{ of } E_i, \\
&j = 1, 2, \ldots, k, \ \text{where the columns are orthogonal:} \\
&u_j(E_i) \perp u_{j'}(E_{i'}), j \neq j', i \neq i'.
\end{aligned}
\tag{10.61}
$$

(ii). A unitary matrix that diagonalizes a normal matrix is constructed as follows: First apply the Gram-Schmidt process, using the rejection algorithm, to the columns of E_1, thereby constructing m_1 orthonormal columns $u_1, \ldots, u_{m_1}$; repeat the procedure for E_2, thereby constructing m_2 orthonormal columns $u_{m_1+1}, \ldots, u_{m_1+m_2}$, each of which is necessarily $\perp$ to the first m_1 columns; $\ldots$; repeat the procedure for E_k, thereby constructing m_k orthonormal columns $u_{m_1+\cdots+m_{k-1}+1}, \ldots, u_{m_1+\cdots+m_k}$, each of which is necessarily $\perp$ to all previously constructed columns. The unitary matrix U defined by $U = (u_1 \ u_2 \ \cdots \ u_n)$ then diagonalizes A.

(iii). Functions of A : If $F(x)$ is any well-defined function of a single variable x, then

$$F(A) = \sum_{i=1}^{k} F(\lambda_i) E_i. \tag{10.62}$$

Special applications of this result are: A matrix U is unitary if and only if there exists a Hermitian matrix H such that $U = \exp(iH)$; a matrix R is real orthogonal if and only if there exists a real skew-symmetric matrix A such that $R = \exp A$.

The idempotent matrices of a normal matrix are very powerful for dealing with many matrix problems that arise in physical systems. The book by Perlis [142] offers a concise and very readable account of their properties.

10.4.2 Inner product on the space of complex matrices

The ring of all complex $m \times n$ matrices can be made into an inner product space by presenting each complex matrix A as a sequence of length mn made up of its rows:

$$A \to (a_{11}, a_{12}, \ldots, a_{1n}; a_{21}, a_{22}, \ldots, a_{2n}; \ldots, a_{m1}, a_{m2}, \ldots, a_{mn}), \tag{10.63}$$

and then using definition (10.33) for the inner product of two matrices A and B, which gives

$$(A,B) = \sum_{i=1}^{m} \sum_{j=1}^{n} a_{ij}^* \, b_{ij} = \text{ trace} \left(A^\dagger B \right). \tag{10.64}$$

Many proofs of results in standard matrix theory can be simplified by introducing such an inner product, but it is not the custom to do so.

10.4.3 Exponentiated matrices

The matrix $Z(t)$ of order n presented as the formal power series

$$Z(t) = e^{tX} = I_n + tX + \frac{t^2}{2!} X^2 + \cdots, \tag{10.65}$$

$$X \in M_{n \times n}(\mathbb{C}), t \in \mathbb{R},$$

is called an *exponentiated matrix* or simply an exponential matrix. It is also called a *matrix curve* through the identity $Z(0) = I_n$ with *infinitesimal element* X defined by

$$\left. \frac{dZ(t)}{dt} \right|_{t=0} = X. \tag{10.66}$$

The matrix $Z(-t)$ is presented as $Z(-t) = e^{-tX}$; it is called the formal inverse of $Z(t)$. The term *formal* means questions of convergence are put aside. The Cayley-Hamilton relation for the matrix X is

$$X^n = \sum_{k=1}^{n} (-1)^{k-1} e_k(x) X^{n-k}, \tag{10.67}$$

in which $e_k(x)$ denotes the elementary symmetric functions in the characteristic roots $x = (x_1, x_2, \ldots, x_n)$ of X; it can be used to determine first X^{n+1}, then X^{n+2}, ..., as linear combinations of $I_n, X, \ldots, X^{n-1}$. Thus, the infinite sum (10.65) can be formally rearranged as a finite sum of the form:

$$Z(t) = e^{tX} = \sum_{k=1}^{n} F_k(x,t) X^{n-k}, \tag{10.68}$$

for some formal functions $F_k(x,t)$ of the roots $x = (x_1, x_2, \ldots, x_n)$ and the parameter t. If the matrix X is diagonable, it is always assured that the functions $F_k(x,t)$ are analytic, since then there exists a nonsingular matrix S such that

$$S^{-1}XS = W = \text{diag}(w_1, w_2, \ldots, w_n). \tag{10.69}$$

The exponentiated matrix $Z(t)$ is always defined, as given by

$$Z(t) = S^{-1}\text{diag}(e^{w_1}, e^{w_2}, \ldots, e^{w_n})S. \tag{10.70}$$

An arbitrary matrix $X \in M_{n\times n}(\mathbb{C})$ with elements $(x_{ij})_{1\le i,j\le n}$ can be written as

$$X = \sum_{k=1}^{n} x_{ij}e_{ij}, \tag{10.71}$$

where e_{ij} denotes the *matrix unit* which has for entries a 1 in row i and column j, and 0 elsewhere. The set $E = \{e_{ij} \mid i,j = 1,2,\ldots,n\}$ of matrix units is thus a basis for the set $M_{n\times n}(\mathbb{C})$ of all complex matrices, and, in particular, of the set $GL(n,\mathbb{C})$ of all nonsingular complex matrices. The matrix units satisfy the multiplication rule

$$e_{ij}e_{kl} = \delta_{j,k}e_{il}, \; i,j,k,l = 1,2,\ldots,n. \tag{10.72}$$

This relation implies the following two useful examples of exponentiated matrix curves:

$$\begin{aligned} Z_{ij}(\alpha) &= e^{\alpha\, e_{ij}} = I_n + \alpha e_{ij}, \; i \neq j; \\ Z_{ii}(\alpha) &= e^{\alpha\, e_{ii}} = I_n + (e^{\alpha} - 1)e_{ii}, \; \alpha \in \mathbb{C}. \end{aligned} \tag{10.73}$$

A principal result for general matrices is the following:

For X an arbitrary complex matrix, the matrix function e^{tX} is defined for all $t \in \mathbb{R}$ by its power series

$$e^{tX} = \sum_{k\ge 0} t^k X^k/k!. \tag{10.74}$$

Given an arbitrary $Z = GL(n,\mathbb{C})$, there exists a unique $X \in M_{n\times n}(\mathbb{C})$ such that

$$Z(t) = e^{tX} \; \textit{and} \; \frac{dZ(t}{dt}\bigg|_{t=0} = X \tag{10.75}$$

for each t in a neighborhood of the origin at $t = 0$.

Thus, in general, the expression $Z(t) = e^{tX}, t \in \mathbb{R}$ has a well-defined meaning as a formal series for each complex matrix X, whose convergence properties can be determined. But, given a nonsingular complex matrix Z of order n, the determination of a complex matrix X such that $Z = e^X$, is not so simple, even as a formal series. We discuss this further below in a formal context, but will not concern ourselves with problems of uniqueness and convergence, since this is not central to the goals of this monograph.

The procedure for determining the relationship between $Z(t) = e^{tX}$ and X for diagonable matrices $Z = Z(t)$ and X is straightforward. In the diagonable case, we have $X = SWS^{-1}$, where S is the nonsingular matrix that diagonalizes Z, that is, $S^{-1}ZS = \operatorname{diag}(z_1, z_2, \ldots, z_n)$, and W is any diagonal matrix defined by $W = \operatorname{diag}(w_1, w_2, \ldots, w_n)$, where $\exp w_i = z_i$. (The branch of z_i to which w_i belongs must be chosen.)

Given a complex, invertible matrix Z of order n, a quite comprehensible method of constructing an X such that $Z = e^X$ is provided by the symmetric orthonormalization procedure described in Sect. 10.3.4, which gives

$$Z = UW,\ U \text{ is unitary},\ W = G_Z^{1/2} \text{ is Hermitian, positive definite.} \tag{10.76}$$

Both U and W are known explicitly from the symmetric orthogonalization construction. Since U and W are normal matrices, they can be diagonalized by unitary transformations. For U, there exists a unitary matrix U_1 such that

$$U_1^\dagger\, U\, U_1 = \operatorname{diag}(e^{i\alpha_1}, e^{i\alpha_2}, \ldots, e^{i\alpha_n}), \tag{10.77}$$

where the $e^{i\alpha_1}, e^{i\alpha_2}, \ldots, e^{i\alpha_n}$ are the complex numbers of unit modulus that are the eigenvalues of U, and the α_i can always be chosen as real numbers in the interval $0 \leq \alpha_i < 2\pi$. For W, there exists a unitary matrix U_2 such that

$$U_2^\dagger\, W\, U_2 = \operatorname{diag}(w_1, w_2, \ldots, w_n), \tag{10.78}$$

where the $w_1, w_2, \ldots, w_n$ are the real positive eigenvalues of the Hermitian, positive definite matrix W, so that $w_i = e^{\beta_i}, i = 1, 2, \ldots, n$, where β_i is real, since each w_i positive. Thus, we can always find a Hermitian matrix H and a real matrix K such that

$$\begin{aligned} U &= e^{iH},\ H = U_1 \operatorname{diag}(\alpha_1, \alpha_2, \ldots, \alpha_n)\, U_1^\dagger, \\ W &= e^{K},\ K = U_2 \operatorname{diag}(\beta_1, \beta_2, \ldots, \beta_n)\, U_2^\dagger. \end{aligned} \tag{10.79}$$

Combining these results, we find that an arbitrary complex nonsingular matrix Z can always be brought to the form of the product of two exponential matrices:

$$Z = e^{iH} e^{K}, \tag{10.80}$$

where the Hermitian matrix H and the Hermitian, positive definite, matrix K can always be determined from Z, although not uniquely, in general. If the eigenvalues of U and those of W are distinct, then conventions can set the values of the α_i and the β_i. But if some eigenvalues among the α_i or among the w_i are equal, then the structure of Z alone admits several solutions of the form (10.80). Nonethless:

Every complex nonsingular matrix of order n can be brought to the form (10.80). Moreover, we can always compute the exponential matrices by the principal idempotent relation (10.62), which for exponentials reads

$$e^{A} = \sum_{i=1}^{k} e^{a_i} E_i, \tag{10.81}$$

where the distinct roots of A are $a_1, a_2, \ldots, a_k$, and the idempotents $E_1, E_2, \ldots, E_k$ refer to those of the normal matrix $A = iH$ and $A = K$ in (10.80). No infinite series expansions (10.65) of the exponentials are required.

Example: It is useful to illustrate the construction (10.80) by a numerical example, which we choose to be a nondiagonable, nonnormal matrix of the form (10.59):

$$Z = \begin{pmatrix} i & 1 \\ 0 & i \end{pmatrix}. \tag{10.82}$$

The Gram matrix is the Hermitian matrix given by

$$G = \begin{pmatrix} 1 & -i \\ i & 2 \end{pmatrix}. \tag{10.83}$$

The eigenvalues of G are given by $\lambda_1 = (3+\sqrt{5})/2$ and $\lambda_2 = (3-\sqrt{5})/2$. The Hermitian idempotent matrices E_1 and E_2 are given by

$$E_1 = \frac{G - \lambda_2 I}{\lambda_1 - \lambda_2} = \frac{1}{\sqrt{5}} \begin{pmatrix} \frac{-1+\sqrt{5}}{2} & -i \\ i & \frac{1+\sqrt{5}}{2} \end{pmatrix}. \tag{10.84}$$

$$E_2 = \frac{G - \lambda_1 I}{\lambda_2 - \lambda_1} = \frac{1}{\sqrt{5}} \begin{pmatrix} \frac{1+\sqrt{5}}{2} & i \\ -i & \frac{-1+\sqrt{5}}{2} \end{pmatrix}, \tag{10.85}$$

These matrices then satisfy $E_i E_j = \delta_{i,j} E_i, E_1 + E_2 = I$, $\text{trace} E_1 =$

trace$E_2 = 1, G = \lambda_1 E_1 + \lambda_2 E_2$. We then calculate

$$G^{1/2} = \sqrt{\lambda_1}E_1 + \sqrt{\lambda_2}E_2 = \frac{1}{\sqrt{5}}\begin{pmatrix} 2 & -i \\ i & 3 \end{pmatrix}, \tag{10.86}$$

$$G^{-1/2} = \frac{1}{\sqrt{\lambda_1}}E_1 + \frac{1}{\sqrt{\lambda_2}}E_2 = \frac{1}{\sqrt{5}}\begin{pmatrix} 2 & i \\ -i & 1 \end{pmatrix}. \tag{10.87}$$

The calculation of the unitary matrix U from the matrix Z and $G^{-1/2}$ now gives

$$U = \frac{1}{\sqrt{5}}\begin{pmatrix} 2i & 1 \\ 1 & 2i \end{pmatrix}. \tag{10.88}$$

We thus obtain $Z = UG^{1/2}$, as verified by

$$\begin{pmatrix} i & 1 \\ 0 & i \end{pmatrix} = \frac{1}{\sqrt{5}}\begin{pmatrix} 2i & 1 \\ 1 & 2i \end{pmatrix} \times \frac{1}{\sqrt{5}}\begin{pmatrix} 2 & -i \\ i & 3 \end{pmatrix}. \tag{10.89}$$

We further have that

$$U = e^{iH},\ H = \begin{pmatrix} \frac{\pi}{2} & \alpha - \frac{\pi}{2} \\ \alpha - \frac{\pi}{2} & \frac{\pi}{2} \end{pmatrix}, \tag{10.90}$$

where α is the angle given by $\cos\alpha = 1/\sqrt{5}, \sin\alpha = 2/\sqrt{5}$. The eigenvalues of U are $e^{i\alpha}, -e^{-i\alpha}$, and the principal idempotents of H are

$$E_1 = \frac{1}{2}\begin{pmatrix} 1 & 1 \\ 1 & 1 \end{pmatrix},\ E_2 = \frac{1}{2}\begin{pmatrix} 1 & -1 \\ -1 & 1 \end{pmatrix}. \tag{10.91}$$

Thus, from the principal idempotent method, the relation $U = e^{iH}$ is verified from

$$U = e^{i\alpha}E_1 - e^{-i\alpha}E_2, \tag{10.92}$$

without the need of the expansion $e^{iH} = I + iH - \frac{1}{2!}H^2 + \cdots$. The calculation of the Hermitian positive definite matrix K in

$$G^{1/2} == \frac{1}{\sqrt{5}}\begin{pmatrix} 2 & -i \\ i & 3 \end{pmatrix} = e^K \tag{10.93}$$

proceeds similarly. The matrix K is given by

$$K = \frac{1}{\sqrt{5}}\begin{pmatrix} k_1\left(\frac{-1+\sqrt{5}}{2}\right) + k_2\left(\frac{1+\sqrt{5}}{2}\right) & -i(k_1 - k_2) \\ i(k_1 - k_2) & k_1\left(\frac{1+\sqrt{5}}{2}\right) + k_2\left(\frac{-1+\sqrt{5}}{2}\right) \end{pmatrix}, \tag{10.94}$$

where k_1 and k_2 are the real logarithms defined by

$$e^{k_1} = \frac{1+\sqrt{5}}{2}, \; e^{k_2} = \frac{-1+\sqrt{5}}{2}. \tag{10.95}$$

The verification of these relations is again most easily carried out by the principal idempotent method. The eigenvalues of K are the real positive numbers k_1 and k_2, from which one calculates, as follows:

$$E_1 = \frac{1}{\sqrt{5}} \begin{pmatrix} \frac{-1+\sqrt{5}}{2} & -i \\ i & \frac{1+\sqrt{5}}{2} \end{pmatrix}, \; E_2 = \frac{1}{\sqrt{5}} \begin{pmatrix} \frac{1+\sqrt{5}}{2} & i \\ -i & \frac{-1+\sqrt{5}}{2} \end{pmatrix}, \tag{10.96}$$

$$e^K = e^{k_1}E_1 + e^{k_2}E_2 = \frac{1}{\sqrt{5}} \begin{pmatrix} 2 & -i \\ i & 3 \end{pmatrix}. \tag{10.97}$$

□

If we wish to bring Z in relation (10.80) to the exponential form $Z = e^X$, we still must deal with the problem of expressing the product of two exponential matrices in terms of a single exponential matrix: $e^A e^B = e^X$. If A and B commute, we have, of course, $X = A + B$ as a solution. In general, we must have $X = A + B + C$, where C takes into account that A and B are noncommuting, and, in general, all three matrices A, B, C are noncommuting. The exponential matrix product problem $e^A e^B = e^X$ of determining X for general complex matrices A and B is quite difficult, although for normal matrices it is always a problem in which X never needs to be presented as an infinite sum of powers of the matrices A and B. Often special pairs of A and B matrices have properties for which the formal infinite sum terminates, or can be summed. Perhaps the most striking feature is that the matrix C is always a sum of multiple commutators of the matrices A and B. We treat the general problem only in a formal sense, without considering questions of convergence. This already leads to very interesting problems, finite and combinatorial in nature, that we present in the next subsection on Lie bracket polynomials.

The series method of solving the relation for $C = C(A, B)$ in the expression

$$e^A e^B = e^{C(A,B)} = I + Z(A, B) \tag{10.98}$$

is to expand into a formal infinite series and obtain

$$Z(A, B) = \sum_{s \geq 1} \frac{1}{s!} \sum_t \binom{s}{t} A^{s-t} B^t, \tag{10.99}$$

$$C = \ln(I + Z) = \sum_{m \geq 1} \frac{(-1)^{m-1}}{m} Z^m, \tag{10.100}$$

where, for notational economy, we sometimes write $C = C(A, B)$ and $Z = Z(A, B)$. Thus, we obtain the following formal infinite series expression for C :

$$C = \sum_{m \geq 1} \frac{(-1)^{m-1}}{m} \sum_{s_1 \geq 1, \ldots, s_m \geq 1} \frac{1}{s_1! \ldots s_m!}$$
$$\times \sum_{t_1, \ldots, t_m} \binom{s_1}{t_1} \cdots \binom{s_m}{t_m} A^{s_1 - t_1} B^{t_1} \ldots A^{s_m - t_m} B^{t_m}. \tag{10.101}$$

The method of solving for C is to write this matrix function as a formal sum of terms C_k, each of which is a polynomial homogeneous of total degree k, which we call the order of C_k; that is,

$$C(A, B) = \sum_{k \geq 1} C_k(A, B). \tag{10.102}$$

We obtain from (10.100)-(10.102) the following result for $k \geq 1$:

$$C_k(A, B) = \sum_{m=1}^{k} \frac{(-1)^{m-1}}{m} \sum_{\substack{s_1 \geq 1, \ldots, s_m \geq 1 \\ s_1 + \cdots + s_m = k}} H_{s_1}(A, B) \cdots H_{s_m}(A, B), \tag{10.103}$$

where the polynomial $H_q(A, B)$, which is homogeneous of degree q, is defined by

$$H_q(A, B) = \sum_{s=0}^{q} \frac{A^{q-s} B^s}{(q-s)! s!}, \; q = 0, 1, \ldots . \tag{10.104}$$

Thus, disregarding all questions of convergence, we obtain a fully explicit formula:

$$e^A e^B = e^{\sum_{k \geq 1} C_k(A, B)}. \tag{10.105}$$

Examples: For small k the matrices $C_k = C_k(A, B)$ can be worked out by hand:

$k = 1$:

$$C_1 = A + B. \tag{10.106}$$

$k = 2:$

$$C_2 = \frac{1}{2}(A^2 + 2AB + B^2) - \frac{1}{2}(A^2 + AB + BA + B^2) = \frac{1}{2}[A, B]. \quad (10.107)$$

$k = 3:$

$$\begin{aligned} C_3 &= \frac{1}{6}(A^3 + 3A^2B + 3AB^2 + B^3) \\ &- \frac{1}{4}(2A^3 + 3A^2B + 3AB^2 + 2ABA + 2B^3 + B^2A + BA^2 + 2BAB) \\ &+ \frac{1}{3}(A^3 + A^2B + AB^2 + ABA + B^3 + B^2A + BA^2 + BAB) \\ &= \frac{1}{12}(A^2B + AB^2 - 2ABA + B^2A + BA^2 - 2BAB) \\ &= \frac{1}{12}[A, [A, B]] + \frac{1}{12}[B, [B, A]]. \end{aligned} \quad (10.108)$$

$k = 4:$

$$C_4 = -\frac{1}{24}[A, [B, [A, B]]]. \quad (10.109)$$

□

A rather remarkable feature is revealed by these examples. In each case, the matrix $C_k(A, B)$, which is always a well-defined matrix of order n, is a sum of multiple commutator matrices of the form

$$\underbrace{[X_1, [X_2, \cdots, [X_{k-1},}_{k-1} X_k \underbrace{]] \cdots]}_{k-1}, \quad (10.110)$$

where each X_i is either an A or a B. This is a general result: *Each matrix $C_k(A, B), k \geq 2$, is a sum of such multiple commutators.* The proof of this property is a combinatorial problem in the theory of matrices of order n and properties of multiple commutators. The appropriate concept for addressing the general case $C_k(A, B)$ is that of a Lie bracket polynomial, which leads to important results for the theory of exponential matrices quite independent of their application to groups and Lie algebras.

10.4.4 Lie bracket polynomials

A *Lie bracket polynomial* in two noncommuting variables x and y is a sum of multiple commutators in x and y. More precisely, a Lie bracket polynomial is defined recursively as follows: Define $L_2(x, y) = \alpha_2[x, y]$, where $[x, y] = xy - yx = -[y, x]$ is the commutator of x and y, and α_2 is an arbitrary complex number. Then, the Lie bracket polynomial

$L_{k+1}(x, y)$ of degree $k+1$ is defined by

$$L_{k+1}(x,y) = \alpha_{k+1}[x, L_k(x,y)] + \beta_{k+1}[y, L_k(x,y)], k = 2, 3, \ldots, \quad (10.111)$$

where α_{k+1} and β_{k+1} are arbitrary complex numbers. We include the polynomial 0 as the Lie bracket polynomial of degree 0, and $L_1(x, y) = \alpha_1 x + \beta_1 y$ as the Lie bracket polynomial of degree 1. All other Lie bracket polynomials are true multiple commutators of x and y, and all Lie bracket polynomials are generated recursively, by degree, in this manner. We often have in mind that x and y are matrices, but this need not be the case, and we never impose the Cayley-Hamilton relation. The results obtained in this section apply to any such pair of noncommuting variables or indeterminates, where we consider only the vector space of polynomials in x and y over the real and complex numbers. Lie bracket polynomials are a subset of all polynomials in x and y and inherit the vector space properties of the parent vector space of polynomials. But, in addition, we have the basic properties of the commutator $[x, y] = xy - yx$ given by

(i). Distributive properties:

$$\begin{aligned} [xy, z] &= x[y, z] + [x, z]y, \\ [x, yz] &= y[x, z] + [x, y]z; \end{aligned} \quad (10.112)$$

(ii). Jacobi identity:

$$[x, [y, z] + [y, [z, x] + [z, [x, y] = 0, \quad (10.113)$$

where x, y, z are any three noncommuting indeterminates.

Consider an arbitrary polynomial $P(x, y)$ in the noncommuting variables or indeterminates x and y, with coefficients $a_{i_1,i_2,\ldots,i_n}$ that are real or complex numbers, of the form

$$P(x,y) = \sum_{i_1,i_2,\ldots,i_n=1}^{n} a_{i_1,i_2,\ldots,i_n} x_{i_1} x_{i_2} \cdots x_{i_n}, \quad (10.114)$$

where each $x_1, x_2, \ldots, x_n$ denotes either an x or a y, and where x and y commute with the coefficients. Then, we define a mapping of each such polynomial into a Lie bracket polynomial by the following rule:

$$P(x,y) \to \langle P(x,y) \rangle = \sum_{i_1,i_2,\ldots,i_n=1}^{n} a_{i_1,i_2,\ldots,i_n} \langle x_{i_1} x_{i_2} \cdots x_{i_n} \rangle, \quad (10.115)$$

where the notation $\langle x_{i_1}x_{i_2}\cdots x_{i_n}\rangle$ denotes the multiple commutator of x and y obtained from the monomial $x_{i_1}x_{i_2}\cdots x_{i_n}$ by the rule:

$$\begin{aligned} x_{i_1}x_{i_2}\cdots x_{i_n} &\rightarrow \langle x_{i_1}x_{i_2}\cdots x_{i_n}\rangle \\ &= \underbrace{[x_{i_1},[x_{i_2},\cdots,[x_{i_{n-1}}}_{n-1},x_{i_n}\underbrace{]]\cdots]}_{n-1}. \end{aligned} \tag{10.116}$$

By definition, the constant polynomial $\alpha_0 \in \mathbb{C}$ is mapped to the zero Lie bracket polynomial; that is, $\langle\alpha_0\rangle = 0$, and the degree 1 polynomial $\alpha_1\, x + \beta_1\, y$ is mapped to itself; that is $\langle\alpha_1\, x + \beta_1\, y\rangle = \alpha_1\, x + \beta_1\, y$.

Examples: We give four examples of the mapping rule (10.115)-(10.116):

1. The monomial x^2y^3x of order 6 is mapped to the Lie bracket polynomial given by $x^2\, y^3\, x \rightarrow [x,[x,[y,[y,[y,x]]]]]$. Thus, if the brackets and the commas are ignored in the bracket form, the Lie bracket polynomial agrees with the original monomial.

2. An interesting feature of the mapping from a polynomial $P(x,y)$ in x,y to the Lie bracket polynomial $\langle P(x,y)\rangle$ is illustrated by the identity
$$\langle(\alpha x+\beta y)^n\rangle = 0,\ n = 2,3,\ldots\,. \tag{10.117}$$
This relation holds because for each term $x_1\cdots x_{n-2}x_{n-1}x_n$ that occurs in the expansion of $(\alpha x+\beta y)^n$ there is a corresponding term $x_1\cdots x_{n-2}x_nx_{n-1}$, for each $n\geq 2$. This is illustrated for $n=2$ by $\langle(\alpha x+\beta y)^2\rangle = \langle\alpha^2x^2+\alpha\beta(xy+yx)+\beta^2y^2\rangle = \alpha^2\langle x^2\rangle+\alpha\beta(\langle xy\rangle + \langle yx\rangle)+\beta^2\langle y^2\rangle = 0$. But, by definition, we have $\langle\alpha x+\beta y\rangle = \alpha\langle x\rangle + \beta\langle y\rangle = \alpha x+\beta y$.

3. The rule for $x^kP(x,y)$ for an arbitrary polynomial $P(x,y)$ is the multiple commutator given by
$$\langle x^kP(x,y)\rangle = \underbrace{[x,[x,\cdots,[x,}_{k}P(x,y)\underbrace{]]\cdots]}_{k}, \tag{10.118}$$
which by definition is $\langle P(x,y)\rangle$ for $k=0$.

4. The identity $e^{b(x,y)}e^x = e^xe^y$, where $b(x,y) = e^xye^{-x}$, follows from
$$e^{b(x,y)} = \sum_{k\geq 0}(b(x,y))^k/k! = \sum_{k\geq 0}e^xy^ke^{-x}/k! = e^xe^ye^{-x}. \tag{10.119}$$
Thus, the Lie bracket polynomial $\langle e^xe^y\rangle$ satisfies the identity
$$\langle e^xe^y\rangle = \langle e^{b(x,y)}e^x\rangle. \qquad \square \tag{10.120}$$

We continue with the development of properties of Lie bracket polynomials. Key relations needed from the previous section are the following:

$$\begin{aligned} e^x e^y &= e^{z(x,y)} = 1 + Z(x,y), \\ 1 + Z(x,y) &= \sum_{q\geq 0} H_q(x,y),\ H_q(x,y) = \sum_{s=0}^{q} \frac{x^{q-s}y^s}{(q-s)!s!}, \quad (10.121) \\ z_k(x,y) &= \sum_{m=1}^{k} \frac{(-1)^{m-1}}{m} \sum_{\substack{s_1\geq 1,\ldots,s_m\geq 1 \\ s_1+\cdots+s_m=k}} H_{s_1}(x,y)\cdots H_{s_m}(x,y). \end{aligned}$$

To these relations, we adjoin the pair of Baker-Campbell-Hausdorff (BCH) relations, in which $P(x,y)$ is an arbitrary polynomial:

$$b(x,y) = e^x y e^{-x} = \sum_{k\geq 0} \frac{\langle x^k y\rangle}{k!}, \tag{10.122}$$

$$\begin{aligned} &e^x P(x,y) e^{-x} = P(e^x x e^{-x}, e^x y e^{-x}) = P(x, e^x y e^{-x}) \\ &= P(x, b(x,y)) = \sum_{k\geq 0} \frac{\langle x^k P(x,y)\rangle}{k!}. \end{aligned} \tag{10.123}$$

The first basic BCH identity is not difficult to prove, and various proofs may be found in many places. Its generalization to $P(x,y)$ is then a direct consequence of the basic one, as shown. These BCH identities can be used to prove a number of properties of the formal polynomials

$$z(x,y) = \sum_{k\geq 1} z_k(x,y),\ z_k(x,y) \text{ homogeneous of degree } k \tag{10.124}$$

that occur in the relation $e^x e^y = e^{z(x,y)}$. These include the following:

1. Functional relation: The functional relation

$$z(b(x,y),x) = z(x,y) \tag{10.125}$$

   is a consequence of $e^x e^y = e^{b(x,y)} e^x$.

2. Symmetry relations:

   (a) The symmetry relation

$$z(-y,-x) = -z(x,y) \tag{10.126}$$

is a consequence of $(e^x e^y)(e^{-y}e^{-x}) = 1$.

(b) The symmetry relation

$$z(x,y) + z(-x,-y) = \sum_{s\geq 1} \frac{(-1)^{s-1}\langle x^s z(x,y)\rangle}{s!}. \tag{10.127}$$

is a consequence of applying (10.123) to $z(y,x)$ and using the symmetry relation (10.126).

The symmetry relations (10.126) and (10.127) preserve the homogeneity of the polynomials $z_k(x,y)$ in relation (10.124). Thus, we also have the following symmetry relations for these polynomials:

$$\begin{aligned}
&z_k(-x,-y) = (-1)^k z_k(x,y),\ k \geq 1,\\
&z_k(y,x) = (-1)^{k-1} z_k(x,y),\ k \geq 1,\\
&(1+(-1)^k) z_k(x,y) = \sum_{s\geq 1}^{k-1} \frac{(-1)^{s-1}\langle x^s z_{k-s}(x,y)\rangle}{s!},\ k \geq 2.
\end{aligned} \tag{10.128}$$

The last of these relations is more than a symmetry relation, as we next discuss.

The last relation (10.128) gives the following results for $k = 2, 3, 4, 5$:

$$\begin{aligned}
z_2 &= \frac{1}{2}[x,y],\\
0 &= \frac{1}{2}[x,[x,y]\,]\,] - [x,z_2],\\
z_4 &= \frac{-1}{24}[x,[x,[x,y]\,]\,] + \frac{1}{2}[x,z_3],\\
0 &= \frac{-1}{24}[x,[x,[x,[x,y]\,]\,]\,] + \frac{1}{2}[x,[x,z_3]\,] - [x,z_4],
\end{aligned} \tag{10.129}$$

where we have used $z_1 = x + y$ and $z_2 = \frac{1}{2}[x,y]$ in obtaining the last two relations. These relations have the following features: The $k = 3$ relation is the commutator with x of the $k = 2$ relation, and the $k = 5$ relation is the commutator with x of the $k = 4$ relation. This is a general feature of the last relation (10.128): *The odd relation is the commutator with x of the even relation preceding it.* This is not a recurrence relation in the usual sense: Knowledge of $z_3(x,y)$ is needed to determine $z_4(x,y)$, then of $z_5(x,y)$ to obtain $z_6(x,y), \cdots$. For example, given that $z_3(x,y) = (\langle x^2 y\rangle + \langle y^2 x\rangle)/12$, we quickly derive $z_4(x,y) = -\langle xyxy\rangle/24$.

Despite the shortcomings of the direct results coming from (10.128), we can derive the following result from these relations:

Let $k \geq 4$ be even, and let all subscripts refer to the homogeneous degree of the polynomial they labeled. Assume that $z_1, z_2, \ldots, z_{k-2}, k \geq 4$, have been shown to be Lie bracket polynomials. Then, necessary and sufficient conditions that z_{k-1} and z_k are Lie bracket polynomials are the following:

(i). The polynomial z_k can be written in the form

$$z_k = \langle x u_{k-1} \rangle, \; u_{k-1} \text{ is a Lie bracket polynomial.} \tag{10.130}$$

(ii). The polynomial z_{k-1} is given by

$$z_{k-1} = 2u_{k-1} - \sum_{s=1}^{k-2} \frac{(-1)^s \langle x^s z_{k-s-1} \rangle}{(s+1)!}. \tag{10.131}$$

Proof. If z_k has the structure given by (i), then it is a Lie bracket polynomial. But then z_{k-1} given by relation (10.131) is just the result of solving the last relation (10.128) for z_{k-1}, where this solution is obtained by removal of x from each of the terms $\langle x u_{k-1} \rangle, \langle x^s z_{k-s} \rangle, s = 1, 2, \ldots, k-1$. This removal is a legitimate operation that leaves behind a Lie bracket polynomial for each term of the form $\langle x v_{k-1} \rangle$, where v_{k-1} is a Lie bracket polynomial, which is the case, by the assumption, for all terms except $\langle x z_{k-1} \rangle = [x, z_{k-1}]$. But if the homogeneous polynomial z_{k-1} contains no term of the form ax^{k-1}, then the factor x can also be removed from the term $[x, z_{k-1}]$, and relation (10.131) obtains. It is a general property of the polynomial $z_k(x, y)$ defined by relation (10.121) for both k even and k odd that it contains no term of the form ax^k in consequence of the multinomial identity (10.146) stated below. □

Example: The case $k = 4$ gives the following relations for (10.130)-(10.131), where the input data is $z_1 = x + y, z_2 = \langle xy \rangle/2$: The form (10.110) (also derived above) for z_4 gives $u_3 = \langle y^2 x \rangle/24$, and then (10.131) gives $z_3 = \langle y^2 x \rangle/12 + \langle xy \rangle/4 - \langle xy \rangle/6 = (\langle x^2 y \rangle + \langle y^2 x \rangle)/12$. □

The above results place the burden of proof that the homogeneous polynomials $z_k(x, y)$ for both k even and odd are Lie bracket polynomials can be given by showing that the even-order polynomials have the form $z_k(x, y) = \langle x u_{k-1} \rangle$, where u_{k-1} is a homogeneous Lie bracket polynomial of degree $k - 1$. Once this is proved, the odd-degree Lie bracket polynomials $z_{k-1}(x, y)$ are given by relation (10.131) from the explicit identification of u_{k-1}.

It is, of course, known from classical results in Lie algebra that the homogeneous polynomials $z_k(x, y)$ are Lie bracket polynomials. If we accept this fact, then relations (10.130)-(10.131) give a practical method for computing these polynomials. But the proof that a finite-order polynomial is a Lie bracket polynomial should be purely combinatorial in methodology. We continue in the next subsection with the presentation of several criteria for determining when a homogeneous polynomial of degree k is a Lie bracket polynomial.

Lie bracket polynomials and homogeneous polynomials

Let

$$\alpha = (\alpha_1, \alpha_1, \ldots, \alpha_m), 1 \leq m \leq k, \tag{10.132}$$

be a composition of $k - m$ into m parts, including 0 as a part. Then, each homogeneous monomial $H_\alpha(x, y)$ of degree k can be written as

$$h_\alpha(x, y) = x^{\alpha_1} y x^{\alpha_2} y \cdots x^{\alpha_m} y. \tag{10.133}$$

Hence, each complex (real) homogeneous polynomial $P_k(x, y)$ can be written in the form

$$P_k(x, y) = \sum_{m=1}^{k} \sum_{\alpha \vdash k-m} \left(a_\alpha h_\alpha(x, y) + b_\alpha h_\alpha(y, x)\right), \tag{10.134}$$

in which a_α and b_α are complex (real) coefficients. The number of terms in $P_k(x, y)$ is given by 2^k, as may be verified by summing the number of compositions of α into $k - m$ parts over m and multiplying by 2 (see relation (1.244), Chapter 1).

It is convenient for the description of the Lie bracket polynomials to use the following abbreviated notations:

$$h_\alpha = h_\alpha(x, y), \ \widetilde{h}_\alpha = h_\alpha(y, x). \tag{10.135}$$

$$P_k = \sum_{m=1}^{k} \sum_{\alpha \vdash k-m} \left(a_\alpha h_\alpha + b_\alpha \widetilde{h}_\alpha\right). \tag{10.136}$$

The Lie bracket polynomial of P_k is given by

$$\langle P_k \rangle = \sum_{m=1}^{k} \sum_{\alpha \vdash k-m} \left(a_\alpha \langle h_\alpha \rangle + b_\alpha \langle \widetilde{h}_\alpha \rangle\right), \tag{10.137}$$

$$\langle h_\alpha \rangle = \langle x^{\alpha_1} y x^{\alpha_2} y \cdots x^{\alpha_m} y \rangle \tag{10.138}$$
$$= \underbrace{[x,[x,\cdots,[x,}_{\alpha_1}[y,\underbrace{[x,[x,\cdots,[x,}_{\alpha_2}[y,\cdots,\underbrace{[x,[x,\cdots,[x,}_{\alpha_m} y \underbrace{]]\cdots]}_{k-1}.$$

The expression for $\langle \widetilde{h}_\alpha \rangle$ is obtained from this result by interchanging x and y. Each such bracket polynomial is of order k for each $m = 1, 2, \ldots, k$.

Examples of notation: At the two extremes of m in (10.138), namely, $m = 1$ and $m = k$, we have the relations:

$$\langle h_{(k-1)} \rangle = \langle x^{k-1} y \rangle,\ \langle h_{(0^k)} \rangle = \langle y^k \rangle = 0,\ k \geq 2. \tag{10.139}$$

For $k = 1, 2, 3, 4,$ we have the following Lie bracket polynomials (10.138)

(i). $k = 1 : \langle h_{(0)} \rangle = \langle y \rangle = y.$

(ii). $k = 2 : \langle h_{(0,0)} \rangle = \langle y^2 \rangle = 0,\ \langle h_{(1)} \rangle = \langle xy \rangle = [x, y].$

(iii). $k = 3 :$

$$\begin{aligned} &\langle h_{(0,0,0)} \rangle = \langle y^3 \rangle = 0,\ \langle h_{(1,0)} \rangle = \langle xy^2 \rangle = 0, \\ &\langle h_{(0,1)} \rangle = \langle yxy \rangle = [y,[x,y]\,],\ \langle h_{(2)} \rangle = \langle x^2 y \rangle = [x,[x,y]\,]. \end{aligned} \tag{10.140}$$

(iv). $k = 4 :$

$$\begin{aligned} &\langle h_{(0,0,0,0)} \rangle = \langle y^4 \rangle = 0,\ \langle h_{(1,0,0)} \rangle = \langle xy^3 \rangle = 0, \\ &\langle h_{(0,1,0)} \rangle = \langle yxy^2 \rangle = 0,\ \langle h_{(0,0,1)} \rangle = \langle y^2 xy \rangle = [y,[y,[x,y]\,]\,], \\ &\langle h_{(2,0)} \rangle = \langle x^2 y^2 \rangle = 0,\ \langle h_{(0,2)} \rangle = \langle y x^2 y \rangle = [y,[x,[x,y]\,]\,], \\ &\langle h_{(1,1)} \rangle = \langle xyxy \rangle = [x,[y,[x,y]\,]\,], \\ &\langle h_{(3)} \rangle = \langle x^3 y \rangle = [x,[x,[x,y]\,]\,]. \end{aligned}$$

The corresponding $\langle \widetilde{h}_\alpha \rangle$ are obtained by interchanging x and y. □

It is meaningful to effect the bracket mapping rule (10.115)-(10.116) on a Lie bracket polynomial simply by writing out the bracket polynomial as a sum of ordinary polynomials. For example, $\langle xy \rangle = xy - yx$, hence,

$$\langle \langle xy \rangle \rangle = \langle xy \rangle - \langle yx \rangle = 2\langle xy \rangle. \tag{10.141}$$

A basic property of the Lie bracket polynomials $\langle h_\alpha \rangle$ defined by (10.138), which are homogeneous of degree k, can be verified to be

$$\langle\, \langle h_\alpha \rangle\, \rangle = k\langle h_\alpha \rangle,\ \langle\, \langle \widetilde{h}_\alpha \rangle\, \rangle = k\langle \widetilde{h}_\alpha \rangle. \qquad (10.142)$$

Since every Lie bracket polynomial $P_k(x, y)$ homogeneous of degree k is a linear combination of the Lie bracket polynomials $\langle h_\alpha \rangle$ and $\langle \widetilde{h}_\alpha \rangle$, it follows that

$$\langle P_k(x, y) \rangle = kP_k(x, y). \qquad (10.143)$$

Conversely, if a polynomial $P_k(x, y)$ homogeneous of degree k satisfies this relation, then it is also true that $\langle\, \langle P_k(x, y) \rangle\, \rangle = k\langle P_k(x, y) \rangle$, so that the polynomial $\langle P_k(x, y) \rangle$ is a Lie bracket polynomial.

Relation (10.143) is necessary and sufficient for a homogeneous polynomial $P_k(x, y)$ of degree k to be a Lie bracket polynomial.

This result is proved in Mielnik and Plebański [134]. It is, of course, not true that every polynomial homogeneous of degree k satisfies relation (10.143). For example, any such polynomial containing the term x^k does not satisfy (10.143).

It is also useful to have conditions for determining when the commutator $[x, P_k(x, y)]$ of x with a homogeneous polynomial $P_k(x, y)$ of degree k is a Lie bracket polynomial. We find the following result:

Let $P_k(x, y)$ be a homogeneous polynomial of degree k. The two relations

$$\begin{aligned} \langle\, [x, P_k(x, y)]\, \rangle &= \frac{k+1}{k}[x, \langle P_k(x, y) \rangle], \\ [x, \langle P_k(x, y) \rangle] &= k[x, P_k(x, y)] \end{aligned} \qquad (10.144)$$

are necessary and sufficient that the commutator $[x, P_k(x, y)]$ is a Lie bracket polynomial.

Proof. If both these relations are true, then $\langle\, [x, P_k(x, y)]\, \rangle = (k+1)$ $[x, P_k(x, y)]$, hence, the commutator $[x, P_k(x, y)]$ is a Lie bracket polynomial. If the commutator $[x, P_k(x, y)]$ is a Lie bracket polynomial, then $\langle\, [x, P_k(x, y)]\, \rangle = (k+1)[x, P_k(x, y)]$, and the second relation is true in consequence of the first. □

Relations (10.143)-(10.144) underlie the proof of the method of construction of the $z_k(x, y)$ given by relations (10.130)-(10.131). Thus, from (10.143), the condition that the homogeneous polynomial $u_{k-1}(x, y)$ in the commutator relation $z_k = [x, u_{k-1}]$ (k even or odd) be a Lie bracket polynomial is that $u_{k-1} = \langle u_{k-1} \rangle/(k-1)$. If this is true, then from (10.144), the relation $\langle z_k \rangle = \langle [x, \langle u_{k-1}] \rangle = k\langle [x, \langle u_{k-1} \rangle] \rangle/(k-1) = k\langle [x, u_{k-1}] \rangle = kz_k$, and z_k is a Lie bracket polynomial.

It is useful to rewrite $z_k(x, y)$ given by (10.121) explicitly in terms of x and y for the application of the above methods:

$$z_k(x,y) = \sum_{m=1}^{k} \frac{(-1)^{m-1}}{m} \sum_{\substack{s_1 \geq 1, \ldots, s_m \geq 1 \\ s_1 + \cdots + s_m = k}} \tag{10.145}$$

$$\times \sum_{t_1=0}^{s_1} \cdots \sum_{t_m=0}^{s_m} \frac{x^{s_1-t_1} y^{t_1} x^{s_2-t_2} y^{t_2} \cdots x^{s_m-t_m} y^{t_m}}{(s_1-t_1)!t_1!(s_2-t_2)!t_2! \cdots (s_m-t_m)!t_m!}.$$

This polynomial has the property that it contains no terms of the form ax^k and ay^k in consequence of the symmetry in relation (10.128) and the following multinomial coefficient identity:

$$\sum_{m=1}^{k} \frac{(-1)^{m-1}}{m} \sum_{s_1 \geq 1, s_2 \geq 1, \ldots, s_m \geq 1} \binom{k}{s_1, s_2, \ldots, s_m} = 0,\ k \geq 2. \tag{10.146}$$

The Lie bracket polynomial of $z_k(x, y)$ is given by

$$\langle z_k(x,y) \rangle = \sum_{m=1}^{k-1} \frac{(-1)^{m-1}}{m} \sum_{\substack{s_1 \geq 1, \ldots, s_m \geq 1 \\ s_1 + \cdots + s_m = k}} \tag{10.147}$$

$$\times \sum_{t_1=0}^{s_1} \cdots \sum_{t_m=0}^{s_m} \frac{\langle\, x^{s_1-t_1} y^{t_1} x^{s_2-t_2} y^{t_2} \cdots x^{s_m-t_m} y^{t_m} \,\rangle}{(s_1-t_1)!t_1!(s_2-t_2)!t_2! \cdots (s_m-t_m)!t_m!}.$$

This form can be simplified greatly because each bracket polynomial under the summation is 0, unless it ends on the right with a single x or a single y; that is, $x^{s_m-t_m} y^t_m = x$ or y. But these relations are too complicated to give the general proof that $\langle z_k(x,y) \rangle = k z_k(x,y)$. For small values of k, they can be used to illustrate the validity of (10.144)-(10.147),as shown by the following examples:

Examples. For $k = 4$, it can be verified from (10.145) and (10.146) that $z_4(x,y) = (2yxyx - yyxx - 2xyxy + xxyy)/24$ and $\langle z_4(x,y) \rangle = 4z_4(x,y) = -\langle xyxy \rangle/6$. Similarly, the method in relations (10.144) uses

$$p_4 = [x, p_3], p_3 = 2yxy - yyx - xyy, \tag{10.148}$$

$$\langle p_3 \rangle = 2\langle yxy \rangle - \langle yyx \rangle - \langle xyy \rangle = 3\langle yxy \rangle = 3p_3. \qquad \square$$

The proof of the known fact that $z_k(x,y)$ is a Lie bracket polynomial, using the above results for Lie bracket polynomials and homogeneous polynomials requires further developments.

Remark. An interesting elementary application of exponential functions to solid state quantum theory is given by the identity

$$\begin{aligned} &\exp(\frac{2i\pi x}{a})\exp(aD_x) = \exp(aD_x)\exp(\frac{2i\pi x}{a}) \\ &= -\exp(\frac{2i\pi x}{a} + aD_x),\ D_x = \frac{d}{dx},\ a \in \mathbb{R}, \end{aligned} \tag{10.149}$$

which asserts that the periodic function $\exp(\frac{2i\pi x}{a})$ commutes with the displacement operator $\exp(aD_x)$ (see Zak [194]). This result proves wrong the often stated result that there exists no function $f(x)$ of the position operator x that commutes with a function $g(p_x)$ of the quantum mechanical linear momentum operator $p_x = -iD_x$. This is an example of two commuting Hermitian operators that can be simultaneously diagonalized, a result of some consequence for periodic physical systems.

10.5 Tensor Product Spaces

The idea underlying tensor product spaces is that of building new vector spaces out of given vector spaces. For physical systems, it is the mathematics that one uses to describe composite systems in terms of more elementary ones. What is surprising is that the linear properties of these mathematical objects allows for composite physical systems to have properties that the individual constituent systems, making up the whole, do not possesses. As is often the case in mathematics, given a collection of mathematical objects, one considers how to build new objects from ordered pairs of the original objects. Here the mathematical objects of interest are vector spaces, in particular, inner product spaces.

A vector space $\mathcal{W}$ is said to be the tensor product of two vector spaces $\mathcal{U}$ and $\mathcal{V}$, written

$$\mathcal{W} = \mathcal{U} \otimes \mathcal{V}, \tag{10.150}$$

if there exists a mapping Φ,

$$\Phi : (u, v) \mapsto w = u \otimes v \in \mathcal{W},\ \text{each}\ u \in \mathcal{U},\ \text{each}\ v \in \mathcal{V}, \tag{10.151}$$

of every ordered pair of vectors (u, v) to a vector $w = u \otimes v \in \mathcal{W}$ such that the following relations hold:

1. The mapping Φ is bilinear; that is, linear in the vectors of $\mathcal{U}$ and

in those of $\mathcal{V}$:

$$(\alpha u + \alpha' u') \otimes v = \alpha(u \otimes v) + \alpha'(u' \otimes v),$$
$$(u \otimes (\beta v + \beta' v') = \beta(u \otimes v) + \beta'(u \otimes v'), \tag{10.152}$$

for all scalars $\alpha, \alpha', \beta, \beta' \in \mathbb{C}$, all vectors $u, u' \in \mathcal{U}$, and all vectors $v, v' \in \mathcal{V}$.

2. Linear independence of the vectors $u, u' \in \mathcal{U}$ and of the vectors $v, v' \in \mathcal{V}$ implies the linear independence of $u \otimes v \in \mathcal{W}$ and $u' \otimes v' \in \mathcal{W}$.

One of the important properties of the tensor product $\mathcal{W} = \mathcal{U} \otimes \mathcal{V}$ of two vector spaces $\mathcal{U}$ and $\mathcal{V}$ is: There are vectors in $\mathcal{W}$ that cannot be obtained as the tensor product of a vector in $\mathcal{U}$ with a vector in $\mathcal{V}$. For example, the vector $\alpha(u \otimes v) + \alpha'(u' \otimes v')$ cannot, in general, be expressed as a tensor product $(au + a'u') \otimes (bv + b'v')$.

If S is an operator acting in $\mathcal{U}$ and T is an operator acting in $\mathcal{V}$, so that $S : u \mapsto Su \in \mathcal{U}$, each $u \in \mathcal{U}$, and $T : v \mapsto Tv \in \mathcal{V}$, each $v \in \mathcal{V}$, the operator $S \otimes T$ is defined to be the operator acting in $\mathcal{U} \otimes \mathcal{V}$ according to the rule: $(S \otimes T)(u \otimes v) = Su \otimes Tv$.

If $\mathcal{H}$ is a separable Hilbert space with inner product $(\ ,\ )_{\mathcal{H}}$ and $\mathcal{K}$ is a separable Hilbert spaces with inner products $(\ ,\ )_{\mathcal{K}}$, then $\mathcal{H} \otimes \mathcal{K}$ is a separable Hilbert space with inner product

$$(\ ,\ )_{\mathcal{H}\otimes\mathcal{K}} = (\ ,\ )_{\mathcal{H}}\ (\ ,\ )_{\mathcal{K}}. \tag{10.153}$$

This definition implies: If $\mathcal{B} = \{u_1, u_2, \ldots\}$ and $\mathcal{C} = \{v_1, v_2, \ldots\}$ are orthonormal bases of $\mathcal{H}$ and $\mathcal{K}$, then $\mathcal{B} \otimes \mathcal{C} = \{u_1 \otimes v_1, u_2 \otimes v_2, \ldots\}$ is an orthonormal basis of $\mathcal{H} \otimes \mathcal{K}$; that is,

$$(u_i \otimes v_j, u_k \otimes v_l)_{\mathcal{B}\otimes\mathcal{C}} = (u_i, u_k)_{\mathcal{H}}\ (v_j, v_l)_{\mathcal{K}} = \delta_{i,k}\delta_{j,l}. \tag{10.154}$$

Let M_S be the $m \times m$ matrix representation of an operator S on a finite-dimensional subspace $\mathcal{H}_m \subset \mathcal{H}$ with orthonormal basis $\mathcal{B}_m$; M_T the $n \times n$ matrix representation of an operator T on a finite-dimensional subspace $\mathcal{K}_n \subset \mathcal{K}$ with orthonormal basis $\mathcal{C}_n$; and $M_S \otimes M_T$ the $mn \times mn$ matrix representation of the operator $(S \otimes T)$ on the finite-dimensional subspace $\mathcal{H}_m \otimes \mathcal{K}_n \subset \mathcal{H} \otimes \mathcal{K}$ with orthonormal basis $\mathcal{B}_m \otimes \mathcal{C}_n$. The action of operators in the respective spaces is the following:

$$Su_k = \sum_{i=1}^{m} s_{ik} u_i,\ \ M_S = (s_{ik})_{1 \le i,k \le m}, \tag{10.155}$$

$$Tv_l = \sum_{j=1}^{n} t_{jl} v_j, \ M_T = (t_{jl})_{1 \le j,l \le n}, \tag{10.156}$$

$$\begin{aligned}(S \otimes T)(u_k \otimes v_l) &= \sum_{i=1}^{m} \sum_{j=1}^{n} s_{ik} t_{jl} (u_i \otimes v_j) \\ &= \sum_{i=1}^{m} \sum_{j=1}^{n} (M_S \otimes M_T)_{ij,kl} (u_i \otimes v_j),\end{aligned} \tag{10.157}$$

where the *Kronecker product matrix* $M_S \otimes M_T$ has its rows enumerated by the pair of row indices i, j of M_S and M_T and the pair of column indices k, l of M_S and M_T :

$$M_S \otimes M_T = \begin{pmatrix} s_{11} M_T & s_{12} M_T & \dots & s_{1m} M_T \\ s_{21} M_T & s_{22} M_T & \dots & s_{2m} M_T \\ & & \vdots & \\ s_{m1} M_T & s_{m2} M_T & \dots & s_{mm} M_T \end{pmatrix}. \tag{10.158}$$

The element $s_{ik} t_{jl}$ in the row indexed by the pair i, j and the column indexed by the pair k, l is actually in row $(i-1)m + j$ and column $(k-1)n + l$ of this matrix of order mn in the usual indexing $1, 2, \dots, mn$.

The Kronecker product $A \otimes B$ extends to rectangular matrices A and B in the obvious way. The Kronecker products of such rectangular matrices A, B, C, D obey the following rules, whenever the matrix multiplication makes sense:

1. Associativity: $(A \otimes B) \otimes C = A \otimes (B \otimes C)$,
2. Multiplication: $(A \otimes B)(C \otimes D) = (AB) \otimes (CD)$, (10.159)
3. Trace: $\mathrm{tr}(A \otimes B) = (\mathrm{tr}A)(\mathrm{tr}B)$ for square matrices.

10.6 Vector Spaces of Polynomials with an Inner Product

We adopt for polynomials the general set-theoretic viewpoint of functions and their sets of values. For a single indeterminate x, this is symbolized by writing $p : x \mapsto \sum_{k=0}^{n} a_k x^k = p(x)$. Thus, coefficients a_k are drawn sequentially from a ring $\mathcal{R}$, multiplied by the power x^k of the indeterminate x, and the summation performed, thus creating the expression

$p(x)$. Again, in the sense of functions, we refer to $p(x)$ as the "value of the function p" at the "point" x. The same procedure is followed for n indeterminates $x = (x_1, x_2, \ldots, x_n)$, where to keep the notation concise, we write

$$i = (i_1, i_2, \ldots, \imath_n), \ x^i = x_1^{i_1} x_2^{i_2} \cdots x_n^{i_n}. \tag{10.160}$$

We next define the inner or scalar product used for polynomials throughout this monograph, unless otherwise specified. The inner product (p, p') of two polynomials p and p' in n indeterminates over the field of complex numbers $\mathbb{C}$ (or $\mathbb{R}$) with values

$$p(x) = \sum_{i \geq 0} a_i x^i \text{ and } p'(x) = \sum_{i \geq 0} a'_i x^i \tag{10.161}$$

is the complex number defined by

$$(p, p') = \sum_{i \geq 0} i! a_i^* a'_i. \tag{10.162}$$

The notations are: $i! = i_1! i_2! \cdots i_n!; i \geq 0$ means $i_1 \geq 0, i_2 \geq 0, \ldots, i_n \geq 0$; and only a finite number of the coefficients a_i and a'_i are nonzero. The notation (p, p') for the inner product, in which the polynomial functions p and p' appear, emphasizes that the mapping is from pairs of polynomials in function space to the complex numbers.

While the proper language of functions and their values is that given above, it is often convenient to abuse this notation and write in place of (p, p') the symbol $(p(x), p'(x))$. This notation is incorrect in that the indeterminates x_i should not appear in the left-hand side of (10.162). Nonetheless, this abuse of notation often makes complicated relations more comprehensible than the insistence on the distinction between functions and their sets of values. We even go further and speak of the polynomial $p(x)$.

An alternative definition of the inner product $(p(x), p'(x))$ of two polynomials over the field of complex or real numbers may be given as follows. The symbol $\frac{\partial}{\partial x}$ denotes the sequence of partial derivatives

$$\frac{\partial}{\partial x} = \left(\frac{\partial}{\partial x_1}, \frac{\partial}{\partial x_2}, \ldots, \frac{\partial}{\partial x_n} \right). \tag{10.163}$$

In analogy to (10.160), we also write

$$\left(\frac{\partial}{\partial x} \right)^i = \left(\frac{\partial}{\partial x_1} \right)^{i_1} \left(\frac{\partial}{\partial x_2} \right)^{i_2} \cdots \left(\frac{\partial}{\partial x_n} \right)^{i_n} \tag{10.164}$$

for each sequence $i = (i_1, i_2, \ldots, i_n)$ of nonnegative integers. The complex conjugate polynomial p^* to p is, of course, defined to be the polynomial with values $p^*(x) = \sum_{i\geq 0} a_i^* x^i$, and, analogously, $p^*(\frac{\partial}{\partial x})$ is defined by

$$p^*\left(\frac{\partial}{\partial x}\right) = \sum_{i\geq 0} a_i^* \left(\frac{\partial}{\partial x}\right)^i. \tag{10.165}$$

The inner product of the two polynomials p and p' is now defined by

$$\begin{aligned}(p, p') &= (p(x), p'(x)) = \text{ constant term of } p^*\left(\frac{\partial}{\partial x}\right) p'(x) \\ &= p^*\left(\frac{\partial}{\partial x}\right) p'(x)\Big|_{x_1=x_2=\cdots=x_n=0}.\end{aligned} \tag{10.166}$$

Almost all the applications of inner product vector spaces of polynomials in this monograph take place in the following context. We are given a set of mn indeterminates $\{z_{ij} \mid i = 1, 2, \ldots, m; j = 1, 2, \ldots, n\}$, which we configure in an $m \times n$ matrix Z :

$$Z = (z_{ij})_{1\leq i\leq m, 1\leq j\leq n} = \begin{pmatrix} z_{11} & z_{12} & \cdots & z_{1n} \\ z_{21} & z_{22} & \cdots & z_{2n} \\ & & \vdots & \\ z_{m1} & z_{m2} & \cdots & z_{mn} \end{pmatrix}. \tag{10.167}$$

We most often deal with polynomials over commuting indeterminates z_{ij}, except as otherwise noted, with coefficients that are complex or real numbers. Such polynomials are denoted by $P(Z)$, where Z signifies that the polynomial P is defined over the **elements** of the matrix Z, and not over the set of all matrices Z. Thus, $P(Z)$ is given by an expression of the form

$$P(Z) = \sum_{A\geq 0} C(A) \prod_{i=1}^{m} \prod_{j=1}^{n} z_{ij}^{a_{ij}}, \tag{10.168}$$

where A is a matrix array of nonnegative integer exponents given by

$$A = (a_{ij})_{1\leq i\leq m, 1\leq j\leq n} = \begin{pmatrix} a_{11} & a_{12} & \cdots & a_{1n} \\ a_{21} & a_{22} & \cdots & a_{2n} \\ & & \vdots & \\ a_{m1} & a_{m2} & \cdots & a_{mn} \end{pmatrix}. \tag{10.169}$$

The coefficients $C(A)$ are functions of the nonnegative integer entries

a_{ij} in the matrix A, and not of the matrix; these coefficients are almost always complex or real numbers. The summation symbol $\sum_{A \geq 0}$ means the summation is to extend over all terms for which each exponent $a_{ij} \geq 0$. We finally bring (10.168) to the abbreviated form

$$P(Z) = \sum_{A \geq 0} C(A)\, Z^A \tag{10.170}$$

by defining, in analogy to (10.160), the symbol Z^A by

$$Z^A = \prod_{i=1}^{m} \prod_{j=1}^{n} z_{ij}^{a_{ij}}. \tag{10.171}$$

The inner product (10.166) takes the following form in terms of polynomials of the form (10.170):

$$(P, P') = \big(P(Z), P'(Z)\big) = P^* \left(\frac{\partial}{\partial Z} \right) P'(Z) \Big|_{Z=0}. \tag{10.172}$$

We leave open the specific nature of the indeterminates z_{ij} that occur as elements of the matrix Z, except that they be commuting. We write $z_{ij} \in \mathcal{Z}$ to specify that the "values" of z_{ij} can range over some set $\mathcal{Z}$. We then define the set of matrices $\mathcal{Z}_{m \times n}$ by

$$\mathcal{Z}_{m \times n} = \{ Z \mid Z \text{ is } m \times n;\ \text{each element } z_{ij} \in \mathcal{Z} \}. \tag{10.173}$$

The entries in the matrix A are always nonnegative integers. We also define the set of all such matrices by

$$\mathbb{M}_{m \times n} = \{ A \mid A \text{ is } m \times n;\ \text{each element } a_{ij} \in \mathbb{N} \}. \tag{10.174}$$

We are also interested in the transformation properties of the polynomials (10.170). Such transformations occur at two levels: transformations of the set of indeterminates $\mathcal{Z}_{m \times n}$ into itself , and transformations of the vector space of polynomials into itself originating from the former.

The transformations of the set $\mathcal{Z}_{m \times n}$ into itself take the form of left transformations and right transformations:

1. Left transformations:

$$L_X : Z \mapsto XZ, \text{ each } Z \in \mathcal{Z}_{m \times n}, \text{ each } X \in \mathcal{Z}_{m \times m}. \tag{10.175}$$

2. Right transformations:

$$R_Y : Z \mapsto ZY^T, \text{ each } Z \in \mathcal{Z}_{m\times n}, \text{ each } Y \in \mathcal{Z}_{n\times n}. \tag{10.176}$$

The product of two left transformations $L_{X'}L_X$ on the set $\mathcal{Z}_{m\times n}$ is defined by

$$\begin{aligned}(L_{X'}L_X)Z &= L_{X'}(L_X Z) = L_{X'}(XZ)\\ &= X'(XZ) = (X'X)Z = L_{X'X}Z.\end{aligned} \tag{10.177}$$

The product of two right transformations $R_{Y'}R_Y$ on the set $\mathcal{Z}_{m\times n}$ is defined by

$$\begin{aligned}(R_{Y'}R_Y)Z &= R_{Y'}(R_Y Z) = R_{Y'}(ZY^T)\\ &= (ZY^T)Y'^T = Z(Y'Y)^T = R_{Y'Y}Z.\end{aligned} \tag{10.178}$$

These product rules are to hold for all $X', X \in \mathcal{Z}_{m\times m}$; all $Y', Y \in \mathcal{Z}_{n\times n}$; and all $Z \in \mathcal{Z}_{m\times n}$.

Two significant properties of left and right transformations on the set $\mathcal{Z}_{m\times n}$ are the following:

1. Commutivity of the product of left and right transformations:

$$L_X R_Y = R_Y L_X \text{ on } \mathcal{Z}_{m\times n}, \text{each } X \in \mathcal{Z}_{m\times m}, \text{ each } Y \in \mathcal{Z}_{n\times n}. \tag{10.179}$$

This relation is true in consequence of the definition of the product action: $(L_X R_Y)Z = L_X(R_Y Z) = L_X(ZY^T) = XZY^T$; $(R_Y L_X)Z = R_Y(L_X Z) = R_Y(XZ) = XZY^T$.

2. Equivalence to Kronecker product action: The transformation $L_X R_Y : Z \mapsto XZY^T = Z'$ of the indeterminates $\{z_{ij}\}$ is identical to the transformation

$$L_X R_Y : \mathrm{col}(Z) \mapsto (X \otimes Y)\mathrm{col}(Z) = \mathrm{col}(Z'), \tag{10.180}$$

where col(Z) denotes the column matrix obtained by arranging the n rows of Z into a single column of length mn :

$$\begin{aligned}\mathrm{col}(Z) = \mathrm{col}(&z_{11}, z_{12}, \ldots, z_{1n}; z_{21}, z_{22}, \ldots, z_{2n};\\ &\ldots; z_{m1}, z_{m2}, \ldots, z_{mn}).\end{aligned} \tag{10.181}$$

Here the Kronecker product $X \otimes Y$ of two matrices of order m and

n emerges naturally as the matrix of order mn defined by

$$X \otimes Y = \begin{pmatrix} x_{11}Y & x_{12}Y & \ldots & x_{1m}Y \\ x_{21}Y & x_{22}Y & \ldots & x_{2m}Y \\ & & \vdots & \\ x_{m1}Y & x_{m2}Y & \ldots & x_{mm}Y \end{pmatrix}. \tag{10.182}$$

The column matrix presentation $\text{col}(Z') = (X \otimes Y)\text{col}(Z)$ and the $m \times n$ matrix presentation $Z' = XZY^T$ encode the same relationship between the indeterminates z_{ij} and z'_{ij}.

The left and right transformations of the inner product space $\mathcal{P}_{mn}$ of polynomials of the form (10.170) in the mn indeterminates z_{ij} that correspond to the left and right transformations L_X and R_Y of the indeterminates themselves are defined by the following actions in the space of polynomials and the space of indeterminates:

1. Left transformations. $\mathcal{L}_X P = P' \in \mathcal{P}_{mn}$:

$$\begin{aligned} &(\mathcal{L}_X P)(Z) = P'(Z) = P(L_{X^T} Z) = P(X^T Z), \\ &\text{each } X \in \mathcal{Z}_{m\times m};\ \text{each } P \in \mathcal{P}_{mn}. \end{aligned} \tag{10.183}$$

2. Right transformations. $\mathcal{R}_Y P = P' \in \mathcal{P}_{mn}$:

$$\begin{aligned} &(\mathcal{R}_Y P)(Z) = P'(Z) = P(R_{Y^T} Z) = P(ZY), \\ &\text{each } Y \in \mathcal{Z}_{n\times n};\ \text{each } P \in \mathcal{P}_{mn}. \end{aligned} \tag{10.184}$$

Again, one has the multiplication properties for the action of left and right transformations in the space $\mathcal{P}_{mn}$ of polynomials:

$$\mathcal{L}_{X'}\mathcal{L}_X = \mathcal{L}_{X'X},\ \mathcal{R}_{Y'}\mathcal{R}_Y = \mathcal{R}_{Y'Y},\ \mathcal{L}_X\mathcal{R}_Y = \mathcal{R}_Y\mathcal{L}_X, \tag{10.185}$$

for all $X, X' \in \mathcal{Z}_{m\times m}$; all $Y, Y' \in \mathcal{Z}_{n\times n}$. In the definitions of the left transformation actions $\mathcal{L}_X$ and the right transformation actions $\mathcal{R}_Y$ in (10.183)-(10.184), it is the transposed matrices X^T and Y that enter into the actions of the left and right transformations of the space $\mathcal{Z}_{m\times n}$ of indeterminates as given by (10.175)-(10.176). This feature is essential for preserving the defining properties (10.8)-(10.10) of operator actions in the separate spaces, polynomial (function) space and indeterminate (variable) space; it is a twist that cannot be avoided.

The vector space of polynomials $\mathcal{P}_{mn}$ is a separable Hilbert space; one in which we have specified a definite inner product (10.172).

10.7 Group Representations

The concept of a representation of a group can be presented in different contexts at various levels of abstraction. In this monograph, the representation of a group always occurs in the context of a group G acting linearly on the set of vectors belonging to a separable Hilbert space $\mathcal{H}$: For each vector $v \in \mathcal{H}$, there is defined a linear mapping $A_g : v \mapsto A_g(v) \in \mathcal{H}$, each $g \in G$. The Hilbert space $\mathcal{H}$ is sometimes said to be a *carrier space* of the group G, or a $G-$space. The Hilbert spaces $\mathcal{H}$ used in this monograph are always a direct sum of a collection of perpendicular subspaces, each of which is invariant under the action of G, and we restrict our considerations to this case. Thus, let $\mathbb{B}$ be an indexing set whose elements are used to label the invariant subspaces of $\mathcal{H}$ under the action of G. Then, the Hilbert space $\mathcal{H}$ may be expressed as the direct sum

$$\mathcal{H} = \sum_{b \in B} \oplus \mathcal{H}_b, \; \mathcal{H}_b \perp \mathcal{H}_{b'}, \quad \text{all pairs } b, b' \in \mathbb{B}. \tag{10.186}$$

The action of the group on each subspace $\mathcal{H}_b$ is to map this space into itself:

$$A_g : \mathcal{H}_b \mapsto \mathcal{H}_b, \text{ each } \mathcal{H}_b \subset \mathcal{H}. \tag{10.187}$$

The action of G on such an *invariant subspace* $\mathcal{H}_b \subset \mathcal{H}$ is implemented by introducing an orthonormal basis $\mathbf{B}$ of $\mathcal{H}_b$:

$$\mathbf{B} = \{| \, b_1\rangle, | \, b_2\rangle, \ldots, | \, b_{\dim \mathcal{H}_b}\rangle\}, \tag{10.188}$$

where the basis is expressed in terms of the Dirac ket notation, for which the orthonormality conditions are given by the bra-ket relations

$$\langle b_i \mid b_j \rangle = \delta_{ij}, \; i, j = 1, 2 \ldots, \dim \mathcal{H}_b. \tag{10.189}$$

Thus, the separable Hilbert space has an inner product, denoted $\langle \mid \rangle$, and the dimension of the subspace $\mathcal{H}_b$, which is invariant under the action of G, is denoted by $\dim\mathcal{H}_b$. We formulate the definitions and properties within this setting.

An *operator representation* of a group G is a set of *linear* $G-$operators A_G

$$A_G = \{A_g \mid g \in G\} \tag{10.190}$$

that acts invariantly in a separable Hilbert space $\mathcal{H}$. A *matrix representation* $M_G(\mathbf{B})$ of a group G corresponding to the action of G on an invariant subspace $\mathcal{H}_b \subset \mathcal{H}$ with basis $\mathbf{B}$ is given by

$$M_G(\mathbf{B}) = \{B(g) \mid g \in G\}, \tag{10.191}$$

where the matrix $B(g)$ is obtained from the action of A_g on the basis $\mathbb{B}$:

$$\begin{aligned} A_g \mid b_j\rangle &= \sum_{i=1}^{\dim\mathcal{H}_b} B_{ij}(g) \mid b_i\rangle,\ j = 1, 2, \ldots, \dim\mathcal{H}_b, \\ B_{ij}(g) &= \langle b_i \mid A_g \mid b_j\rangle. \end{aligned} \tag{10.192}$$

Thus, the correspondence

$$A_g \mapsto B(g),\ \text{each } g \in G \tag{10.193}$$

between $G-$operators and $G-$matrices gives the matrix representation of the action of A_g on the basis $\mathbf{B}$ of $\mathcal{H}_b \subset \mathcal{H}$: the set of matrices $M_G(\mathbf{B})$ is a matrix representation of G. In consequence of the multiplication properties of the $G-$operators, the matrices in the set $M_G(\mathbf{B})$ satisfy the same rules as those for the group G itself:

$$\begin{aligned} & B(g')B(g) = B(g'g), \\ & B(e) = I_{\dim\mathcal{H}_b},\ B(g^{-1}) = (B(g))^{-1}, \\ & B(g'')\left(B(g')B(g)\right) = \left(B(g'')B(g')\right)B(g), \end{aligned} \tag{10.194}$$

where these properties hold for all $g, g', g'' \in G$, where e is the identity element of the group, and $I_{\dim\mathcal{H}_b}$ is the identity matrix of order $\dim\mathcal{H}_b$.

Examples. We give two examples illustrating these concepts for finite groups and one example for a continuous group:

1. Let G be any finite group of order n with elements given by

$$G = \{g_1 = e, g_2, \ldots, g_n\}. \tag{10.195}$$

Select any set of n orthonormal vectors from $\mathbb{R}^n$, and label the orthonormal vectors in this basis, denoted $\mathbf{B}_n$, by the elements of the group itself:

$$\mathbf{B}_n = \{\mid g_1\rangle, \mid g_2\rangle, \ldots, \mid g_n\rangle\}, \langle g_i \mid g_j\rangle = \delta_{ij},\ i, j = 1, 2, \ldots, n. \tag{10.196}$$

For each $g_i \in G$, define the left and right actions, respectively, of G on $\mathbf{B}_n$ by

$$L_{g_i} \mid g_k\rangle = \mid g_i g_k\rangle,\ R_{g_i} \mid g_k\rangle = \mid g_i g_k^{-1}\rangle, k = 1, 2, \ldots, n. \tag{10.197}$$

The following properties of the left and right action on the basis

$\mathbf{B}_n$ are now verified directly from the definitions (10.197):

$$L_{g_i}L_{g_j} = L_{g_ig_j},\ R_{g_i}R_{g_j} = R_{g_ig_j},\ L_{g_i}R_{g_j} = R_{g_j}L_{g_i}, \qquad (10.198)$$

for all pairs $g_i, g_j \in G$. The matrices corresponding to these linear transformations of the basis $\mathbf{B}_n$ are given by

$$L_{g_i} \mapsto B_L(g_i),\ (B_L(g_i))_{hk} = \delta_{g_h, g_ig_k}, \qquad (10.199)$$

$$R_{g_i} \mapsto B_R(g_i), (B_R(g_i))_{hk} = \delta_{g_h, g_ig_k^{-1}}, \qquad (10.200)$$

where $1 \leq h, k \leq n$ denote the elements in row h and column k of the matrix $B_L(g_i)$ and $B_R(g_i)$.

These matrices satisfy the same multiplication rules (10.198) as the left and right actions L_{g_i} and R_{g_i} :

$$\begin{aligned} &B_L(g_i)B_L(g_j) = B_L(g_ig_j),\ B_R(g_i)B_R(g_j) = B_R(g_ig_j), \\ &B_L(g_i)B_R(g_j) = B_R(g_j)B_L(g_i), \end{aligned} \qquad (10.201)$$

for all pairs $g_i, g_j \in G$. The set of matrices $M_L(\mathbf{B}) = \{B_L(g_i) \mid g_i \in G\}$ is called a *left regular representation* of G, and the set of matrices $M_R(\mathbf{B}) = \{B_R(g_i) \mid g_i \in G\}$ is called a *right regular representation* of G. The fact that these two representations commute can be used to construct all inequivalent irreducible representations of any finite group (the definitions of the terms *inequivalent* and *irreducible* are given below). The left and right regular matrix representations are $0-1$ matrices; that is, matrices containing only 0 or 1. The distribution of 0 and 1 into elements of the matrices is intrinsic to the group structure itself, as determined by the Kronecker delta functions in (10.199)-(10.200).

2. As a second example, let S_n denote the group of permutations of the integers 1,2,. . . ,n, as described in Sect. 10.1. Let $\mathbf{E}_n$ denote the set of basis vectors of $\mathbb{R}^n$ given by the unit column vectors

$$\mathcal{E}_n = \{e_1, e_2, \ldots, e_n\}, \qquad (10.202)$$

where the $1 \times n$ column matrix e_i has 1 in row i and $0's$ in all other rows. The action of each permutation

$$\pi = \begin{pmatrix} 1 & 2 & \cdots & n \\ \pi_1 & \pi_2 & \cdots & \pi_n \end{pmatrix} \in S_n \qquad (10.203)$$

on the basis $\mathcal{E}_n$ is defined by

$$A_\pi e_j = e_{\pi_j} = \sum_{i=1}^{n} \delta_{i,\pi_j} e_i, \; j = 1, 2, \ldots, n. \tag{10.204}$$

This gives the matrix corresponding to the linear operator transformation A_π as

$$A_\pi \mapsto P_\pi, \; (P_\pi)_{ij} = \delta_{i,\pi_j}, \tag{10.205}$$

so that P_π is the matrix with unit column matrices given by

$$P_\pi = (e_{\pi_1} \; e_{\pi_2} \; \ldots \; e_{\pi_n}). \tag{10.206}$$

Thus, we obtain the representation of the symmetric group S_n by the set of $n \times n$ matrices given by

$$\mathbf{P}_{S_n} = \{P_\pi \mid \pi \in S_n\}. \tag{10.207}$$

The matrices in this set are called *permutation matrices.*The set of permutation matrices coincides with the subset of all $n \times n$ $0-1$ matrices, in which it is further specified that all row and column sums are equal to 1. Alternatively, the set of permutation matrices can be obtained as the set of matrices obtained by effecting all possible permutations of the columns of the unit matrix of order n. The above procedure could just as well have been carried out on unit row vectors to obtain still another method of obtaining the set of permutation matrices of order n.

3. Let Z_i, each $i = 1, 2, \ldots, k$, be an arbitrary nonsingular complex matrix of order n; that is, an element of the group $GL(n, \mathbb{C})$ of all nonsingular complex matrices of order n, where multiplication of group elements is matrix multiplication. Then, the Kronecker product

$$Z = Z_1 \otimes Z_2 \otimes \cdots \otimes Z_k, \text{ each } Z_i \in GL(n, \mathbb{C}) \tag{10.208}$$

is a matrix representation of the group $GL(n, \mathbb{C})$. $\square$

In examples 1 and 2 above, the positive integer n occurs in two contexts. In the first example, the group G is any group with n elements, and the left and right regular matrix representations are by matrices of size equal to the order of the group. In the second example, the symmetric (permutation) group S_n of order $|S_n| = n!$ is represented by the permutation matrices of order n. The second example can also be used to obtain a representation of any finite group G by $n \times n$ $0-1$ matrices by using the fact that *every finite group is a subgroup of a permutation group S_n for some positive integer n.*

Given an action of a group G on an orthonormal basis $\mathbf{B}$ of an invariant subspace $\mathcal{H}_b \subset \mathcal{H}$, one can always determine the action of G on an arbitrary basis $\mathbf{A}$, not necessarily orthonormal, of $\mathcal{H}_b$:

$$\mathbf{A} = \{| a_1\rangle, | a_2\rangle, \ldots, | a_n\rangle\}, \tag{10.209}$$

where $n = \dim\mathcal{H}_b$. Each basis vector $| a_j\rangle \in \mathbf{A}$ can be written as a linear combination

$$| a_j\rangle = \sum_{i=1}^{n} s_{ij} | b_i\rangle, \; s_{ij} \in \mathbb{C}, \tag{10.210}$$

where the transformation matrix $S = (s_{ij})_{1\leq i,j\leq n}$ is nonsingular. Since each operator A_g is linear, the action of G on the basis $\mathbf{A}$ is given by

$$\begin{aligned} A_g | a_j\rangle &= \sum_{i=1}^{n} s_{ij} A_g | b_i\rangle \\ &= \sum_{k=1}^{n} (S^{-1} M_g S)_{kj} | a_k\rangle, \; j = 1, 2, \ldots, n. \end{aligned} \tag{10.211}$$

Thus, the matrix representation $M_G(\mathbf{A})$ of G on the basis $\mathbf{A}$ is given by

$$M_G(\mathbf{A}) = S^{-1} M_G(\mathbf{B}) S = \{A(g) = S^{-1} B(g) S \mid g \in G\}. \tag{10.212}$$

The matrices $A(g) = S^{-1} M_g S$ satisfy, of course, all the rules required of a matrix representation.

If A and B are complex square matrices for which there exists a nonsingular matrix S such that $A = S^{-1}BS$, then A is said to be *similar* to B under S. Similarity is an equivalence relation on the set $\mathbb{M}_n$ of all $n \times n$ complex matrices; that is, if we write $A \sim B$ to mean that there exists an S such that $A = S^{-1}BS$, then $A \sim A$ (reflexive), $B \sim A$ (symmetric), and the two relations $A \sim B, B \sim C$ imply $A \sim C$ (transitive). Thus, the set $\mathbb{M}_n$ of all complex matrices of order n can be partitioned into disjoint equivalence classes under the similarity rule $\sim$.

A matrix $A \in \mathbb{M}_n$ is said to be *diagonable* if A is similar to a diagonal matrix. Since each matrix $A \in \mathbb{M}_n$ is either diagonable or nondiagonable, the set $\mathbb{M}_n$ is partitioned into two disjoint subsets, the subset of diagonable matrices and the subset of nondiagonable matrices. Diagonability is an equivalence relation. The set of nondiagonable matrices is nonempty for all $n \geq 2$. For example, for $n = 2$, each Jordan matrix of the upper diagonal form

$$H = \left\{ h(z) = \begin{pmatrix} 1 & z \\ 0 & 1 \end{pmatrix} \middle| z \in \mathbb{C} \right\} \tag{10.213}$$

satisfies the rules of an additive abelian group, as given by

$$h(0) = I, \; h(z')h(z) = h(z' + z), \; h(-z)h(z) = h(0), \tag{10.214}$$

and this group of matrices is nondiagonable. In general, each matrix $A \in \mathbb{M}_n$ is either similar to a diagonal matrix or similar to a Jordan matrix..

10.7.1 Irreducible representations of groups

Representations of groups can be constructed in many ways, as illustrated by the above examples. The direct construction of the class of representations called *irreducible* is not so immediate. In the context of this monograph, this concept is the following: Let G act invariantly on an orthonormal basis set $\mathbf{B}$ of $\mathcal{H}_b \subset \mathcal{H}$. The set $A_G = \{A_g \mid g \in G\}$ of $G-$operators is said to be *reducible* on the space $\mathcal{H}_b$ if there exists a nonsingular transformation S of the orthonormal basis $\mathbf{B}$ of the form (10.212) to a basis $\mathbf{A}$ such that the matrix representation $A(g)$ of G has the form

$$A(g) = S^{-1}B(g)S = \begin{pmatrix} A'(g) & C(g) \\ 0 & A''(g) \end{pmatrix}, \text{ for all } g \in G \;, \tag{10.215}$$

where $A'(g)$ and $A''(g)$ are square matrices, and S is the same for all $g \in G$. In this case, each of the sets of matrices $M'_G(\mathbf{A}) = \{A'(g) \mid g \in G\}$ and $M''_G(\mathbf{A}) = \{A''(g) \mid g \in G\}$ is a matrix representation of G, as may be shown directly from the fact that $M_G(\mathbf{A})$ is a matrix representation. If relation (10.215) holds, the carrier space $\mathcal{H}_b$ is said to be *reducible,* as also is the corresponding matrix representation $M_G(\mathbf{B})$. If no transformation S to a new basis $\mathbf{A}$ of $\mathcal{H}_b$ such that (10.215) holds for all $g \in G$, the carrier space $\mathcal{H}_b$ of G is said to be *irreducible,* as also is the corresponding matrix representation $M_G(\mathbf{B})$.

In this monograph, we will encounter only Hilbert spaces $\mathcal{H}$ such that each invariant subspace that carries a finite-dimensional representation of G under the action of G is either irreducible, or possesses a basis such that it may be expressed as a direct sum of two perpendicular subspaces, each of which is invariant under the action of G. Thus, the subspace $\mathcal{H}_b \subset \mathcal{H}$ of G in (10.215) must satisfy the following conditions:

$$\begin{aligned}
&\mathcal{H}_b = \mathcal{H}'_b \oplus \mathcal{H}''_b,\ \mathcal{H}'_b \perp \mathcal{H}''_b,\\
&A_G : \mathcal{H}'_b \mapsto \mathcal{H}'_b;\ A_G : \mathcal{H}''_b \mapsto \mathcal{H}''_b.
\end{aligned} \tag{10.216}$$

This property is alternatively expressed as follows: A nonsingular transformation of the basis of $\mathcal{H}_b$ exists such that the matrix representation $M_G(\mathbf{B})$ has one or the other of the following two properties:

(i). $M_G(\mathbf{B})$ is irreducible.

(ii). $M_G(\mathbf{B})$ is reducible, and there exists a transformation S to a basis $\mathbf{A}$ of $\mathcal{H}_b$ such that

$$\begin{aligned}
A(g) &= S^{-1}B(g)S = \begin{pmatrix} A'(g) & 0 \\ 0 & A''(g) \end{pmatrix}\\
&= A'(g) \oplus A''(g),\ \text{for all } g \in G\ .
\end{aligned} \tag{10.217}$$

This property is referred to as *complete reducibility,* in contrast to the form in (10.215).

In the case of complete reducibility described in Item (ii), it is also assumed that the subspaces, $\mathcal{H}'_b$ and $\mathcal{H}''_b$ are also completely reducible, etc., so that by continuing the reduction process, the subspace $\mathcal{H}_b$ can be written as a direct sum of perpendicular subspaces, each of which is invariant and irreducible under the action of G. We may summarize these results on reducibility and irreducibility, as implemented in this monograph, as follows: The separable Hilbert spaces $\mathcal{H}$ that occur are those for which an indexing set $\mathbb{J}$ exists such that for each $j \in \mathbb{J}$ the space $\mathcal{H}$ is a direct sum of perpendicular subspaces, each of which is finite, invariant, and irreducible under the action of G :

$$\mathcal{H} = \sum_{j \in \mathbb{J}} \oplus \mathcal{H}_j,\ \mathcal{H}_j \perp \mathcal{H}_{j'},\ \text{all pairs } j, j' \in \mathbb{J}. \tag{10.218}$$

Some of the invariant and irreducible subspaces in the direct sum may carry equivalent irreducible matrix representations of G. Matrix representations related by similarity transformations are said to be *equivalent,* otherwise, inequivalent.

A *unitary action* $\mathcal{U}_g = A_g$ of a group G on a Hilbert space $\mathcal{H}$ having the direct sum structure given by (10.218) is one in which each operator $\mathcal{U}_g, g \in G$ satisfies the inner product relation

$$\langle \psi, \mathcal{U}_{g^{-1}} \psi' \rangle = \langle \psi', \mathcal{U}_g \psi \rangle^*,\ \text{all pairs } \psi, \psi' \in \mathcal{H}, \tag{10.219}$$

where $\langle\ ,\ \rangle$ is the inner product on $\mathcal{H}$. This property is expressed by

$$\langle\psi \mid \mathcal{U}_{g^{-1}} \mid \psi'\rangle = \langle\psi' \mid \mathcal{U}_g \mid \psi\rangle^* \tag{10.220}$$

in terms of the Dirac bra-ket notation, which we use frequently.

A unitary action $\mathcal{U}_G = \{\mathcal{U}_g \mid g \in G\}$ of a group G on $\mathcal{H}$ is represented by a group

$$U_G = \{U(g) \mid g \in G, U(g) \text{ unitary}\} \tag{10.221}$$

of unitary matrices on any orthonormal basis of each of the invariant, irreducible subspaces $\mathcal{H}_j \subset \mathcal{H}, j \in \mathbb{J}$:

$$U^*_{j'j}(g) = \langle j' \mid \mathcal{U}_g \mid j\rangle^* = \langle j \mid \mathcal{U}_{g^{-1}} \mid j'\rangle = U_{jj'}(g^{-1}); \tag{10.222}$$

that is, the matrix $U(g)$ with rows and columns indexed, respectively, by the pairs $j, j' \in \mathbb{J}$ is unitary:

$$\begin{aligned} U^\dagger(g) &= (U^T(g))^* = U(g^{-1}) = (U(g))^{-1}; \\ U^\dagger(g)U(g) &= U(g)U^\dagger(g) = I. \end{aligned} \tag{10.223}$$

We will mostly be interested in unitary actions and the associated inequivalent, unitary, irreducible matrix representations of groups carried by a separable Hilbert space of the sort described in this section.

10.7.2 Schur's lemmas

In physical applications, a principal goal of representation theory is to find all inequivalent, unitary, irreducible matrix representations of a given group G, which is a symmetry group of the system. The irreducible representations of such a symmetry group is a principal method of classifying the quantum states of the system.

A criterion for determining whether a given matrix representation M_G of G is irreducible is given by

Schur's Lemma (matrix form): *If the only nonsingular matrix C that commutes with all matrices of a finite-dimensional matrix representation M_G of a group G is a multiple of the identity matrix, then M_G is irreducible; otherwise, it is reducible.*

To prove this lemma, let us first consider the following problem: Let $M_G = \{M_g \mid g \in G\}$ be an $n \times n$ matrix representation of a group G, and let C be an $n \times n$ nonsingular complex matrix that is not a multiple of the unit matrix and commutes with the representation M_G. The determination of the properties of the matrices in the set M_G is intrinsically a problem in matrix theory. It does not depend on the origin of the matrices, whether they come from a group action on a Hilbert space, or are just presented outright with no hint as to their origin.

The matrix C is either diagonable or it is nondiagonable. The class of diagonable matrices includes the class of all normal matrices, where a normal matrix N is any matrix that commutes with its Hermitian conjugate: $NN^\dagger = N^\dagger N$. The class of normal matrices includes the following types: Hermitian, skew-Hermitian, real symmetric, real skew-symmetric, unitary, orthogonal, diagonal, and all matrices unitarily similar to normal matrices. Indeed, a complex matrix is unitarily similar to a diagonal matrix if and only if it is normal.

Let us next prove the following result from which Schur's lemma is an immediate corollary:

Let $C \neq \lambda I_n$ be an $n \times n$ nonsingular matrix that commutes with an $n \times n$ matrix representation M_G of a group G. Then M_G is one of two types: (i) If the matrix C is diagonable, then M_G is completely reducible; (ii) if the matrix C is nondiagonable, then M_G is reducible, but not completely reducible.

Proof. (i). If the matrix $C \neq \lambda I$ is diagonable, then there exists a nonsingular matrix S such that $S^{-1}CS = D$, where D is diagonal. It then follows from $M_g C = C M_g$ that $(S^{-1} M_g S) D = D(S^{-1} M_g S)$. A diagonal matrix D has the direct sum form $D = c_1 I_1 \oplus c_2 I_2 \oplus \cdots \oplus c_k I_k$, where the $c_i, i = 1, 2, \ldots, k \geq 2$, are the distinct eigenvalues of C, the multiplicity of c_i being equal to $\dim I_i$, the I_i being unit matrices such that $I_1 \oplus I_2 \oplus \cdots \oplus I_k = I$. But then $M_g C = C M_g$ implies that $M_g = M_g^{(1)} \oplus M_g^{(2)} \oplus \cdots \oplus M_g^{(k)}$, where $\dim M_g^{(i)} = \dim I_i$. This relation holds for each $g \in G$, since C is independent of g. Thus, the representation M_G is completely reducible. A simpler proof may also be given for the case where the matrix representation is by unitary matrices $U(g)$. If $U(g)C = CU(g)$, all $g \in G$, then also $U(g)C^\dagger = C^\dagger U(g)$, all $g \in G$, so that $U(g)H = HU(g)$ and $U(g)K = KU(g)$, all $g \in G$, where $H = C + C^\dagger$ and $K = i(C - C^\dagger)$ are both Hermitian. One, or both, of H and K are Hermitian, and nonzero. But Hermitian matrices are diagonable, in fact, normal, hence, the unitary representation U_G is completely reducible by a unitary similarity transformation. (ii). If the matrix C is nondiagonable, it has at least one eigenvalue λ and one eigenvector v; that is, $Cv = \lambda v$. But then $CM_g = M_g C$, all $g \in G$ implies that $CM_g v = \lambda M_g v$, all $g \in G$. Thus, each vector in the set $Y_G = \{y_g = M_g v \mid g \in G\}$ is an eigenvector of C with eigenvalue λ. Moreover, the action of the matrix M_g on the set of vectors Y_G is invariant:

$$M_g : Y_G \to Y_G \ \text{ with } M_g y_{g'} = y_{gg'} \in Y_G. \tag{10.224}$$

Each of the vectors $y_g \in Y_G$ is also a column vector of length n; hence, at most n of them can be linearly independent. But the number must be less than n, since if equal to n, then C would be diagonable: The number k of linearly independent vectors in the set Y_G satisfies $1 \leq k < n$, since $y_{g_1} = y_{\text{identity}} = v \in Y_G$. Thus, there must exist a set $\mathbf{V}_k$ of linearly independent vectors given by $\mathbf{V}_k = \{y_{g_1} = v, y_{g_2}, \ldots, y_{g_k}\}$, for some $g_i \in G$, $i = 1, 2, \ldots, k$. The action of M_g on the set $\mathbf{V}_k$ is given by the linear transformations

$$M_g y_{g_j} = y_{gg_j} = \sum_{i=1}^{k} C_{ij}(g) y_{g_i}, j = 1, 2, \ldots, k. \tag{10.225}$$

The set $\mathbf{V}_k$ of vectors can be extended, in infinitely many ways, to a basis of the space of column vectors of length n by adjoining a set $\{y_1, y_2, \ldots, y_{n-k}\}$ of $n-k$ linearly independent vectors, each of which is linearly independent of the vectors in the set $\mathbf{V}_k$. Thus, the set of vectors

$$\{y_1, y_2, \ldots, y_{n-k}, y_{g_1}, y_{g_2}, \ldots, y_{g_k}\} \tag{10.226}$$

spans the space of column vectors of length n. The action of M_g on this basis set of vectors is given by linear transformations of the form

$$\begin{aligned} M_g y_j &= \sum_{i=1}^{n-k} S_{ij}(g) y_i + \sum_{i=1}^{k} R_{ij}(g) y_{g_i},\ j = 1, 2, \ldots, n-k, \\ M_g y_{g_j} &= y_{gg_j} = \sum_{i=1}^{k} C_{ij}(g) y_{g_i}, j = 1, 2, \ldots, k. \end{aligned} \tag{10.227}$$

Thus, the nonsingular matrix S

$$S = (y_1\ y_2\ \ldots\ y_{n-k}\ y_{g_1}\ y_{g_2}\ \cdots\ y_{g_k}) \tag{10.228}$$

with columns given by the column vectors in the set (10.226) has the property

$$S^{-1} M(g) S = \begin{pmatrix} S(g) & R(g) \\ 0 & C(g) \end{pmatrix}. \tag{10.229}$$

This relation is true whenever the matrix C commuting with the matrix representation M_G is nondiagonable. Since $1 \leq k < n$, the representation M_G is reducible, but not completely reducible. □

An example of a reducible group that is not completely reducible is the group H given by relation (10.213): Each element $h(z) \in H$ commutes with all elements of H; that is, $h(z)H = Hh(z)$, and $h(z)$ has eigenvalue 1 and only one linearly independent eigenvector col(1 0); hence, $h(z)$ is nondiagonable. This result generalizes to the abelian group of complex upper triangular matrices T given by

$$\begin{aligned} &T = \left\{ t(a,b) = \begin{pmatrix} a & b \\ 0 & a \end{pmatrix} \middle| a, b \in \mathbb{C}; a \neq 0 \right\}, \\ &t(1,0) = I, t(a',b')t(a,b) = t(a'a, a'b + b'a), \qquad (10.230) \\ &t(a^{-1}, -ba^{-2})t(a,b) = t(1,0). \end{aligned}$$

Each matrix in the subgroup H commutes with each matrix in the group T. The group T is reducible, but not completely reducible.

We have found the following references to be very useful for our synopsis of topics in this compendium:

1. Linear algebra and Hilbert space: Berberian [10], Fraleigh [56], Gelfand [59], Halmos [74, 75], Hungerford [81], Perlis [142].
2. Analysis: Boothby [29], Helgason [79], Simmons [162].
3. Lie algebra and Lie groups: Gelfand *et al.* [63], Fulton and Harris [57], Goodman and Wallach [67], Hamermesh [76], Humphreys [80], Jacobsen [84], Littlewood [105], Michel [132], Sattinger and Weaver [159], Sternberg [166], Weyl [177, 178], Wigner [181], Wybourne [189].
4. General: Roman [152], van der Waerden [172, 173].

Chapter 11

Compendium B. Basic Combinatorial Objects and Structures

This Compendium introduces objects from combinatorics that we need in order to relate the theory of the representations of the unitary group $U(n)$, and more generally the integer representations of general linear group $GL(n,\mathbb{C})$, as closely as possible to combinatorics. Much of the focus of this work is not on the groups themselves, but rather on the class of multivariable $D^{\lambda}-$polynomials that yield all inequivalent, irreducible representations of $U(n)$ (and the integer representations of $GL(n,\mathbb{C})$) when the indeterminates z_{ij} are suitably specialized. It is accordingly not unexpected that the present work encompasses many of the basic combinatorial tools on which present-day approaches to combinatorial theory are built. We pay special tribute to the books by Andrews [1], Kerber [88], Macdonald [126], Rota [154] (and his many lectures, notes, and review articles), and Stanley [163]) for making combinatorics accessible, and also to colleague and collaborator, Bill Chen, for help in the interpretation. These topics include the concepts of partitions, Young-Weyl tableaux, Gelfand-Tsetlin patterns, graph theory, expecially binary trees, digraphs, and cubic graphs, generating functions, the umbral calculus, and group actions. For example, it turns out that MacMahon's master theorem is a unifying combinatorial relation underlying the quantum theory of angular momentum of composite systems. Many books, in addition to those mentioned, are devoted to detailed and rigorous development of many of these topics. Of necessity, our treatment is very uneven: some topics are presented very briefly; others are given great discussion, depending on our needs. The properties of the Littlewood-Richardson numbers, which are so important for physics, falls among the latter topics. We list additional reading at the end of this Compendium.

11.1 Partitions and Tableaux

Partitions are sequences of positive integers $(\lambda_1, \lambda_2, \ldots, \lambda_n)$ satisfying the conditions

$$\lambda_1 \geq \lambda_2 \geq \ldots \geq \lambda_n > 0,\ n \in \mathbb{P}, \tag{11.1}$$

where the length $l(\lambda) = n$ of the sequence is any positive integer. The positive integer λ_i is called the $i-$th *part* of the partition. If $\sum_{i=1}^{n} \lambda_i = k$ and $l(\lambda) = n$, then λ is said to be a partition of k into n parts. This set of partitions is denoted by

$$\mathrm{Par}_n(k) = \{\lambda = (\lambda_1, \lambda_2, \ldots, \lambda_n) \mid \lambda \vdash k\},\ 1 \leq n \leq k, \tag{11.2}$$

where $\lambda \vdash k$ denotes that the parts of λ sum to k. The cardinality of the set $\mathrm{Par}_n(k)$ is denoted by

$$p_n(k) = |\mathrm{Par}_n(k)|. \tag{11.3}$$

By convention, we take $p_n(k) = 0$, for all $k < n$. The number $p_n(k)$ has the special values $p_1(k) = p_k(k) = 1, k \geq 1$, and $p_{k-1}(k) = 1, k \geq 2$, since the corresponding partitions are $(k), (1^k)$, and $(2, 1^{k-2}),\ k \geq 2$. By convention, The notation α^h for h a nonnegative integer means that the symbol α is to be written sequentially h times with $h = 0$ denoting no occurrence. For all other values of the pair (k, n), the number $p_n(k)$ can be calculated from the recurrence relation (Stanley [163], p. 28):

$$p_n(k) = p_{n-1}(k-1) + p_n(k-n), 2 \leq n \leq k, \tag{11.4}$$

where, by convention, $p_n(k-n) = 0$, for $k < 2n$. The set $\mathrm{Par}(k)$ of all partitions of k is then given by

$$\mathrm{Par}(k) = \bigcup_{n=1}^{k} \mathrm{Par}_n(k),\ k \geq 1. \tag{11.5}$$

For example, the set Par(4) is given by

$$\mathrm{Par}(4) = \{(4), (3,1), (2,2), (2,1,1), (1,1,1,1)\}. \tag{11.6}$$

The cardinality of the set $\mathrm{Par}(k)$ is given by

$$|\mathrm{Par}(4)| = \sum_{n=1}^{k} p_n(k). \tag{11.7}$$

The set of all nonempty partitions of arbitrary length is given by

$$\mathrm{Par} = \bigcup_{k \geq 1} \mathrm{Par}(k), \tag{11.8}$$

By definition, Par(0) is the empty partition, which we denote by $\lambda_\emptyset$, and could be included in (11.8). (By convention, the empty set is assigned cardinality 0.)

In the theory of partitions, the integer 0 is often not admitted as a part. *Our applications of partitions are most conveniently carried out by allowing zeros.*

A *partition with zero as part* is a partition λ to which a sequence of zeros $(0, 0, \ldots)$ has been adjoined at the right-hand end:

$$(\lambda, 0^{n-l}) = (\lambda_1, \lambda_2, \ldots, \lambda_l, \underbrace{0, 0, \ldots, 0}_{n-l}), \tag{11.9}$$

where $l = l(\lambda)$ denotes the number of nonzero parts. This situation corresponds to modifying definition (11.1) of a partition to the following: A partition with zero as a part is a sequence of nonnegative integers $(\lambda_1, \lambda_2, \ldots, \lambda_n)$ satisfying the conditions

$$\lambda_1 \geq \lambda_2 \geq \ldots \geq \lambda_n \geq 0, \; n \in \mathbb{P}. \tag{11.10}$$

We use notations analogous to those above for the following sets of partitions with zero as a part:

$$\mathbb{P}\mathrm{ar}_n(k) = \{\lambda = (\lambda_1, \ldots, \lambda_n) | \lambda_1 \geq \ldots \geq \lambda_n \geq 0, \lambda \vdash k\}, \tag{11.11}$$

$$\mathbb{P}\mathrm{ar}_n = \bigcup_{k \geq 0} \mathbb{P}\mathrm{ar}_n(k), \tag{11.12}$$

$$\mathbb{P}\mathrm{ar} = \bigcup_{n \geq 1} \mathbb{P}\mathrm{ar}_n. \tag{11.13}$$

The sets $\mathbb{P}\mathrm{ar}_n(k)$ and $\mathbb{P}\mathrm{ar}_n$ of partitions always has n parts, where we count each repeated zero as a part; $\mathbb{P}\mathrm{ar}_n(k)$ contains a finite number of elements, but $\mathbb{P}\mathrm{ar}_n$ is countably infinite, as is $\mathbb{P}\mathrm{ar}$. The set $\mathbb{P}\mathrm{ar}_n$ includes $\mathbb{P}\mathrm{ar}_n(0) = (0^n)$. In general, the number of zero parts is unspecified. For example, the set $\mathbb{P}\mathrm{ar}_3(5)$ is given by

$$\mathbb{P}\mathrm{ar}_3(5) = (5, 0, 0), (4, 1, 0), (3, 2, 0), (3, 1, 1), (2, 2, 1). \tag{11.14}$$

In order to avoid the awkward phrase "partitions with zero as a part," we refer to such sequences simply as "partitions," occasionally amplifying the description of the parts for clarity, but generally letting the notation $\mathbb{P}$ carry the burden that such sets of partitions can have 0 as a part.

The relation between the two kinds of partitions, those with no 0 parts and those with 0 parts is

$$\mathrm{Par}(k) = \hat{\mathbb{P}}\mathrm{ar}_k(k), \tag{11.15}$$

where $\hat{\mathbb{P}}\mathrm{ar}_k(k)$ is the set of partitions obtained from $\mathbb{P}\mathrm{ar}_k(k)$ by deleting all zeros from each partition in the set. Similarly, the set $\mathbb{P}\mathrm{ar}_n(k)$ is given by

$$\mathbb{P}\mathrm{ar}_n(k) = \check{\mathrm{P}}\mathrm{ar}(k), \tag{11.16}$$

where $\check{\mathrm{P}}\mathrm{ar}(k)$ is the set of partitions obtained from $\mathrm{Par}(k)$ by deleting all partitions of length greater than n, and adjoining $0's$, as necessary, to the remaining ones so that all partitions have n parts.

It is useful to remark further on partitions with no zero parts and those with zero parts. The viewpoint that one adopts depends on the information that one wishes to encode in a particular study. In the description of Young tableaux (see Sect. 11.2), a Young frame or *shape* is fully determined by a partition $\lambda \in \mathrm{Par}(k)$. Thus, $(2, 1)$ gives the shape with two boxes in the first row and one in the second, left-justified, as depicted by

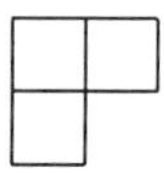 . (11.17)

In the case of partitions with zero parts, each partition $(2, 1, 0, 0, \ldots)$ containing arbitrarily many zeros is associated with the same Young frame (11.17).

Associated with each Young frame of shape $\lambda \in \mathrm{Par}(k)$, there is a set of "filled-in" shapes corresponding to the assignment of nonnegative integers to the boxes. In particular, such a rule of assignment is that each row of boxes contains only sequences of nonincreasing integers and each column only sequences of strictly increasing integers. Thus, in the case of the shape $(2, 1)$, the filled-in shapes are countably infinite in number as given by

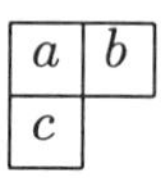

$$a \geq b, a > c, \tag{11.18}$$

where a, b, c are nonnegative integers. Each such filled-in frame constitutes what is called a *semistandard Young tableau (SSY tableau).*

The Young frame (10.17) of shape $(2,1)$ is associated with every partition $(2,1,0^{n-2}) \in \mathbb{P}\mathrm{ar}_n(3), n \geq 2$. The rule for filling in the boxes to obtain a semistandard Young tableau is still that each row of boxes contains only sequences of nonincreasing integers and each column only sequences of strictly increasing integers, but now we assign a further role to n, which counts the number of parts, including 0 parts: *The integers assigned to the rows and columns are to be drawn from the set* $\{1,2,\ldots,n\}$. We define the finite set of semistandard Young tableaux $\mathbb{T}_\lambda$, where $\lambda \in \mathbb{P}\mathrm{ar}_n$, by

$$\mathbb{T}_{(\lambda_1,\lambda_2,\ldots,\lambda_n)} = \left\{ \begin{array}{l} \text{all semistandard tableaux of shape } \lambda \\ \text{filled in with integers } 1,2,\ldots,n \end{array} \right\}. \tag{11.19}$$

Examples. Sets of semistandard tableaux $\mathbb{T}_{(2,1,0^{n-2})}$ for $n = 2, 3$:

$\mathbb{T}_{(2,1)}$:

$$\begin{array}{|c|c|}\hline 1 & 1 \\ \hline 2 \\ \cline{1-1}\end{array} \qquad \begin{array}{|c|c|}\hline 1 & 2 \\ \hline 2 \\ \cline{1-1}\end{array} \tag{11.20}$$

$\mathbb{T}_{(2,1,0)}$:

$$\begin{array}{cccc} \begin{array}{|c|c|}\hline 1 & 1 \\ \hline 2 \\ \cline{1-1}\end{array} & \begin{array}{|c|c|}\hline 1 & 2 \\ \hline 2 \\ \cline{1-1}\end{array} & \begin{array}{|c|c|}\hline 1 & 1 \\ \hline 3 \\ \cline{1-1}\end{array} & \begin{array}{|c|c|}\hline 1 & 2 \\ \hline 3 \\ \cline{1-1}\end{array} \\ \\ \begin{array}{|c|c|}\hline 2 & 2 \\ \hline 3 \\ \cline{1-1}\end{array} & \begin{array}{|c|c|}\hline 1 & 3 \\ \hline 2 \\ \cline{1-1}\end{array} & \begin{array}{|c|c|}\hline 1 & 3 \\ \hline 3 \\ \cline{1-1}\end{array} & \begin{array}{|c|c|}\hline 2 & 3 \\ \hline 3 \\ \cline{1-1}\end{array} \end{array} \tag{11.21}$$

□

We refer to the Young frame itself as having shape λ, it being understood that there are no rows in the shape corresponding to the 0 parts of the partition. For reasons given below, we refer to a semistandard tableau $T_\lambda \in \mathbb{T}_\lambda$ as a semistandard Young-Weyl (SSYW) tableau: *A SSYW tableau is a collection of boxes of shape λ such that each row of boxes contains sequences of nonincreasing integers and each column sequences of strictly increasing integers, these integers being drawn from the set* $\{1,2,\ldots,n\}$.

The cardinality of the set $\mathbb{T}_\lambda$ is given by the famous Weyl dimension formula derived in Sect. 11.8.1. For example, the cardinality of the sets illustrated in (11.20)-(11.21) can be shown directly, or calculated from the Weyl formula for general n, to be $|\mathbb{T}_{(2,1,0^{n-2})}| = 2\binom{n+1}{3}$. The enumeration of semistandard Young tableaux based on the set of partitions $\mathbb{P}\mathrm{ar}_n$ contain partitions with 0 parts is motivated by our use of what are known as *Gelfand-Tsetlin patterns.* These patterns, which are defined in Sect. 11.3 have a principal role not only in the enumeration of the inequivalent irreducible representations of the unitary group $U(n)$, but also in the enumeration of a class of polynomials, called D^λ−polynomials, treated extensively in Chapters 5 and forward. Since Weyl introduced partitions with 0 parts, we refer to the corresponding semistandard Young tableaux as semistandard Young-Weyl (SSYW) tableaux (see Sect. 11.3.2).

The utility of the enumeration of polynomials by the use of partitions with zero parts can be illustrated quite directly. In applications of partitions, one often encounters functions of the partitions themselves, or their occurrence in an ancillary role of indexing other mathematical objects. Let us illustrate this. The function of partitions of three parts given by

$$\mathrm{Dim}(\lambda_1,\lambda_2,\lambda_3) = (\lambda_1-\lambda_2+1)(\lambda_1-\lambda_3+2)(\lambda_2-\lambda_3+1)/2. \quad (11.22)$$

is the Weyl formula for the number of semistandard Young-Weyl tableaux of shape $(\lambda_1,\lambda_2,\lambda_3)$. This function may be viewed as being defined over all real values of $(\lambda_1,\lambda_2,\lambda_3)$, and in particular over all nonnegative integral sequences $(\lambda_1,\lambda_2,\lambda_2)$ that satisfy $\lambda_1 \geq \lambda_2 \geq \lambda_3 \geq 0$. Thus, the dimensions of the inequivalent unitary matrix representations of the unitary group $U(3)$, and, equivalently, of the number of semistandard Young-Weyl tableaux of shape $(\lambda_1,0,0),(\lambda_1,\lambda_2,0)$, and $(\lambda_1,\lambda_2,\lambda_2)$ is synthesized into a single formula. Another example shows how functions of several variables are also unified into a single formula:

$$\begin{aligned}
&F_{(\lambda_1)}(x_1) = x_1^{\lambda_1},\\
&F_{(\lambda_1,\lambda_2)}(x_1,y_1,x_2,y_2) = x_1^{\lambda_1-\lambda_2}\begin{vmatrix} x_1 & x_2\\ y_1 & y_2\end{vmatrix}^{\lambda_2}, \qquad (11.23)\\
&F_{(\lambda_1,\lambda_2,\lambda_3)}(x_1,y_1,z_1,x_2,y_2,z_2,x_3,y_3,z_3)\\
&= x_1^{\lambda_1-\lambda_2}\begin{vmatrix} x_1 & x_2\\ y_1 & y_2\end{vmatrix}^{\lambda_2-\lambda_3}\begin{vmatrix} x_1 & x_2 & x_3\\ y_1 & y_2 & y_3\\ z_1 & z_2 & z_3\end{vmatrix}^{\lambda_3}.
\end{aligned}$$

Not only are the first two formulas special cases of the last one, but there is a reciprocity between values of the variables and values of the partitions: If we set $x_3 = y_3 = z_1 = z_2 = z_3 = 0$ in the third function, we obtain 0, unless $\lambda_3 = 0$, in which case we obtain the second function; and if we next set $y_1 = x_2 = y_2 = 0$, we obtain 0, unless $\lambda_2 = 0$, in which case we obtain the first function. Such hierarchical relations and reciprocities are essential for the explicit construction of irreducible representations of the unitary groups. Such structures are captured most economically by using partitions with zero parts.

The development of the properties of partitions is a fairly sophisticated subject with a long history (see Andrews [1]). The cardinality of various classes of partitions is often available from generating functions. The simplest case is

$$\prod_{j\geq 1}(1-t^j)^{-1} = \sum_{k\geq 0} |\mathrm{Par}(k)|t^k. \tag{11.24}$$

If constraints are imposed on the parts of a partition, the situation can become quite intricate. A renewed interest in generating functions based on interpretations and extensions of MacMahon's Partition Analysis (MacMahon [129]) has been carried out by Andrews *et al.* [4] (many reference to the published literature are found here).

An example of this, going back to Gauss and amenable to the new methods, is given by the subset of $\mathrm{Par}(k)$ defined as follows: Let k be an arbitrary positive integer, and let m and l be any integers in the set $\{1, 2, \ldots, k\}$. For each triplet of such integers, define the set of partitions $\mathrm{Par}(k; l, m) \subset \mathrm{Par}(k)$ by

$$\begin{aligned}&\mathrm{Par}(k; l, m) = \\ &\{\lambda \in \mathrm{Par}(k) \mid \lambda \vdash k \text{ has at most } l \text{ parts with each part } \leq m\}.\end{aligned} \tag{11.25}$$

The generating function for $|\mathrm{Par}(k; l, m)|$ is the Gaussian polynomial $G(l, m; t) = G(m, l; t)$, which is a polynomial of degree lm in the indeterminate t :

$$\begin{aligned} G(l, m; t) &= \begin{bmatrix} l+m \\ m \end{bmatrix}_t \\ &= \frac{(1-t^{l+m})(1-t^{l+m-1})\cdots(1-t^{m+1})}{(1-t^l)(1-t^{l-1})\cdots(1-t)} \\ &= \sum_{k=0}^{lm} |\mathrm{Par}(k; l, m)|t^k, \end{aligned} \tag{11.26}$$

where $G(l, m; 0) = 1$. It is nontrivial to effect the division of the numerator factors by the denominator factors in (11.26) and to multiply together the resulting polynomial factors to obtain the coefficients $|Par(k;l,m)|$ on the right-hand side. We refer to Andrews [1], Chapter 3, for properties of Gaussian polynomials and additional properties of the numbers $|\mathrm{Par}(k;l,m)|$.

11.1.1 Restricted compositions

It is quite interesting that the counting of the partitions in the set $P(k;l,m)$ given by the generating function (11.26) can be generalized to that of the counting of compositions (see (1.244), Chapter 1) with restricted parts, where the set of restricted compositions is defined by

$$\mathbb{C}_{k,n}(h) = \{\alpha = (\alpha_1, \alpha_2, \ldots, \alpha_n) \,|\, \alpha \vdash k; 0 \leq \alpha_i \leq h_i, i = 1, \ldots, n\}, \tag{11.27}$$

where $h = (h_1, h_2, \ldots, h_n)$ is an arbitrary sequence of nonnegative integers. The cardinality of this set, $c_{k,n}(h) = |\,\mathbb{C}_{k,n}(h)\,|$, can be given in terms of the Clebsch-Gordan (GG) numbers $N_j(j_1, j_2, \ldots, j_n), j = j_{\min}, j_{min}+1, \ldots, j_{\max}$, defined and discussed in Sect. 2.1.1 and Sect. 2.1.3, Chapter 2. They are obtained as follows: Define recursively the multisets $\langle j_1, j_2, j_3\rangle, \langle j_1, j_2, j_3, j_4\rangle, \ldots$ by

$$\langle j_1, j_2, \ldots, j_{i+1}\rangle = \{\langle j_{12\cdots i}, j_{i+1}\rangle \,|\, j_{12\cdots i} \in \langle j_1, j_2, \ldots, j_i\rangle\}, \tag{11.28}$$

where $i = 2, 3, \ldots,$ and

$$\langle j_1, j_2\rangle = \{|j_1 - j_2|, |j_1 - j_2| + 1, \ldots, j_1 + j_2\}. \tag{11.29}$$

is the set with elements given by the well-known CG series. Then, the CG number $N_j(j_1, j_2, \ldots, j_n)$ is the number of times j occurs in the multiset $\langle j_1, j_2, \ldots, j_n\rangle$; they are fully determined from the multisets generated by (11.28)-(11.29) for all $n \geq 2$. Relations (2.36)-(2.39) give

$$c_{k,n}(h) = \sum_{j=j_{\max}-k}^{j_{\max}} N_j(j_1, j_2, \ldots, j_n), \tag{11.30}$$

where $j_i = h_i/2, i = 1, 2, \ldots, n$. Remarkably, for all $h_i = s$, we have:

$$|\mathrm{Par}(k;n,s)| = c_{k,n}(s, s, \ldots, s) = \sum_{j=\mathrm{J}_{\max}-k}^{j_{\max}} N_j(s/2, s/2, \ldots, s/2), \tag{11.31}$$

where $j_{\max} = \frac{1}{2}ns$. All coefficients in the expansion (11.26) of the Gaussian polynomial $G(n, s; t)$ are expressed in terms of CG numbers by using the symmetry relation

$$|\mathrm{Par}(k;n,s)| = |\mathrm{Par}(ns-k;n,s)|. \tag{11.32}$$

While the above relations follow a different presentation, we first learned of the relation between CG numbers for equal angular momenta and Gaussian polynomials from the preprint by Šunko and Svrtan [168].

11.1.2 Linear ordering of sequences and partitions

For the applications considered in this monograph, we require a linear ordering on the set of all *compositions* of length N, which is the set of all sequences of nonnegative integers of the same length. Thus, we define the set $\mathbb{A}_N$ of compositions for each $N = 1, 2, \ldots$ by

$$\mathbb{A}_N = \{\alpha = (\alpha_1, \alpha_2, \ldots, \alpha_N) \mid \text{each } \alpha_i \in \mathbb{N}\}. \tag{11.33}$$

We require a linear ordering of all sequences in this set. Notice that sequences in this set may have 0 as a part. This ordering is defined as follows: Let $\alpha, \beta \in \mathbb{A}_N$. Form the difference sequence $(\alpha_1 - \beta_1, \alpha_2 - \beta_2, \ldots, \alpha_N - \beta_N)$. If the first nonzero difference is positive, write $\alpha > \beta$; if negative, write $\alpha < \beta$; if all zeros, write $\alpha = \beta$. This ordering is called *lexicographic (page) ordering,* which we denote by $\mathcal{L}_N$. The first page is 0^N, where to obtain a page number the parenthesis pair and comma separators are removed from the sequence notation.

We use this same lexicographic order $\mathcal{L}_n$ on the set $\mathbb{P}\mathrm{ar}_n$ of partitions with 0 as a part. For example, the partitions in the set $\mathbb{P}_7(6)$ are ordered as follows:

$$\begin{aligned}
&(6,0,0,0,0,0,0) \;>\; (5,1,0,0,0,0,0) > (4,2,0,0,0,0,0)\\
&> (4,1,1,0,0,0,0) \;>\; (3,3,0,0,0,0,0) > (3,2,1,0,0,0,0)\\
&> (3,1,1,1,0,0,0) \;>\; (2,2,2,0,0,0,0) > (2,2,1,1,0,0,0)\\
&> (2,1,1,1,1,0,0) \;>\; (1,1,1,1,1,1,0).
\end{aligned} \tag{11.34}$$

If all $0's$ are deleted from the partitions in this set $\mathbb{P}\mathrm{ar}_7(6)$, the mapping $\mathbb{P}\mathrm{ar}_7(6) \mapsto \mathrm{Par}(6)$ is obtained and the partitions in the set $\mathrm{Par}(6)$ can be considered as inheriting the order rule. In general, an ordering rule for the partitions in $\mathrm{Par}(k)$ is obtained by adjoining zeros to the right end of a given partition as necessary to obtain the set of partitions $\mathbb{P}\mathrm{ar}_n(k)$, for any $n \geq k$, and then applying the lexicographic order rule to the latter set, which is then carried back to $\mathrm{Par}(k)$.

In the theory of partitions (with no zero parts), one encounters the notation $\lambda \supseteq \mu$. This notation refers to any pair of partitions $\lambda, \mu \in \mathrm{Par}$

such that

$$\lambda_i \geq \mu_i, \text{ all } i = 1, 2, \ldots . \tag{11.35}$$

By design, this notation and condition express the property that the Young frame corresponding to the partition μ is contained within (or coincides with) the Young frame corresponding to the partition λ (see Sect. 11.2 below). The relation $\lambda \supseteq \mu$ is a partial order relation on the set of all partitions. This relation is only a partial order because, for example, sequences such as $\lambda = (5,4,2)$ and $\mu = (4,2,1,1)$, where the length of μ exceeds that of λ, are not comparable. The relation $\lambda \supseteq \mu$ can be applied to all sequences in the set $\mathbb{P}\mathrm{ar}_n$ of partitions having the same number of parts. For this set of partitions, the relation $\lambda \supseteq \mu$ is a total order relation; that is, either $\lambda \supset \mu$, $\mu \supset \lambda$, or $\lambda = \mu$; moreover, it agrees with the lexicographic order given above. We use the partial order symbol $\lambda \supseteq \mu$ only in the context that the shape μ is contained in the shape λ in the sense of Young frames.

Another important partial order on the set of partitions is known as *dominance order.* It is defined only for partitions in $\mathrm{Par}(k)$, or, equivalently, for partitions in $\mathbb{P}\mathrm{ar}_n(k)$. For each pair $\lambda, \mu \in \mathbb{P}\mathrm{ar}_n(k)$, one writes $\lambda \geq \mu$, for all partitions that satisfy the relations

$$\begin{aligned} &\lambda_1 \geq \mu_1, \lambda_1 + \lambda_2 \geq \mu_1 + \mu_2, \\ &\qquad \ldots, \lambda_1 + \cdots + \lambda_n \geq \mu_1 + \cdots + \mu_n. \end{aligned} \tag{11.36}$$

This is a linear order for $k \leq 5$, but is only a partial order for $k > 5$. For example, the partitions $(3,1,1,1), (2,2,2) \in \mathrm{Par}(6)$, or, equivalently, $(3,1,1,1,0^{n-4}), (2,2,2,0^{n-3}) \in \mathbb{P}\mathrm{ar}_n(6), n \geq 4$, are not comparable.

Our use of order relations is for the purpose of placing objects labeled by partitions and compositions into matrices, for which we require a total order relation: lexicographic order is quite adequate.

11.2 Young Frames and Young-Weyl Tableaux

A partition $\lambda \in \mathbb{P}\mathrm{ar}_n$ is sometimes called a *shape.* This terminology originates from the figure depicting a set of planar boxes arranged in n rows that are left-justified with λ_i boxes being placed in row $i, 1 \leq i \leq n$, with row 1 at the top:

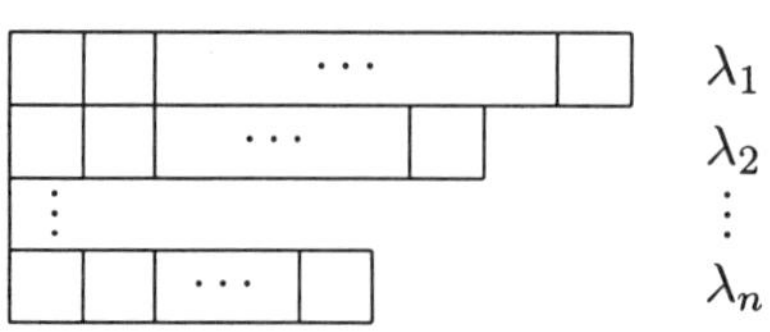

(11.37)

Altogether there are $\lambda_1 + \lambda_2 + \cdots + \lambda_n = |\lambda|$ boxes in the shape λ. If $\lambda = (\lambda_1, \lambda_2, \ldots, \lambda_l, 0, \ldots, 0)$ has l nonzero parts, there are, of course, no boxes in rows $l + 1, \ldots, n$. Nonetheless, we shall still say that the shape is λ, and ignore the 0 parts in the depiction of its true shape, but, for general description, still use the generic Young frame as shown in (11.37). We use the terms *partition* and *shape* interchangeably for λ when referring to Young frames.

The shape obtained from $\lambda = (\lambda_1, \lambda_2, \ldots, \lambda_n)$ by interchanging rows and columns gives the conjugate shape λ^c or conjugate partition

$$\lambda^c = (n^{\lambda_n}, (n-1)^{\lambda_{n-1}-\lambda_n}, \ldots, 2^{\lambda_2-\lambda_3}, 1^{\lambda_1-\lambda_2}), \tag{11.38}$$

where the notation i^m denotes that integer i is to be repeated m times with $m = 0$ meaning no occurrence. One use of the conjugate partition is that it allows us to see that the number of boxes occurring in column j of the shape λ is given by λ^c_j, each $j = 1, 2 \ldots, \lambda_1$. Square shapes are self-conjugate: $\lambda^c_j = \lambda_j, j = 1, 2, \ldots, n$.

A *Young-Weyl tableau* is a shape $\lambda \in \mathbb{P}\mathrm{ar}_n$ in which the integers $1, 2, \ldots, n$ have been distributed among the $|\lambda|$ boxes, one integer in each box. The "filled-in" shape is called a *semistandard tableau* if the sequence of integers appearing in each row is a weakly increasing (non-decreasing) sequence as read from left-to-right, and if the sequence of integers appearing in each column is strictly increasing as read from top-to-bottom. Because Weyl admitted zero parts in the definition of a partition and filled in the boxes with repeated integers selected from $\{1, 2, \ldots, n\}$, we refer to such a filled-in Young frame as a semistandard Young-Weyl (SSYW) tableau.

An important quantity associated with a tableau is it weight: The *weight* of a given SSYW tableau of shape λ is the sequence of nonnegative integers $\alpha = (\alpha_1, \alpha_2, \ldots, \alpha_n)$, where α_i equals the number of times integer i appears in the tableau. Since the assigned integers fill in all the boxes, we necessarily have that $\alpha_1 + \alpha_2 + \ldots + \alpha_n = |\lambda|$, a relation denoted by $\alpha \vdash |\lambda|$. A *standard tableau* is a special semistandard tableau in which each integer $1, 2, \ldots, n$ appears exactly once, hence, has weight $(1, 1, \ldots, 1)$ and $|\lambda| = n$. For example, the set $\mathbb{T}_{(3,0,0)}$ contains one standard tableau, $\mathbb{T}_{(2,1,0)}$ contains two, and $\mathbb{T}_{(1,1,1)}$ contains one:

$$\begin{array}{|c|c|c|}\hline 1 & 2 & 3 \\ \hline \end{array}\;; \qquad \begin{array}{|c|c|}\hline 1 & 2 \\ \hline 3 \\ \cline{1-1} \end{array}\;, \qquad \begin{array}{|c|c|}\hline 1 & 3 \\ \hline 2 \\ \cline{1-1} \end{array}\;; \qquad \begin{array}{|c|}\hline 1 \\ \hline 2 \\ \hline 3 \\ \hline \end{array} \tag{11.39}$$

Of course, the sets $\mathbb{T}_{(3,0,0)}$ and $\mathbb{T}_{(2,1,0)}$ contain semistandard tableaux as well. For $\lambda = (1, 0^{n-1}))$, there is only one box, which may be filled in with any of the integers $1, 2, \ldots, n$, corresponding to weights

$$\alpha = (1, 0, \ldots, 0), (0, 1, 0, \ldots, 0), \ldots, (0, \ldots, 0, 1). \tag{11.40}$$

We next give an incomplete list of some of the terminology and properties arising in describing SSYW tableaux of shape $\lambda \in \mathbb{P}\mathrm{ar}_n$, repeating in the list the definition of weight or *content,* as a weight is also called.

1. The *weight* or *content* of a given SSYW tableau of shape λ is the sequence of nonnegative integers $\alpha = (\alpha_1, \alpha_2, \ldots, \alpha_n)$, where α_i equals the number of times integer i appears in the tableau. Since the assigned integers fill in all the boxes, we necessarily have that $\alpha_1 + \alpha_2 + \cdots + \alpha_n = |\lambda|$, a relation we also denote by $\alpha \vdash |\lambda|$. This definition of weights may be summarized as follows: The set of weights $\mathbb{W}_\lambda$ of the set $\mathbb{T}_\lambda$ of SSYW tableau of shape $\lambda \in \mathbb{P}\mathrm{ar}_n$ is given by

$$\mathbb{W}_\lambda = \left\{ (\alpha_1, \alpha_2, \ldots, \alpha_n) \;\middle|\; \begin{array}{l} \alpha \text{ is the weight of a} \\ \text{SSYW tableau of shape } \lambda \end{array} \right\}. \tag{11.41}$$

The implementation of this result to obtain explicit sets of weights $\mathbb{W}_\lambda, \lambda \in \mathbb{P}\mathrm{ar}_n$, is a intricate process in this initial form; it requires further development. For such purposes, it is useful to define the subset of $\mathbb{T}_\lambda$ having a prescribed weight $\alpha \in \mathbb{W}_\lambda$ by $\mathbb{T}_\lambda(\alpha)$.

2. Let the partition $\lambda \in \mathbb{P}\mathrm{ar}_n$, and let α be a weight of a SSYW tableau of shape λ. In general, there are several distinct SSYW tableaux of shape λ having the same weight α. The number of such SSYW tableaux is denoted $K(\lambda, \alpha)$ and is called a *Kostka number:*

$$K(\lambda, \alpha) = |\mathbb{T}_\lambda(\alpha)|. \tag{11.42}$$

An example of this relation is given by the partition $(2, 1, 0)$ of 3 into 3 parts, for which there are two standard tableaux having weight $(1, 1, 1)$, as given above in (11.39).

3. The action of a permutation

$$\pi = \begin{pmatrix} 1 & 2 & \cdots & n \\ \pi_1 & \pi_2 & \cdots & \pi_n \end{pmatrix} \in S_n \tag{11.43}$$

on a weight $\alpha = (\alpha_1, \alpha_2, \ldots, \alpha_n)$ is defined by

$$\pi : \alpha \mapsto (\alpha_{\pi_1}, \alpha_{\pi_2}, \ldots, \alpha_{\pi_n}) = \pi(\alpha). \tag{11.44}$$

A principal property of weights is that if α is the weight of a SSYW tableau of shape λ, then every permutation of α is also a weight of a SSYW tableau of shape λ, and $\pi(\alpha)$ is of the same multiplicity as that of α, that is, the Kostka numbers are invariant under permutations:

$$K(\lambda, \pi(\alpha)) = K(\lambda, \alpha), \text{ each } \pi \in S_n, \text{ each weight } \alpha. \qquad (11.45)$$

4. The cardinality of the set $\mathbb{T}_\lambda$ is given by the formula

$$|\mathbb{T}_\lambda| = \operatorname{Dim}\lambda = \frac{\prod_{1\le i<j\le n}(p_i - p_j)}{1!2!\cdots(n-1)!}, \; p_i = \lambda_i + n - i. \qquad (11.46)$$

This is the so-called *Weyl dimension formula.* This important formula is proved algebraically in Sect. 11.8, where basic properties of Vandermonde's determinant are given. The notation $\operatorname{Dim}\lambda$ arises in the context of group representation theory, where this function of a partition occurs as the dimensionality of a vector space.

5. The Kostka numbers and the Weyl dimension are related by

$$\operatorname{Dim}\lambda = \sum_{\alpha\in\mathbb{W}_\lambda} K(\lambda, \alpha). \qquad (11.47)$$

A formula for the Kostka numbers is unknown, except for $n = 2, 3$ (see relations (11.140) and (11.144) below).

The Weyl formula for the cardinality of the set $\mathbb{T}_\lambda$ of SSYW tableaux of shape $\lambda \in \mathbb{P}\mathrm{ar}_n$ can also be expressed in terms of the *hooks* associated with the shape λ. A hook is a special shape of the form $h = (l, 1^{m-l}), m \ge l \ge 1$, having l boxes in the first row and 1 box in each of $m - l$ rows:

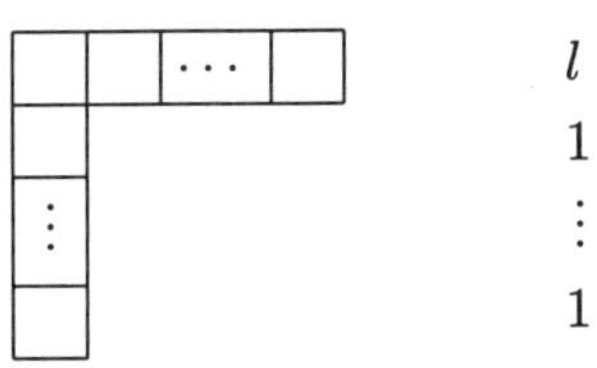

(11.48)

The length of a hook equals the number of boxes it contains and is given by $|h| = m$. An arbitrary partition of shape λ contains a hook associated to each of its boxes, defined as follows: Let (i, j) denote the box in row i and column j of a Young frame of shape λ, hence, $i = 1, 2, \ldots, \lambda_i; j =$

$1, 2, \ldots, \lambda_i^c$. The hook $h(i,j)$ associated with box (i,j) is the partition

$$\begin{aligned} h(i,j) &= (\lambda_i - j + 1, 1^{\lambda_j^c - i}) \text{ of length} \\ |h(i,j)| &= \lambda_i - j + 1 + \lambda_j^c - i. \end{aligned} \tag{11.49}$$

The Weyl dimension formula expressed in terms of hook lengths is

$$\text{Dim}\lambda = \frac{\prod_{1 \le i < j \le n} (\lambda_i - \lambda_j + j - i)}{1!2! \cdots (n-1)!} = \prod_{i,j} \frac{(n + j - i)}{|h(i,j)|}, \tag{11.50}$$

where the product is over all $|\lambda|$ boxes (i,j) contained in the shape λ.

Two other useful functions defined on the set of partitions $\mathbb{P}\text{ar}_n$ that can be written in terms of the set of hooks of $\lambda \in \mathbb{P}\text{ar}_n$ are given by

$$M(\lambda) = \frac{\prod_{i=1}^{n} (\lambda_i + n - 1)!}{\prod_{i \le j = 1}^{n} (\lambda_i - \lambda_j + j - i)} = \prod_{i,j} |h(i,j)|, \tag{11.51}$$

$$K(\lambda, \alpha)|_{\alpha = (1^n)} = \dim \lambda = \frac{n!}{\prod_{i,j} |h(i,j)|}, \quad \lambda \vdash n. \tag{11.52}$$

The first of these is the normalization factor for D^λ–polynomials introduced in Chapter 5; the second is the Kostka number for the multiplicity of the weight $\alpha = (1^n)$ in the set of standard Young tableaux, hence, $\dim \lambda$ is the number of **standard** Young tableaux of shape $\lambda \vdash n$. (No such formula for the general Kostka number is known.) The function $\dim \lambda$ is also the order of the irreducible matrix representations of the symmetric group S_n, all such matrix representations being enumerated by partitions $\lambda \in \mathbb{P}\text{ar}_n(n)$. The notation dim is chosen in parallel to that for the Weyl dimension formula.

An important concept for the polynomials and group representations considered in this monograph is the Cartesian product $\mathbb{T}_\lambda \times \mathbb{T}_\lambda$ of two sets of SSYW tableaux of the same shape λ, which is called a *double* SSYW tableau of shape λ. Let $T_\lambda(\alpha) \in \mathbb{T}_\lambda(\alpha)$. The elements of $\mathbb{T}_\lambda(\alpha) \times \mathbb{T}_\lambda(\alpha')$ are the ordered pairs $(T_\lambda(\alpha), T_\lambda(\alpha'))$ of SSYW tableaux. The ordered pair of weights (α, α') specifies the weight of the first and second SSYW tableau; it is called a weight of the double SSYW tableau. We now also have the following set:

$$\mathbb{T}_\lambda(\alpha, \alpha') = \left\{ \begin{array}{l} \text{subset of all double } SSYW \text{ tableau} \\ \text{of shape } \lambda \text{ having a given weight } (\alpha, \alpha') \end{array} \right\}. \tag{11.53}$$

The cardinality of this set is given by the product of Kostka numbers $K_\lambda(\alpha, \alpha') = K(\lambda, \alpha)K(\lambda, \alpha')$; that is,

$$|\mathbb{T}_\lambda(\alpha, \alpha')| = K_\lambda(\alpha, \alpha'). \tag{11.54}$$

There exists a remarkable formula relating these numbers to the cardinality of another set of combinatorial objects.

The second set of combinatorial objects is a matrix array of nonnegative integers, where the row and column sums are specified nonnegative integers. To be precise, we first introduce a matrix array of nonnegative integers $(a_{ij})_{1\leq i\leq n, 1\leq j\leq m}$ having n rows and m columns:

$$A = \begin{pmatrix} a_{11} & a_{12} & \cdots & a_{1m} \\ a_{21} & a_{22} & \cdots & a_{2m} \\ \vdots & \vdots & \cdots & \vdots \\ a_{n1} & a_{n2} & \cdots & a_{nm} \end{pmatrix}. \tag{11.55}$$

Next, we introduce a composition $\alpha = (\alpha_1, \alpha_2, \ldots, \alpha_n)$ of a nonnegative integer k into n nonnegative parts; that is, $|\alpha| = \sum_{i=1}^n \alpha_i = k$, each $\alpha_i \in \mathbb{N}$. We introduce, similarly, a second composition $\alpha' = (\alpha'_1, \alpha'_2, \ldots, \alpha'_m)$, $|\alpha'| = k$ of k into m nonnegative parts. We now impose the constraints that row i of A should sum to α_i, $i = 1, 2, \ldots, n$, and column j of A should sum to α'_j, $j = 1, 2, \ldots, m$. In this context, α is called the *row-sum sequence* of A, and α' *the column-sum sequence* of A. We thus arrive at the notion of the set $\mathbb{M}^k_{n\times m}(\alpha, \alpha')$ of $n \times m$ matrix arrays with fixed row-sum sequence α and fixed column-sum sequence α', which, in turn, are compositions of k into n and m nonnegative parts:

$$\mathbb{M}^k_{n\times m}(\alpha, \alpha') = \left\{ A = (a_{ij}) \;\middle|\; \begin{array}{l} A \text{ is } n \times m; a_{ij} \in \mathbb{N}, \\ A \text{ has row-sum sequence } \alpha, \\ A \text{ has column-sum sequence } \alpha' \end{array} \right\}. \tag{11.56}$$

The Kostka numbers (11.54) and the number of matrix arrays (11.56) are related by

$$\sum_{\lambda\vdash k} K_\lambda(\alpha, \alpha') = |\mathbb{M}^k_{n\times n}(\alpha, \alpha')|. \tag{11.57}$$

This relation is known as the Robinson-Schensted-Knuth (RSK) identity.

Knuth's [91] combinatorial technique for proving this relation has many other applications, as discussed by Stanley [163]. For us, this relation is an important combinatorial verification of the equality of two different bases of a vector space, as discussed in Sect. 5.1, Chapter 5.

11.2.1 Skew tableaux

Let $\lambda, \mu \in \mathbb{P}\mathrm{ar}_n$ be partitions, hence, both λ and μ have corresponding diagrams depicting Young frames. Suppose now that $\lambda \supseteq \mu$, so that $\lambda_i \geq \mu_i, i = 1, 2, \ldots, n$, and the shape μ "fits inside" the shape λ, as discussed in Sect. 11.1.2. The *skew frame* $\lambda - \mu$ refers to the shape of the "staggered" set of boxes that remains after deleting all the boxes of shape μ from the shape λ :

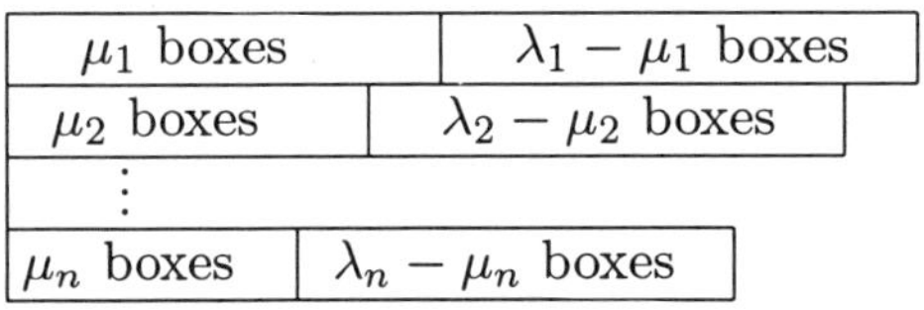

(11.58)

Both λ and μ can have 0 parts, but, of course, 0 parts are ignored in the mapping to Young frames.

Example. We have for $\lambda = (6, 5, 3, 0)$ and $\mu = (2, 2, 0, 0)$ that $\lambda - \mu = (4, 3, 3, 0)$ so that the shape is given by

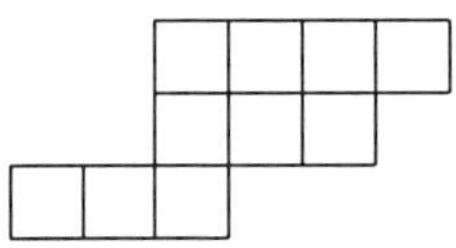

(11.59)

The presence of the 0 parts and the common length 4 carries the information that this skew tableau is to be filled in with integers selected from $\{1, 2, 3, 4\}$, as we next describe for the general case. $\square$

Let $\lambda, \mu \in \mathbb{P}\mathrm{ar}_n$ with $\lambda \supseteq \mu$. A *semistandard skew tableau* is a skew shape $\lambda - \mu$ filled in with the integers $1, 2, \ldots, n$ such that each row is a sequence of weakly increasing integers and each column is a sequence of strictly increasing integers. Such a filled-in shape is denoted by $T_{\lambda/\mu}$. The set of all such semistandard skew tableaux is denoted by $\mathbb{T}_{\lambda/\mu}$.

The weight α of a skew tableaux $T_{\lambda/\mu}$ is defined in the same way as for SSYW tableau: $\alpha = (\alpha_1, \alpha_2, \ldots, \alpha_n)$, where α_i is the number of times

i appears, hence, $|\alpha| = \sum_{i=1}^{n} \alpha_i = |\lambda| - |\mu|$. The subset of $\mathbb{T}_{\lambda/\mu}$ having the same weight α is denoted by $\mathbb{T}_{\lambda/\mu}(\alpha)$, and its cardinality is the skew Kostka number, denoted $K(\lambda/\mu, \alpha)$. For $\mu = p_\emptyset$, the empty shape or the shape with no boxes, skew shapes become ordinary shapes. Many concepts for ordinary tableaux transfer directly to skew tableaux.

We note, however, that, while the removal from a semistandard tableau T_λ of all boxes of integers in shape μ always gives a semistandard skew tableau $T_{\lambda/\mu}$, the converse is not true. For example, for $\lambda = (6,5,3,0)$ and $\mu = (3,2,0,0)$, the following picture shows a semistandard skew tableau:

$T_{(6,5,3,0)/(3,2,0,0)} =$ 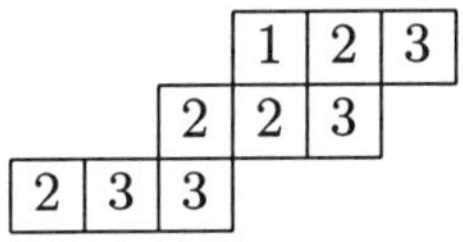

(11.60)

This skew tableau cannot be obtained by striking some semistandard tableau $T_{(3,2,0,0)}$ from any semistandard tableau in $\mathbb{T}_{(6,5,3,0)}$. This example brings us to the notion of a *horizontal strip.*

Horizontal strips

An important special skew shape is known as a *horizontal strip.* Let $\lambda, \mu \in \mathbb{P}\mathrm{ar}_n$. *A skew shape $\lambda - \mu$ each of whose λ_1 columns contains no boxes or one box is called a horizontal strip of order n.* This situation may be illustrated by the following two pictures in terms of the Young frames λ and μ. First, we have $\mu \subseteq \lambda$, as show by

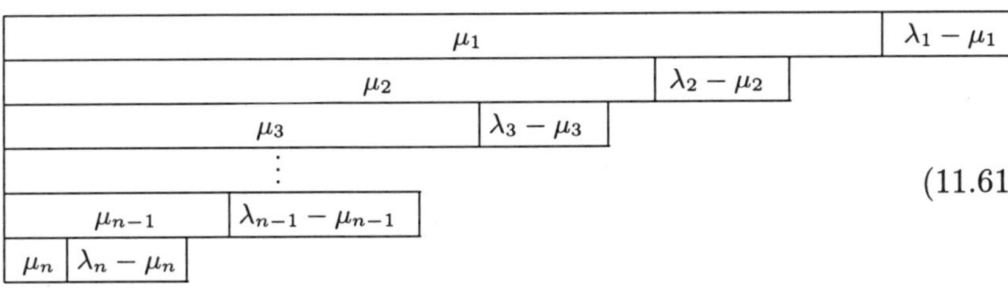

(11.61)

Second, after the removal of the shape μ, we are left with the shape $\lambda - \mu$ in which there is **no overlapping** between any of the staggered rows of boxes, as depicted by the following diagram of the horizontal strip, which is denoted $\mathrm{Sh}_{\lambda/\mu}$:

$$\mathrm{Sh}_{\lambda/\mu} =$$

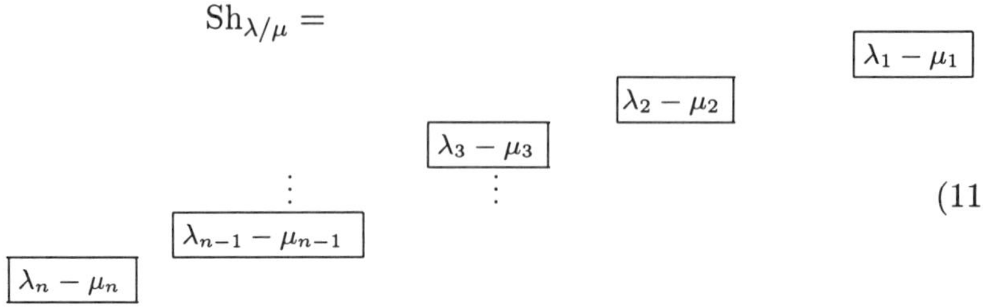

(11.62)

In these two diagrams (11.61) and (11.62), the partitions $\lambda, \mu \in \mathbb{P}\mathrm{ar}_n$ satisfy the relation

$$\lambda_1 \geq \mu_1 \geq \lambda_2 \geq \mu_2 \geq \cdots \geq \lambda_{n-1} \geq \mu_{n-1} \geq \lambda_n \geq \mu_n \geq 0. \tag{11.63}$$

This relation is equivalent to the two conditions

$$\lambda \supseteq \mu \text{ and } \mu \supseteq (\lambda_2, \ldots, \lambda_n, 0). \tag{11.64}$$

The additional condition $\mu \supseteq (\lambda_2, \ldots, \lambda_n, 0)$ assures that there are no columns containing two boxes. However, some columns can be empty (contain no boxes). The vertical lines at the left edge of $\lambda_1 - \mu_1$ and the right edge of $\lambda_2 - \mu_2$ align only when $\mu_1 = \lambda_2$; the vertical lines between the left edge of $\lambda_2 - \mu_2$ and the right edge of $\lambda_3 - \mu_3$ align only when $\mu_2 = \lambda_3$; $\ldots$; the vertical lines between the left edge of $\lambda_{n-1} - \mu_{n-1}$ and the right edge of $\lambda_n - \mu_n$ align only when $\mu_{n-1} = \lambda_n$.

It is sometimes convenient when working with skew SSYW tableaux to keep the full picture (11.61) showing the parts of the partition $\mu \in \mathbb{P}\mathrm{ar}_n$ and to leave all these boxes blank, but to fill-in the remaining boxes with the integers $1, 2, \ldots, n$ by the prescribed rule of nondecreasing rows and strictly increasing columns. This keeps visible at each step exactly the effect of removing boxes, a practice we follow below.

The main result for horizontal strips is known an Pieri's rule (see Stanley [163, Vol. 2, p. 339]). It may be stated as follows: Let $T_{\lambda/\mu}$ be a skew SSYW tableau with $\lambda, \mu \in \mathbb{P}\mathrm{ar}_n$ that is filled in with integers selected from $1, 2, \ldots, n$. Define $m_{n-1} \in \mathbb{P}\mathrm{ar}_n$ to be the shape of the full SSYW tableau that remains when all boxes containing $n's$ are deleted; $m_{n-2} \in \mathbb{P}\mathrm{ar}_n$ the full shape that remains when all boxes containing $(n-1)'s$ are deleted;$\ldots$; $m_1 \in \mathbb{P}\mathrm{ar}_n$ the full shape that remains when all boxes containing $2's$ are deleted; and $m_0 \in \mathbb{P}\mathrm{ar}_n$ the full shape that remains when all boxes containing $1's$ are deleted, where $m_j = (m_{1,j}, m_{2,j}, \ldots, m_{n,j}), j = n, n-1, \ldots, 0$, in which $\lambda = m_n$ and

$\mu = m_0$. Then, Pieri's rule is

$$T_{\lambda/\mu} = \sum_{j=1}^{n} \boxplus T_{m_j/m_{j-1}}(j), \tag{11.65}$$

where $T_{m_j/m_{j-1}}(j)$ denotes the semistandard horizontal strip in which all boxes are filled in with j, and the symbol $\boxplus$ designates that the filled-in horizontal strips are simply to be superposed, one over the other.

Proof: The proof of this result is a straightforward application of the manner in which the integers $n, n-1, \ldots, 1$ are distributed into a SSYW tableau. The proof is also clear from the following example.

Example. Relation (11.65) has the following form for the skew SSYW tableau (11.60):

$$\begin{aligned} T_{(6,5,3,0)/(3,2,0,0)} &= T_{(6,5,3,0)/(5,4,1,0)}(3) \boxplus T_{(5,4,1,0)/(4,2,0,0)}(2) \\ &\boxplus \ T_{(4,2,0,0)/(3,2,0,0)}(1). \end{aligned} \tag{11.66}$$

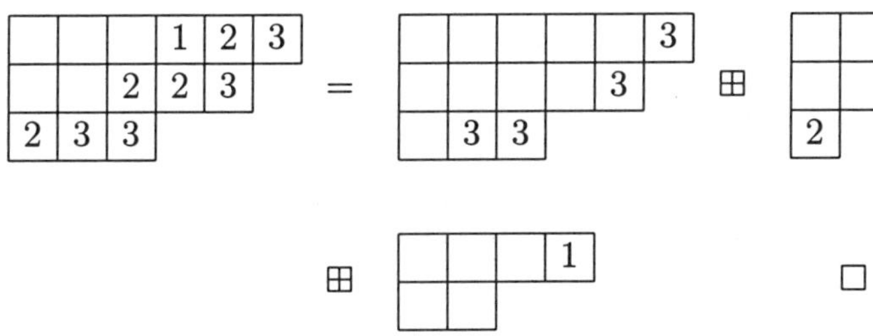

(11.67)

We return to further consequences of (11.65) in Sect. 11.3.3 below.

11.3 Gelfand-Tsetlin Patterns

11.3.1 Combinatorial origin of Gelfand-Tsetlin patterns

Gelfand-Tsetlin patterns containing n rows were introduced (see Gelfand and Tsetlin [62]) for the purpose of enumerating the integer irreducible representations of the general linear group, and, in particular, the unitary irreducible representations of $U(n)$. This result can also be understood from Pieri's rule, which applies to SSYW tableaux as well as to SSYW skew tableaux. It is only necessary to choose the partition $\mu = (0^n)$, in which case $T_{\lambda/(0^n)} = T_\lambda$. We then also have

$$\begin{aligned} &m_j = (m_{1,j}, m_{2,j}, \ldots, m_{j,j}, 0^{n-j}),\ j = 1, 2, \ldots, n, \\ &(m_{1,j}, m_{2,j}, \ldots, m_{j,j}) \in \mathbb{P}\mathrm{ar}_j, \\ &\lambda = m_n = (\lambda_1, \lambda_2, \ldots, \lambda_n) \in \mathbb{P}\mathrm{ar}_n. \end{aligned} \tag{11.68}$$

Since the zeros past the j–th place in m_j have no role, they are dropped. Pieri's rule becomes:

$$\begin{aligned} T_\lambda &= \sum_{j=1}^{n} \boxplus T_{m_j/m_{j-1}}(j), \\ m_j &= (m_{1,j}, m_{2,j}, \ldots, m_{j,j}) \in \mathbb{P}\mathrm{ar}_j;\ m_1/m_0 = m_1 = m_{1,1}. \end{aligned} \tag{11.69}$$

Pieri's rule in this form is the combinatorial basis for the origin of GT patterns, which were originally based on the Weyl group reduction law as explained in Sect. 11.3.2 below.

The relation to GT patterns comes about in the following way. We introduce a two-rowed pattern as follows:

$$\begin{pmatrix} m_j \\ m_{j-1} \end{pmatrix} = \begin{pmatrix} m_{1,j} & & m_{2,j} & & \cdots & & m_{j-1,j} & & m_{j,j} \\ & m_{1,j-1} & & m_{2,j-1} & & \cdots & & m_{j-1,j-1} & \end{pmatrix}, \tag{11.70}$$

where the partitions in this relation are to satisfy

$$\begin{aligned} m_j &= (m_{1,j}, m_{2,j}, \ldots, m_{j,j}) \in \mathbb{P}\mathrm{ar}_j, \\ m_{j-1} &= (m_{1,j-1}, m_{2,j-1}, \ldots, m_{j-1,j-1}) \in \mathbb{P}\mathrm{ar}_{j-1}, \\ m_{1,j} &\geq m_{1,j-1} \geq m_{2,j} \geq m_{2,j-1} \geq \cdots \geq m_{j-1,j} \geq m_{j-1,j-1} \\ &\geq m_{j,j} \geq 0, \qquad\qquad \text{each } j = 1, 2, \ldots, n. \end{aligned} \tag{11.71}$$

These patterns are called two-rowed GT patterns. The inequalities in (11.71) are called betweenness conditions, which may also be written in the closed interval form:

$$m_{i,j-1} \in [m_{i+1,j}, m_{i,j}],\ i = 1, 2, \ldots, j-1. \tag{11.72}$$

The placement of the parts of m_{j-1} between the part of m_j in (11.70) is intended to indicate geometrically the betweenness conditions, which we also write as $m_j \succ m_{j-1}$.

We next stack the collection of $n-1$ two-rowed GT patterns (11.70) corresponding to $j = 1,,2,\ldots,n$, but do not repeat the common rows, to obtain the n−rowed triangular pattern as follows:

$$\binom{\lambda}{m} = \left(\begin{array}{ccccccccccc} \lambda_1 & & \lambda_2 & \cdots & \lambda_j & & \cdots & & \lambda_{n-1} & & \lambda_n \\ & m_{1,n-1} & & m_{2,n-1} & \cdots & m_{j,n-1} & & \cdots & & m_{n-1,n-1} & \\ & & & & & \vdots & & & & & \\ & & & m_{1,j} & & m_{2,j} & \cdots & m_{j,j} & & & \\ & & & & & \vdots & & & & & \\ & & & & m_{1,2} & & m_{2,2} & & & & \\ & & & & & m_{1,1} & & & & & \end{array}\right). \tag{11.73}$$

The rows of this pattern are to satisfy the betweenness conditions

$$\lambda = m_n \succ m_{n-1} \succ \cdots \succ m_2 \succ m_1. \tag{11.74}$$

Each pattern of the form (11.73) satisfying the conditions (11.74) is called a *Gelfand-Tsetlin pattern;* it is written in the notation of the original paper by Gelfand and Tsetlin [62]. We use the convention of writing row n (the top row) in the Greek notation for a partition:

$$\lambda = (\lambda_1, \lambda_2, \ldots, \lambda_n) = (m_{1,n}, m_{2,n}, \ldots, m_{n,n}) \in \mathbb{P}\mathrm{ar}_n. \tag{11.75}$$

The single letter m on the left-hand side of (11.73) subsumes the entire $(n-1)$−rowed GT pattern on the right-hand side. Such an abbreviated notation is required to present the many relations involving these symbols in manageable form.

The weight α_j of the two-rowed GT pattern (11.70) is defined by

$$\begin{aligned} \alpha_j &= |m_j| - |m_{j-1}| = \sum_{i=1}^{j} m_{i,j} - \sum_{i=1}^{j-1} m_{i,j-1}, \\ &\qquad \text{each } j = 1,2,\ldots,n; \alpha_1 = m_{1,1}. \end{aligned} \tag{11.76}$$

Thus, α_j is the number of times that j appears in the SSYW tableau T_λ of shape λ. To each horizontal strip $T_{m_j/m_{j-1}}(j)$, whose boxes are all filled in with j, there corresponds a unique two-rowed GT pattern (11.70) of weight α_j, and the n−rowed triangular GT pattern is the unique pattern corresponding to the presentation of a SSYW tableau in terms of its semistandard horizontal strips (11.73). The weight of the full GT pattern is given by

$$\alpha = (\alpha_1, \alpha_2, \ldots, \alpha_n), \tag{11.77}$$

where the significance of α_j is the number of times j appears in the SSYW tableau. We conclude that the set $\mathbb{T}_\lambda$ of SSYW tableau of shape λ is bijective with the set of all GT patterns of shape λ defined by

$$\mathbb{G}_\lambda = \left\{ \begin{pmatrix} \lambda \\ m \end{pmatrix} \middle| \, m \text{ is a lexical pattern} \right\}. \tag{11.78}$$

We also refer to GT patterns corresponding to the partition $\lambda \in \mathbb{P}\mathrm{ar}_n$ as having (triangular) shape λ. As noted above, m itself in (11.73) denotes an $(n-1)$–rowed GT pattern. The statement *"m is a lexical pattern"* means that the betweenness conditions are all fulfilled. The statement *"m is a nonlexical pattern"* means these conditions are violated. Thus, from the concept of a horizontal strip and Pieri's rule, we have arrived at the weight preserving bijection

$$\mathbb{T}_\lambda \leftrightarrow \mathbb{G}_\lambda. \tag{11.79}$$

It is then the case that the cardinality of these sets are both equal to the Weyl dimension:

$$|\mathbb{T}_\lambda| = |\mathbb{G}_\lambda| = \mathrm{Dim}\lambda. \tag{11.80}$$

We have used the most elemental combinatorial concept, the horizontal strip, to establish the bijection between the set of SSYW tableaux and the set of GT patterns of shape λ. It is also useful to give two alternative ways of making this correspondence.

Given a SSYW tableau $T_\lambda \in \mathbb{T}_\lambda$, the corresponding GT pattern $G_\lambda \in \mathbb{G}_\lambda$ is read off by the sequential removal of boxes containing $n, \ldots, 2$ as follows: First, the partition $\lambda \in \mathbb{P}\mathrm{ar}_n$ is read off the shape of the tableau; then the partition $m_{n-1} \in \mathbb{P}\mathrm{ar}_{n-1}$ is read off the shape that remains after all boxes containing n are removed;...; the partition $m_2 \in \mathbb{P}\mathrm{ar}_2$ is read off the shape that remains after all boxes containing 3 are removed; and the partition $m_1 \in \mathbb{P}\mathrm{ar}_1$ is read off the shape that remains after all boxes containing 2 are removed.

Given a GT pattern $G_\lambda \in \mathbb{G}_\lambda$, the corresponding SSYW tableau $T_\lambda \in \mathbb{T}_\lambda$ is read off in the following way: The entries in row j of the SSYW tableau are read off the j–th down-diagonal

$$\begin{matrix} m_{j,n} & & \\ & m_{j,n-1} & \\ & \searrow & \\ & & m_{j,j} \end{matrix} \qquad m_{j,n} = \lambda_j \tag{11.81}$$

of the GT pattern (11.73) by the following rule:

$m_{j,j}$ $j's$	$m_{j,j+1} - m_{j,j}$ $(j+1)'s$	$\cdots$	$m_{j,n} - m_{j,n-1}$ $n's$

(11.82)

This rule applies to each $j = 1, 2, \ldots, n$, where $m_{n,n-1} = 0$.

Some caution must be exercised in implementing these rules, since it follows from the definition of the weight $\alpha = (\alpha_1, \alpha_2, \ldots, \alpha_n), \alpha_j =$ number of times j appears, of T_λ that some integers from among $1, 2, \ldots, n$ may not appear. Also, the requirement that $m_j \in \mathbb{P}\mathrm{ar}_j$ must be carefully observed. The rules are made clear by several examples.

Examples. Four simple examples illustrate the rules for determining the GT pattern corresponding to a SSYW tableau:

$n = 3$: Member of the family $\mathbb{G}_{(3,0,0)}$:

$$\boxed{3}\boxed{3}\boxed{3} \qquad \longleftrightarrow \qquad \begin{pmatrix} 3 & & 0 & & 0 \\ & 0 & & 0 & \\ & & 0 & & \end{pmatrix} \tag{11.83}$$

$n = 4$: Member of the family $\mathbb{G}_{(3,0,0,0)}$:

$$\boxed{3}\boxed{3}\boxed{3} \qquad \longleftrightarrow \qquad \begin{pmatrix} 3 & & 0 & & 0 & & 0 \\ & 3 & & 0 & & 0 & \\ & & 0 & & 0 & & \\ & & & 0 & & & \end{pmatrix} \tag{11.84}$$

$n = 4$: Member of the family $\mathbb{G}_{(6,5,3,0)}$:

$$\begin{array}{|c|c|c|c|c|c|} \hline 1 & 1 & 1 & 3 & 3 & 4 \\ \hline 3 & 3 & 3 & 4 & 4 & \\ \cline{1-5} 4 & 4 & 4 & & & \\ \cline{1-3} \end{array} \qquad \longleftrightarrow \qquad \begin{pmatrix} 6 & & 5 & & 3 & & 0 \\ & 5 & & 3 & & 0 & \\ & & 3 & & 0 & & \\ & & & 3 & & & \end{pmatrix} \tag{11.85}$$

$n = 5$: Member of the family $\mathbb{T}_{(6,5,3,0,0)}$:

$$\begin{array}{|c|c|c|c|c|c|}\hline 1&1&1&3&3&4\\\hline 3&3&3&4&4\\\cline{1-5} 4&4&4\\\cline{1-3}\end{array} \quad \longleftrightarrow \quad \begin{pmatrix} 6 & & 5 & & 3 & & 0 & & 0 \\ & 6 & & 5 & & 3 & & 0 & \\ & & 5 & & 3 & & 0 & & \\ & & & 3 & & 0 & & & \\ & & & & 3 & & & & \end{pmatrix} \tag{11.86}$$

□

11.3.2 Group theoretical origin of Gelfand-Tsetlin patterns

Gelfand and Tsetlin [62] introduced the triangular patterns (11.73) having n rows and shape λ for the purpose of indexing an orthonormal basis of an abstract Hilbert space $\mathcal{H}_\lambda$ on which the Lie algebra of the general linear group $GL(n, \mathbb{C})$ has an irreducible linear action. This notation was based on a famous result proved by Weyl [177, p. 369]:

Branching law for c_n.

On restricting c_n to the sub-group of linear transformations of an (n-1)-dimensional sub-space the irreducible representation $(f_1, f_2, \cdots)$ of c_n reduces into the sum of all those representations $(f'_1, f'_2, \cdots)$ of c_{n-1} for which

$$f_1 \geq f'_1 \geq f_2 \geq f'_2 \geq \cdots \geq f'_{n-1} \geq f_n;$$

each of these constituents appears exactly once.

While not stated explictly in the above branching law for $c_n = GL(n, \mathbb{C})$, it is the case that $f_n \geq 0$ (Weyl [177, p. 381]. This result then gives the two-rowed symbol (11.70) for $j = n$ with $(f_1, f_2, \cdots, f_n) = \lambda = m_n$ and $(f'_1, f'_2, \cdots, f'_{n-1}) = m_{n-1}$. (See King [90] for a modern treatment of the character formulas leading to the above branching law.)

Gelfand-Tsetlin patterns were introduced in the context of vector spaces, Lie algebras, and groups, but these patterns are precisely those originating from SSYW tableaux of shape λ, as discussed above in the context of two-rowed patterns and Pieri's rule.

11.3.3 Skew Gelfand-Tsetlin patterns

The entire analysis of Sect. 11.3.1 can now be repeated to obtain the set of skew GT patterns based on the concept of horizontal strips. The exposition of results follows, step-by-step, the earlier procedures, and we simply highlight the results.

The two-rowed skew GT pattern is defined as follows:

$$\begin{bmatrix} m_j \\ m_{j-1} \end{bmatrix} = \begin{bmatrix} m_{1,j} & & m_{2,j} & & \cdots & & m_{j,j-1} & & m_{n,j} & \\ & m_{1,j-1} & & m_{2,j-1} & & \cdots & & m_{n-1,j-1} & & m_{n,j-1} \end{bmatrix}, \tag{11.87}$$

where the partitions in this relation satisfy

$$\begin{gathered} m_j = (m_{1,j}, m_{2,j}, \ldots, m_{n,j}) \in \mathbb{P}\mathrm{ar}_n, \\ m_{j-1} = (m_{1,j-1}, m_{2,j-1}, \ldots, m_{n,j-1}) \in \mathbb{P}\mathrm{ar}_n, \\ m_{1,j} \geq m_{1,j-1} \geq m_{2,j} \geq m_{2,j-1} \geq \cdots \geq m_{n,j} \geq m_{n,j-1} \geq 0, \\ \text{each } j = 1, 2, \ldots, n. \end{gathered} \tag{11.88}$$

The weight of this pattern is defined by

$$\alpha_j = \sum_{i=1}^{n} m_{i,j} - \sum_{i=1}^{n} m_{i,j-1}. \tag{11.89}$$

The set of two-rowed skew GT patterns for $j = 1, \cdots, n$ are in one-to-one correspondence with the set of filled-in horizontal strips $T_{m_j/m_{j-1}}(j)$ in relation (11.65).

The geometric placement of the parts of m_{j-1} in (11.87) fall between those of m_j, which also may be stated as the following closed interval conditions hold:

$$m_{i,j-1} \in [m_{i+1,j}, m_{i,j}],\ i = 1, 2, \ldots, n-1;\ m_{n,j-1} \in [0, m_{n,j}]. \tag{11.90}$$

We denote these betweenness relations by $m_j \sqsupseteq m_{j-1}$.

The $(n+1)-$ rowed skew GT pattern $\begin{bmatrix} \lambda/\mu \\ m \end{bmatrix}$ of shape $\lambda - \mu$ is now obtained as the stacked collection of n two-rowed $(n+1)-$rowed skew GT patterns (11.87):

$$\begin{bmatrix} \lambda/\mu \\ m \end{bmatrix} = \begin{bmatrix} \lambda_1 & & \lambda_2 & & \cdots & & \lambda_n & & & \\ & m_{1,n-1} & & m_{2,n-1} & & \cdots & & m_{n,n-1} & & \\ & & & & \vdots & & & & & \\ & & m_{1,j} & & m_{2,j} & & \cdots & & m_{n,j} & \\ & & & & & \vdots & & & & \\ & & & m_{1,1} & & m_{2,1} & & \cdots & & m_{n,1} \\ & & & & \mu_1 & & \mu_2 & & \cdots & & \mu_n \end{bmatrix}, \tag{11.91}$$

in which the betweenness conditions hold between the rows; that is,

$$\lambda = m_n \sqsupseteq m_{n-1} \sqsupseteq \ldots \sqsupseteq m_1 \sqsupseteq m_0 = \mu. \tag{11.92}$$

Each pattern of this form, where the partitions that make up its rows satisfy the betweenness conditions (11.92), is called a *skew Gelfand-Tsetlin pattern.*We use the convention of writing row n (the top row) in the Greek notation λ for a partition and row 0 in the Greek notation μ for a partition:

$$\begin{aligned} \lambda &= (\lambda_1, \lambda_2, \ldots, \lambda_n) = (m_{1,n}, m_{2,n}, \ldots, m_{n,n}) \in \mathbb{P}\mathrm{ar}_n, \\ \mu &= (\mu_1, \mu_2, \ldots, \mu_n) = (m_{1,0}, m_{2,0}, \ldots, m_{n,0}) \in \mathbb{P}\mathrm{ar}_n. \end{aligned} \tag{11.93}$$

The single letter m on the left-hand side of (11.91) subsumes the entire $(n-1)$–rowed skew GT pattern on the right-hand side.

We denote the set of all skew GT patterns $G_{\lambda/\mu}$

$$\mathbb{G}_{\lambda/\mu} = \left\{ \left[\begin{array}{c} \lambda/\mu \\ m \end{array} \right] \middle| m \text{ is a lexical pattern} \right\}. \tag{11.94}$$

The statement *"m is a lexical pattern"* means that the betweenness conditions (11.92) are all fulfilled. The statement *"m is a nonlexical pattern"* means these conditions are violated.

We have a weight preserving bijection

$$\mathbb{T}_{\lambda/\mu} \leftrightarrow \mathbb{G}_{\lambda/\mu} \tag{11.95}$$

between the set of skew SSYW tableaux of shape λ/μ and the set of skew GT patterns of shape λ/μ, where the weight is given by

$$\begin{aligned} \alpha &= (\alpha_1, \alpha_2, \ldots, \alpha_n),\ |\alpha| = |\lambda| - |\mu|, \\ \alpha_j &= |m_j| - |m_{j-1}| = \sum_{i=1}^{n} m_{i,j} - \sum_{i=1}^{n} m_{i,j-1}. \end{aligned} \tag{11.96}$$

The multiplicity of a given weight α is denoted by $K(\lambda/\mu, \alpha)$; it is called a *skew Kostka number.*

The above procedure establishes the bijection (11.95) between skew SSYW tableau and skew GT patterns by using the elemental combinatoric concept of a horizontal strip. Equivalent methods of establishing this bijection parallel those stated for SSYW tableau and GT patterns, which we repeat.

Given a skew SSYW tableau $T_{\lambda/\mu} \in \mathbb{T}_{\lambda/\mu}$, the corresponding skew GT pattern $G_{\lambda/\mu} \in \mathbb{G}_{\lambda/\mu}$ is read off by the sequential removal of boxes containing $n, \ldots, 2$ as follows: First, the skew shape λ/μ with $\lambda, \mu \in \mathbb{P}\mathrm{ar}_n$ is read off the shape of the skew SSYW tableau, which gives both $\lambda = m_n$ and $\mu = m_0$; then the skew shape m_{n-1}/μ is read off the skew shape that remains after all boxes containing n have been removed, which gives the partition $m_{n-1} \in \mathbb{P}\mathrm{ar}_n$; the skew shape m_{n-2}/μ is read off the skew shape that remains after all boxes containing $n-1$ have been removed, which gives the partition $m_{n-2} \in \mathbb{P}\mathrm{ar}_n$; ...; the skew shape m_2/μ is read off the skew shape that remains after all boxes containing 3 have been removed, which gives the partition $m_2 \in \mathbb{P}\mathrm{ar}_n$; and the skew shape m_1/μ is read off the skew shape that remains after all boxes containing 2 have been removed, which gives the partition $m_1 \in \mathbb{P}\mathrm{ar}_n$.

Given a skew GT pattern $G_{\lambda/\mu} \in \mathbb{G}_{\lambda/\mu}$ with $\lambda, \mu \in \mathbb{P}\mathrm{ar}_n$, the corresponding skew SSYW tableau $T_{\lambda/\mu} \in \mathbb{T}_{\lambda/\mu}$ is read off in the following way: The entries in row j of the skew SSYW tableau are read off the j-th down-diagonal

$$\begin{matrix} \lambda_j & & & \\ & m_{j,n-1} & & \\ & \searrow & & \\ & & m_{j,1} & \\ & & & \mu_j \end{matrix} \tag{11.97}$$

of the skew GT pattern (11.91) by the following rule:

$$\boxed{m_{j,1} - m_{j,0}\ 1's} \boxed{m_{j,2} - m_{j,1}\ 2's} \boxed{\cdots} \boxed{m_{j,n} - m_{j,n-1}\ n's}\,, \tag{11.98}$$

where $\lambda_j = m_{j,n}$ and $\mu_j = m_{j,0}$, with $j = 1, 2, \ldots, n$.

The set $\mathbb{G}_\lambda$ of n-rowed triangular patterns is obtained from the relation

$$\left[\begin{matrix} \lambda/(0^n) \\ m \end{matrix} \right] = \left(\begin{matrix} \lambda \\ m \end{matrix} \right), \tag{11.99}$$

where we drop the inverted triangle of n rows of $0's$ that occurs in the skew array (11.91).

Examples.

$\lambda = (6, 5, 3, 0), \mu = (3, 2, 1, 0)$:

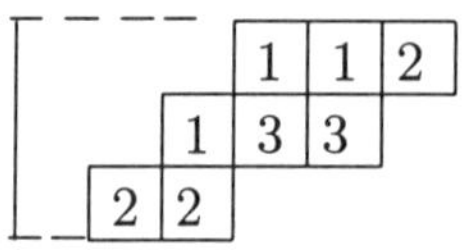

$$\longleftrightarrow \quad \begin{array}{ccccccccccccc} 6 && 5 && 3 && 0 &&&&&& \\ & 6 && 5 && 3 && 0 &&&&& \\ && 6 && 3 && 3 && 0 &&&& \\ &&& 5 && 3 && 1 && 0 &&& \\ &&&& 3 && 2 && 1 && 0 && \end{array} \tag{11.100}$$

$\lambda = (6, 5, 3, 0), \mu = (2, 2, 0, 0)$:

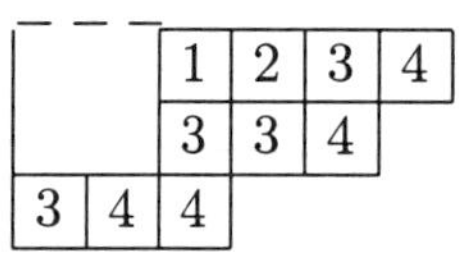

$$\longleftrightarrow \quad \begin{array}{ccccccccccccc} 6 && 5 && 3 && 0 &&&&&& \\ & 5 && 4 && 1 && 0 &&&&& \\ && 4 && 2 && 0 && 0 &&&& \\ &&& 3 && 2 && 0 && 0 &&& \\ &&&& 2 && 2 && 0 && 0 && \end{array}$$

□

11.3.4 Notations

It is useful to give a summary of the many special symbols introduced:

NOTATIONS FOR GT PATTERNS

$$\begin{aligned}
\mathbb{G}_\lambda &= \text{all GT patterns of shape } \lambda\text{, with elements } \begin{pmatrix} \lambda \\ m \end{pmatrix}; \\
\mathbb{G}_{\lambda/\mu} &= \text{all skew GT patterns of shape } \lambda/\mu\text{, with} \\
&\quad \text{elements } \begin{bmatrix} \lambda/\mu \\ m \end{bmatrix}; \\
\mathbb{G}_\lambda(\alpha) &= \text{all GT patterns of shape } \lambda \text{ and weight } \alpha \\
&= \left\{ \begin{pmatrix} \lambda \\ m \end{pmatrix} \middle| \alpha = (\alpha_1, \alpha_2, \ldots, \alpha_n) = W\begin{pmatrix} \lambda \\ m \end{pmatrix} \right\}; \\
\mathbb{G}_{\lambda/\mu}(\alpha) &= \text{all skew GT patterns of shape } \lambda/\mu \text{ and weight } \alpha \\
&= \left\{ \begin{bmatrix} \lambda/\mu \\ m \end{bmatrix} \middle| \alpha = W\begin{bmatrix} \lambda/\mu \\ m \end{bmatrix} = (\alpha_1, \alpha_2, \ldots, \alpha_n) \right\}; \\
\mathbb{W}_\lambda &= \text{all weights of GT patterns of shape } \lambda;
\end{aligned}$$

$$\begin{aligned}
\mathbb{SW}_{\lambda/\mu} &= \text{all weights of skew GT patterns of shape } \lambda/\mu;\\
|\mathbb{G}_\lambda(\alpha)| &= K(\lambda,\alpha) = \text{Kostka number};\\
|\mathbb{G}_{\lambda/\mu}(\alpha)| &= K(\lambda/\mu,\alpha) = \text{skew Kostka number.}
\end{aligned}$$

NOTATIONS FOR SSYW TABLEAUX

$$\begin{aligned}
\mathbb{T}_\lambda &= \text{all SSYW tableaux of shape } \lambda \text{, with elements } T\begin{pmatrix}\lambda\\ m\end{pmatrix};\\
\mathbb{T}_{\lambda/\mu} &= \text{all skew SSYW tableaux of shape } \lambda/\mu \text{, with}\\
&\quad\ \text{elements } T\begin{bmatrix}\lambda/\mu\\ m\end{bmatrix};\\
\mathbb{T}_\lambda(\alpha) &= \text{all SSYW tableaux of shape } \lambda \text{ and weight } \alpha\\
&= \left\{T\begin{pmatrix}\lambda\\ m\end{pmatrix}\ \middle|\ \alpha=(\alpha_1,\alpha_2,\ldots,\alpha_n)=W\begin{pmatrix}\lambda\\ m\end{pmatrix}\right\};\\
\mathbb{W}_\lambda &= \text{all weights of SSYW tableaux of shape } \lambda;\\
\mathbb{W}_{\lambda/\mu} &= \text{all weights of skew SSYW tableaux of shape } \lambda/\mu;\\
|\mathbb{T}_\lambda(\alpha)| &= K(\lambda,\alpha) = \text{Kostka number};\\
|\mathbb{T}_{\lambda/\mu}(\alpha)| &= K(\lambda/\mu,\alpha) = \text{skew Kostka number.}
\end{aligned}$$

11.3.5 Triangular and skew Gelfand-Tsetlin patterns

It is a rather remarkable property that the set of skew GT patterns $\mathbb{G}_{\lambda/\mu}$ of shape λ/μ can be identified with a subset of the ordinary triangular GT patterns in the set $\mathbb{G}_{(\lambda,0^n)}, \lambda \in \mathbb{P}\mathrm{ar}_n$. Thus, consider the triangular GT pattern with $2n$ rows and partition $(\lambda, 0^n), \lambda \in \mathbb{P}\mathrm{ar}_n$, as follows:

$$\begin{pmatrix}
\lambda_1 \quad \lambda_2 \quad \cdots \quad \lambda_n \quad 0 \quad \cdots \quad 0\\
m_{1,n-1} \quad m_{2,n-1} \quad \cdots \quad m_{n,n-1} \quad 0 \quad \cdots \quad 0\\
\vdots \qquad\qquad \vdots\\
m_{1,2} \quad m_{2,2} \quad \cdots \quad m_{n,2} \quad 0 \quad 0\\
m_{1,1} \quad m_{2,1} \quad \cdots \quad m_{n,1} \quad 0\\
\mu_1 \quad \mu_2 \quad \cdots \quad \mu_n\\
\\
(l)
\end{pmatrix}$$

$$= \left(\begin{array}{cc} \left[\begin{array}{c} \lambda/\mu \\ m \end{array} \right] & (0) \\ \left(\begin{array}{c} \mu \\ l \end{array} \right) & \end{array} \right), \tag{11.101}$$

where $\left(\begin{array}{c} \mu \\ l \end{array} \right)$ is an ordinary triangular GT pattern with n rows. In the abbreviated notation on the right, the partition μ is included both as the bottom row in the skew pattern $\left[\begin{array}{c} \lambda/\mu \\ m \end{array} \right]$ and as the top row in $\left(\begin{array}{c} \mu \\ l \end{array} \right)$. We can obtain a one-to-one correspondence between skew GT patterns $\left[\begin{array}{c} \lambda/\mu \\ m \end{array} \right]$ and triangular patterns (11.101) in which we choose the entries $l_{i,j}$ in the triangular pattern $\left(\begin{array}{c} \mu \\ l \end{array} \right)$ to be maximal; that is, we choose $l_{i,j} = \mu_i, j = i, i+1, \ldots, n-1$; each $i = 1, 2, \ldots, n-1$. Thus, we obtain the one-to-one correspondence given by

$$\left[\begin{array}{c} \lambda/\mu \\ m \end{array} \right] \longleftrightarrow \left(\begin{array}{cc} \left[\begin{array}{c} \lambda/\mu \\ m \end{array} \right] & (0) \\ \left(\begin{array}{c} \mu \\ max \end{array} \right) & \end{array} \right). \tag{11.102}$$

A consequence of this correspondence is the following relation between skew Kostka numbers for the weights of skew GT patterns having $n+1$ rows and the Kostka numbers for the weights of the special triangular patterns (11.102) having $2n$ rows:

$$K(\lambda/\mu, \alpha) = K(\Lambda, W). \tag{11.103}$$

The partition Λ is that of the GT pattern on the left-hand side of (11.101),

$$\Lambda = (\lambda, 0^n), \ \lambda \in \mathbb{P}ar_n, \tag{11.104}$$

and the weight W is that of this GT patterns for $l = (max)$, which we write as

$$W = (\mu, \alpha) \in \mathbb{W}_\Lambda, \ \alpha = W \left[\begin{array}{c} \lambda/\mu \\ m \end{array} \right]. \tag{11.105}$$

These relations were pointed out in Ref. [120]. Relation (11.101) also implies that the following formula between the number of skew SSYW tableaux and the number of GT patterns of shape Λ must hold:

$$\sum_{\mathrm{all}\, \mu \subseteq \lambda} \mathrm{Dim}(\lambda/\mu)\mathrm{Dim}\mu = \mathrm{Dim}\Lambda \,, \tag{11.106}$$

where we have defined

$$\mathrm{Dim}(\lambda/\mu) = |\mathbb{T}_{\lambda/\mu}| = |\mathbb{G}_{\lambda/\mu}| \,. \tag{11.107}$$

11.3.6 Words and lattice permutations

Words of skew SSYW tableaux and skew GT patterns and the subset of words called *lattice permutations* are very important for this monograph. This is because the number of such lattice permutations is equal to the Littlewood-Richardson numbers. The Littlewood-Richardson numbers $c^{\lambda}_{\mu\,\nu}$ are among the most important quantities in combinatorics, and arise in many different contexts (see Stanley [163]). In the context of the subject of this book, they arise when one considers Kronecker products $D^{\mu}(Z) \otimes D^{\nu}(Z)$ of a certain class of matrices with polynomial elements; the $c^{\lambda}_{\mu\,\nu}$ are the multiplicity numbers for counting the number of times the matrix $D^{\lambda}(Z)$ appears in the reduction of the Kronecker product. This feature then carries over to the reduction of the Kronecker product of two unitary irreducible representations of the general unitary group. The properties of the Littlewood-Richardson numbers are essential to a comprehensive theory of tensor operators in the unitary groups. Thus, for symmetries of many physical systems going beyond $SU(2)$, the study of the properties of the Littlewood-Richardson numbers is basic. It is for these reasons that we focus on this subject, in-depth.

Let $\gamma = (\gamma_1, \gamma_2, \cdots, \gamma_n)$ be a sequence of nonnegative integers with $|\gamma| = k$ (composition of k into n nonnegative parts).The weakly increasing sequence defined by

$$(1, 2, \cdots, n)^{\gamma} = 1^{\gamma_1} 2^{\gamma_2} \cdots n^{\gamma_n} \tag{11.108}$$

is called the *standard form* or *type* of any sequence

$$A_k = A_k(\gamma) = a_1 a_2 \cdots a_k, \;\; k = \gamma_1 + 2\gamma_2 + \cdots + n\gamma_n \tag{11.109}$$

containing $\gamma_1\; 1's, \gamma_2\; 2's, \ldots, \gamma_n\; n's$. Such a sequence A_k is also called a *word* in the alphabet of "letters" $\{1, 2, \ldots, n\}$. We denote the set of all such words of type γ by $\mathbb{A}_k(\gamma)$. In writing out words, it is the custom to omit all separating commas in sequences such as (11.108).

The *word of a semistandard skew tableau*

$$T\left[\begin{array}{c}\lambda/\mu\\ m\end{array}\right] \in \mathbb{T}_{\lambda/\mu} \tag{11.110}$$

is the sequence defined by

$$L\left[\begin{array}{c}\lambda/\mu\\ m\end{array}\right] = L_{\lambda/\mu}^{(1)}\, L_{\lambda/\mu}^{(2)} \cdots L_{\lambda/\mu}^{(i)} \cdots L_{\lambda/\mu}^{(n)}, \tag{11.111}$$

where

$$L_{\lambda/\mu}^{(i)} = n^{l_{i,n}} \cdots 2^{l_{i,2}}\, 1^{l_{i,1}}, \tag{11.112}$$

in which the nonnegative integer $l_{i,j} = m_{i,j} - m_{i,j-1}$ equals the number of times integer j appears in row i of the semistandard skew tableau:

$$\text{row } i = \boxed{m_{i,1} - m_{i,0}\ 1's} \boxed{m_{i,2} - m_{i,1}\ 2's} \boxed{\cdots} \boxed{m_{i,n} - m_{i,n-1}\ n's}\,. \tag{11.113}$$

Thus, the partial word $L_{\lambda/\mu}^{(i)}$ is just the sequence of integers that appears in row i of the skew SSYW tableau $T\left[\begin{array}{c}\lambda/\mu\\ m\end{array}\right]$, as read from right-to-left, and the full word $L\left[\begin{array}{c}\lambda/\mu\\ m\end{array}\right]$ defined by (11.111) is the product of the n partial words from rows $i = 1, 2, \ldots, n$. The word $L\left[\begin{array}{c}\lambda/\mu\\ m\end{array}\right]$ is a sequence of type $(1, 2, \ldots, n)^{\alpha}$, where $\alpha \in \mathbb{W}_{\lambda/\mu}$ is the weight of $T\left[\begin{array}{c}\lambda/\mu\\ m\end{array}\right]$ and

$$L\left[\begin{array}{c}\lambda/\mu\\ m\end{array}\right] \in \mathbb{A}_{|\lambda|-|\mu|}(\alpha). \tag{11.114}$$

In this way, we be obtain a one-to-one correspondence between skew SSYW tableaux and words:

$$T\left[\begin{array}{c}\lambda/\mu\\ m\end{array}\right] \longleftrightarrow L\left[\begin{array}{c}\lambda/\mu\\ m\end{array}\right]. \tag{11.115}$$

The partial word $L_{\lambda/\mu}^{(i)}$ given by (11.112) can also be read off from the corresponding skew GT pattern $\begin{bmatrix} \lambda/\mu \\ m \end{bmatrix}$ by mapping the i–th down-diagonal ($i = 1, 2, \ldots, n$) of this pattern to this word, as follows:

$$\begin{matrix} m_{i,n} & & \\ & \ddots & \\ & & m_{i,1} \\ & & & m_{i,0} \end{matrix} \quad \mapsto L^{(i)}_{\lambda/\mu} = n^{l_{i,n}} \cdots 2^{l_{i,2}}\, 1^{l_{i,1}}, \tag{11.116}$$

where the nonnegative integers $l_{i,j} = m_{i,j} - m_{i,j-1}, j = 1, 2, \ldots, n$, are the successive differences, from bottom-to-top, of the entries along the down-diagonal. The product of all such words then gives the word $L\begin{bmatrix} \lambda/\mu \\ m \end{bmatrix}$ of the skew GT pattern. In this way, we obtain a one-to-one correspondence between skew GT patterns and words:

$$\begin{bmatrix} \lambda/\mu \\ m \end{bmatrix} \longleftrightarrow L\begin{bmatrix} \lambda/\mu \\ m \end{bmatrix}. \tag{11.117}$$

We define the set of all words associated with the set $\mathbb{T}_{\lambda/\mu}$ of skew SSYW tableaux, or, equivalently, with the set $\mathbb{G}_{\lambda/\mu}$ of skew GT patterns by

$$\mathbb{L}_{\lambda/\mu} = \left\{ L\begin{bmatrix} \lambda/\mu \\ m \end{bmatrix} \middle|\, m \in \mathbb{G}_{\lambda/\mu} \right\}. \tag{11.118}$$

We clearly have the following bijections of sets:

$$\mathbb{L}_{\lambda/\mu} \longleftrightarrow \mathbb{G}_{\lambda/\mu} \longleftrightarrow \mathbb{T}_{\lambda/\mu}. \tag{11.119}$$

All the preceding results for skew SSYW tableaux, skew GT patterns, and words apply also to ordinary SSYW tableaux and GT patterns by choosing $\mu = (0^n)$.

We have obtained three distinct sets of mathematical objects, mutually bijective. Properties of one set can be transferred to the others. Often, characterizing objects is more naturally effected from one perspective than in another. This appears to be so for the notion of a *lattice permutation,* which falls naturally into a subset of words of type $(1, 2, \ldots, n)^\alpha$, as we next discuss.

11.3.7 Lattice permutations and Littlewood-Richardson numbers

A word

$$A_k(\alpha) = a_1 a_2 \cdots a_j \cdots a_k \in \mathbb{A}_k(\alpha) \tag{11.120}$$

of type $(1, 2, \ldots, n)^\alpha$ is a lattice permutation if and only if Rule L as follows is true:

> **Rule** L : *In each left factor* $A_j = a_1 a_2 \cdots a_j, 1 \le j \le k$, of A_k, the number of $i's$ is greater than or equal to the number of $(i+1)'s$.

It follows from this rule that the sequence α is a partition, which we henceforth denote by ν.

We observe that each partition $\alpha = \nu \in \mathbb{W}_{\lambda/\mu}$ satisfies $\lambda \supseteq \nu \in \mathbb{P}\mathrm{ar}_n$. Therefore, for $\lambda, \mu, \nu \in \mathbb{P}\mathrm{ar}_n$ with $\mu, \nu \subseteq \lambda, |\lambda| = |\mu| + |\nu|$, we can define the subsets of words

$$\mathbb{L}^{\lambda}_{\mu,\nu} \subseteq \mathbb{L}_{\lambda/\mu}(\nu) \subseteq \mathbb{L}_{\lambda/\mu} \tag{11.121}$$

as follows:

$$\begin{aligned} \mathbb{L}_{\lambda/\mu}(\nu) &= \left\{ L \left[\begin{array}{c} \lambda/\mu \\ m \end{array} \right] \middle| \, W \left[\begin{array}{c} \lambda/\mu \\ m \end{array} \right] = \nu \right\}, \\ \mathbb{L}^{\lambda}_{\mu,\nu} &= \left\{ L \left[\begin{array}{c} \lambda/\mu \\ m \end{array} \right] \in \mathbb{L}_{\lambda/\mu}(\nu) \middle| \, L \left[\begin{array}{c} \lambda/\mu \\ m \end{array} \right] \text{ is a lattice permutation} \right\}. \end{aligned} \tag{11.122}$$

Similarly, we define the sets of skew GT patterns and skew SSYW tableaux whose words are those in these sets:

$$\begin{aligned} \mathbb{G}_{\lambda/\mu}(\nu) &= \{\text{subset of } \mathbb{G}_{\lambda/\mu} \text{ of weight } \nu \}, \\ \mathbb{G}^{\lambda}_{\mu,\nu} &= \{\text{subset of } \mathbb{G}_{\lambda/\mu}(\nu) \text{ with words that} \\ &\qquad \text{are lattice permutations}\}, \\ \mathbb{T}_{\lambda/\mu}(\nu) &= \{\text{subset of } \mathbb{T}_{\lambda/\mu} \text{ of weight } \nu \}, \\ \mathbb{T}^{\lambda}_{\mu,\nu} &= \{\text{subset of } \mathbb{T}_{\lambda/\mu}(\nu) \text{ with words that} \\ &\qquad \text{are lattice permutations}\}. \end{aligned} \tag{11.123}$$

The significance of these sets for the Littlewood-Richardson numbers is the following result:

Let $\lambda, \mu, \nu \in \mathbb{P}ar_n$. The Littlewood-Richardson numbers $c^{\lambda}_{\mu\nu}$ are given by

$$c^{\lambda}_{\mu\nu} = \begin{cases} |\mathbb{L}^{\lambda}_{\mu,\nu}|, & \text{for } \mu, \nu \subseteq \lambda,\ |\lambda| = |\mu| + |\nu| \\ 0, & \text{otherwise} \end{cases} \tag{11.124}$$

(See Macdonald [126], especially, the second edition.)

A basic problem is:

Give a constructive method for finding the set $\mathbb{L}^{\lambda}_{\mu,\nu} \subseteq \mathbb{L}_{\lambda/\mu}(\nu)$ of lattice permutations, hence, the Littlewood-Richardson numbers.

We can make some progress in this direction by narrowing the possibilities of selecting the lattice permutations from among the words of the skew GT pattern (11.91).

We introduce the following $(n+1)-$rowed triangular GT pattern:

$$\begin{pmatrix} \lambda/\mu \\ m \end{pmatrix} = \begin{pmatrix} \lambda_1 \quad \mu_1 \quad\quad \mu_2 \quad\quad \mu_3 \quad \cdots \quad\quad \mu_n \\ \lambda_1 \quad m_{1,n-1} \quad m_{2,n-1} \quad \cdots \quad m_{n-1,n-1} \\ \lambda_2 \quad m_{1,n-2} \quad m_{2,n-2} \cdots m_{n-2,n-2} \\ \vdots \\ \lambda_{n-1} \quad m_{1,1} \\ \lambda_n \end{pmatrix}, \tag{11.125}$$

which is to satisfy all betweenness relations in the standard way. This pattern is unusual only in our perspective of it, where the first down-diagonal is take to be the partition λ, with an extra λ_1 adjoined at the top in row $n+1$, and row $n+1$ itself continues with the partition μ, as shown. The notation on the left indicates that we consider the partitions $\lambda, \mu \in \mathbb{P}\mathrm{ar}_n$ to be specified, but otherwise all the $m_{i,j}$ entries giving a lexical pattern are allowed. The lexical conditions on the full pattern automatically require that

$$\lambda_1 \geq \mu_1 \geq \lambda_2 \geq \mu_2 \geq \cdots \geq \lambda_n \geq \mu_n \geq 0, \tag{11.126}$$

which has previously been denoted $\lambda \sqsupseteq \mu$. Because of the convention of fixing λ and μ, we call the pattern a modified Gelfand-Tsetlin (MGT) pattern.

The relationship of a MGT pattern to the skew GT pattern (11.91) is described as follows:

(i). The skew GT pattern (11.91) can be split by the up-diagonal of the parallelogram into two triangular patterns containing the same number of entries, as follows:

$$\begin{bmatrix} \lambda/\mu \\ m \end{bmatrix} = \left[\begin{pmatrix} \lambda \\ m' \end{pmatrix} \Big/ \begin{pmatrix} \widehat{m} \\ \mu \end{pmatrix}\right]. \tag{11.127}$$

(ii). The pattern $\binom{\lambda}{m'}$ in (11.127) can be chosen to be maximal; that is,

$$\begin{pmatrix} \lambda \\ m' \end{pmatrix} = \begin{pmatrix} \lambda_1 \quad \lambda_2 \quad \cdots \quad \lambda_n \\ \lambda_1 \quad \lambda_2 \quad \cdots \quad \lambda_{n-1} \\ \vdots \\ \lambda_1 \quad \lambda_2 \\ \lambda_1 \end{pmatrix}. \tag{11.128}$$

(iii). The right-most up-diagonal in the maximal pattern can be retained together with the inverted pattern $\binom{\widehat{m}}{\mu}$, and the whole array inverted to arrive at the pattern

$$\begin{pmatrix} X \quad \mu_1 \quad \mu_2 \quad \mu_3 \quad \cdots \quad \mu_n \\ \lambda_1 \quad m_{1,n-1} \quad m_{2,n-1} \quad \cdots \quad m_{n-1,n-1} \\ \lambda_2 \quad m_{1,n-2} \quad m_{2,n-2} \cdots \quad m_{n-2,n-2} \\ \vdots \\ \lambda_{n-1} \quad m_{1,1} \\ \lambda_n \end{pmatrix}, \tag{11.129}$$

in which the place marked by X has no entry. The entries $m_{i,j}$ in this array are given in terms of the entries $\widehat{m}_{k,l} = m_{k,l}$, $1 \leq l < k \leq n$, in the skew GT pattern (11.91) by $m_{i,j} = m_{n+i-j,n-j}$, $1 \leq i \leq j \leq n-1$.

We next take the definition (11.96) of the weight of the skew GT pattern (11.91) and apply it to the special MGM patterns (11.127)-(11.128). This gives the following rule: The weight α of the MGT pattern (11.125) is defined by

$$\begin{aligned} \alpha &= (\alpha_1, \alpha_2, \ldots, \alpha_n), \\ \alpha_i &= \lambda_i - \beta_{n-i+1},\ i = 1, 2, \ldots, n, \\ \beta &= (\beta_1, \beta_2, \cdots, \beta_n) = W\begin{pmatrix} \mu \\ m \end{pmatrix}. \end{aligned} \tag{11.130}$$

We also take the definition (11.111)-(11.113) of the word of the skew GT pattern (11.91) and apply it to the special MGM patterns (11.127)-(11.128). This gives the following rule: The word of the MGT pattern

(11.125) is read off as the successive differences from bottom-to-top of the entries in the $i-$th up-diagonal of the pattern (11.125) given by

$$\begin{matrix} & \mu_1 \\ \lambda_1 & \end{matrix}\ , \ i=1; \qquad \begin{matrix} & & & \mu_i \\ & & m_{i-1,n-1} & \\ & / & & \\ & m_{1,n-i+1} & & \\ \lambda_i & & & \end{matrix}\ , \ i=2,3,\ldots,n. \tag{11.131}$$

Thus, the word of the MGT pattern (11.125) is defined by

$$\begin{aligned} L\begin{pmatrix} \lambda/\mu \\ m \end{pmatrix} &= L^{(1)}\begin{pmatrix} \lambda/\mu \\ m \end{pmatrix} L^{(2)}\begin{pmatrix} \lambda/\mu \\ m \end{pmatrix} \cdots L^{(n)}\begin{pmatrix} \lambda/\mu \\ m \end{pmatrix}, \\ L^{(i)}\begin{pmatrix} \lambda/\mu \\ m \end{pmatrix} &= i^{l_{i,i}}(i-1)^{l_{i,i-1}} \cdots 1^{l_{i,1}}, \\ l_{i,j} &= m_{i-j,n-j} - m_{i-j+1,n-j+1},\ j=i,i-1,\ldots,1, \end{aligned} \tag{11.132}$$

in which $m_{0,n-i} = \lambda_i$ and $m_{i,n} = \mu_i$.

We introduce the following notations to describe the various sets introduced above:

$$\begin{aligned} \mathbb{MG}_{\lambda/\mu} &= \left\{ \begin{pmatrix} \lambda/\mu \\ m \end{pmatrix} \middle|\ m \text{ is lexical} \right\}, \\ \mathbb{MW}_{\lambda/\mu} &= \left\{ W\begin{pmatrix} \lambda/\mu \\ m \end{pmatrix} \middle|\ \text{m is lexical} \right\}, \\ \mathbb{MG}_{\lambda/\mu}(\alpha) &= \left\{ \begin{pmatrix} \lambda/\mu \\ m \end{pmatrix} \middle| W\begin{pmatrix} \lambda/\mu \\ m \end{pmatrix} = \alpha \right\}, \\ \mathbb{ML}_{\lambda/\mu}(\nu) &= \left\{ \mathbb{MG}_{\lambda/\mu}(\nu) \middle|\ \nu \in \mathbb{P}\mathrm{ar}_n \right\}. \end{aligned} \tag{11.133}$$

Not all the words in the set $\mathbb{ML}_{\lambda/\mu}(\nu)$ need be lattice permutations. We introduce the following notation for the subset of lattice permutations associated with MGT patterns:

$$\mathbb{ML}^{\lambda}_{\mu,\nu} = \{\text{subset of } \mathbb{MG}_{\lambda/\mu}(\nu) \text{ that are lattice permutations}\}. \tag{11.134}$$

We can now justify the introduction of MGT patterns (11.125) by the following main result:

Let $\lambda, \mu, \nu \in \mathbb{P}ar_n$. The Littlewood-Richardson numbers $c^{\lambda}_{\mu\nu}$ are given by

$$c^{\lambda}_{\mu\nu} = \begin{cases} |\mathrm{M}\mathbb{L}^{\lambda}_{\mu,\nu}|, & \text{for } \mu,\nu \subseteq \lambda,\ |\lambda| = |\mu| + |\nu| \\ 0, & \text{otherwise} \end{cases} \tag{11.135}$$

Proof: The proof is given by demonstrating that each lattice permutation in the set of words of the SSYW skew GT patterns $\mathbb{G}_{\lambda/\mu}$ defined by (11.111) and (11.116) is also contained in the set of words of the MGT patterns $\mathrm{M}\mathbb{G}_{\lambda/\mu}$ defined by (11.131)-(11.132); that is,

$$L^{\lambda}_{\mu,\nu} \in \mathbb{L}^{\lambda}_{\mu,\nu} \text{ implies } L^{\lambda}_{\mu,\nu} \in \mathrm{M}\mathbb{L}^{\lambda}_{\mu,\nu}. \tag{11.136}$$

While tedious, it may be shown directly that this is true, and we omit these details. □

The lattice permutations encoded in the skew GT patterns are also fully encoded in the modified GT patterns (11.129). The unspecified entry X in (11.129) can be arbitrarily chosen such that the pattern is lexical, such as $X = \lambda_1$ in (11.125). The search for lattice permutations among the words of the MGT patterns (11.125) is substantially reduced from that for the skew GT patterns (11.91).

Examples. It is useful to give examples of the above results for the Littlewood-Richardson numbers for $n = 2, 3$:

$n = 2$: Let $\mu = (\mu_1, \mu_2), \nu = (\nu_1, \nu_2), \lambda = (\lambda_1, \lambda_1) \in \mathbb{P}\mathrm{ar}_2$, with $\lambda_1 + \lambda_2 = \mu_1 + \mu_2 + \nu_1 + \nu_2$. The modified GT pattern is

$$\begin{pmatrix} \lambda_1 & & \mu_1 & & \mu_2 \\ & \lambda_1 & & \lambda_2 - \nu_2 & \\ & & \lambda_2 & & \end{pmatrix}, \quad \begin{matrix} \text{weight } = (\alpha_1, \alpha_2) = (\nu_1, \nu_2) = \\ (\lambda_1 + \lambda_2 - \mu_1 - \mu_2 - \nu_2, \nu_2) \end{matrix} \quad . \tag{11.137}$$

The word of this MGT pattern is

$$\text{word} = 1^{\lambda_1 - \mu_1}\ 2^{\nu_2}\ 1^{\lambda_2 - \mu_2 - \nu_2}, \tag{11.138}$$

which is a lattice permutation if and only if $\lambda_1 - \mu_1 \geq \nu_2$. This condition for a lattice permutation and the betweenness relations for the MGT pattern require that

$$\lambda_1 \geq \mu_1,\ \lambda_1 \geq \mu_1 + \nu_2;\ \mu_2 + \nu_2 \leq \lambda_2 \leq \mu_1 + \nu_2. \tag{11.139}$$

Thus, we find that the modified GT pattern (11.137) corresponds to a lattice permutation, that is,

$$\begin{aligned} c^{\lambda}_{\mu\nu} &= 1, \text{ if and only if} \\ (\lambda_1, \lambda_2) \in &\left\{ (\mu_1 + \nu_1 - k, \mu_2 + \nu_2 + k) \,\middle|\, k = 0, 1, \ldots, \bar{k} \right\}; \qquad (11.140) \\ c^{\lambda}_{\mu\nu} &= 0, \text{ otherwise,} \end{aligned}$$

where $\overline{k}$ is the greatest integer such that $2\overline{k} \leq \mu_1 + \nu_1 - \mu_2 - \nu_2$.

$n = 3$: Let $\mu = (\mu_1, \mu_2, \mu_3), \nu = (\nu_1, \nu_2, \nu_3), \lambda = (\lambda_1, \lambda_2, \lambda_3) \in \mathbb{P}ar_3$, with $\lambda_1 + \lambda_2 + \lambda_3 = \mu_1 + \mu_2 + \mu_3 + \nu_1 + \nu_2 + \nu_3$. Again, we can obtain an explicit formula for the Littlewood-Richardson numbers, this result, in turn, being a consequence of an explicit formula for the Kostka numbers.

We first give the formula for the Kostka numbers. For $n = 3$, the general GT pattern is

$$\begin{pmatrix} \mu_1 & & \mu_2 & & \mu_3 \\ & m_{1,2} & & m_{2,2} & \\ & & m_{1,1} & & \end{pmatrix} \qquad (11.141)$$

and has weight

$$\begin{aligned} \beta &= (\beta_1, \beta_2, \beta_3) \qquad (11.142) \\ &= (m_{1,1}, m_{1,2} + m_{2,2} - m_{1,1}, \mu_1 + \mu_2 + \mu_3 - m_{1,2} - m_{2,2}). \end{aligned}$$

Here we specify the weight β and enumerate the *GT* patterns having this given weight. It is not difficult to verify that $\beta \in \mathbb{W}_\mu$ if and only if $|\beta| = |\mu|$ and $\mu_1 \geq \beta_i \geq \mu_3, i = 1, 2, 3$. We next define the step function σ_i by

$$\sigma_i = \max(0, \mu_2 - \beta_i), \text{ each } i = 1, 2, 3. \qquad (11.143)$$

Then, the Kostka number $K(\mu, \beta)$ is given by

$$K(\mu, \beta) = (\mu_2 - \mu_3 + 1) - (\sigma_1 + \sigma_2 + \sigma_3), \qquad (11.144)$$

and the explicit set of GT patterns in $\mathbb{G}_\mu(\beta)$ having weight β are

$$\begin{pmatrix} \mu_1 & & \mu_2 & & \mu_3 \\ & \beta_1 + \beta_2 - \mu_3 - \sigma_3 - s + 1 & & \mu_3 + \sigma_3 + s - 1 & \\ & & \beta_1 & & \end{pmatrix},$$
$$s = 1, 2, \ldots, K(\mu, \beta). \qquad (11.145)$$

Proof. The proof is by direct enumeration: Write out all possible eight cases corresponding to the σ_i and verify that in each case the corresponding GT patterns corresponding to $s = 1, 2, \ldots, K(\mu, \beta)$ give exactly all patterns satisfying the betweenness conditions. $\square$

The Littlewood-Richardson numbers for $n = 3$ are obtained by determining the number of lattice permutations that occur among the words associated with the MGT patterns as follows:

$$\begin{pmatrix} \nu_1+\beta_1 & & \mu_1 & & & & \mu_2 & & & & \mu_3 \\ & \nu_1+\beta_1 & & \beta_2+\beta_3-\mu_3-\sigma_1-s+1 & & & & \mu_3+\sigma_1+s-1 & & & \\ & & \nu_2+\beta_2 & & & & \beta_3 & & & & \\ & & & & \nu_3+\beta_3 & & & & & & \end{pmatrix},$$

$$s = 1, 2, \ldots, K(\mu, \beta). \tag{11.146}$$

These MGT patterns encode the same information on lattice permutations as the skew GT patterns (11.91) for $n = 3$. We have set $\lambda = \nu + \beta$, since there always exists a weight $\beta \in \mathrm{W}_\mu$ such that this is possible, for all $\lambda \supseteq \mu, \nu$ (see (11.156) and (11.158) below). We obtain the Littlewood-Richardson numbers $c^\lambda_{\mu\nu}$ by counting the number of lattice permutations among the words of the MGT patt'ern (11.146).

The weight (11.130) of the MGT pattern (11.146) is

$$\alpha_1 = \nu_1,\ \alpha_2 = \nu_2,\ \alpha_3 = \nu_3. \tag{11.147}$$

The word (11.132) of the MGT pattern (11.146) is

$$\begin{aligned} &\text{word} = 1^{l_{1,1}} \quad 2^{l_{2,2}}1^{l_{2,1}} \quad 3^{l_{3,3}}2^{l_{3,2}}1^{l_{3,1}}, \\ &l_{1,1} = \nu_1 + \beta_1 - \mu_1, \qquad\qquad (11.148) \\ &l_{2,2} = \nu_2 - \beta_3 + \mu_3 + \sigma_1 + s - 1,\ l_{2,1} = \mu_1 - \beta_1 - \sigma_1 - s + 1, \\ &l_{3,3} = \nu_3,\ l_{3,2} = \beta_3 - \mu_3 - \sigma_1 - s + 1,\ l_{3,1} = \sigma_1 + s - 1. \end{aligned}$$

We have used $\beta_1 + \beta_2 + \beta_3 = \mu_1 + \mu_2 + \mu_3$ in obtaining $l_{2,1}$. Necessary and sufficient conditions that the word (11.146) be a lattice permutation are

$$l_{1,1} \geq l_{2,2},\ l_{1,1} + l_{2,1} \geq l_{2,2} + l_{3,2},\ l_{2,2} \geq l_{3,3}. \tag{11.149}$$

Application of these conditions gives

$$\begin{aligned} s &\leq (\nu_1 - \nu_2 + 1) + (\mu_2 - \beta_2 - \sigma_1), \\ s &\leq (\nu_1 - \nu_2 + 1) - \sigma_1, \\ s &\geq (\beta_3 - \mu_3 - \sigma_3 + 2) - (\nu_2 - \nu_3 + 1). \end{aligned} \tag{11.150}$$

The first and second of these relations can be written as the single form $s \leq (\nu_1 - \nu_2 + 1) + (\mu_2 - \beta_2 - \sigma_1 - \sigma_2)$. Thus, we obtain the result:

Necessary and sufficient conditions that the MGT pattern (11.146) of weight $\alpha = (\alpha_1, \alpha_2, \alpha_3) = (\nu_1, \nu_2, \nu_3) \in \mathbb{P}ar_3$ is a lattice permutation are that the index s satisfies the following three conditions:

$$\begin{aligned} &(i). \quad s \in \{1, 2, \ldots, K(\mu, \beta)\}; \\ &(ii). \quad s \leq (\nu_1 - \nu_2 + 1) + (\mu_2 - \beta_2 - \sigma_1 - \sigma_2); \\ &(iii). \quad s \geq -(\nu_2 - \nu_3 + 1) + (\beta_3 - \mu_3 - \sigma_3 + 2). \end{aligned} \tag{11.151}$$

These relations can be solved to obtain the Littlewood-Richardson numbers.

For the description the solutions of relations (11.151), it is convenient to rewrite these conditions in terms of lattice points belonging to the Möbius coordinate description of the Cartesian plane $\mathbb{R}^2$. In this description, the points are assigned the three real numbers (coordinates) (x_1, x_2, x_3) obtained by perpendicular projection onto the three axes with positive directions oriented at 120°. Accordingly, the three numbers add to zero: $x_1 + x_2 + x_3 = 0$. The positive axes are also the directions of the vertices of an equilateral triangle positioned at the origin $(0, 0, 0)$, the three axis being labeled counterclockwise. This arrangement of axes and perpendicular projections for the assignment of points is shown in Fig. 9.1 of Chapter 9, where a very detailed description of the Littlewood-Richardson numbers is carried out because of their importance for the null space properties of unit tensor operators in $U(3)$. Here we give the algebraic solutions of relations (11.151) presented in terms of the lattice points (points with integral coordinates) of the Möbius plane, and refer to Figs. 9.1-9.4 of Chapter 9 for the geometric portrayal of the Littlewood-Richardson numbers.

The lattice points in the Möbius plane of interest are those defined by the set of all partitions $\nu = (\nu_1, \nu_2, \nu_3) \in \mathbb{P}\mathrm{ar}_3$, as defined by the *lattice point variables*

$$\begin{aligned} x_1 &= \nu_2 - \nu_3 + 1, \\ x_2 &= \nu_3 - \nu_1 - 2, \\ x_3 &= \nu_1 - \nu_2 + 1. \end{aligned} \tag{11.152}$$

The solutions of (11.151) can now be described as follows: For each given Kostka number (11.144), abbreviated to $K = K(\mu, \beta)$, we define the following fixed Möbius coordinates $x(K) = (x_1(K), x_2(K), x_3(K))$:

$$\begin{aligned} x_1(K) &= \beta_3 - \mu_3 - \sigma_1 + 1, \\ x_2(K) &= -K + \beta_1 - \mu_1 - \sigma_2 - 1, \\ x_3(K) &= \beta_2 - \mu_3 - \sigma_3 + 1. \end{aligned} \tag{11.153}$$

For each value of $L \in \{0, 1, \ldots, K\}$, we define the following Möbius coordinates $x'(L) = (x'_1(L), x'_2(L), x'_3(L))$:

$$\begin{aligned} x'_1(L) &= x_1(K) + L - K, \\ x'_2(L) &= x_2(K) + 2K - 2L, \\ x'_3(L) &= x_3(K) + L - K, \end{aligned} \tag{11.154}$$

In terms of these discrete Möbius plane variables, we now formulate (11.151) as follows:

Necessary and sufficient conditions that the modified GT pattern (11.146) determines a lattice permutation are

$$\max(1, x_1(K) - x_1 + 1) \leq s \leq \min(K, x_3 - x_3(K) + K). \tag{11.155}$$

These inequalities can be solved to give uniquely the values of the index s that enumerate all modified GT patterns (11.146) that correspond to the L lattice permutations for which the value of Littlewood-Richardson number is L. We have the following result:

The Littlewood-Richardson number has the value

$$c_{\mu\nu}^{\nu+\beta} = L,\ 1 \leq L \leq K, \tag{11.156}$$

if and only if the lattice points (x_1, x_2, x_3) in the Möbius plane satisfy Conditions 1, 2, or 3, as given by

$$\begin{aligned} &1.\ x_1 = \max(1, x'_1(L)),\ x_3 \geq x_3(K), \\ &2.\ x_2 = x'_2(L),\ x_3(L) \leq x_3 \leq x_3(K),\ x_1(L) \leq x_1 \leq x_1(K), \\ &3.\ x_3 = \max(1, x'_3(L)),\ x_1 \geq x_1(K). \end{aligned} \tag{11.157}$$

Proof. The values of s for which the MGT patterns (11.140) correspond to lattice permutations are: (1) $s = K - L + 1, K - L + 2, \ldots, K$; (2) $s = x_1(K) - x_1 + 1, x_1(K) - x_1 + 2, \ldots, K - x_1 - x_2(L) - x_3(K)$; and (3) $s = 1, 2, \ldots, L$. In each of these three case, there are L patterns. □

Relations (11.151) also apply to the case where there are no values of s for which the inequalities are true; that is, there are no lattice permutations among the words of the MGT patterns (11.146), in which case the Littlewood-Richardson numbers are zero. We find:

The Littlewood-Richardson number has the value

$$c_{\mu\nu}^{\nu+\beta} = 0, \tag{11.158}$$

if and only if the lattice points (x_1, x_2, x_3) in the Möbius plane satisfy the following conditions (a),(b), or (c):

$$\begin{aligned} &\text{(a). } 1 \leq x_1 \leq x_1'(0) = x_1(K) - K;\\ &\text{(b). } 1 \leq x_3 \leq x_3'(0) = x_3(K) - K;\\ &\text{(c). } x_1'(0) + 1 \leq x_1 \leq x_1(K) - 1,\\ &\quad\;\; x_3'(0) + 1 \leq x_3 \leq x_3(K) - 1,\\ &\quad\;\; x_2'(0) - K \leq x_2 \leq -x_2'(0) - 2. \end{aligned} \tag{11.159}$$

Since the lattice points corresponding to $L = 0, 1, \ldots, K$ obtained above cover all lattice points in the Möbius plane satisfying $x_1 \geq 1, x_2 \leq -2, x_3 \geq 1$, all lattice permutations originating from the MGT pattern (11.146) have been found. The nature of these solutions is more vividly portrayed by the Möbius plane pictures in Figs. 9.1-9.4, Chapter 9. □

A number of properties of Littlewood-Richardson numbers can be deduced from the general MGT pattern (11.125), such as

Let $\lambda, \mu, \nu \in \mathbb{P}ar_n$, with $\lambda \supseteq \mu, \nu$, and $|\lambda| = |\mu| + |\nu|$. Then, the maximum value of the Littlewood-Richardson number $c_{\mu\nu}^{\lambda}$ is given by a Kostka number, namely,

$$c_{\mu\nu}^{\lambda} \leq \min\{K(\mu, \lambda - \nu), K(\nu, \lambda - \mu)\}. \tag{11.160}$$

Proof: For $\lambda_n \geq \mu_1$ in the MGT pattern (11.125), every lexical pattern $\binom{\mu}{m}$ is allowed, and the total number of patterns for given λ and μ is equal to Dimμ. Thus, there are $K(\mu, \beta)$ patterns of weight β. But the weight β is given in terms of the weight α of the full MGT pattern by $\beta = (\beta_1, \beta_2, \ldots, \beta_n)$, where, from (11.130), we have $\beta_i = \lambda_{n-i+1} - \alpha_{n-i+1}$. But the Kostka number $K(\mu, \beta)$ is invariant under permutations of the parts β_i of β. Thus, the number of MGT patterns with weight $\alpha = \nu \in \mathbb{P}\text{ar}_n$ is given by $K(\mu, \lambda - \nu)$. Since the number of lattice permutations cannot exceed this number, we have $c_{\mu\nu}^{\lambda} \leq K(\mu, \lambda - \nu)$. This applies to the case where $\lambda_n \geq \mu_1$, but any other choice of λ restricts the value of $c_{\mu\nu}^{\lambda}$ to an even lesser value, hence, $c_{\mu\nu}^{\lambda} \leq K(\mu, \lambda - \nu)$ holds in all cases. Using the symmetry $c_{\mu\nu}^{\lambda} = c_{\nu\mu}^{\lambda}$, proves the full property (11.160) (see (11.178) and (11.186) below). □

An immediate consequence of (11.160) is:

Let $\lambda, \mu, \nu \in \mathbb{P}ar_n$ with $|\lambda| = |\mu| + |\nu|$ and $\mu, \nu \subseteq \lambda$. The Littlewood-Richardson number $c^{\lambda}_{\mu\nu}$ is 0 unless the following relations are valid:

$$c^{\lambda}_{\mu\nu} = 0, \text{ unless } (\lambda - \nu) \in \mathbb{W}_{\mu} \text{ and } (\lambda - \mu) \in \mathbb{W}_{\nu}. \tag{11.161}$$

These relations are stronger than the ones usually stated, which are : $c^{\lambda}_{\mu\nu} = 0$, unless $\mu \subseteq \lambda$ and $\nu \subseteq \lambda$. For example, for $\mu = (6,2,0), \nu = (2,1,0), \lambda = (6,5,0)$, both $\mu \subseteq \lambda$ and $\nu \subseteq \lambda$ are satisfied, yet $c^{\lambda}_{\mu\nu} = 0$ because $\lambda - \mu = (0,3,0)$ is not a weight in $\mathbb{W}_{2,1,0}$.

Summary: The set of all MGT patterns $\mathbb{MG}_{\lambda/\mu}$, $\lambda, \mu \in \mathbb{P}\mathrm{ar}_n, \lambda \supseteq \mu$, is the collection of all patterns $\binom{\lambda/\mu}{m}$ of the form (11.125) filled-in in all possible ways with the integers $1, 2, \ldots, n$ such that the pattern is lexical. While, in appearance, such a pattern looks like an ordinary $(n+1)$−rowed triangle GT pattern, it codifies quite different information: It inherits its weight and its word from the parent special skew GT pattern of shape λ/μ, as defined by relations(11.130)-(11.132). The set of MGT patterns $\mathbb{MG}_{\lambda/\mu}$ splits into those patterns having words that are lattice permutations, and those that are not. These MGT patterns are called *lattice* MGT *patterns* and *nonlattice* MGT *patterns,* respectively. The number of lattice MGT patterns of weight ν equals the Littlewood-Richardson numbers $c^{\lambda}_{\mu\nu}$, where the partition $\nu \in \mathbb{P}\mathrm{ar}_n$ is uniquely identified as the weight of the lattice MGT pattern. The set of all lattice MGT permutations and their weights ν thus gives the set of all Littlewood-Richardson numbers that occur in the reduction of the Kronecker product $\mu \otimes \nu$; that is, confirms the relation

$$\mu \otimes \nu = \sum_{\lambda} \oplus\, c^{\lambda}_{\mu\nu}\, \lambda. \tag{11.162}$$

The split of the set $\mathbb{MG}_{\lambda/\mu}$ of MGT patterns into lattice MGT patterns and nonlattice MGT patterns can be carried in the general case by using the explicit expressions for weights, words, and the conditions for a lattice permutation. We give the results, but not the details.

We define the $\binom{n}{2}$ linear forms $L_{j,k}$ on the entries of the GT pattern $\binom{\mu}{m}$ contained in the MGM pattern (11.125) as follows:

$$\begin{aligned} L_{j,k} &= \sum_{i=1}^{k-j+1} (m_{i,n-j+1} - m_{i,n-j}) - \sum_{i=1}^{k-j} (m_{i,n-j} - m_{i,n-j-1}), \\ &\qquad 1 \leq j \leq k \leq n-1, \end{aligned} \tag{11.163}$$

where $m_{j,n} = \mu_j$, and the second summation term is 0 for $j = k$. We next define the $n-1$ integers M_j by

$$M_j = \max\{L_{j,k} \,|\, k = j, j+1, \ldots, n-1\}, \qquad \text{each } j = 1, 2, \ldots, n-1. \tag{11.164}$$

Then, the set of MGT patterns $\mathbb{MG}_{\lambda/\mu}$ that are lattice MGT patterns is the set for which the following $n-1$ conditions hold:

$$\lambda_j - \lambda_{j+1} \geq M_j,\ j = 1, 2, \ldots, n-1. \tag{11.165}$$

We conclude:

Let $\lambda, \mu, \nu \in \mathbb{P}ar_n$ be arbitrary partitions such that $\lambda \supseteq \mu$, $\lambda \supseteq \nu$, $|\lambda| = |\mu| + |\nu|$. Select the subset of MGT patterns $\mathbb{MG}_{\lambda/\mu}$ that has weight ν. Then, the Littlewood-Richardson number $c^{\lambda}_{\mu\nu}$ is the number of these MGT patterns of weight ν that are lattice MGT patterns; that is, the number that satisfy the $n-1$ conditions (11.165).

An alternative way of formulating this result synthesizes the full content of the Kronecker product relation (11.162):

Algorithm: Let $\mu, \nu \in \mathbb{P}ar_n$ be arbitrarily chosen partitions. Enumerate all lexical MGT patterns $\binom{\lambda/\mu}{m}$ corresponding to all λ given by $\lambda_i = \nu_i - \beta_{n-i+1}, i = 1, 2, \ldots, n$, where the weight $\beta = (\beta_1, \beta_2, \ldots, \beta_n)$ takes on all values $\beta \in \mathbb{W}_\mu$. Then, the number of times the partition λ occurs in the direct sum (11.162) is given by the Littlewood-Richardson number $c^{\lambda}_{\mu\nu}$, which equals the number of MGT patterns that satisfy the $n-1$ conditions

$$\lambda_j - \lambda_{j+1} \geq M_j,\ j = 1, 2, \ldots, n-1. \tag{11.166}$$

Example. Let $\mu = (2,1,0)$ and $\nu = (1,1,0)$. There are only three patterns in the set of MGT patterns (11.125) having weight $\nu = (1,1,0)$, as given by

$$\begin{pmatrix} 3 \quad 2 \quad 1 \quad 0 \\ 3 \quad 1 \quad 0 \\ 2 \quad 0 \\ 0 \end{pmatrix}, \begin{pmatrix} 3 \quad 2 \quad 1 \quad 0 \\ 3 \quad 1 \quad 0 \\ 1 \quad 1 \\ 1 \end{pmatrix}, \begin{pmatrix} 2 \quad 2 \quad 1 \quad 0 \\ 2 \quad 1 \quad 1 \\ 2 \quad 1 \\ 1 \end{pmatrix}. \tag{11.167}$$

Each of these patterns is a lattice MGT pattern (satisfies (11.165)). Thus, $c^{\lambda}_{(2,1,0)(1,1,0)} = 1$, for $\lambda = (3,2,0), (3,1,1), (2,2,1)$, which gives

$$(2,1,0) \otimes (1,1,0) = (3,2,0) \oplus (3,1,1) \oplus (2,2,1). \qquad \square \tag{11.168}$$

The algorithm given above is a fully implementable procedure for determining the Littlewood-Richardson numbers $c^{\lambda}_{\mu\nu}$ and the full Kronecker reduction rule (11.162) for arbitrary n. A geometrical interpretation remains an open question for $n \geq 4$ (see, however, Ref. [120]).

11.3.8 Kostka and Littlewood-Richardson numbers

There are many other relations between weights $\alpha \in \mathbb{W}_\lambda$, their multiplicities, the Kostka numbers $K(\lambda, \alpha)$, and the Littlewood-Richardson numbers $c^{\lambda}_{\mu\nu}$. We summarize in this section some of these relations, giving short proofs of those that are less well-known. One of the more important relations is (11.178) below, which expresses the Littlewood-Richardson numbers as a summation over Kostka numbers.

1. Generation of weights and their multiplicities.

   Let $\mathbb{W}^*_\mu$, $\mu \in \mathbb{P}\mathrm{ar}_{n-1}$, denote the multiset of all weights of the GT patterns $\mathbb{G}_\mu$ at level $n-1$. Thus, the set $\mathbb{W}^*_\mu$ contains each weight $\beta \in \mathbb{W}_\mu$ a number of times equal to the Kostka number $K(\mu, \beta)$, so that the number of partitions in the multiset is given by the Weyl dimension formula, $|\mathbb{W}^*_\mu| = \mathrm{Dim}\mu$. Let $\lambda \in \mathbb{P}\mathrm{ar}_n$ denote a selected partition and $\mathbb{G}_\lambda$ the set of GT patterns for which we wish to calculate all weights and their multiplicities, the Kostka numbers $K(\lambda, \alpha)$, at level n. It is evident from the two-rowed GT pattern

   $$\begin{pmatrix} \lambda_1\ \lambda_2\ \ldots\ \lambda_{n-1}\ \lambda_n \\ \mu_1\ \mu_2\ \ldots\ \mu_{n-1} \end{pmatrix},\ \mu \prec \lambda \tag{11.169}$$

   that each weight $\alpha \in \mathbb{W}^*_\lambda$ is included $K(\lambda, \alpha)$ times in the multiset of weights $\mathbb{W}^*_\lambda$ at level n defined by

   $$\mathbb{W}^*_\lambda = \{(\beta, |\lambda| - |\beta|) \mid \beta \in \mathbb{W}^*_\mu,\ \text{all}\ \mu \prec \lambda\}. \tag{11.170}$$

   Thus, if we have already generated all multisets of weights $\mathbb{W}^*_\mu$, each $\mu \prec \lambda$, at level $n-1$, the multiset $\mathbb{W}^*_\lambda$ gives all weights α and their multiplicities $K(\lambda, \alpha)$ at level n. This gives the following recursion relation for the Kostka numbers at level n in terms of those at level $n-1$:

   $$K(\lambda, \alpha) = \sum_{\substack{\mu \prec \lambda \\ |\mu| = |\lambda| - \alpha_n}} K(\mu, \alpha'),$$
   $$\alpha' = (\alpha_1, \alpha_2, \ldots, \alpha_{n-1}). \tag{11.171}$$

Examples.

$$\begin{aligned}
&K((4,2,0),(2,2,2)) = K((4,0),(2,2)) + K((3,1),(2,2)) \\
&+K((2,2),(2,2)) = 3, \\
&K((4,2,0,0),(2,2,1,1)) = K((4,1,0),(2,2,1)) \qquad (11.172)\\
&+K((3,2,0),(2,2,1)) = K((4,0),(2,2)) \\
&+2K((3,1),(2,2)) + K((2,2),(2,2)) = 1+2+1 = 4. \qquad \square
\end{aligned}$$

2. Symmetric group invariance of Kostka numbers.

   If $\alpha \in \mathbb{G}_\lambda$, then $\pi(\alpha) \in \mathbb{W}_\lambda$, each $\pi \in S_n$. Thus, the Kostka numbers $K(\lambda, \alpha)$ are symmetric functions of α, as also are the skew Kostka numbers $K(\lambda/\mu, \alpha)$. This result follows by induction on n from (11.171).

3. Partitions as weights (proved below).

   A partition $\mu \vdash |\lambda|$, with $\mu, \lambda \in \mathbb{P}ar_n$, is a weight in $\mathbb{W}_\lambda$ if and only if $\mu \leq \lambda$. It then follows that the set of weights $\mathbb{W}_\lambda$ is given by

$$\mathbb{W}_\lambda = \{\pi(\mu) \mid \mu \in \mathbb{P}\mathrm{ar}_n, \mu \vdash |\lambda|, \mu \leq \lambda, \pi \in S_n\}. \qquad (11.173)$$

4. Conditions for a composition $\alpha \vdash |\lambda|$ to belong to $\mathbb{W}_\lambda$.

   It follows from Item 3 that a necessary and sufficient condition that a composition $\alpha \vdash k$ be a weight in $\mathbb{W}_\lambda$, $\lambda \vdash k$, is that $\alpha_+ \leq \lambda$, where α_+ is the partition obtained from α by ordering its parts so that $(\alpha_+)_1 \geq (\alpha_+)_2 \geq \cdots \geq (\alpha_+)_n$.

5. The Baird-Biedenharn Kronecker product rule (proved below).

   Let $\mu, \nu \in \mathbb{P}\mathrm{ar}_n$. The Kronecker product $\mu \otimes \nu$ is the direct sum of partitions $\lambda \in \mathbb{P}\mathrm{ar}_n$ and $\lambda \supseteq \mu, \nu$, with $|\lambda| = |\mu| + |\nu|$, given by

$$\begin{aligned}
\mu \otimes \nu &= \sum_{m \in \mathbb{G}_\mu}^{\text{restricted}} \oplus\, \varepsilon_\pi \left(\pi \left(\nu + W\binom{\mu}{m} + \delta \right) - \delta \right) \\
&= \sum_\lambda \oplus\, c^\lambda_{\mu\nu}\, \lambda, \qquad (11.174)
\end{aligned}$$

where the quantities in this relation have the following definitions:

(i). $W\binom{\mu}{m}$ denotes the weight of the GT pattern $\binom{\mu}{m} \in \mathbb{G}_\mu$ and δ is the sequence $\delta = (n-1, n-2, \ldots, 1, 0)$.

(ii). If the parts of the sequence $\nu + W\binom{\mu}{m} + \delta$ are distinct, then $\pi \in S_n$ is the unique permutation $\pi \in S_n$ such that $\pi(\nu + W\binom{\mu}{m} + \delta) = (\nu + W\binom{\mu}{m} + \delta)_+$, and ε_π is the signature of the permutation π; if two or more parts of $\nu + W\binom{\mu}{m} + \delta$ are equal, the term $\pi(\nu + W\binom{\mu}{m} + \delta) - \delta$ is omitted, which is the meaning of the restricted summation.

(iii). When the middle restricted summation is effected, according to the rule in Items (ii), the summation comes to the form of the right-hand side; hence, the Littlewood-Richardson number is uniquely determined.

Example. For $\mu = (2,1,0)$ and $\nu = (1,1,0)$, the terms corresponding to $m = \binom{2\ 1}{1}$, $m = \binom{1\ 0}{1}$, $m = \binom{1\ 0}{0}$ all have sequences $\nu + W\binom{\mu}{m} + \delta = (3,2,0) + W\binom{2\ 1\ 0}{m}$ with two parts equal; these terms are omitted from the restricted summation. The remaining five terms corresponding to $m = \binom{2\ 1}{2}$, $m = \binom{2\ 0}{2}$, $m = \binom{2\ 0}{1}, m = \binom{2\ 0}{0}, m = \binom{1\ 1}{1}$ all have sequences $\nu + W\binom{\mu}{m} + \delta = (3,2,0) + W\binom{2\ 1\ 0}{m}$ that are distinct, and the corresponding permutations are $\pi = (1,2,3), (1,2,3), (1,2,3), (2,1,3), (1,2,3)$, with $\varepsilon_\pi = 1, 1, 1, -1, 1$. The five terms corresponding terms in the middle summation in (11.174) are $(3,2,0) \oplus (3,1,1) \oplus (2,2,1) \ominus (2,2,1) \oplus (2,2,1) = (3,2,0) \oplus (3,1,1) \oplus (2,2,1)$. Thus, we find

$$(2,1,0) \otimes (1,1,0) = (3,2,0) \oplus (3,1,1) \oplus (2,2,1), \tag{11.175}$$

from which the Littlewood-Richardson numbers can be identified in the right-hand side of (11.174), The entire procedure just carried out is somewhat simpler if μ and ν are interchanged. □

The Baird-Biedenharn rule (11.174) (Baird and Biedenharn [7]) is very useful from a calculational viewpoint, but other forms are interesting for revealing the inter-relations between Kostka numbers, the symmetric group, and Littlewood-Richardson numbers (Ref. [120]).

It is notationally economical to introduce the action of a permutation $\pi = (\pi_1, \pi_2, \ldots, \pi_n) \in S_n$ on the set of all sequences of the form $a = (a_1, a_2, \ldots, a_n)$ by

$$a \circ \pi = (a_{\pi_1} - \pi_1 + 1, a_{\pi_2} - \pi_2 + 2, \ldots, a_{\pi_n} - \pi_n + n) = \pi(a + \delta) - \delta. \tag{11.176}$$

This group action satisfies the rules for the action of a group on a set.

The restricted summation over GT patterns $m \in \mathbb{G}_\mu$ in relation (11.174) can be eliminated in favor of a summation over all permutations $\pi \in S_n$. We first observe that the restricted summation can be written as

$\sum_{\alpha \in \mathbb{W}_\mu} K(\mu, \alpha)\left((\nu+\alpha) \circ \pi\right)$, where the summation is over all $\alpha \in \mathbb{W}_\mu$ such that the parts of $\nu+\alpha$ are distinct. But then $(\nu+\alpha) \circ \pi = (\nu+\alpha)_+ = \lambda$ is a partition satisfying $\lambda \in \mathbb{P}\mathrm{ar}_n, \lambda \supseteq \mu, \nu, |\lambda| = |\mu| + |\nu|$: these are exactly the partitions that occur in the right-hand side of(11.174). But now the weight α in the summation may be expressed as $\alpha = \lambda \circ \pi^{-1} - \nu$; hence, the restricted summation over $\alpha \in \mathbb{W}_\mu$ can now be taken as a sum over a corresponding restricted set of permutations $\pi^{-1} \in S_n$. But we define

$$K(\mu, \beta) = 0, \text{ for } \beta \notin \mathbb{W}_\mu. \tag{11.177}$$

Thus, the sum can be extended to all permutations $\pi^{-1} \in S_n$, or, equivalently, all $\pi \in S_n$, noting that $\varepsilon_\pi = \varepsilon_{\pi^{-1}}$. Thus, replacing π^{-1} by π, we obtain the result

$$\begin{aligned} \mu \otimes \nu &= \sum_\lambda \oplus\, c^\lambda_{\mu\nu}\, \lambda, \\ c^\lambda_{\mu\nu} &= \sum_{\pi \in S_n} \varepsilon_\pi K(\mu, \lambda \circ \pi - \nu) = \sum_{\pi \in S_n} \varepsilon_\pi K(\nu, \lambda \circ \pi - \mu). \end{aligned} \tag{11.178}$$

The second relation in (11.178) for $c^\lambda_{\mu\nu} = c^\lambda_{\nu\mu}$ is similarly derived. These relations are nice results, relating three fundamental combinatorial entities.

We can go one step further and prove the more detailed relation:

$$c^\lambda_{\mu\nu} = \begin{cases} 0, \text{ unless } \lambda - \nu = \alpha \in \mathbb{W}_\mu; \\ \displaystyle\sum_{\pi \in S_n} \varepsilon_\pi\, K(\mu, (\nu+\alpha) \circ \pi - \nu), \\ \qquad \text{for } \lambda - \nu = \alpha \in \mathbb{W}_\mu, \end{cases} \tag{11.179}$$

with similar expressions for $c^\lambda_{\mu\nu}$, where μ and ν are interchanged in the right-hand side. We refer to Ref. [120] for the proof of this result (see also Kostant [94]). The proofs of Items 3 and 5 above are given next.

Proof of Item 3: We use GT patterns of shape λ. That every partition $\mu \vdash |\lambda|$ that is a weight of λ has the property $\mu \leq \lambda$ is a consequence of the fact that **every** weight α of λ satisfies $\alpha \leq \lambda$. The proof of the converse is by induction on n, where we note that the result is true for $n = 2$. We assume the property (11.173) is true for $n-1$ and prove it for n, thus closing the induction loop. Consider the two-rowed GT pattern

$$\begin{pmatrix} \lambda_1 & & \lambda_2 & & \cdots & & \lambda_{n-1} & & \lambda_n \\ & \nu_1 & & \nu_2 & & \cdots & & \nu_{n-1} & \end{pmatrix}, \ \nu \prec \lambda. \tag{11.180}$$

By the induction hypothesis, each partition $\rho \vdash |\nu|$, $\rho \leq \nu$, is a weight of ν. Define the subset $\mathbb{P}\mathrm{ar}_{n-1}(\lambda)$ of partitions in $\mathbb{P}\mathrm{ar}_{n-1}$ by

$$\mathbb{P}\mathrm{ar}_{n-1}(\lambda) = \{\rho = (\rho_1, \rho_2, \ldots, \rho_{n-1}) \mid \rho \in \mathbb{P}\mathrm{ar}_{n-1},\ \rho \leq \nu,\ \text{all } \nu \prec \lambda\}. \tag{11.181}$$

Also define the set of weights $\mathbb{W}'_\lambda$ by

$$\mathbb{W}'_\lambda = \{(\rho, |\lambda| - |\rho|) \mid \rho \in \mathbb{P}\mathrm{ar}_{n-1}(\lambda)\}. \tag{11.182}$$

That $\mathbb{W}'_\lambda \subset \mathbb{W}_\lambda$ is a consequence of Item 1. We must show that every partition $\mu \vdash |\lambda|, \mu \leq \lambda$ occurs in $\mathbb{W}'_\lambda$. By the induction hypothesis and Item 1, we see that the set $\mathbb{P}\mathrm{ar}_{n-1}(\lambda)$ contains the subset given by

$$\mathbb{P}\mathrm{ar}_{n-1}(\lambda) = \{\rho \in \mathbb{P}\mathrm{ar}_{n-1} \mid (\lambda_2, \ldots, \lambda_n) \leq \rho \leq (\lambda_1, \ldots, \lambda_{n-1})\}. \tag{11.183}$$

But now the set $\mathbb{W}'_\lambda$, which is obtained by adjoining (part n) $= |\lambda| - |\rho|$ to each $(\rho_1, \rho_2, \cdots, \rho_{n-1}) \in \mathbb{P}\mathrm{ar}_{n-1}(\lambda)$ contains all partitions $\mu = (\rho_1, \rho_2, \cdots, \rho_{n-1}, |\lambda| - |\rho|)$ of $|\lambda|$ that are less than or equal to λ. This result closes the induction loop and proves that every partition $\mu \in \mathbb{P}\mathrm{ar}_n, \mu \vdash \lambda, \mu \leq \lambda$ is a weight of λ. □

Proof of Item 5. For the proof of the Baird-Biedenharn rule (11.174), we refer ahead to the definition and properties of *extended Schur functions* given in Sect. 11.6.4. For each pair of partitions $\mu, \nu \in \mathbb{P}\mathrm{ar}_n$, the extended Schur functions satisfy the relation

$$S_\mu(x)S_\nu(x) = \sum_{\alpha \in \mathbb{W}_\mu} K(\mu, \alpha) S_{\nu+\alpha}(x) = \sum_{\beta \in \mathbb{W}_\nu} K(\nu, \beta) S_{\mu+\beta}(x), \tag{11.184}$$

which expresses the basis property of extended Schur functions. They are defined for arbitrary compositions $\alpha = (\alpha_1, \alpha_2, \ldots, \alpha_n)$ and satisfy:

$$S_\alpha(x) = \begin{cases} 0, & \text{if at least two parts of } \alpha + \delta \text{ are equal,} \\ \varepsilon_\pi s_{\alpha \circ \pi}(x), & \text{if } \alpha + \delta \text{ has distinct parts,} \end{cases} \tag{11.185}$$

where $\pi \in S_n$ is the unique permutation such that $\pi(\alpha + \delta) = (\alpha + \delta)_+$. Thus, if $\alpha = \lambda \in \mathbb{P}\mathrm{ar}_n$, then $S_\lambda(x) = s_\lambda(x)$. Simple examples for α not a partition are $S_{1,2,0}(x) = 0, S_{1,0,2}(x) = -s_{1,1,1}(x)$. The term $S_{\nu+\alpha}(x)$ and $S_{\mu+\beta}(x)$ in the right-hand side of relation (11.184) are, in general, extended Schur functions, while those on the left are ordinary Schur functions. Ordinary Schur functions satisfy the relation

$$s_\mu(x)s_\nu(x) = \sum_\lambda c^\lambda_{\mu\nu}\, s_\lambda(x), \tag{11.186}$$

which expresses the basis property of Schur functions. We use (11.186) and substitute $S_\mu(x) = s_\mu(x), S_\nu(x) = s_\nu(x)$, in the left-hand side of relation (11.184) to obtain

$$\sum_\lambda c^\lambda_{\mu\nu}\, s_\lambda(x) = \sum_{\alpha\in\mathbb{W}_\mu} K(\mu,\alpha)S_{\nu+\alpha}(x). \qquad (11.187)$$

Using (11.185) in this result now gives the first relation for $c^\lambda_{\mu\nu}$ in (11.178). The second form (11.178) is similarly proved. □

Summary of basic relations

Let $\mu,\nu,\lambda \in \mathbb{P}\mathrm{ar}_n$; $\mu,\nu \subseteq \lambda$; $|\mu| + |\nu| = |\lambda|$. Then, we have the following relations between skew and ordinary Kostka numbers, Littlewood-Richardson numbers, and Weyl dimension formulas:

$$K(\lambda/\mu,\alpha) = \sum_{\nu\subseteq\lambda} c^\lambda_{\mu\nu} K(\nu,\alpha), \qquad (11.188)$$

$$K(\lambda/\mu,\alpha) = K((\lambda,0^n),(\mu,\alpha)), \qquad (11.189)$$

$$\mathrm{Dim}(\lambda/\mu) = \sum_{\nu\subseteq\lambda} c^\lambda_{\mu\nu}\mathrm{Dim}\nu, \qquad (11.190)$$

$$\mathrm{Dim}(\lambda,0^n) = \sum_{\mu\subseteq\lambda} \mathrm{Dim}\mu\,\mathrm{Dim}(\lambda/\mu), \qquad (11.191)$$

$$\mathrm{Dim}(\lambda,0^n) = \sum_{\mu,\nu\subseteq\lambda} c^\lambda_{\mu\nu}\mathrm{Dim}\mu\,\mathrm{Dim}\nu, \qquad (11.192)$$

$$\mathrm{Dim}\mu\,\mathrm{Dim}\nu = \sum_\lambda c^\lambda_{\mu\nu}\,\mathrm{Dim}\lambda. \qquad (11.193)$$

The first relation is proved in Macdonald [126, p. 69], and the second in Sect. 11.3.5 above. Relation (11.190) is obtained from (11.188) by summing over all weights α; relation (11.191) from (11.106) in Sect. 11.3.5; relation (11.192) from (11.191) by substitution of (11.190); and relation (11.193) from the dimensionality of the irreducible representations of $U(n)$ appearing in the Kronecker product reduction relation (11.162).

The number of GT patterns of shape $(\lambda,0^n)$ is related to the number of GT patterns of shape λ by

$$\frac{\mathrm{Dim}(\lambda,0^n)}{\mathrm{Dim}(\lambda)} = \frac{1!2!\ldots(n-1)!}{n!(n+1)!\ldots(2n-1)!}\prod_{i=1}^n(\lambda_i+n-i+1)_n. \qquad (11.194)$$

We also collect together four important relations between combinatorial quantities that occur in this monograph:

$$|\mathbb{M}^p_{n\times n}(\alpha,\beta)| = \sum_{\lambda\vdash p} K(\lambda,\alpha)K(\lambda,\beta), \tag{11.195}$$

$$|\overline{\mathbb{M}}^p_{n\times n}(\alpha,\beta)| = \sum_{\lambda\vdash p} K(\lambda,\alpha)K(\lambda^c,\beta), \tag{11.196}$$

$$K(\lambda,\alpha) = \sum_{\pi\in S_n} \varepsilon_\pi |\mathbb{M}^p_{n\times n}(\lambda\circ\pi,\alpha)|, \tag{11.197}$$

$$K(\lambda/\nu,\alpha) = \sum_{\pi\in S_n} \varepsilon_\pi |\mathbb{M}^p_{n\times n}(\lambda\circ\pi-\nu,\alpha)|. \tag{11.198}$$

We recall that $\mathbb{M}^p_{n\times n}(\alpha,\beta)$ denotes the set of $n\times n$ matrix arrays with nonnegative entries having line sums $\alpha=(\alpha_1,\ldots,\alpha_n)\vdash p$ and column sums $\beta=(\beta_1,\ldots,\beta_n)\vdash p$. Relation (11.195) is the Robinson-Schensted-Knuth identity (5.16). The notation $\overline{\mathbb{M}}^p_{n\times n}(\alpha,\beta)\subset\mathbb{M}^p_{n\times n}(\alpha,\beta)$ denotes the subset of $n\times n$ matrix arrays with only $0's$ and $1's$ as entries. The symbol λ^c denotes the Young frame with shape conjugate to λ. In the last two relations, it is implicit that $\lambda\vdash p$.

The collection of relations obtained in this section (and others) is placed under the full purview of invariant theory, the symmetric group, and modern combinatorics in the articles by Désarménien *et al.* [47], Doubilet [48], Doubilet *et al.* [50], and in the context of tensor operators in the unitary groups by Louck and Biedenharn [120]. Gustafson and Milne [73] have developed the inter-relations still further.

11.4 Generating Functions and Relations

The concept of a generating functions is often given a formal definition in mathematics (Doubilet *et al.* [49], Riordin [149], Wilf [186]). Relations for generating mathematical objects come in a bewildering variety of contexts and forms, and we must acknowledge an attitude of informality: Many relations do not fit the formal definition, and yet convey essential structural properties of objects of interest for this monograph.

We require a number of standard notations in order to present a reasonably concise description of some aspects of the fascinating subject of generating functions. Some of the notations are designed to give the same simplicity of form to multivariable objects as that of single variable objects. For clarity, we sometimes sacrifice this compactness.

11.4.1 Notations

Rising factorials:

$$(x)_n = x(x+1)\cdots(x+n-1), n \geq 1,\ (x)_0 = 1. \tag{11.199}$$

Falling factorials:

$$\begin{aligned} [x]_n &= x(x-1)\cdots(x-n+1) \\ &= (-1)^n(-x)_n,\ n \geq 1,\ [x]_0 = 1. \end{aligned} \tag{11.200}$$

Binomial functions:

$$\binom{x}{n} = \frac{[x]_n}{n!} = (-1)^n \frac{(-x)_n}{n!} = \frac{(x-n+1)_n}{n!}. \tag{11.201}$$

Addition theorem for binomial functions:

$$\binom{x+y}{n} = \sum_{k \geq 0} \binom{x}{k} \binom{y}{n-k}. \tag{11.202}$$

Binomial coefficient:

$$\binom{n}{k} = \begin{cases} \frac{n!}{k!(n-k)!} = \frac{[n]_k}{k!},\ 0 \leq k \leq n, \\ 0,\ \text{otherwise} \end{cases} \tag{11.203}$$

Factorial of a sequence of nonnegative integers $\alpha = (\alpha_1, \alpha_2, \ldots, \alpha_n)$:

$$\alpha! = \alpha_1!\alpha_2!\cdots\alpha_n!. \tag{11.204}$$

Power of a sequence of indeterminates $x = (x_1, x_2, \ldots, x_n)$:

$$x^\alpha = x_1^{\alpha_1} x_2^{\alpha_2} \cdots x_n^{\alpha_n}. \tag{11.205}$$

Multinomial coefficients:

$$\binom{k}{\alpha} = \begin{cases} \frac{k!}{\alpha_1!\alpha_2!\cdots\alpha_n!},\ \text{for } \sum_i \alpha_i = k, \\ 0, \text{otherwise} \end{cases} \tag{11.206}$$

where α is a composition of a nonnegative integer k into n nonnegative parts, which is often written as $\alpha \vdash k$.

Factorial of an $n \times m$ matrix array of nonnegative integers $A = (a_{ij})$:

$$A! = \prod_{i=1}^{n}\prod_{j=1}^{m} a_{ij}. \tag{11.207}$$

Power of an $n \times m$ matrix array of indeterminates $X = (x_{ij})$:

$$X^A = \prod_{i=1}^{n}\prod_{j=1}^{m} x_{ij}^{a_{ij}}. \tag{11.208}$$

11.4.2 Counting relations

Counting functions. This type of generating function is often of the form $f(t) = \sum_{k\geq 0} a_k t^k$, or $g(t) = \sum_{k\geq 0} a_k t^k/k!$, where $f(t)$ and $g(t)$ are known functions of a parameter (indeterminate) and the sequence of numerical coefficients $(a_0, a_1, a_2, \ldots)$ is a finite or infinite sequence. Often it is the case that this sequence has a combinatoral interpretation and may be obtained by combinatorial arguments and perhaps generated by recurrence relations. Often this leads to a simple expression for the corresponding function $f(t)$ or $g(t)$. Conversely, it may be these functions of the parameter that are known, and this may lead to a combinatorial interpretation of the expansion coefficients, and, in favorable cases, to a combinatorial explanation of their significance.

Binomial coefficients:

$$(1+t)^n = \sum_{k\geq 0}\binom{n}{k}t^k = \sum_{k\geq 0}[n]_k\frac{t^k}{k!}, \tag{11.209}$$

$$(1+t+t^2+\cdots)^n = \frac{1}{(1-t)^n} = \sum_{k\geq 0}\binom{n+k-1}{k}t^k. \tag{11.210}$$

Multinomial coefficients:

$$(x_1+x_2+\cdots+x_n)^k = \sum_{\alpha\vdash k}\binom{k}{\alpha}x^\alpha. \tag{11.211}$$

Fubonacci numbers f_n :

$$\begin{aligned}\frac{1}{1-t-t^2} &= \sum_{k\geq 0} f_k t^k, &(11.212)\\ f_{n+1} &= f_n + f_{n-1},\ n\geq 1,\ f_0 = f_1 = 1,\\ \begin{pmatrix} 1 & 1\\ 1 & 0\end{pmatrix}^n &= \begin{pmatrix} f_{n+1} & f_n\\ f_n & f_{n-1}\end{pmatrix},\ n\geq 1.\end{aligned}$$

Bernoulli numbers B_n :

$$\frac{t}{e^t-1} = \sum_{k\geq 0} B_k \frac{t^k}{k!}. \qquad (11.213)$$

11.4.3 Generating relations of functions

This type of generating function is similar to counting generating functions, except that now the generating function depends not only on a parameter t, but also on a variable x. The expansion "coefficients" are now the values $a_k(x)$ of functions defined on the variable x. Such generating functions are invaluable for developing a variety of properties of the sequence of functions $(a_0(x), a_1(x), a_2(x), \ldots)$. Treatises on special functions abound with examples of this sort. The examples included here occur in numerous contexts of physical applications.

Bernoulli polynomials $B_n(x)$:

$$\frac{2e^{tx}}{e^t-1} = \sum_{k\geq 0} B_k(x)\frac{t^k}{k!}, \qquad (11.214)$$

where the first three functions are: $B_0(x) = 1, B_(x) = x - \frac{1}{2}, B_2(x) = x^2 - x + \frac{1}{6}, \ldots$. The Bernoulli numbers are given by $B_k = B_k(0)$. One of the important applications of the Bernoulli functions is the formula for the sum of the $k-$th power of successive integers:

$$\sum_{x\in[a,b-1]} x^k = \frac{B_{k+1}(b) - B_{k+1}(a)}{k+1}, \qquad (11.215)$$

where $[a, b-1]$ denotes the set of nonnegative integers $a, a+1, \ldots, b-1$, which is defined to be empty for $b \leq a$.

Hermite polynomials:

$$\begin{aligned} e^{2tx-t^2} &= \sum_{k\geq 0} H_k(x)\frac{t^k}{k!}, &(11.216)\\ \frac{H_n(x)}{n!} &= \sum_{k\geq 0} \frac{(-1)^k}{k!}\frac{(2x)^{n-2k}}{(n-2k)!}, \\ H_n(x+y) &= \frac{1}{2^{n/2}}\sum_{k\geq 0}\binom{n}{k} H_k(\sqrt{2}x)H_{n-k}(\sqrt{2}y), \\ \frac{(2x)^n}{n!} &= \sum_{k\geq 0}\frac{1}{k!}\frac{H_{n-2k}(x)}{(n-2k)!}. \end{aligned}$$

Associated Laguerre polynomials:

$$\begin{aligned} \frac{1}{(1-t)^{\alpha+1}}e^{(-tx/(1-t))} &= \sum_{k\geq 0} L_k^{(\alpha)}(x)t^k, &(11.217)\\ L_n^{(\alpha)}(x) &= \sum_{k\geq 0}\binom{n+\alpha}{n-k}\frac{(-x)^k}{k!}, \alpha > -1/2, \\ L_n^{(\alpha+\beta+1)}(x+y) &= \sum_{k=0}^{n} L_k^{(\alpha)}(x)L_{n-k}^{(\beta)}(y), \\ \frac{x^n}{n!} &= \sum_{k\geq 0}(-1)^k\binom{n+\alpha}{n-k}L_k^{(\alpha)}(x). \end{aligned}$$

Jacobi polynomials:

$$\frac{2^{\alpha+\beta}}{(1-2tx+t^2)^{1/2}[1-t+(1-2tx+t^2)^{1/2}]^{\alpha}} \times\frac{1}{[1+t+(1-2tx+t^2)^{1/2}]^{\beta}} = \sum_{k\geq 0} P_k^{(\alpha,\beta)}(x)t^k, \quad (11.218)$$

$$P_n^{(\alpha,\beta)}(x) = \sum_{k\geq 0}\binom{n+\alpha}{s}\binom{n+\beta}{n-s}\left(\frac{x+1}{2}\right)^s\left(\frac{x-1}{2}\right)^{n-s},$$

$$\left(\frac{1-x}{2}\right)^n$$
$$= n!\sum_{k=0}^{n}(-1)^k \frac{(\alpha+\beta+2k+1)(\alpha+k+1)_{n-k}}{(n-k)!(\alpha+\beta+k+1)_{n+1}} P_k^{(\alpha,\beta)}(x).$$

Chebyshev polynomials of the first kind:

$$\begin{aligned}
\frac{1-tx}{(1-2tx+t^2)^{1/2}} &= \sum_{k\geq 0} T_k(x)t^k, \qquad (11.219)\\
T_n(x) &= \cos(n \arccos x) = \frac{1}{2^{2n}}\binom{2n}{n} P_n^{(-1/2,-1/2)}(x)\\
&= \frac{1}{2}\sum_{k\geq 0}(-1)^k \frac{n}{n-k}\binom{n-k}{k}(2x)^{n-2k},\\
T_n(\cos\theta) &= \sin n\theta.
\end{aligned}$$

Chebyshev polynomials of the second kind:

$$\begin{aligned}
\frac{1-tx}{(1-2tx+t^2)^{1/2}} &= \sum_{k\geq 0} T_k(x)t^k, \qquad (11.220)\\
U_n(x) &= \frac{1}{2}\sum_{k\geq 0}(-1)^k \binom{n-k}{k}(2x)^{n-2k},\\
\begin{pmatrix} 2x & -1 \\ 1 & 0 \end{pmatrix}^n &= \begin{pmatrix} U_n(x) & -U_{n-1}(x) \\ U_{n-1}(x) & -U_{n-2}(x) \end{pmatrix}, \ n \geq 1,\\
U_n(\cos\theta) &= \frac{\sin(n+1)\theta}{\sin\theta}.
\end{aligned}$$

11.4.4 Operator generated functions

Lie algebras and groups are perhaps the greatest source for what might be called *operator generated functions.* The general idea is that operators, usually differential operators, act on rather simple functions to produce other functions of considerable complexity. The examples given here arise in quantum mechanics from operator similarity transformations used by King [89] to transform the Schrödinger equation for simple systems to easily solved first-order linear differential equations. Such operator methods fall under the formal finite operator calculus presented in G.-C. Rota [154] and Rota *et al.* [157]. Recent work on this subject by Penson *et al.* [141] extends these results.

Displacement operator:

$$e^{aD}f(x) = f(x+a), \; D = \frac{d}{dx}. \tag{11.221}$$

Hermite polynomials:

$$H_n(x) = e^{-\frac{1}{4}D^2}(2x)^n = (2x - D)^n\,(1), \tag{11.222}$$

where the notation $\Omega(1)$ denotes the action of a differential operator Ω on the function $x^0 = 1$.

Associated Laguerre polynomials:

$$L_k^\alpha(x) = \frac{(-1)^k}{k!}e^{-\Omega_\alpha}x^k, \; \Omega_\alpha = xD^2 + (\alpha+1)D. \tag{11.223}$$

Baker-Campbell-Hausdorff identities:

$$\begin{aligned} e^{tA}Be^{-tA} &= \sum_{k\geq 0}\frac{t^k}{k!}[A,B]_k, \\ e^{tA}B^ne^{-tA} &= \left(e^{tA}Be^{-tA}\right)^n, \end{aligned} \tag{11.224}$$

where the multiple commutators $[A,B]_k$ are defined by

$$\begin{aligned} &[A,B]_0 = B, \; [A,B]_1 = AB - BA, \; [A,B]_2 = [A,[A,B]], \\ &\qquad \ldots, \; [A,B]_k = [A,[A,B]_{k-1}], \; \ldots . \end{aligned} \tag{11.225}$$

If A is a differential operator such that A $(1) = 0$, then,

$$e^AB\,(1) = \sum_{k\geq 0}\frac{1}{k!}[A,B]_k\,(1). \tag{11.226}$$

Examples.

Displacement operator:

$$\begin{aligned} e^{aD}xe^{-aD} &= x+a, \\ e^{aD}x^n e^{-aD} &= (x+a)^n, \\ e^{aD}f(x)e^{-aD} &= f(x+a). \end{aligned} \tag{11.227}$$

Rotation operator:

$$e^{-i\phi A_3}A_1 e^{i\phi A_3} = \cos(2\phi)A_1 + \sin(2\phi)A_2, \tag{11.228}$$

for either $A_i = \frac{1}{2}\sigma_i$, where the σ_i are the Pauli matrices,

$$\begin{aligned} \sigma_1 &= \begin{pmatrix} 0 & 1 \\ 1 & 0 \end{pmatrix}, \ \sigma_2 = \begin{pmatrix} 0 & -i \\ i & 0 \end{pmatrix}, \\ \sigma_3 &= \begin{pmatrix} 1 & 0 \\ 0 & -1 \end{pmatrix}; \end{aligned} \tag{11.229}$$

or the differential operators given by

$$\begin{aligned} A_1 &= \frac{1}{4}\left(z_1\frac{\partial}{\partial z_2} + z_2\frac{\partial}{\partial z_1}\right), \ A_2 = \frac{-i}{4}\left(z_1\frac{\partial}{\partial z_2} - z_2\frac{\partial}{\partial z_1}\right), \\ A_3 &= \frac{1}{2}\left(z_1\frac{\partial}{\partial z_1} - z_2\frac{\partial}{\partial z_2}\right). \end{aligned} \tag{11.230}$$

The **commutators** are the same for either $A_i = \frac{1}{2}\sigma_i$ or for the differential operators A_i in the complex variables (z_1, z_2, z_3) : The Baker-Campbell-Hausdorff relations depend only on the commutation relations of mathematical objects, and not on their specific realization, thus giving the same form to the right-hand side of relation (11.228). □

11.5 Multivariable Special Functions

In this section, we summarize a number of multivariable functions that are related to topics in this monograph. These descriptions are very brief. References to the literature for further details are given. Such functions are very important for applications to the physics and chemistry of complex composite systems of many particles.

11.5.1 Solid harmonics

Solid harmonics are polynomials of degree l in $(x_1, x_2, x_3) \in \mathbb{R}^3$ that solve Laplace's equation; they are also eigenfunctions of the operator $L_3 = -i(x_1\partial/\partial x_2 - x_2\partial/\partial x_1)$ with eigenvalue m (Sect. 1.3, Chapter 1):

$$\mathcal{Y}_{lm}(x_1, x_2, x_3) = \left[\frac{2l+1}{4\pi}(l+m)!(l-m)!\right]^{1/2} \times \sum_k \frac{(-x_1 - ix_2)^{m+k}(x_1 - ix_2)^k x_3^{l-m-2k}}{2^{m+2k}(m+k)!k!(l-m-2k)!}, \quad (11.231)$$

each $l \in \mathbb{N}$, each $m = -l, -l+1, \ldots, l$, where the summation is over all nonnegative integers k such that the exponents $k+m, k, l-m-2k$ are nonnegative. They are orthogonal in the inner product $(\,,\,)$ (Sect. 1.3.1) and orthonormal on the unit sphere $\mathbb{S}^2$:

$$(\mathcal{Y}_{lm}, \mathcal{Y}_{l'm'}) = \delta_{l,l'}\delta_{m,m'}\frac{(2l+1)!}{2^l l!(4\pi)}, \quad (11.232)$$

$$\int_{\mathbb{S}^2} \mathcal{Y}^*_{lm}(x)\mathcal{Y}_{l'm'}(x)dS = \delta_{l,l'}\delta_{m,m'}. \quad (11.233)$$

The definition of the solid harmonic can be extended to homogeneous polynomials of degree l in three arbitrary complex variables $z = (z_1, z_2, z_3) \in \mathbb{C}^3$, or three arbitrary commuting indeterminates. They remain eigenfunctions of the operator $L_3 = -i(z_1\partial/\partial z_2 - z_2\partial/\partial z_1)$ with eigenvalue m. The definition is made with a normalization factor:

$$\mathcal{C}_{lm}(z) = \left[\frac{4\pi}{(2l+1)(l+m)!(l-m)!}\right]^{1/2} \mathcal{Y}_{lm}(z). \quad (11.234)$$

Then, these functions satisfy the remarkably simple addition rule (see Biedenharn and Louck [21, p. 314]):

$$\mathcal{C}_{lm}(z+z') = \sum_{\substack{l_1+l_2=l \\ m_1+m_2=m}} \mathcal{C}_{l_1m_1}(z)\mathcal{C}_{l_2m_2}(z'). \quad (11.235)$$

This relation generalizes to an arbitrary number of commuting indeterminates $z^k = (z_1^k, z_2^k, z_3^k)$, $k = 1, 2, \ldots, n$:

$$\mathcal{C}_{lm}(z^1 + \cdots + z^n) = \sum_{\substack{l_1+l_2+\cdots+l_n=l \\ m_1+m_2+\cdots+m_n=m}} \mathcal{C}_{l_1m_1}(z^1)\cdots\mathcal{C}_{l_nm_n}(z^n). \quad (11.236)$$

11.5.2 Double harmonic functions in $\mathbb{R}^3$

Define $a = (a_1, a_2, a_3) \in \mathbb{R}^3, b = (b_1, b_2, b_3) \in \mathbb{R}^3$, and the dot product $a \cdot b$ by $a \cdot b = a_1 b_1 + a_2 b_2 + a_3 b_3$. Then, for points

$$x = (x_1, x_2, x_3) \in \mathbb{R}^3 \text{ and } y = (y_1, y_2, y_3) \in \mathbb{R}^3, \tag{11.237}$$

and for gradient operators defined by

$$\nabla_x = (\frac{\partial}{\partial x_1}, \frac{\partial}{\partial x_2} \frac{\partial}{\partial x_2}) \text{ and } \nabla_y = (\frac{\partial}{\partial y_1}, \frac{\partial}{\partial y_2}, \frac{\partial}{\partial y_3}), \tag{11.238}$$

we have the generator relations given by

$$\begin{aligned} &\frac{1}{((x_1 - ty_1)^2 + (x_2 - ty_2)^2 + (x_3 - ty_3)^2)^{1/2}} \\ &= e^{-ty \cdot \nabla_x} \frac{1}{(x \cdot x)^{1/2}} = \sum_{l \geq 0} \frac{I_l(x, y)}{(x \cdot x)^{(2l+1)/2}} t^l, \\ &I_l(x, y) = \frac{1}{2^l} \sum_{k \geq 0} \binom{l}{k} \binom{2l - 2k}{l} (x \cdot y)^{l-2k} (x \cdot x)^k (y \cdot y)^k, \end{aligned} \tag{11.239}$$

where these relations hold for all real $t \leq \left(\frac{x \cdot x}{y \cdot y}\right)^{1/2}$. The function $I_l(x, y)$ is harmonic in both x and y; that is, satisfies both the Laplace's equations:

$$\nabla_x^2 I_l(x, y) = 0, \ \nabla_y^2 I_l(x, y) = 0. \tag{11.240}$$

These results are proved in Biedenharn and Louck [21, p. 304].

11.5.3 Hypergeometric functions

The hypergeometric functions ${}_pF_q$ in p numerator parameters, q denominator parameters, and a single indeterminate z, all of which are commuting indeterminates, is the formal series defined by

$${}_pF_q \left(\begin{matrix} a_1, a_2, \ldots, a_p \\ b_1, b_1, \ldots, b_q \end{matrix} ; z \right) = \sum_{k \geq 0} \frac{(a_1)_k (a_2)_k \cdots (a_p)_k}{(b_1)_k (b_2)_k \cdots (b_q)_k} \frac{z^k}{k!}. \tag{11.241}$$

We do not concern ourselves with convergence: the form (11.241) is a formal power series in z. Of principal interest for this monograph are the hypergeometric polynomials in z that are obtained by taking the numerator parameter a_p to be the negative integer $a_p = -n \in \mathbb{N}$ so that the infinite series (11.241) terminates to a polynomial of degree n in z :

$$ {}_pF_q\left(\begin{matrix} a_1, a_2, \ldots, a_{p-1}, -n \\ b_1, b_1, \ldots, b_q \end{matrix} ; z\right) = \sum_{k=0}^{n} \frac{(a_1)_k(a_2)_k \cdots (a_{p-1})_k}{(b_1)_k(b_2)_k \cdots (b_q)_k} (-1)^k \binom{n}{k} z^k . \quad (11.242) $$

We may view relation (11.242) as defining a family of polynomials in z of degree i. Thus, we write

$$ {}_pF_q^{(i)}(z) = \sum_{j=0}^{i} a_{ij} z^j, i = 0, 1, 2, \ldots, n, \quad (11.243) $$

with coefficients

$$ a_{ij} = \frac{(a_1)_j(a_2)_j \cdots (a_{p-1})_j}{(b_1)_j(b_2)_j \cdots (b_q)_j} (-1)^j \binom{i}{j}, \ 0 \leq j \leq i \leq n. \quad (11.244) $$

Relation (11.243) is now expressed as the following column matrix multiplication for each $n = 0, 1, \ldots$:

$$ \mathrm{col}({}_pF_q^{(0)}(z),\ {}_pF_q^{(1)}(z),\ \ldots,\ {}_pF_q^{(n)}(z)), = A\,\mathrm{col}(1, z, \ldots, z^n), \quad (11.245) $$

where A is the lower triangular matrix with elements in row i and column j given by (11.244). The inversion of (11.245) is given by

$$ \mathrm{col}(1, x, \ldots, x^n) = A^{-1}\,\mathrm{col}({}_pF_q^{(1)}(z),\ {}_pF_q^{(2)}(z),\ \ldots,\ {}_pF_q^{(n)}(z)), \quad (11.246) $$

which gives the general inversion of relation (11.243) as

$$ z^i = \sum_{j=0}^{n} (A^{-1})_{ij}\ {}_pF_q\left(\begin{matrix} a_1, a_2, \ldots, a_{p-1}, -j \\ b_1, b_1, \ldots, b_q \end{matrix} ; z\right), \quad (11.247) $$
$$ i = 0, 1, \ldots, n. $$

Polynomials having such inversion formulas as this one, and the earlier ones in (11.216)-(11.218) imply, of course, the basis property for such polynomials: Every polynomial can be expressed in terms of polynomials having such inversions. There are many such formulas for familiar classical polynomials in terms of hypergeometric polynomials and other such invertible polynomials (see, for example, Andrews *et al.* [3]).

There is still another feature of hypergeometric polynomials defined by (11.242) that is important in unitary group theory: Using the identity $(x)_n/(x)_k = (x+k)_{n-k}, k \leq n$, we define the polynomial (see p. 136):

$$
{}_pP_q\left(\begin{matrix} x_1, x_2, \ldots, x_{p-1}, -n \\ y_1, y_1, \ldots, y_q \end{matrix} ; z\right)
$$

$$
\begin{aligned}
&= (y_1)_n(y_2)_n \cdots (y_q)_n \, {}_pF_q\left(\begin{matrix} x_1, x_2, \ldots, x_{p-1}), -n \\ y_1, y_1, \ldots, y_q \end{matrix} ; z\right) \\
&= \sum_{k=0}^{n} (-1)^k \binom{n}{k} z^k \prod_{i=1}^{p-1} (x_i)_k \prod_{j=1}^{q} (y_j + k)_{n-k} \, . \qquad (11.248)
\end{aligned}
$$

The polynomial ${}_pP_q$ is a well-defined polynomial of $p+q$ indeterminates $(x_1, x_2, \ldots, x_{p-1}; y_1, y_2, \ldots, y_q; z)$; it is invariant under all permutations of $(x_1, x_2, \ldots, x_{p-1})$ and all permutations of $(y_1, y_2, \ldots, y_q)$; that is, it is an invariant polynomial under the action of the direct product group $S_{p-1} \times S_q$.

There are many ways of extending the definition of hypergeometric functions to several variables $z_1, z_2, \ldots$. The ones defined below in terms of partitions have a significant role in the properties of $U(n)$ tensor operators (see Biedenharn and Louck [22]) and Ref. [112]. Let μ be a partition $\mu \in \mathbb{P}\mathrm{ar}_n$, $a = (a_1, a_2, \ldots, a_p)$ a sequence of p numerator parameters, and $b = (b_1, b_2, \ldots, b_q)$ a sequence of q denominator parameters. We first define hypergeometric coefficients in terms of partitions, and then hypergeometric Schur functions as follows:

$$
\begin{aligned}
\langle {}_p\mathcal{F}_q(a;b)|\mu\rangle &= \frac{1}{M(\mu)} \prod_{k=1}^{n} \frac{\prod_{i=1}^{p}(a_i - s + 1)_{\mu_k}}{\prod_{j=1}^{q}(b_j - s + 1)_{\mu_k}}; \qquad (11.249) \\
{}_p\mathcal{F}_q(a;b;z) &= \sum_{\mu \in \mathbb{P}\mathrm{ar}_n} \langle {}_p\mathcal{F}_q(a;b)|\mu\rangle \, s_\mu(z). \qquad (11.250)
\end{aligned}
$$

The factor $M(\mu)$ is the combinatorially defined normalization factor defined by relation (11.51). *We could replace the Schur functions s_μ in this definition by any of the symmetric functions labeled by partitions μ (see Sect. 11.6). Indeed, even the D^μ-polynomials whose definition and study constitutes much of this monograph could be used.* The properties of generalized hypergeometric functions are quite interesting (see Beyer *et al.* [12], Biedenharn and Louck [22], Louck and Biedenharn [119], Gustafson [72], Shulka [161]).

11.5.4 MacMahon's master theorem

MacMahon's master theorem states:

The coefficient of y^α (notation that of (10.21), Compendium A) in the expansion of

$$\left(\sum_{j=1}^{n} z_{1j}y_j\right)^{\alpha_1}\left(\sum_{j=1}^{n} z_{2j}y_j\right)^{\alpha_2}\cdots\left(\sum_{j=1}^{n} z_{nj}y_j\right)^{\alpha_n} \tag{11.251}$$

is equal to the coefficient of y^α in the expansion of the reciprocal of the determinant given by

$$\frac{1}{\det(I_n - D(y)Z)}, \tag{11.252}$$

where we have the following definitions of the diagonal matrix $D(y)$ and the matrix Z :

$$D(y) = diag(y_1, y_2, \ldots, y_n),\ Z = (z_{ij})_{1\le i,j\le n}\,. \tag{11.253}$$

MacMahon's master theorem has two generalizations, which are given in Sect. 1.6.2, Chapter 1. MacMahon's classical theorem is the basis of Schwinger's [160] treatment of angular momentum theory. We use this theorem in several different contexts to synthesis the so-called "quantum theory of angular momentum of many-particle systems."

11.5.5 Power of a determinant

The expression of the power of a determinant of an $n \times n$ matrix $Z = (z_{ij})_{1\le i,j\le n}$ as a polynomial in the n^2 variables is a quite interesting result, which also relates to the topics of this monograph. It is described by the following relations:

$$\det Z = \sum_{\pi\in S_n} \varepsilon_\pi z_{1,\pi_1} z_{2,\pi_2}\cdots z_{n,\pi_n}, \tag{11.254}$$

$$(\det Z)^k = \sum_{A\in \mathbb{M}_n(k)} C_k(A) Z^A, \tag{11.255}$$

$$C_k(A) = \sum_{\alpha\in \mathbb{C}_{n,k}(A)} (-1)^{\sigma(\alpha)}\binom{k}{\alpha}, \tag{11.256}$$

where the notations in these relations have the following definitions:

1. The summation in (11.254) is over all permutations $\pi \in S_n$, and ε_π is the signature of the permutation, which is 1 for even permutations and -1 for odd permutations.

2. The symbol $\mathbb{M}_n(k)$ denotes the set of magic squares of order n and line-sum k; that is, a matrix of order n in which the entries are nonnegative integers such that the row and column sums are all equal to the positive integer k. The summation in (11.255) is over all magic square matrices $A \in \mathbb{M}_n(k)$.

3. The summation over α in (11.256) is over all special compositions of k into $n!$ nonnegative parts that are also members of the set $\mathbb{C}_{n,k}(A)$ by

$$\mathbb{C}_{n,k}(A) = \left\{ \alpha_\pi \,\middle|\, A = \sum_{\pi \in S_n} \alpha_\pi P_\pi \right\}, \tag{11.257}$$

where P_π denotes a *permutation matrix* of order n with elements given by $(P_\pi)_{ij} = \delta_{i,\pi_j}$, where π_j is the $j-$th part of the permutation $\pi = (\pi_1, \ldots, \pi_j, \cdots, \pi_n)$. Thus, for each magic square $A \in \mathbb{M}_n(k)$, the compositions in the set $\mathbb{C}_{n,k}(A)$ must satisfy the n^2 relations:

$$a_{ij} = \sum_{\pi \in S_n^{(i,j)}} \alpha_\pi, \; 1 \le i, j \le n, \tag{11.258}$$

where $S_n^{(i,j)}$ is the subgroup of the symmetric group S_n defined by

$$S_n^{(i,j)} = \{\pi \in S_n \mid \pi_j = i\}. \tag{11.259}$$

These relations are such that the conditions $\sum_{\pi \in S_n} \alpha_\pi = k$ in (11.256) for a nonzero multinomial coefficient $\binom{k}{\alpha}$ are always satisfied. The sign factor $\sigma(\alpha)$ in (11.256) is given by

$$\sigma(\alpha) = \sum_{\pi\,\text{odd}} \alpha_\pi. \tag{11.260}$$

Example. The expansion coefficients $C_3(A), A \in \mathbb{M}_3(k)$, are given by

$$\begin{aligned} C_k(A) &= \sum_{\alpha \in \mathbb{C}_{3,k}(A)} (-1)^{\alpha_{132}+\alpha_{213}+\alpha_{321}} \\ &\times \binom{k}{\alpha_{123}, \alpha_{132}, \alpha_{231}, \alpha_{213}, \alpha_{312}, \alpha_{321}}, \end{aligned} \tag{11.261}$$

where the set $\mathbb{C}_{3,k}(A)$ is the set of all nonnegative integers $\alpha_{\pi_1\pi_2\pi_3}$ that satisfy the following Diophantine relations for each magic square $A \in \mathbb{M}_3(k)$:

$$\begin{array}{lll} \alpha_{123}+\alpha_{132}=a_{11} & \alpha_{312}+\alpha_{213}=a_{12} & \alpha_{231}+\alpha_{321}=a_{13} \\ \alpha_{231}+\alpha_{213}=a_{21} & \alpha_{123}+\alpha_{321}=a_{22} & \alpha_{312}+\alpha_{132}=a_{23} \\ \alpha_{312}+\alpha_{321}=a_{31} & \alpha_{231}+\alpha_{132}=a_{32} & \alpha_{123}+\alpha_{213}=a_{33} \end{array} \quad (11.262)$$

These constraints imply that $A \in \mathbb{M}_3(k)$. □

The exponential generating function for the power of a determinant is given by

$$e^{t\det X} = \sum_{k\geq 0} (\det X)^k \frac{t^k}{k!}. \quad (11.263)$$

These results for the power of a determinant with further properties are given in Ref. [114]. The formulation given here is modified somewhat to bring in more clearly the role of the symmetric group and to relate the result to the expression of magic square matrices and doubly stochastic matrices as sums over permutation matrices with coefficients that are nonnegative integers (see Birkoff [26], Brualdi and Ryser [34], Carter and Louck [36, 37], Louck [110, 116]).

11.6 Symmetric Functions

11.6.1 Introduction

There is, perhaps, no arena in which generating functions play a more profound and unifying role than in the expansive subject of symmetric functions. The term "symmetric functions" means that this class of functions is invariant under all permutations of the variables; that is, invariant under the standard action of the symmetric group. Much of the theory concerns itself with developing various basis functions for the space of all symmetric functions, and the relationships between such bases. There is also a corresponding effort and goal of bringing the subject under the full purview of combinatorial interpretations. Despite more than a hundred years of development, the subject still inspires new insights and directions. This subject is also of great interest for this monograph because it is the model for generalization, and its role in defining the general D^λ–polynomials is fundamental. *Indeed, the $D^\lambda(Z)$ matrices could be called matrix Schur functions.*

11.6.2 Four basic symmetric functions

The following list of the various symmetric functions is given in which it is always the case that $x = (x_1, x_2, \dots, x_n)$ are commuting indeterminates, and all partitions that appear are elements of $\mathbb{P}\mathrm{ar}_n$, unless otherwise noted. Vandermonde determinants and Schur functions are given their own subsections, since their properties are so vital to this monograph.

Monomial symmetric functions $m_\lambda(x)$:

$$m_\lambda(x) = \sum_{\substack{\text{distinct permutations}\\ \alpha \text{ of the parts of } \lambda}} x_1^{\alpha_1} x_2^{\alpha_2} \cdots x_n^{\alpha_n}. \tag{11.264}$$

Elementary symmetric functions $e_k(x)$:

$$\begin{aligned} e_k(x) = m_{(1,1,\dots,1)}(x) &= \sum_{1 \le i_1 < i_2 < \cdots < i_k \le n} x_{i_1} x_{i_2} \cdots x_{i_n}, \\ &\qquad e_0(x) = 1, \\ \prod_{i \ge 1} (1 + x_i t) &= \sum_{k \ge 0} e_k(x) t^k, \\ e_{(\lambda_1, \lambda_2, \dots, \lambda_n)}(x) &= e_{\lambda_1}(x) e_{\lambda_2}(x) \cdots e_{\lambda_n}(x). \end{aligned} \tag{11.265}$$

Power sum symmetric functions $p_k(x)$:

$$\begin{aligned} p_k(x) = m_{(k,0,\dots,0)}(x) &= \sum_{i=1}^{n} x_i^k, \; p_0(x) = n, \\ \sum_{i \ge 1} \frac{x_i}{1 - x_i t} &= \sum_{k \ge 0} p_{k+1}(x) t^k, \\ p_{(\lambda_1, \lambda_2, \dots, \lambda_n)}(x) &= p_{\lambda_1}(x) p_{\lambda_2}(x) \cdots p_{\lambda_n}(x). \end{aligned} \tag{11.266}$$

Complete homogeneous symmetric functions $h_k(x)$:

$$\begin{aligned} h_k(x) &= \sum_{\lambda \vdash k} m_\lambda(x), \\ \prod_{i=1}^{n} \frac{1}{1 - x_i t} &= \sum_{k \ge 0} h_k(x) t^k. \end{aligned} \tag{11.267}$$

11.6.3 Vandermonde determinants

Vandermonde's determinant $V_n(x)$ is defined as follows:

$$\begin{aligned} V_n(x) &= \prod_{1\le i<j\le n} (x_i - x_j) \qquad (11.268)\\ &= \det \begin{pmatrix} x_1^{n-1} & \cdots & x_1 & 1 \\ x_2^{n-1} & \cdots & x_2 & 1 \\ \vdots & \cdots & \vdots & \vdots \\ x_n^{n-1} & \cdots & x_n & 1 \end{pmatrix}, \quad V_2(x_1, x_2) = x_1 - x_2. \end{aligned}$$

The product of two Vandermonde determinants satisfies the relation

$$V_m(x)V_n(y) = V_{m+n}(x;y) \Big/ \prod_{i=1}^{m} \prod_{j=1}^{n} (x_i - y_j). \qquad (11.269)$$

Determinants with elements $a_{ij} = x_i^{a_j}$ in row i and column j are powers a_j of a single variable x_i are often written in abbreviated form:

$$\det A = \left(x_i^{a_j}\right)_{1\le i,j\le n}. \qquad (11.270)$$

The expansion of such a determinant is then written

$$\det \left(x_i^{a_j}\right)_{1\le i,j\le n} = \sum_{\pi\in S_n} \varepsilon_\pi x^{a_\pi}, \qquad (11.271)$$

where $\pi = (\pi_1, \pi_2, \ldots, \pi_n) \in S_n$ and

$$x^{a_\pi} = x_1^{a_{\pi 1}} x_2^{a_{\pi 2}} \cdots x_n^{a_{\pi n}}, \text{ for } a_\pi = (a_{\pi_1}, a_{\pi_2}, \ldots, a_{\pi_n}). \qquad (11.272)$$

In terms of these notations, the Vandermonde determinant is written

$$V_n(x) = \det \left(x_i^{\delta_j}\right)_{1\le i,j\le n} = \sum_{\pi\in S_n} \varepsilon_\pi x^{\delta_\pi}, \qquad (11.273)$$

where $\delta = (n-1, n-2, \ldots, 0)$, $\delta_\pi = (\delta_{\pi_1}, \delta_{\pi_2}, \ldots, \delta_{\pi_n})$, $\delta_{\pi_i} = n - \pi_i$.

The Vandermonde determinant and the Weyl dimension $\mathrm{Dim}\lambda$ are related in the following interesting way:

$$(\mathrm{Dim}\lambda)\, V_n(x) = \sum_{\alpha\in \mathbb{W}_\lambda} K(\lambda,\alpha) V_n(x+\alpha). \qquad (11.274)$$

Proof: The right-hand side is skew-symmetric in the interchange of the every pair (x_i, x_j) in consequence of the fact that for each weight $\alpha \in \mathbb{W}_\lambda$ every permutation of that weight also belongs to $\mathbb{W}_\lambda$. From the degree of $V_n(x+\alpha)$ in each x_i, we conclude that the left-hand side is a multiple of $V_n(x)$. From $V_n(x+\alpha) = V_n(x)+$(terms of degree less than $n-1$ in each x_i), and $\sum_{\alpha\in\mathbb{W}_\lambda} K(\lambda,\alpha) = \mathrm{Dim}\lambda$, we obtain (11.274). □

Relation (11.274) can also be written in the eigenvector-eigenvalue form given by

$$\Delta_\lambda V_n(x) = (\mathrm{Dim}\lambda)\, V_n(x), \tag{11.275}$$

where Δ_λ is the differential operator defined by

$$\Delta_\lambda = \sum_{\alpha\in\mathbb{W}_\lambda} K(\lambda,\alpha)\exp(\alpha\cdot\nabla), \quad \alpha\cdot\nabla = \sum_{i=1}^{n}\alpha_i\frac{\partial}{\partial x_i}. \tag{11.276}$$

Thus, *the Vandermonde determinant is an eigenfunction of the displacement operator Δ_λ with eigenvalue equal to the Weyl dimension.*

11.6.4 Schur functions

The Schur functions $s_\lambda(x) = s_{(\lambda_1,\lambda_2,\ldots,\lambda_n)}(x_1,x_2,\ldots,x_n)$, each $\lambda \in \mathbb{P}\mathrm{ar}_n$, are often defined by the Jacobi-Trudi formula as follows:

$$s_\lambda(x) = \det\left(x_i^{\lambda_j+\delta_j}\right)_{1\le i,j\le n} \Big/ \det\left(x_i^{\delta_j}\right)_{1\le i,j\le n}, \tag{11.277}$$

where the Vandermonde determinant in the denominator divides the numerator determinant to give a homogeneous polynomial of degree n. The division is nontrivial to effect directly.

The combinatorial definition of the Schur function is based on SSYW tableaux of shape λ, or, equivalently, on GT patterns of shape λ (see Sects. (11.2) and (11.3)). The homogeneous monomial of degree $|\lambda|$ given by

$$x^\alpha = x_1^{\alpha_1}x_2^{\alpha_2}\cdots x_n^{\alpha_n},\ \text{each } \alpha \in \mathbb{W}_\lambda, \tag{11.278}$$

is associated with each filled-in SSYW tableau (or GT pattern). These monomials are then summed over all such patterns, including those of the same weight, to obtain the Schur function $s_\lambda(x)$ defined by

$$s_\lambda(x) = \sum_{\alpha\in\mathbb{W}_\lambda} K(\lambda,\alpha)x^\alpha, \tag{11.279}$$

where $K(\lambda,\alpha)$ is the Kostka number giving the multiplicity of weight $\alpha \in \mathbb{W}_\lambda$.

The Schur functions are invariant under the action of the symmetric group as expressed by permutations of the variables as given by

$$\sum_{\alpha \in \mathbb{W}_\lambda} K(\lambda, \alpha) x^{\alpha_\pi} = s_\lambda(x), \text{ each } \pi \in S_n. \tag{11.280}$$

Schur functions are among a class of symmetric polynomials that are referred to as a Z−basis for symmetric functions. This aspect of symmetric function theory is developed in detail by Macdonald [126] and Stanley [63, Vol. 2]. The invariance of the Kostka numbers under all permutations of the parts of the weight α may be used to prove (11.280) from (11.279), and conversely. We conclude:

For $\lambda \in \mathbb{P}ar_n$ and $x = (x_1, x_2, \ldots, x_n)$, the Schur function $s_\lambda(x)$ is a symmetric, homogeneous polynomial of degree $|\lambda|$.

One of the most significant properties of Schur functions is the multiplication rule:

$$s_\mu(x) s_\nu(x) = \sum_{\lambda \supseteq \mu, \nu} c^\lambda_{\mu\nu}\, s_\lambda(x), \tag{11.281}$$

where the $c^\lambda_{\mu\nu}$ are the Littlewood-Richardson numbers. Two other relations of the Schur functions to objects of interest in this monograph are:

$$s_\lambda(1, 1, \ldots, 1) = \sum_{\alpha \in \mathbb{W}_\lambda} K(\lambda, \alpha) = \text{Dim}\lambda, \tag{11.282}$$

$$\sum_{\alpha \in \mathbb{W}_\lambda} K(\lambda, \alpha) \det \left(x_i^{\alpha_j}\right)_{1 \le i,j \le n} = 0. \tag{11.283}$$

Relation (11.282), applied to the product rule for Schur functions, gives the important Weyl dimension product rule:

$$\text{Dim}\mu\, \text{Dim}\nu = \sum_{\lambda \supseteq \mu, \nu} c^\lambda_{\mu\nu}\, \text{Dim}\lambda. \tag{11.284}$$

The iteration of relation (11.281) is also relevant to this monograph in view of the relation (see relation (5.53), Chapter 5):

$$\text{trace} D^\lambda(\text{diag}(x_1, x_2, \ldots, x_n)) = s_\lambda(x_1, x_2, \ldots, x_n), \ \lambda \in \mathbb{P}\text{ar}_n. \tag{11.285}$$

Thus, we have the two relations:

$$s_{\mu^{(1)}}(x) s_{\mu^{(2)}}(x) \cdots s_{\mu^{(k)}}(x) = \sum_\lambda c^\lambda_{\mu^{(1)}\, \mu^{(2)} \ldots \mu^{(k)}}\, s_\lambda(x), \tag{11.286}$$

$$c^{\lambda}_{\mu^{(1)}\,\mu^{(2)}\ldots\mu^{(k)}} = \sum_{\nu^{(1)},\nu^{(2)},\ldots,\nu^{(k-2)}} c^{\lambda}_{\mu^{(1)}\,\nu^{(1)}} c^{\nu^{(1)}}_{\mu^{(2)}\,\nu^{(2)}} c^{\nu^{(2)}}_{\mu^{(3)}\,\nu^{(3)}} \cdots c^{\nu^{(k-2)}}_{\mu^{(k-1)}\,\nu^{(k-1)}},$$

in which $\nu^{(0)} = \lambda$, $\nu^{(k-1)} = \mu^{(k)}$, and $k = 2, 3, \ldots$ (There is no summation in the second relation for $k = 2$). For $n = 2$, this relation gives

$$\begin{aligned} c^{\lambda}_{\mu^{(1)},\mu^{(2)},\ldots,\mu^{(k)}} &= N_j(j_1, j_2, \ldots, j_k), \\ \lambda &= (2j, 0), \mu^{(i)} = (2j_i, 0),\ i = 1, 2, \ldots, k, \end{aligned} \tag{11.287}$$

where $N_j(j_1, j_2, \ldots, j_k)$ is the Clebsch-Gordan number for the coupling of k angular momenta (p. 87, Chapter 2). The generalized Littlewood-Richardson numbers (11.286) give the dimension of the multiplicity space that occurs in the reduction of the $k-$fold Kronecket product of $D^{\lambda}-$polynomials into the Kronecker direct sum. The product (11.286) of Schur functions unifies the role of CG and Littlewood-Richardson numbers.

Skew Schur functions

The definition and properties of skew Schur functions parallel those of Schur functions, with skew SSYW tableaux replacing SSYW tableaux (skew GT patterns replacing GT patterns). The definition is

$$s_{\lambda/\mu}(x) = \sum_{\alpha\in\mathbb{W}_{\lambda/\mu}} K(\lambda/\mu, \alpha) x^{\alpha}, \tag{11.288}$$

where $K(\lambda/\mu, \alpha)$ is the skew Kostka number giving the multiplicity of the weight $\alpha \in \mathbb{W}_{\lambda/\mu}$. For $\mu = (0, 0, \ldots, 0)$, skew Schur functions reduce to ordinary Schur functions. Again, the skew Schur functions are a $Z-$basis for symmetric functions.

Skew Schur functions are invariant under the action of the symmetric group S_n as expressed by permutations of the variables given by

$$\sum_{\alpha\in\mathbb{W}_{\lambda/\mu}} K(\lambda/\mu, \alpha) x^{\alpha_{\pi}} = s_{\lambda/\mu}(x), \text{ each } \pi \in S_n. \tag{11.289}$$

This relation is proved by using the invariance of the skew Kostka numbers under permutations of the parts of the weight α.

Skew Schur functions are expressed in terms of ordinary Schur functions by the relations (see Stanley [163, Vol. 2, p. 338]:

$$s_{\lambda/\mu}(x) = \sum_{\nu\subseteq\lambda} c^{\lambda}_{\mu\nu}\, s_{\nu}(x), \qquad s_{\lambda/\nu}(x) = \sum_{\mu\subseteq\lambda} c^{\lambda}_{\mu\nu}\, s_{\mu}(x). \tag{11.290}$$

These two relations shows that $c^\lambda_{\mu\nu} = c^\lambda_{\nu\mu}$. The evaluation of the first relation in (11.290) at x = (1,1,...,1) gives

$$\mathrm{Dim}(\lambda/\mu) = \sum_{\nu\subseteq\lambda} c^\lambda_{\mu\nu}\,\mathrm{Dim}\nu\,; \tag{11.291}$$

the second relation gives this relation with μ and ν interchanged. Thus:

For $\lambda, \mu \in \mathbb{P}ar_n$ and $x = (x_1, x_2, \ldots, x_n)$, the skew Schur function $s_{\lambda/\mu}(x)$ is a symmetric, homogeneous polynomial of degree $|\lambda| - |\mu|$.

We can now show an important property of the Littlewood-Richardson numbers. Define the matrix C^λ to be the matrix with elements given by

$$\left(C^\lambda\right)_{\mu,\nu} = c^\lambda_{\mu\nu},\ \text{all}\ \mu,\nu \subseteq \lambda. \tag{11.292}$$

We call this matrix the Littlewood-Richardson matrix, for which we have the result:

The Littlewood-Richardson matrix C^λ is a nonsingular, real, symmetric matrix of finite order.

Proof: The symmetric property already follows from the two relations (11.290). These relations must also be invertible, since the set of skew Schur functions is also a basis of symmetric functions:

$$s_\nu(x) = \sum_{\mu\subseteq\lambda} d^\lambda_{\nu\mu}\, s_{\lambda/\mu}(x)\,; \tag{11.293}$$

$$d^\lambda_{\nu\mu} = d^\lambda_{\mu\nu} = \left((C^\lambda)^{-1}\right)_{\nu,\mu}. \tag{11.294}$$

The order of the matrix C^λ is equal to the number of partitions whose shape is contained in shape λ, including the partitions $(0,0,\ldots,0)$ and λ. □

Schur functions and skew Schur functions also satisfy the relation

$$s_{\lambda,0^n}(x;y) = \sum_{\mu\subseteq\lambda} s_\mu(x)\, s_{\lambda/\mu}(y) = \sum_{\mu,\nu\subseteq\lambda} c^\lambda_{\mu\nu}\, s_\mu(x)\, s_\nu(y), \tag{11.295}$$

for $\lambda, \mu, \nu \in \mathbb{P}\mathrm{ar}_n$ where $x = (x_1, x_2, \ldots, x_n)$, $y = (y_1, y_2, \ldots, y_n)$. The evaluation of these relations at $x = y = (1, \ldots, 1)$ gives (see (11.106)):

$$\mathrm{Dim}(\lambda, 0^n) = \sum_{\mu\subseteq\lambda} \mathrm{Dim}\mu\, \mathrm{Dim}(\lambda/\mu) = \sum_{\nu\subseteq\lambda} \mathrm{Dim}\nu\, \mathrm{Dim}(\lambda/\nu). \tag{11.296}$$

The concept of *extended* Schur functions generalizes that of Schur functions in another direction; namely, by replacing the partition λ in $s_\lambda(x)$ by an arbitrary composition $\alpha = (\alpha_1, \alpha_2, \dots, \alpha_n)$ of nonnegative integers (see Littlewood [105]). Extended Schur functions are defined by a formula analogous to the Jacobi-Trudi formula for ordinary Schur functions:

$$S_\alpha(x) = \det\left(x_i^{\alpha_j+\delta_j}\right)_{1\le i,j\le n} \Big/ \det\left(x_i^{\delta_j}\right)_{1\le i,j\le n}, \tag{11.297}$$

where, if at least two parts of the sequence $\alpha+\delta$ are equal, then $S_\alpha(x) = 0$, since the numerator determinant is 0. Otherwise, we have the following property in direct consequence of the determinantal definition (11.297):

$$S_\alpha(x) = \varepsilon_\pi s_\lambda(x), \text{ if all parts of } \alpha+\delta \text{ are unequal}, \tag{11.298}$$

where π is the unique permutation in S_n that brings the sequence $\alpha+\delta$ to the ordered form $\pi(\alpha+\delta) = (\alpha+\delta)_+$, in which $\lambda \in \mathbb{P}\mathrm{ar}_n$ and $\lambda \vdash |\alpha|$.

Extended Schur functions satisfy the product rule:

$$S_\mu(x)S_\nu(x) = \sum_{\alpha\in\mathbb{W}_\mu} K(\mu,\alpha)S_{\nu+\alpha}(x) = \sum_{\beta\in\mathbb{W}_\nu} K(\nu,\beta)S_{\mu+\beta}(x), \tag{11.299}$$

where in the left-hand side $S_\mu(x) = s_\mu(x), S_\nu(x) = s_\nu(x)$, since extended and ordinary Schur functions agree for all partitions $\mu, \nu \in \mathbb{P}\mathrm{ar}_n$.

Proof: The multiplication rule (11.299) is equivalent to showing that

$$s_\mu(x)\det\left(x_i^{h_j}\right)_{1\le i,j\le n} = \sum_{\alpha\in\mathbb{W}_\mu} K(\mu,\alpha)\det\left(x_i^{h_j+\alpha_j}\right)_{1\le i,j\le n},$$
$$h_j = \mu_j + n - j. \tag{11.300}$$

From the definition of $\det\left(x_i^{h_j+\alpha_j}\right)_{1\le i,j\le n}$, we have the identity

$$\det\left(x_i^{h_j+\alpha_j}\right)_{1\le i,j\le n} = \sum_{\pi\in S_n} \varepsilon_\pi x^{\alpha_\pi} x^{h_\pi}. \tag{11.301}$$

Multiplying this relation by the Kostka number $K(\mu,\alpha)$, summing over all weights $\alpha \in \mathbb{W}_\mu$, using relation (11.280) with λ replaced by μ, and the fact that the summation in (11.280) is independent of the permutation $\pi \in S_n$, gives relation (11.299). □

It is interesting that the following discrete version of relations (11.275)-(11.276) may be formulated. Define the shift operator T_α by its action in the set of all compositions $\alpha, \beta, \ldots$ of length n by $\Delta_\alpha : \beta \to \beta + \alpha$, and define the corresponding action of T_α on the set of functions $\{F_\beta(x)\}$ enumerated by such compositions β by

$$T_\alpha F_\beta(x) = F_{\beta+\alpha}(x). \tag{11.302}$$

In particular, we have that

$$T_\alpha \det\left(x_i^{h_j}\right)_{1\le i,j\le n} = \det\left(x_i^{h_j+\alpha_j}\right)_{1\le i,j\le n}. \tag{11.303}$$

Extending the definition (11.302) by linearity, we can then define a new shift operator by

$$\Omega_\mu = \sum_{\alpha\in\mathbb{W}_\mu} K(\mu,\alpha) T_\alpha. \tag{11.304}$$

Relation (11.300) can now be formulated in terms of these shift operators as

$$\Omega_\mu \det\left(x_i^{h_j}\right)_{1\le i,j\le n} = s_\mu(x) \det\left(x_i^{h_j}\right)_{1\le i,j\le n}, \; h_j = \mu_j + n - j. \tag{11.305}$$

The Schur functions $s_\mu(x)$ are the eigenvalues of the shift operator Ω_μ with eigenvectors given by $\det\left(x_i^{h_j}\right)_{1\le i,j\le n}$.

11.6.5 Dual bases for symmetric functions

The inner product of pairs of symmetric functions plays an important role is determining inter-relations between symmetric functions. The inner product (,) defined and discussed in Sect. 10.6, Compendium A, is such that the monomial symmetric polynomials are orthogonal, but not orthonormal:

$$(m_\lambda, m_\mu) = m_\lambda(\partial/\partial x) m_\mu(x)\Big|_{x=0} = \delta_{\lambda,\mu} N_\lambda, \tag{11.306}$$

where N_λ is the number of distinct permutations of the parts of λ. This inner product, which is applicable to the vector space of arbitrary polynomials, is not the one introduced for symmetric functions. Instead, an inner product $\langle \, , \, \rangle$ is defined such that the set $\{m_\lambda\}$ of monomial symmetric polynomials and the set of complete symmetric functions $\{h_\mu\}$ are dual bases:

$$\langle m_\lambda, h_\mu \rangle = \delta_{\lambda,\mu}. \tag{11.307}$$

(See Macdonald [126], Stanley [163].) The motivation for such a definition of inner product comes from the following relation (Stanley [163, Vol. 2, p. 307]:

Let $\{u_\lambda\}$ and $\{v_\lambda\}$ be bases of Λ such that for all $\lambda \vdash k$ we have $u_\lambda, v_\lambda \in \Lambda^k$. Then, $\{u_\lambda\}$ and $\{v_\lambda\}$ are dual bases if and only if

$$\frac{1}{\prod_{i,j=1}^{n}(1-x_iy_j)}=\sum_{k\geq 0}\sum_{\lambda\in\mathbb{P}ar_k} u_\lambda(x)v_\lambda(y). \tag{11.308}$$

Thus, for all bases $\{u_\lambda\}$ and $\{v_\lambda\}$ satisfying this relation, we have that

$$\langle u_\lambda, v_\mu\rangle = \delta_{\lambda,\mu}. \tag{11.309}$$

Three identities leading to dual bases are (Macdonald [124, p. 33]:

$$\begin{aligned}\frac{1}{\prod_{i,j=1}^{n}(1-x_iy_j)} &= \sum_{k\geq 0}\sum_{\lambda\in\mathbb{P}ar_k} z_\lambda^{-1}p_\lambda(x)p_\lambda(y) \qquad (11.310)\\ &= \sum_{k\geq 0}\sum_{\lambda\in\mathbb{P}ar_k} m_\lambda(x)h_\lambda(y) = \sum_{k\geq 0}\sum_{\lambda\in\mathbb{P}ar_k} s_\lambda(x)s_\lambda(y),\end{aligned}$$

in which z_λ is defined by $z_\lambda = \prod_{i\geq 1} i^{h_i}\, h_i!$, where h_i denotes the number of parts of λ equal to i. Thus, the following three pairs of basis sets of symmetric polynomials give dual bases: $\{p_\lambda\}$ and $\{z_\lambda^{-1}p_\lambda\}$, $\{m_\lambda\}$ and $\{h_\lambda\}$, $\{s_\lambda\}$ and $\{s_\lambda\}$, the Schur function basis being self-dual. In particular, the definition of dual bases then leads to the result that the Schur functions are an orthonormal basis of Λ :

$$\langle s_\lambda, s_\mu\rangle = \delta_{\lambda,\mu}. \tag{11.311}$$

As examples of the inner product $\langle\ ,\ \rangle$ introduced above, we have the following relations:

$$\begin{aligned} c_{\mu\nu}^{\lambda} &= \langle s_\lambda, s_\mu s_\nu\rangle = \langle s_{\lambda/\mu}, s_\nu\rangle,\\ \langle s_\lambda, m_\mu\rangle &= K(\lambda,\mu)N_\mu, \qquad (11.312)\\ \langle s_\lambda, s_\mu\rangle &= \sum_{\nu\subseteq\lambda,\mu} K(\lambda,\nu)K(\mu,\nu)N_\nu.\end{aligned}$$

We also note the form of (11.310) (right-most identity) given by

$$\frac{1}{\prod_{i=1}^{m}\prod_{j=1}^{n}(1-x_iy_j)} = \sum_{\lambda\in\mathbb{P}\mathrm{ar}_n} s_\lambda(x)\, s_{(\lambda,0^{n-m})}(y),\ m\leq n. \tag{11.313}$$

This relation, in turn, implies all of the following: The first obtains by setting all $x_i = t$ (and changing y_j to x_i); the second from the first by setting all $x_i = 1$; and the third by equating powers of t in the second.

$$\begin{aligned}\frac{1}{\prod_{i=1}^{n}(1-tx_i)^m} &= \sum_{k\geq 0} t^k \sum_{\lambda\vdash k} \mathrm{Dim}\lambda\, s_{(\lambda,0^{n-m})}(x),\\ \frac{1}{\prod_{i=1}^{n}(1-t)^m} &= \sum_{k\geq 0} t^k \sum_{\lambda\vdash k} \mathrm{Dim}\lambda\, \mathrm{Dim}(\lambda,0^{n-m}), \\ \binom{mn+k-1}{k} &= \sum_{\lambda\vdash k} \mathrm{Dim}\lambda\, \mathrm{Dim}(\lambda,0^{n-m}).\end{aligned} \tag{11.314}$$

The right-most relation in (11.310) is a special case of a generalization of MacMahon's master theorem given by

$$\frac{1}{\det(I_{n^2}-tX\otimes Y)} = \sum_{k\geq 0} t^k \sum_{\lambda\vdash k} \mathrm{Tr}D^\lambda(X)\,\mathrm{Tr}D^\lambda(Y), \tag{11.315}$$

in which X and Y are arbitrary matrices of order n of commuting indeterminates, and $X\otimes Y$ is their Kronecker product. The relation in (11.310) is obtained by choosing X and Y to be diagonal matrices and using the following relation of the D^λ–matrices to Schur polynomials:

$$\mathrm{Tr}D^\lambda(\mathrm{diag}(x_1,x_2,\ldots,x_n)) = s_\lambda(x). \tag{11.316}$$

The structure and properties of the matrices $D^\lambda(X)$ is one of the main themes of this monograph, and relation (11.315) is an important example of their properties. This relation was conjectured to be true by Louck [108] and proved by Méndez [130, 131]. Still other generalizations of MacMahon's master theorem may prove useful for applications (see Konvalinka and Pak [92]).

11.7 Sylvester's Identity

A relationship between symmetric functions known as Sylvester's identity has a fundamental role in our development of the properties of the

$D^\lambda-$polynomials in Chapters 5-8. Recurrence relations for these polynomials depend in an essential way on Sylvester's identity, as also does the Lie algebra of the general unitary group. The history of Sylvester's identity and its proof is surveyed by Bhatnagar [13].

We next state Sylvester's identity together with some of its properties:

Sylvester's identity:

$$\sum_{j=1}^{n} x_j^{k+n-1} \Big/ \prod_{\substack{i=1\\ i\neq j}}^{n} (x_j - x_i) = \begin{cases} 0, & k = -1, \ldots, -n+1, \\ h_k(x), & k = 0, 1, \ldots . \end{cases} \tag{11.317}$$

The $h_k(x)$ are the complete homogeneous symmetric functions, which are expressed in terms of the elementary symmetric functions by

$$h_k(x) = \sum_{\alpha_1+2\alpha_2+\cdots+n\alpha_n=k} (-1)^{k-|\alpha|} \binom{|\alpha|}{\prod \alpha_i} e_1^{\alpha_1}(x) e_2^{\alpha_2}(x) \cdots e_n^{\alpha_n}(x),$$
$$\text{for } k = 1, 2, \ldots; \; h_0(x) = 1. \tag{11.318}$$

The homogeneous symmetric functions extend naturally to all negative integers from (11.317) for $x_i \neq 0, i = 1, \ldots, n:$

$$h_k(x) = 0, \text{ for } k = -1, -2, \ldots -n+1; \tag{11.319}$$

$$\begin{aligned} h_k(x) &= \sum_{\alpha_1+2\alpha_2+\cdots+n\alpha_n=-n-k} (-1)^{k-|\alpha|-1} \binom{|\alpha|}{\prod \alpha_i} \\ &\quad \times \frac{e_{n-1}^{\alpha_1}(x) e_{n-2}^{\alpha_2}(x) \cdots e_1^{\alpha_{n-1}}(x)}{e_n^{|\alpha|+1}(x)}, \end{aligned} \tag{11.320}$$
$$\text{for } k = -n-1, -n-2, \ldots; \; h_{-n}(x) = (-1)^{n-1}/e_n(x).$$

Sylvester's identity implies the following relation for all $k \geq 0:$

$$\begin{aligned} \sum_{j=1}^{n} \frac{x_j^k V_n(x - e_n(j))}{V_n(x)} &= \sum_{j=1}^{n} x_j^k \prod_{i=1}^{n} (x_j - x_i - 1) \Big/ \prod_{\substack{i=1\\ i\neq j}}^{n} (x_j - x_i) \\ &= \sum_{j=0}^{k+1} (-1)^j e_j(x+1) h_{k-j+1}(x), \end{aligned} \tag{11.321}$$

where $V_n(x)$ is Vandermonde's determinant, $e_n(j)$ is the unit row vector of length n with j–th part 1 and 0 for all other parts. The argument $x+1$ of the elementary symmetric function $e_j(x+1)$ denotes $(x_1+1, x_2+1, \ldots, x_n+1)$. The right-most form in (11.321) is obtained by expanding the factor $\prod_{i=1}^{n}(x_j - x_i - 1)$ and using Sylvester's identity (11.317).

Sylvester's identity is an important relation, not only for its use in this monograph, but also to other topics such as relativistic wave equations (Fischback *et al.* [55]) and epidemiology (Hyman and Stanley [83]).

11.8 Algebraic Derivation of Weyl's Dimension Formula

11.8.1 Vandermonde determinants, Bernoulli polynomials, and Weyl's formula

Let $y = (y_1, y_2, \ldots, y_{n+1})$ denote a set of n positive integers $(y_1, y_2, \ldots, y_n)$ and one nonnegative integer y_{n+1} satisfying $y_1 > y_2 > \cdots > y_{n+1} \geq 0$. Define disjoint closed intervals of integers by

$$\begin{aligned}[y_{i+1}, y_i - 1] &= \{x_i \,|\, x_i = y_{i+1}, y_{i+1}+1, \ldots, y_i - 1\}, \quad (11.322)\\ &\quad \text{each } i = 1, 2, \ldots, n \geq 2.\end{aligned}$$

Define also the sequence $[y] = [y_1, y_2, \ldots, y_{n+1}]$ of these intervals by

$$[y] = ([y_2, y_1 - 1], [y_3, y_2 - 1], \ldots, [y_{n+1}, y_n - 1]). \quad (11.323)$$

The notation $x = (x_1, x_2, \ldots, x_n) \in [y]$ designates the family of interval conditions: $x_i \in [y_{i+1}, y_i - 1], i = 1, 2, \ldots, n$. With these definitions and notations, we then have the following result:

The following identity between Vandermonde determinants is true for these discrete integral variables:

$$n! \sum_{x \in [y]} V_n(x) = V_{n+1}(y),\ n \geq 2. \quad (11.324)$$

This relation uniquely determines $V_{n+1}(y)$ in terms of the $V_n(x)$.

This identity is a special case of the following family of relations between the elementary symmetric functions $e_k(x)$, the Vandermonde determinant $V_n(x)$, and the Bernoulli polynomials $B_k(x)$:

$$\frac{1}{(n+1)!}\det\begin{pmatrix} (n+1)t^n & nt^{n-1} & \cdots & 1 \\ \Delta_{n+1}(y_1,y_{n+1}) & \Delta_n(y_1,y_{n+1}) & \cdots & \Delta_1(y_1,y_{n+1}) \\ \Delta_{n+1}(y_2,y_{n+1}) & \Delta_n(y_2,y_{n+1}) & \cdots & \Delta_1(y_2,y_{n+1}) \\ \vdots & \vdots & & \vdots \\ \Delta_{n+1}(y_n,y_{n+1}) & \Delta_n(y_n,y_{n+1}) & \cdots & \Delta_1(y_n,y_{n+1}) \end{pmatrix}$$

$$\begin{aligned} &= t^n \sum_{x\in[y]} V_n(x) - t^{n-1} \sum_{x\in[y]} e_1(x)V_n(x) \qquad (11.325) \\ &+ t^{n-2} \sum_{x\in[y]} e_2(x)V_n(x) + \cdots + (-1)^n \sum_{x\in[y]} e_n(x)V_n(x), \end{aligned}$$

$$\Delta_j(y_i,y_{n+1}) = B_j(y_i) - B_j(y_{n+1}), \qquad (11.326)$$
$$i = 1,2\ldots,n;\, j = 1,2,\ldots n+1\ .$$

Proof: First, we prove (11.325), then (11.324). We start with the $(n+1)\times(n+1)$ determinant with row 1 given by $(t^n,\ t^{n-1},\cdots,1)$ and row i by $(x_i^n,\ x_i^{n-1},\cdots,1), i = 1,2,\ldots,n$. This determinant is expanded by the row 1 cofactor rule to obtain:

$$V_n^{(k)}(x) = e_{k-1}(x)V_n(x), \qquad (11.327)$$

where $V_n^{(k)}(x)$ denotes the subdeterminant obtained by striking row 1 and column k. We next carry out the summation $\sum_{x\in[y]}$, noting that $\sum_{x_i\in[y_{i+1},y_i-1]}$ can be taken inside the determinantal form of $V_n(x)$ in front of every power $x_i^j, j = n, n-1, \ldots 0$, in row i, this being valid for each $i = 1,2,\ldots,n$. This step is followed by using the identity:

$$\sum_{x_i\in[y_{i+1},y_i-1]} x_i^j = \left(B_{j+1}(y_i) - B_{j+1}(y_{i+1})\right)\Big/(j+1). \qquad (11.328)$$

Finally, we use elementary row and column operations to bring the determinant to the form given by (11.325), which completes the proof of that identity.

We continue by using (11.325) to prove (11.324). Equating the coefficient of t^n on both sides of (11.325) gives:

$$\sum_{x\in[y]} V_n(x) \tag{11.329}$$

$$= \frac{1}{n!}\det\begin{pmatrix} \Delta_n(y_1,y_2) & \Delta_{n-1}(y_1,y_2) & \cdots & \Delta_1(y_1,y_2) \\ \Delta_n(y_2,y_3) & \Delta_{n-1}(y_2,y_3) & \cdots & \Delta_1(y_2,y_3) \\ \vdots & \vdots & & \vdots \\ \Delta_n(y_n,y_{n+1}) & \Delta_{n-1}(y_n,y_{n+1}) & \cdots & \Delta_1(y_n,y_{n+1}) \end{pmatrix}$$

$$= \frac{1}{n!}\det\begin{pmatrix} \Delta_n(y_1,y_{n+1}) & \Delta_{n-1}(y_1,y_{n+1}) & \cdots & \Delta_1(y_1,y_{n+1}) \\ \Delta_n(y_2,y_{n+1}) & \Delta_{n-1}(y_2,y_{n+1}) & \cdots & \Delta_1(y_2,y_{n+1}) \\ \vdots & \vdots & & \vdots \\ \Delta_n(y_n,y_{n+1}) & \Delta_{n-1}(y_n,y_{n+1}) & \cdots & \Delta_1(y_n,y_{n+1}) \end{pmatrix}.$$

This relation shows that $y_i - y_j$ is a factor of the right-hand side for every $1 \leq i < j \leq n$. Since the degree in each y_i is n, the determinant must be of the form $a_n V_{n+1}(y)$. For the values $y_1 = n, y_2 = n-1, \ldots, y_n = 1, y_{n+1} = 0$, the left-hand side of relation (11.329) reduces to $V_n(n-1, n-2, \ldots, 0) = 1!2!\cdots(n-1)!$, while $a_n V_{n+1}(y) = 1!2!\ldots n! a_n$, which gives $a_n = 1/n!$. Thus, the determinant on the right-hand side of (11.324) is $V_{n+1}(y)$. (This method was developed in another context in Ref. [106]). □

Weyl's dimension formula

It follows from the definition of GT patterns of shape λ and the betweenness conditions $\mu \prec \lambda$ that Weyl's dimension formula is obtained recursively by

$$\mathrm{Dim}\lambda = \sum_{\mu\in[\lambda]} \mathrm{Dim}\mu, \tag{11.330}$$

where the sequence of intervals $[\lambda]$ is defined for a partition $\lambda \in \mathbb{P}\mathrm{ar}_n$ by

$$[\lambda] = ([\lambda_2, \lambda_1], [\lambda_3, \lambda_2], \ldots, [\lambda_n, \lambda_{n-1}]). \tag{11.331}$$

This relation uniquely determines $\mathrm{Dim}\lambda$ in terms of the $\mathrm{Dim}\mu$, $\mu \prec \lambda$.

We can now prove the formula for the Weyl dimension:

Proof: We replace n by $n-1$ in relation (3.324) and set

$$\begin{aligned} V_{n-1}(x) &= 1!2!\cdots(n-2)!\mathrm{Dim}\mu, \\ x_j &= \mu_j + n - j - 1;\ j = 1, 2, \ldots, n-1. \end{aligned} \tag{11.332}$$

But then the left-hand side of relation (3.324) becomes exactly the right-hand side of (11.329), so that we obtain

$$V_n(y) = 1!2!\cdots(n-1)!\mathrm{Dim}\lambda,\ y_i = \lambda_i + n - i,\ i = 1, 2, \ldots, n, \quad (11.333)$$

provided the initial conditions agree. This is the case, since $V_2(x_1, x_2) = x_1 - x_2$ and $\mathrm{Dim}(\mu_1, \mu_2) = \mu_1 - \mu_2 + 1 = V(\mu_1 + 1, \mu_2)$. □

11.9 Other Topics

11.9.1 Alternating sign matrices

We present in this section some recent results establishing a bijection between a certain class of Gelfand-Tsetlin patterns and a class of matrices known as alternating sign matrices. The counting formula for the number of alternating sign matrices has been derived by Zeilberger [195], with contributions from many others (see Andrews [2], Lascoux [102], Robbins *et al.* [150], the rederivation by Kuperberg [99], and the nice survey by Bressoud and Propp [33]).

Gelfand-Tsetlin patterns have a major role in this monograph, and closed counting formulas for their number, such as the Weyl formula for the case when all betweenness relations are fulfilled, are of considerable interest. It is for this reason that we address the problem of counting alternating sign matrices from the perspective of counting GT patterns. If restrictions are imposed on the betweenness conditions, then the counting problem of the associated GT patterns may or may not be tractable, but it is solvable for the alternating sign matrix problem, as given by the Zeilberger formula. *Here, we do not present a new solution of the problem, but offer instead a different focus on its various structural aspects.*The interest is in what are called *strict* Gelfand-Tsetlin patterns.

The Gelfand-Tsetlin pattern (11.73) is called a *strict* GT pattern, if, in addition to all the betweenness conditions, it is further restricted by the conditions that its shape λ is a strict partition, that is, $\lambda_1 > \lambda_2 > \cdots > \lambda_n > 0$, and, moreover, that each of its remaining $n-1$ rows is also a strict partition. We denote the set of all strict Gelfand-Tsetlin patterns of shape λ by $\mathrm{Str}\,\mathbb{G}_\lambda$. The number of patterns in the general set $\mathrm{Str}\,\mathbb{G}_\lambda$ is unknown, but for the special partition $\lambda = (n, \cdots, 2, 1)$, it is given by the Andrews-Zeilberger number d_n :

$$|\mathrm{Str}\,\mathbb{G}_{(n,\cdots,2,1)}| = d_n = \prod_{j=0}^{n-1} \frac{(3j+1)!}{(n+j)!},\ n \geq 1. \quad (11.334)$$

We call this the Andrews-Zeilberger number because Andrews [2] had earlier proved a conjecture of Mills *et al.* [135] that gives the same number d_n as the number of totally symmetric, self-complementary plane partitions. As yet, a bijection between such plane partitions and the set of alternating sign matrices has not been found (We do not concern ourselves here with this problem).

We first define the set of alternating sign matrices (AS matrices) and then demonstrate the bijection with strict GT patterns of shape $(n, \cdots, 2, 1) = \lambda$.

An alternating sign matrix of order n is a matrix $A = (a_{ij})_{1 \le i,j \le n}$ with elements $a_{ij} \in \mathcal{A} = \{-1, 0, 1\}$ such that when a given row (column) is read from left (top) to right (bottom), and all zeros in that row (column) are ignored, the row (column) contains either a single 1, or is an alternating series $1, -1, 1, -1, \ldots, 1, -1, 1$ of odd length; hence, each row and column adds to 1. We denote the set of alternating sign matrices of order n by $\mathbb{A}_n$.

This definition includes all the $n!$ permutation matrices $P_\pi, \pi \in S_n$, of order n. For $n = 1, 2, 3$, the alternating sign matrices are the following:

$$\begin{gathered} n=1: \ (1); \qquad n=2: \ \begin{pmatrix} 1 & 0 \\ 0 & 1 \end{pmatrix}, \begin{pmatrix} 0 & 1 \\ 1 & 0 \end{pmatrix}; \\ n=3: \ \text{all six } P_\pi, \pi \in S_3, \text{ and } \begin{pmatrix} 0 & 1 & 0 \\ 1 & -1 & 1 \\ 0 & 1 & 0 \end{pmatrix}. \end{gathered} \tag{11.335}$$

Thus, the number of alternating sign matrices for $n = 1, 2, 3$ is $1, 2, 7$, which is in agreement with relation (11.334).

The bijection between $\operatorname{Str} \mathbb{G}_{n,\cdots,2,1}$ and $\mathbb{A}_n$ is given by the following mapping rule (see the GT pattern (11.73) for notation):

$$\begin{pmatrix} n & \cdots & 1 \\ & m & \end{pmatrix} \mapsto A = (A_1\, A_2\, \cdots\, A_n), \tag{11.336}$$

where the A_i are the columns of the alternating sign matrix A given by

$$\begin{aligned} &A_1 = e_{m_{1,1}}, \\ &A_i = (e_{m_{1,i}} + e_{m_{2,i}} + \cdots + e_{m_{i,i}}) \\ &-(e_{m_{1,i-1}} + e_{m_{2,i-1}} + \cdots + e_{m_{i-1,i-1}}),\ i = 2, 3, \ldots, n, \end{aligned} \tag{11.337}$$

in which e_k denotes the unit column matrix of length n with 1 in position k and 0 elsewhere. The mapping (11.336)-(11.337) is a bijection because the explicit reverse mapping for each $A \in \mathbb{A}_n$ is obtained as follows: The sum of the first i columns of A is given by

$$A_1 + A_2 + \cdots + A_i = \sum_{j=1}^{i} e_{m_{j,i}}, \tag{11.338}$$

from which row i of the GT pattern is read off as $m_i = (m_{1,i}\, m_{2,i} \cdots m_{1,i})$, for each $i = 1, 2, \ldots, n$. Robbins *et al.* [150] first pointed out the bijection between strict GT patterns and AS matrices.

Examples. It is useful to give examples of the bijection between strict GT patterns and AS matrices:

$n = 4:$

$$\begin{pmatrix} 4 & & 3 & & 2 & & 1 \\ & 4 & & 3 & & 1 & \\ & & 4 & & 2 & & \\ & & & 3 & & & \end{pmatrix} \mapsto \begin{pmatrix} 0 & 0 & 1 & 0 \\ 0 & 1 & -1 & 1 \\ 1 & -1 & 1 & 0 \\ 0 & 1 & 0 & 0 \end{pmatrix}. \tag{11.339}$$

$n = 5:$

$$\begin{pmatrix} 5 & & 4 & & 3 & & 2 & & 1 \\ & 5 & & 4 & & 2 & & 1 & \\ & & 4 & & 3 & & 1 & & \\ & & & 3 & & 2 & & & \\ & & & & 2 & & & & \end{pmatrix} \mapsto \begin{pmatrix} 0 & 0 & 1 & 0 & 0 \\ 1 & 0 & -1 & 1 & 0 \\ 0 & 1 & 0 & -1 & 1 \\ 0 & 0 & 1 & 0 & 0 \\ 0 & 0 & 0 & 1 & 0 \end{pmatrix}. \quad \square \tag{11.340}$$

Recurrence relation for strict GT patterns

It is quite easy to give a recurrence relation (see (11.330)) for counting the set of all strict GT patterns of general strict shape λ. We define the cardinality $\mathrm{Str}_n(\lambda)$ of a strict partition $\lambda \in \mathbb{P}\mathrm{ar}_n$ by

$$\mathrm{Str}_n(\lambda) = |\mathrm{Str}\mathbb{G}_\lambda|. \tag{11.341}$$

Then, the recurrence relation for these quantities follows directly from (11.330) by taking the general GT pattern (11.73) to be strict:

$$\mathrm{Str}_n(\lambda) = \sum_{\substack{\mu_i \in [\lambda_{i+1}, \lambda_i] \\ \mu_1 > \mu_2 > \cdots > \mu_{n-1}}} \mathrm{Str}_{n-1}(\mu), \tag{11.342}$$

where the starting point for the iteration is

$$\mathrm{Str}_2(\lambda_1, \lambda_2) = \lambda_1 - \lambda_2 + 1. \tag{11.343}$$

Example. By direct enumeration of the restricted GT patterns, we have that $\mathrm{Str}_3(2, 1, 0) = 7$, since the only pattern in the full set $\mathbb{G}_{(2,1,0)}$ of eight that fails to quality is $\begin{pmatrix} 2\,1\,0 \\ 1\,1 \\ 1 \end{pmatrix}$. Relation (3.342) correspondingly gives $\mathrm{Str}_3(2,1,0) = \mathrm{Str}_2(2,1) + \mathrm{Str}_2(2,0) + \mathrm{Str}_2(1,0) = 2+3+2 = 7.$ □

The important property of relation (11.342) is:

Relation (11.342) determines uniquely all strict GT patterns starting with the initial values (11.343), iterating first to $n = 3$, then to $n = 4$, etc.

Example. The calculation of $\mathrm{Str}_4(4,3,2,1)$.

$$\begin{aligned} \mathrm{Str}_4(4,3,2,1) &= \sum_{\substack{\mu_1 \in [3,4];\, \mu_2 \in [2,3],\, \mu_3 \in [1,2] \\ \mu_1 > \mu_2,\, \mu_2 > \mu_3}} \mathrm{Str}_3(\mu_1,\mu_2,\mu_3) \\ &= \mathrm{Str}_3(3,2,1) + \mathrm{Str}_3(4,2,1) + \mathrm{Str}_3(4,3,1) + \mathrm{Str}_3(4,3,2), \end{aligned} \tag{11.344}$$

where the terms in this summation are given by

$$\begin{aligned} \mathrm{Str}_3(3,2,1) &= \sum_{\substack{\mu_1 \in [2,3];\, \mu_2 \in [1,2] \\ \mu_1 > \mu_2}} (\mu_1 - \mu_2 + 1) = 7, \\ \mathrm{Str}_3(4,2,1) &= \sum_{\substack{\mu_1 \in [2,4];\, \mu_2 \in [1,2] \\ \mu_1 > \mu_2}} (\mu_1 - \mu_2 + 1) = 14, \\ \mathrm{Str}_3(4,3,1) &= \sum_{\substack{\mu_1 \in [3,4];\, \mu_2 \in [1,3] \\ \mu_1 > \mu_2}} (\mu_1 - \mu_2 + 1) = 14, \\ \mathrm{Str}_3(4,3,2) &= \sum_{\substack{\mu_1 \in [3,4];\, x_2 \in [2,3] \\ \mu_1 > \mu_2}} (\mu_1 - \mu_2 + 1) = 7. \end{aligned}$$

Thus, we obtain

$$\mathrm{Str}_4((4,3,2,1)) = 42. \qquad \square \tag{11.345}$$

The above example shows clearly the importance of the general relation (11.342) for determining the variety of strict GT patterns that occur in the expression for the cardinality $\mathrm{Str}_n(n,\ldots,2,1)$ of strict GT patterns in which only adjacent integers appear.The general formula for the cardinality of these special strict GT patterns is an application of (11.342):

$$\begin{aligned} &\mathrm{Str}_n(n,\ldots,2,1) \\ &= \sum_{\substack{\mu_n \in [1,2];\, \mu_{n-1} \in [2,3];\ldots,\, \mu_1 \in [n-1,n] \\ \mu_1 > \mu_2,\, \mu_2 > \mu_3,\, \ldots,\, \mu_{n-2} > \mu_{n-1}}} \mathrm{Str}_n(\mu_1,\mu_2,\ldots,\mu_{n-1}). \end{aligned} \tag{11.346}$$

This relation, with all of its restrictions, simplifies to the form:

$$\begin{aligned}\mathrm{Str}_n(n,\ldots,2,1) &= \sum_{i=1}^{n}\mathrm{Str}_{n-1}(\mu^{(i)}),\\ \mu^{(i)} &= (n,n-1,\ldots,\widehat{i},\ldots,2,1),\end{aligned} \qquad (11.347)$$

where the notation $\widehat{i}$ for the $i-$th strict partition $\mu^{(i)}$ designates that integer i is missing from the partition $(n, n-1, \ldots, 1)$. For example, for $n = 4$, we have $\mu^{(1)} = (4,3,2), \mu^{(2)} = (4,3,1), \mu^{(3)} = (4,2,1), \mu^{(4)} = (3,2,1)$, as in the example above. The meaning of the set of strict partitions $\mu^{(i)}$, $i = 1,2,\ldots,n$, is that they are the partitions that enter into row $n-1$ of the set $\mathbb{G}_{n,\ldots,2,1}$ of strict GT patterns. But a principal hurdle is: *Relation (11.347) does not stand on its own as a recurrence relation.* It is necessary to revert back to the complete recurrence relation (11.342) to determine the strict partitions that can occur in row $n-2$ of a GT pattern in $\mathbb{G}_{n,\ldots,2,1}$, etc. This becomes increasing difficult when rows $n-2$, $n-3,\ldots,1$ are considered, in turn.

We observe next two important properties that preserve the counting of ordinary GT patterns as well as strict GT patterns:

(i). Translation invariance: The number of patterns in $\mathbb{G}_\lambda$ is invariant to the shift of all entries in the pattern by the same integer $h \in \mathbb{Z}$, which implies

$$|\mathbb{G}_{\lambda+h}| = |\mathbb{G}_\lambda|, \qquad (11.348)$$

where $\lambda + h$ denotes $(\lambda_1 + h, \lambda_2 + h, \ldots, \lambda_n + h)$.

(ii). Sign-reversal shift conjugation: The number of patterns in $\mathbb{G}_\lambda$ is invariant to the operation of writing each row of the pattern backwards with minus signs and translating by $h = \lambda_1 + \lambda_n$. This operation effects the transformation of shape λ to shape λ^*, where $\lambda_i^* = \lambda_1 - \lambda_{n-i+1} + \lambda_n, i = 1,2,\ldots,n$. Thus, we have the identity

$$|\mathbb{G}_\lambda| = |\mathbb{G}_{\lambda^*}|. \qquad (11.349)$$

This operation is called sign-reversal shift conjugation; it is an *involution:* repetition of the operation restores the original pattern.

It is evident that the properties of shift invariance and sign-reversal shift conjugation apply as well to strict GT patterns for all $n \geq 2$:

$$\begin{aligned}\mathrm{Str}_n(\lambda_1+h,\ldots,\lambda_n+h) &= \mathrm{Str}_n(\lambda_1,\ldots,\lambda_n),\ h\in\mathbb{Z},\\ \mathrm{Str}_n(\lambda^*) &= \mathrm{Str}_n(\lambda).\end{aligned} \qquad (11.350)$$

Since relations (11.350) hold for all n, we have the result:

The general recurrence formula (11.342) giving the number of strict GT patterns of arbitrary order n is invariant under the operations of translations and sign-reversal shift conjugation.

It is, perhaps, an rather unexpected result that the positive integer $\mathrm{Str}_{n-1}(\mu^{(i)})$, which is uniquely determined by the recurrence relation (11.342), is given by the value $Z_n(i)$ of a polynomial $Z_n(x)$ of degree $2n-2$ that satisfies the sign-reversal shift conjugation rule:

$$Z_n(x) = Z_n(-x+n+1). \tag{11.351}$$

This value of $\mathrm{Str}_{n-1}(\mu^{(i)})$ is given by

$$\mathrm{Str}_{n-1}(\mu^{(i)}) = Z_n(i),\ n \geq 2,\ i = 1, 2, \ldots, n, \tag{11.352}$$

as we next show, by examples.

Examples. It is a fascinating exercise to calculate directly from the general recurrence relation (11.342) the values of $\mathrm{Str}_{n-1}(\mu^{(i)})$ for $n = 2, 3, 4, 5$, and verify that they are the values of the polynomials:

$$\begin{aligned}
Z_1(x) &= 1, \\
Z_2(x) &= \frac{1}{2}\binom{x}{1}\binom{-x+3}{1}, \\
Z_3(x) &= \frac{1}{3}\binom{x+1}{2}\binom{-x+5}{2}, \\
Z_4(x) &= \frac{7}{20}\binom{x+2}{3}\binom{-x+7}{3}, \\
Z_5(x) &= \frac{3}{5}\binom{x+3}{3}\binom{-x+9}{3}.
\end{aligned} \tag{11.353}$$

These polynomials have the following zeros:

$$\begin{aligned}
&n = 2 : x = 0, 3;\ n = 3 : x = -1, 0, 4, 5; \\
&n = 4 : x = -2, -1, 0, 5, 6, 7;\ n = 5 : x = -3, -2, -1, 0, 6, 7, 8, 9.
\end{aligned} \tag{11.354}$$

These polynomials satisfy the sign-reversal shift conjugation rule $Z_n(x) = Z_n(-x+n+1)$, which upon setting $x = y + (n+1)/2$ becomes

$$Z_n(\frac{n+1}{2} - y) = Z_n(\frac{n+1}{2} + y),\ y \geq 0. \tag{11.355}$$

The polynomial $Z_n(x)$ is a symmetric polynomial under reflection through the vertical line containing the point $a = (n+1)/2$.

The simplicity of the structural pattern in the examples (11.353) suggests immediately the general result: *The polynomial $Z_n(x)$ is given by*

$$Z_n(x) = a_n \binom{x+n-2}{n-1}\binom{-x+2n-1}{n-1}, \; n \geq 1, \tag{11.356}$$

where a_n is an undetermined positive constant.

But the inference of the form (11.356), which satisfies sign-reversal shift conjugation, does not, of course, constitute a proof of its validity. Even though the n numbers $Z_n(i), i = 1, 2, \ldots, n$, in relation (11.352) are uniquely determined by the general recurrence relation (11.342), it is not easy to implement this calculation (see (11.357) below). Nonetheless, the simplicity of the polynomial (11.356) is appealing, and challenges a direct proof.

That the polynomial $Z_n(x)$ is correct is a consequence of results obtained by Zeilberger [196] in his second paper, where $Z_n(x)$ appears in terms of the binomial coefficients in the value $Z_n(i)$ at $x = i$, together with the explicit multiplying factor a_n. Nonetheless, it is informative to explore further the properties of the polynomials (11.356) without assuming Zeilberger's result.

As noted above, the partitions that occur in row $n-1$ in the set $\mathbb{G}_\lambda$, $\lambda = (n, \ldots, 2, 1)$ of strict GT patterns are $\mu^{(1)}, \mu^{(2)}, \ldots, \mu^{(n)}$; that is, the set $\mathbb{G}_\lambda$ of GT patterns is the union of the subsets $\mathbb{G}^{(i)}_\lambda$, $i = 1, 2, \ldots, n$, of strict GT patterns defined by

$$\mathbb{G}^{(i)}_\lambda = \left\{ \begin{pmatrix} \lambda \\ \mu^{(i)} \\ (m)_{n-2} \end{pmatrix} \;\middle|\; \begin{array}{l} \text{the } n-2 \text{ rows of the lexical GT} \\ \text{pattern } (m)_{n-2} \text{ are strict partitions} \end{array} \right\}. \tag{11.357}$$

By definition, the positive integer $Z_n(i)$ gives the number of GT patterns in the set $\mathbb{G}^{(i)}_\lambda$. The sum $\sum_{i=1}^{n} Z_n(i) = |\mathbb{G}_{n,\ldots,2,1}|$ is the number of strict GT patterns sought. In the statement of these properties of the GT patterns (11.357), no use is made of (11.356). We now make the additional assumption:

Assumption: *The numbers $Z_n(i)$ are the values of the polynomials $Z_n(x)$ of the form (11.356) at $x = i$, $i = 1, 2, \ldots, n$, where a_n is undetermined; that is,*

$$\mathrm{Str}_{n-1}(\mu^{(i)}) = Z_n(i), \; n \geq 2; \; i = 1, 2, \ldots, n. \tag{11.358}$$

We remark further on this assumption below, but now use it to prove: *The following relations hold:*

$$\begin{aligned} Z_n(x) &= d_n \frac{\binom{x+n-2}{n-1}\binom{-x+2n-1}{n-1}}{\binom{3n-2}{n-1}},\ n \geq 1, &(11.359)\\ d_n &= \prod_{j=0}^{n-1} \frac{(3j+1)!}{(n+j)!}. &(11.360)\end{aligned}$$

Proof: The polynomial (11.356), evaluated at $x = i$, gives

$$Z_n(i) = a_n \binom{i+n-2}{n-1}\binom{-i+2n-1}{n-1}, \tag{11.361}$$

where a_n is an undetermined constant. We next carry out the following summation of the polynomials $Z_n(x)$ evaluated at $x = i$:

$$\sum_{i=1}^{n} Z_n(i) = a_n \sum_{i=1}^{n} \binom{i+n-2}{n-1}\binom{-i+2n-1}{n-1}. \tag{11.362}$$

The summation on the right-hand side of this relation can be effected by using the binomial function identities:

$$\begin{aligned} \binom{x+y}{n-1} &= \sum_{s+t=n-1} \binom{x}{s}\binom{y}{t},\\ \binom{i+n-2}{n-1} &= (-1)^{i-1}\binom{-n}{i-1}, &(11.363)\\ \binom{-i+2n-1}{n-1} &= (-1)^{n-i}\binom{-n}{n-i}.\end{aligned}$$

Thus, we obtain:

$$\begin{aligned} \sum_{i=1}^{n} \binom{i+n-2}{n-1}\binom{-i+2n-1}{n-1} &= (-1)^{n-1}\sum_{i=1}^{n}\binom{-n}{i-1}\binom{-n}{n-i}\\ &= (-1)^{n-1}\binom{-2n}{n-1} = \binom{3n-2}{n-1}, &(11.364)\end{aligned}$$

$$\sum_{i=1}^{n} Z_n(i) = a_n \binom{3n-2}{n-1} = d_n, \tag{11.365}$$

where d_n is not yet defined. But it is also the case that $Str_{n-1}(\mu^{(n)}) = \mathrm{Str}_{n-1}(n-1,\ldots,2,1) = Z_n(n)$, which gives

$$d_{n-1} = a_n \binom{2n-2}{n-1}. \tag{11.366}$$

Thus, we obtain the recurrence relation

$$d_n = d_{n-1} \frac{\binom{3n-2}{n-1}}{\binom{2n-2}{n-1}},\ n \geq 2,\ d_1 = 1. \tag{11.367}$$

This relation is easily iterated to obtain

$$d_n = \prod_{j=0}^{n-1} \frac{(3j+1)!}{(n+j)!}, \tag{11.368}$$

which, in turn, gives the value of a_n in relation (11.356):

$$a_n = \frac{d_n}{\binom{3n-2}{n-1}}. \qquad \square \tag{11.369}$$

Remarks.

1. It is appropriate to use the term Zeilberger polynomials for the fully determined polynomials $Z_n(x)$. It is the multiplying coefficient $a_n = d_n \big/ \binom{3n-2}{n-1}$ in these polynomials that gives them a nontrivial structure. The proof of (11.359) for the Zeilberger polynomial $Z_n(x)$ shows that this polynomial is fully defined by its degree $2n-2$ and its set of zeros:

   $$\{-n+2, -n+3, \ldots, -1, 0; n+1, n+2, \ldots, 2n-1\}. \tag{11.370}$$

   The split of zeros into these two sets is a consequence of sign-reversal shift conjugation; the values $Z_n(i)$ at the integers $i = 1, 2, \ldots, n$ between the sets of zeros counts the number of strict GT patterns in the set (11.357). Such an exquisite characterization of the Zeilberger polynomials suggests a corresponding simplicity of origin (see Remark 3 below).

2. There is no freedom in the definition of the polynomials $Z_n(x)$. The general recurrence relation (11.342), shift invariance, and sign-reversal shift conjugation uniquely determine $Z_n(x)$, but a simple, direct proof seems difficult. An alternative approach makes use of still another property of GT patterns, as follows: The removal of the left-most down-diagonal from an arbitrary GT pattern (11.73) having n rows leaves behind a (lexical) GT pattern having $n-1$ rows. This rule applies to strict GT patterns as well. In particular, when applied to strict partitions in the set $\mathrm{Str}\mathbb{G}_{n,\ldots,1}$, the result is:

$$d_n = \sum_{\mu^{(2)}} \Big(\sum_{\mu^{(1)}} R_{\mu^{(1)}/\mu^{(2)}}\Big) d_{n-1}(\mu^{(2)}), \tag{11.371}$$

where: $\mu^{(1)}$ and $\mu^{(2)}$ are, respectively, the first down-diagonal in a strict GT patterns in $\mathrm{Str}\mathbb{G}_{n,\ldots,1}$ and the first down-diagonal in a strict GT pattern $\mathrm{Str}\mathbb{G}_{n-1,\ldots,1}$:

$$\mu^{(1)} = (n, \mu_2^{(1)}, \ldots, \mu_n^{(1)}) \supseteq \mu^{(2)} = (n-1, \mu_2^{(2)}, \ldots, \mu_{n-1}^{(2)}). \tag{11.372}$$

The number $d_{n-1}(\mu^{(2)})$ is the number of GT patterns in $\mathrm{Str}\mathbb{G}_{n-1,\ldots,1}$ that have the fixed down-diagonal $\mu^{(2)}$. The summation is, first of all, over all down-diagonals $\mu^{(1)}$ for fixed $\mu^{(2)}$ such that

$$\mu_i^{(1)} \geq \max\{\mu_{i-1}^{(2)}, \mu_i^{(2)} + 1\},\ i = 1, 2, \ldots, n, \tag{11.373}$$

where $\mu_0^{(2)} = n-1$ and $\mu_n^{(2)} = 0$. Finally, the sum is over all down-diagonals $\mu^{(2)}$ that arise in $\mathrm{Str}\mathbb{G}_{n-1,\ldots,1}$. Only GT patterns of the type $\mathbb{G}_{k,k-1,\ldots,1}$, $k = n, n-1, \ldots, 1$, enter into the iteration of (11.371) through the associated family of down-diagonals. It is a nice result that the iteration entails restricted partitions for which each part satisfies a constraint based on the parts of the contiguous down-diagonal. A formula similar to (11.371) can be derived for $Z_n(i)$ by applying this same method to the GT pattern (11.357). The implications of these relations remains to be investigated.

3. A completely different approach to the zeros problem uses a special class of magic squares of order n that is bijective with the set of alternating sign matrices (see Ref. [116]). It is quite intriguing that the polynomial $H_n(x)$ of degree $(n-1)^2$ that gives the number $H_n(r)$ of all magic squares of order n with row and column sums equal to r has the zeros $x = -n+1, -1, \ldots, -2, -1$ (see Stanley [163, Vol. 1, p. 232]), which are the left half of the zeros in (11.370) shifted to the left by one unit. The polynomials $H_n(x)$ also have the symmetry $H_n(-x-n) = (-1)^{n-1} H_n(x)$. The zeros of $H_n(x)$ give, in

general, only a partial determination of the polynomial. The theory of magic squares is a complex subject that is of considerable interest for this monograph because it is a special case of the ubiquitous matrix arrays $\mathbb{M}^{p}_{n\times n}(\alpha,\beta)$ (see Refs. [28, 36, 37].)

4. The simplest meaning of the number d_n can be stated in terms of SSYW tableaux themselves: Consider the set $\mathbb{T}_{(n,\ldots,2,1)}$ of SSYW tableaux obtained by filling in the shape in all possible ways with the integers $1,2,\ldots,n$. The number of such tableaux is $2^{\binom{n}{2}}$. Select from this set the subset obtained by the following rule: If the shape of the tableau is a strict partition after removing all boxes containing n, keep the tableau, otherwise discard it; if the shape of the tableau is a strict partition after removing all boxes containing $n-1$, keep the tableau, otherwise discard it; $\ldots$; if the shape of the tableau is a strict partition after moving all boxes containing 2, keep the tableau, otherwise discard it. Denote the set of all such tableaux having strict shapes by $\mathrm{Str}\,\mathbb{T}_{(n,\ldots,2,1)}$. Then, $d_n = |\mathrm{Str}\,\mathbb{T}_{(n,\ldots,2,1)}|$.

5. The results obtained in this section illustrate further the versatility of Gelfand-Tsetlin patterns in enumeration problems, going well beyond their original purpose. Such GT type patterns are also known for other groups, with corresponding tableaux (Baclawski [5, 6] and Krillov [95]). Generalizations also turn up in other applications as developed by Krillov and Berenstein [96].

11.9.2 Binary trees, graphs, and digraphs

The subjects of binary trees and related graphs are so important for the main themes of this monograph that they are developed in the main text in the context of their usage. In this connection, we note an undeveloped aspect of binary trees: The Weyl formula, applied to patterns having $2n$ rows with partition $\lambda = (n,n,0,\cdots,0)$ of weight $(1,1,\ldots,1)$ (length $2n$), gives the famous Catalan numbers $a_n = \frac{1}{n+1}\binom{2n}{n}$ (see Stanton and White [165]). Thus, the theory of binary trees could also be formulated in terms of these specialized GT patterns.

11.9.3 Umbral calculus and double tableau calculus

The foundations of combinatorics set forth in Roman and Rota [153], Rota [154, 155, 156], Rota *et al.* [157], Kung and Rota [97], and Kung [98] are the basis for a more combinatorial-oriented treatment of much of the subject matter of this monograph. Several such applications are pointed out in the main text, but a more detailed combinatorial exposition of the relationship of the umbral calculus to coupled angular momentum theory with its $3n-j$ coefficients, to the shift operator approach to the properties of the general $D^\lambda-$polynomials, and to the fundamental role of Sylvester's identity in the Lie algebra of the general linear group remains a challenge, not fully realized in the present volume. In particular, the bracket algebra developed by Kung and Rota [97] is intrinsically related to the methods presented in this volume.

11.9.4 Special functions

Relations of the $D^\lambda-$polynomials and their Kronecker products to special functions of many variables, but in the spirit of Wigner [182], Talman [169], and Vilenkin [170], have not been included in this volume because of space and a rather random development. Results in that direction can be found in Louck and Biedenharn [118], Biedenharn *et al.* [15], Biedenharn and Louck [22, 23, 24], and Chen and Louck [40, 41]. Recent results on Hermite polynomials and their relation to the totally symmetric D^k-polynomials by Yang [191] are also a beginning for even more complex relations originating from the harmonic oscillator creation operator realization of the indeterminates in the general $D^\lambda-$polynomials. The paper by Gelfand and Graev [61] places some aspects of this subject in the broader context of general hypergeometric functions, as also does that of Milne [136] in a different direction.

11.9.5 Other generalizations

The discovery of two other noteworthy combinatorial based functions have been by-products of the approach to the calculation of explicit $SU(3)$ CG coefficients suggested by the properties of canonical tensor operators: The first is the concept of the factorial Schur function with properties presented and discussed in Ref. [23, 24, 41, 109]; the second is the notion of the generalized hypergeometric series presented and discussed in Ref. [15, 112]. The richness of physical theory in suggesting new mathematical objects of interest is well-illustrated by these examples. But a quite different perspective led Goulden and Hamel [69] to a similar theory of the factorial Schur function.

Additional reading in combinatorics

We have found the following books (in addition to those already cited) to be helpful in preparing this Compendium: Berge [11], Godsil [65], Graham *et al.* [70], Riordin [148], Robinson [151], Stanton [164], Stanton and White [165], Wilf [186].

Bibliography

[1] G. E. Andrews, The Theory of Partitions, Encyclopedia of Mathematics and Its Applications, Vol. 2, edited by G.-C. Rota, Cambridge University Press, Cambridge, 1976. [1, 92, 505, 511, 512] [1]

[2] G. E. Andrews, Plane partitions V: The TSSCPP conjecture, J. Combin. Theory Ser. A **66** (1994) 28-39. [585, 586]

[3] G. E. Andrews, R. A. Askey, and R. Roy, Special Functions, Encyclopedia of Mathematics and Its Applications, Vol. 71, edited by G.-C. Rota, Cambridge University Press, Cambridge, 1999. [preface, 566]

[4] G. E. Andrews, P. Paule, and A. Reise, Omega package for Mathematica, http://www.risc.unilinz.ac.at/research/combinat/risc/software. [511]

[5] K. Baclawski, Recursive algorithm for unitary and symplectic group representations, SIAM J. Alg. Disc. Meth. **3** (1982) 592-605. [595]

[6] K. Baclawski, A new rule for computing Clebsch-Gordan series, Adv. Appl. Math. **5** (1984) 416-432. [403, 589, 595]

[7] G. E. Baird and L. C. Biedenharn, On the representations of the semisimple Lie groups. II, J. Math. Phys. **5** (1963) 1449–1466; III, The explicit conjugation operation, *ibid* (1964) 1723–1730; IV, A canonical classification of tensor operators in SU_3, *ibid* **5** (1964) 1730-1747. [363, 394, 552]

[8] V. Bargmann, On the representations of the rotation group, Rev. Mod. Phys. **34** (1962) 829-845. [246, 248, 257]

[9] J. S. Bell, Speakable and unspeakable in quantum mechanics, Cambridge University Press, Cambridge, New York, 1987. [12]

[10] S. K. Berberian, Introduction to Hilbert Space, Oxford University Press, New York, 1961. [504]

[11] C. Berge, Principles of Combinatorics, Academic Press, New York, 1971. [596]

[1]The page numbers where the reference is made are enclosed in brackets.

[12] W. A. Beyer, J. D. Louck, and P. R. Stein, Group theoretical basis for some identities for the generalized hypergeometric series, J. Math. Phys. **28** (1987) 497-508. [567]

[13] G. Bhatnagar, A short proof of an identity of Sylvester, Int. J. Math. and Math. Sci. **22** (1999) 431-440. [367, 581]

[14] L.C. Biedenharn, Resolving the problem posed by multiplicity in the unitary groups, Group Theory and Special Symmetries in Nuclear Physics, Proceedings of the International Symposium in Honor of K. T. Hecht, editors, J. P. Draayer and J. Jänecke, World Scientific, Singapore, 1992, pp. 38-48. [398]

[15] L. C. Biedenharn, A. M. Bincer, M. A. Lohe, and J. D. Louck, New relations and identities for generalized hypergeometric series, Adv. Appl. Math. **13** (1992) 62-121. [596]

[16] L. C. Biedenharn, W. Y. C. Chen, M. A. Lohe, and J. D. Louck, The role of $3n-j$ coefficients in $SU(3)$, editors, T. Lulek, W. Florek, B. Lulek, World Scientific, Singapore, 1996, pp. 150-182. [257, 446]

[17] L. C. Biedenharn and M. A. Lohe, Quantum Group Symmetry and q-Tensor Operator Algebras, World Scientific, Singapore, 1995. [446]

[18] L. C. Biedenharn, M. A. Lohe, and J. D. Louck, On the denominator function for canonical tensor $SU(3)$ tensor operators, J. Math. Phys. **26** (1985) 1458-1492. [421, 429, 431, 433, 446]

[19] L. C. Biedenharn and J. D. Louck, A pattern calculus for tensor operators in the unitary groups, Commun. Math. Phys. **8** (1968) 89-131. [421, 446]

[20] L. C. Biedenharn and J. D. Louck, On the structure of the canonical tensor operators in the unitary groups. II. The tensor operators in U(30 characterized by maximal null space, J. Math. Phys. **13** (1972) 1985-2001. [394, 421, 429, 446]

[21] L. C. Biedenharn and J. D. Louck, Angular Momentum in Quantum Physics, Vol.8: The Racah-Wigner Algebra in Quantum Theory, Vol.9: Encyclopedia of Mathematics and Its Applications, edited by G.-C. Rota, Cambridge University Press, Cambridge, 1981. [15, 21, 31, 34, 44, 46, 47, 53, 58, 59, 129, 130, 134, 142, 143, 164, 202, 216, 258, 259, 334, 440, 446, 564, 565]

[22] L. C. Biedenharn and J. D. Louck, Some properties of generalized hypergeometric coefficients, Adv. Appl. Math. **9** (1988) 447-464. [567, 596]

[23] L. C. Biedenharn and J. D. Louck, A new class of symmetric polynomials defined in terms of tableaux, Adv. Appl. Math. **10** (1989) 396-438. [596]

[24] L. C. Biedenharn and J. D. Louck, Inhomogeneous basis set of symmetric polynomials defined by tableaux, Proc. Natl. Acad. Sci. **87** (1990) 1441-1445. [596]

[25] L. C. Biedenharn and H. van Dam, Quantum Theory of Angular Momentum, Academic Press, New York, 1965. [67]

[26] G. Birkhoff, Tres ovservaciones sobre el algebra lineal, Univ. Nac. Tucuman Rev. Ser. A (1946) 147-151. [570]

[27] D. J. Bohm, Quantum Theory, Prentice-Hall, New York, 1951. [12]

[28] M. Bóna and H-K Ju, Enumerating solutions of a system of linear inequalities related to magic squares, Annal of Comb. **10** (2006) 179-191. [595]

[29] W. M. Boothby, An Introduction to Differentiable Manifolds and Riemannian Geometry, Academic Press, New York, 1986. [504]

[30] M. Born and P. Jordan, Zur Quantenmechnik, Z. Phys. **34** (1925) 858-888. [29]

[31] M. Born, W. Heisenberg, and P. Jordan, Zur Quantenmechnik II, Z. Phys. **35** (1926) 557-615. [29]

[32] M. Born and P. Jordan, Elementare Quantenmechanik, Springer, Berlin, 1930. [29]

[33] D. Bressoud and J. Propp, How the alternating sign matrix conjecture was solved, Notices of the AMS **46** (1999) 637-646. [585]

[34] R. A. Brualdi and H. J. Ryser, Combinatorial Matrix Theory, Encyclopedia of Mathematics and Its Applications, edited by G.-C. Rota, Cambridge University Press, Cambridge, 1991. [226, 230, 570]

[35] E. Cartan, Thesis, Sur la Structure des Groupes Finis and Continus, Paris, Nony, 1894, Ouevres Complète, Part 1, Gauthier-Villars, Paris (1952) 137-287. [14]

[36] J. H. Carter and J. D. Louck, to appear in Annals of Combinatorics. [570, 595]

[37] J. H. Carter and J. D. Louck, Magic squares: symmetry and combinatorics, Mol. Phys. **102** (2004) 1243-1267. [570, 595]

[38] H. B. G. Casimir, Thesis, Rotation of a Rigid Body in Quantum Mechanics, University of Leyden, Wolters, Groningen,1931, Koninkl. Ned. Akad. Wetenschap, Proc. **34** (1931) 844. [44]

[39] A. Cayley, On the analytical forms called trees, with application to the theory of chemical combinations, Report of the British Associaton for the Adv. of Science (1875) 257-305. [163]

[40] W. Y. C. Chen and J. D. Louck, The combinatorics of a class of representation functions, Adv. Math. **14** (1998) 207-236. [63, 65, 68, 70, 230, 232, 595, 596]

[41] W. Y. C. Chen and J. D. Louck, The factorial Schur function, J. Math. Phys. **34** (1993) 4144-4160. [596]

[42] W. Y. C. Chen, and J. D. Louck, Enumeration of cubic graphs by inclusion-exclusion, J. Comb. Theory (1999) 151-164. [226]

[43] A. Clebsch, "Theorie der binären algebraischen Formen," Teubner, Leipzig (1872). [37]

[44] L. Comtet, Advanced Combinatorics, D. Reidel Publishing Company,, Dordrecht, 1974. [99, 106, 217, 182, 217]

[45] E. U. Condon and G. H. Shortley, The Theory of Atomic Spectra, Cambridge University, London, 1935. [preface]

[46] M. W. Crofton, On operative symbols in the differential calculus, Proceedings of the London Mathematical Society **12** (1881) 12. [8]

[47] P. Désarménien, J. P. S. Kung, and G.-C. Rota, Invariant theory, Young bitableaux, and combinatorics, Adv. Math. **27** (1978) 63-92. [556]

[48] P. Doubilet, On the foundations of combinatorial theory. VII. Symmetric functions through the theory of distributions and occupancies. Stud. Appl, Math **51** (1972) 377-396. [556]

[49] P. Doubilet, G-C. Rota, R. P. Stanley, On the foundations of combinatorial theory. VI. The idea of a generating function. Sixth Berkeley Symposium on Statistics and Probability Theory, Vol II, University of California, 1972, 267-318. (Reprinted in G.-C Rota, Finite Operator Calculus, Academic Press, 1975). [556]

[50] P. Doubilet, G.-C. Rota, and J. Stein, On The Foundations of Combinatorial Theory: IX. Combinatorial methods in invariant theory, Vol. **53** (1974) 185-216. [556]

[51] P. A. M. Dirac, The fundamental equations of quantum mechanics, Proc. Roy. Soc. A **109** (1925) 642-653. [29]

[52] P. A. M. Dirac, The Principles of Quantum Mechanics, Oxford University Press, 4th ed., London, 1958. [29]

[53] G. W. F. Drake, Editor, Springer Handbook of Atomic, Molecular, and Optical Physics, Springer, Berlin, 2006, pp. 9-72. [31, 250, 254, 257]

[54] C. Eckart, The application of group theory to the quantum mechanics of monatomic systems, Rev. Mod. Phys. **2** (1930) 305-380. [358]

[55] E. Fischback, J. D. Louck, M. M. Nieto, and C. F. Scott, The Lie algebra $so(n)$ and the Duffin-Kemmer-Peteau Ring, J. Math. Phys. **15** (1974) 60-64. [582]

[56] J. B. Fraleigh, A First Course in Abstract Algebra, Addison-Wesley, Reading, Massachusetts, 1993. [504]

[57] W. F. Fulton and J. Harris, Representation Theory, Springer-Verlag, New York, 1991. [504]

[58] I. M. Gelfand, The center of an infinitesimal group ring, Mat. Sb. **26** (1950) 103-112. [368, 369]

[59] I. M. Gelfand, Lectures on Linear Algebra, Interscience, New York, 1961. [504]

[60] I. M. Gelfand and M. L. Graev, Finite-dimensional irreducible representations of the unitary and full linear groups, and related special functions, Izv. Akad. Nauk SSSR **29** (1965) 1329-1356. (Am. Math. Soc. Trans. **64** (1967) 116-146. [363]

[61] I. M. Gelfand and M. L. Graev, Louck polynomials and their relation to general hypergeometric functions and GG-functions, Doklady Mathematics **61** (2000) 336-339. [64, 596]

[62] I. M. Gelfand and M. L. Tsetlin, Matrix elements for the unitary group, Dokl. Akad. Nauk SSSR **71** (1950) 825-828, 1017-1020. (Reprinted in I. M. Gelfand *et al.*, Representations of the Rotation and Lorentz Group, Pergamon, 1963.) [30, 285, 363, 523, 525, 528]

[63] I. M. Gelfand, R. A. Minlos, and Z. Ya. Shapiro, Representations of the Rotation and Lorentz Groups and Their Applications, Pergamon, New York, 1963. [504]

[64] J. W. L. Glaisher, Note on certain symbolic operators, and their application to the solution of certain partial differential equations, Proceedings of the London Mathematical Society **12** (1881) 201-207. [8]

[65] C. D. Godsil, Algebraic Combinatorics, Chapman and Hall, New York, 1993. [596]

[66] I. J. Good, A short proof of a conjecture of Dyson, J. Math. Phys. **11** (1970) 1884. [8, 367]

[67] R. Goodman and N. R. Wallach, Representations and Invariant Theory of the Classical Groups, Encyclopedia of Mathematics and Its Applications, Vol. 68, edited by G.-C. Rota, Cambridge University Press, Cambridge, 1998. [504]

[68] P. Gordan, "Über das Formensystem binärer Formen," Teubner, Leipzig, 1875. [37]

[69] I. P. Goulden and A. M. Hamel, Shift operators and factorial symmetric functions, J. Comb. Theory A **69** (1995) 51-60. [596]

[70] R. L. Graham, M. Grötschel, L. Lovász, Editors, Handbook of Combinatorics, Editors, The MIT Press, Cambridge, 1995, Elsevier Science B. V., Amsterdam, The Netherlands. [596]

[71] R. B. Griffiths, Letters, Physics Today, March 2007, pp. 8-9. [12]

[72] R. A. Gustafson, A Whipple's transformation for hypergeometric series in $U(n)$ and multivariable hypergeometric orthogonal polynomials, SIAM J. Math. Anal. **18** (1987) 495-430. [567]

[73] R. A. Gustafson and S. C. Milne, Schur functions and the invariant polynomial characterizing $U(n)$ tensor operators, Adv. in Appl. Math. **4** (1983) 422-488. [556]

[74] P. R. Halmos, Introduction to Hilbert Space and the Theory of Specral Multiplicity, D. Van Nostrand, Princeton, New Jersey, 1957. [504]

[75] P. R. Halmos, Finite-Dimensional Vector Spaces. Appendix on Hilbert Space, D. Van Nostrand, Princeton, New Jersey, 1958. [504]

[76] M. Hamermesh, Group Theory and Its Applications to Physical Problems, Addison-Wesley, 1662. [504]

[77] F. Harary, Graph Theory, Addison-Wesley, 1969. [164, 216]

[78] F. Harary and E. M. Palmer, Graphical Enumeration, Academic Press, 1773. [164, 226, 227]

[79] S. Helgason, Differential Geometry and Symmetric Spaces, Academic Press, New York,1962. [504]

[80] J. E. Humphreys, Introduction to Lie Algebras and Representation Theory, Springer-Verlag, New York, 1972. [504]

[81] T. W. Hungerford, Algebra, Springer-Verlag, New York, 1974. [504]

[82] W. Heisenberg, Über quantentheoretische Umdeutung kinematischer und mechanischer Beziehungen, Z.Physik **33** (1925) 879-893. [29]

[83] J. M. Hyman and E. A. Stanley, private communication, 1984. [582]

[84] N. Jacobsen, Lectures in Abstract Algebra Vols. 1-3, D. Van Nostrand, New York, 1953. [504]

[85] B. R. Judd, Operator Techniques in Atomic Spectroscopy, McGraw-Hill, New York, 1963. [260]

[86] B. R. Judd, Second Quantization and Atomic Spectroscopy, Johns Hopkins Press, Baltimore, 1967. [260]

[87] B. R. Judd and G. M. S. Lister, A class of generalized $9-j$ symbols for $Sp(2n)$, J. Phys. A: Math. Gen. **20** (1987) 3159-3169. [260]

[88] A. Kerber, Applied Finite Group Actions. Algorithms and and Combinatorics, Vol. 19, Second Edition, Springer, 1999. [505]

[89] C. M. King, The Quantum Mechanical Solutions of Several Problems by Means of Operator Similarity Transformations, Thesis, Auburn University, Alabama, 1963. [562]

[90] R. C. King, Weyl's denominator identity and its deformation, Symmetry and Structural Properties of Condensed Matter, editors., T. Lulek, B. Lulek, A. Wal, World Scientific, Singapore, 2001, pp. 37-48. [528]

[91] D. E. Knuth, Permutations, matrices, and generalized Young tableaux, Pac. J. Math. **34** (1970) 709-727. [265, 520]

[92] M. Konvalinka and I. Pak, Non-commutative extensions of the MacMahon master theorem, Adv. Math. **216** (2007) 29-61. [580]

[93] R. R. Korfhage, Discrete Computational Structures, Academic Press, New York, 1974. [216]

[94] B. Kostant, A formula for the multiplicity of a weight, Trans. Amer. Math. Soc. **93** (1959) 53-73. [553]

[95] A. N. Kirillov, A remark on the Gelfand-Tsetlin patterns for symplectic groups, Geometry and Physics, Vol. 5. Essays in honour of I. M. Gelfand, editors, S. Gindikin and I. M. Singer, North-Holland, Amsterdam, 1989, pp. 483-481. [595]

[96] A. N. Kirillov and A. D. Berenstein, Groups generated by involutions, Gelfand-Tsetlin patterns, and combinatoric of Young tableau, Algebra i Analiz **7** (1995) 92-152; translation in St. Petersburg Math. J. **7** (1996) 77-127. [595]

[97] J. P. S. Kung and G. C. Rota, The invariant theory of binary forms, Bull. Amer. Math. Soc. **10** (1984) 27-85. [595, 596]

[98] J. P. S. Kung, Editor, Gian-Carlo Rota on Combinatorics, Birhaäuser, Bostin, 1995. [375, 595]

[99] G. Kuperberg, Another proof of the alternating sign matrix conjecture, Internat. Math. Res. Notices (3) (1996) 139-150. [585]

[100] G. Lakoff and R. E. Núñez, Where Mathematics Comes From, Basic Books, Perseus, New York, 2000. [3]

[101] A. Lascoux and M.-P. Schützenberger, Formulaire raisonné symétriques, Publication LIPT, Universiteé Paris 7, 1995. [448]

[102] A. Lascoux, Square-ice enumeration, The Andrews Festschrift: Seventeen Papers on Classical Number Theory and Combinatorics, editors, D. Foata and G. N, Han. Springer-Verlag, 2001, pp. 335-348. [585]

[103] A. J. Leggett, The Quantum Measurement Problem, Science **307** (2005) 871-872. [5,12]

[104] D. E. Littlewood and A. R. Richardson, Group characters and algebras, Phil. Trans. Royal Society A (London) **233** (1934) 99-141. [269]

[105] D. E. Littlewood, The Theory of Group Characters, Oxford University Press, London, 1950, reprinted 1958. [504, 577]

[106] J. D. Louck, Quantum Theory of Angular Momentum in N-Dimensional Space, Los Alamos Scientific Laboratory Report, LA-2451, 1960. [584]

[107] J. D. Louck, Recent progress toward a theory of tensor operators in the unitary groups, Am. J. Phys. **38** (1970) 3-42. [368, 394, 408, 446]

[108] J. D. Louck, MacMahon's master theorem, double tableau polynomials, and representations of groups, Adv. Appl. Math. **17** (1996)143-168. [67, 446, 580]

[109] J. D. Louck, Unitary group theory and the discovery of the factorial Schur functions, Annals Comb. **4** (2000) 413-432. [446, 596]

[110] J. D. Louck, Doubly stochastic matrices in quantum mechanics, Found. of Phys. **27** (1997) 1085-1104. [570]

[111] J. D. Louck, Combinatorial aspects of the representations of the unitary group, Symmetry and Structural Properties of Condensed Matter, editors, T. Lulek, W. Florek, B. Lulek, World Scientific, Singapore, 1997, pp. 231-252. [257, 258, 446]

[112] J. D. Louck, Unitary symmetry, combinatorics and special functions, Physical Applications and Mathematical Aspect of Geometry, Groups, and Algebras, Vol. 1, editors, H.-D. Doebner, P. Nattermann, and W. Scherer, World Scientific, Singapore, 1997, pp. 50-70. [433, 446, 567, 596]

[113] J. D. Louck, Survey of zeros of $3j$ and $6j$ coefficients by Diophantine equation methods, Group Theory and Special Symmetries in Nuclear Physics, Proceedings of the International Symposium in Honor of K. T. Hecht, editors, J. P. Draayer and J. Jänecke, World Scientific, Singapore, 1992, pp. 38-48. [136, 398]

[114] J. D. Louck, Power of a determinant with two physical applications, J. Math. and Math. Sci. **22** (1999) 745-759. [570]

[115] J. D. Louck, New perspectives on the unitary group and its tensor operators, Symmetry and Structural Properties of Condensed Matter, editors, T. Lulek, B. Lulek, and A. Wal, World Scientific, 2001, pp. 23-36. [349, 446]

[116] J. D. Louck, Alternating sign matrices and Gelfand-Tsetlin polynomials, Symmetry, Spectroscopy and SCHUR, Proceedings of the Professor Brian G. Wybourne Commemorative Meeting Toruń, edited by R. C. King, M. Bylicki, and J. Karowski, Nicolaus Copernicus University Press, Toruń, 2006, pp. 180-186. [570, 594]

[117] J. D. Louck and L. C. Biedenharn, Canonical unit adjoint tensor operators in $U(n)$, J. Math. Phys. **11** (1970) 2368-2414. [377, 394, 395, 421, 443, 446]

[118] J. D. Louck and L. C. Biedenharn, Further developments of the boson polynomials and their implications: III, J. Math. Phys. **14** (1973) 1336-1357. [265, 399, 446, 534, 596]

[119] J. D. Louck and L. C. Biedenharn, A generalization of the Gauss hypergeometric series, J. Math. Analysis and Appl. **59** (1977) 423-431. [446, 567]

[120] J. D. Louck and L. C. Biedenharn, Some properties of the intertwining number of the general linear group, Science and Computers. Adv. Math. Suppl. Studies, editor, G.-C. Rota, Vol. 10, Academic Press, New York, 1986, pp. 265-311. [446, 534, 550, 552, 553, 556]

[121] J. D. Louck and L. C. Biedenharn, Canonical and noncanonical methods in applications of group theory to physical problems, Adv. Quant. Chem. **23** (1992) 127-204. [446]

[122] J. D. Louck, W. Y. C. Chen, and H. W. Galbraith, Combinatorics of $3n-j$ coefficients, in Proc. VIII International Conf. on Symmetry Methods in Physics, July, 1997, Dubna, Russia, Special issue of Physics of Atomic Nuclei **61** (1998) 1868-1873. [254, 349, 446]

[123] J. D. Louck, W. Y. C. Chen, and H. W. Galbraith, Generating functions for $SU(2)$ binary recoupling coefficients, Symmetry and Structural Properties of Condensed Matter, editors, T. Lulek, A. Wal, and B. Lulek, World Scientific, Singapore, 1999, pp. 112-137. [254, 349, 446]

[124] J.D. Louck and H. W. Galbraith, Application of orthogonal and unitary group methods to the $N-$body problem, Rev. Mod. Phys. **44** (1972) 540-601. [50]

[125] T. Lulek, Bethe ansatz, Young tableaux, and the spectrum of Jucys-Murphy operators, Symmetry and Structural Properties of Condensed Matter, editors, T. Lulek, W. Florek, B. Lulek, World Scientific, Singapore, 1903, pp. 279-291. [448]

[126] I. Macdonald, Symmetric Functions and Hall Polynomials, Oxford University Press, Oxford, 1979, second edition, 1995. [preface, 317, 505, 539, 555, 574, 579]

[127] G. W. Mackey, Mathematical Foundations of Quantum Mechanics, Benjamin, New York,1963. [12]

[128] S. Mac Lane. Mathematics: Form and Function, Springer-Verlag, New York, 1986. [1, 2, 3, 4]

[129] P. A. MacMahon, Combinatory Analysis, Chelsea, New York, 1915. [preface, 64, 511]

[130] M. A. Méndez, Directed graphs and the combinatorics of the polynomial representations of $GL(n, \mathbb{C})$, Annals of Comb. **5** (2001) 459-478. [580]

[131] M. A. Méndez, Towards a combinatorial description of the matrices corresponding to irreducible representations of the unitary and general linear groups, Symmetry and Structural Properties of Condensed Matter, editors, T. Lulek, A. Wal, and B. Lulek, World Scientific, Singapore, 2003, pp. 265-273. [580]

[132] L. Michel, Lecture Notes in Physics: Group Representations in Mathematics and Physics, Battelle Recontres, edited by V. Bargmann, Springer, Berlin, 1970, pp. 36–143. [504]

[133] L. Michel and B. I. Zhilinskii, Symmetry, Invariants, Topology, Physics Reports, North-Holland, Amsterdam **341** (2001) 11- 84. [67]

[134] B. Mielnik and J. Plebański, Combinatorial approach to Baker-Campbell-Hausdorff exponents, Ann. Inst. Henri Poincaré **12** (1970) 215-254. [484]

[135] W. H. Mills, D. P. Robbins, and H. Rumsey, Self-complementary, totally symmetric plane partitions, J. Combin Theory Ser. A **42** (1986) 277–292. [586]

[136] S. C. Milne, Basic hypergeometric series very well-poised in $SU(n)$, J. Math. Anal. Appl. **122** (1987) 223-256. [596]

[137] T. Molien, Über die Invarianten der linearen Substitutionsgruppen, Sitzungber. Konig. Preuss. Akad. Wiss. **52** (1897) 1152-1156. [67]

[138] M. Moshinsky, Bases for the irreducible representations of the unitary groups and some applications, J. Math. Phys. **4** (1963) 1128-1139. [363]

[139] L. Page, Introduction to Theoretical Physics, Van Nostrand, New York, 1935. [4]

[140] W. Pauli, Zur Quantenmechanik des magnetischen Elektrons, Z. Physik **37** (1927) 601–623. [13]

[141] K. A. Penson, P. Blasiak, G. Dattoli, G. H. E. Duchamp, A. Horzela, and A. I. Solomon, Monomiality principle, Sheffer-type polynomials and the normal ordering problem, Symmetry and Structural Properties of Condensed Matter, editors, T. Lulek, B. Lulek, and A. Wal, J. Phys. Conf. Series, Vol. 30, Institute of Phys. 2006. pp. 86-97. [562]

[142] S. Perlis, Theory of Matrices, Addison-Wesley, Cambridge, Massachusetts, 1952. [465, 467, 468, 504]

[143] G. Racah, Theory of complex spectra. I, Phys. Rev. **61** (1942) 186-197; II, *ibid.* **62** (1942) 438-462; III,*ibid.* **63** (1943) 367-382. [30, 129, 131]

[144] G. Racah, Group Theory and Spectroscopy, Ergebnisse der exakten Naturwissenschaften **37** (1965) 28-84, Springer-Verlag, Berlin (based on Princeton lectures, 1951). [30, 318]

[145] G. Racah, in Group Theoretical Concepts and Methods in Elementary Particle Physics, edited by F. Gürsey, Gordon and Breach, New York, 1962, pp. 31. [316, 378]

[146] R. C. Read, The enumeration of locally restricted graphs. I, J. London Math. Soc. **34** (1959) 417-436; II, *ibid.* **35** (1960) 344-351. [226]

[147] T. Regge, Symmetry properties of Clebsch-Gordan coefficients, Nuovo Cimento **10** (1958) 544-545; Symmetry properties of Racah's coefficients, *ibid.* **11** (1959) 116-117. [58, 60, 246]

[148] J. Riordin, An Introduction to Combinatorial Analysis, Wiley, New York, 1958. [596]

[149] J. Riordan, Generating functions, Applied Combinatorial Mathematics, editor E. F. Beckenback, Wiley, New York,1964, pp. 67-90. [556]

[150] D. P. Robbins, W. H. Mills, and H. Rumsey, Jr, Alternating sign matrices and descending plane partitions, J. Comb. Theory Ser. A **34** (1983) 340-359. [585, 587]

[151] G. de B. Robinson, Representation Theory of the Symmetric Group, University of Toronto Press, 1961. [596]

[152] P. Roman, Some Modern Mathematics for Physicists and Other Outsiders, Vol. 1, Pergamon Press, New York, 1975. [504]

[153] S. M. Roman and G.-C Rota, The umbral calculus, Advances in Math. **27** (1978) 95-188. [ii, 54, 595]

[154] G.-C. Rota, Finite Operator Calculus, Academic Press, 1975. [54, 505, 562, 595]

[155] G.-C. Rota, Hopf algebra methods in combinatorics, Colloques Internationaux, C.N.R.S., N° 260—Pròblemes Combinatores et Theéorie des Graphes, 363-365 (Reprinted in Ref. [98]). [54, 595]

[156] G.-C. Rota, Invariant theory, old and new, The American Math. Soc. Colloquium Series, Lecture 2, Baltimore, 1998. [54, 595]

[157] G-C. Rota, D. Kahaner, and A. Odlyzko, Finite operator calculus, J. Math. Anal. Appl. **42** (1973) 685-760 (Reprinted in Ref. [154]). [54, 562, 595]

[158] D. J. Rowe and J. Repka, An algebraic algorithm for calculating Clebsch-Gordan coefficients in $SU(2)$ and $SU(3)$, J. Math. Phys. **38** (1997) 4363-4388. [349]

[159] D. H. Sattinger and O. L. Weaver, Lie Groups and Algebras with Applications to Physics, Geometry, and Mechanics, Springer-Verlag, New York, 1986. [504]

[160] J. Schwinger, On Angular Momentum, U.S. Atomic Energy Commission Report NYO-3071, 1952 (Reprinted in Ref. [25]). [preface, 64, 230, 234, 568]

[161] D. P. Shulka, Certain transformations of generalized hypergeometric series, Indian J. Pure Appl. Math. **12** (1981) 994-1000. [567]

[162] G. F. Simmons, Topology and Modern Analysis, McGraw-Hill, New York, 1963. [504]

[163] R. P. Stanley, Enumerative Combinatorics, Vols. 1,2, Cambridge University Press, Cambridge, 1997. [preface, 60, 182, 317, 505, 506, 520, 522, 535, 574, 575, 579, 594]

[164] D. Stanton, Editor, Invariant Theory and Tableau, The IMA Volumes in Mathematics and Its Applications, Vol. 19, Springer-Verlag, New York, 1988. [596]

[165] D. Stanton and D. White, Constructive Combinatorics, Springer-Verlag, New York, 1986. [100, 595, 596]

[166] S. Sternberg, Group Theory and Physics, Cambridge Univ. Press, Cambridge, 1994. [355, 356, 357, 504]

[167] H. Henry Stroke, Editor, The Physical Review: The First Hundred Years, AIP Press, New York, 1995. [13, 14]

[168] D. K. Sunko and D. Svrtan. A generating function for angular momentum multiplicities, (Univ. of Zagreb, Yugoslavia, preprint, 1984). [92, 513]

[169] J. D. Talman, Special Functions: A Group Theoretic Approach, Benjamin, New York, 1968. [596]

[170] N. Vilenkin, Special Functions and the Theory of Group Representations,Vol. 22, translation in Amer. Math. Soc., Providence, 1968. [596]

[171] B. L. van der Waerden, Die Gruppentheoretische Methode in der Quantenmechanik, Springer, Berlin, 1932. [21]

[172] B. L. van der Waerden, Sources of Quantum Mechanics, North-Holland, Amsterdam, 1967. [504]

[173] B. L. Van der Waerden, Algebra, Frederick Ungar Publishing Co., New York, 1970. [453, 504]

[174] J. von Neumann, Mathematical Foundations of Quantum Mechanics, Princeton University Press, 1955. [12]

[175] L. Wei, Unified approach for calculation of angular momentum coupling and recoupling coefficients, Comput. Phys. Commun. **120** (1999) 222-230. [245]

[176] H. Weyl, Space—Time—Matter, Dover, New York, 1922. [3]

[177] H. Weyl, The Theory of Groups and Quantum Mechanics, Methuen, London, 1931; reissued by Dover, New York, 1949. [3, 30, 358, 504, 528]

[178] H. Weyl, The Classical Continuous Groups, Princeton University Press, Princeton, New Jersey, 1946. [356, 504]

[179] J. A. Wheeler and W. H. Zurek, Editors, Quantum Theory and Measurement, Princeton University Press, Princeton, New Jersey,1983. [12]

[180] E. T. Whittaker, History of the Theories of the Aether and Electricity, 1900-1926, Nelson, London, 1953. [14]

[181] E. P. Wigner, Group Theory and Its Application to the Quantum Mechanics of Atomic Spectra. Academic Press, New York, 1959 (Original German edition: Gruppentheorie und ihre Anwendung auf die Quanten-mechanik der Atomspektren, Braunschweig, 1931). [15, 30, 358, 504]

[182] E. P. Wigner, Application of Group Theory to the Special Functions of Mathematical Physics, Lecture notes, 1955. [360, 596]

[183] E. P. Wigner, On the matrices which reduce the Kronecker products of representations of S. R. groups, 1940. (Reprinted in Quantum Theory of Angular Momentum, editors, L. C. Biedenharn and H. van Dam, Academic Press, New York, 1965, pp. 87-133.) [143]

[184] E. P. Wigner, Einige Folgerungen aus der Schrödingerschen Theorie für die Termstrukuren, Z. Phys. **43** (1927) 624-652. [37]

[185] E. P. Wigner, Berichtigung zu der Arbiet: Einige Folgerungen aus der Schrödingerschen Theorie für die Termstrukuren, Z. Phys. **44** (1927) 601-602. [37]

[186] H. F. Wilf, Generatingfunctionolgy, Academic Press, New York, 1994. [556, 596]

[187] S. Wolfram, A New Kind of Science, Wolfram Media, Champaign, Illinois, 2002. [preface]

[188] A. C. T. Wu, Structure of the Wigner $9j$ coefficients in the Bargmann approach, J. Math. Phys. **13** (1972) 84-90. [248, 254, 257]

[189] B. G. Wybourne, Symmetry Principles and Atomic Spectroscopy, Wiley-Interscience, New York, 1970. [504]

[190] B. G. Wybourne, Symmetric functions and their application to physics, Symmetry and Structural Properties of Condensed Matter, editors, W. Florek, D. Lipskiński, T. Lulek, World Scientific, Singapore, 1993, pp. 79–100. [448]

[191] Lingling Yang, Louck polynomials, Hermite polynomials, and harmonic oscillators, Master's Thesis, Center for Combinatoric, Nankai University,Tianjin, P. R. China, 2002. [596]

[192] A. P. Yutsis, I. B. Levinson, and V. V. Vanagas, The Theory of Angular Momentum, Vilnius, 1960. [202]

[193] A. P. Yutsis and A. A. Bandzaitis, Angular Momentum Theory in Quantum Physics, Moksias, Vilnius 1977. [202, 258]

[194] J. Zak, Finite transitions in solid-state physics, Phys. Rev. Lettrs. **19** (1967) 1385-1387. [486]

[195] D. Zeilberger, Proof of the alternating sign matrix conjecture, Electronic J. of Combinatorics **3** (2) (1996). [585]

[196] D. Zeilberger, Proof of the refined alternating sign matrix conjecture, J. Math. **2,** New York (1996) 59-68. [591]

Index

A

abstract angular momentum theory 28–43
  bra-ket vectors 32
  compounding of 29
  coupled basis 37, 83
  irreducible unitary representation 34
  orthonormality of basis vectors 84–85
  parametrizations of $SU(2)$ 34
  standard action of components 31–33, 37, 84, 86
  standard unitary action 31–33, 38
  uncoupled basis 36
action of a group 449–450
  left and right 23
  of generators of $SU(2)$ 23–28
adjacent points of a graph 163
alternating sign matrices 585–593
  Andrews-Zeilberger numbers 586
  and strict Gelfand-Tsetlin patterns 586–591
  and strict semistandard tableaux 594–595
arc digraphs 295–300
associated Laguerre polynomials 560, 562

B

Baker-Campbell-Hausdorff identities 357
basis of space of irreducible tensor operators 382
Bernoulli polynomials 559, 583
Biedenharn-Elliott identity 143–147
Biedenharn operator patterns 398,403, 445
  origin 445
binary coupling theory 97–162
  associated generalized WCG coefficients 109–111
  binary bracketings 97–104
  binary state vectors 117–119
  binary tree(s) 100–104, 188
  Catalan number 99
  coupled state vector 112–114,
  problem of parentheses 97
  shape of a binary tree 100
  structure of $C^\lambda-$coefficients 335
binary recoupling matrices 115–119
binary recoupled state vectors 112–119
binary reduction of Kronecker products 114–115
binary tree(s) 100–104, 188
  internal and external points 100
  fork 102–104
  fork matrix 102–104
  four types of forks 109
  labeled and unlabeled 107–108
  map to trivalent tree 163
  polynomial $T_n(x)$ 104–106
  shape 100
  skew-symmetric matrix 232–237
  standard labeling 107–108

binomial coefficients and functions 43
  sum formula 43, 592

C

Cartesian 3-space 2
Catalan numbers 99, 187
Cayley-Hamilton theorem 355
C−coefficient mapping rule 80
characteristic null space 389–425
Chebyshev polynomials 561
classical angular momentum 12–13
Clebsch-Gordon numbers 87–89, 171, 207, 512
  combinatorial significance 91-92
  multiset generation 88
Clebsch-Gordon series 88–89
  and Gaussian polynomials 91–92, 512–513
  cardinality 88
  evaluation 54–55
combinatorial definition of Wigner-Clebsch-Gordan coefficients 51–55
combinatorial formula for the $C^{\lambda}(A)$ matrices 325–329
combinatorial proof of multiplication of D^k−polynomials 67–70
common fork 142
commutator 25, 357
complete set of Hermitian operators 368
  coupled basis transformation 90
  state vectors 83–87
  uncoupled basis transformation 90
  unitary frame rotations 89
composite systems 83–162
compositions of integers 65, 512
compounding of angular momenta 29
coupling rules for irreducible tensor operators 433–440
cubic graphs 158, 161, 198–227
  adjacent points 198
  and trivalent trees 189–198
  angular momentum type 202, 209, 212, 219–220
  conservation of adjacency of triangles 197
  cut and join properties 213–220
  cutting and rejoining of lines
  factoring properties 199, 213
  isomorphisms 190–192, 199
  labeled cubic graphs 226–228
  matchings 225
  matrices 221–226
  of angular momentum type 200, 202, 209–210
  of non angular momentum type 206, 208
  of $6-j$ and $9-j$ coefficients
  planar graphs 198
  points as triangles 193
  points in Cartesian 3-space 198
  symbolic relations 193–196
  tetrahedron and Racah coefficients 193
  Yutsis factor theorem 200

D

D^{λ}−polynomials, form 261–284
  abbreviated notations 262–264
  basic properties 271, 273–277
  defining relations 268–272
  fundamental shift operators 270
  Gelfand-Tsetlin patterns 262
  group property 275
  in terms of Maclaurin polynomials 263
  invariant normalizing factor $M(\lambda)$ 264
  inversion 264
  Kronecker product 274
  multiplication property 274
  order relation 266–267
  orthogonality 264–265

over $l \times n$ variables 278–280
partitions with two nonzero parts 342
proof of multiplication property 67–70, 277
reduction relation 276
relation to Robinson-Schensted-Knuth identity 265
relation to Schur functions 275
structural relations 282–283
transpositional symmetry 274
vector space aspects 280–282
D^{λ}–polynomials, structure 325–354
binary tree structure of C^{λ}–coefficients 335–342
combinatorial formula 325–329
differential operator actions 372–373
in angular momentum notation 77, 350
partitions with two nonzero parts 342–350
Pieri's rule 335, 339–340, 522–523
recursion relation 351–354
reduction of Kronecker product $D^{p} \otimes D^{\lambda}$ 329–335
determinants 568
power of 568–570
Vandermonde 572
Dirac bra-ket 32
domain of definition 10
of a function 10
of a wavefunction 14
of angular momentum quantum numbers 32–37, 41, 84–89, 94–95
doubly stochastic matrices 570
as sums of permutation matrices 570

E

eigenvalues of Gelfand invariants 369, 371–378
Euclidean geometry 1
and Cartesian space 1–5
Euler operator 47
exponential matrices 355, 469–476

F

factorial Schur function 596
factorization lemma 315, 329–335, 397, 399
finite groups 447
left and right regular representations 495–496
of permutations 448–449
fork matrix 102
fundamental results of angular momentum theory: first, second, third, fourth, fifth, sixth 35, 35, 42, 96, 119, 154
fundamental shift operators 285–295, 420
algebra 301–304
factorization 302
orthogonality relations 301–302
fundamental triangle coefficients 126, 153, 157
and no common fork 153,157

G

Gelfand-Tsetlin (GT) patterns 76, 523–550
betweenness conditions 76, 524
combinatorial origin 523–524
double 76
group theoretical origin 528
lexical 526
linear order 266–267
of modified type 539
of skew type 528–532
two-rowed 524
Gelfand invariants 368–371
eigenvalues 369–378
eigenvalues as symmetric functions 373

generating functions in angular momentum theory 229–260
  double Pfaffian of the $9-j$ cubic graph 256
  double Pfaffian of the tetrahedron 255
  elements of recoupling matrix 244
  integer property 245
  $9-j$ cubic graph rule 257
  of coupled wave function 239–240
  of general recoupling coefficients 241–245
  of the $9-j$ coefficients 248–254
  of the $6-j$ coefficients 246–248
  of the WCG coefficients 239–240
  Pfaffians and double Pfaffians 230–232
  skew-symmetric matrix of a binary tree 232–237
  tetrahedral rule 257
  triangle monomials 238
  triangle monomial of a binary tree 238
generating functions of classical special functions 558–563
generator of rotations 8
generator of translations 8
geometric vector 3
geometry of level sets 317
Gram-Schmidt procedure 461–464
  Gram matrix 461
graph isomorphism 163
group(s) 447
  abelian 448
  action 449–450
  defining properties 447–448
  diagonal subgroup 39, 63–64
  direct product 38–39
  invariant subset under group action 481
  irreducible representations 499–500
  isotropy 451
  matrix representation 494–496
  one-parameter subgroup 358
  operator representation 494–496
  permutation 448–449, 496–497
  unitary action in a Hilbert space 500–501
  unitary action in composite systems 89–90
group representations 494–504
  diagonal subgroup of $SU(2)$ 63
  irreducible 499–500
  Kronecker product 497
  matrix 494–495
  operator 494
  Schur's lemma 501
  unitary 500–501

H

Hermite polynomials 562, 569, 596
Hilbert space(s) 456
  isomorphic 459
  model Hilbert space 410-411
  orthogonal basis 458
  product of operators 459–460
  properties of linear operators 458–461
hypergeometric functions 565–567
  classical definition 565
  relation of ${}_4F_3$ to Racah coefficients 59
  relation of ${}_3F_2$ to WCG coefficients 59
  terminating type as polynomials 136, 567

I

indeterminates 452–453
inner product(s) 49–51
  of bosons polynomials 49–51
inversion of reference frame 6

irreducible tensor operators of $U(n)$ 389–446
basis 382
canonical tensor operators 394–398
characteristic null space 394
coupling rules 433–435
limit relations 440–445
ordered set 395–396,402–403
phase rule 445
Racah invariant operators and coefficients 435–440
reduced matrix elements 403–410
subgroup conditions 403–404
unit tensor operators 386–394
Wigner-Eckart theorem 406
isomorphism of graphs 163
isotropy group 326–327

J

Jacobi identity 358
Jacobi polynomials 560

K

Kostka numbers 516
for $n = 3$ 543
invariance under permutations 517
recurrence relation 550–551
relation to Littlewood-Richardson numbers 552-553
relation to Weyl dimension 517
Kronecker product 39, 62, 79, 312, 488, 492–493
multiplicity space 316
of solid harmonics 61,79–81
reduction of 90–91, 529–335
shift operator-valued polynomials 312–315
standard direct sum 91, 93-94, 312–313

L

Laplace's equation 47
lattice permutations 327, 535, 537
and Littlewood-Richardson numbers 538–543
of modified GT patterns 539–543
of skew SSYW tableaux and skew GT patterns 538–539
Lie algebra 358
and Sylvester's identity 364, 367
differential operator representation of $gl(n, \mathbb{C})$ 360
Jacobi identity 358
left and right actions 360
matrix elements of generators 363–369
matrix representation 361
matrix representations of $u(n)$ 359–362
of $GL(n, \mathbb{C})$ 358–360
Lie bracket polynomials 476–486
limit properties 440–445
of reduced matrix elements 440–443
of $U(n)$ CG coefficients 444–445
linear operators 458–461
action in Hilbert space 458–461
linear ordering 513–514
compositions 65, 512–513
Gelfand-Tsetlin patterns 266–267
lexicographic 65, 396, 266, 513
operator patterns 395–397, 402
partitions 513–514
Littlewood-Richardson matrix 398–401
Littlewood-Richardson numbers 538–543
as basic objects in Kronecker products 548
Baird-Biedenharn rule 551
for $n = 3$ 544–547
in product of Schur functions 554, 574

level sets 395
range of definition 317
range of values 395
relation to lattice permutations 538–544
relation to Schur functions 538–544
role in characteristic null space 395–403
role in factorization lemma 397–403
shift invariance 421
sum of Kostka numbers 553

M

Maclaurin monomials 71
MacMahon master theorem 66–67, 229–230, 568
and generating functions of recoupling coefficients 241–259
basic form 67
generalization 580
Molien group form 67
Schwinger form 66
magic squares 56
number of order three 60
relation to angular momentum 57
zeros of the counting polynomial 594
many-particle systems 22–28
matrices 465
Cayley-Hamilton theorem 466
characteristic equation 466
diagonal and nondiagonable 465, 498
exponentiated 469–486
matrix unit 359
normal 465, 467–468
permutation 497
representation of linear operators 464–465
matrix arrays $\mathbb{M}^p_{n\times m}(\alpha,\beta)$ 263
matrix curve through the identity 359
matrix representation of Lie algebra $gl(n,\mathbb{C})$ 359
measure factor $M(\lambda)$ 264
in terms of hooks 518
method of Cartan 15
multiplication of unit tensor operators
multivariable special functions 563–567

N

Newtonian physics 9
null space of tensor operators 389–425
of canonical tensor operators 394–398
of operator-valued D^Λ−polynomials 392–394
of $U(3)$ canonical tensor operators 425–433
null tensor operators 389

O

operator actions 449–450
operator-valued polynomials 308–311, 382–384, 400–403
orthonormalization methods 461–464
Gram-Schmidt 461–462
rejection rule 462
symmetric 463–464

P

parametrizations of $SU(2)$ solid harmonics 44
partitions 506–511
path of a recoupling matrix 148, 158
associations and commutations 148
pattern calculus 295–301, 415–419
arrow pattern 416
labeled arc multigraph 416, 419
shift-weight function 417–418
permutation(s)
lattice 327, 535, 537

matrices 349, 497, 586
matrices of order three 57
perpendicular projection 2
Pfaffians and double Pfaffians 230–232
in generating functions 241–259
matchings 230
phase factor rule 445
Pieri's rule 335, 339–340, 522–523
and binary trees 338–339
planar matrix arrays $\mathbb{M}^{p}_{n\times m}(\alpha,\beta)$ 67–75, 263
power of a determinant 55–57, 568–570
Maclaurin polynomial 56–57
WCG coefficients 55–58

Q

quaternion representations of $SU(2)$ 34

R

Racah coefficients 129–136
in terms of paths 148
in terms of recoupling matrices 139
in terms of 6−j coefficients 133
in terms of ${}_4F_3$ hypergeometirc coefficients 135
orthogonality relations 133
Racah's sum rule 139–140, 148
real number line 2
recoupling matrices 94–96, 125, 137–162
action of symmetric group 117
and triangle coefficients 125–131,
156–162
association and commutation rules 153
classification 156–162
for $n = 3$ 137–140
for $n = 4$ 140–152
fundamental 126, 153, 157
isomorphic 162
multiplication property 119, 142
reduced coupling matrix 96
reduced matrix elements 405–410
as matrix elements of unit projective operators 410–415
reference frame 3
inversion 6
rotation 7
translation 6
regular union 189,191,202
right-handed triad (frame) of unit vectors 8, 12
rings 454
of polynomials 452–454
Robinson-Schensted-Knuth identity 265, 384, 556
rotation of reference frame 7

S

Schur functions 573–580
extended 577–578
factorial 596
Schur's lemma 501–504
Schwinger master theorem 66–67
shape of a binary tree100
shape of a Gelfand-Tsetlin pattern 526
shape of a Young frame 514
shape transformation rule between binary trees 153
shift operator(s) 285–295
action 286–291
algebra 301–304
arc digraph interpretation 295–300
fundamental 285–295, 381
numerator and denominator factors 298–299
polynomials 308–312, 382–384
triangular shift pattern 291–292
skew Gelfand-Tsetlin patterns 528–532
skew tableaux 520–523

solid harmonics 43–48, 64–74, 81
as matrices 65
coordinate and derivative actions on $U(2)$ solid harmonics 81–82
left and right transformations 45
multiplication property 45
of $SO(3,\mathbb{R})$ 46
of $SU(2)$ 43–44
of $SU(n)$ 64
of $U(2)$ 75–78
parametrizations 44
transposition property 45
special functions 559–567
generating functions 556–565
invertible 566,
related to $SU(2)$ representations 63
spin states 20–21
spinor harmonics (polynomials) 44, 48, 65
standard $C-$coefficient mapping rule 80
standard reduction of Kronecker product 90-91, 93–94
superposition of state functions 11
Sylvester's's identity 364, 366, 367, 580–582
and the $GL(n,\mathbb{C})$ Lie algebra 364–367
Symmetric functions 570–578
binomial type 374–375
complete homogeneous 571
dual basis 578–580
elementary 571
extended Schur 577–578
monomial 371
power sum 571
Schur 573–575
skew Schur 575–576

T

tensor operators 379–446
as operator-valued polynomials 383
basis 382
canonical 386–403
coupling rules 433–440
definition 380
irreducible 380
null space 425
unit 386–403
tensor product space(s) 486–488
and Kronecker product of matrices 488
of separable Hilbert spaces 487
spatial and spin 22
totally symmetric CG coefficients 335
translation of reference frame 6
triangle coefficients 125–131, 156–162
adjacency diagram 157–161
and recoupling matrices 152, 156–162
fundamental 126, 153, 157
phase-equivalent 126–130
triangle monomials 238
triangle patterns 119
equivalence classes 121–124
left and right 126
representative 124, 129
standard labeling 119–120
trivalent trees 162–188
binary tree skeleton 165
complete sets of nonisomophic trivalent trees 186
compound fork 167
counting formulas 169, 188
internal lines and external lines, points 168
labeled and unlabeled 164
local scaling, reflections, rotations, Ω operations 174–180
nonisomorphic 172–184
occupation numbers 181
regular union 189,191,202

root points 168
shape labels 184
upper and lower branches 168
word labeling 165

U

umbral calculus 54, 595
unit projective operators 410–415
unitary action of a group on a Hilbert space 500–501
unitary action on $SU(2)$ 33–34
unitary frame rotations 17–19
unitary group representations 358–368
unitary group $U(3)$ 422–433
canonical tensor operators 422–425
characteristic null space 425–432
unit projective operators 432–433
unitary similarity transformation 90–91

V

vector spaces 455–458,
and linear operators 458–461
model Hilbert spaces 285–286, 379–380, 410–415
of polynomials 455–456, 488–494
separable Hilbert spaces 456–458
with an inner product 456–457, 488–491
Vandermonde determinants 572
and Bernoulli polynomials 582–584
and Sylvester's identity 582
and Weyl's dimension formula 572, 584–585

W

Waring's formulas 466
Wedderburn-Etherington numbers 182
weight of SSYW tableau 54
Weyl's dimension formula 572, 584–585
and Bernoulli polynomials 583
and Vandermonde's determinant 583–584
Wigner's $9-j$ coefficient 143, 149
in terms of paths 148
in terms of recoupling matrices 147–148
orthogonality relations 151
symmetries 151
Wigner-Clebsch-Gordan (WCG) coefficients 37
as an evaluation 54–55
domain of definition 39
in terms of ${}_3F_2$ hypergeometric coefficients 134
inter-relations with D^{λ}−polynomials 63
orthogonality relations 40
Regge generating function 58, 61
relation to $3-j$ coefficients 58
symmetries 58, 59
words 536–539, 541–542
as labels of binary trees 165
of modified Gelfand-Tsetlin patterns 539–542
of SSYW tableaux and skew GT patterns 536–537

X

$\times$ as direct product symbol 518
$\times$ as multiplication 451

Y

Young standard tableaux 515, 518
Young-Weyl semistandard tableaux (YWSS) 515

Z

Zeilberger polynomials 591, 593–594